AF342109

EDITORIAL COPY

Heme Oxygenase in Biology and Medicine

Heme Oxygenase in Biology and Medicine

Edited by

Nader G. Abraham

New York Medical College
Valhalla, New York

and

Associate editors

Jawed Alam

Alton Ochsner Medical Foundation
New Orleans, Louisiana

Karl Nath

Mayo Clinic/Foundation
Rochester, Minnesota

EDITORIAL COPY

Kluwer Academic / Plenum Publishers
New York, Boston, Dordrecht, London, Moscow

Library of Congress Cataloging-in-Publication Data

Hemeoxygenase in biology and medicine/edited by Nader G. Abraham.
p. cm.
"The chapters include several important discoveries introduced at the First International Symposium on Heme Oxygenase, which was held in New York City at the World Trade Center, Marriott Hotel, July 17–20, 2000."—Pref.
Includes bibliographical references and index.
ISBN 0-306-47264-3
1. Heme oxygenase—Physiological effect. 2. Heme oxygenase—Pathophysiology.
I. Abraham, Nader G. II. International Symposium on Heme Oxygenase
(1st: 2000: New York, N.Y.)

QP603.H45 H45 2002
612′.015191—dc21

2002066907

ISBN 0-306-47264-3

© 2002 Kluwer Academic / Plenum Publishers
233 Spring Street, New York, NY 10013

http://www.wkap.nl/

10 9 8 7 6 5 4 3 2 1

A C.I.P. record for this book is available from the Library of Congress

All rights reserved

No part of this book may be reproduced, stored in a retrieval system, or transmitted in any form or by any means, electronic, mechanical, photocopying, microfilming, recording, or otherwise, without written permission from the Publisher, with the exception of any material supplied specifically for the purpose of being entered and executed on a computer system, for exclusive use by the purchaser of the work.

Printed in the United States of America

To the many international scientists and clinicians for their contributions to the field of carbon monoxide and cellular stress responses and antioxidants as presented at the First International Symposium on Heme Oxygenase, which was held in New York City at the World Trade Center, Marriott Hotel, July 17–20, 2000.

This volume is, in a personal way, also dedicated to my wife, Miki, Professor of Pharmacology at New York Medical College, to my parents, and to my children, Richard and Oriane, who have been loving supporters of all my professional endeavors

CONTRIBUTORS

Atallah Kappas
Laboratory of Pharmacology
The Rockefeller University Hospital
New York, New York

Stefan W. Ryter
Department of Medicine
Division of Pulmonary, Allergy, and
  Critical Care Medicine, The
  University of Pittsburgh
Medical Center, Pittsburgh, PA

Augustine M. K. Choi
Department of Medicine, Division of
  Pulmonary
Allergy, and Critical Care Medicine,
  The University of Pittsburgh Medical
  Center, Pittsburgh, PA

Rui Wang
Department of Physiology, University
  of Saskatchewan, Saskatoon,
  Canada

Robert Peterson-Wakeman
Department of Physiology, University
  of Saskatchewan, Saskatoon,
  Canada

Toshisuke Morita
Fourth Department of Internal
  Medicine
Saitama Medical School, Saitama,
  Japan

Daniel S. Seidman
Department of Obstetrics and
  Gynecology, Sheba Medical Center,
  Tel-Hashomer, Sackler
School of Medicine, Tel-Aviv
  University, Israel

Charles W. Leffler
Department of Physiology
University of Tennessee Health Science
  Center
Memphis, Tennessee

Mutsuo Yamaya
Department of Geriatric and
  Respiratory Medicine
Tohoku University School of Medicine,
  Sendai, Japan

Hidetada Sasaki
Department of Geriatric and
  Respiratory
Medicine, Tohoku University School of
  Medicine, Sendai, Japan

Jian-Xion Chen
Department of Pathology
Center for Lung Research
Vanderbilt University Medical Center
Nashville, Tennessee

Chen-Ching Lai
Southern Illinois University, School of
  Medicine,
Southern Illinois

Józef Dulak
Division of Cardiology
Innsbruck University, Austria

Stella Kourembanas
Department of Medicine, Division of
    Newborn Medicine
Children's Hospital & Department of
    Pediatrics, Harvard Medical School,
    Boston, Massachusetts

Camille Taille
Institut National pour la Santé et la
    Recherche
Médicale, Unité 408, Faculté de
    Médecine Xavier
Bichat, Paris, France

Jorge Boczkowski
Institut National pour la Santé et la
    Recherche
Médicale, Unité 408, Faculté de
    Médecine Xavier
Bichat, Paris, France

P. O. Berberat
Immunobiology Research Center
Department of Surgery
Beth Israel Deaconess Medical Center
Harvard Medical School, Boston, MA

S. Brouard
Immunobiology Research Center
Department of Surgery
Beth Israel Deaconess Medical Center
Harvard Medical School, Boston, MA

M. P. Soares
Immunobiology Research Center
Department of Surgery
Beth Israel Deaconess Medical
    Center
Harvard Medical School, Boston, MA

F. H. Bach
Immunobiology Research Center
Department of Surgery

Beth Israel Deaconess Medical Center
Harvard Medical School, Boston, MA

Mark A. Perrella
Program of Develop. Cardiovascular
    Biology
The Cardiovascular Division
The Pulmonary and Critical Care
    Division
Brigham & Women's Hospital, Boston,
    MA

Jesus Araujo
Department of Medicine
University of California-Los Angeles,
    Los Angeles, California

Shaw-Fang Yet
Program of Develop. Cardiovascular
    Biology
The Cardiovascular Division
The Pulmonary and Critical Care
    Division
Brigham & Women's Hospital, Boston,
    MA

Aldons J. Lusis
Department of Medicine
University of California-Los Angeles,
    Los Angeles, California

German Camejo
Wallenberg Laboratory for
    Cardiovascular Research
University of Gothenburg, Sahlgrenska
    Hospital
Gothenburg, Sweden

Rafael Apitz-Castro
Laboratorio de Trombosis
    Experimental, Centro de Biofisica y
    Bioquimica, Instituto Venezolano de
    Investigaciones
Cientificas
Caracas, Venezuela

Kazunobu Ishikawa
The First Department of Internal
   Medicine
Fukushima Medical University
Fukushima-city, Fukushima, Japan

Amel F. Khelifi
Gray Cancer Institute
Mount Vernon Hospital, Middlesex,
   UK

James E. Clark
Department of Surgical Research
Northwick Park Institute for Medical
   Research
Harrow, Middlesex UK

Roberto Motterlini
Department of Surgical Research
Northwick Park Institute for Medical
   Research
Harrow, Middlesex UK

Gillian M. Tozer
Gray Cancer Institute
Mount Vernon Hospital, Middlesex,
   UK

David Sacerdoti
Department of Clinical and
   Experimental Medicine
University and Azienda Ospedaliera of
   Padova, Padova, Italy

Angelo Gatta
Department of Clinical and
   Experimental Medicine
University and Azienda Ospedaliera of
   Padova, Padova, Italy

Scapagnini G.
Section of Biochemistry & Molecular
   Biology Faculty of Medicine
Department of Chemistry
University of Catania, Italy

V. Calabrese
Section of Biochemistry & Molecular
   Biology Faculty of Medicine
Department of Chemistry
University of Catania, Italy

Atsushi Takeda
Department of Neurology
Tohoku University School of Medicine
Sendai, Japan

R. Krishnan Kutty
National Eye Institute
National Institutes of Health, LRCMB-
   NEI
Bethesda, MD

George Perry
Institute of Pathology
Case Western Reserve University
Cleveland, Ohio

Mark A. Smith
Institute of Pathology
Case Western Reserve University
Cleveland, Ohio

Yogesh Mawal
Bloomfield Centre for Research in
   Aging, Lady Davis Institute for
   Medical Research, Sir Mortimer B.
   Davis Jewish General Hospital,
   McGill University, Montreal, Quebec,
   Canada

Hyman M. Schipper
Bloomfield Centre for Research in
   Aging, Lady Davis Institute for
   Medical Research, Sir Mortimer B.
   Davis Jewish General Hospital,
   McGill University, Montreal, Quebec,
   Canada

A. Yachie
Department of Lab. Sci.
Kanazawa Univ., Kanazawa, Japan

K. Maruhashi
Department of Lab. Sci.
Kanazawa Univ., Kanazawa, Japan

Y. Kasahara
Department of Lab. Sci.
Kanazawa Univ., Kanazawa, Japan

S. Koizumi
Department of Pediatrics
Kanazawa Univ., Kanazawa, Japan

Alvin I. Goodman
Division of Nephrology and
    Department of Pharmacology, New
    York Medical College, Valhalla, NY

Giovanni Li Volti
Department of Biomedical Sciences
University of Catania, Italy

Lucia Malaguarnera
Department of Biomedical Sciences
University of Catania, Italy

Stephan Immenschuh
Istitute of Clinical Chemistry and
    Pathobiochemistry, Justus-Liebig-
    University, Giessen, Germany

Thomas Kietzmann
Institute of Biochemistry
Georg-August-University
Göttingen, Germany

Daniel Stewart
Department of Molecular Genetics
Alton Ochsner Medical Foundation
New Orleans, Louisiana

Jawed Alam
Department of Molecular Genetics
Alton Ochsner Medical Foundation
New Orleans, Louisiana

Prasun K. Datta
Department of Medicine, Division of
    Nephrology
University of Medicine and Dentistry,
    RWJMS
New Brunswick, New Jersey

Elias A. Lianos
Department of Medicine, Division of
    Nephrology, University of Medicine
    and Dentistry, RWJMS
New Brunswick, New Jersey

Karl Nath
Division of Nephrology
Mayo Clinic/Foundation
Rochester, Minnesota

Toru Takahashi
Department of Anesthesiology and
    Resuscitology
Okayama University Medical School,
    Okayama, Japan

Masahisa Hirakawa
Department of Anesthesiology and
Resuscitology, Okayama University
    Medical School, Okayama, Japan

Shigeru Sassa
Laboratory of Biochemical Hematology
The Rockefeller University Hospital
New York, New York

Maria Alfonsina Desiderio
Istituto di Patologia Generale-
Università degli Studi di Milano, e
    Centro di
Studio sulla Patologia Cellulare del
    C.N.R., Milano-Italy

Lorenza Tacchini
Istituto di Patologia Generale-
    Università degli Studi di Milano, e
    Centro di Studio sulla Patologia
    Cellulare del C.N.R., Milano-Italy

Gerben J. Schaaf
Department of Veterinary
  Pharmacology, Pharmacy and
  Toxicology (VFFT), Faculty of
  Veterinary Medicine, Utrecht
  University, Utrecht, The Netherlands.

Johanna Fink-Gremmels
Department of Veterinary
  Pharmacology,
Pharmacy and Toxicology (VFFT),
  Faculty of Veterinary Medicine,
  Utrecht University, Utrecht, The
  Netherlands.

Elba Vazquez
Centro de Investigaciones sobre
  Porfirinas
y Porfirias (CIPYP) (CONICET-
  Facultad de
Ciencias Exactas y Naturales, UBA),
  Ciudad Universitaria
Buenos Aires, Argentina

Maria Lujan Tomaro
Departamento de Quimica Biologica,
  Facultad de Farmacia y Bioquimica,
  UBA, Buenos Aires, Argentina

Alcira Batlle
Centro de Investigaciones sobre
  Porfirinas
y Porfirias (CIPYP) (CONICET-
  Facultad de
Ciencias Exactas y Naturales, UBA),
  Ciudad Universitaria
Buenos Aires, Argentina

Kenneth Maiese
Division of Cellular and Molecular
  Cerebral
Ischemia
Wayne State University School of
  Medicine
Detroit, Michigan

Zhao Zhong Chong
Division of Cellular and Molecular
  Cerebral Ischemia
Wayne State University School of
  Medicine
Detroit, Michigan

Martin H. Deininger
Institute of Brain Research
University of Tuebingen, Medical
  School
Tuebingen, Germany

Tobias Polte
University of Bath, Bath, UK

Henning Schroder
School of Pharmacy
Martin Luther University
Halle (Saale), Germany

Florence Favatier
Laboratoire de Physiologie Respiratoire
UFR Cochin Port-Royal
Paris, France

Barbara S. Polla
Laboratoire de Physiologie Respiratoire
UFR Cochin Port-Royal
Paris, France

Boon-Seng Wong
Institute of Pathology
Case Western Reserve University
Cleveland, Ohio

David R. Brown
Institute of Pathology
Case Western Reserve University
Cleveland, Ohio

Alfredo Vannacci
Department of Preclinical and Clinical
  Pharmacology
University of Florence, Florence Italy

Pier Francesco Mannaioni
Department of Preclinical and Clinical
   Pharmacology
University of Florence, Florence Italy

Liming Yang
Department of Pharmacology
New York Medical College
Valhalla, NY

Shuo Quan
Department of Pharmacology
New York Medical College
Valhalla, NY

Shu-Hui Juan
Institute of Biomedical Sciences
Academia Sinica, Taipei
Taiwan, ROC

Lee-Young Chau
Institute of Biomedical Sciences
Academia Sinica, Taipei
Taiwan, ROC

Xiao-Ming Liu
VA Medical Center
Houston, Texas

William Durante
Departments of Medicine and
   Pharmacology

Baylor College of Medicine
Houston, Texas

Michael Dunn
Department of Ophthalmology
New York Medical College
Valhalla, NY

Michal Laniado-Schwartzman
Department of Pharmacology
New York Medical College
Valhalla, NY

Maivel H. Ghattas
Department of Pharmacology
New York Medical College
Valhalla, NY

Peter Hewett
Department of Reproductive and
   Vascular Biology
The Medical School
University of Birmingham,
Edgbaston, Birmingham, UK

Asif Ahmed
Department of Reproductive and
   Vascular Biology The Medical School
University of Birmingham, Edgbaston,
   Birmingham, UK

PREFACE

Heme oxygenase is rapidly taking its place as the centerpiece of multiple interacting metabolic systems. Only 25 years ago heme oxygenase and its metabolic products appeared to be merely a simple metabolic system—one substrate, heme; one enzyme, heme oxygenase; and one set of products, iron to be recycled, and bilirubin and carbon monoxide to be disposed. From a group of about 25 people in 1974, as judged by attendance at various Gordon conferences, heme oxygenase has, in the year 2000, attracted working scientists—and clinicians I might add—by the hundreds and has produced referenced publications by the thousands. It is well-deserved attention. Heme oxygenase system is now similar to the metabolic networks surrounding glucose in those complex maps of glycolytic and non-glycolytic metabolic pathways, which we had to memorize as students.

The relevance of heme oxygenase to regulatory biology was recognized many years ago, but the work conducted over the past five years has created a new wave of emphasis focusing on genetic manipulation to alter heme oxygenase gene expression, the regulatory actions of heme oxygenase products including carbon monoxide, and the significance of changes in the heme oxygenase system. The physiological and pathological relevance of heme oxygenase in the brain, heart, liver, bone marrow, organ transplant, lung and kidney, opens many areas of investigation in various disciplines. Advances in the pharmacology of bilirubin and its ability as an antioxidant have provided a new avenue in clinical research. The altered function and cellular levels of bilirubin and carbon monoxide have been related to normal cell cycle progression in many organs, including those affected by cardiovascular and neurological disease. The resurgence of interest in the heme oxygenase system and its products, carbon monoxide and bilirubin, has led to the development of several clinical trials in the neonatal jaundice and cardiovascular disease.

This book is the product of collaboration by many respected scientists around the world. It presents concise up-to-date reviews that describe current developments regarding the biochemical properties of heme oxygenase, the regulation of its activity by endogenous and exogenous factors, and its involvement in pathological processes and the molecular biological aspects of the control mechanism for regulating heme oxygenase activity. For the first time, we are describing the significance of the heme oxygenase system and its products in clinical medicine and biology. The chapters include several important discoveries introduced at the First International Symposium on Heme Oxygenase, which was held in New York City at the World Trade Center, Marriott Hotel, on July 17–20, 2000. In addition, a few important papers that were not presented at the meeting have been included, because they highlight the role of HO/CO in multidiscipline areas. We offer our sincerest apologies for not being able to include papers submitted after the publication deadline.

Many thanks to all the contributors and to the Scientific Committee and the abstract reviewers for their help and support. Finally, we thank The Honorable Mayor of New York City, Rudolph W. Guiliani, who honored us by presenting the Welcome Address and for making New York City the ideal environment for holding this meeting.

Nader G. Abraham
Meeting Chairman

TABLE OF CONTENTS

SECTION II.   PHYSIOLOGICAL FUNCTION OF HEME OXYGENASE AND THE CENTRAL NERVOUS SYSTEM

SECTION III.   CLINICAL IMPLICATIONS OF HEME OXYGENASE SYSTEM IN INFLAMMATION

SECTION IV.   HEME OXYGENASE AND CARDIOVASCULAR SYSTEM

SECTION V.   HEME OXYGENASE SYSTEM AND OXIDATIVE STRESS RESPONSE

SECTION VI.   THE NETWORK OF HEME OXYGENASE AND PROGRAM CELL GROWTH AND DEATH

PHYSIOLOGY/PATHOLOGY OF HEME OXYGENASE AND ITS PRODUCTS, CARBON MONOXIDE AND BILIRUBIN

DEVELOPMENT OF HEME OXYGENASE INHIBITORS FOR THE PREVENTION OF SEVERE JAUNDICE IN INFANTS

Studies from Laboratory Bench to Newborn Nursery*

Attallah Kappas

Laboratory of Pharmacology
The Rockefeller University Hospital
New York, New York

INTRODUCTION

"Study of the biological properties of synthetic metalloporphyrins represents a potentially fruitful area of research and the results may have significant value for basic as well as clinical disciplines." This statement, made in a Perspectives article published in the Journal of Clinical Investigation in 1986,[1] presaged the sustained effort my laboratory group would undertake to bring to clinical application, for the prevention of severe newborn jaundice, the findings of our studies on the regulatory actions of synthetic heme analogues on the activity and expression of heme oxygenase, the rate-limiting enzyme in the catabolism of heme to bilirubin.[2-9]

The goal of this research program has been achieved. The success of this effort represents an important demonstration of the potential ways in which knowledge gained from laboratory studies on the regulation of heme metabolism can lead to practical applications in patients.

---

* This report was presented as The First Annual Lang Research Lecture, The New York Hospital Medical Center, Queens-Weill Cornell Medical College, New York City.

The clinical problem of severe newborn jaundice results from the disparity between the rate of bilirubin production and the rate of bilirubin disposal in newborns during the immediate post-natal period.[10] The magnitude of this problem is reflected in the large number of babies (several hundred thousand out of a birth-rate of nearly 4,000,000 annually in the United States), treated for hyperbilirubinemia to pre-empt the possible development of brain damage caused by bilirubin. The costs of this therapeutic effort are substantial and the difficulties of predicting which infants will be at risk from progressive jaundice are considerable.[11]

Bilirubin-induced encephalopathy (i.e., "kernicterus") is increasing in frequency, in part because of the current practice of discharging mother and baby from the hospital much earlier (24–48 hours after birth) than was the practice in the past. Further, the failure to regularly measure plasma bilirubin levels in babies before hospital discharge and the fact that, generally, plasma bilirubin levels peak at about 96 hours after birth, results in infants being denied close medical supervision during a period of potential risk for them.

When hyperbilirubinemia is recognized as threatening to newborns, two principal treatment methods are available to manage the problem—phototherapy and exchange transfusion. Both methods are effective in lowering plasma bilirubin levels, but have drawbacks; moreover, both suffer the disadvantage that they are based on attempts to dispose of bilirubin after this potential neurotoxin has already been formed and reached dangerous levels in the bloodstream of newborns. Decisions for treatment are made at this point with some medical urgency and carry with them a significant element of subjectivity.

It seemed to us more logical to address this problem at an earlier stage in the clinical course of the infant by attempting to interdict the production of bilirubin before its level in the bloodstream becomes threatening. To this end we decided to embark on a major and sustained effort to develop inhibitors of heme oxygenase which would have a pharmacological and toxicological profile permitting their clinical use in human newborns for the prevention of severe hyperbilirubinemia thus eliminating, for practical purposes, concern about the problem of bilirubin-induced brain damage.

From the demonstration, first made in this laboratory,[2] that an inhibitor of heme oxygenase could significantly decrease plasma bilirubin levels in newborn animals this research effort progressed steadily, ultimately leading us to the conduct of extensive controlled, clinical trials involving the use of heme oxygenase inhibitors to interdict the development of severe jaundice in infants. The results of these studies[12–19] have established a sound basis for the adoption of a new, effective preventive approach to the management of newborn hyperbilirubinemia.

The chronology of this research program is outlined in the studies from this laboratory summarized below. These studies describe the translation, into practical application in patients, of basic knowledge concerning the biochemistry and regulation of heme oxygenase and the interactions of synthetic heme analogues with this enzyme; and they affirm the important clinical potential of developing pharmacologic, genetic and other methods for controlling heme metabolism in humans.

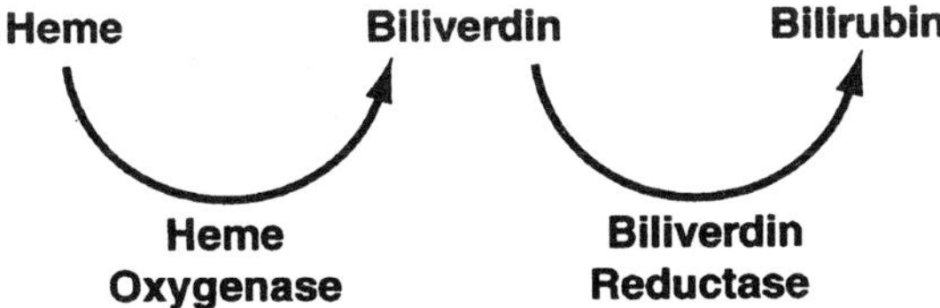

Figure 1.

METHODS

The catabolism of heme to bilirubin is mediated by the two enzymes shown in Fig. 1; heme oxygenase and biliverdin reductase. Heme oxygenase is rate-controlling in this process. Our goal was to develop synthetic analogues of heme which would competitively inhibit heme oxygenase, since blocking its activity would be the most direct way to favorably alter the balance between bilirubin production and bilirubin disposal after birth.

Many analogues of heme were synthesized and tested for their ability to act as competitive substrates for heme oxygenase *in vitro* and to inhibit bilirubin production and lower plasma bilirubin levels *in vivo* in animals and in humans. It should be emphasized that in such studies, interactions of synthetic metalloporphyrins with heme oxygenase demonstrated solely *in vitro*, without proof that they can lower plasma bilirubin levels in the whole animal,[20] is not a sufficient basis for concluding that such compounds would be effective in controlling hyperbilirubinemia in newborns. Studies with heme analogues such as the cobalt porphyrins which inhibit heme oxygenase activity *in vitro* but powerfully induce the enzyme *in vivo* and the dual effects of synthetic metalloporphyrins on the synthesis as well as the activity of heme oxygenase, which we earlier demonstrated,[3,6,21] make this point clear.

Our studies showed that certain analogues of heme in which the central iron atom is replaced by tin, such as stannic protoporphyrin (SnPP) and stannic mesoporphyrin (SnMP), are especially potent competitive inhibitors of heme oxygenase *in vitro* as well as *in vivo*.[4,22] Other analogues, such as the zinc and chromium porphyrins which also inhibit heme oxygenase *in vitro* display significant cellular or whole animal toxicities[23] and would not be suitable for use in newborns.

SnMP acts in the manner depicted in Fig. 2. The metalloporphyrin has an affinity for the catalytic site of heme oxygenase much greater than that of heme, the natural substrate of the enzyme. Heme is thus displaced from its binding site on the enzyme and the production of bilirubin rapidly decreases. SnMP cannot be degraded to bilirubin itself since its central metal does not bind the molecular oxygen required for this process. There is, in fact, no known physiological mechanism by which the tetrapyrrole ring of SnMP is cleaved with release of its contained metal and the formation of bile pigments. Unmetabolized heme is excreted from the liver into the biliary tract,[24] as is a fraction of the unaltered inhibitor.

The efficacy and safety of stannic heme analogues have been amply demonstrated in animals and in humans. In every experimental or naturally occurring form of jaundice in animals and in man listed in Table 1, heme oxygenase inhibition by

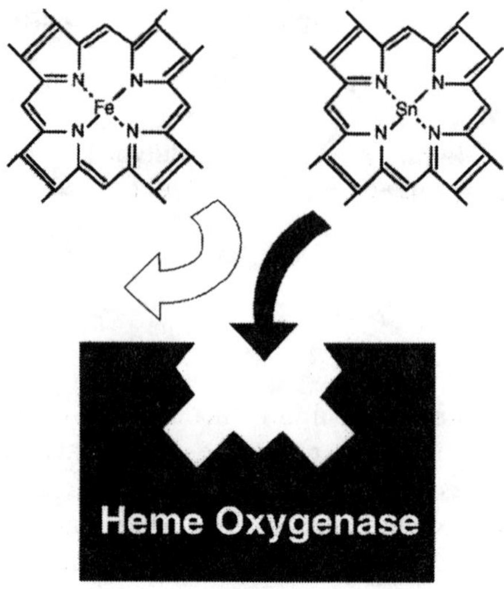

Figure 2.

SnPP or SnMP has significantly diminished bilirubin production and reduced plasma levels of bilirubin.

Early studies with SnPP in normal volunteers, in adult patients with liver disorders and in pre-term babies, were followed by a series of controlled clinical trials in newborns using SnMP which is presently the inhibitor of choice for human studies. The structural differences between the synthetic compound, SnMP, and natural heme are indicated by the arrows in the right panel of Fig. 3.

The ratio of efficacy to toxicity of SnMP is very high and the favorable

Table 1. Studies with Heme Oxygenase Inhibitors

Laboratory	Clinical
Jaundice in newborn rats	Normal subjects
Jaundice in newborn monkeys	Patients with primary biliary cirrhosis
Jaundice induced by the heme precursor, ALA	Patients with Gilbert's syndrome
Jaundice induced by bile duct ligation	Patients with hemochromatosis
Jaundice in mice associated with severe hereditary hemolytic anemia	Pre-term and neartenn newborns
Jaundice produced by the injection of heme or heat-damaged BBC	Term newborns with GGPD deficiency
Jaundice induced by starvation in squirrel monkeys	Term newborns with ABO incompatibility
Jaundice in EHBR/Eis rats (conjugated hyperbilirubinemia)	Term newborns with high bilirubin levels (15–18 mg/dL.)
Jaundice in Gunn rats (unconjugated hyperbilirubinemia)	Patients with the Crigler-Najjar Type 1 syndrome

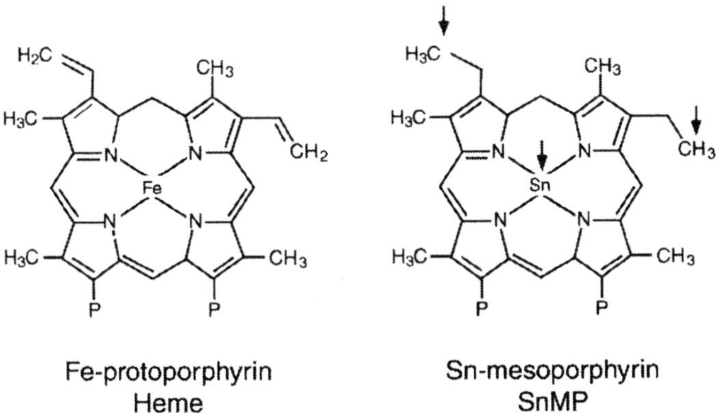

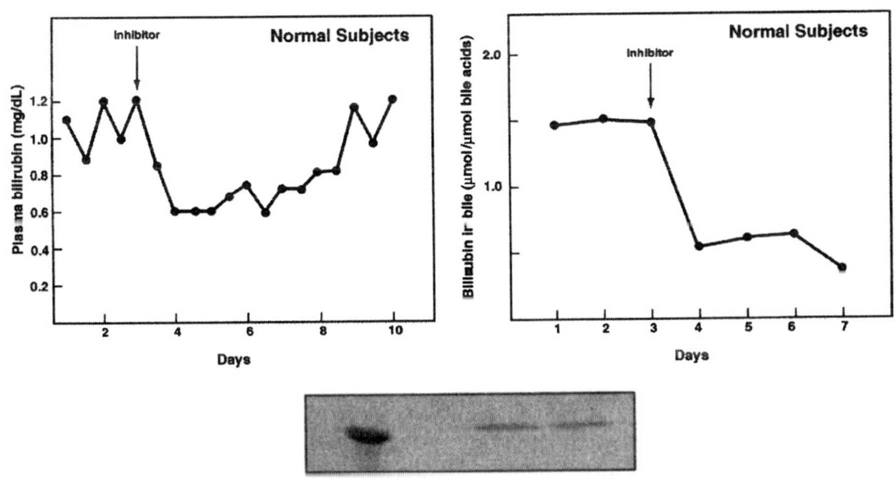

Fe-protoporphyrin
Heme

Sn-mesoporphyrin
SnMP

Figure 3.

results of extensive pharmacological and toxicological studies of the inhibitor led to FDA approval for its use in clinical trials in pre-term, near-term and term newborns.

RESULTS

The effect of a single, small dose of heme oxygenase inhibitor, administered intramuscularly to 10 normal volunteers,[25] is depicted in Fig. 4.

Figure 4.

Table 2. Newborns Treated with a Single Dose of SnMP and Phototherapy (PT) Requirement in Each Group

	Control		SnMP	
	n	PT	n	PT
Pre-term (210–250 days GA)	56	35	55	13
Near-term/Term	42	42*	44	0
Term with high (15–18 mg/dL) plasma bilirubin levels	86	19	80	0
ABO-incompatibility	37	15	39	0
G6PD-deficiency	58	18	225	0

*By study design newborns requiring PT were randomized to either the control group or the SnMP group.

The values shown are the averages for the group. Plasma bilirubin levels and bilirubin content of bile decreased promptly after the inhibitor was administered, as indicated. The chromatogram in the bottom panel shows a heme standard for comparison at the left; the absence of heme in normal bile is shown in the adjacent panel; and in the two panels to the right, the excretion of unmetabolized heme into bile after the administration of the inhibitor is demonstrated.

These findings indicate that a single dose of heme oxygenase inhibitor will significantly interdict even normal levels of bilirubin production and they confirm in humans our earlier finding in animals that heme which cannot be degraded by heme oxygenase is excreted unaltered from the liver into the bile.[24]

A summary of the results of all clinical trials we have conducted to-date with use of SnMP in a dose of 6 μmol/Kg birth weight in newborns,[26–31] is shown in Table 2.

In the series of controlled, clinical trials carried out using this dose of SnMP, a total of 443 newborns were administered the inhibitor to assess its effects on the development of newborn jaundice. There were 279 control infants in these trials of whom 129 ultimately required phototherapy to control progressive hyperbilirubinemia. Of the 443 infants receiving the single dose of SnMP, only 13 pre-term infants required a small amount of supplemental phototherapy to control hyperbilirubinemia.

The over-all findings from the trials with SnMP, at the dose indicated, confirmed the potent ability of this inhibitor to greatly diminish or entirely eliminate the need for phototherapy to control the development of severe hyperbilirubinemia in the infant populations studied. A brief summary of the results of specific clinical trials is presented below. In these trials, a decreased requirement for phototherapy is directly related to the moderating effect of SnMP on plasma bilirubin levels in the treated newborns.

The cumulative use of phototherapy to control developing hyperbilirubinemia in the dose-ranging study with SnMP in pre-term infants during the 9–10 days after birth is depicted in Fig. 5. This and other principal studies in newborns (except as noted) were carried out in the Metera Maternity Hospital, Athens, Greece. The patient population served by this institution is stable, homogeneous, and largely middle class. The institution allowed mother and baby to remain in the hospital for the follow-up periods indicated which was considered essential for these studies.

The administration of increasing doses of SnMP, from 1–6 μmol/Kg birth weight, to these babies exerted a progressively greater effect in lowering plasma bilirubin levels in treated infants compared with controls; this was reflected in the markedly diminished need for phototherapy (reduced by 76%) in babies receiving the highest dose of inhibitor. Since a plateauing of the SnMP effect was not fully achieved even at the highest dose studied, it is likely that use of a somewhat increased dose of inhibitor would have entirely eliminated the need for supplemental light treatment in the treated infants.

A comparison of SnMP versus phototherapy for control of newborn hyperbilirubinemia in term and near-term babies was also made. Newborns requiring phototherapy, based on age-related progressive hyperbilirubinemia, were randomized to a phototherapy group or to a group receiving a single dose of SnMP (6 μmol/Kg/birth weight) instead of light treatment. The results are shown in Fig. 6.

The newborns in the phototherapy group ultimately required an average of 40 hours of light per treated infant before the episode of hyperbilirubinemia subsided. None of the SnMP-treated infants required supplemental light treatment. Further, hyperbilirubinemia in the SnMP-treated newborns subsided more quickly than it did in light-treated babies; their requirement for follow-up plasma bilirubin determinations was significantly less than for babies receiving phototherapy; and there was no "rebound" hyperbilirubinemia in the SnMP-treated infants as was observed in a significant number of babies receiving light treatment. Thus a single, small dose of SnMP proved to be at least as effective as prolonged phototherapy in controlling hyperbilirubinemia in these newborns.

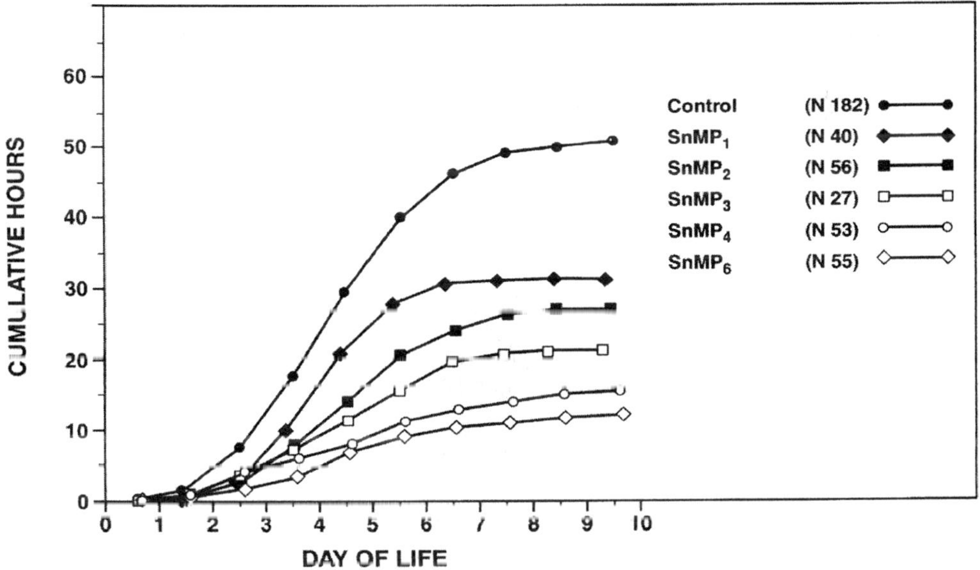

Figure 5. Effect of Sn-Mesoporphyrin on Cumulative Phototherapy Time.

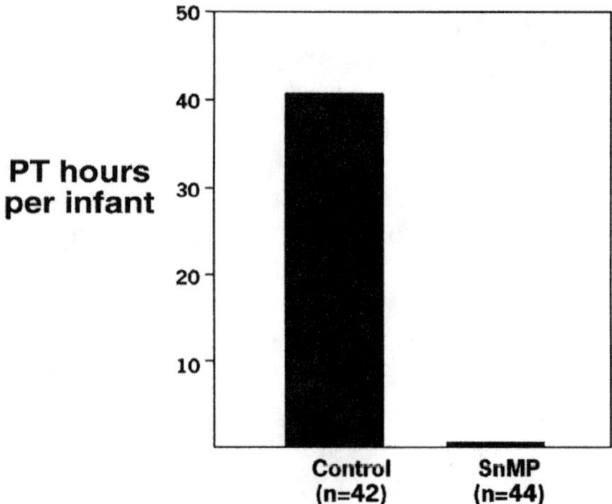

Figure 6. Phototherapy Requirements in Control and Sn-Mesoporphyrin-Treated Term and Near-term Infants.

Studies were also carried out (in Buenos Aires, Argentina) in infants whose plasma bilirubins had already reached a high risk level—that is, newborns whose bilirubin levels were in the range of 15–18 mg/dL. About one-half of the infants were treated with a single dose of SnMP; the remainder received light treatment, if and when the plasma bilirubin concentration reached the threshold level for phototherapy of 19.5 mg/dL or greater. The results are shown in Fig. 7.

None of the SnMP-treated infants in this study reached the threshold level for initiation of phototherapy. Nearly a quarter of the untreated infants did reach this level and required light treatment. Thus SnMP prevented progression of hyperbilirubinemia in all of the high-risk infants treated and entirely eliminated the need for use of phototherapy in them.

In a further study to examine the effectiveness of SnMP in preventing the development of severe hyperbilirubinemia in newborns, the inhibitor was administered as

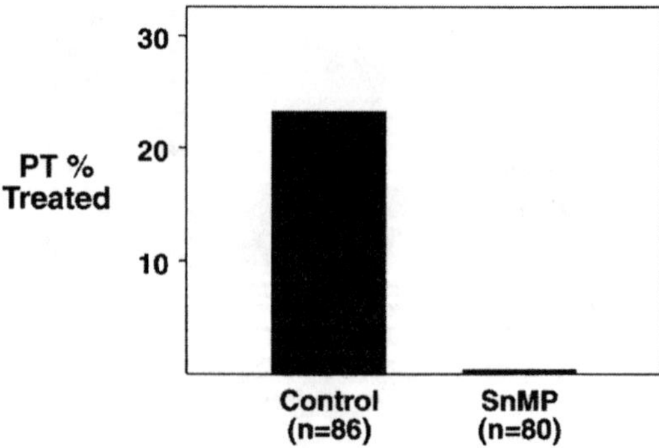

Figure 7. Phototherapy Requirements in Control and Sn-Mesoporphyrin-Treated Term Infants with High (15–18 mg/dL) Plasma Bilirubin Levels.

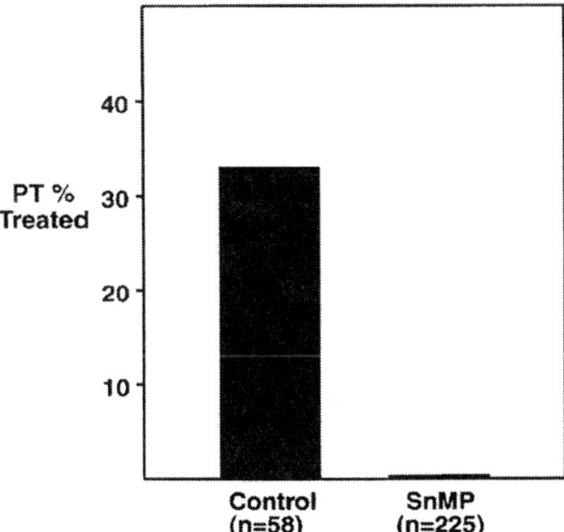

Figure 8. Phototherapy Requirements in Control and Sn-Mesoporphyrin Treated Infants with G6PD Deficiency.

a single dose (6 μmol/Kg birth weight) to 225 G6PD-deficient infants within about 24 hours of birth, after the diagnosis of the enzyme deficiency had been confirmed. The results are shown in Fig. 8.

None of the SnMP-treated infants required supplemental light treatment for hyperbilirubinemia. In the control group of newborns, more than 30% required phototherapy to control progressive jaundice.

The mean rate of increase of plasma bilirubin levels in the SnMP-treated group slowed significantly after administration of the inhibitor compared with the rate of increase in controls, as shown in Fig. 9.

This differential in incremental plasma bilirubin levels continued throughout the course of the study so that none of the babies in the SnMP group reached a plasma bilirubin level that would dictate use of phototherapy. Thus in this large population of G6PD-deficient newborns a single dose of SnMP prevented the development of severe hyperbilirubinemia and entirely eliminated the need for phototherapy in them.

We have also studied the use of SnMP in normal volunteers, in adults with liver diseases including hereditary hepatic porphyria[25,32,33] and in children with the genetic liver disease, Crigler-Najjar Type I syndrome.[34,35] In the latter disorder, bilirubin cannot be conjugated and plasma levels of the pigment may exceed 20–25 times normal. Bilirubin encephalopathy can develop at any time in affected children but especially so during episodes of stress (i.e., infection, trauma, etc.) when hyperbilirubinemia becomes even more severe. The possibility of moderating hyperbilirubinemia in these children on a long-term basis is under study. Figure 10 shows the response to six successive weekly injections of SnMP (3 μmol/Kg/body weight) in a 4 year old child with this syndrome.

This treatment schedule with SnMP substantially lowered the patient's plasma bilirubin levels for a sustained period of time. It is possible that periodic, preventive

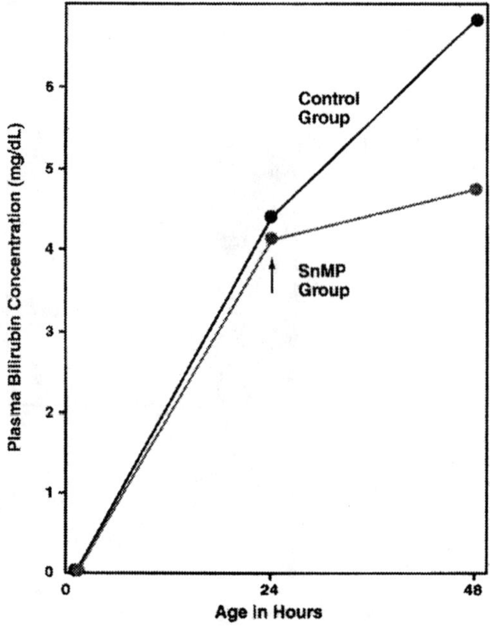

Figure 9. Incremental Change in Plasma Bilirubin Concentration in G6PD-Deficient Newborns.

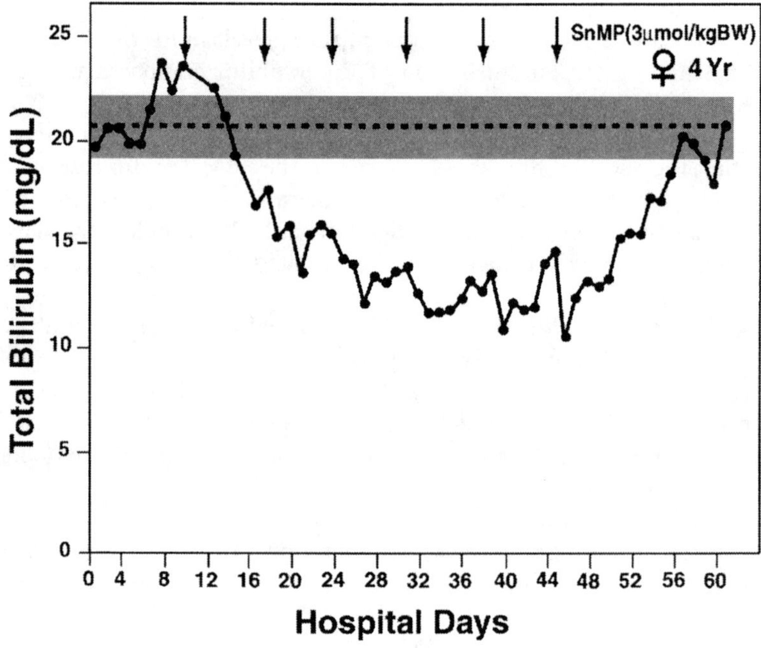

Figure 10.

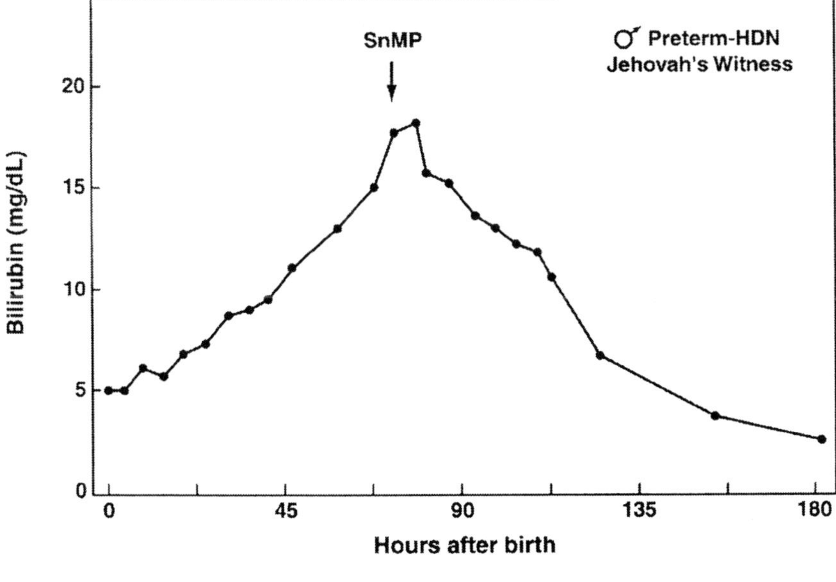

Figure 11.

use of heme oxygenase inhibition in this way can benefit these children by temporarily moderating the acute, severe exacerbations of hyperbilirubinemia which occur during stresses and which produce irreversible brain damage, or may prove lethal, in them.

Finally, heme oxygenase inhibition can also be used as an alternative to exchange transfusion in the management of severe hyperbilirubinemia which has proved unresponsive to phototherapy.[36] In the pre-term infant whose data are shown in Fig. 11, progressive, severe jaundice associated with hemolytic disease of the newborn (HDN) was unresponsive to 75 hours of intensive light treatment.

The parents of the baby were Jehovah's Witnesses and rejected, on religious grounds, the exchange transfusion considered necessary by their physician. Pending legal action to force the use of this procedure, a single dose of SnMP was administered to the baby at the time indicated by the arrow; the intensity of phototherapy was reduced simultaneously.

The effect of the inhibitor was prompt and dramatic as shown. The rate of increase in plasma bilirubin levels rapidly subsided and was followed by a steady regression of hyperbilirubinemia. Exchange transfusion proved unnecessary. We have had three similar experiences with use of the inhibitor in this population of newborns. The administration of a single dose of SnMP to interdict progression of hyperbilirubinemia in all three babies resolved the moral, medical and legal dilemmas faced by physicians and parents in these unusual clinical circumstances and demonstrated that SnMP can serve as an effective alternative to exchange transfusion in the management of severe newborn jaundice.

DISCUSSION

Our goal in this research effort was to develop a method, based on the principle of competitive substrate inhibition of heme oxygenase, for interdicting the production of bilirubin in newborns before blood levels of this potential neurotoxin become threatening. Such a method would eliminate the need for cumbersome and costly therapeutic procedures such as light treatment and exchange transfusion to control hyperbilirubinemia and minimize the subjectivity involved in deciding when and for how long to use these treatment modalities in individual clinical situations. Further, a simple, safe and effective pharmacologic method for blocking bilirubin production would make available to economically deprived communities a means for controlling severe jaundice in newborns who might otherwise not have available to them other medical resources for the management of this problem.

SnMP achieves this goal. This synthetic heme analogue is a potent and rapidly-acting competitive inhibitor of heme oxygenase, the enzyme which controls the rate-limiting step in the production of bilirubin. The inhibitor has a highly favorable pharmacological and toxicological profile; follow-up of treated infants for up to 5 years has shown no detectable long-term side effects of SnMP treatment. A single, small dose of the inhibitor is sufficient to protect newborns considered vulnerable to brain damage as a consequence of severe hyperbilirubinemia and effectively substitutes for phototherapy or exchange transfusion in managing this problem. Finally, SnMP can be used for either preventive or therapeutic purposes as the physician chooses and the clinical circumstances warrant, thus offering a high degree of flexibility in the management of progressive hyperbilirubinemia in the newborn.

SUMMARY

These studies constitute the first successful application of heme oxygenase inhibitors for the control of a prevalent and serious medical problem and they lay the basis for a new, preventive approach to the management of progressive, severe hyperbilirubinemia in the newborn. In a broader sense, they represent the effective translation into the clinical arena of basic biochemical knowledge relating to the regulation of heme metabolism by pharmacologic means. It can be predicted that there will be more examples of this process in the future.

The attention now being focused on heme oxygenase is well-deserved. This enzyme is rapidly taking its place as the centerpiece of multiple, interacting metabolic systems having major biomedical significance.[37] The potential importance of developing means for regulating this enzyme at the clinical level is becoming increasingly evident. Methods now being devised for the control of heme oxygenase such as induction, inhibition and genetic techniques,[38–41] are likely to provide physicians access to a new therapeutic armamentarium for regulating those aspects of human physiology and pathology which are influenced by, or dependent upon, heme and its metabolic products.

ACKNOWLEDGMENTS

I thank the many younger colleagues in this laboratory group, especially Dr. George S. Drummond, who have made important contributions to the work reported here. Clinical trials with SnMP in newborns were carried out in collaboration with Dr. Timos Valaes in Athens and Dr. Jorge Martinez in Buenos Aires. Other clinical studies were conducted with Dr. Bo Angelin and Dr. Lars Berglund, Stockholm; and Dr. Maria Vogiatzi and Dr. Maria New in New York. Dr. Nader Abraham has provided valuable counsel in our ongoing studies of heme oxygenase.

The author is indebted to the National Institute of Child Health and Human Development (contract N01-HD53234) and to the Ablon and Renfield Foundations for sustained support of this research. This research was also supported in part by a General Clinical Research Center grant (M01RR00102) to The Rockefeller University and a Childrens' Clinical Research Center Grant (M01RR06020) to the Department of Pediatrics, Weill-Cornell Medical College.

REFERENCES*

1. Kappas, A. and Drummond, G.S.: Control of heme metabolism with synthetic metalloporphyrins. *Journal of Clinical Investigation, 77*:335–339, 1986.
2. Drummond, G.S. and Kappas, A.: Prevention of neonatal hyperbilirubinemia by Sn-protoporphyrin IX, a potent competitive inhibitor of heme oxidation. *Proceedings of the National Academy of Sciences, 78*:6466–6470, 1981.
3. Yoshinaga, T., Sassa, S., and Kappas, A.: Purification and properties of bovine spleen heme oxygenase: Amino acid composition and sites of action of inhibitors of heme oxidation. *Journal of Biological Chemistry, 257*:7778–7785, 1982.
4. Drummond, G.S. and Kappas, A.: Chemoprevention of neonatal jaundice: Potency of tin-protoporphyrin in an animal model. *Science, 217*:1250–1252, 1982.
5. Kappas, A. and Drummond, G.S.: Control of heme and cytochrome P-450 metabolism by inorganic metals, organometals and synthetic metalloporphyrins. *Environmental Health Perspectives, 57*:301–306, 1984.
6. Sardana, M.K. and Kappas, A.: Dual control mechanism for heme oxygenase; Sn-protoporphyrin potently inhibits the enzyme activity while markedly increasing content of the enzyme protein in liver. *Proceedings of the National Academy of Sciences USA, 84*:2464–2468, 1987.
7. Kappas, A. and Drummond, G.S.: Synthetic metalloporphyrins: A class of compounds of pharmacological interest. *BioEssays, 3*:256–259, 1985.
8. Simionatto, C.S., Anderson, K.E., Drummond, G.S., and Kappas, A.: Studies on the mechanism of Sn-protoporphyrin suppression of hyperbilirubinemia: Inhibition of heme oxidation and bilirubin production. *Journal of Clinical Investigation, 75*:513–521, 1985.
9. Landaw, S.A., Sassa, S., Drummond, G.S., and Kappas, A.: Proof that Sn-protoporphyrin inhibits the enzymatic catabolism of heme in vivo: Suppression of ^{14}CO generation from radiolabeled endogenous and exogenous heme sources. *Journal of Experimental Medicine, 165*:1195–1200, 1987.
10. Dennery, P.A., Seidman, D.S., and Stevenson, D.K.: Neonatal hyperbilirubinemia. *New England Journal of Medicine, 344*:581–590, 2001.
11. Valaes, T.: Problems with prediction of neonatal hyperbilirubinemia. *Pediatrics, 108*:1–3, 2001.

---

* References are limited principally to selected studies from this laboratory which reflect the sequential development and clinical application of inhibitors of heme oxygenase.

12. Drummond, G.S. and Kappas, A.: Suppression of hyperbilirubinemia in the rat neonate by chromium-protoporphyrin: Interactions of metalloporphyrins with microsomal heme oxygenase of human spleen. *Journal of Experimental Medicine, 156*:1878–1883, 1982.

13. Sassa, S., Drummond, G.S., Bernstein, S.E., and Kappas, A.: Tin-protoporphyrin suppression of hyperbilirubinemia in mutant mice with severe hemolytic anemia. *Blood, 61*:1011–1013, 1983.

14. Drummond, G.S. and Kappas, A.: An experimental model of postnatal jaundice in the suckling rate: Suppression of induced hyperbilirubinemia by Sn-protoporphyrin. *Journal of Clinical Investigation, 74*:142–149, 1984.

15. Kappas, A., Drummond, G.S., Simionatto, C.S., and Anderson, K.E.: Control of heme oxygenase and plasma levels of bilirubin by a synthetic heme analogue, tin-protoporphyrin. *Hepatology, 4*:336–341, 1984.

16. Sassa, S., Drummond, G.S., Bernstein, S.E., and Kappas, A.: Long-term administration of massive doses of Sn-protoporphyrin in anemic mutant mice (sph^{ha}/sph^{ha}). *Journal of Experimental Medicne, 162*:864–876, 1985.

17. Drummond, G.S. and Kappas, A.: Sn-protoporphyrin inhibition of fetal and neonatal brain heme oxygenase: Transplacental passage of the metalloporphyrin and prenatal suppression of hyperbilirubinemia in the newborn animal. *Journal of Clinical Investigation, 77*:971–976, 1986.

18. Sisson, T.P.C., Drummond, G.S., Samonte, D., Calabia, R., and Kappas, A.: Sn-protoporphyrin blocks the increase in serum bilirubin levels which develops postnatally in homozygous Gunn rats. *Journal of Experimental Medicine, 167*:1247–1252, 1988.

19. Wissel, P., Galbraith, R.A., Sassa, S., and Kappas, A.: Tin-protoporphyrin inhibits heme oxygenase and prevents the decline in hepatic heme and cytochrome P450 contents produced in nude mice by tumor transplantation. *Biochemical and Biophysical Research Communications, 150*:822,827, 1988.

20. Maines, M.D.: Zinc-Protoporphyrin is a selective inhibitor of heme oxygenase activity in the neonatal rat. *Biochem. Biophys. Acta, 673*:339–350, 1981.

21. Drummond, G.S. and Kappas, A.: The cytochrome P450-depleted animal: An experimental model for in vivo studies in chemical biology. *Proceedings of the National Academy of Sciences, 79*:2384–2388, 1982.

22. Drummond, G.S., Galbraith, R.A., Sardana, M.K., and Kappas, A.: Reduction of the C_2 and C_4 vinyl groups of Sn-protoporphyrin to form Sn-mesoporphyrin markedly enhances the ability of the metalloporphyrin to inhibit in vivo heme catabolism. *Archives of Biochemistry and Biophysics, 255*:64–74, 1987.

23. Lutton, J.D., Abraham, N.G., Drummond, G.S., Levere, R.D., and Kappas, A.: Zinc porphyrins: Potent inhibitors of hematopoiesis in animal and human bone marrow. *Proceedings of National Academy of Science, 94*:1432–1436, 1997.

24. Kappas, A., Simionatto, C.S., Drummond, G.S., Sassa, S., and Anderson, K.E.: The liver excretes large amounts of heme into bile when heme oxygenase is inhibited competitively by Sn-protoporphyrin. *Proceedings of the National Academy of Sciences, 82*:896–900, 1985.

25. Berglund, L., Angelin, B., Blomstrand, R., Drummond, G.S., and Kappas, A.: Sn-protoporphyrin lowers serum bilirubin levels, decreases biliary bilirubin output, enhances biliary heme excretion and potently inhibits microsomal heme oxygenase activity in normal human subjects. *Hepatology, 8*:625–631, 1988.

26. Kappas, A., Drummond, G.S., Manola, T., Petmezaki, S., and Valaes, T.: The use of Sn-protoporphyrin in the management of hyperbilirubinemia in newborn infants with direct Coombs positive ABO-incompatibility. *Pediatrics, 81*:485–497, 1988.

27. Valaes, T., Petmezaki, S., Henschke, C., Drummond, G.S., and Kappas, A.: Control of jaundice in preterm newborns by an inhibitor of bilirubin production; Studies with tin-mesoporporphyrin. *Pediatrics, 93*:1–11, 1994.

28. Kappas, A., Drummond, G.S., Henschke, C.I., and Valaes, T.: Direct comparison of Sn-mesoporphyrin, an inhibitor of bilirubin production, and phototherapy in controlling hyperbilirubinemia in term and near-term newborns. *Pediatrics, 95*:468–474, 1995.

29. Valaes, T., Drummond, G.S., and Kappas, A.: Control of hyperbilirubinemia in glucose-6-phosphate dehydrongenase deficient newborns using an inhibitor of bilirubin production, Sn-mesoporphyrin. *Pediatrics, 101*:1–7, 1998.

30. Martinez, J.C., Garcia, H.O., Otheguy, L., Drummond, G.S., and Kappas, A.: Control of severe hyperbilirubinemia in full-term newborns with the inhibitor of bilirubin production Sn-mesoporphyrin. *Pediatrics, 103*:1–5, 1999.

31. Kappas, A., Drummond, G.S., and Valaes, T.: A single dose of Sn-mesoporphyrin prevents development of significant hyperbilirubinemia in glucose-6-phosphate dehydrogenase-deficient newborns. *Pediatrics, 108*:25–30, 2001.

32. Berglund, L., Angelin, B., Hultcrantz, R., Einarsson, K., Emtestam, L., Drummond, G., and Kappas, A.: Studies with the haem oxygenase inhibitor Sn-protoporphyrin in patients with primary biliary cirrhosis and idiopathic haemochromatosis. *Gut, 31*:899–904, 1990.

33. Galbraith, R.A. and Kappas, A.: Pharmacokinetics of tin-mesoporphyrin in man and the effects of tin-chelated porphyrins on hyperexcretion of heme pathway precursors in patients with acute inducible porphyria. *Hepatology*, 9:882–888, 1989.

34. Galbraith, R.A., Drummond, G.S., and Kappas, A.: Suppression of bilirubin production in the Crigler-Najjar Typer I syndrome: Studies with the heme oxygenase inhibitor, tin-mesoporphyrin. *Pediatrics, 89*:175–182, 1992.

35. Kappas, A., Drummond, G.S., and Galbraith, R.A.: Prolonged clinical use of a heme oxygenase inhibitor: Hematological evidence for an inducible but reversible iron-deficiency state. *Pediatrics, 91*:537–539, 1993.

36. Kappas A., Drummond, G.S., Munson, D.P., and Marshall, J.R.: Sn-Mesoporphyrin interdiction of severe hyperbilirubinemia in Jehovah's Witness newborns as an alternative to exchange transfusion. *Pediatrics, 108*:1374–1377, 2001.

37. Abraham, N.G., Drummond, G.S., Lutton, J.D., and Kappas, A.: The biological signifance and physiological role of heme oxygenase. *Cell Physiology and Biochemistry*, 6:129–168, 1996.

38. Abraham, N.G., Lavrovsky, Y., Schwartzman, M.L., Stoltz, R.A., Levere, R.D., Gerritsen, E., Shibahara, S., and Kappas, A. Transfection of the human heme oxygenase gene into rabbit coronary microvessel endothelial cells: Protective effect against heme and hemoglobin toxicity. *Proceedings of the National Academy of Sciences*, 92:6798–6802, 1995.

39 Laniado-Schwartzman, M., Abraham, N.G., Conners, M., Dunn, M., Levere, R.D., Kappas, A.: Heme oxygenase induction with attenuation of experimentally-induced corneal inflammation. *Biochemical Pharmacology*, 53:1069–1075, 1997.

40. Quan, S., Yang, L., Abraham, N.G., and Kappas, A.: Regulation of human heme oxygenase, in endothelial cells by using sense and antisense retroviral constructs. *Proceedings of the National Academy of Sciences*, 98:12203–12208, 2001.

41. Abraham, N.G., Quan, S., Shenouda, S., Kappas, A.: Selective increase in human heme oxygenase-1 gene expression attenuates development of hypertension and increase body growth in spontaneously hypertensive rats. In *Carbon Monoxide and Cardiovascular Functions*, Wang, Rui (ed.), CRC Press, Boca Raton, FL, Chapter 13, 233–245, 2002.

2

CARBON MONOXIDE

A Potential Anti-Inflammatory Agent and Mediator of Lung Anti-Oxidant Defenses

Stefan W. Ryter, Leo E. Otterbein, Danielle Morse,
and Augustine M. K. Choi

The Department of Medicine
Division of Pulmonary
Allergy, and Critical Care Medicine
The University of Pittsburgh Medical Center
MUH628NW, 3459 Fifth Ave, Pittsburgh
PA, 15213

1. INTRODUCTION

Heme oxygenase [EC 1:14.99.3, heme, hydrogen donor: oxygen oxidoreductase, (α-methene hydroxylating, decyclizing)] catalyzes the rate-limiting step in the oxidative catabolism of heme. In a coupled reaction with NADPH: cytochrome p-450 reductase, HO generates equimolar carbon monoxide (CO), ferrous iron, and biliverdin IXα per heme oxidized. Cytosolic NAD(P)H: biliverdin reductase (BVR) [E.C. 1:3:1:24] subsequently converts biliverdin-IXα to bilirubin-IXα (Tenhunen et al., 1968; Tenhunen et al., 1969). In addition to HO 1, the inducible form, two genet ically distinct isozymes exist (HO-2, and HO-3). Many mammalian tissues constitutively express HO-2 including the brain, testes, and vascular endothelium. HO-1 and HO-2 catalyze an identical reaction with different kinetic parameters, whereas HO-3 displays little enzymatic activity (Maines et al., 1986; McCoubrey et al., 1997). Taken together, HO enzymes serve an essential physiological function in regulating intracellular heme turnover (Maines, 1992).

The transcriptional up-regulation of the HO-1 gene follows cellular exposure to agents that generate intracellular reactive oxygen species (ROS), including hydrogen peroxide (H_2O_2), ultraviolet-A (UVA: 320–380 nm) radiation (Keyse and Tyrrell, 1987;

Keyse and Tyrrell, 1989), as well as deviations in normal physiological oxygen (O_2) tension (ie., hyperoxia or hypoxia) (Lee et al., 1996; Lee et al., 1997). Furthermore, HO-1 activation responds to thiol (-SH)-reactive substances that complex and effectively deplete intracellular reduced glutathione (GSH), including such compounds as sodium *m*-arsenite, diethylmaleate (DEM), heavy metal salts, and nitric oxide (NO) (Keyse and Tyrrell, 1989; Saunders et al., 1991; Maines and Kappas, 1977; Foresti et al., 1997; Hartsfield et al., 1997). The prior chemical depletion of intracellular GSH enhances the activation of HO-1 by oxidants (Lautier et al., 1992). Intracellular chelatable iron levels also influence HO-1 gene expression in pro-oxidative or hypoxic states (Ryter et al., 2000; Panchenko et al., 2000; Fogg et al., 1999; Keyse et al., 1990).

HO-1 elevation may occur under pathophysiological conditions associated with increased ROS production, including ischemia/reperfusion injury, inflammation and sepsis (Sharma et al., 1996; Rizzardini et al., 1993; Rizzardini et al., 1994). HO-1 responds to cell stimulation with pro-inflammatory mediators, including cytokines (interleukin 1: IL-1, interleukin-6: IL-6, tumor necrosis factor alpha: TNFα) (Rizzardini et al., 1993), bacterial endotoxins (lipopolysaccharide: LPS) (Camhi et al., 1995), and tumor promoters (12-*O*-tetradecanoylphorbol-13-acetate) (Kageyama et al., 1988).

A large body of evidence has accumulated *in vivo* and *in vitro* that HO-1 participates in cellular and systemic defenses against oxidative stress, which includes possible anti-apoptotic and anti-inflammatory functions. Among numerous examples, the adenoviral mediated gene transfer of HO-1 into rat lungs protected against the development of lung apoptosis and inflammation during hyperoxia (Otterbein et al., 1999). Furthermore, transgenic mice overexpressing HO-1 specifically in the lung, displayed resistance to the inflammatory and hypertensive effects of hypoxia (Minamino et al., 2001). Conversely, transgenic HO-1 (–/–) mouse embryo fibroblasts displayed hypersensitivity to the toxic effects of hemin and H_2O_2 and generated increased intracellular ROS production in response to these agents (Poss and Tonegawa, 1997). HO-1 (–/–) mice developed right ventricular dilation and right myocardial infarction, during chronic hypoxia (10% O_2), relative to wild type mice that sustained the treatment (Yet et al., 1999). In earlier *in vitro* studies, the overexpression of HO-1 in endothelial cells conferred protection against heme and hemoglobin mediated toxicity (Abraham et al., 1995). The *in vitro* application of HO-1 antisense oligonucleotides blocked the UVA-stimulated cytoprotection against subsequent lethal UVA challenge in human skin fibroblasts (Vile et al., 1994).

The underlying mechanism(s) of HO-mediated protection remain incompletely understood, but may pertain to the biological activities of the heme metabolites (reviewed in Ryter and Tyrrell, 2000; Otterbein and Choi, 2000; Choi and Alam, 1996). HO fulfills a theoretical anti-oxidant function by removing heme, whose intercellular accumulation may increase cellular pro-oxidant burden (Keyse and Tyrrell, 1989). The reactive iron released from heme by HO activity may stimulate the expression of the iron sequestration protein ferritin, promoting a secondary cellular desensitization to oxidative stress (Vile et al., 1993; Vile et al., 1994). The potent *in vitro* antioxidant properties of the bile pigments, biliverdin-IXα, and bilirubin-IXα, suggest a possible anti-oxidative function for HO activity (Stocker et al., 1987). The oxidation of the α-methene bridge carbon of heme by HO, releases carbon monoxide (CO), a heme ligand. CO has demonstrated effects on vascular function, including the regulation of vessel tone, platelet aggregation, and smooth muscle proliferation (Durante and Schafer, 1998; Morita and Kourembanas 1995; Morita et al., 1995; Morita et al., 1997). Studies of HO-1 localization in the brain have also implied CO as a potential neurotransmitter (Verma et al., 1993). These previously described roles of CO have been associated with the direct activation of soluble guanylate cyclase (sGC) by binding to its heme iron, stimulating the production of guanosine 3′, 5′-cyclic monophosphate (cGMP), a second messenger molecule (Verma et al., 1993; Morita et al., 1995; Morita and Kourembanas 1995, Morita *et al.* 1997; Maines, 1997).

Recent studies from this laboratory, have implicated CO in the modulation of mitogen activated protein kinase (MAPK)-signaling cascades (Otterbein et al., 2000, Otterbein et al., 2001, *in press*). These effects of CO on MAPK apparently bypass classical small gas signal transduction pathways involving sGC activation and cGMP production; however the proximal target (ie., CO receptor) in this case remains unknown. This chapter will highlight recent work that demonstrates novel roles for HO derived CO as an anti-inflammatory mediator, and also as a possible underlying mechanism for HO-1 mediated tissue protection in oxidative lung injury.

2. HEME OXYGENASE-DERIVED CARBON MONOXIDE (CO): AN ANTI-INFLAMMATORY EFFECTOR

2.1. Carbon Monoxide Inhibits Pro-Inflammatory Cytokine Production *In Vitro* and *In Vivo* via a cGMP and Nitric Oxide-Independent Pathway

This laboratory has recently discovered a novel anti-inflammatory effect of HO-1 mediated by carbon monoxide (CO) generated in the HO reaction (Otterbein et al., 2000). A mouse macrophage cell line (RAW 264.7) overexpressing HO-1 protein served as an *in vitro* model to test the effect of HO-1 on the inflammatory response. In two independent HO-1 positive clones, the ability of bacterial lipopolysacharide (LPS) to stimulate the production of the pro-inflammatory cytokine tumor necrosis factor-alpha (TNFα), was markedly diminished compared to that in control (*Neo*) transfected cells not expressing HO-1. To test the hypothesis that CO, a reaction product of HO activity, contributed to the anti-inflammatory effect of HO-1 overexpression, untransfected RAW 264.7 cells were assayed for LPS inducible TNFα production in the absence or presence of CO (250 ppm). Indeed, exogenously administered CO inhibited the production of TNFα in the media of RAW 264.7 cells after LPS treatment (1 μg/ml), independently of HO-1 overexpression. CO also inhibited LPS-inducible TNFα protein levels (by Western analysis), but did not affect LPS-inducible TNFα mRNA levels. These results indicated that, in macrophages, CO inhibits the post-transcriptional expression of TNFα.

Exposure of RAW 264.7 cells to exogenous CO (250 ppm) also inhibited the expression of other pro-inflammatory cytokines including IL-1β, and the macrophage inflammatory protein-β (MIP-1β). Conversely, CO in the same concentration range stimulated the production of an anti-inflammatory cytokine interleukin-10 (IL-10).

The effects of CO on cytokine expression were also observed in an *in vivo* model of inflammation. Mice received injections of LPS (1 mg/kg) with or without prior exposure (1 hr) in a CO enriched environment (10–500 ppm). CO pretreatment dose-dependently inhibited LPS-inducible serum TNFα levels (EC_{50} = 69.9 ppm). Conversely, CO (250 ppm) increased LPS-inducible IL-10 production. These observed effects of CO did not depend on hypoxia occurring secondary to the CO treatment, since exposure of mice to hypoxia alone (10% O_2) did not affect the ability of LPS to stimulate TNFα production.

CO exposure (250 ppm, 2 h) did not significantly modulate cGMP production in RAW 264.7 macrophages. In contrast, cultured smooth muscle cells increased their cGMP levels up to 16 fold in response to a similar CO treatment. Pretreatment of the macrophages with a non-hydrolysable cGMP analog 8-Bromo-cGMP did not affect LPS-inducible TNFα production. These results confirmed that the observed anti-inflammatory effects of CO in macrophages likely did not involve activation of the guanylyl cyclase-cGMP pathway. A possible role for endogenous NO generated secondary to the CO treatment was also excluded. Pretreatment with the inhibitor of nitic oxide synthase, N^G-Nitro-L-arginine-methyl ester (L-NAME) in combination with CO, did not compromise the ability of CO to inhibit LPS-inducible TNFα production in RAW 264.7 cells. Furthermore CO did not significantly modulate nitrate or nitrite levels in RAW 264.7 cells within an hour following LPS treatment. In conclusion, the anti-inflammatory effects of CO likely did not involve stimulation of iNOS activity or NO generation.

2.2. CO Exerts Anti-Inflammatory Effects by Modulating MAP Kinases

The exclusion of classical signaling pathways involving sGC/cGMP from the anti-inflammatory action of CO, led to the search for alternate mechanisms of action. The LPS mediated stimulation of pro-inflammatory cytokines in macrophages involves the activation of mitogen activated protein kinase (MAPK) signaling pathways (Chow et al., 1999; Hambelton et al., 1996; Han et al., 1994; Raingeaud et al., 1995). Experiments in this laboratory by Otterbein *et al.* (2000) confirmed that LPS treatment activated the p38, ERK1/ERK2 and JNK pathways in RAW 264.7 macrophages. CO treatment (250 ppm) significantly increased LPS inducible p38 MAPK activation, but had no effect on the LPS mediated stimulation of ERK1/ERK2 and JNK. The p38 MAPK may be activated in turn by three MAP kinase kinases (MKK): MKK3, MKK4, and MKK6 (Derijard et al., 1995; Raingeaud et al., 1996). Of these, CO enhanced the LPS-mediated stimulation of MKK3 and MKK6 in RAW 264.7 cells, relative to LPS treatment alone.

Since CO treatment selectively and positively affected the p38 kinase and its corresponding kinase kinases (MKK), but not other MAPK involved in the LPS mediated pro-inflammatory response, it was hypothesized that the anti-inflammatory effects of CO would be compromised in mice deficient in the p38 activation pathway. Mice with an MKK3 (–/–) genotype or matched wild type controls were treated with either LPS (1 mg/kg) alone, or LPS in combination with CO (250 ppm). The TNFα response to LPS treatment was downregulated, but not abolished, in MKK3 (–/–) mice compared to WT (+/+) mice that displayed a strong induction response. As described above, CO treatment diminished the LPS inducible TNFα response in WT (+/+) mice. However, CO failed to downregulate LPS-inducible TNFα levels in MKK3 (–/–) mice.

The addition of CO further increased LPS-inducible IL-10 levels in the WT (+/+) mice. However, CO failed to modulate serum IL-10 levels in LPS treated MKK3 (–/–) mice. In control studies, MKK3 (–/–) or WT (+/+) mice treated with CO or air in the absence of LPS did not modulate either TNFα or IL-10.

IL-10, an anti-inflammatory cytokine, may limit the expression of pro-inflammatory cytokines, such as TNFα (Howard et al., 1993). However, IL-10 induction was excluded from the possible mechanism by which CO inhibits LPS-inducible TNFα production. CO (250 ppm) inhibited TNFα levels observed within 1 hr of LPS treatment to a similar extent in either WT (–/–) mice or in IL 10 (–/–) mice completely deficient in IL-10.

These results taken together demonstrate that CO exerts anti-inflammatory effects by limiting the synthesis of pro-inflammatory cytokines. The mechanism underlying these altered cytokine profiles under inducing conditions depends on the CO mediated super-induction of the MKK3/p38 MAPK pathway. Although the classical sGC/cGMP pathway was excluded from the events leading to p38/MAPK activation, the direct target of CO (possibly also a hemoprotein) remains obscure.

3. CO PROTECTS THE LUNG IN A MODEL OF OXIDATIVE INJURY

Recent work from this laboratory demonstrates that CO, by virtue of anti-inflammatory effects, also protects the lung in a model of oxidative lung injury (Otterbein et al., 2001, *in press*). In this model, mice sustained exposure to an atmosphere of high O_2 partial pressure (hyperoxia, >95% O_2). Hyperoxia generates an oxidative stress in the lung, presumably by elevating mitochondrial ROS production relative to normoxia (Freeman and Crapo, 1981). Mice exposed to hyperoxia develop a condition similar to human acute respiratory distress syndrome (ARDS), displaying signs of lung injury by 64–72 h, and generally dying within 90–100 h of continuous exposure (Clark and Lambertson, 1971).

To test the effects of CO on oxidative lung injury, mice were exposed either to hyperoxia alone, or to hyperoxia in the presence of CO (250 ppm). The hyperoxia alone killed 100% of the mice between 90–100 h of exposure. However, at 95 h of exposure to hyperoxia in the presence of 250 ppm CO, 95% of the mice remained alive. Furthermore, in the presence of CO, 50% of the animals survived past 128 h of continuous hyperoxia.

The presence of CO prevented the manifestation of histological markers of tissue injury, with the lungs appearing microscopically normal after 84 h hyperoxia. In contrast, lungs from mice treated with hyperoxia alone displayed visible hemorrhage, edema, and fibrin deposition upon microscopic examination. Furthermore, hyperoxia alone (84 h) increased measurable markers of lung injury including pulmonary edema and protein accumulation in the airway (as estimated by wet/dry tissue ratio, and protein concentration in broncoalveolar lavage (BAL) fluid, respectively). The presence of CO throughout the treatment (250 ppm) improved lung injury parameters measured at 84 h of hyperoxia. CO also inhibited hyperoxia-inducible lipid peroxidation in the lung, a marker of oxidative damage. In control studies, neither air nor CO alone in the absence of hyperoxia affected these markers. Hyperoxia treatment (84 h) also triggered an influx of inflammatory neutrophils into the airways, which mediates in part, the manifestation of oxidative lung injury. Significantly elevated neutrophil content was detected in the BAL fluid of animals exposed to hyperoxia. On the other hand, animals exposed to a hyperoxia in combination with CO did not demonstrate significant neutrophil influx relative to air treated controls.

3.1. The Cytoprotective and Anti-Inflammatory Effects of CO in Hyperoxia Involve the MKK3/p38 MAP Kinase Pathway

Hyperoxia induced the expression of numerous proinflammatory cytokines including TNFα, IL-1β, and IL-6, in lung tissues by 84 h of exposure. Comparable to the inhibitory effects of CO described in an LPS-induced model of inflammation (*See section 2.*), CO also inhibited the expression of pro-inflammatory cytokines (TNFα, IL-1β and IL-6) in lung tissue following hyperoxia exposure. Hyperoxia treatment activated the major stress kinases in lung tissue including ERK1/2, JNK, P38/MAPK and MKK3/MKK6. To test the relative importance of these kinases in the hyperoxic shock response mice genetically deficient in stress kinase genes were employed, with the JNK (–/–) and MKK3(–/–) genotypes. Other genotypes of interest (ie., p38 (–/–), Erk1 (–/–)), Erk2 (–/–) were unavailable for study due to embryonic lethality. The MKK(–/–) mice displayed increased sensitivity to the lethal effects of hyperoxia, dying within the 65–72 h of exposure relative to 90–100 h for lethality in WT (+/+) mice. MKK3(–/–) exposed to hyperoxia also displayed the accelerated manifestation of tissue damage markers relative to WT (+/+) mice. Specifically, MKK3 (–/–) exposed to hyperoxia displayed similar BAL protein accumulation and histological damage at 60–65 h as evident in WT (+/+) mice at 80–95 h of hyperoxic exposure. Neutrophil influx in the lung, as measured in BAL, however did not appear in MKK3 (–/–) as an accelerated damage marker within 60–65 h of hyperoxic exposure.

CO treatment (250 ppm) protected against the lethal effects of hyperoxia (>95% O_2) in WT (+/+) mice, and afforded similar protection in JNK (–/–) mice, thus excluding a role for JNK in the underlying mechanism. CO, however, failed to confer protection or extend survival against hyperoxia in MKK3 (–/–) mice. The selective chemical inhibitor of the α and β isoforms of p38 (SB203580) was administered to mice (20 mg/kg) by intraperitoneal injection. CO (250 ppm) failed to significantly protect against hyperoxia in mice preinjected with SB203580, relative to control mice that did not receive the inhibitor. In the absence of CO, mice receiving injections of

SB203580 succumbed earlier to the lethal effects of hyperoxia (72–80 h) relative to mice that did not receive the inhibitor pretreatment (95–100 h). These experiments taken together, point to a critical role for the MKK3/p38 pathway in mediating the protective effects of CO against hyperoxic stress.

The CO-induced protection against hyperoxia in WT (+/+) mice correlated with the inhibited expression of the pro-inflammatory cytokines TNFα, IL-1β and IL-6. Interestingly, in this model, the CO attenuated the expression of the mRNA corresponding to these cytokines, in contrast to clearly post-transcriptional effects of CO on cytokine expression observed in *in vitro* with LPS-activated macrophages.

Cytokine mRNA (TNFα, IL-1β and IL-6) expression in response to hyperoxia appeared earlier in the MKK3 (–/–) mice when compared to the WT (+/+) exposed to 65 h continuous hypoxia. Finally, CO (250 ppm) failed to inhibit the expression of the pro-inflammatory cytokines in the MKK3 (–/–) mice.

In vitro experiments were performed in lung epithelial cells (A549) to further investigate these findings. The CO treatment (250 ppm) of A549 cells increased p38 and MKK3 activation, with a maxima at 16–24 h exposure. Immunoprecipitation experiments confirmed that CO selectively activates the β-isoform of p38. The presence of CO (250 ppm) increased the survival (51%) of cells grown in continuous hyperoxia ($>$95% O_2), relative to cells exposed to hyperoxia alone (20%). Finally treatment with the chemical inhibitor of p38 (SB203580) or transient transfection with dominant negative mutants of p38β or MKK3 abolished the cytoprotective effect of CO against hyperoxia. These experiments taken together demonstrate that CO protects against the lethal and inflammatory effects of hyperoxia by downregulating the expression of pro-inflammatory cytokines, through a mechanism dependent on the p38β/MKK3 pathway.

4. SUMMARY

The recent studies described in this chapter have shown that carbon monoxide downregulates the inflammatory response by limiting the expression of pro-inflammatory cytokines *in vitro*, and *in vivo*, in two models of inflammation stimulated by either endotoxin (LPS) treatment or hyperoxic shock. The observed protection afforded by CO against hyperoxia-induced oxidative lung injury, also related to downregulation of the inflammatory response. These studies have uncovered a novel target for CO, the modulation of the MKK3/p38 pathway, which mediates the CO-dependent downregulation of inflammatory mediators. These results suggest a possible physiological anti-inflammatory role for the heme oxygenase enzyme system (HO-1, HO-2), one of the principle known sources of endogenously occurring biological carbon monoxide. Furthermore, the potent anti-inflammatory properties of exogenous CO suggest medical applications for limiting the progression of inflammatory states, by either employing controversial inhalation therapy or heme oxygenase gene therapy approaches. Recently the anti-inflammatory properties of CO have been successfully applied to decrease xenograft rejection during experimental organ transplantation in rodents (Sato et al., 2001).

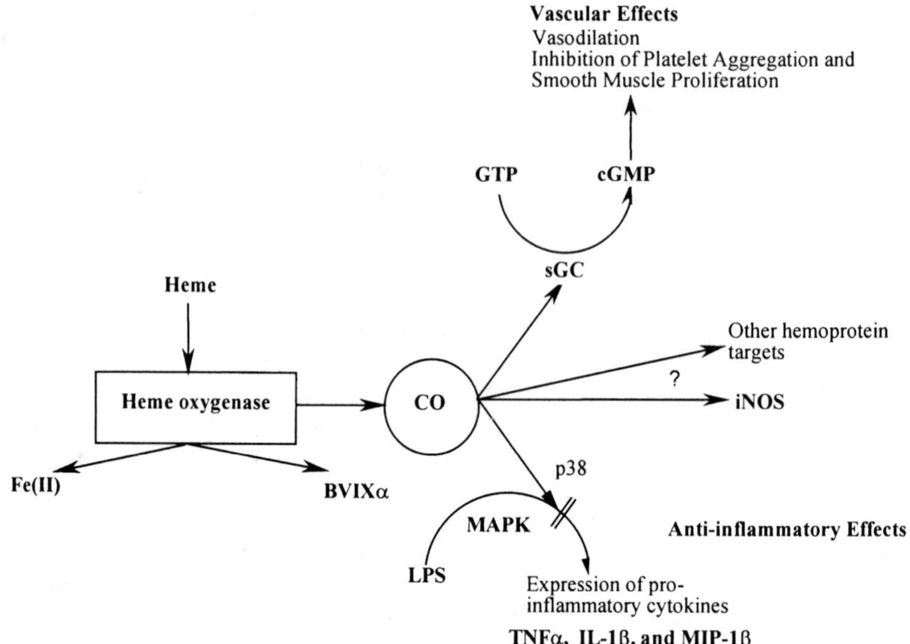

Figure 1. The schemata shows possible consequences of carbon monoxide (CO) generated in the heme oxygenase (HO) reaction. CO may bind to the heme of soluble guanylate cyclase (sGC), stimulating the enzyme to produce guanosine 3′, 5′-cyclic monophosphate (cGMP). Activation of the sGC/cGMP pathway has been associated with the known vascular effects of CO. CO potentially binds to the heme iron of other hemoproteins, including that of nitric oxide synthase (NOS). In studies described in this chapter, CO exerts a novel anti-inflammatory effect in two models of experimental inflammation (hyperoxic shock and bacterial lipopolysaccharide exposure). In both models, CO downregulates the expression of pro-inflammatory cytokines by modulating the p38/mitogen activated protein kinase (MAPK) pathway. Abbreviations include: BVIXa: Biliverdin IX-alpha, Fe (II): Ferrous iron, sGC: Soluble guanylate cyclase, cGMP: guanosine 3′, 5′-cyclic monophosphate, GTP: Guanosine triphosphate, IL-1β: Interleukin-1-beta, iNOS: inducible nitric oxide synthase, LPS: Lipopolysaccharide, MAPK: Mitogen activated protein kinase, MIP-β: Macrophage inflammatory protein-beta, p38: p38 mitogen activated protein kinase, TNFα: Tumor necrosis factor-alpha.

REFERENCES

Abraham, N.G., Lavrovsky, Y., Schwartzman, M.L., Stoltz, R.A., Levere, R.D., Gerritsen, M.E., Shibahara, S., and Kappas, A., 1995, Transfection of the human heme oxygenase gene into rabbit coronary microvessel endothelial cells: protective effect against heme and hemoglobin toxicity, *Proc. Natl. Acad. Sci. USA* **92**:6798–6802.

Camhi, S.L., Alam, J., Otterbein, L., Sylvester, S.L., and Choi, A.M., 1995, Induction of heme oxygenase-1 gene expression by lipopolysaccharide is mediated by AP-1 activation, *Am. J. Respir. Cell Mol. Biol.* **13**:387–398.

Choi, A.M. and Alam, J., 1996, Heme oxygenase-1: function, regulation, and implication of a novel stress-inducible protein in oxidant-induced lung injury, *Am. J. Respir. Cell Mol. Biol.* **15**:9–19.

Chow, J.C., Young, D.W., Golenbock, D.T., Christ, W.J., and Gusovsky, F., 1999, Toll-like receptor-4 mediates lipopolysaccharide-induced signal transduction, *J. Biol. Chem.* **274**:10689–10692.

Clark, J.M. and Lambertson, C.J., 1971, Pulmonary oxygen toxicity: a review, *Pharmacol. Rev.* **23**:37–133.

Derijard, B., Raingeaud, J., Barrett, T., Wu, I.H., Han, J., Ulevitch, R.J., and Davis, R.J., 1995, Independent human MAP-kinase signal transduction pathways defined by MEK and MKK isoforms, *Science* **267**:682–685.

Durante, W. and Schafer, A.I., 1998, Carbon monoxide and vascular cell function (Review), *Int. J. Mol. Med.* **2**:255–262.

Fogg, S., Agarwal, A., Nick, H.S., and Visner, G.A., 1999, Iron regulates hyperoxia-dependent human heme oxygenase 1 gene expression in pulmonary endothelial cells, *Am. J. Respir. Cell Mol. Biol.* **20**: 797–804.

Foresti, R., Clark, J.E., Green, C.J., and Motterlini, R., 1997, Thiol compounds interact with nitric oxide in regulating heme oxygenase-1 induction in endothelial cells. Involvement of superoxide and peroxynitrite anions, *J. Biol. Chem.* **272**:18411–18417.

Freeman, B.A. and Crapo, J.D., 1981, Hyperoxia increases oxygen radical production in rat lungs and lung mitochondria, *J. Biol. Chem.* **256**:10986–10992.

Hambleton, J., Weinstein, S.L., Lem, L., and DeFranco, A.L., 1996, Activation of c-Jun N-terminal kinase in bacterial lipopolysaccharide- stimulated macrophages, *Proc. Natl. Acad. Sci. USA* **93**:2774–2778.

Han, J., Lee, J.D., Bibbs, L., and Ulevitch, R.J., 1994, A MAP kinase targeted by endotoxin and hyperosmolarity in mammalian cells, *Science* **265**:808–811.

Hartsfield, C.L., Alam, J., Cook, J.L., and Choi, A.M., 1997, Regulation of heme oxygenase-1 gene expression in vascular smooth muscle cells by nitric oxide, *Am. J. Physiol.* **273**:L980–L988.

Howard, M., Muchamuel, T., Andrade, S., and Menon, S., 1993, Interleukin 10 protects mice from lethal endotoxemia, *J. Exp. Med.* **177**:1205–1208.

Kageyama, H., Hiwasa, T., Tokunaga, K., and Sakiyama, S., 1988, Isolation and characterization of a complementary DNA clone for a Mr 32,000 protein which is induced with tumor promoters in BALB/c 3T3 cells, *Cancer Res.* **48**:4795–4798.

Keyse, S.M. and Tyrrell, R.M., 1987, Both near ultraviolet radiation and the oxidizing agent hydrogen peroxide induce a 32-kDa stress protein in normal human skin fibroblasts, *J. Biol. Chem.* **262**:14821–14825.

Keyse, S.M. and Tyrrell, R.M., 1989, Heme oxygenase is the major 32-kDa stress protein induced in human skin fibroblasts by UVA radiation, hydrogen peroxide, and sodium arsenite, *Proc. Natl. Acad. Sci. USA* **86**:99–103.

Keyse, S.M. and Tyrrell, R.M., 1990, Induction of the heme oxygenase gene in human skin fibroblasts by hydrogen peroxide and UVA (365 nm) radiation: evidence for the involvement of the hydroxyl radical, *Carcinogenesis* **11**:787–791.

Lautier, D., Luscher, P., and Tyrrell, R.M., 1992, Endogenous glutathione levels modulate both constitutive and UVA radiation/hydrogen peroxide inducible expression of the human heme oxygenase gene, *Carcinogenesis* **13**:227–232.

Lee, P.J., Alam, J., Sylvester, S.L., Inamdar, N., Otterbein, L., and Choi, A.M., 1996, Regulation of heme oxygenase-1 expression *in vivo* and *in vitro* in hyperoxic lung injury, *Am. J. Respir. Cell Mol. Biol.* **14**:556–568.

Lee, P.J., Jiang, B.H., Chin, B.Y., Iyer, N.V., Alam, J., Semenza, G.L., and Choi, A.M., 1997, Hypoxia-inducible factor-1 mediates transcriptional activation of the heme oxygenase-1 gene in response to hypoxia, *J. Biol. Chem.* **272**:5375–5381.

Maines, M.D. and Kappas, A., 1977, Metals as regulators of heme metabolism, *Science* **198**:1215–1221.

Maines, M.D., 1992, *Heme Oxygenase: Clinical Applications and Functions.* CRC Press, Boca Raton, FA.

Maines, M.D., Trakshel, G.M., and Kutty, R.K., 1986, Characterization of two constitutive forms of rat liver microsomal heme oxygenase. Only one molecular species of the enzyme is inducible, *J. Biol. Chem.* **261**:411–419.

Maines, M.D., 1997, The heme oxygenase system: a regulator of second messenger gases, *Annu. Rev. Pharmacol. Toxicol.* **37**:517–554.

McCoubrey, W.K., Jr., Huang, T.J., and Maines, M.D., 1997, Isolation and characterization of a cDNA from the rat brain that encodes hemoprotein heme oxygenase-3, *Eur. J. Biochem.* **247**:725–732.

Minamino, T., Christou, H., Hsieh, C.M., Liu, Y., Dhawan, V., Abraham, N.G., Perrella, M.A., Mitsialis, S.A., and Kourembanas, S., 2001, Targeted expression of heme oxygenase-1 prevents the pulmonary inflammatory and vascular responses to hypoxia, *Proc. Natl. Acad. Sci. USA* **98**:8798–8803.

Morita, T., Perrella, M.A., Lee, M.E., and Kourembanas, S., 1995, Smooth muscle cell-derived carbon monoxide is a regulator of vascular cGMP, *Proc. Natl. Acad. Sci. USA* **92**:1475–1479.

Morita, T. and Kourembanas, S., 1995, Endothelial cell expression of vasoconstrictors and growth factors is regulated by smooth muscle cell-derived carbon monoxide, *J. Clin. Invest.* **96**:2676–2682.

Morita, T., Mitsialis, S.A., Koike, H., Liu, Y., and Kourembanas, S., 1997, Carbon monoxide controls the proliferation of hypoxic vascular smooth muscle cells, *J. Biol. Chem.* **272**:32804–32809.

Otterbein, L.E., Kolls, J.K., Mantell, L.L., Cook, J.L., Alam, J. and Choi, A.M., 1999, Exogenous administration of heme oxygenase-1 by gene transfer provides protection against hyperoxia-induced lung injury, *J. Clin. Invest.* **103**:1047–1054.

Otterbein, L.E. and Choi, A.M., 2000, Heme oxygenase: colors of defense against cellular stress, *Am. J. Physiol. Lung Cell Mol. Physiol.* **279**:L1029–L1037.

Otterbein, L.E., Bach, F.H., Alam, J., Soares, M., Tao, L.H., Wysk, M., Davis, R.J., Flavell, R.A., and Choi, A.M., 2000, Carbon monoxide has anti-inflammatory effects involving the mitogen- activated protein kinase pathway, *Nat. Med.* **6**:422–428.

Otterbein, L.E., Otterbein, S.L., Morse, D.E., Fearns, C., Ulevitch, R.J., Knickelbein, R., Davis R.J., Flavell, R.A., and Choi, A.M., 2001, MKK3 mitogen activated protein kinase pathway mediates carbon monoxide-induced protection against oxidant-induced lung injury, *in press.*

Panchenko, M.V., Farber, H.W., and Korn, J.H., 2000, Induction of heme oxygenase-1 by hypoxia and free radicals in human dermal fibroblasts, *Am. J. Physiol. Cell Physiol.* **278**:C92–C101.

Poss, K.D. and Tonegawa, S., 1997, Reduced stress defense in heme oxygenase 1-deficient cells, *Proc. Natl. Acad. Sci. USA* **94**:10925–10930.

Raingeaud, J., Gupta, S., Rogers, J.S., Dickens, M., Han, J., Ulevitch, R.J., and Davis, R.J., 1995, Pro-inflammatory cytokines and environmental stress cause p38 mitogen- activated protein kinase activation by dual phosphorylation on tyrosine and threonine, *J. Biol. Chem.* **270**:7420–7426.

Raingeaud, J., Whitmarsh, A.J., Barrett, T., Derijard, B., and Davis, R.J., 1996, MKK3 and MKK6 regulation of gene expression is mediated by the p38 MAP kinase signal transduction pathway, *Mol. Cell Biol.* **16**:1247–1255.

Rizzardini, M., Terao, M., Falciani, F., and Cantoni, L., 1993, Cytokine induction of haem oxygenase mRNA in mouse liver. Interleukin 1 transcriptionally activates the haem oxygenase gene, *Biochem. J.* **290 (Pt 2)**: 343–347.

Rizzardini, M., Carelli, M., Cabello Porras, M.R., and Cantoni, L., 1994, Mechanisms of endotoxin-induced haem oxygenase mRNA accumulation in mouse liver: synergism by glutathione depletion and protection by N- acetylcysteine, *Biochem. J.* **304 (Pt 2)**:477–483.

Ryter, S.W. and Tyrrell, R.M., 2000, The heme synthesis and degradation pathways: role in oxidant sensitivity. Heme oxygenase has both pro- and antioxidant properties, *Free Radic. Biol. Med.* **28**:289–309.

Ryter, S.W., Si, M., Lai, C.C., and Su, C.Y., 2000, Regulation of endothelial heme oxygenase activity during hypoxia is dependent on chelatable iron, *Am. J. Physiol. Heart Circ. Physiol.* **279**:H2889–H2897.

Sato, K., Balla, J., Otterbein, L., Smith, R.N., Brouard, S., Lin, Y., Csizmadia, E., Sevigny, J., Robson, S.C., Vercellotti, G., Choi, A.M., Bach, F.H., and Soares, M.P., 2001, Carbon monoxide generated by heme oxygenase-1 suppresses the rejection of mouse-to-rat cardiac transplants, *J. Immunol.* **166**:4185–4194.

Saunders, E.L., Maines, M.D., Meredith, M.J., and Freeman, M.L., 1991, Enhancement of heme oxygenase-1 synthesis by glutathione depletion in Chinese hamster ovary cells, *Arch. Biochem. Biophys.* **288**:368–373.

Sharma, H.S., Maulik, N., Gho, B.C., Das, D.K., and Verdouw, P.D., 1996, Coordinated expression of heme oxygenase-1 and ubiquitin in the porcine heart subjected to ischemia and reperfusion, *Mol. Cell Biochem.* **157**:111–116.

Stocker, R., Yamamoto, Y., McDonagh, A.F., Glazer, A.N., and Ames, B.N., 1987, Bilirubin is an antioxidant of possible physiological importance, *Science* **235**:1043–1046.

Tenhunen, R., Marver, H.S., and Schmid, R., 1968, The enzymatic conversion of heme to bilirubin by microsomal heme oxygenase, *Proc. Natl. Acad. Sci. USA* **61**:748–755.

Tenhunen, R., Marver, H.S., and Schmid, R., 1969, Microsomal heme oxygenase. Characterization of the enzyme, *J. Biol. Chem.* **244**:6388–6394.

Verma, A., Hirsch, D.J., Glatt, C.E., Ronnett, G.V., and Snyder, S.H., 1993, Carbon monoxide: a putative neural messenger, *Science* **259**:381–384.

Vile, G.F. and Tyrrell, R.M., 1993, Oxidative stress resulting from ultraviolet A irradiation of human skin fibroblasts leads to a heme oxygenase-dependent increase in ferritin, *J. Biol. Chem.* **268**:14678–14681.

Vile, G.F., Basu-Modak, S., Waltner, C., and Tyrrell, R.M., 1994, Heme oxygenase 1 mediates an adaptive response to oxidative stress in human skin fibroblasts, *Proc. Natl. Acad. Sci. USA* **91**:2607–2610.

Yet, S.F., Perrella, M.A., Layne, M.D., Hsieh, C.M., Maemura, K., Kobzik, L., Wiesel, P., Christou, H., Kourembanas, S., and Lee, M.E., 1999, Hypoxia induces severe right ventricular dilatation and infarction in heme oxygenase-1 null mice, *J. Clin. Invest.* **103**:R23–R29.

3

CORRELATION OF THE ALTERED VASCULAR EFFECTS OF CARBON MONOXIDE AND THE CARDIOVASCULAR COMPLICATIONS OF DIABETES

Rui Wang[a], Xianfeng Sun[a], Lingyun Wu[c], Zunzhe Wang[b],
Salma Toma Hanna[a], and Robert Peterson-Wakeman[a]

[a]Department of Physiology
University of Saskatchewan
Saskatoon, Canada S7N 5E5
[b]Laboratory of Cellular Morphology
Weifang Medical College
Weifang, P.R. China
[c]Department of Anatomy and Cell Biology
University of Saskatchewan
Saskatoon, Canada S7N 5E5

1. THE VASCULAR EFFECT OF CO AND THE UNDERLYING MECHANISMS

Carbon monoxide (CO) is not only a hypoxia agent resulting from environmental pollution, but also an endogenously generated biological gas. In as early as 1991, Marks et al. proposed that CO might possess physiologically important vasoactive functions. That pioneer hypothesis has gained more and more support. The vasorelaxing effects of CO have been observed in many vascular preparations. Several features of the vasorelaxant effects of CO have also been described. (1) The CO-induced muscle relaxation is not mediated by stimulation of adrenergic receptors, by adenosine, or by prostaglandins. (2) The vasorelaxation induced by CO is not due to hypoxic hypoxia or functional hypoxia since CO relaxed aorta smooth muscles under conditions where oxygen uptake had been completely inhibited and since CO induced a greater vasorelaxation than hypoxic hypoxia did. (3) The vascular effect of CO was independent of the presence of endothelium in vascular tissues. Isolated aortic smooth muscle cells (SMCs) could also be relaxed by CO (Ramos et al., 1989).

Among the mechanisms underlying the vascular effect of CO an increase in cGMP level in vascular SMCs has been emphasized mostly. CO may increase cGMP content via its stimulatory interaction with the heme in the regulatory subunit of guanylyl cyclase. Increased cGMP would consequently decrease $[Ca^{2+}]_i$ in SMCs through the inhibition of IP_3 formation, the activation of Ca^{2+}-ATPase, and the inhibition of Ca^{2+} channels. Finally, the relaxation of smooth muscles would occur. In non-vascular preparations, such as human neutrophils (Morita et al., 1995), rat thyomocytes (Alvarez et al., 1992), and human platelets (Alonso et al., 1991), calcium influx evoked by different stimuli was inhibited by CO. CO also significantly reduced high-K^+ induced increase in intracellular calcium concentration (^{45}Ca uptake) in rat aortic rings (Lin et al., 1988). This effect of CO was similar to that of verapamil, indicating that CO may inhibit voltage-dependent calcium entry. The possibility is thus raised that CO may inhibit Ca^{2+} channels by increasing the content of cGMP in SMCs, although direct electrophysiological evidence on the modulation of Ca^{2+} channels by CO in single SMCs is lacking.

In vascular SMCs, voltage-gated K^+ channels, ATP-sensitive K^+ channels (K_{ATP}), and Ca^{2+}-activated K^+ channels (K_{Ca}) have been identified. The opening of K^+ channels will lead to membrane hyperpolarization, which in turn inhibits the agonist-induced increase in IP_3, reduces Ca^{2+} sensitivity and resting Ca^{2+} level, and relaxes SMCs. The distribution of K^+ channels and their modulation by CO are different in different tissue preparations. CO transiently increased a TEA-insensitive K channel current and induced membrane potential oscillation in human jejunal SMCs (Farrugia et al., 1993). In cultured urinary bladder SMCs, CO inhibited a whole-cell K_{Ca} channel current (Trischmann et al., 1991). Our recent studies have demonstrated that CO directly acts on K channels in rat tail artery SMCs via a cGMP-independent mechanism (Wang et al., 1997; Wang and Wu, 1997).

It has been suggested that CO could relax certain types of blood vessels, especially the ductus arteriosus, by inhibiting the cytochrome P-450 dependent monooxygenase reaction (Coceani et al., 1996; 1997). The cytochrome P-450 comprises a family of hemoproteins present in many organs, including a variety of blood vessels (Liu et al., 1993). Levels of cytochrome P-450 are regulated by the availability of cellular heme which in turn is controlled by the level of heme oxygenase (Levere et al., 1990). Our results have shown that the CO-induced relaxation of adult rat tail artery tissues was not modulated by cytochrome P-450 inhibition (Wang, 1998). Therefore, the role played by cytochorme P-450 in the CO-induced vasorelaxation was likely related to specific developmental stage of the investigated subjects.

2. THE REGULATION OF VASCULAR TONE IN DIABETES MELLITUS AND THE INVOLVEMENT OF CO MECHANISMS

Vascular complications of diabetes mellitus are largely manifested as coronary heart disease, hypertension (Christlieb, 1973; Epstein and Soweres, 1992), peripheral vessel occlusion, cerebral ischemia and stroke (Mayhan, 1989). These vascular complications are responsible for most of the morbidity and mortality of patients with diabetes. Experimental studies support these clinical observations by revealing the occurrence of hypertension (Shah et al., 1995), enhanced vascular reactivity to agonists (Bodmer et al., 1995; Christ et al., 1992; Kamata et al., 1992), and increased contractility of blood vessels (Jackson and Carrier, 1981; Macleod and McNeill, 1985; Weidmann et al., 1979; White and Carrier, 1988; White and Carrier, 1990) in both insulinopenic (Reddy et al., 1990) and insulin-resistant rats (Shehin et al., 1989). Although dysfunction of endothelium in diabetes (Tesfamariam et al., 1995) has been extensively studied (Tesfamariam, 1994; Vallance et al., 1992), the mechanisms responsible for the altered vascular contractility in diabetes, especially the structural and functional changes in vascular SMCs, remain unknown.

The involvement of CO in the etiology of diabetes has been implicated as CO up-regulated, whereas nitric oxide (NO) down-regulated insulin secretion from pancreatic islets. The altered metabolism of CO in cardiac tissues from streptozotocin-induced diabetic rats has also been shown recently. Whether the vascular effects of CO and the endogenous generation of CO are altered in diabetes had not been reported. The roles played by cGMP and K_{Ca} channels in the vascular complications of diabetes were also unknown. We have recently studied the vasorelaxant effect of CO on tail artery tissues from streptozotocin-induced diabetic rats in comparison to that from normal control rats. The relative contributions of cGMP pathway and K_{Ca} channels to the putatively altered effect of CO on the contractility of diabetic vascular tissues were investigated. The direct effect of CO on single K_{Ca} channel activity was determined. The role of glycation of K_{Ca} channels in the altered effect of CO was further analyzed. Finally, the vascular responses to the endogenously generated CO in diabetic rat tail artery tissues were assayed. Our results, as presented below, provide evidence that corroborates the altered vascular effect of CO in diabetes and unravels the underlying cellular mechanisms.

3. RESEARCH APPROACHES

3.1. Animal Model of Diabetes

Male Sprague-Dawley adult rats weighing 150–180 g were maintained on standard rat chow and tap water ad libitum with 12 h light/dark cycles in a quiet environment. Diabetes was induced by a single injection via the lateral tail vein or penis vein of streptozotocin (STZ, 60 mg/kg body weight), dissolved in sodium citrate buffer (pH 4.5), after the rats were anaesthetized by intraperitoneal injection of sodium pentobarbital (60 mg/kg body weight). Age-matched control rats were injected of equal volume of vehicle (sodium citrate buffer). The STZ-injected rats were used in the present study 1 month after the induction of diabetes. The "Principles of laboratory animal care" (NIH publication no. 85–23, revised 1985) were followed, and the Committee on Animal Care and Supply of the University of Saskatchewan approved animal experimental protocols.

3.2. Measurement of Isometric Tension Development of Isolated Rat Tail Artery Tissues

Briefly, rat tail artery tissues were mounted in 10 ml organ baths filled with Krebs' bicarbonate solution (bubbled with 95% O_2 and 5% CO_2) which was composed of (in mM): NaCl 115, KCl 5.4, $MgSO_4$ 1.2, NaH_2PO_4 1.2, $NaHCO_3$ 25, glucose 11, and $CaCl_2$ 1.8. The tail artery strips were mechanically stretched to achieve a basal tension of approximately 0.7 g, and were allowed to equilibrate for 1 h before the start of experiments. Since the vascular effect of CO was not dependent on the presence of an intact endothelium (Wang et al., 1997), endothelium was removed from vascular strips by a rubbing procedure. The isometric tension development was measured with FT 03 force displacement transducers (Grass Ins. Co., Quincy). Data acquisition and analysis were accomplished using a Biopac system (Biopac System, Inc., Golata).

There are several reasons for using rat tail artery in this study. (1) This artery has been widely used for pharmacology assays. It is a representative peripheral blood vessel, unlike aorta being a conduit vessel. (2) The innervation and receptor distribution on this artery is relatively clear. (3) We have established the methods to measure tissue contraction and to isolate single cells from this preparation. Thus, we can easily link the functional changes from the tissue level to the cellular/ionic level. It is noticed that, however, this artery plays a minor role in thermoregulation, does not contribute much to peripheral resistance, and generates small number of single SMCs.

3.3. Cell Preparation

Single SMCs were dispersed enzymatically following our established procedure (Wang et al., 1989) with modifications. Rats were anesthetized by intraperitoneal injection of sodium pentobarbital (60 mg/kg body weight). Tail arteries were isolated and connective tissue removed under a dissecting microscope. The arteries were cut open longitudinally and immersed in a Ca^{2+}- and Mg^{2+}-free Hanks' buffered saline solution (HBSS, Gibco) at 4 °C. The arterial strips were then digested with various enzymes and single cells were dispersed. The dispersed cells were plated onto 35 mm Petri dishes in Dulbecco's modified Eagle's medium (Gibco) containing penicillin (100 U/ml, Sigma) and streptomycin (0.1 mg/ml, Sigma), and maintained at 4 °C for at least 4 h. These freshly isolated cells were used in electrophysiological recording within 8–24 h of isolation.

3.4. Whole-Cell Recording

The Petri dish with attached cells was mounted on the stage of an inverted phase contrast microscope (Olympus, Japan). Pipettes with tip resistances of 2–4 MΩ were used. Transmembrane currents were recorded using an Axopatch-1D patch-clamp amplifier, controlled by a Digidata 1200 interface and a pClamp software (6.01, Axon Instruments, Inc.). Current-voltage (I-V) curves were constructed using the sustained current amplitudes at the end of 800 ms test pulses. The bath solution contained (in mM) NaCl 130, KCl 5, $CaCl_2$ 1.8, $MgCl_2$ 1.2, HEPES 5, and glucose 5. The pipette solution was composed of (in mM) K-aspartate 110, KCl 20, HEPES 5, ATP-Na$_2$ 1, MgCl$_2$ 1, EGTA 0.1. Adequate CaCl$_2$ was added to the pipette solution with the pCa adjusted to 7.7.

3.5. Single Channel Recording

The inside-out and outside-out configurations of the patch-clamp technique were used to record single K_{Ca} channel currents as described previously (Wang and Wu, 1997; Wang et al., 2001). Pipettes with a resistance of 6–8 MΩ were used and the seal resistance was usually greater than 10 GΩ. For each concentration of CO tested at least 60s of channel activity was recorded directly on the hard disk of a computer. The open probability (NPo), with N representing the number of single channels in one patch, and the unit amplitude of K_{Ca} channels were determined from all point histograms using a Fetchan program (Axon Instruments, Inc.). NPo of K_{Ca} channels was averaged over 2–5 minute recording to describe the changes in channel activity following different treatments. Membrane patches with unstable NPo over time were excluded from further analysis. The external surface of membrane patches was bathed in a solution containing (in mM): KCl 145, HEPES 10, and glucose 10. The internal surface of membrane patches was exposed to a solution containing (in mM): KCl 145, HEPES 10, MgCl$_2$ 1.2, glucose 10, EGTA 1, and 0.5 µmM of free Ca^{2+}, $[Ca^{2+}]_i$. $[Ca^{2+}]_i$ of the recording solution was calculated using a computer program (EQCAL, Biosoft, USA).

3.6. Data Process

The data were expressed as means ± S.E.M. The comparison of EC50 under different conditions was performed by analysis of variance (ANOVA) followed by Student's t test in conjunction with the Newman-Keuls test where applicable. The significant difference between treatments was defined at a level of $p < 0.05$.

4. ALTERED VASCULAR EFFECTS OF CO IN DIABETES AND THE UNDERLYING MECHANISMS

Diabetic rats after 1 month STZ injection lost body weight, had glycosuria, and developed hyperglycemia with the fasting glucose concentration of plasma elevated to 32.2 ± 2.2 mM (n = 10), while the control non-diabetic rats had a fasted plasma glucose concentration of 7.3 ± 0.9 mM (n = 7). CO induced a concentration dependent relaxation of the phenylephrine (PHE)-precontracted endothelium-free tail artery tissues. This vasorelaxant effect of CO was significantly reduced in diabetic vascular tissues. EC50 of the vasorelaxant effect of CO was 58 ± 24 µM in normal tissues (n = 8) but 131 ± 38 µM in diabetic tissues (n = 8, p < 0.05).

To explore the mechanisms underlying the reduced vasorelaxant effect of CO on diabetic tail artery tissues, the effect of CO on cGMP levels was determined. In tail artery tissues from both normal rats and diabetic rats, CO significantly enhanced the levels of cGMP. However, the effect of CO on cGMP level was significantly reduced in diabetic vascular tissues. CO elevated the tissue level of cGMP by about 310% (taking the basal level as 100%) in normal tissue but only about 130% in diabetic tissues (n = 8, p < 0.05 vs. normal tissues). Whether this reduced production of cGMP could fully account for the decreased CO effect on diabetic vascular tissues was further examined by incubating vascular tissues with 10 μM ODQ, a specific inhibitor for the soluble guanylyl cyclase. Without pretreating vascular tissues with ODQ, CO (300 μM) induced a 60 ± 7% relaxation of normal vascular tissues (n = 8). In the presence of ODQ, CO only induced a 38 ± 8% relaxation of normal vascular tissues (n = 8). This represents a 63% inhibition of CO effect by ODQ. Further prolonging the ODQ incubation time from 10 min to 20 min (n = 4) or increasing the concentration of ODQ from 10 to 30 μM (n = 4) did not induce additional inhibition of the CO effect. In contrast, 10 min incubation of diabetic tail artery tissues with 10 μM ODQ completely abolished the vasorelaxant effect of CO (Wang et al., 2001).

The interaction of CO and tetraethylammonium (TEA) on vascular tone was examined. The vascular tissues were pretreated with TEA (30 mM). Subsequently, the CO (300 μM)-induced relaxation of the vascular strips precontracted with PHE (1 μM) was examined. TEA had no effect on either the resting tension level or the PHE-induced tonic contraction forces of normal tissue (Wang, 1998) or diabetic tissue (Fig. 1A). In the presence of TEA, the CO-induced vasorelaxation was reduced significantly from 45.9 ± 9% to 19 ± 14% (n = 8, p < 0.05) in normal tissues (Wang, 1998). In diabetic tail artery tissues, CO induced an oscillated vasorelaxation in the presence of TEA (Fig. 1A). The mean relaxation amplitude was not different with or without TEA pre-treatment (n = 6, p > 0.05).

Since CO relaxes normal rat tail artery by stimulating cGMP and high-conductance K_{Ca} channels (Wang, 1998; Wang et al., 1997; Wang and Wu, 1997; Wang et al., 1997), the complete blockade of CO effect by ODQ and the failure of TEA to reduce the relaxant potency of CO in diabetic tail artery hinted a diminished role of K_{Ca} channels in mediating the CO effect. Therefore, the subsequent study focused on the characteristics of K_{Ca} channels and their modulation by CO in diabetic tail artery SMCs.

It was found that the whole-cell outward K^+ channel currents in vascular SMCs were enhanced by CO. CO (30 μM) enhanced K^+ channel currents with more prominent effect at more depolarized potentials (Fig. 1B). At a membrane potential of +30 mV, the amplitude of whole-cell K^+ channel currents was increased by 276 ± 74% by CO in normal tail artery SMCs (n = 7, p < 0.05), but only 65 ± 42% in diabetic SMCs (p < 0.05, n = 6) (Fig. 1C).

Extracelluarly or intracellularly applied CO increased the open probability of single big-conductance K_{Ca} channels in a concentration-dependent fashion without affecting the single channel conductance. This effect of CO could be explained by the CO-induced increase in the calcium sensitivity of single K_{Ca} channels. Furthermore, the absence of cGMP-dependent protein kinase or the stimulation of G proteins (Gi/Go or Gs) in excised cell membrane patches did not affect the activities of single K_{Ca} channels. Our results indicate that CO directly modulated big-conductance K_{Ca} channels in vascular SMCs. The characteristics of this K_{Ca} channel were not different between normal and diabetic tail artery SMCs. The single channel conductances were 239 ± 8 pS (n = 8) in normal tail artery SMCs, and 230 ± 6 pS (n = 6) in diabetic

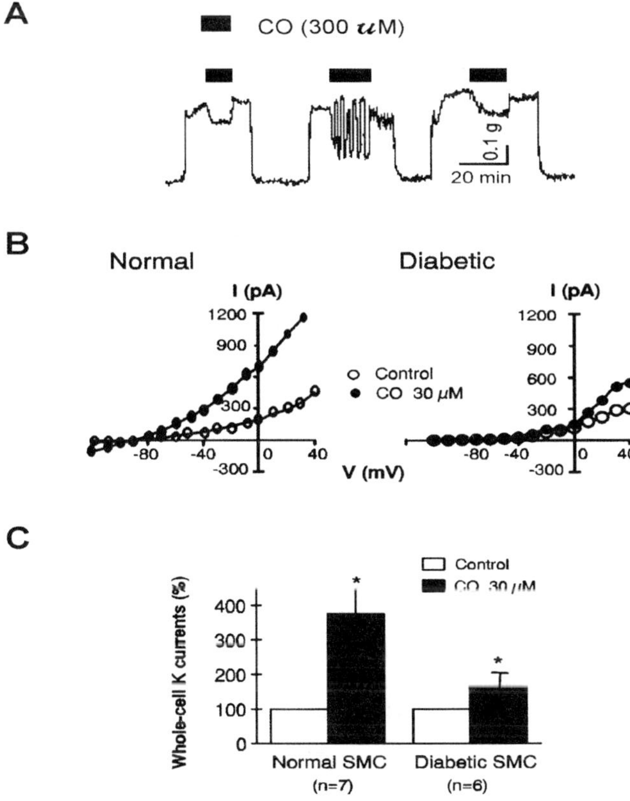

Figure 1. Altered effects of CO on vascular contractility and K^+ channel currents in diabetes. **A.** CO induced an oscillated vasorelaxation of the PHE-precontracted tail artery tissues from a diabetic rat in the presence of TEA. **B.** The I-V relationships of the whole-cell outward K^+ channels show that the stimulatory effect of CO was significantly reduced in a diabetic SMC (right panel) as compared to its effect on a normal SMC (left panel). HP = –80 mV. **C.** Summary of the effect of CO on the whole-cell outward K^+ current in normal or diabetic rat tail artery SMCs. HP = –80 mV, test potential = +30 mV, *p < 0.05.

SMCs (p > 0.05). The open probability (NPo) of K_{Ca} channels was decreased by charybdotoxin (100 nM) or iberiotoxin (100 nM), but not by apamin (100 nM) in both normal and diabetic tail artery SMCs. The concentration-dependent stimulatory effect of CO (3–30 μM) was greatly reduced in diabetic artery SMCs. For instance, the mean NPo over 3 min of recording was increased by CO (10 μM) by 81 ± 24% in normal SMCs (n = 6, p < 0.05). At the same concentration, CO had no effect on the mean NPo of K_{Ca} channels in diabetic SMCs. When the concentration of CO was increased to 30 μM, the mean NPo of single K_{Ca} channels over 3 min of recording was increased by 173 ± 14% in normal SMCs (n – 5), but only by 48 ± 30% in diabetic SMCs (n = 4, p < 0.05 vs. the effect of CO on normal SMCs).

The decreased K_{Ca} channel sensitivity to CO might be explained by the altered glycation status of K_{Ca} channel proteins or hyperosmolality-induced structural and/or functional alterations in diabetic SMCs. To test the influence of glycation on K_{Ca}

channels, tail artery SMCs were isolated and incubated *in vitro* under different conditions. After culturing normal vascular SMCs for 8 days with 5 mM glucose in the culture medium, the single channel conductance and the sensitivity to charybdotoxin or iberiotoxin of high-conductance of K_{Ca} channels recorded in the cell-free patches were similar to those of freshly isolated normal SMCs. CO (10 μM) increased the mean NPo over 3 mins of the recording of the single K_{Ca} channels by $95 \pm 23\%$, taking the NPo level before the application of CO as 100% ($p < 0.05$). After culturing SMCs isolated from diabetic rats for 8 days with 25 mM glucose, high-conductance K_{Ca} channels had similar characteristics as those of freshly isolated diabetic SMCs. In these 8-day cultured diabetic cells, K_{Ca} channels were also not sensitive to CO with the 3 min mean NPo showing no change after the application of CO (10 μM). In another set of experiments, normal SMCs were cultured with 25 mM glucose for 8 days. Single K_{Ca} channels in these cells lost their responsiveness to CO (10 μM) as the 3 min mean NPo was not changed by CO. Culturing normal SMCs for 8 days with 25 mM mannitol did not alter the effect of CO on K_{Ca} channels. A $73 \pm 9\%$ increase in the mean NPo of K_{Ca} channels in these cells was observed in the presence of CO (10 μM) ($p < 0.05$), indicating that osmolality change cannot explain the effect of hyperglycemia on the diminished sensitivity of K_{Ca} channels to CO.

Whether the vascular responses to the endogenous CO were altered in diabetic vascular tissues was further studied. Tail artery tissues were pre-incubated in the dark for 6 h with hemin (20 μM). The enhanced expression of heme oxygenase-1 by hemin incubation would promote the endogenous CO production from vascular tissues, and subsequently decreased the PHE-induced vasoconstriction. In agreement with our previous studies, a 6-h incubation of tail artery tissues from normal rats or diabetic rats without hemin added did not alter the resting tension level or the PHE-induced concentration-dependent vasoconstriction. The concentration-dependent vasoconstriction induced by PHE was significantly inhibited by hemin incubation. The EC50 of PHE effects was 0.24 ± 0.03 μM and 1.19 ± 0.12 μM without or with hemin incubation, respectively ($p < 0.05$). In contrast, the vasoconstrictive effect of PHE on diabetic tail artery tissues was not affected by hemin incubation. This result suggests that the endogenous CO production in response to hemin incubation was suppressed in diabetic vascular tissues.

5. CONCLUSION

The following novel results were obtained from our present study. (i) The CO concentration-dependent relaxation of tail artery tissues from streptozotocin-induced diabetic rats was significantly decreased as compared to that of non-diabetic control rats. (ii) The vasorelaxant effect of CO in diabetes was solely mediated by cGMP, and the CO effect on cGMP production in diabetic vascular tissues was decreased. (iii) The sensitivity of K_{Ca} channels in diabetic vascular SMCs to CO was significantly reduced. (iv) The glycation of K_{Ca} channel proteins may account for its reduced sensitivity to CO. (v) The vascular relaxation induced by endogenous CO appears to be significantly reduced in diabetes. To reiterate, the vascular functions of CO were altered in diabetes, which may contribute to certain types of vascular complications of diabetes.

As demonstrated in our previous studies, CO relaxes rat tail artery mainly via its effects on the cGMP-dependent pathway and the cGMP-independent modulation of K_{Ca} channels (Wang, 1998). The decreased vasorelaxant effect of CO in diabetes may be caused by the changed responsiveness of one or both of these two independent signalling pathways in diabetic SMCs. The STZ-induced experimental diabetes is a well-established animal model of diabetes mellitus, characterized with hyperglycemia and hypoinsulinemia (Wang Z et al., 1998; Wang et al., 2000; Wang et al., 1998). In these experimental diabetic rats, the CO-evoked generation of cGMP in diabetic vascular tissues was significantly lower as compared to normal tissues. This could partially explain the reduced vasorelaxant effect of CO in diabetes. More importantly, our tissue contraction study showed that the effect of CO on diabetic vascular tissues was completely eliminated by pretreating the tissues with ODQ that blocked the cGMP pathway. Furthermore, a significantly lower sensitivity of K_{Ca} channels of diabetic SMCs to CO was illustrated, indicating that another leg of CO reaction system, i.e. K_{Ca} channels, was malfunctioning in diabetes. Taken together, our results suggested that the decreased vasorelaxant effect of CO in diabetic rats could be related to both a decreased CO-induced cGMP production and a diminished sensitivity of K_{Ca} channels in the diabetic SMCs to CO.

ACKNOWLEDGMENTS

This study was supported by research grants from Canadian Institutes of Health Research (CIHR), and from Smokeless Tobacco Research Council Inc., USA. Rui Wang is a CIHR Scientist. Lingyun Wu is supported by a post-doctoral fellowship from CIHR/Heart and Stroke Foundation of Canada. The excellent technical assistance from Ginger Beal and Koleen Safiniuk is greatly appreciated.

REFERENCES

Alonso, M.T., Alvarez, J., Montero, M., Sanchez, A., and Garcia-Sancho, J., 1991, Agonist-induced Ca^{2+} influx into human platelet is secondary to the emptying of intracellular calcium stores. *Biochem. J.* **280**:783–789.

Alvarez, J., Montero, J., and Garcia-Sancho, J., 1992, Cytochrome P450 may regulate plasma membrane Ca^{2+} permeability according to the filling state of the intracellular Ca^{2+} stores. *FASEB J.* **6**:786–792.

Bodmer, C.W., Schaper, N.C., Janssen, M., de Leeuw, P.W., and Williams, G., 1995, Selective enhancement of alpha 2-adrenoceptor-mediated vasoconstriction in insulin-dependent diabetic patients with microalbuminuria. *Clin. Sci.* **88**:421–426.

Christ, G.J., Schwartz, C.B., Stone, B.A., Parker, M., Janis, M., Gondre, M., Valcic, M., and Melman, A., 1992, Kinetic characteristics of alpha 1-adrenergic contractions in human corpus cavernosum smooth muscle. *Am. J. Physiol.* **263**:H15–H19.

Christlieb, A.R., 1973, Diabetes and hypertensive vascular disease. Mechanisms and treatment. *Am. J. Cardiol.* **32**:592–606.

Coceani, F., Kelsey, L., and Seidlitz, E., 1996, Carbon monoxide-induced relaxation of the ductus arteriosus in the lamb: evidence against the prime role of guanylyl cyclase. *Br. J. Pharmacol.* **118**: 1689–1696.

Coceani, F., Kelsey, L., Seidlitz, E., Marks, G.S., McLaughlin, B.E., Vreman, H.J., Stevenson, D.K., Rabinovitch, M., and Ackerley, C., 1997, Carbon monoxide formation in the ductus arteriosus in the lamb: implications for the regulation of muscle tone. *Br. J. Pharmacol.* **120**:599–608.

Epstein, M. and Soweres, J.R., 1992, Diabetes mellitus and hypertension. *Hypertension* **19**:403–418.

Farrugia, G., Irons, W.A., Rae, J.L., Sarr, M.G., and Szurszewski, J.H., 1993, Activation of whole cell currents in isolated human jejunal circular smooth muscle cells by carbon monoxide. *Am. J. Physiol.* **264**:G1184–G1189.

Jackson, C.V. and Carrier, G.O., 1981, Supersensitivity of isolated mesenteric arteries to noradrenaline in the long-term experimental diabetic rat. *J. Auton. Pharmacol.* **1**:399–405.

Kamata, K., Miyata, N., Abiru, T., and Kasuya, Y., 1992, Functional changes in vascular smooth muscle and endothelium of arteries during diabetes mellitus. *Life Sci.* **50**:1379–1387.

Levere, R.D., Martasek, P., Escalante, B., Schwartzman, M.L., and Abraham, N.G., 1990, Effect of heme arginate administration on blood pressure in spontaneously hypertensive rats. *J. Clin. Invest.* **86**:213–219.

Lin, H. and McGrath, J.J., 1988, Carbon monoxide effects on calcium levels in vascular smooth muscle. *Life Sci.* **43**:1813–1816.

Liu, Z., Brien, J.F., Marks, G.S., McLaughlin, B.E., and Nakatsu, K., 1993, Lack of evidence for the involvement of cytochrome P-450 or other hemoproteins in metabolic activation of glyceryl trinitrate in rabbit aorta. *J. Pharmacol. Exp. Ther.* **264**:1432–1439.

Macleod, K.M. and McNeill, J.H., 1985, The influence of chronic experimental diabetes on contractile responses of rat isolated blood vessels. *Can. J. Physiol. Pharmacol.* **63**:52–57.

Magyar, J., Rusznak, Z., Szentesi, P., Szucs, G., and Kovacs, L., 1992, Action potentials and potassium currents in rat ventricular muscle during experimental diabetes. *J. Mol. Cell. Cardiol.* **24**:841–853.

Marks, G.S., Brien, J.F., Nakatsu, K., and McLaughlin, B.E., 1991, Does carbon monoxide have a physiological function? *Trends Pharmacol. Sci.* **12**:185–188.

Mayhan, W.G., 1989, Impairment of endothelium-dependent dilatation of cerebral arterioles during diabetes mellitus. *Am. J. Physiol.* **256**:H621–H625.

Morita, T., Perrella, M.A., Lee, M.E., and Kourembanas, S., 1995, Smooth muscle cell-derived carbon monoxide is a regulator of vascular cGMP. *Proc. Natl. Acad. Sci. USA.* **92**:1475–1479.

Ramos, K.S., Lin, H., and McGrath, J.J., 1989, Modulation of cyclic guanosine monophosphate levels in cultured aortic smooth muscle cells by carbon monoxide. *Biochem. Pharmacol.* **38**:1368–1370.

Reddy, S., Shehin, S., Sowers, J.R., Dardas, G., and Zemel, M.B., 1990, Aortic ^{45}Ca^{2+} flux and blood pressure regulation in streptozotocin-induced diabetic rats. *J. Vasc. Med. Biol.* **2**:47–51.

Shah, T.S., Satia, M.C., Gandhi, T.P., Bangaru, R.A., and Goyal, R.K., 1995, Effects of chronic nifedipine treatment on streptozotocin-induced diabetic rats. *J. Cardiovas. Pharmacol.* **26**:6–12.

Shehin, S.E., Sowers, J.R., and Zemel, M.B., 1989, Impaired vascular smooth muscle Ca^{2+} efflux and hypertension in Zucker obese rats. *J. Vasc. Med. Biol.* **1**:278–282.

Shimoni, Y., Firek, L., Severson, D., and Giles, W.R., 1994, Short-term diabetes alters K$^+$ currents in rat ventricular myocytes. *Circ. Res.* **74**:620–628.

Stoffel, M., Tokuyama, Y., Trabb, J.B., German, M.S., Tsaar, M.L., Jan, L.Y., Polonsky, K.S., and Bell, G.I., 1995, Cloning of rat KATP-2 channel and decreased expression in pancreatic islets of male Zucker diabetic fatty rats. *Biochem. Biophy. Res. Commun.* **26**:894–899.

Tesfamariam, B., Brown, M.L., and Cohen, R.A., 1995, 15-Hydroxyeicosatetraenoic acid and diabetic endothelial dysfunction in rabbit aorta. *J. Cardiovas. Pharmacol.* **25**:748–755.

Tesfamariam, B., 1994, Free radicals in diabetic endothelial cell dysfunction. *Free Radical Biology & Medicine.* **16**:383–391.

Trischmann, U., Klockner, U., Isenberg, G., Utz, J., and Ullrich, V., 1991, Carbon monoxide inhibits depolarization-induced Ca rise and increases cyclic GMP in visceral smooth muscle cells. *Biochem. Pharmacol.* **41**:237–241.

Vallance, P., Calver, A., and Collier, J., 1992, The vascular endothelium in diabetes and hypertension. *J. Hypertens.* **10**:S25–S29.

Wang, R., Wang, Z.Z., Wu, L., Hanna, S.T., and Peterson-Wakeman, R., 2001, Reduced vasorelaxant effect of carbon monoxide in diabetes and the underlying mechanisms. *Diabetes* **50**:166–174.

Wang, R., Wu, Y.J., Wu, L., Tang, G., and Hanna, S., 2000, Altered L-type voltage-dependent calcium channels in diabetic vascular smooth muscle cells. *Am. J. Physiol.* **278**:H714–H722.

Wang, R., 1998, Resurgence of carbon monoxide: an endogenous gaseous vasorelaxing factor. *Can. J. Physiol. Pharmacol.* **76**:1–15.

Wang, R., Liu, Y., Sauvé, R., and Anand-Srivastava, M.B., 1998, Diabetes-related abnormal calcium mobilization in smooth muscle cells are induced by hyperosmolality. *Mol. Cell. Biochem.* **183**:79–85.

Wang, R., Wu, L., and Wang, Z.Z., 1997. The direct effect of carbon monoxide on K_{Ca} channels in vascular smooth muscle cells. *Pflügers Arch.* **434**:285–291.

Wang, R. and Wu, L., 1997, The chemical modification of K_{Ca} channels by carbon monoxide in vascular smooth muscle cells. *J. Biol. Chem.* **272**:8222–8226.

Wang, R., Wang, Z.Z., and Wu, L., 1997, Carbon monoxide-induced vasorelaxation and the underlying mechanisms. *Br. J. Pharmacol.* **121**:927–934.

Wang, Z., Wu, L., and Wang, R., 1998, Kinin B2 receptor-mediated contraction of tail artery from normal or streptozotocin-induced diabetic rats. *Br. J. Pharmacol.* **125**:143–151.

Weidmann, P., Beretta-Piccoli, C., Keusch, G., Gluck, Z., Mujagic, M., Grimm, M., Meier, A., and Zeigler, W.H., 1979, Sodium-volume factor, cardiovascular reactivity and hypotensive mechanism of diuretic therapy in mild hypertension associated with diabetes mellitus. *Am. J. Med.* **67**:779–784.

White, R.E. and Carrier, G.O., 1988, Enhanced vascular alpha-adrenergic neuroeffector system in diabetes: importance of calcium. *Am. J. Physiol.* **255**:H1036–H1042.

White, R.E. and Carrier, G.O., 1990, Vascular contraction induced by activation of membrane calcium ion channels is enhanced in streptozotocine-diabetes. *J. Pharmacol. Exp. Ther.* **253**:1057–1062.

4

ENDOGENOUS CARBON MONOXIDE HAS PROTECTIVE ROLES IN NEOINTIMAL DEVELOPMENT ELICITED BY ARTERIAL INJURY

Toshisuke Morita[a], Yuko Togane[b], Makoto Suematsu[c],
Jun-ichi Yamazaki[b], and Shigehiro Katayama[a]

[a]Fourth Department of Internal Medicine
Saitama Medical School, Saitama 350-0495
[b]First Department of Internal Medicine
Toho University, School of Medicine
Tokyo 143-8540
[c]Department of Biochemistry and Integrative Medical Biology
School of Medicine, Keio University
Tokyo 160-8582, Japan

INTRODUCTION

The accumulation of vascular smooth muscle cells (VSMCs) in neointima resulting from the migration and proliferation of medial VSMCs in response to endothelial damage is believed to be one of the main events involved in the initiation of atherosclerosis. Although various types of growth factor and cytokines, including endothelin-1 (ET-1), platelet-derived growth factor-B (PDGF-B) and angiotensin II (Ang II), have been acknowledged to contribute generally to the development of atherosclerosis, recent studies have indicated that many species of oxidants can be considered to be early growth signals.[1,2]

Correspondence to Toshisuke Morita, MD, Ph.D., Fourth Department of Internal Medicine, Saitama Medical School, 38 Morohongo Moroyama Iruma-gun Saitama, 350-0495, Japan. Phone: 81-492-76-1204. Fax: 81-492-94-9752. E-mail: *toshijpn@saitama-med.ac.jp*

Recently, we demonstrated that cultured VSMCs expressed both heme oxygenase (HO) HO-1 and HO-2 and released CO into their conditioned media.[3] Furthermore, VSMCs-derived CO was found to suppress VSMC proliferation through the inhibition of ET-1, PDGF-B and the E2F-1 gene, in a manner similar to nitric oxide (NO).[4,5] It is thus not unreasonable to hypothesize that induction of HO could lead to inhibition of excessive VSMCs proliferation in atherosclerotic lesions through the biological action of CO as a reaction product.

In view of potential roles of HO in pathophysiological conditions, this study aimed to examine perturbation of HO expression in the carotid artery and its functional consequence on VSMC proliferative responses after denudation by balloon injury. We have also attempted to address whether such inhibitory effects of the HO-1 induction on VSMC proliferation are ascribable to the biological action of endogenous CO.

METHODS

Balloon Injury

Male Sprague Dawley rat (400–450 g) obtained from the Charles River Corporation was anesthetized with an intraperitoneal injection of pentobarbital sodium at 40 mg/kg, and endothelial denudation of the carotid artery was performed by three passages of a Forgaty 2F balloon catheter inflated to 2 atm (measured by manometer) as described elsewhere.[6]

Histological Examinations

At the required time after balloon injury, rats were sacrificed with a lethal dose of pentobarbital sodium, and their vascular systems were perfused via the left ventricle with phosphate buffer saline (PBS) for 5 minutes at 100 mmHg, and then fixed with neutral formaldehyde for 10 minutes at 100 mmHg. The left common carotid arteries were removed and 3 vessel rings (5 mm long) were cut and embedded in paraffin. Sections of each 5-mm ring were cut and stained with hematoxylin and eosin, as well as Azan-Mallory stain. Histological micrographs were captured and processed digitally by a computer-assisted 8-bit image analyzer (Power Macintosh 8800/NIH Image 1.58), and cross-sectional areas of medial and neointimal areas were quantified by the software at three times and the mean values of the area of interests were calculated for each section.

HO-1 and HO-2 were detected in tissues using monoclonal antibodies against rat HO-1 and HO-2.[7] Briefly, sections were trypsinized with 1% v/v trypsin in PBS at 37°C for 60 minutes, and washed with PBS. Nonspecific protein binding was blocked with 0.1% v/v horse normal serum at room temperature for 30 minutes. The sections were then incubated overnight at 4°C with the required monoclonal antibody, and the bound primary antibody was detected using an avidin-biotinylated horseradish peroxidase complex (Vectastain ABC kits, Vector Laboratories, Burlingame, CA). Finally the tissues were lightly counter-stained with hematoxylin. Positive staining with DAB appeared as a brownish-black color. The staining of uninjured portions of the external carotid artery of each rat was used as a control.

Treatment of Rats

Hemin was dissolved in dimethyl sulfoxide (DMSO) as described previously.[8] ZnPP was dissolved in 50 mmol/L Na_2CO_3 solution as described previously.[9] SnPP was dissolved in distilled water, which was adjusted to pH 11 with 0.1 N NaOH as described previously.[10] To block production of NO in vivo, N^w-nitro—L-arginine (L-NNA) was dissolved in drinking water to give a final dose of 60 mg/kg/day and given from 14 days before until 14 days after balloon injury.[11] L-NNA-untreated or -treated rats were treated with intraperitoneal injection of hemin (15 mg/kg), ZnPP (40 μmol/kg) or SnPP (50 μmol/kg) IP every other day from 3 days before until 14 days after balloon injury (n = 8–10 per group). Control rats were treated with equivalent volumes of vehicles.

Cell Culture

Primary cultures of rat aortic VSMCs were grown in Dulbecco's modified Eagle's medium (Gibco Laboratories, Grand Island, NY) with 10% v/v newborn calf serum as described previously.[3] When the culture reached a 70% confluence, the medium was changed to Dulbecco's modified Eagle's medium with supplemented 0.2% v/v newborn calf serum, and the cells were cultured for further 48 hrs prior to experiment.

Cell Proliferation

Cell proliferation was assessed by counting cells. 48 hrs after exposure to ET-1 (10 nM) or Ang II (100 nM), in the presence or absence of reagents, the cells were washed twice with ice-cold PBS, harvested and centrifuged. The cell pellets were resuspended in ice-cold PBS and the cells were counted with a Coulter counter (Coulter Corp., Hialeah, FL). The values are shown as percentages of the number versus that of control VSMC at the start of culture (n = 8).

RNA Analysis

Total tissue RNA was prepared from carotid arteries by guanidinium isothiocyanate extraction from carotid artery, 15 μg/lane was separated by electrophoresis on 1% w/v agarose gels containing formaldehyde, transferred to nitrocellulose membranes by blotting, and the filters were hybridized with cDNA probes specific for rat HO-1 and HO-2 as described previously.[3]

Determination of Heme Oxygenase Activity

Heme oxygenase activity in microsomes was determined in rat carotid artery at different time points after balloon injury and compared with the activity in control rats as described elsewhere.[7]

Determination of cGMP Levels

Levels of cGMP were measured in extracts of carotid artery with use of a commercial ELISA kit (Amersham). 14 days after balloon injury, carotid artery was harvested from rats. At the end of experiments, samples were immediately frozen in liquid nitrogen and stored at −80°C until the cGMP assay was carried out as described previously.[3]

Determination of concentrations of NOx in the urine

Concentrations of NOx were measured in urine samples collected before, 14 and 28 days after L-NNA treatment in the presence or absence of ZnPP, SnPP or hemin with a commercial kit (Nitrate/Nitrite assay kit Cat#780001, Cayman Chemical Co, MI). In these experiments, rats were placed in individual metabolic cages for 24 hrs and urine samples were collected for assay. The values were normalized by the levels of creatinine in the urine (Creatinine-Testwako assay kit 275-10502, Wako Pure Chemical Co. Osaka, Japan).

Drug Preparation

SnPP and ZnPP were purchased from Porphyrin Products, Inc. (Logan, UT). All other reagents used were obtained from Sigma, unless otherwise specified. Pure hemoglobin (Hb) and MetHb were prepared as described previously.[3,7] L-NNA, ODQ, 8-bromo-cGMP and clotrimazole were prepared as described elsewhere.[4,12,13]

Data Analysis

Significant differences were determined by one-way ANOVA and $p < 0.05$ was considered statistically significant.

RESULTS

Balloon Injury Induced HO-1 Expression in the Carotid Artery with the Increase in HO Activity

Figure 1a shows representative changes in HO gene expression demonstrated with Northern blot analysis, after balloon injury. Northern blot analysis revealed that low levels of HO-1 transcripts, whereas HO-2 mRNA expression was demonstrated clearly, in normal cartid artery, and that marked HO-1 mRNA expression was induced as early as 1 day after balloon injury, and declined thereafter. In contrast, the levels of HO-2 gene expression did not change as clearly as HO-1 during the 14 days after injury. Typical histological assessments of HO isozymes at 1, 4, 7 and 14 days after balloon injury are shown in Fig. 1b. As seen in the upper panel, the HO-1 expression was very low in the normal artery in medial VSMCs and increased markedly as early as 1 day after injury. By 14 days after injury, the HO-1-associated immunoreactivities

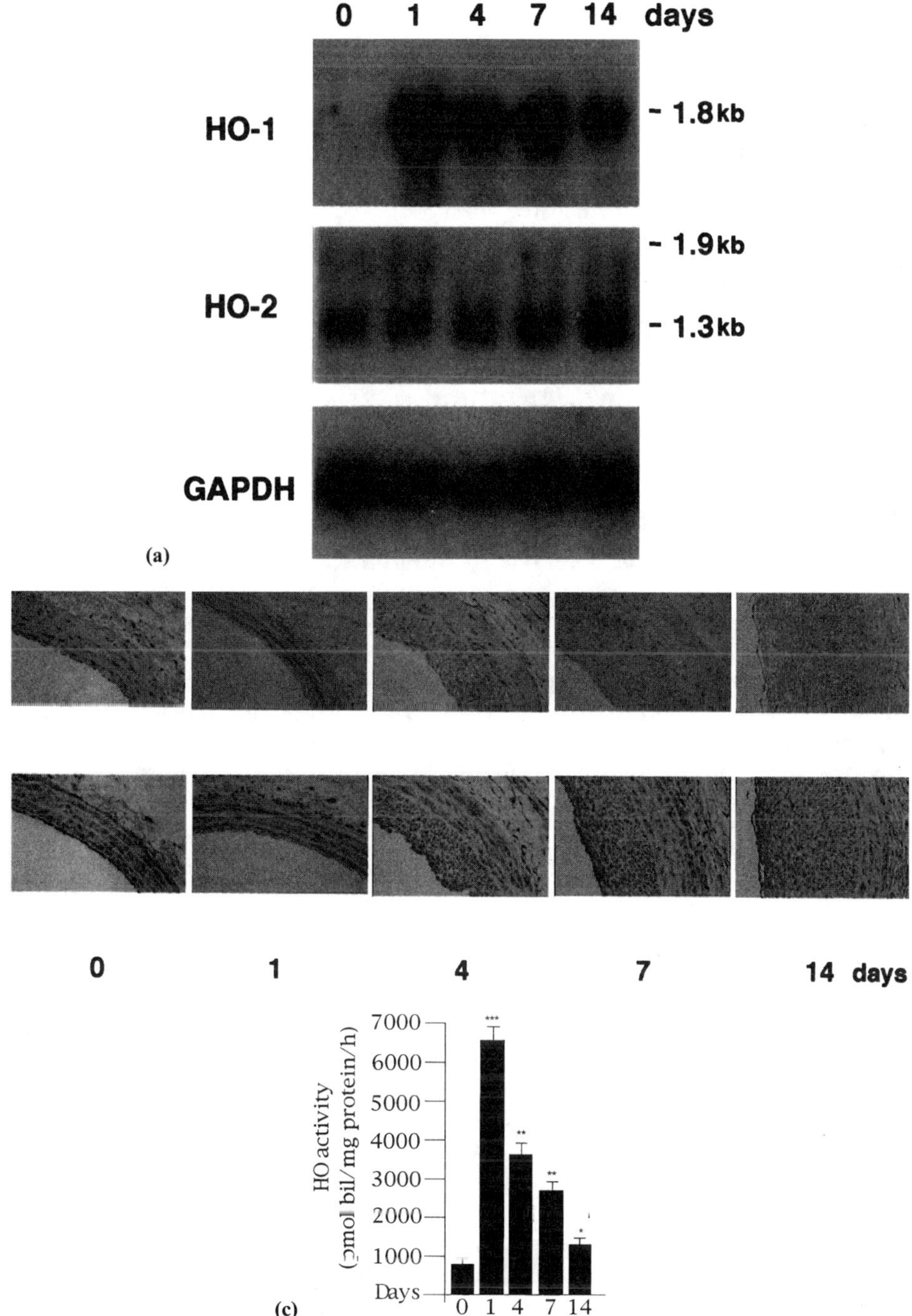

Figure 1. Changes in HO expression after balloon injury. (a) HO-1 gene expression was induced as early as 1 day after balloon injury and declined subsequently. HO-2 expression gene did not change significantly after injury. (b) Upper panels show that HO-1 protein expression was induced initially in the media, and then relocalized in the intima in parallel with neointimal development (×400). Lower panels show that HO-2 protein expression in the media and adventitia did not change, but it was detected in the neointima, in a manner similar to that of HO-1 (×400). (c) Summary of balloon-elicited changes in the HO activity in denuded artery. Days 0 shows intact carotid artery. HO activity increased after balloon injury, in a similar manner to those of HO-1 expression. Mean values ± S.E. are shown. *, p < 0.05; **, p < 0.01; ***, p < 0.001 versus HO activity of intact carotid artery.

were decreased, a similar time history to that of Northern blot analysis, and confined to the neointima. In contrast, the HO-2 immunoreactivities shown in lower panel in medial VSMCs and advential cells did not change after injury, and HO-2 was detected in the intima, its expression increasing in parallel with neointimal development.

We then inquired whether the injury-induced HO gene could result in an increase in the HO activity, we determined the HO activity in carotid artery from rats subjected to balloon injury. Figure 1c shows the time course of the HO activity in denuded artery at the indicated time periods after balloon injury. Upon balloon injury, the HO activity in denuded artery significantly increased as compared with the control after balloon injury; the activity became maximal at 1 day after injury and declined subsequently. The changes in the HO activity in denuded artery is essentially similar to those in the HO-1 expression.

The Enzymatic Product of HO, Presumably CO, Reduces of the Proliferative Response of VSMCs to Balloon Injury

We examined effects of alterations in the HO activity in the denuded carotid artery, by treating rats with ZnPP or SnPP to block or hemin to induce HO activity further. Figure 2a shows the HO activity in denuded artery collected at 14 days after balloon injury. The HO activity in denuded artery of the HO inhibitor-treated rats was significantly suppressed, while hemin-treated rats showed markedly elevated HO activity, compared with control rats. Figure 2b shows the typical neointimal development 14 days after balloon injury in the normal and HO-modulated rat carotid artery. The HO inhibitor-treated rats exhibited markedly augmented neointimal formation, as compared with that observed in untreated rats. In contrast, treating rats with hemin suppressed neointimal formation.

The inhibitors of metalloporphyrin are known to affect NOS activity.[14,15] To eliminate the role of NO on neointimal formation, we therefore performed parallel experiments using rats undergoing the 4-week L-NNA treatment. We found that the HO modulators changed intima / media ratio in a similar manner to those observed in L-NNA-untreated rats (data not shown). We then investigated whether cGMP levels in denuded artery is changed by the treatment of HO modulators. Figure 2c shows the changes in cGMP levels in denuded artery from four-week L-NNA-treated rats in the presence or absence of the HO modulators. The HO modulators regulate cGMP levels in a similar manner to those of HO activity. These findings indicate that the induction of HO expression leads to an actual increase in CO and results in cGMP elevation in denued artery independently of NO.

To establish further whether HO activity regulate the proliferative response of VSMCs to ET-1 or Ang II, we performed cell proliferation assay, as production of these mitogens has been reported to be induced in the vascular wall by balloon injury.[16,17] Figure 3a shows the proliferative response of serum-deprived VSMCs to ET-1 or Ang II in the presence of L-NNA at the concentrations previously reported,[3] since NO is expected to have a suppressive effect on VSMC proliferation. Incubation with ZnPP or SnPP enhanced the mitogen-induced increase in VSMCs proliferation significantly. Hb, which captures CO from cultures, also augmented this response in

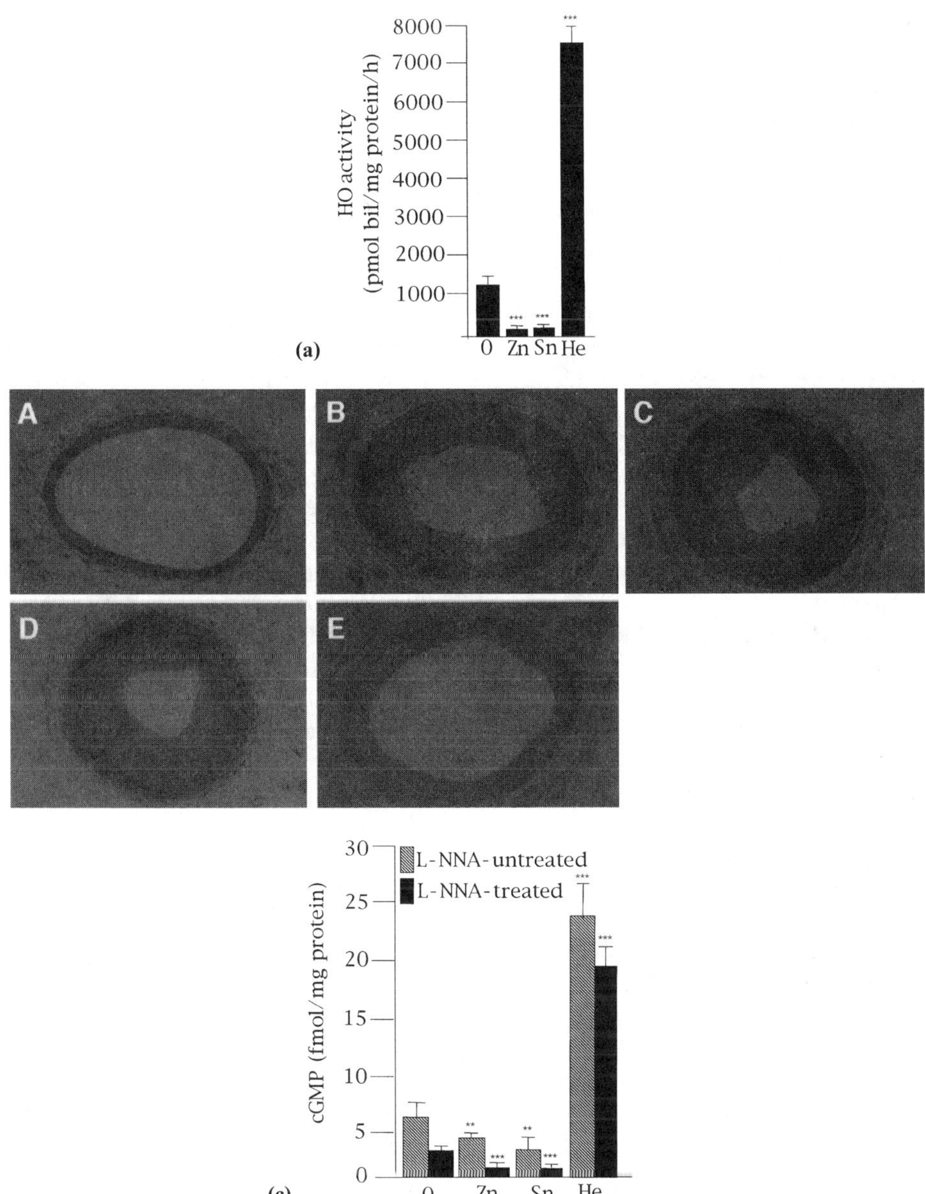

Figure 2. The modulators of HO regulate neointimal development 14 days after balloon injury accompanied with changes in HO activity and cGMP levels. (a) HO activity was lowered with ZnPP or SnPP, and elevated with hemin, respectively. 0: L-NNA treatment, Zn: ZnPP, Sn: SnPP, He: hemin. Mean values ± S.E. are shown ***, $p < 0.001$ versus HO activity of denuded carotid artery of control rats. (b) Histological examination (hematoxylin-eosin staining) (×80) revealed that the HO inhibitors or an HO inducer augmented or reduced neointimal formation. (A) normal, (B) untreated, (C) ZnPP-treated, (D) SnPP-treated and (E) hemin-treated rat. (c) The HO modulators changed cGMP levels in denuded artery in L-NNA-treated rats, in a similar manner to those in L-NNA-untreated rats. 0: no treatment with HO modulators, Zn: ZnPP, Sn: SnPP, and He: hemin. Mean values ± S.E. are shown. **, $p < 0.01$; ***, $p < 0.001$; versus cGMP levels of denuded carotid artery of corresponding control rats.

a similar manner to the HO blockers. In contrast, the addition of metHb, which can not bind to CO, did not stimulate mitogen-induced increase in VSMCs proliferation significantly. Furthermore, addition of hemin to the cultures significantly reduced the proliferative response of VSMCs. These findings suggest that the enzymatic product, that is CO, regulates VSMC proliferation.

Observation suggesting that suppression or elevation of CO alters proliferative responses of VSMCs in parallel with cellular cGMP levels led us to examine whether supplements of cGMP could mimic the inhibitory action of CO on the cell proliferation in vitro. As seen in Fig. 3b, administration of Hb cancelled out and further enhanced the proliferative responses. Application of ODQ mimicked the Hb-induced enhancement of the response, but its effect appeared to be in part. The Hb-elicited changes were suppressed by supplement with 8-bromo-cGMP, but its effect was also in part. These results raised a possibility that the stimulatory effect of the CO scavenger such as Hb involves both cGMP-dependent and -independent mechanisms. Among such cGMP-independent mechanisms for the CO-mediated signaling events, cytochrome P 450 monooxygenase constitute a putative candidate receptor besides soluble guanylate cyclase, since CO could bind to heme enzymes which possess ferrous heme as a prosthetic molecule for the enzyme reaction under steady-state conditions.[13] We have thus tested effects of clotrimazole on the Hb-induced changes. This reagent is known to bind the prosthetic heme of cytochrome P 450 monooxygenases and shares the inhibitory action on the enzyme with CO. As seen, pretreatment with the reagent significantly repressed the Hb-induced changes. Furthermore, co-application of this reagent with 8-bromo-cGMP additively suppressed the Hb-induced enhancement of the proliferation, suggesting the presence of cGMP-dependent and— independent mechanisms.

DISCUSSION

In this study, we found that the HO expression patterns of normal and denuded carotid arteries differed. Only little expression of HO-1 was detected if any, except for a small extent of the expression in the medial VSMCs layer, while HO-2 was expressed prominently in the endothelium, medial VSMCs and adventitial cells of the rat normal carotid artery. As early as 1 day after balloon injury, the HO-1 expression increased rapidly and site-specifically in medial VSMCs layers and then HO-1 relocalized to the neointima in parallel with the development of intimal thickening. On the other hand, the levels of HO-2 gene expression in denuded artery determined as a whole tissue did not change significantly throughout 14 days after injury, suggesting that the transient elevation of the HO activity apparently occurs mainly through upregulation of HO-1. However, careful examination of the regional protein expression by immunohistochemistry revealed that HO-2 became detectable in the neointima formed after the vascular insult, to a similar extent to that of HO-1. Upregulation of protein expression of these HO isozymes are accompanied by the enzyme activity in the same tissue. Considering that CO generated through HO in the extravascular space could lead to alterations in vascular functions such as a reduction of vascular tone,[7,18] it is not unreasonable to suggest that these newly upregulating HO

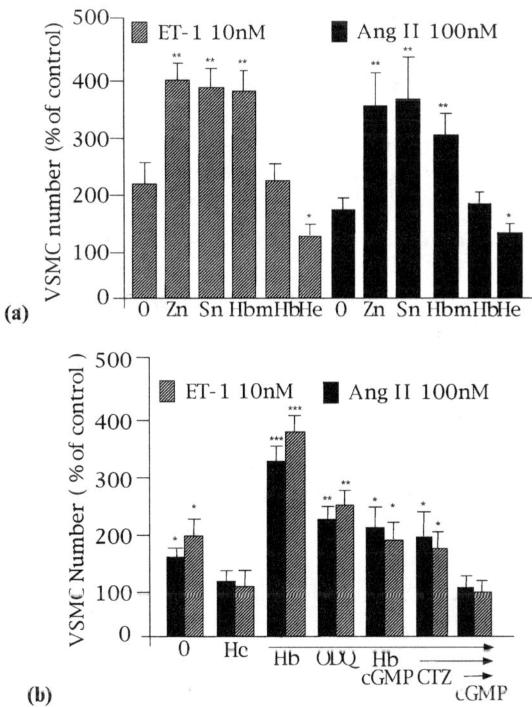

Figure 3. (a) Endogenous CO levels regulate mitogen-induced VSMCs proliferation. The numbers of VSMCs were determined 48 h after exposure to ET-1 or Ang II in the presence of L-NNA. Cultures were treated with the indicated reagents for the last 12 h of the exposure to ET-1 or Ang II in the presence of L-NNA to modulate the endogenous CO levels. SnPP (Sn) (100 μM), ZnPP (Zn) (10 μM), HbO2 (Hb) (50 μM), metHb (mHb) (50 μM) and hemin (He) (5 μM). Each experiment was repeated four times in quadruplicate; mean values ± S.E. (n = 6) are shown. *, $p < 0.05$; **, $p < 0.01$ versus numbers in the absence of HO modulators at each mitogen. (b) 8-bromo-cGMP and clotrimazole mimic the inhibitory effect of the enzymatic products of HO on VSMCs proliferation. The numbers of VSMCs were determined 48 h after exposure to ET-1 or Ang II. Cultures were treated with the indicated reagents for the last 12 h of the exposure to ET-1 or Ang II in the presence of L-NNA. ODQ (1 μM), cGMP (8-bromo-cGMP) (1 mM), and clotrimazole (CTZ) (1 μM). Each experiment was repeated four times in quadruplicate; mean values ± S.E. (n = 6) are shown. *, $p < 0.05$; **, $p < 0.01$; ***, $p < 0.001$ versus numbers of VSMCs treated with hemin.

isozymes contribute to modulation of proliferative responses in VSMCs. The current study suggests that endogenous CO derived mainly from HO-1 and partly from HO-2 functions as an inhibitory regulator preventing excessive VSMC proliferation in the denuded carotid artery.

Balloon injury has been reported to induce the production of many vasoactive factors, including ET-1,[17] renin-angiotensin system[17] and reactive oxygen species,[19] resulting in changes in the phenotype of VSMCs. Furthermore, after the denudation of endothelial cells, the VSMCs layer is exposed directly to red blood cells in the blood stream and this may change the shear stress and redox state in the vascular wall. As HO-1 expression is known to be upregulated by mechanical stress,[20] as well as by chemical factors, the rapid induction of HO-1 expression in VSMCs may be mediated by these changes in the vascular wall. We observed that HO-1 expression translocated from the media to the intima during the first week post-injury, and during 14 days, it was virtually restricted to the neointima, although expression levels immediately after balloon injury were high in the medial. The time history of the relocalization of HO-1 immunostaining after balloon injury are very similar to that of VSMCs: Neointimal formation following balloon injury of the rat carotid artery consists of four steps.[21] First, replication of VSMCs in the media, which occurs 0 to 3 days after balloon injury. Secondly, VSMCs exhibit migration from the media to the intima, which occurs 3 to 14 days after balloon injury. The third and fourth steps are VSMCs proliferation and extracellular matrix deposition in the neointima, respectively, which begin 7 days after balloon injury. HO-1 was reported to be expressed by VSMCs in human atherosclerotic lesion, but not in normal vessels.[25] The phenotype of VSMCs in the intima and atherosclerotic lesions is known to differ from that of those in the media. Taken together, these findings suggest that HO-1 expression and VSMCs differentiation are linked to each other. In fact, the levels of HO-1 in cultured VSMCs after passaging are increased, as compared with those observed in primary cultures of VSMCs (unpublished observation).

Taken together with our previous studies, the current results suggest that CO derived from the HO reaction serves as a potentially inhibitory mediator for stimulus-elicited cell proliferation in vitro and in vivo. As was suggested previously, activation of soluble guanylate cyclase appears to be involved in the CO-mediated signaling mechanisms in the current study, inasmuch as ODQ restored and 8-bromo-cGMP mimicked the hemin-mediated reduction of mitogen-induced cell proliferation at least in part. At the same time, the current result raised an important possibility that the guanylate cyclase-independent mechanism could be involved in mechanism for CO-mediated regulation of cell proliferation, since sufficient amounts of the membrane-permeable cGMP analogue did not fully mimic the effect of the HO-1 induction by hemin. Such a notion against the involvement of soluble guanylate cyclase in the CO-mediated signaling events has recently been emphasized in that CO is not so potent as NO to activate soluble guanylate cyclase.[23] We therefore examined whether inhibition of cytochrome P 450 monooxygenase could mimic the effects of CO on VSMCs proliferation, since the blockade of this enzyme reaction is one of well-known biological actions of CO.[24] As expected, the enzyme inhibitor such as clotrimazole exhibited additive suppressive effects with 8-bromo-cGMP on mitogen-induced proliferative responses, suggesting the involvement of cGMP-independent mechanisms. Detailed mechanisms for cytochrome P 450-mediated regulation of cell proliferation ha not fully been addressed in the current study. However, the CO-mediated signal reception by this enzyme system has recently paid much attention in that a variety of cytochrome -P450-derived products are biologically active to modulate potassium and calcium concentrations.[25,26,27] Further investigation is obviously required to fully address whether endogenous CO could actually modulate generation of specific cytochrome P450-derived products and thereby cause alterations in cell fuction as a result of HO-1.

In this study, we demonstrated clearly that HO-1 expression is markedly increased in denuded carotid artery and that the increased levels of CO via HO in the vascular wall results in inhibition of balloon injury-induced neointimal development. Therefore, the HO/CO system in vascular wall may protect against the excessive VSMCs proliferation associated with vascular diseases.

ACKNOWLEDGMENTS

The authors thank Dr. Stella Kourembanas for her critical review, and Ms. Junko Tatebe and Ms. Machi Ohno for her technical assistance.

This work was supported in part by a grant-in-aid for Scientific Research from the Ministry of Education of Japan (09770496), a grant from Maruki Memorial Foundation (B 95003), and by the Japan Foundation of Cardiovascular Research.

REFERENCES

1. A.S. Baas and B.C. Berk, Differential activation of mitogen-activated protein kinases by H_2O_2 and O_2- in vascular smooth muscle cells, *Circ. Res.* 77, 29–36 (1995).
2. G.N. Rao and B.C. Berk, Active oxygen species stimulate vascular smooth muscle cell growth and proto-oncogene expression, *Circ. Res.* 70, 593–599 (1992).
3. T. Morita, M.A. Perrella, M.E. Lee, and S. Kourembanas, Smooth muscle cell-derived carbon monoxide is a regulator of vascular cGMP, *Proc. Natl. Acad. Sci. USA.* 92, 1475–1479 (1995).
4. T. Morita, S.A. Mitsialis, Y. Liu, and S. Kourembanas, Carbon monoxide controls the proliferation of hypoxic vascular smooth muscle cells, *J. Biol. Chem.* 52, 32804–32809 (1997).
5. T. Morita and S. Kourembanas, Endothelial expression of vasoconstrictors and growth factors is regulated by smooth muscle cell-derived carbon monoxide, *J. Clin. Invest.* 96, 2676–2682 (1995).
6. A.W. Clowes, M.A. Reidy, and M.M. Clowes, Kinetics of cellular proliferation after arterial injury, I: smooth muscle growth in the absence of endothelium, *Lab. Invest.* 49, 327–333 (1983).
7. N. Goda, K. Suzuki, M. Naito, S. Takeoka, E. Tsuchida, Y. Ishimura, T. Tamatani, and M. Suematsu, Distribution of heme oxygenase isozymes in rat liver, *J. Clin. Invest.* 101, 604–612 (1998).
8. R.D. Levere, P. Martasek, B. Escalante, M.L. Schwartzmannn, and N.G. Abranam, Effect of heme arginate administration on blood pressure in spontaneously hypertensive rats, *J. Clin. Invest.* 86, 213–219 (1990).
9. R.A. Johnson, M. Lavesa, D. Askari, N.G. Abraham, and A. Nasjletti, A heme oxygenase product, presumably carbon monoxide, mediates a vasodepressor function in rats, *Hypertension.* 25, 166–169 (1995).
10. M. Maines and D. Trakshel, Differential regulation of heme oxygenase isozymes by Sn- and Zn-protoporphyrins: possible relevance to suppression of hyperbilirubinemia, *Biochem. Biophys. Acta.* 1131, 166–174 (1992).
11. N.L. Kanagy, Increased vascular responsiveness to alpha 2-adrenergic stimulation during NOS inhibition induced hyperonoion, *Am. J. Physiol.* 273, H2756 H2761 (1997).
12. E. Nisoli, E. Clementi, C. Tonello, C. Sciorati, L. Briscini, and M.O. Carruba, Effects of nitric oxide on proliferation and differentiation of rat brown adipocytes in primary cultures, *Br. J. Pharmacol.* 125, 888–894 (1998).
13. Y. Shinoda, M. Suematsu, Y. Wakabayashi, T. Suzuki, N. Goda, S. Saito, T. Yamaguchi, and Y. Ishimura, Carbon monoxide as a regulator of bile canalicular contrctility in cultured rat hepatocytes, *Hepatology.* 28, 286–295 (1998).
14. D. Luo and S.R. Vincent, Metalloporphyrins inhibit nitric oxide-dependent cGMP frmation in vivo, *Eur. J. Pharmacol.* 267, 253–262 (1994).
15. M.K. Meffert, J.E. Haley, E.M. Schuman, H. Schulman, D.V. Madison, Inhibition of Hippocampal heme oxygenase, nitric oxide synthese, and long-term potentiation by metalloporphyrins, *Neuron.* 13, 1225–1233 (1994).

16. T. Naruko, K. Ueda, K. Haze, A.C. van der Wal, C.M. van der Loos, K. Tanzawa, and A.E. Becker, Endothelin and endothelin-converting enzyme expression at the site of percutaneus transluminal coronary angioplasty in humans, *Circulation.* 96, I-348 (1997).

17. H. Rakugi, H.J. Jacob, J.E. Krieger, J.R. Ingelfinger, and R.E. Pratt, Vascular injury induces angiotensinogen gene expression in the media and neointima, *Circulation.* 87, 283–290 (1993).

18. R. Zakhary, S.P. Gaine, J.L. Dinerman, M. Ruat, N.A. Flavahaan, and S.H. Snyder, Heme oxygenase 2: Endothelial and neuronal localization and role in endothelium-dependent relaxation, *Proc. Natl. Acad. Sci. USA.* 93, 795–798 (1996).

19. G.A. Ferns, L. Forster, L.A. Stewart, M. Konneh, Z.J. Nouroos, and E.E. Anggard, Probucol inhibits neointimal thickening and macrophage accumulation after balloon injury in the cholesterol-fed rabbit, *Proc. Natl. Acad. Sci. USA.* 89, 11312–11316 (1992).

20. C.T. Wagner, W. Durante, N. Christodoullides, J.D. Hellums, and A.I. Schafer, Hemodynamic forces induced the expression of heme oxygenase in cultured vascular smooth muscle cells, *J. Clin. Invest.* 100, 589–596 (1997).

21. S.M. Schwartz, D. deBlois, and E.R. O'Brien, The intima:soil for atherosclerosis and restenosis, *Circ. Res.* 77, 445–465 (1995).

22. L.J. Wang, T.S. Lee, F.Y. Lee, R.C. Pai, and L.Y. Chau, Expression of heme oxygenase-1 in atherosclerotic lesions, *Am. J. Pathol.* 152, 711–720 (1998).

23. F. Coceani, L. Kelsey, and E. Seidlitz, Carbon monoxide-induced relaxation of the ductus arteriosus in the lamb: evidence against the prime role of guanylyl cyclase, *Br. J. Pharmacol.* 118, 1689–1696 (1996).

24. F. Coceani, C.A. Breen, J.G. Lees, J.R. Falck, and P.M. Olley, Further evidence implicating a cytochrome P 450-mediated reaction in the contractile tension of lamb ductus arteriosus, *Cir. Res.* 62, 471–477 (1988).

25. H. Lin and J.J. McGrath, Carbon monoxide effects on calcium levels in vascular smooth muscle, *Life Sci.* 43, 1813–1816 (1988).

26. H. Lin and J.J. McGrath, Is carbon monoxide a calcium blocking agent? I. Effect of carbon monoxide on mechanical tension in isolated thoracic aorta, *Fed. Proc.* 2, A 372 (1988).

27. R. Wang, Z.Z. Wang, and L. Wu, Carbon monoxide-induced vasorelaxation and underlying mechanisms, *Br. J. Phramacol.* 121, 927–934 (1997).

5

END TIDAL BREATH CARBON MONOXIDE (ETCO) LEVELS IN PREGNANT WOMEN

Daniel S. Seidman[a], Micha Baum[a], Doron Kreiser[a], Israel Hendler[a],
Eyal Schiff[a], Maurice Druzin[b], David K. Stevenson[c], Dinesh Shah[d],
Avery A. Fanaroff[d], and Phyllis A. Dennery[c]

Department of Obstetrics and Gynecology[a]
Sheba Medical Center, Tel-Hashomer
Sackler School of Medicine
Tel-Aviv University
Department of Ob/Gyn[b], Pediatrics[c]
Stanford University School of Medicine
Stanford, CA, and Department of Pediatrics[d]
University Hospitals of Cleveland
Case Western Reserve University
Cleveland, OH

INTRODUCTION

Carbon monoxide (CO) is a simple diatomic gas molecule that shares some of the physiochemical properties of nitric oxide (NO).[1] Carbon monoxide, like NO, stimulates soluble guanylyl cyclase and thereby raises intracellular levels of cyclic guanosine monophosphate (cGMP). Endogenous CO production under physiological conditions serves to regulate smooth muscle tone,[2] and may play a role in neurotransmission.[3] Exogenously added CO has been shown to increase the smooth muscle intracellular levels of cGMP and to cause endothelium-dependant vascular relaxation.[4]

Heme oxygenase (HO) is the enzyme responsible for the conversion of heme to bilirubin and CO. Two isoforms of this enzyme are known to exist. The inducible form, HO-1, is regulated by various forms of oxidative stress[5] whereas the constitutive form, HO-2, has been shown to vary only in response to glucocorticoids.[6,7] HO activity is found in most tissues but is not evenly distributed.[8] The HO pathway may affect several physiological functions including vascular tone, oxidative stress and cellular proliferation.[9–11]

HEME OXYGENASE AND PREGNANCY

Both known isoforms of HO were found to be expressed in the human placenta.[12] The distribution of the HO immunoreactivity was shown to be wide, including the syncytiotrophoblast layer of placenta villi, the endothelium and smooth muscle cells of the umbilical-placental blood vessels, and in all layers of the fetal membranes.[13] Immunostaining for HO-2 has been noted in healthy women to be prominent in the syncytiotrophoblast in the first trimester and to be significantly reduced by term.[12] Endothelial immunostaining for HO-2 was weak in the first trimester, but significantly increased toward term.[13] It appears that in the umbilical cord and myometrial tissues predominance of a particular isoenzyme is not readily observed.[14]

Impaired nitric oxide (NO) synthesis was recently implicated in the pathogenesis pregnancy induced hypertension. This may reflect a state due to failure of the vasodilatation and decreased vascular reactivity characteristic of normal pregnancy, which is at least in part regulated by NO.[15–17] In an animal model, inhibition of NO synthesis during pregnancy causes a long-lasting blood pressure increase.[18] It was therefore implied that a reduction in the synthesis of NO might contribute to the pathogenesis of preeclampsia.[17,19,20] However, contradicting studies have demonstrated an increase in NO production in preeclampsia. These findings implied a compensatory response to improve blood flow in the placenta and/or a role for NO in limiting platelet adhesion and aggregation.[21–23]

The similarities in the mechanisms of action of NO and CO in endothelium-dependent arterial relaxation suggest that CO formation could also have a contributory role in the pathogenesis of pregnancy. The possible effect of the HO/CO pathway in pregnancy has only recently been investigated. It has been suggested that HO plays a role in human placental throphoblast invasion[12] and vascular function control.[13] However, little is known regarding endogenous CO production during pathophysiological conditions of pregnancy.

The aim of our investigation was therefore to estimate endogenous CO formation in pregnant women experiencing complications of pregnancy by monitoring their breath end tidal CO (ETCO) levels.

END TIDAL CO (ETCO) MEASUREMENTS

Increased ETCO measurements can reflect exposure to both exogenous and endogenous sources of CO. The major endogenous source of CO is due to increased CO production when hemolysis occurs, because hemoglobin degradation is the only metabolic pathway that produces significant amounts of this gas.[24] ETCO was accordingly shown to accurately identify hemolysis.[25] Smoking is the most important environmental source of CO. Measurement of CO levels in the maternal breath has been shown in late pregnancy to accurately assess the exposure of women to tobacco smoke.[26] Therefore, once hemolysis and exposure to smoking can be ruled out, it is assumed that endogenous CO formation in pregnant women can be reliably assessed by measuring their breath ETCO levels corrected for CO in the ambient air (ETCOc).

ETCOc measurements were obtained using a portable automated CO analyzer (The Natus® CO-Stat® End Tidal Breath Analyze, Natus Medical Inc., San Carlos, CA). This device can perform automatic sampling and analysis of end expiratory air without the requirement for laboratory testing or highly trained personnel. Thereby such highly accurate devices can provide on the spot immediate results in a noninvasive and straightforward manner.[27]

Measurements were obtained from all pregnant mothers enrolled in our studies following instruction by one of the investigators and at least a 15 minutes rest. Each participant gave a 90 seconds breath exhale sample, collected and simultaneously analyzed, via a 5F catheter placed 0.5 cm into the anterior nares.

PREECLAMPSIA AND PREGNANCY INDUCED HYPERTENSION

Preeclampsia and pregnancy induced hypertension (PIH) are major complications of pregnancy. As the pathogenesis of these disorders is poorly understood, prediction and prevention remains an elusive goal.

Preeclampsia, and especially its severe variant the HELLP (Hemolysis, Elevated Liver enzymes and Low Platelets) syndrome, are frequently misdiagnosed at initial presentation.[28,29] Early diagnosis is critical because the morbidity and mortality rates associated with the severe forms of this disease have been reported to be as high as 25 percent.[29] Neonatal morbidity and mortality is also high in these pregnancies.[30]

Since delay in diagnosis of preeclampsia can be associated with mortal consequences,[31] better means of early recognition of women destined to develop preeclampsia is an urgent need. An accurate predictive test may allow timely transfer of patients to tertiary centers where adequate intervention and treatment can be promptly provided. So far, no test exit that satisfies the criteria for an ideal screening test.[32]

As we described above it was recently postulated that preeclampsia might reflect a state of impaired NO synthesis, due to failure of the vasodilatation and decreased vascular reactivity. The similarity in the mechanisms of action of NO and CO in endothelium-dependent arterial relaxation suggest that CO formation could also have a contributory role in the pathogenesis of preeclampsia. This is supported by our recent observation that HO is expressed more intensely in the placenta, umbilical cord, and myometrium of preeclamptic patients compared with normal pregnant patients.[14] Furthermore, It has been shown that CO may play a role in the regulation of placental hemodynamics.[12]

We therefore performed a prospective multicenter study in order to estimate the endogenous CO production via ETCOc measurements in women with PIH and preeclampsia compared with control pregnant women. We prospectively obtained ETCOc measurements in three tertiary medical centers (Lucile Salter Packard Children's Hospital at Stanford University, MacDonald Women's Hospital at Cleveland and Sheba Medical Center, Tel-Hashomer, Israel).[33,34] A Natus® CO-Stat® End Tidal

Breath Analyzer (Natus Medical Inc., San Carlos, CA) was used. The study group included 52 women with PIH/PET. The control groups included 42 first and 63 third trimester normotensive pregnant women and 46 non-pregnant women.

The mean ± SD ETCOc measurements were significantly lower (p < 0.0001) in the PIH/PET group compared to first and third trimester normotensive pregnant women and non pregnant women, 1.29 ± 0.35 vs. 1.72 ± 0.46, 1.77 ± 0.48, and 1.72 ± 0.54 ppm, respectively. The ETCOc values were <1.6 ppm in 86.5% of PIH/PET women compared with, respectively, only 52.4%, 46%, 47.8% of first and third trimester normotensive pregnant women and non-pregnant women (p < 0.0001). In the three centers the controls had a similar mean ETCOc and the differences found remained significant when results for each center were analyzed separately. The ETCOc concentrations were also found not to be effected by gestational age and ethnic origin.[34]

It was therefore concluded that ETCOc levels might be significantly lower in women with PIH/PET. Further investigation is required to determine if the lower CO levels reflect a deficient compensatory response to the increase in blood pressure or whether these are primary changes of significance to our understanding of the obscure pathogenesis of this common and severe disorder of pregnancy. As we discussed in the introduction, CO may act as an endogenous vasodilator[1,2,4] and can inhibit platelet aggregation.[11] It is therefore plausible that CO might influence the regulation of uterine and feto-placental hemodynamics in normal pregnancy, and decreased generation of CO in preeclampsia may point at an inadequate compensatory response. This is consistent with a close relationship found in animal models between heme metabolism and blood pressure regulation.[35,36]

It has been proposed that the poorly perfused fetoplacental unit in preeclamptic patients is the origin of oxygen free radicals and lipid peroxides.[37–39] Furthermore, a common hypothesis suggest that in preeclampsia placental pathologic mechanisms result in endothelial injury,[39] leading in turn to an increase in fetal-placental vascular resistance, associated with altered shear stress, nuetrophil activation and generation of free radicals.[40,41] Both reactions can potentially induce HO, generating the heme catabolic pathway products CO and bilirubin. Carbon monoxide can decrease placental vascular resistance through the production of cGMP thereby protecting against PIH.[2] In turn, the bilirubin generated by heme breakdown through HO, is an endogenous antioxidant, and can interact with the excess free radicals[40,41] thereby modifying the effects of preeclampsia. The low levels of ETCO found in our study among patients with PIH or preeclampsia may point at an improper activation of the important HO/CO pathway in patients developing this severe disease of pregnancy.

SMOKING AND HYPERTENSION DURING PREGNANCY

Cigarette smoking during pregnancy has been recognized for over three decades to be associated with reduced risks of preeclampsia.[42] A recent systematic review of the existing evidence found that the risk of preeclampsia in pregnant women who smoked was 32% lower than that among nonsmoking pregnant women.[44] Furthermore, pooled data from cohort and case-control studies showed that this inverse association was dose-related and remarkably consistent across studies conducted in various populations and countries.[44] A current study that measured urinary cotinine in order to assess tobacco exposure confirmed the reduced risk of developing preeclampsia with cigarette smoking.[45] Using data from the Collaborative Perinatal Project, it was recently demonstrated that smoking is associated not only with a reduced risk of hypertension during pregnancy, but also with a protective effect that appeared to continue even after cessation of smoking.[46]

Maternal smoking in pregnancy is a well-recognized health hazard to both the mother and her newborn.[47–49] The paradoxical effect of exposure to tobacco during pregnancy, where smoking reduces the incidence of preeclampsia, but increases the perinatal morbidity and mortality, has long puzzled investigators.[46,50]

Many hypotheses have been put forward to explain the association between maternal smoking and reduced risk of preeclampsia, but none have been substantiated by much basic research. One explanation speculates that smokers, compared with nonsmokers, have a reduced incidence of preeclampsia, due to lesser expansion of their plasma volume.[50] Another explanation suggests that the decreased risk of PIH among smokers might be attributed to nicotine inhibition of thromboxane A2 production.[51] Others have speculated that chronic smoking desensitizes the endothelium, thereby reducing its responsivity to acute perturbations that might develop from the uteroplacental vascular injury characteristics of preeclampsia. Such down-regulation of endothelial sensitivity by chronic endothelial injury may have a protective effect during pregnancy.[52] It has also been indicated that tobacco exposure may provide protection against the development of preeclampsia by reducing the immune response[45] or by inducing liver enzymes that could metabolize endothelial toxins produced by the placenta.[46]

Carbon monoxide has been implicated as responsible for some of the detrimental effects of smoking. This may be, at least in part, attributed to the high affinity of CO to fetal hemoglobin. Maternal CO levels in the exhaled air were shown to be associated with a linear decrease in birth weight.[25,53] However, the physiological role of CO has only recently been recognized. It is now known that the heme degradation pathway and CO formation could have a contributory role in the regulation of placental hemodynamics.[54] We therefore speculated that in pregnancy CO, generated by smoking, may have in addition to its familiar adverse effect also a direct influence on the fetal-placental vasculature, resulting in the so called "protective" effect against preeclampsia.[34]

The results of our study, showing a significantly lower level of CO in women with preeclampsia,[34] support the concept that moderately elevated levels of CO may offer an advantage in terms of the risk for preeclampsia. We therefore suggest that CO might at least in part explain the lower incidence of preeclampsia among women who smoke.

PREMATURE LABOR

Premature labor remains the most important problem facing modern care of pregnant women, demanding tremendous costs for care of prematurely born infants.

Preterm birth is the major cause of perinatal morbidity and mortality in the world. Prematurity is responsible for 75% of infant deaths and 50% of the long-term neurologic handicaps, including cerebral palsy, blindness, deafness, and developmental delays. The survival rate of neonates is improved by 2% per day from the 23rd to the 26th week of pregnancy (i.e., from 16% at 23 weeks to 57% at 26 weeks), reaching 80% survival at 28 weeks and >90% by 30 weeks of gestation. Therefore any treatment that prevents premature birth will profoundly reduce neonatal mortality and morbidity rates. However, the utilization of premature delivery depends on accurate early identification of preterm contractions.[55]

Recent attention has focused on the part that NO might play in maintaining myometrial (uterine smooth muscle) contractility during pregnancy.[56–58] It has been suggested that the L-arginine-NO system may contribute to uterine quiescence during gestation and the initiation of labor at term,[57,58] although not all authors have confirmed this proposed physiologic role for NO.[59] The involvement of a NO in control of human uterine contractility during pregnancy is presumed to be through the stimulation of soluble guanylyl cyclase, thereby raising intracellular levels of cyclic guanosine monophosphate (cGMP).[56] Carbon monoxide can similarly activate the cyclic guanosine monophosphate pathway in smooth muscle to produce relaxation.[2] It was thus hypothesized that CO may similarly suppress myometrial contractility during pregnancy.

The expression of HO was demonstrated in the human myometrium.[14,54] Furthermore, it was recently shown that induction of HO produces CO that limits uterine contractility in pregnant myometrium indicating a role for the HO-CO-cGMP pathway in the maintenance of the quiescent state of the uterus during pregnancy.[60] However, a subsequent study could not support either an up-regulation of HO-1 and HO-2 during pregnancy or a consistent role for CO in human myometrial quiescence.

We proposed that ETCOc measurements could be helpful in detecting women at risk of developing premature labor, thereby allowing early initiation of preventive treatment.[62] Even if CO does not serve as the major endogenous inhibitor of myometrial contractility during pregnancy, very early changes in the quiescent state of the uterus during pregnancy may offset alterations in the HO-CO-cGMP pathway.

We therefore undertook a prospective study where ETCOc was measured using the CO-Stat™ End Tidal Breath Analyzer (Natus Medical Inc., San Carlos, CA) in 10 women with premature uterine contractions (PMC) during the second half of their pregnancy, 13 women in active labor at term and 32 pregnant mothers at matched gestational ages not experiencing uterine contractions. We found that the mean ± SD ETCOc measurements were significantly lower (p < 0.001) in women with PMC and in women actively delivering at term compared with control women, 0.99 ± 0.38 and 1.15 ± 0.41 vs. 1.70 ± 0.52 ppm, respectively.[62] There was no difference in mean ± SD ETCOc measurements between women with PMC or active contractions during labor at term. The ETCOc values were lower or equal to 1.6 ppm in all women with PMC and in 92.3% of women in active delivery compared with only about a third (37.5%) of the control pregnant women (p < 0.001). The ETCOc values were lower than 1.3 ppm in two thirds (65.2%) of women with PMC or active delivery compared with only 15.6% of the control pregnant women (p < 0.0001).[62]

We therefore concluded that the ETCOc levels might be significantly higher in pregnant women with a relaxed myometrium. Measurement of ETCOc may therefore be of value in the clinical assessment of PMC. At present we are undertaking a larger longitudinal study in an attempt to confirm our observation in a bigger patient population and based on serial measurements.

INTRAUTERINE GROWTH RETARDATION

Intrauterine growth retardation (IUGR) is an important cause of perinatal morbidity and mortality. The pathophysiology that precedes the development of IUGR remains incompletely understood. The importance of the placental blood flow to the growing fetus is obvious. The possible role of HO and its by-product CO in the regulation of blood pressure and blood flow has only been realized over the last few years.

Recently a case of HO-1 deficiency was presented.[63] The patient had a complete loss of exon-2 of the maternal allele and a two-nucleotide deletion within exon-3 of the paternal allele. This child had severe growth retardation, hemolytic anemia, low bilirubin levels, elevated thromomodulin and Von Wilebrand factor as well as iron deposition in the liver and kidney. This presentation was very similar to that observed in the HO-1 null mutant mice. In normal gestation the HO-1 enzyme is seen at high levels in the neonatal and fetal rat lung and liver, compared to adults.[64,65]

A current study found that HO expression in human placenta and placental bed implies a role in regulation of trophoblast invasion and placental function.[66] Furthermore, their results suggested a role for CO in placental function, trophoblast invasion, and spiral artery transformation.[66]

We therefore presumed that, in light of these observations, CO formation could have a contributory role in the pathogenesis of IUGR. Our preliminary ETCOc measurements in pregnant women with suspected IUGR fetuses, suggest that such measurements can be helpful in detecting women at risk of developing IUGR, thereby allowing early initiation of follow-up and treatment. We are currently continuing to collect additional data on endogenous CO production in pregnancies complicated by IUGR by measuring ETCOc levels.

SUMMARY

The possible role of CO in pregnancy has only recently been recognized However, all previous studies have been based on laboratory analysis of tissue samples. We therefore undertook a series of on-going investigations in an attempt to assess endogenous CO production in pathological conditions of pregnancy using ETCOc measurements.

We found significantly lower ETCOc levels in women with PIH, preeclampsia and premature contractions. We are therefore currently trying to determine whether ETCOc measurements may identify, directly or indirectly, a fundamental disturbance in normal regulation of placental and myometrial function. Thereby allowing early detection of various severe disorders of normal physiology associated with the major morbidities of pregnancy. ETCOc measurements may also offer a valuable clinical tool, since they are very easily performed, safe, non-invasive, relatively inexpensive, and provide on-spot results. These characteristics make this tool suitable for daily office use. Moreover, ETCOc measurements avoid the hazards associated with the handling of blood samples. Additional data is needed in order to determine the role of ETCOc measurements in improving our understanding of the pathophysiology of the most common disorders of pregnancy, including PIH, preeclampsia, premature contractions and IUGR.

REFERENCES

1. Vedernikov Y.P., Graser T., and Vanin A.F. Similar endothelium-dependent arterial relaxation by carbon monoxide and nitric oxide. Biomed Biochim Acta 1989;48:601–603.
2. Ramos K.S., Lin H., and McGrath J.J. Modulation of cyclic guanosine monophosphate levels in cultured aortic smooth muscle cells by carbon monoxide. Biochemical Pharmacology 1989; 38:1368–1370.
3. Verma A., Hirsch D.J., Glatt C.E., Ronnett G.V., and Snyder S.H. Carbon monoxide: a putative neural messenger. Science 1993;259:381–384.
4. Graser T., Vedernikov Y.P., and Li D.S. Study on the mechanism of carbon monoxide induced endothelium-independent relaxation in porcine coronary artery and vein. Biomedica Biochimica Acta 1990;49:293–296.
5. Maines M.D. Heme oxygenase: function, multiplicity, regulatory mechanisms, and clinical applications. FASEB Journal 1988;2:2557–2568.
6. Abraham N.G., Lavrovsky Y., Schwartzman M.L., et al. Transfection of the human heme oxygenase gene into rabbit coronary microvessel endothelial cells: protective effect against heme and hemoglobin toxicity. Proc Natl Acad Sci USA 1995;92:6798–6802.
7. Dennery P., Sridhar K., Lee C., et al. Heme oxygenase-mediated resistance to oxygen toxicity in hamster fibroblasts. J Biol Chem 1997;272:14937–14942.
8. Clark J.E., Green C.J., and Motterlini R. Involvement of the heme oxygenase-carbon monoxide pathway in keratinocyte proliferation. Biochemical & Biophysical Research Communications 1997; 241:215–220.
9. Liu Y., Christou H., Morita T., Laughner E., Semenza G.L., and Kourembanas S. Carbon monoxide and nitric oxide suppress the hypoxic induction of vascular endothelial growth factor gene via the 5′ enhancer. Journal of Biological Chemistry 1998;273:15257–15262.
10. Suttner D.M., Sridhar K., Lee C.S., Tomura T., Hansen T.N., and Dennery P.A. Protective effects of transient HO-1 overexpression on susceptibility to oxygen toxicity in lung cells. American Journal of Physiology 1999;276:L443–L451.
11. Brune B. and Ullrich V. Inhibition of platelet aggregation by carbon monoxide is mediated by activation of guanylate cyclase. Molecular Pharmacology 1987;32:497–504.
12. Lyall F., Barber A., Myatt L., Bulmer J.N., and Robson S.C. Hemeoxygenase expression in human placenta and placental bed implies a role in regulation of trophoblast invasion and placental function. FASEB Journal 2000;14:208–219.
13. McLean M., Bowman M., Clifton V., Smith R., and Grossman A.B. Expression of the heme oxygenase-carbon monoxide signalling system in human placenta. Journal of Clinical Endocrinology & Metabolism 2000;85:2345–2349.
14. Seidman D.S., Hallak M., Kelly D.K., et al. The role of carbon monoxide (CO) in the pathogenesis of pre-eclampsia. Am J Obstet Gynecol 1997;176:S101.
15. Pinto A., Sorrentino R., Sorrentino P., et al. Endothelial-derived relaxing factor released by endothelial cells of human umbilical vessels and its impairment in pregnancy-induced hypertension. Am J Obstet Gynecol 1991;164:507–513.
16. Ahokas R.A., Mercer B.M., and Sibai B.M. Enhanced endothelium-derived relaxing factor activity in pregnant, spontaneously hypertensive rats. Am J Obstet Gynecol 1991;165:801–807.

17. Seligman S.P., Buyon J.P., Clancy R.M., Yonng B.K., and Abramson S.B. The role of nitric oxide in the pathogenesis of preeclampsia. Am J Obstet Gynecol 1994;171:944–948.

18. Yallampalli C. and Garfield R. Inhibition of nitric oxide synthesis in rats during pregnancy produces signs similar to those of preeclampsia. Am J Obstet Gynecol 1993;169:1316–1320.

19. Morris N.H., Sooranna S.R., Learmont J.G., Poston L., Ramsey B., Pearson J.D., and Steer P.J. Nitric oxide synthase activities in placental tissue from normotensive, pre-eclamptic and growth retarded pregnancies. Br J Obstet Gyaecol 1995;102:711–714.

20. Rees D.D., Palmer R.M.J., and Moncada S. Role of endothelium-derived nitric oxide in the regulation of blood pressure. Proc Natl Acad Sci USA 1989;86:3375–3378.

21. Lyall F., Young A., and Greer I.A. Nitric oxide concentrations are increased in the feto-placental circulation in preeclampsia. Am J Obstet Gynecol 1995;173:714–718.

22. Davidge S.T., Baker P.N., and Roberts J.M. NOS expression is increased in endothelial cells exposed to plasma from women with preeclampsia. Am J Physiol 1995;269:H1106–H1112.

23. Baker P.N., Davidge S.T., and Roberts J.M. Plasma from women with preeclampsia increases endothelial cell nitric oxide production. Hypertension 1995;26:244–248.

24. Rodgers P.A., Vreman H.J., Dennery P.A., and Stevenson D.K. Sources of carbon monoxide in biological systems and applications of CO detection technologies. Seminar Peinatol 1994;18:2–10.

25. Seidman D.S., Shiloh M., Stevenson D.K., Vreman H.J., and Gale R. Role of hemolysis in neonatal jaundice associated with glucose-6 phosphate dehydrogenase deficiency. J Pediatr 1995;127: 804–806.

26. Seidman D.S., Paz I., Merlet-Aharoni I., Vreman H.J., Stevenson D.K., and Gale R. Noninvasive validation of tobacco smoke exposure in late pregnancy using end-tidal carbon monoxide measurements. J Perinatol 1999,19.1–4.

27. Vreman H.J., Baxter L.M., Stone R.T., and Stevenson D.K. Evaluation of a fully automated end-tidal carbon monoxide instrument for breath analysis. Clinical Chemistry 1996;42:50–56.

28. Isler C.M., Rinehart B.K., Terrone D.A., Martin R.W., Magann E.F., and Martin J.N. Jr. Maternal mortality associated with HELLP (hemolysis, elevated liver enzymes, and low platelets) syndrome. Am J Obstet Gynecol 1999;181:924–928.

29. Padden M.O. HELLP syndrome: recognition and perinatal management. Am Fam Physician 1999;60:829–836, 839.

30. Dotsch J., Hohmann M., and Kuhl P.G. Neonatal morbidity and mortality associated with maternal haemolysis elevated liver enzymes and low platelets syndrome. Eur J Pediatr 1997;156:389–391.

31. Sibai B.M. Hypertension in Pregnancy. In: Gabbe S.G., Niebyl J.R., and Simpson J.L. (eds.). Obstetrics: Normal & Problem Pregnancies. New York, Churchill Livingstone, 1996.

32. Mattar F. and Sibai B.M. Prediction and prevention of preeclampsia/eclampsia. Gynec Forum 1999;4:16–21.

33. Baum M., Schiff E., Kreiser D., Dennery P.A., Stevenson D.K., Rosenthal T., and Seidman D.S. End tidal carbon monoxide measurements in women with pregnancy induced hypertension and preeclampsia. Am J Obstet Gynecol 2000:183:900–903.

34. Kreiser D., Druzin M., Hendler D., Baum M., Schiff E., Stevenson D.K., Dennery P.A., and Seidman D.S. End tidal carbon monoxide levels are lower in women with gestational hypertension and preeclampsia. Am J Obstet Gynecol 2001:184:S69.

35. Martasek P., Schwartzman M.L., Goodman A.I., Solangi K.B., Levere R.D., and Abraham N.G. Hemin and l-arginine regulation of blood pressure in spontaneous hypertensive rats. J Am Soc Nephrol 1991;2:1078–1084.

36. Levere R.D., Martasek P., Escalante B., Schwartzman M.L., and Abraham N.G. Effect of heme arginate administration on blood pressure in spontaneously hypertensive rats. J Clin Invest 1990;86:213–219.

37. Tsukimori K., Maeda H., Ishida K., Nagata H., Koyanagi T., and Nakano H. The superoxide generation of neutrophils in normal and pre-eclamptic pregnancies. Obstet Gynecol 1993;81:536–540.

38. Hubel C.A., Roberts J.M., Taylor R.N., Musci T.J., Rodgers G.M., and McLaughlin M.K. Lipid peroxidation in pregnancy: New perspectives on pre-eclampsia. Am J Obstet Gynecol 1989;161:1025–1034.

39. Roberts J.M., Taylor R.N., and Goldfien A. Clinical and biochemical evidence of endothelial cell dysfunction in the pregnancy syndrome preeclampsia. Am J Hypertens 1991;4:700–708.

40. Stocker R., Yamamoto Y., McDonagh A.F., Glazer A.N., and Ames B.N. Bilirubin is an antioxidant of possible physiologic importance. Science 1987;235:1043–1046.

41. Chandra L., Lali P., and Jain A. Role of bilirubin in pregnancy-induced hypertension. Int J Gynecol Obstet 1996;53:267–268.

42. Underwood P.B., Kesler K.F., O'Lane J.M., and Callagan D.A. Parental smoking empirically related to pregnancy outcome. Obstet Gynecol 1967;29:1–8.

43. Duffus G.M. and MacGillivray I. The incidence of pre-eclamptic toxaemia in smokers and non-smokers. Lancet 1968;1:994–995.

44. Conde-Agudelo A., Althabe F., Belizan J.M., and Kafury-Goeta A.C. Cigarette smoking during pregnancy and risk of preeclampsia: a systematic review. Am J Obstet Gynecol 1999;181:1026–1035.

45. Lain K.Y., Powers R.W., Krohn M.A., Ness R.B., Crombleholme W.R., and Roberts J.M. Urinary cotinine concentration confirms the reduced risk of preeclampsia with tobacco exposure. Am J Obstet Gynecol 1999;181:1192–1196.

46. Zhang J., Klebanoff M.A., Levine R.J., Puri M., and Moyer P. The puzzling association between smoking and hypertension during pregnancy. Am J Obstet Gynecol 1999;181:1407–1413.

47. Andres R.L. and Larrabee K. The perinatal consequences of smoking and alcohol use. Curr Probl Obstet Gynecol Fertil 1996;19:167–206.

48. Seidman D.S. and Mashiach S. Involuntary smoking and pregnancy. Eur J Obstet Gynecol Reprod Biol 1991;41:105–116.

49. Ananth C.V., Smulian J.C., and Vintzileos A.M. Incidence of placental abruption in relation to cigarette smoking and hypertensive disorders during pregnancy: a meta-analysis of observational studies. Obstet Gynecol 1999;93:622–628.

50. Cnattingius S., Mills J.L., Yuen J., Eriksson O., and Salonen H. The paradoxical effect of smoking in preeclamptic pregnancies: smoking reduces the incidence but increases the rates of perinatal mortality, abruptio placentae, and intrauterine growth restriction. Am J Obstet Gynecol 1997; 177:156–161.

51. Marcoux S., Brisson J., and Fabia J. The effect of cigarette smoking on the risk of preeclampsia and gestational hypertension. Am J Epidemiol 1989;130:950–957.

52. Salafia C. and Shiverick K. Cigarette smoking and pregnancy II: vascular effects. Placenta 1999; 20:273–279.

53. Secker-Walker R.H., Vacek P.M., Flynn B.S., and Mead P.B. Smoking in pregnancy exhaled carbon monoxide, and birth weight. Obstet Gynecol 1997;89:648–653.

54. Odrcich M.J., Graham C.H., Kimura K.A., McLaughlin B.E., Marks G.S., Nakatsu K., and Brien J.F. Heme oxygenase and nitric oxide synthase in the placenta of the guinea-pig during gestation. Placenta 1998;19:509–516.

55. Lu G.C. and Goldenberg R.L. Current concepts on the pathogenesis and markers of preterm births. Clin Perinatol 2000;27:263–283.

56. Buhimschi I., Yallampalli C., Dong Y.L., and Garfield R.E. Involvement of a nitric oxide-cyclic guanosine monophosphate pathway in control of human uterine contractility during pregnancy. Am J Obstet Gynecol 1995;172:1577–1584.

57. Kaya T., Cetin A., and Sarioglu Y. Changes in the nitric oxide system of rat myometrium during midgestation and delivery at term. Pharmacol Res 1998;37:403–408.

58. Ekerhovd E., Weidegard B., Brannstrom M., and Norstrom A. Nitric oxide-mediated effects on myometrial contractility at term during prelabor and labor. Obstet Gynecol 1999;93:987–994.

59. Mirabile C.P. Jr, Massmann G.A., and Figueroa J.P. Physiologic role of nitric oxide in the maintenance of uterine quiescence in nonpregnant and pregnant sheep. Am J Obstet Gynecol 2000; 183:191–198.

60. Acevedo C.H. and Ahmed A. Hemeoxygenase-1 inhibits human myometrial contractility via carbon monoxide and is upregulated by progesterone during pregnancy. J Clin Invest 1998; 101:949–955.

61. Barber A., Robson S.C., and Lyall F. Hemoxygenase and nitric oxide synthase do not maintain human uterine quiescence during pregnancy. Am J Pathol 1999;155:831–840.

62. Hendler I., Baum M., Kreiser D., Schiff E., Druzin M., Stevenson D.K., Dennery P.A., Mashiach S., and Seidman D.S. Pregnant women with uterine contractions have lower end tidal carbon monoxide levels. Isr J Obstet Gynecol 2000;11;86–87.

63. Yachie A., Niida Y., Wada T., Igarashi N., Kaneda H., Toma T., et al. Oxidative stress causes enhanced endothelial cell injury in human heme oxygenase-1 deficiency. J Clin Invest 1999; 103:129–135.
64. Abraham N.G., Lin J.H., Mitrione S.M., Schwartzman M.L., Levere R.D., and Shibahara S. Expression of heme oxygenase gene in rat and human liver. Biochem Biophys Res Commun 1988; 150:717–722.
65. Rodgers P.A., Lee C.S., Stevenson D.K., and Dennery P.A. Ontogeny of lung heme oxygenase in the neonatal Wistar rat. Pediatr Res 1995.
66. Fiona L., Barber A., Myatt L., Bulmer J.N., and Robson S.C. Hemeoxygenase expression in human placenta and placental bed implies a role in regulation of trophoblast invasion and placental function. The FASEB Journal. 2000;14:208–219.

6

THE ROLE OF HEME OXYGENASE IN PREGNANCY

Peter Hewett and Asif Ahmed

Department of Reproductive and Vascular Biology
The Medical School
University of Birmingham
Edgbaston, Birmingham, B15 2TT, UK

1. INTRODUCTION

The human placenta performs essential transport, metabolic and endocrine functions to support fetal development and serves as a physical and immunological barrier between the maternal and fetal blood. The placenta develops following embryo implantation when the syncytiotrophoblast invade the vasculature of the decidualised endometrium and migrate along the spiral arterioles transforming them into large low resistance vessels. Throughout pregnancy there is progressive remodelling of the placental vasculature to increase diffusion capacity and meet the gaseous and nutritive demands of the growing foetus. This is achieved through the constant elaboration of the terminal villi and reduction in the thickness of the villous membrane separating the maternal and fetal circulations. The regulation of this process is complex and still poorly understood involving both physical forces such as stretch and blood flow, and chemical stimuli including local oxygen tension. The essential requirement of placental adaptation is highlighted by the maternal hypertensive condition of preeclampsia and intrauterine growth restriction (IUGR) that result from a failure of trophoblast invasion and transformation of the maternal spiral arterioles. Preeclampsia results in increased placental resistance which may be due to physical changes in vascular anatomy or altered sensitivity and/or production of vasoactive factors, local hypoxia and systemic endothelial dysfunction.[1]

The maintenance of uterine tone is also essential for the successful progression of pregnancy to term. A key feature of this is the ability of the uterus to maintain a relaxed state (uterine quiescence) despite having to accommodate the developing fetus.

Disturbances of normal myometrial contractility may result in spontaneous pre-term (prior to 37 weeks gestation) birth the most significant factor contributing to both perinatal mortality and morbidity.[2,3]

The microsomal enzyme heme oxygenase (HO) catalyses the oxidative degradation of intracellular heme to biliverdin generating carbon monoxide (CO).[4] The biliverdin generated is subsequently reduced by biliverdin reductase to bilirubin, which acts as a potent anti-oxidant[5] and also inhibits complement activation.[6] CO, like nitric oxide (NO), is an endogenous chemical messenger that activates soluble guanylate cyclase to generate intracellular cyclic GMP (cGMP) and is involved in the regulation of many organ systems. It is implicated in the control of vascular tone acting as a smooth muscle relaxant.[7–9] Three homologous HO isozymes encoded by different genes have been identified. HO-1 (heat shock protein 32 kDa is inducible and is highly expressed in spleen and liver.[4,10] Induced by heme, oxidative stress, metals and hypoxia, HO-1 forms part of cells stress-response mechanism to oxidative injury. In contrast, HO-2 expression appears to be constitutive and widely distributed in the body with the highest concentration detected in the brain.[11] The recently identified HO-3 has very low catalytic activity and may be involved in heme binding.[12] Our laboratory has investigated the expression, distribution of HO-1 and HO-2 and functional activity of HO in pregnancy.[13,14] In this chapter we review the current evidence supporting the involvement of HO in the regulation of normal placental and uterine function and its association with the common complications of pregnancy; preeclampsia, IUGR and pre-term birth.

2. HEME OXYGENASE PROMOTES VESSEL RELAXATION AND VASCULAR PROTECTION IN THE HUMAN PLACENTA

The feto-placental circulation is unique in that it lacks autonomic innervation[15] and the vasomotor control of placental blood flow is mediated via the action of humoral and/or autocrine/paracrine factors on the smooth muscle cells surrounding the stem villous arterioles. Vasodilators such as Bradykinin, histamine and acetyl-choline, which induce NO release in other systems do not appear to be active in the isolated perfused placenta. However, NO has been shown to maintain low vascular tone in the isolated perfused placenta *in vitro*.[16,17] Nitric oxide synthestase (NOS) and HO-2 appear to play complementary roles in other tissues.[18] Both endothelial NO synthase (eNOS) and NO production are actually down-regulated by hypoxia[19–21] and vessel tone may then be modulated by other factors such as CO, as HO-1 is induced rapidly under conditions oxidative stress such as hypoxia.[8,22] Based on this evidence our laboratory and others have investigated both the expression and functional activity of HO in placenta.

2.1. Expression and Distribution of HO-1 and HO-2 in Human Placenta throughout Gestation

2.1.1. Detection of HO-1 and HO-2 Expression in Placental Extracts. We examined both the expression and activity of HO in full thickness human placental tissue obtained following termination (first trimester and second trimester) and elective caesarean section (third trimester).[14] In placental tissue extracts HO-1 mRNA and protein were detected in placentae, and fetal membranes by RT-PCR, ribonuclease (RNase) protection assay and Western blotting respectively. Quantitative analysis revealed low abundance HO-1 transcripts relative to β-actin mRNA in early pregnancy (7–12 weeks gestation) which increased throughout gestation and were higher by the third trimester (27–34 weeks gestation). Consistent with the mRNA data, HO-1 protein expression was significantly higher (2.6-fold) in full thickness term placenta compared with first trimester placenta suggesting a role in placental vascular development/function. Levels of HO-1 protein in the term placenta were similar to those detected in the amnion and choriodecidua. However, no difference in HO-2 protein expression was detected in first trimester and term placentae, or in extraembryonic tissues. In agreement with these findings McLean et al.[23] detected HO-1 mRNA by RT-PCR and protein by immunoblotting in normal human placentae. In another recent study, HO-1 and HO-2, mRNA and protein were detected in chorionic villous extracts of first trimester and term placentae from 10 women by RT-PCR and Western blotting. Although they did not fully quantify their results, Yoshiki and colleagues[24] reported greater expression of HO-2 in term chorionic villous extracts compared with those from the first trimester. Surprisingly, two studies by Lyall colleagues[25,26] failed to detect HO-1 protein by Western blotting in full thickness placental biopsies taken throughout gestation. However, these authors report the constitutive expression of HO-2 in term human placenta[25,26] that is consistent with our findings.[14]

2.1.2. Immunolocalization of HO-1 and HO-2 in Placenta. The immuno-localisation of HO-1 and HO-2 in the placentae of various species has been reported by several groups[14,23–27] and most observe distinct spatial patterns of HO-1 and HO-2 expression in placental biopsies throughout gestation. HO-1 and HO-2 have been detected in the trophoblast of rat placenta.[28] In the guinea pig HO was detected in the adventitial layer of large fetal blood vessels of the placenta[27] and similarly in the adventitial layer and the stem villi in a preliminary study of human placentae. In our study of human placenta, HO-1 was localised predominantly in the extravascular connective tissue that forms the perivascular contractile sheath surrounding the developing blood vessels[14] (Fig. 1 and Table 1). The intensity of HO-1 immunoreactivity was greatest in the chorionic plate and stem villi arising from the plate in term placentae but was low or absent in villi devoid of muscularized vessels. In contrast, Yoshiki and colleagues[24] observed HO-1 primarily in the cells of the syncytiotrophoblast layer that are in direct contact with the maternal blood. Similarly, Lyall et al[25,26] only detected occasional HO-1 immunostaining on the syncytiotrophoblast layer, which did not vary with gestation. McLean and colleagues[23] report the presence of HO in the syncitiotrophoblast and endothelium but due to the cross-reactivity of their antibody could not conclusively differentiate between HO-1 and HO-2.

HO-2 is expressed in the capillaries as well as the villous stroma, with weak staining of trophoblast and was similar throughout gestation[14] (Fig. 1 and Table 1). In general agreement with our findings, Yoshi and colleagues[24] report HO-2 in the endothelium and smooth muscle cells of the blood vessels in the placental villi and this pattern of distribution remained constant from the first trimester to term. Lyall and colleagues[25,26] observed HO-2 expression primarily in the trophoblast layer with

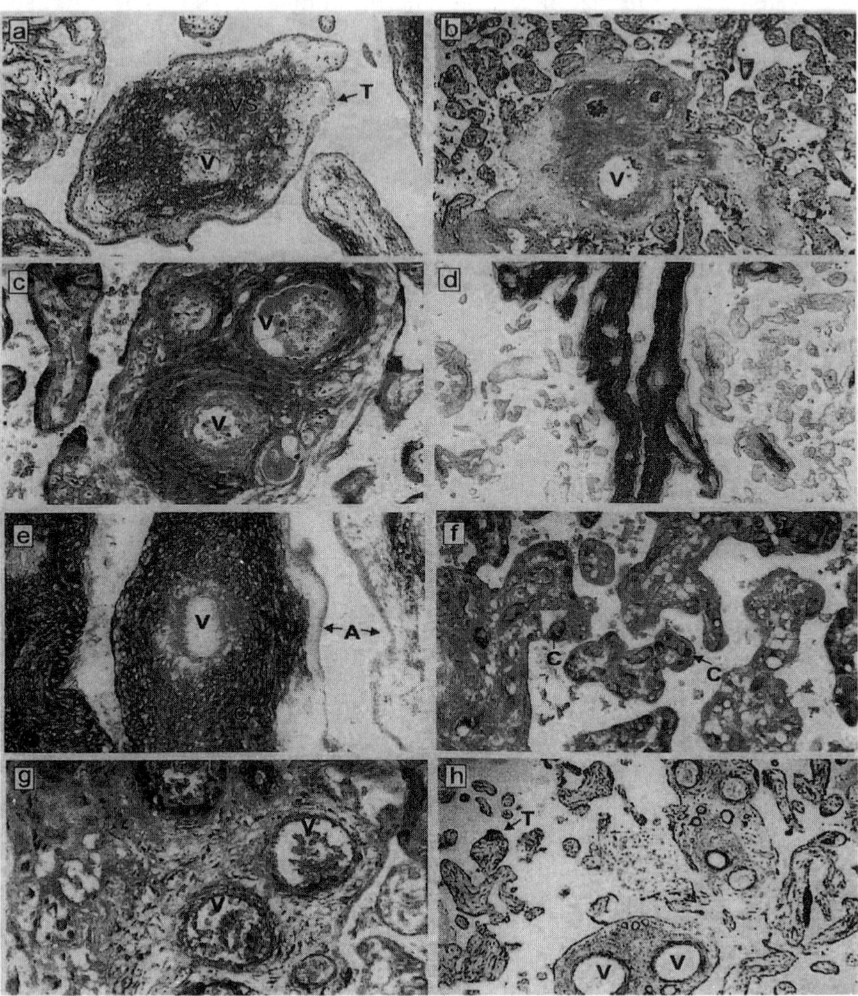

Figure 1. Immunolocalization of HO-1 and HO-2 in placental villi and feto-placental membranes. Immonohistochemical staining of HO-1 in (a) first trimester villi (7 weeks) and (b) placental villi at term (37 weeks). Strong HO-1 staining is localized in the perivascular villous stromal (VS) with weaker staining of the trophoblast layer (T) in panel (a). Intense staining in the large blood vessels (V) of the stem villi detected in term placenta (panel c). Very strong HO-1 staining (panels f–g) was also observed in the chorio-decidual cells (C) and weak staining in the amnion (A). Examples of HO-2 immunoreactivity are shown in panels f and g. Intense HO-2 staining of the endothelium of the terminal villous capillary bed (C) was observed at term (panel f). In the endothelium of large blood vessels (g) and the media of the muscularised vessels there is strong staining for HO-2. Panel (h) shows staining with α-actin in term placenta. Magnifications were ×350 for panels c, e, g; and ×284 for panels a, b, d, f and ×142 for panel h. *Adapted from Ahmed* et al.[14]

Table 1. Distribution of immunoreactive HO-1 and HO-2 in human placental during gestation*

	HO-1			HO-2		
	Weeks of gestation			Weeks of gestation		
	7	10	39	7	10	39
Trophoblast brush border	+	−	−	+	+	+
Syncytiotrophpoblast	+/−	−	−	−	−	−
Cytotrophoblast	−	−	−	++	+	−
Villous stromal cells	+	+	+	+	+/−	−
Perivascular stromal cells	++	+++	++++	+	++	+
Stem vessel media	−	−	−	−	−	−
Endothelium/intima	−	−	+/−	+/−	+/−	+++
Decidua	+++	+++	++	++	++	++
Chorion	++	+++	+++	+	+	+
Amnion	+/−	−	+	+	+/−	+

*Immunoreactivity is expressed as weak (+), moderate (++), strong (+++), intense (++++), variable or equivocal (+/−) and no staining (−).

occasional staining of the endothelium in first trimester placentae and noted a progressive decrease in trophoblast and increase in endothelial cell HO-2 staining in these tissues to term. CO like NO, prevent platelet aggregation[29] and the consistently reported observation of HO-2 in the endothelium of the placental villi suggests that it may serve to inhibit the adhesion of platelets in the utero-placental circulation.

The discrepancies in the expression and distribution of HO isozymes in these studies are probably due to the use of the different antibodies to detect HO-1 and HO-2, and the region of the placental tissue examined. Highlighting this fact Dr Lyall's group[26] report the different activities of three commercially available anti-HO-1 antibodies in human and rat tissue. Moreover different batches (904409 and 708405) of rabbit anti-HO-1 polyclonal antibodies purchased from StressGen (OSA-100, StressGen Biotechnologies, Victoria, Canada) produced different patterns of immunostaining.[26] In our study we also used the StressGen anti-HO-1 antibody (OSA-100) but from different batches (706414 and 702414) to Barber et al.[26] However, in a similar manner to Barber et al.,[26] we confirmed the binding specificity of these antibodies by running control sections incubated with antibodies pre-absorbed with purified HO-1 and HO-2 protein. A clearer picture may emerge following *in situ* hybrididization studies that are underway.

2.2. Functional Activity of HO in Human Placenta

Our finding that HO-1 expression is significantly higher in term compared with first trimester placentae and localisation to the perivascular contractile sheath of developing vessels and media of large stem vessels suggests it may play a role in placental vascular development/function.[14] We therefore investigated the functional activity of HO in human placentae.

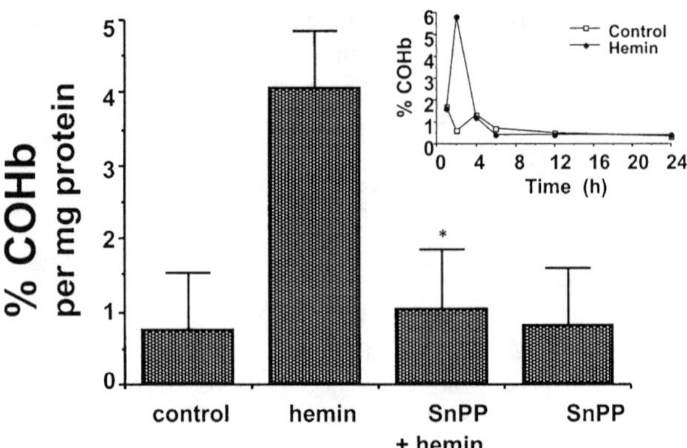

Figure 2. Hemin-mediated release of carbon monoxide. Placental tissues and cells were exposed to hemin and the carbon monoxide (CO) generated determined as a measure of the percentage of carboxyhemoglobin (COHb) in the medium. The inset shows a peak of CO production in term villous fragments at 2 h following the addition of hemin (5 μM). The effect of hemin and the HO inhibitor, tin protoporphyrin IX (SnPP) on CO production in placental explants exposed to 20 μM SnPP or vehicle for 1 h and then 5 μM hemin or vehicle for 2 h. CO release is expressed as the percent carboxyhemoglobin (%COHb) in conditioned media per mg/ml of total protein. Data are the mean (±1 SEM) of three separate experiments performed in triplicate. *$p < 0.01$ vs. hemin; **$p < 0.001$ vs. + vehicle. *Adapted from Ahmed* et al.[14]

2.2.1. Production of CO in Placenta and Fetal Membranes.

To demonstrate functional activity of the HO-CO system in human placenta, villous fragments from term placentae were exposed to hemin, an inducer of HO activity, for up to 24 hours and CO production was determined by measuring the percent carboxyhaemoglobin (COHb) in the conditioned medium.[14] Hemin (5 μM) induced a transient CO production in the placental villous fragments and fetal membranes that peaked after 2 hours and returned to baseline levels after 4 hours incubation. The HO inhibitor, tin protoporphyri IX (SnPP) significantly attenuated hemin-induced CO production in these villous fragments demonstrating the specificity of this CO production (Fig. 2).

2.2.2. Effect of the HO-CO System on Placental Vascular Contractility.

As the feto-placental circulation lacks autonomic innervation[15] the vasomotor regulation of the smooth muscle cells surrounding the villous arterioles play a pivotal role in maintaining the blood flow within the placenta.[30] NO acts as a vasodilator in the placenta[16,30] and generation of NO appears to attenuate the effect of vasoconstrictors such as thromboxane A_2 and endothelin-1 in the placental circulation.[30] While NO contributes to the physiological regulation of vessel tone under normoxia, the HO-CO pathway was proposed as a local regulator of feto-placental circulation under conditions of hypoxia[14] when NO levels are down-regulated.[31,32] To determine placental vessel contractility in response to up-regulated HO activity, villous arterial rings were mounted under tension and monitored using isometric recording equipment.[14] Rapid and prolonged contraction of the arterial rings was achieved using the thromboxane A_2 mimetic, U46619. Addition of hemin (5 μM) to the pre-constricted arterial rings caused a rapid decrease in tension compared with the vehicle alone. The specificity of this response was demonstrated by incubation of the tissue with the HO inhibitor SnPP, which prevented the hemin-induced reduction in tension in the pre-constricted placental vessels. It is noteworthy that hemin induced a 15% reduction in preconstricted vessel tension even in the presence of SnPP, which alone had no effect on these vessels. Although NO is a recognised regulator of resting vascular tone[16,17,33] the potential involvement of the NO pathway in HO-mediated vessel relaxation was ruled out in these experiments by the use of the NOS inhibitor N^G-methyl-L-arginine (L-NMA), which had no effect on HO-1-mediated vessel relaxation.

The contribution of the HO-CO system to the maintenance of basal vascular tone in the isolated dual perfused human cotyledon has also been demonstrated by Lyall and co-workers[25] using the HO inhibitor zinc protoporphyrin IX (ZnPP). To avoid any confounding effects of ZnPP, which may also affect NOS activity and prostaglandin release, N^{ω}-nitro-L-arginine methyl ester (L-NAME) and meclofena-mate respectively were used to block these pathways prior to stimulation with ZnPP. These authors showed a significant concentration-dependent vasoconstrictive effect of ZnPP in human placenta. Collectively, these studies provide evidence for the mod-ulatory role of the HO-CO system in placental blood vessel relaxation indicating that placental and fetal membrane-derived CO may play a pivotal role in the control of feto-placental vascular tone.

The distribution of HO activity was recently examined in the microsomal fractions of tissue homogenates obtained from different regions of human term pla-centa.[34] By gas chromatography the CO produced was found to be significantly greater in the chorionic plate, chorionic villi, basal plate and chorio-decidua compared with the amnion. This is consistent with our earlier observation that chorio-decidua generated the highest levels of CO and may play a role in parturition.

2.2.3. Cytoprotective Effect of HO in Normal Human Placenta. The metabolites of HO-mediated breakdown of heme are thought to represent potent endogenous inhibitors of stress-induced inflammatory injury[6,35] and HO activation inhibits the expression of oxidative-stress-induced adhesion molecules.[36,37] Indeed, in human HO-1 deficiency there is severe and persistent endothelial damage characterised by increased circulating thrombomodulin and von Willerand factor.[38] Bilirubin acts as an anti-oxidant[5] suppressing hydrogen peroxide-mediated endothelial cell death *in vitro*,[39] and HO is thought to protect against both endotoxin and hyperoxia-induced lung injury in rats.[40,41] In addition, the protective effect of NO in ischemia reperfu-sion injury and tumor necrosis factor α (TNF α) cytotoxicity is in part dependent on HO activation and HO inhibition attenuates the cytoprotective effect of NO.[42,43] We therefore investigated the role of HO-1 in preventing TNF-α-induced cytotoxicity in term placental villous explants.[14] Explants were incubated with TNFα (50 ng/ml) overnight in the presence or absence of hemin (5 μM) which stimulates HO expres-sion. In the presence TNF-α, hemin increased HO-1 expression compared with control or hemin alone. TNF-α-induced plasma membrane damage was assessed by measurement lactate dehydrogenase (LD) leakage into the medium. Hemin-induced activation of HO-1 significantly reduced TNF-α-mediated LD leakage from placen-tal villous explants. Moreover, this effect was specifically blocked by the HO-1

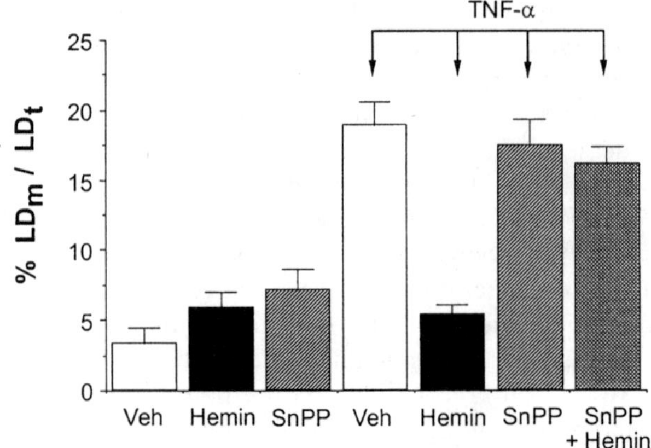

Figure 3. The cytoprotective effect of hemin in TNF-α-treated placental extracts. Villous fragments from normal term placenta were exposed to either vehicle (veh), 5 μM hemin, 20 μM tin protoporphyrin IX (SnPP) or 20 μM SnPP for 1 h prior to the addition of 5 μM hemin, for 2 h. TNF-α (50 ng/ml) or vehicle were then added to these samples for 1 h and lactate dehydrogenase release into the medium measured as a marker of plasma membrane perturbation. SnPP completely blocked the protective effect of Hemin completely inhibited TNF-α-induced cytotoxicity and this effect was totally abolished by SnPP. The cytotoxicity index is expressed as a percentage ratio of lactate dehydrogenase (LD) in the medium/LD in tissue (LD_m/LD_t). Data represents the mean (±1 SEM) of eight separated experiments performed in duplicate. *Adapted from Ahmed* et al.[14]

inhibitor, SnPP (Fig. 3) clearly establishing a protective role for the HO-CO system against TNF-α-induced cytotoxicity in the normal human placenta.[14]

3. IMPAIRMENT OF HEME OXYGENASE ACTIVITY IN PREECLAMPSIA AND IUGR

Preeclampsia is a hypertensive condition that affects 7–10% of all first pregnancies and is a major cause of maternal death and peri-natal morbidity.[1,44] In addition women with a history of preeclampsia commonly suffer hyperinsulinemia and are at two-to-three-fold greater risk of developing hypertension and isch emic heart disease in later life.[45,46] Although the precise underlying causes of preeclampsia remain unknown central to this disorder is the failure of adequate extravillous trophoblast transformation of the spiral arterioles. results in fetal growth retardation, compromised utero-placental circulation/perfusion and subsequent local hypoxia that leads to the secretion of cytotoxic factors into the maternal blood and circulatory disturbances including hypertension and proteinuria.[1] This is clearly demonstrated by the ability of sera of pre-eclamptic women to inhibit the bradykinin-stimulated relaxation of pre-constricted vessels.[47] Although the circulatory factor(s) responsible for the systemic effects have yet to be fully elucidated, increasing circulating levels of the inflammatory cytokine TNF-α are associated with placental lipid peroxidation and implicated in the general endothelial dysfunction observed in Preeclampsia.[48]

Abnormalities in spiral artery transformation are also observed in IUGR, which occurs in the absence of maternal hypertension and is defined according to a birth weight below the 10th centile. Although less is known of the aetiology of this disorder it may result from defective placentation and some maternal vessels may undergo changes in the absence of invasive cytotrophoblasts.[49–51]

3.1. Expression and Distribution of HO-1 and HO-2 in Preeclampsia and IUGR

Acute hypoxia and inflammation may impair eNOS/NO production in the trophoblast contributing to the underlying pathologies of gestational diseases.[52] Indeed, under acute fetal hypoxia in the rat placenta, HO-1 is down-regulated.[28] In studies of pregnancies complicated by preeclampsia and IUGR in our laboratory, HO-1 mRNA levels were not found to be significantly different from those in gestationally matched normal third trimester placentae.[14] However, HO-1 protein expression was significantly reduced ($p < 0.01$) in preeclamptic, but not IUGR placentae, compared with control placenta ($p < 0.01$). While HO-1 mRNA expression was similar in preeclamptic and normal placentae, the significant decrease in the level of HO-1 in preeclampsia suggests translational differences between these tissues.[14]

There was no apparent change in the expression of HO-2 protein in preeclamptic and IUGR pregnancies compared with normal placentae.[14] This is consistent with the recent findings of Lyall and colleagues.[25,26] who similarly reported no significant changes in total HO-2 expression in third-trimester placental homogenates from normal, preeclamptic and IUGR pregnancies by Western blotting. However, while HO-2 immunostaining in the extravillous trophoblasts of the placental bed was comparable in preeclampsia, IUGR and normal pregnancies these authors[26] report the specific down-regulation of HO-2 in the villous endothelial cells in both preeclampsia and IUGR. This down-regulation of HO-2 was particularly marked in a subset of patients with extremely abnormal laser doppler blood flow measurements.[26] In contrast eNOS is up-regulated in the endothelium of the stem and terminal villous in PE and IUGR[53] which may represent an adaptive response to the increased vascular resistance in these conditions.

3.2. Reduced Placental HO-Mediated Cytoprotection in Preeclampsia

It has been proposed that preeclampsia represents an exaggeration of an inflammatory process characteristic of normal pregnancy due to an impairment of the normal compensatory mechanisms.[1,44] HO-1 may counteract stress-induced inflammatory injury[35,54–56] and thus our observed down-regulation of HO-1 in preeclampsia[14] supports this theory. Consistent with our HO expression data,[14] the leakage of lactate dehydrogenase from preeclamptic villous explants was three times higher than in normal villous tissue under basal conditions and addition of the hemin did not reduce, nor TNF-α exacerbate, lactate dehydrogenase leakage in these explants. Preeclampsia is associated with poor placental perfusion and subsequent local hypoxia and secretion of cytotoxic factors into the maternal circulation leading to widespread circulatory disturbances secondary to endothelial dysfunction.[1,45] There is evidence to support the notion that impaired HO-1 activity may contribute to the endothelial activation and cytotoxicity observed in preeclampsia. For example lipid peroxidation products such as malonlidialdehyde are increased, and the activity of antioxidant enzymes, superoxide dismutase and glucose 6-phosphate-dehydrogenase decreased in preeclamptic placentae.[57] However, bilirubin which acts as a physiological antioxidant preventing lipid peroxidation[5] is significantly reduced in pregnancy-induced hypertension compared with normal pregnancies[58] suggesting that it may in part, result from the decreased HO-1 activity in preeclamptic placenta. Moreover, in HO-1 deficiency there is chronic severe endothelial damage[38] and induction of HO-1 directly down-regulates endothelial cell activation blocking the oxidative stress-induced expression of adhesion molecules.[36,37] Taken together these studies provide evidence that HO acts as a cytoprotective pathway counteracting pathophysiological insults during pregnancy.

4. THE ROLE OF HEME OXYGENASE IN MYOMETRIAL CONTRACTILLITY DURING PREGNANCY

The endogenous NO-cGMP pathway has been proposed as a key regulator of uterine smooth muscle relaxation during pregnancy and may be an important in the initiation of labor.[59–62] Although iNOS was recently reported not to increase in pregnant myometrium[63] studies from this laboratory have previously demonstrated a marked increase in the levels of immunoreactive NOS in pregnant compared with non-pregnant myometrium[64] consistent with studies linking the involvement of NO to the maintenance of myometrial quiescence during pregnancy.[61] However, myometrial cGMP is reported to increase during pregnancy by a NO-independent pathway in the guinea-pig[65] suggesting that cGMP is not under the sole control of the NO-cGMP pathway in the placenta. Levels of NOS detected in laboring and non-laboring human myometrium[64] are in general agreement with pharmacological data.[19] Below we review the evidence that CO can induce smooth muscle relaxation in the pregnant human uterus.

4.1. Expression and Distribution of HO-1 and HO-2 in Human Myometrium

Our laboratory was the first to report the expression of HO-1 and HO-2 isozymes in human myometrium collected from pre-menopausal women by hysterectomy and pregnant women during cesarean section.[13] Both HO-1 and HO-2 mRNA and protein were detected in myometrial tissue by Western blotting. The expression of HO-1 and HO-2 was markedly higher (16-fold and 17-fold respectively) in pregnant myometrium obtained at cesarean section compared with non-pregnant myometrium. However, there was no significant difference ($p > 0.5$) in HO expression between non-laboring and laboring groups obtained from elective and emergency cesarean sections respectively. In contrast to our findings and similar to their data from placental tissue, Barber and colleagues[63] were unable to detect HO-1 protein in myometrium despite the presence of low level HO-1 mRNA, again this may be down to the source of the primary antibodies used (see Section 2.1.). In addition they did not observe a difference in HO-2 expression between pregnant and non-pregnant myometrium.[63] However, consistent with our studies they observed no differences in HO-2 expression in laboring and non-laboring myometrium.[63]

In our study immunostaining for both HO isozymes was localised in the smooth muscle and endothelium of the myometrial blood vessels.[13] Although inter-individual variation was observed, the distribution of HO was similar in both pregnant (non-laboring and laboring) and non-pregnant human myometrium.[13] In agreement with our findings Barber and colleagues describe a comparable pattern of HO-2 distribution in smooth muscle and endothelial cells.[63]

4.2. The Induction of HO by Sex Steroids in Uterine Tissue

The onset of labor in non-primate mammals is characterised by a shift from progesterone to oestrogen dominance.[39,66] Although there is no such change in circulating oestradiol and progesterone signalling human parturition, Mitchell and Wong[67] suggested that local increases in the oestradiol/progesterone ratio within uterine tissue and fetal membranes may initiate labor. More recently, Karalis and co-workers[68] proposed a model for functional progesterone withdrawal at the end of human pregnancy. The shift from progesterone to oestrogen dominance leads to the activation of multiple pathways that contribute to the onset of labor, including stimulation of oxytocin and cognate receptors, and prostaglandin synthesis.[66] Myometrial cGMP increases with advancing gestational age and the greatest increase corresponds to the period of maximal fetal growth.[65]

As the shift from progesterone to oestrogen dominance appears to be a key regulator of parturition we have examined the influence of sex steroids on HO expression in myometrial explants. Tissue from non-pregnant and non-laboring patients at term were exposed to increasing concentrations of oestradiol-17β (10^{-8}–10^{-6} M) and progesterone (10^{-8}–10^{-4} M) for 24 hours in phenol red indicator-free medium. In unstimulated myometrium mRNA encoding HO-1 and HO-2 was surprisingly not readily detected possibly due to the withdrawal of sex steroids during 48-hour period in culture. However, following 24 hour incubation with progesterone but not oestradiol-17β, HO-1 (10^{-8} M) and HO-2 (10^{-4} M) mRNA was detected in non laboring pregnant but not in non pregnant myometrium. Confirming these results HO-1 protein was significantly induced in a concentration-dependent manner by progesterone but not by oestradiol-17β in non-laboring pregnant myometrium. The induction of HO-1 protein was significant when the myometrial tissue was exposed to relatively low concentrations of progesterone (10^{-8} M) and when progesterone (10^{-6} M) and oestradiol-17β (10^{-7} M) were used in combination ($p < 0.05$). HO-2 was also induced by oestradiol-17β and the highest concentration of progesterone (10^{-6} M).

In aggreement with the reported finding that estradiol activates guanylate cyclase in macaque myometrium[68] our studies show that oestradiol-17β up-regulates

HO-2 protein expression in the non-laboring myometrium and HO-1 and HO-2 mRNA in non-pregnant myometrium.[13]

4.3. Effect of Sex Steroids on HO-Induced Myometrial Contractility

To assess the functional activity of HO in pregnant myometrium, COHb production was measured in response to hemin, progesterone or the vehicle (control) in term non-laboring myometrial explants.[13] Maximal CO release (approximately 6-fold) occurred after 2 hours incubation with hemin ($10 \mu M$) or progesterone ($10^{-6} M$) and returned to baseline levels after 12 hours. The specificity of this effect was demonstrated using SnPP ($20 \mu M$) which significantly blocked hemin-induced CO production in myometrial strips.

Myometrial contractility was examined using human myometrial strips from non-laboring uteri isometric recording under tension and equilibrated in Krebs Henseleit physiological solution maintained at 37° in an organ bath as described by Morrison et al.[69] The stimulation of HO activity with hemin ($10 \mu M$) completely inhibited spontaneous myometrial required for hemin to induce HO activity to generate sufficient CO (see above). In addition, lower concentrations hemin (~$3 \mu M$) significantly inhibited (by 45%) oxytocin-induced contractions in this system (Fig. 4).

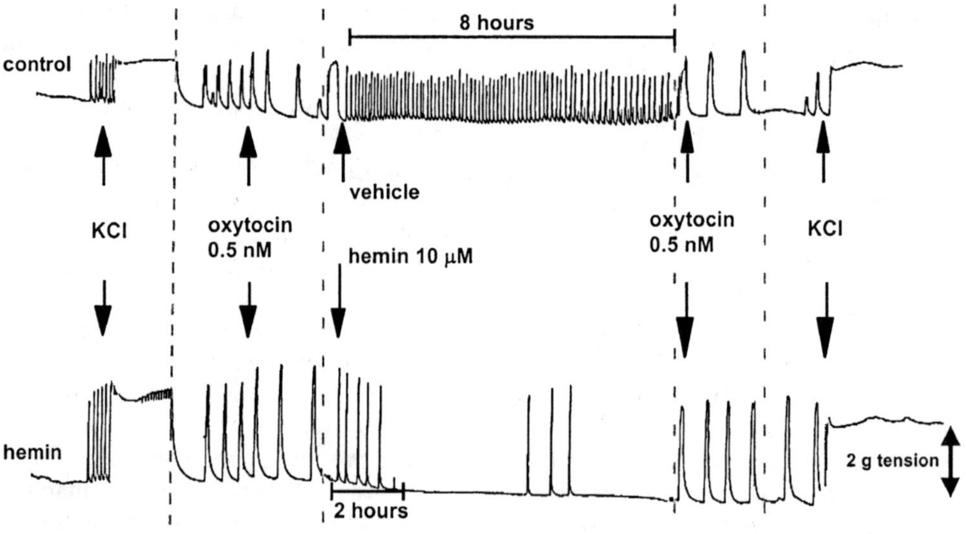

Figure 4. The effect of hemin on oxytocin-induced contractions in human non-laboring pregnant myometrium at term. Myometrial strips were equilibrated for 2 h under 2 g tension prior to the isometric assessment of contractile activity. Oxytocin (0.5 nM) and KCL (45 mM) were used to assess the responsiveness of the myometrial tissue before and after the addition of hemin and the tissue was washed as indicated by the dashed lines. Hemin ($10 \mu M$) completely inhibited the spontaneous contractions (vehicle) in the myometrial tissue after 2 h incubation. The trace is a representative example from three similar experiments and the speed of the recording was slowed during the hemin incubation. *Adapted from Acevedo and Ahmed.*[13]

Coupled with the higher expression of HO protein in term pregnant uterus, the finding that contractions following a 2 to 4 hour incubation. The 2 to 4 hour delay is due to the time progesterone up-regulates HO gene and protein expression and CO production in non-laboring term myometrium *in vitro* indicates that the HO-CO-cGMP system promotes uterine quiescence during pregnancy. In the rat brain and spleen, HO activity is modulated by inhibitors and donors of NO[20] and NO donor (SIN-1) increases rat liver HO-1 activity.[56] The inhibition of uterine contractions by hemin may be direct through the CO-cGMP pathway and/or as a result of a positive feedback mechanism between NOS-NO and HO-CO production in the myometrium during pregnancy.

5. SUMMARY

HO activity appears to be central to the maintenance of normal pregnancy to term and perturbations in its expression may exacerbate the underlying pathologies of preeclampsia and IUGR or lead to premature birth. The specific induction of HO activity leads to the production of CO from villous placental fragments and fetal membranes and significantly relaxes pre-constricted placental arteries. HO expression is significantly reduced in placenta from pregnancies complicated with preeclampsia compared with gestationally-matched normal pregnancies suggesting that the impairment of HO activation may predispose placenta to cellular injury and subsequent maternal endothelial cell activation. Collectively these findings identify the HO-CO pathway as a modifier of feto-placental circulation and indicate that HO-1 may offer protection against cytotoxic damage in the placenta. Furthermore the HO-CO system under the control of progesterone appears to act as an endogenous tocolytic agent in human myometrium during pregnancy indicating the importance of HO in the maintenance of normal pregnancy.

REFERENCES

1. C.W. Redman and I.L. Sargent, The pathogenesis of pre-eclampsia, *Gynecol. Obstet. Fertil.*, **29**(7–8), 518–522 (2001).
2. R.W. Rush, M.J. Keirse, P. Howat, J.D. Baum, A.B. Anderson, and A.C. Turnbull, Contribution of preterm delivery to perinatal mortality, *Br. Med. J.*, **2**(6042), 965–968 (1976).
3. G.S. Berkowitz and E. Papiernik, Epidemiology of preterm birth, *Epidemiol. Rev.*, **15**(2), 414–443 (1993).
4. M.D. Maines, Heme oxygenase: function, multiplicity, regulatory mechanisms, and clinical applications, *FASEB J.*, **2**(10), 2557–2568 (1988).
5. R. Stocker, A.N. Glazer, and B.N. Ames, Antioxidant activity of albumin-bound bilirubin, *Proc. Natl. Acad. Sci. USA*, **84**(16), 5918–5922 (1987).
6. D. Willis, A.R. Moore, R. Frederick, and D.A. Willoughby, Heme oxygenase: a novel target for the modulation of the inflammatory response, *Nat. Med.*, **2**(1), 87–90 (1996).
7. S.J. McFaul and J.J. McGrath, Studies on the mechanism of carbon monoxide induced vasodilation in the isolated perfused rat heart, *Toxicol. Appl. Pharmacol.*, **87**(3), 464–473 (1987).
8. T. Morita, M.A. Perrella, M.E. Lee, and S. Kourembanas, Smooth muscle cell-derived carbon monoxide is a regulator of vascular cGMP, *Proc. Natl. Acad. Sci. USA*, **92**(5), 1475–1479 (1995).

9. R.A. Johnson, M. Lavesa, B. Askari, N.G. Abraham, and A. Nasjletti, A heme oxygenase product, presumably carbon monoxide, mediates a vasodepressor function in rats, *Hypertension*, **25**(2), 166–169 (1995).

10. T. Yoshida, P. Biro, T. Cohen, R.M. Muller, and S. Shibahara, Human heme oxygenase cDNA and induction of its mRNA by hemin, *Eur. J. Biochem.*, **171**(3), 457–461 (1988).

11. W.K. McCoubrey, Jr., J.F. Ewing, and M.D. Maines, Human heme oxygenase-2: characterization and expression of a full-length cDNA and evidence suggesting that the two HO-2 transcripts may differ by choice of polyadenylation signal, *Arch Biochem. Biophys.*, **295**(1), 13–20 (1992).

12. W.K. McCoubrey, Jr., T.J. Huang, and M.D. Maines, Isolation and characterization of a cDNA from the rat brain that encodes hemoprotein heme oxygenase-3, *Eur. J. Biochem.*, **247**(2), 725–732 (1997).

13. C.H. Acevedo and A. Ahmed, Hemeoxygenase-1 inhibits human myometrial contractility via carbon monoxide and is upregulated by progesterone during pregnancy, *J. Clin. Invest.*, **101**(5), 949–955 (1998).

14. A. Ahmed, M. Rahman, X. Zhang, C.H. Acevedo, S. Nijjar, I. Rushton, B. Bussolati, and J. St John, Induction of placental heme oxygenase-1 is protective against TNFalpha-induced cytotoxicity and promotes vessel relaxation, *Mol. Med.*, **6**(5), 391–409 (2000).

15. R.D. Reilly and P.T. Russell, Neurohistochemical evidence supporting an absence of adrenergic and cholinergic innervation in the human placenta and umbilical cord, *Anat. Rec.*, **188**(3), 277–286 (1977).

16. N.M. Gude, R.G. King, and S.P. Brennecke, Role of endothelium-derived nitric oxide in maintenance of low fetal vascular resistance in placenta, *Lancet*, **336**(8730), 1589–1590 (1990).

17. L. Myatt, A. Brewer, and D.E. Brockman, The action of nitric oxide in the perfused human fetal-placental circulation, *Am. J. Obstet. Gynecol.*, **164**(2), 687–692 (1991).

18. R. Zakhary, S.P. Gaine, J.L. Dinerman, M. Ruat, N.A. Flavahan, and S.H. Snyder, Heme oxygenase 2: endothelial and neuronal localization and role in endothelium-dependent relaxation, *Proc. Natl. Acad. Sci. USA*, **93**(2), 795–798 (1996).

19. G.D. Jones and L. Poston, The role of endogenous nitric oxide synthesis in contractility of term or preterm human myometrium, *Br. J. Obstet. Gynaecol.*, **104**(2), 241–245 (1997).

20. R.D. Lees, P. Slater, and S.W. D'Souza, Nitric oxide does not modulate kainate receptor binding in human brain, *Neurosci. Lett.*, **233**(2–3), 133–136 (1997).

21. C. Lees, S. Campbell, E. Jauniaux, R. Brown, B. Ramsay, D. Gibb, S. Moncada, and J.F. Martin, Arrest of preterm labour and prolongation of gestation with glyceryl trinitrate, a nitric oxide donor, *Lancet*, **343**(8909), 1325–1326 (1994).

22. B.J. Murphy, K.R. Laderoute, S.M. Short, and R.M. Sutherland, The identification of heme oxygenase as a major hypoxic stress protein in Chinese hamster ovary cells, *Br. J. Cancer*, **64**(1), 69–73 (1991).

23. M. McLean, M. Bowman, V. Clifton, R. Smith, and A.B. Grossman, Expression of the heme oxygenase-carbon monoxide signalling system in human placenta, *J. Clin. Endocrinol. Metab.*, **85**(6), 2345–2349 (2000).

24. N. Yoshiki, T. Kubota, and T. Aso, Expression and localization of heme oxygenase in human placental villi, *Biochem. Biophys. Res. Commun.*, **276**(3), 1136–1142 (2000).

25. F. Lyall, A. Barber, L. Myatt, J.N. Bulmer, and S.C. Robson, Hemeoxygenase expression in human placenta and placental bed implies a role in regulation of trophoblast invasion and placental function, *FASEB J.*, **14**(1), 208–219 (2000).

26. A. Barber, S.C. Robson, L. Myatt, J.N. Bulmer, and F. Lyall, Heme oxygenase expression in human placenta and placental bed: reduced expression of placenta endothelial HO-2 in preeclampsia and fetal growth restriction, *FASEB J.*, **15**(7), 1158–1168 (2001).

27. M.J. Odrcich, C.H. Graham, K.A. Kimura, B.E. McLaughlin, G.S. Marks, K. Nakatsu, and J.F. Brien, Heme oxygenase and nitric oxide synthase in the placenta of the guinea-pig during gestation, *Placenta*, **19**(7), 509–516 (1998).

28. N. Ihara, R. Akagi, K. Ejiri, T. Kudo, K. Furuyama, and H. Fujita, Developmental changes of gene expression in heme metabolic enzymes in rat placenta, *FEBS Lett.*, **439**(1–2), 163–167 (1998).

29. B. Brune and V. Ullrich, Inhibition of platelet aggregation by carbon monoxide is mediated by activation of guanylate cyclase, *Mo.l Pharmacol.*, **32**(4), 497–504 (1987).

30. L. Myatt, Control of vascular resistance in the human placenta, *Placenta*, **13**(4), 329–341 (1992).

31. L.P. McQuillan, G.K. Leung, P.A. Marsden, S.K. Kostyk, and S. Kourembanas, Hypoxia inhibits expression of eNOS via transcriptional and posttranscriptional mechanisms, *Am J Physiol*, **267**(5 Pt 2), H1921–1927 (1994).

32. D.V. Faller, Endothelial cell responses to hypoxic stress, *Clin Exp Pharmacol Physiol*, **26**(1), 74–84 (1999).

33. S. Moncada, Nitric oxide in the vasculature: physiology, and pathophysiology, *Ann. NY Acad. Sci.*, **811**, 60–67; discussion 67–69 (1997).

34. B.E. McLaughlin, J.M. Hutchinson, C.H. Graham, G.N. Smith, G.S. Marks, K. Nakatsu, and J.F. Brien, Heme oxygenase activity in term human placenta, *Placenta*, **21**(8), 870–873 (2000).

35. K.A. Nath, G. Balla, G.M. Vercellotti, J. Balla, H.S. Jacob, M.D. Levitt, and M.E. Rosenberg, Induction of heme oxygenase is a rapid, protective response in rhabdomyolysis in the rat, *J. Clin. Invest.*, **90**(1), 267–270 (1992).

36. Wagener, E. Feldman, T. de Witte, and N.G. Abraham, Heme induces the expression of adhesion molecules ICAM-1, VCAM-1, and E selectin in vascular endothelial cells, *Proc. Soc. Exp. Biol. Med.*, **216**(3), 456–463 (1997).

37. W.W. Hancock, R. Buelow, M.H. Sayegh, and L.A. Turka, Antibody-induced transplant arteriosclerosis is prevented by graft expression of anti-oxidant and anti-apoptotic genes, *Nat. Med.*, **4**(12), 1392–1396 (1998).

38. A. Yachie, Y. Niida, T. Wada, N. Igarashi, H. Kaneda, T. Toma, K. Ohta, Y. Kasahara, and S. Koizumi, Oxidative stress causes enhanced endothelial cell injury in human heme oxygenase-1 deficiency, *J. Clin. Invest.*, **103**(1), 129–135 (1999).

39. R. Motterlini, R. Foresti, M. Intaglietta, and R.M. Winslow, NO-mediated activation of heme oxygenase: endogenous cytoprotection against oxidative stress to endothelium, *Am. J. Physiol.*, **270**(1 Pt 2), H107–114 (1996)

40. L. Otterbein, S.L. Sylvester, and A.M. Choi, Hemoglobin provides protection against lethal endotoxemia in rats: the role of heme oxygenase-1, *Am. J. Respir. Cell. Mol. Biol.*, **13**(5), 595–601 (1995).

41. L.E. Otterbein, L.L. Mantell, and A.M. Choi, Carbon monoxide provides protection against hyperoxic lung injury, *Am. J. Physiol.*, **276**(4 Pt 1), L688–694 (1999).

42. N. Maulik, D.T. Engelman, M. Watanabe, R.M. Engelman, and D.K. Das, Nitric oxide–a retrograde messenger for carbon monoxide signaling in ischemic heart, *Mol. Cell. Biochem.*, **157**(1–2), 75–86 (1996).

43. T. Polte, S. Oberle, and H. Schroder, Nitric oxide protects endothelial cells from tumor necrosis factor-alpha-mediated cytotoxicity: possible involvement of cyclic GMP, *FEBS Lett.*, **409**(1), 46–48 (1997).

44. J.M. Roberts and C.W. Redman, Pre-eclampsia: more than pregnancy-induced hypertension, *Lancet*, **341**(8858), 1447–1451 (1993).

45. L.S. Jonsdottir, R. Arngrimsson, R.T. Geirsson, H. Sigvaldason, and N. Sigfusson, Death rates from ischemic heart disease in women with a history of hypertension in pregnancy, *Acta. Obstet. Gynecol. Scand.*, **74**(10), 772–776 (1995).

46. P. Hannaford, S. Ferry and S. Hirsch, Cardiovascular sequelae of toxaemia of pregnancy, *Heart*, **77**(2), 154 158 (1997).

47. J.R. Ashworth, A.Y. Warren, I.R. Johnson, and P.N. Baker, Plasma from pre-eclamptic women and functional change in myometrial resistance arteries, *Br. J. Obstet. Gynaecol.*, **105**(4), 159 161 (1998).

48. Y. Wang and S.W. Walsh, TNF alpha concentrations and mRNA expression are increased in preeclamptic placentas, *J. Reprod. Immunol.*, **32**(2), 157–169 (1996).

49. B.L. Sheppard and J. Bonnar, An ultrastructural study of utero placental spiral arteries in hypertensive and normotensive pregnancy and fetal growth retardation, *Br. J. Obstet. Gynaecol.*, **88**(7), 695–705 (1981).

50. R. Pijnenborg, J. Anthony, D.A. Davey, A. Rees, A. Tiltman, L. Vercruysse, and A. van Assche, Placental bed spiral arteries in the hypertensive disorders of pregnancy, *Br. J. Obstet. Gynaecol.*, **98**(7), 648–655 (1991).

51. I.R. McFadyen, A.B. Price, and R.T. Geirsson, The relation of birthweight to histological appearances in vessels of the placental bed, *Br. J. Obstet. Gynaecol.*, **93**(5), 476–481 (1986).

52. H. Kiss, C. Schneeberger, W. Tschugguel, H. Lass, J.C. Huber, P. Husslein, and M. Knofler, Expression of endothelial (type III) nitric oxide synthase in cytotrophoblastic cell lines: regulation by hypoxia and inflammatory cytokines, *Placenta*, **19**(8), 603–611 (1998).

53. L. Myatt, A.L. Eis, D.E. Brockman, I.A. Greer, and F. Lyall, Endothelial nitric oxide synthase in placental villous tissue from normal, pre-eclamptic and intrauterine growth restricted pregnancies, *Hum. Reprod.*, **12**(1), 167–172 (1997).

54. J. Balla, H.S. Jacob, G. Balla, K. Nath, and G.M. Vercellotti, Endothelial cell heme oxygenase and ferritin induction by heme proteins: a possible mechanism limiting shock damage, *Trans. Assoc. Am. Physicians*, **105**, 1–6 (1992).

55. N.G. Abraham, Y. Lavrovsky, M.L. Schwartzman, R.A. Stoltz, R.D. Levere, M.E. Gerritsen, S. Shibahara, and A. Kappas, Transfection of the human heme oxygenase gene into rabbit coronary microvessel endothelial cells: protective effect against heme and hemoglobin toxicity, *Proc. Natl. Acad. Sci. USA*, **92**(15), 6798–6802 (1995).

56. D. Willis, A. Tomlinson, R. Frederick, M.J. Paul-Clark, and D.A. Willoughby, Modulation of heme oxygenase activity in rat brain and spleen by inhibitors and donors of nitric oxide, *Biochem. Biophys. Res. Commun.*, **214**(3), 1152–1156 (1995).

57. A.K. Poranen, U. Ekblad, P. Uotila, and M. Ahotupa, Lipid peroxidation and antioxidants in normal and pre-eclamptic pregnancies, *Placenta*, **17**(7), 401–405 (1996).

58. P.C. Chandra, H.J. Schiavello, S.L. Briggs, and J.D. Samuels, Heterotopic pregnancy with term delivery after rupture of a first- trimester tubal pregnancy. A case report, *J. Reprod. Med.*, **44**(6), 556–558 (1999).

59. S.M. Sladek, R.R. Magness, and K.P. Conrad, Nitric oxide and pregnancy, *Am. J. Physiol.*, **272**, R441–463 (1997).

60. H. Izumi and R.E. Garfield, Relaxant effects of nitric oxide and cyclic GMP on pregnant rat uterine longitudinal smooth muscle, *Eur. J. Obstet. Gynecol. Reprod. Biol.*, **60**(2), 171–180 (1995).

61. I. Buhimschi, M. Ali, V. Jain, K. Chwalisz, and R.E. Garfield, Differential regulation of nitric oxide in the rat uterus and cervix during pregnancy and labour, *Hum. Reprod.*, **11**(8), 1755–1766 (1996).

62. C. Yallampalli, I. Buhimschi, K. Chwalisz, R.E. Garfield, and Y.L. Dong, Preterm birth in rats produced by the synergistic action of a nitric oxide inhibitor (NG-nitro-L-arginine methyl ester) and an antiprogestin (onapristone), *Am. J. Obstet. Gynecol.*, **175**(1), 207–212 (1996).

63. A. Barber, S.C. Robson, and F. Lyall, Hemoxygenase and nitric oxide synthase do not maintain human uterine quiescence during pregnancy, *Am. J. Pathol.*, **155**(3), 831–840 (1999).

64. D.H. Howe, R. Sangha, M.D. Kilby, M.J. Whittle, and A. Ahmed, Identification and expression of nitric oxide synthase isoforms in human myometrium before and after the onset of labor., *J. Soc. Gynaecol. Invest.*, **3**(Suppl), A326 (1996).

65. C.P. Weiner, R.G. Knowles, S.E. Nelson, and L.D. Stegink, Pregnancy increases guanosine 3',5'-monophosphate in the myometrium independent of nitric oxide synthesis, *Endocrinology*, **135**(6), 2473–2478 (1994).

66. J.N. Anderson, E.J. Peck, Jr., and J.H. Clark, Estrogen-induced uterine responses and growth: relationship to receptor estrogen binding by uterine nuclei, *Endocrinology*, **96**(1), 160–167 (1975).

67. B.F. Mitchell, and S. Wong, Changes in 17 beta,20 alpha-hydroxysteroid dehydrogenase activity supporting an increase in the estrogen/progesterone ratio of human fetal membranes at parturition, *Am. J. Obstet. Gynecol.*, **168**(5), 1377–1385 (1993).

68. K. Karalis, G. Goodwin, and J.A. Majzoub, Cortisol blockade of progesterone: a possible molecular mechanism involved in the initiation of human labor, *Nat. Med.*, **2**(5), 556–560 (1996).

69. J.J. Morrison, S.R. Dearn, S.K. Smith, and A. Ahmed, Activation of protein kinase C is required for oxytocin-induced contractility in human pregnant myometrium, *Hum. Reprod.*, **11**(10), 2285–2290 (1996).

7

INCREASED CARBON MONOXIDE IN EXHALED AIR IN PATIENTS WITH INFLAMMATORY RESPIRATORY DISEASES

Mutsuo Yamaya[a], Shoji Okinaga[a], Kiyohisa Sekizawa[b], Mizue Monma[a], and Hidetada Sasaki[a]

[a]Department of Geriatric and Respiratory Medicine
Tohoku University School of Medicine
Sendai 980-8574, and
[b]Department of Pulmonary Medicine
Institute of Clinical Medicine
University of Tsukuba
Tsukuba 306-8575, Japan

1. INTRODUCTION

Carbon monoxide (CO), like nitric oxide (NO), has been reported to have biologic actions such as smooth muscle relaxation[1] or inhibition of platelet aggregation,[2] and to act as a neural messenger in the brain.[3,4] CO is produced endogenously in many tissues of the body by the class of enzymes known collectively as heme oxygenase (HO).[5] Two forms of HO have been characterized. Of these, HO-1 is present in the pulmonary vascular endothelium,[6] alveolar macrophages[7] and human airway epithelium,[8] and is induced by oxidative stress,[6,9] inflammatory cytokines,[10,11] and NO.[12] HO-2 is not inducible and is widely distributed throughout the body, with high concentrations in the brain.[5] CO can be detected in exhaled air in smokers and non-smokers.[13] The pathogenesis of inflammatory respiratory diseases is associated with several factors including oxidative stress and inflammatory cytokines. Therefore, we studied whether the levels of exhaled CO increase in patients with inflammatory respiratory diseases.

2. INCREASED CARBON MONOXIDE IN EXHALED AIR OF ASTHMATIC PATIENTS

Exhaled CO was measured on a portable Bedfont EC50 analyzer (Bedfont Tehnical Instruments Ltd., Sittingbourne, UK) using the method described by Jarvis et al.[13] in which subjects are asked to exhale fully, inhale deeply, and hold their breath for 20s before exhaling rapidly into a disposable mouthpiece. This procedure was repeated three times, with 1 min of normal breathing between each repetition, and mean values was used for analysis. The exhaled CO concentration was determined by subtracting the background level from the observed reading as previously described.[13] To avoid analysis with a value of exhaled CO concentration below 1.0 ppm, the background level was subtracted from the average value obtained from three sequential maneuvers in each patient. The exhaled CO concentration values were always above 1.0 ppm before subtracting the background level throughout the experiments. Prior to the start of the study, the analyzer was calibrated with a mixture of 50 ppm CO in air.[13]

First, we examined whether asthmatic patients produce more CO than healthy control subjects and if the levels of the exhaled CO concentration are reduced in asthmatic patients receiving regular inhaled corticosteroids, which control inflammation in the asthmatic airways.[14] Asthma was defined as a clinical history of intermittent wheeze, cough, chest tightness, or dyspnea, and documented reversible airflow limitation either spontaneously or with treatment during the preceding year.[15] All the asthmatic subjects were nonsmokers and their airway obstruction was stable for at least 2 wk before the study. One group received inhaled β_2-agonists only and the others received regular inhaled corticosteroids (beclomethasone dipropionate 400 to 1,200 μg daily). In order to further investigate the effect of inhaled corticosteroids on exhaled CO concentration, 12 patients with symptomatic asthma, which was being treated by β_2-agonists alone and which was considered severe enough to require prophylactic treatment for disease control, were followed before and 4 wk after the initiation of inhaled corticosteroid treatment (beclomethasone dipropionate 400 μg daily). We also examined eosinophil cell counts in sputum.

The mean exhaled CO concentration was 1.5 ± 0.1 ppm in nonsmoking control subjects. In asthmatic patients not receiving corticosteroids the exhaled CO concentration was significantly higher (5.6 ± 0.6 ppm, $p < 0.001$), whereas in asthmatic patients receiving inhaled corticosteroids the exhaled CO concentration did not differ significantly from that in nonsmoking control subjects (1.7 ± 0.1 ppm, $p > 0.20$) (Fig. 1). The exhaled CO concentration before and after the initiation of inhaled corticosteroid treatment is shown in Fig. 2. There was a significant relation between changes in the exhaled CO concentration and those in eosinophil cell counts in sputum ($p < 0.001$) (Fig. 3). Likewise, changes in the exhaled CO concentration significantly correlated with those in FEV1 ($r = 0.71$, $p < 0.01$). This study has shown that exhaled CO can be reliably measured in healthy control subjects and asthmatic patients; the latter has an elevated exhaled CO concentration.[16] The improved lung function was accompanied by concomitant decreases in exhaled CO concentration and eosinophil cell counts in sputum in patients who needed inhaled corticosteroid therapy. These

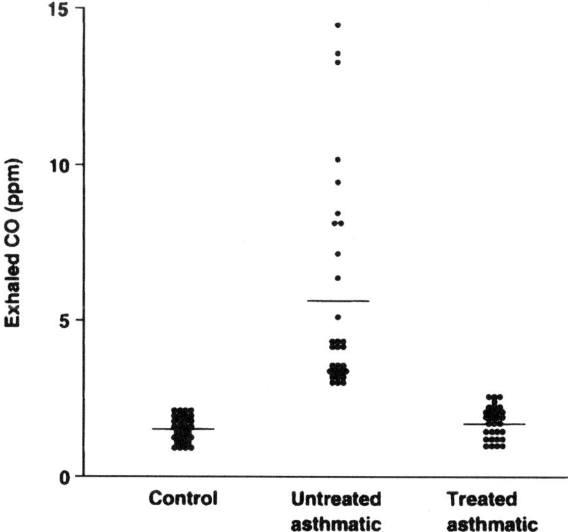

Figure 1. Carbon monoxide (CO) concentrations in exhaled air in non-smoking subjects (n = 30), untreated asthmatics (n = 30), and treated asthmatics (n = 30). Untreated = no inhaled corticosteroids; treated = regular inhaled corticosteroids. Bar = mean value.

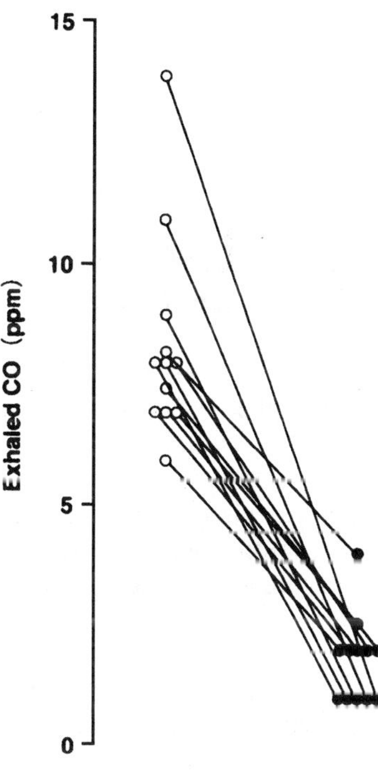

Figure 2. Carbon monoxide (CO) concentrations in exhaled air from patients with symptomatic asthma before (open circles) and after (closed circles) inhaled corticosteroid treatment.

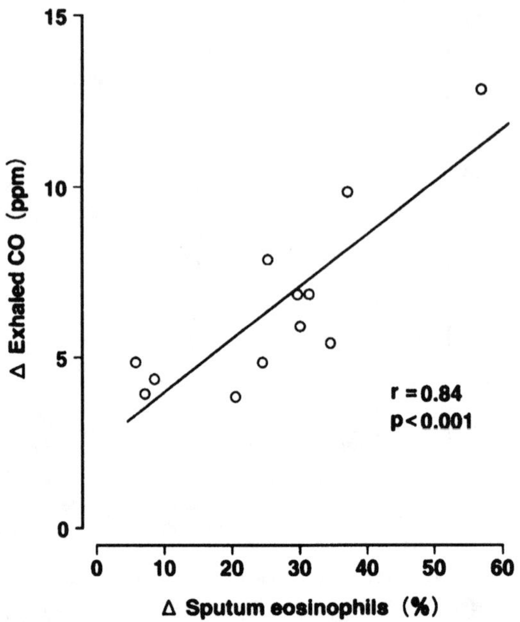

Figure 3. Relation between changes in exhaled carbon monoxide (CO) concentrations and those in eosinophil cell counts in sputum from patients with symptomatic asthma after inhaled corticosteroid treatment; r is the correlation coefficient; the *line* and p value correspond to the fitted regression equation.

findings showed the possibility that an increase in exhaled CO concentration may reflect inflammation of the asthmatic lung.[16]

3. EXHALED CARBON MONOXIDE LEVELS DURING TREATMENT OF ACUTE ASTHMA

As described above, CO has been recovered in exhaled air from normal subjects and at higher levels from the exhaled air of patients with asthma.[16] Although significantly lower CO concentrations have been reported in exhaled air from asthmatic patients receiving glucocorticoid treatment than from asthmatic patients not receiving this therapy,[16] the relationship between changes in exhaled CO concentrations and a therapeutic effect of glucocorticoids on the airways has not been shown in patients with acute asthma exacerbations. The time course changes in exhaled CO concentrations after treatment of acute asthma with oral glucocorticoids were therefore examined in a group of patients presenting with an asthma exacerbation.[17] The patients were taking regular treatment with inhaled glucocorticoids (beclomethasone dipropionate 200 μg daily), along with β-adrenergic agonists on demand, and were categorized as moderate asthma.[15] All the asthmatic patients were non-smokers, and were usually in a stable condition before exacerbations of asthma.

The time course changes in exhaled CO and PEFR after treatment with oral corticosteroids in all patients are shown in Fig. 4. Acute asthma exacerbations caused increases in exhaled CO concentrations and decreases in PEFR in all patients. The mean exhaled CO concentration in asthmatic subjects at the baseline condition was 1.4 ± 0.2 ppm. The exhaled CO was significantly higher at the start of glucocorticoid treatment for acute asthma (4.6 ± 0.4 ppm, $p < 0.01$, two-way ANOVA). Although there were some individual variations, treatment with oral glucocorticoids decreased exhaled CO concentrations in association with increases in PEFR ($p < 0.05$, in each case by two-way ANOVA). Decreases in exhaled CO after treatment with oral corticosteroids were coincident with an improvement of PEFR ($p < 0.05$, Student's t-test). Although the maximal exhaled CO concentration did not correlate with maximal fall in PEFR ($p > 0.20$, Student's t-test), a significant relationship was observed between the maximal exhaled CO concentration and recovery time of PEFR after treatment of acute asthma exacerbations with oral glucocorticoids ($p < 0.01$, Student's t-test). All asthmatic patients had a total symptom score for URTIs of >5 (7.9 ± 0.5, $n = 20$).

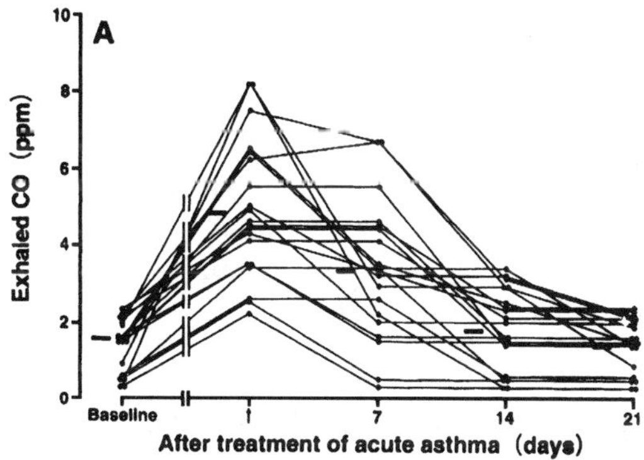

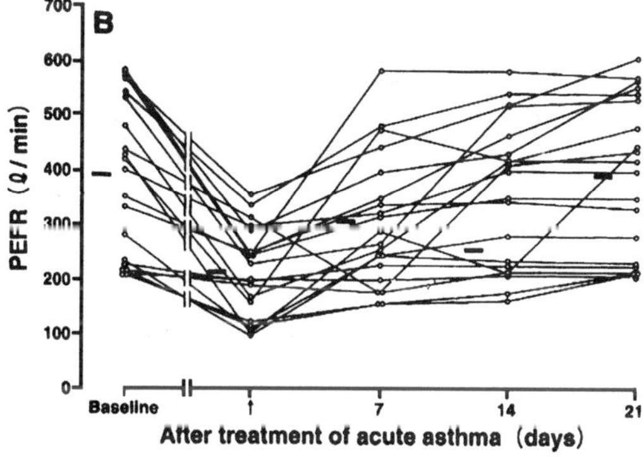

Figure 4. Time course changes in exhaled carbon monoxide concentrations (A) and peak expiratory flow rate (PEFR; B) in asthmatic patients before acute asthma exacerbations (baseline) and after treatment with oral glucocorticoids. ppm: parts per million. ↑: the start of treatment of acute asthma exacerbations with oral glucocorticoids. Horizontal bars indicate mean values for each time-point.

Influenza type A viruses were identified in 14 patients, and no viruses were detected in the other 6 patients at the onset of exacerbations of asthma.

Glucocorticoid therapy is commonly employed to treat acute asthma exacerbations. In this study, an improved PEFR was found to be accompanied by a concomitant decrease in exhaled CO concentration in patients who need emergency treatment for acute asthma. Furthermore, the maximal exhaled CO concentration correlates with recovery time of PEFR after treatment of acute asthma exacerbations with oral glucocorticoids. In the present study, all the patients had symptoms of URTIs and wheeze during thr exacerbations of bronchial asthma. Viral infections may induce HO in a variety of cell types, including airway epithelial cells and macrophages[7] via the induction of pro-inflammatory cytokines[10] and NO.[12] These cytokines, including interleukin-1, interleukin-6 and tumor necrosis factor, are involved in asthmatic inflammation. Furthermore, asthmatic airways produce high levels of NO.[18] Therefore, both viral infection and asthmatic inflammation may upregulate HO-1 activity. Viral infection may induce airway inflammation and exacerbation of bronchial asthma, resulting in a rise in exhaled CO and a fall in PEFR in the present study. The present study suggests that exhaled CO may be a useful means of monitoring the control of airway inflammation of asthma. CO analysers are not widely used at the present time, but they are portable and not expensive. Therefore, it is possible to conceive the use of personal monitors in conjunction with home peak flow meters. Although an exhaled CO elevation was demonstrated in patients who need emergency treatment for acute asthma, it is expected that exhaled CO levels may be elevated in even quite mild asthma with airway inflammation. Thus, exhaled CO measurements may be useful in monitoring the control of asthma and the response to antiinflammatory treatments in individual asthmatic patients.

4. RELATION BETWEEN EXHALED CARBON MONOXIDE LEVELS AND CLINICAL SEVERITY OF ASTHMA

Histological examinations of specimens from bronchial biopsy and autopsy in asthma patients revealed airway inflammation including increases in eosinophils, lymphocytes, mast cells, and neutrophils in airways.[19,20] Furthermore, levels of eosinophils and eosinophil cationic protein are increased in the bronchoalveolar lavage fluid from asthmatic patients and correlate with the clinical severity.[21] However, it is uncertain whether exhaled CO concentrations are related to the severity of asthma. To determine whether the level of CO is related to the severity of asthma, exhaled CO concentration was measured in asthmatic patients for 12 months when the patients attended the hospital for regular treatment or for treatment of asthma exacerbations.[22]

A classification of asthma severity was performed according to clinical features before treatment, lung function and regular medication usually required to maintain control,[15] and patients were categorized as mild (20 patients), moderate (20 patients) and severe asthma (31 patients), respectively.[15] Furthermore, severe asthmatic patients were divided into two groups: stable and unstable severe asthma. Patients with unstable severe asthma had been admitted to the hospital (1 to 4 times/year) or had received

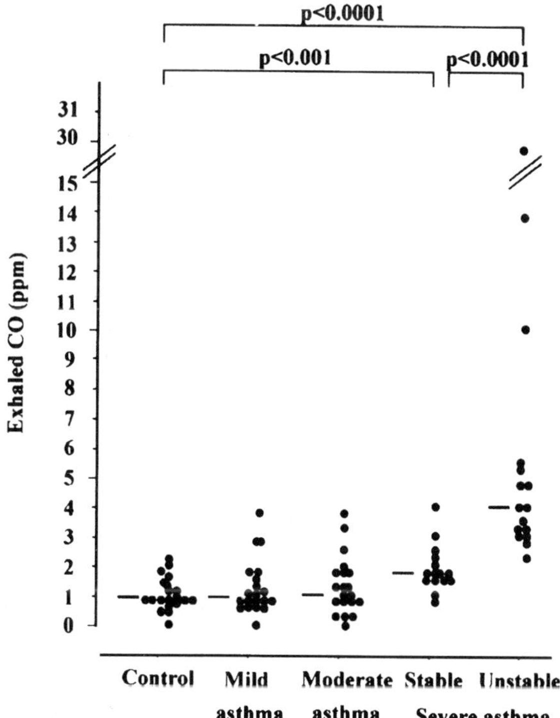

Figure 5. Mean exhaled carbon monoxide (CO) concentrations over 12 months in individual non-smoking control subjects (n = 20) and individual patients with mild (n = 20), moderate (n = 20), stable severe (n = 15) and unstable severe (n = 16) asthma. Horizontal bars indicate median values for each group.

additional oral (prednisolone 80 to 560 mg/year) or intravenous (dexamethasone sodium phosphate 4 to 22 mg/year) corticosteroids because of asthma exacerbations. Exhaled CO concentrations were measured for 12 months from December of 1997 to November of 1998, when the patients attended the hospital for regular treatment or for the treatment of asthma exacerbations. Exhaled CO concentrations were also measured in 16 unstable severe asthmatics who visited the hospital every 7 or 14 days for treatment. The mean value of exhaled CO concentration in each season (winter; December–February, spring; March–May, summer; June–August, autumn; September–November) was calculated. Mild asthmatic patients free from any symptoms for several weeks were asked to visit the hospital every month for measurements of exhaled CO. Exhaled CO concentrations were measured 3 weeks after treatment of acute exacerbations of asthma for the analysis in severe asthmatic patients.

At the start of treatment with corticosteroids, exhaled CO concentrations in severe asthma (7.6 ± 8.9 ppm) were significantly reduced 4 weeks after treatment with corticosteroids (4.3 ± 5.7 ppm, n = 31, p < 0.05). However, the mean exhaled CO concentrations were significantly higher in severe asthmatic patients than those in mild and moderate asthmatic patients and non-smoking control subjects (Fig. 5).[22]

Furthermore, exhaled CO concentrations in unstable severe asthma were significantly higher than those in stable severe asthma (Fig. 5). The mean CO concentrations over the 12 months were significantly higher in three cases of mild asthma (3.3, 2.8, and 3.5 ppm, respectively). However, mean exhaled CO concentrations in mild asthmatic patients did not differ significantly from those in non-smoking control subjects ($P > 0.50$) (Fig. 5). Likewise, exhaled CO concentrations in moderate asthmatic patients did not differ significantly from those in non-smoking control subjects ($P > 0.50$) (Fig. 5). There was no seasonal change in mean exhaled CO concentrations in asthmatic patients ($P > 0.10$). The CO values of severe asthma patients in winter season appear to be higher than those in summer season, but the differences were not significant ($P > 0.20$). There was a significant correlation between mean exhaled CO concentration over 12 months and FEV_1 in asthmatic patients ($P < 0.01$, r = 0.55). Even when the data of two subjects with extremely higher CO concentrations (10.3 and 14.0 ppm) were excluded, there was a significant correlation between mean exhaled CO concentration over 12 months and FEV_1 in asthmatic patients ($P < 0.0001$). We have shown that exhaled CO concentration in severe asthmatic patients increased even after treatment with corticosteroids.[22] In contrast, exhaled CO concentration in mild and moderate asthmatic patients did not differ from that in non-smoking control subjects. In the present study, there was a significant correlation between mean exhaled CO concentration over 12 months and FEV_1 in asthmatic patients. Structural changes occur in asthma in both the large and small airways, including subepithelial collagen deposition, smooth muscle growth and thickening of the basement membrane.[19,23,24] Therefore, the persistent airway inflammation detected by increases in exhaled CO concentrations may cause structural changes in airways, thereby the decreases in FEV_1 in severe asthma, although we did not perform histological examinations of the lung in the present study.

5. INCREASED CARBON MONOXIDE IN EXHALED AIR OF SUBJECTS WITH UPPER RESPIRATORY TRACT INFECTIONS

Because upper respiratory tract infections (URTIs) are reported to induce increases in IL-1, IL-6 and TNF-α in nasal lavage fluid[25] as well as increases in exhaled NO,[18] it is likely that viral respiratory tract infection increases CO production. We therefore studied whether URTIs increase the concentration of CO in the exhaled air of normal persons. Twenty subjects with URTIs and 10 control subjects volunteered for this study.[26] None of the subjects in either group were smokers, ex-smokers, or passive smokers. All subjects with URTIs were studied at the time of symptomatic URTIs. Subjects were interviewed regarding the presence and severity of the following 10 symptoms: sneezing, nasal discharge, nasal congestion, malaise, headache, chills, feverishness, sore throat, hoarseness, and cough. Symptoms were rated for severity on a scale from 0 to 3. URTIs were defined if they had a total symptom score more than 5.[27] In the 10 control subjects there was no history of URTIs for at least 4 wk prior to the study, and no history of respiratory or cardiovascular disease. Ten smokers were recruited from volunteers and were studied within 1 to 3 h after the last cigarette.

Mean exhaled CO concentration was 1.2 ± 0.3 ppm in non-smoking control subjects. In subjects with URTIs, the exhaled CO (5.7 ± 0.4 ppm) was significantly higher during the acute phase of URTIs than in non-smoking control subjects (1.0 ± 0.1 ppm, $p < 0.001$). As expected, smoking control subjects had the highest CO concentration among the three groups (18.5 ± 2.3 ppm). The exhaled CO during the acute phase of URTIs (5.7 ± 0.4 ppm) decreased after 3 wk of recovery (1.2 ± 0.2 ppm, $p < 0.001$) and the exhaled CO values in subjects who had recovered from URTIs did not differ significantly from those in non-smoking control subjects (1.0 ± 0.1 ppm, $p > 0.20$). There was a significant relation between changes in the exhaled CO concentration and those in symptom scores from the time of the acute phase to after 3 wk of recovery in subjects with URTIs ($r = 0.72$, $p < 0.01$, $n = 20$). Influenza type A viruses were identified using the method described previously, from all subjects and other viruses were not detected in any of the 20 subjects with URTIs. This study has shown that exhaled CO can be reliably measured in healthy control subjects and subjects during the acute phase of URTIs.[26] We have demonstrated that URTIs are associated with an increase in exhaled CO in normal persons in the acute phase when symptoms are present, and that there is a reduction in exhaled CO after recovery to values that are similar to those in age-matched normal subjects. This suggests that URTIs, presumably caused by influenza viral infection, increases the production of CO in the respiratory tract.

6. INCREASED CARBON MONOXIDE IN EXHALED AIR OF PATIENTS WITH SEASONAL ALLERGIC RHINITIS

The infiltration of T lymphocytes, eosinophils and mucosal type mast cells is observed in the nasal specimens obtained from symptomatic seasonal allergic rhinitis.[28] Likewise, a number of proinflammatory cytokines and chemokines such as interleukin (IL)-1, IL-5 and IL-6 have been identified as possible mediators in rhinitis.[29,30] However, no measures of CO in nasal exhaled air have been made in rhinitis, in which there is a common form of inflammation similar to that described in asthma in the lower airways. Thus to determine whether levels of CO are increased in patients with seasonal allergic rhinitis, measurements of CO in exhaled air were made during nasal and oral exhalation in patients with seasonal allergic rhinitis, and the results were compared with the same measurements made in nonatopic volunteers.[31] We studied 30 normal nonsmoking control subjects and 86 patients with seasonal cedar pollen-sensitive rhinitis who did not have asthma. In the 30 control subjects there was no history of respiratory or cardiovascular disease. None of these 30 control subjects were receiving long-term medication. All patients had a history of seasonal allergic rhinitis for at least 2 years, with upper airway but not lower airway symptoms. A positive cutaneous response (skin prick test that produced a 3 mm wheal) to cedar pollen was identified in all patients but none of the controls had positive skin test responses to a battery of standard antigens including cedar pollen. The study was performed in February during the cedar pollen season in Japan. To assess the clinical symptoms, subjects were instructed to record their clinical symptoms every day over 1 week before the study. The presence of symptoms (nasal obstruction, rhinorrhea and sneezing)

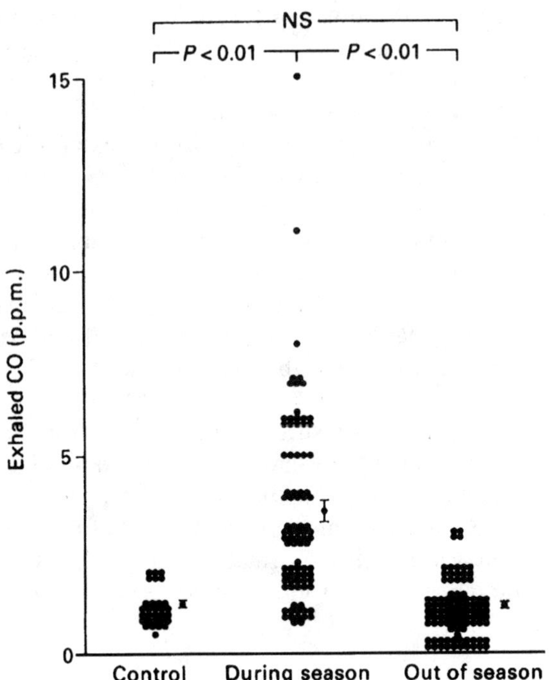

Figure 6. Carbon monoxide (CO) concentrations in exhaled air in control subjects (n = 30) and patients with allergic rhinitis (n = 86) during and out of the cedar pollen season. Mean values ± SE are indicated by closed circles with error bars.

were recorded and the intensity of each symptom was evaluated every day. The intensity of symptoms were scored according to a 4-point scale: 0, no symptoms; 1, mild symptoms; 2, moderate symptoms ; and 3, severe symptoms.[32] Mean value of the daily total symptom score of the three symptoms over 1 week before the study was used to define two subgroups: patients with a mean value of a daily total symptom score below 1 (called without symptoms) and patients with a mean value of a daily total symptom score above 2 (called with symptoms).[33] All patients with seasonal allergic rhinitis were also studied in May, out of the cedar pollen season, at a time when they were free of symptoms.

Mean CO concentration exhaled orally was 1.2 ± 0.1 ppm in healthy control subjects. In patients with allergic rhinitis, the exhaled CO was significantly higher during the cedar pollen season than that in control subjects (Fig. 6).[35] However, the exhaled CO values out of the season in patients with allergic rhinitis did not differ significantly from those in control subjects ($P > 0.20$) (Fig. 6). Orally exhaled CO values during the cedar pollen season were significantly higher in patients with symptoms than in those without symptoms. However, exhaled CO values did not differ significantly among oral and nasal exhalation, and oral exhalation with an expiratory resistance in both control subjects and patients with allergic rhinitis ($P > 0.20$). We have shown that seasonal cedar pollen-sensitive rhinitis is associated with an increase in exhaled CO in patients during the cedar pollen season, and that levels of exhaled CO are higher in patients with symptoms than in those without symptoms. Furthermore, there is a reduction in exhaled CO out of the cedar pollen season to values that are similar to those in nonatopic control subjects. This suggests that allergic rhinitis increases the production of CO in the respiratory tract. The exhaled CO concentration was similar between oral and nasal exhalation in both control subjects and patients with allergic rhinitis, suggesting that the increase in CO may be derived from the lower respiratory tract. This is consistent with the results that the exhaled CO concentration did not differ between oral exhalation with and without an expiratory resistance used to close the velum during expiration to exclude nasal CO that may leak throughout expiration in the presence of an open velum.[34] In contrast to symptomatic rhinitis, lower airway symptoms were not present in any patients. Even in the absence of asthmatic symptoms, however, subclinical inflammation has been reported in patients with allergic rhinitis with evidence of eosinophil accumulation[35,36] and enhanced collagen deposition in the lower airways.[36] It is therefore possible that in patients with seasonal allergic rhinitis, inflammatory cell accumulation occurs within the lower airways but does not reach a sufficient threshold to induce symptoms.

7. SUMMARY

Carbon monoxide (CO) is known to be present in measurable quantities in the exhaled air of normal subjects. We have shown that exhaled CO is increased in patients with inflammatory pulmonary and airway diseases such as bronchial asthma, upper respiratory tract infections, and seasonal allergic rhinitis. Treatments with inhaled and oral corticosteroids, which have been shown to reduce airway inflammation, are associated with a reduction in the exhaled levels of CO in asthma. Furthermore, exhaled CO is increased in the exacerbations of bronchial asthma induced by respiratory virus infections. Based on these findings, measurements of CO may serve as an indirect marker of airway inflammation.

REFERENCES

1. M. Suematsu, N. Goda, T. Sano, S. Kashiwagi, T. Egawa, Y. Shinoda, and Y. Ishimura, Carbon monoxide, an endogenous modulator of sinusoidal tone in the perfused rat liver, J Clin Invest 96:2431–2437 (1995).
2. B. Brune and J. Ullrich, Inhibition of platelet aggregation by carbon monoxide is mediated by activation of guanylate cyclase, Mol Pharmacol 32:497–504 (1987).
3. A. Verma, D.J. Hirsch, C.E. Glatt, and G.V. Ronnett, and S.H. Snyder, Carbon monoxide: a putative neural messenger, Science 259:381–384 (1993).
4. M. Zhuo, S.A. Small, E.R. Kandel, and R.D. Hawkins, Nitric oxide and carbon monoxide produce activity dependent long term synaptic enhancement in hippocampus, Science 260:1946–1950 (1993).
5. M.D. Maines, Heme oxygenase: function, multiplicity, regulatory mechanisms, and clinical applications, FASEB J 2:2557–2568 (1988).
6. L. Otterbein, S.L. Sylvester, and A.M.K. Choi, Hemoglobin provides protection against lethal endotoxin in rats: the role of heme oxygenase-1, Am J Respir Cell Mol Biol 13:595–601 (1995).

7. T. Fukushima, S. Okinaga, K. Sekizawa, T. Ohrui, M. Yamaya, and H. Sasaki, The role of carbon monoxide in lucigenin-dependent chemiluminescence of rat alveolar macrophages, Eur J Pharmacol **289**:103–107 (1995).

8. N. Yamada, M. Yamaya, S. Okinaga, R. Lie, T. Suzuki, K. Nakayama, A. Takeda, T. Yamaguchi, Y. Itoyama, K. Sekizawa, and H. Sasaki, Protective effects of heme oxygenase-1 against oxidant-induced injury in the cultured human tracheal epithelium. Am J Respir Cell Mol Biol **21**:428–435 (1999).

9. S.L. Camhi, J. Alam, L. Otterbein, S.L. Sylvester, and A.M.K. Choi, Induction of heme oxygenase-1 gene expression by lipopolysaccharide is mediated by AP-1 activation, Am J Respir Cell Mol Biol **13**:387–398 (1995).

10. L. Cantoni, C. Rossi, M. Rizzardini, M. Gardina, and P. Ghezzi, Interleukin-1 and tunor necrosis factor induce hepatic heme oxygenase: feedback regulation by glucocorticoids, Biochem J **279**: 891–94 (1991).

11. Y. Lavrovsky, G.S. Drummond, and N.G. Abraham, Downregulation of the human heme oxygenase gene by glucocorticoids and identification of 56b regulatory elements, Biochem Biophys Res Commun **218**:759–765 (1996).

12. Y.M. Kim, H.A. Bergonia, C. Muller, B.R. Pitt, W.D. Watkins, and J.R. Lancaster, Loss and degradation of enzyme-bound heme induced by cellular nitric oxide synthesis, J Biol Chem **270**:5710–5713 (1995).

13. M.J. Jarvis, M. Belcher, C. Vesey, and D.C.S. Hutchison, Low cost carbon monoxide monitors in smoking assessment, Thorax **41**:886–887 (1986).

14. P.J. Barnes, Anti-inflammatory therapy for asthma, Annu Rev Med **44**:229–242 (1993).

15. A.L. Sheffer, International consensus report on the diagnosis and management of asthma, Clin Exp Allergy **22**:1–72 (1992).

16. K. Zayasu, K. Sekizawa, S. Okinaga, M. Yamaya, T. Ohrui, and H. Sasaki, Increased carbon monoxide in exhaled air of asthmatic patients, Am J Respir Crit Care Med **156**:1140–1143 (1997).

17. M. Yamaya, K. Sekizawa, S. Ishizuka, M. Monma, and H. Sasaki, Exhaled carbon monoxide levels during treatment of acute asthma, Eur Respir J **13**:757–760 (1999).

18. S.A. Kharitonov, D. Yates, and P.J. Barnes, Increased nitric oxide in exhaled air of normal human subjects with upper respiratory tract infections, Eur Respir J **8**:295–297 (1995).

19. P.M. O'Byrne, and D.S. Postma, The many faces of airway inflammation. Asthma and chronic obstructive pulmonary disease, Am J Respir Crit Care Med **159**:S41–S66 (1999).

20. R. Djukanovic, W.R. Roche, J.W. Wilson, C.R.W. Beasley, O.P. Twentyman, P.H. Howarth, and S.T. Holgate, Mucosal inflammation in asthma, Am Rev Respir Dis **142**:434–457 (1990).

21. J. Bousquet, P. Chanez, J.Y. Lacoste, G. Barneon, N. Ghavanian, I. Enander, P. Venge, S. Ahlstedt, J. Simony-Lafontaine, and P. Godard, Eosinophilic inflammation in asthma, N Engl J Med **323**: 1033–1039 (1990).

22. M. Yamaya, M. Hosoda, S. Ishizuka, M. Monma, T. Matsui, T. Suzuki, K. Sekizawa, and H. Sasaki, Relation between exhaled carbon monoxide levels and clinical severity of asthma, Clin Exp Allergy **31**:417–722 (2001).

23. P.D. Pare and T.R. Bai, The consequences of chronic allergic inflammation, Thorax **50**:328–332 (1995).

24. N. Carroll, J. Elliot, A. Morton, and A. James A, The structure of large and small airways in non-fatal and fatal asthma, Am Rev Resspir Dis **147**:405–410 (1993).

25. L.C.K. Lau, J.M. Corne, S.J. Scott, R. Davies, E. Friend, and P.H. Howarth, Nasal cytokines in common cold (abstract), Am J Respir Crit Care Med **153**:A697 (1996).

26. M. Yamaya, K. Sekizawa, S. Ishizuka, M. Monma, K. Mizuta, and H. Sasaki H, Increased carbon monoxide in exhaled air of subjects with upper respiratory tract infections, Am J Respir Crit Care Med **158**:311–314 (1998).

27. G.G. Jackson, H.F. Dowling, I.G. Spiesman, and A.V. Board, Transmission of the common cold to volunteers under controlled conditions: I. The common cold as a clinical entity, Arch Intern Med **101**:267–278 (1958).

28. Y. Igarashi, M.S. Goldrich, M.A. Kaliner, A.M.A. Irani, L.B. Schwartz, and M.V. White, Quantitation of inflammatory cells in the nasal mucosa of patients with allergic rhinitis and normal subjects. J Allergy Clin Immunol **95**:716–725 (1995).

29. S. Ying, S.R. Durham, J. Barkans, K. Masuyama, M. Jacobson, S. Rak, O. Lowhagen, R. Moqbel, A.B. Kay, and Q.A. Hamid, T cells are the principal source of interleukin-5 mRNA in allergen-induced rhinitis, Am J Respir Cell Mol Biol **9**:356–360 (1993).
30. T.C. Sim, L.M. Reece, K.A. Hilsmeier, J.A. Grant, and R. Alam, Secretion of chemokines and other cytokines in allergen-induced nasal responses: Inhibition by topical steroid treatment. Am J Respir Crit Care Med **152**:927–933 (1995).
31. M. Monma, M. Yamaya, K. Sekizawa, K. Ikeda, N. Suzuki, T. Kikuchi, T. Takasaka, and H. Sasaki, Increased carbon monoxide in exhaled air of patients with seasonal allergic rhinitis, Clin Exp Allergy **29**:1537–1541 (1999).
32. J. Day and T. Carrillo, Comparison of the efficacy of budesonide and fluticasone propionate aqueous nasal spray for once daily treatment of perennial allergic rhinitis, J Allergy Clin Immunol **102**:902–908 (1998).
33. J.F. Arnal, A. Didier, J. Rami, C. M'Rini, J.P. Charlet, E. Serrano, and J.P. Besombes, Nasal nitric oxide is increased in allergic rhinitis, Clin Exp Allergy **27**:358–362 (1997).
34. P.E. Silkoff, P.A. McClean, A.S. Slutsky, H.G. Furlott, E. Hoffstein, S. Wakita, K.R. Chapman, J.P. Szalai, and N. Zamel, Marked flow-dependence of exhaled nitric oxide using a new technique to exclude nasal nitic oxide, Am J Respir Crit Care Med **155**:260–267 (1997).
35. P.H. Howarth, J. Wilson, R. Djukanovic, S. Wilson, K. Britten, A. Walls, W.R. Roche, and S.T. Holgate, Airway inflammation and atopic asthma: a comparative bronchoscopic investigation. Int Arch Allergy Appl Immunol **94**:266–269 (1991).
36. J. Chakir, M. Laviolette, M. Boutet, R. Laliberte, J. Dube, and L.P. Boulet, Lower airways remodeling in nonasthmatic subjects with allergic rhinitis, Lab Invest **75**:735–744 (1996).

8

CARBON MONOXIDE AND IRON, BY-PRODUCTS OF HEME OXYGENASE, MODULATE VASCULAR ENDOTHELIAL GROWTH FACTOR SYNTHESIS IN VASCULAR SMOOTH MUSCLE CELLS

Józef Dulak[a,*], Roberto Motterlini[b], Ihor Huk[c], Otmar Pachinger[a], Franz Weidinger[a], and Alicja Józkowicz[c]

[a]Division of Cardiology, Innsbruck University, Austria
[b]Vascular Biology Unit,
Department of Surgical Research,
Northwick
Park Institute for Medical Research,
Harrow, United Kingdom
[c]Department of Vascular Surgery,
University of Vienna, Austria

1. INTRODUCTION

Vascular endothelial growth factor (VEGF; VEGF-A) is a major player in angiogenesis (for review see ref. 1). Enhancement of VEGF expression is mediated by hypoxia, inflammatory cytokines and some growth factors.[1]

Besides VEGF, two gaseous molecules are also important determinants of angiogenesis: nitric oxide (NO) and carbon monoxide (CO). NO is continuously synthesized by endothelial cells and its production can also be markedly increased under inflammatory conditions when inducible NO synthase (iNOS) is activated in numerous cell types (for review see ref. 2). Recently we have demonstrated that NO enhances VEGF synthesis in VSMC.[3–5]

* J. Dulak, and A. Jozkowicz contributed equally to this work; send correspondence to: Dr. J. Dulak, Department of Cell Biochemistry, Institute of Molecular Biology, Jagiellonian University, Gronostajowa 7, 30-387 Krakow, Poland.

CO is generated by heme oxygenases (HO) during the degradation of heme to iron and biliverdin, the latter being converted to bilirubin by a cytosolic biliverdin reductase (for review see ref. 6 and 7). Of the three known different HO isoforms (HO-1, HO-2 and HO-3), the stress inducible enzyme HO-1 is highly activated both under hypoxic conditions and by a variety of inflammatory stimuli. Recently it has been demonstrated that transfer of the HO-1 gene resulted in enhancement of endothelial cell proliferation suggesting an involvement of this enzyme in angiogenesis.[8] It has also been described that the number of HO-1 expressing macrophages positively correlated with the extent of angiogenesis in glioma.[9]

The mechanisms of HO-dependent angiogenesis has not been, however, investigated. Therefore, we aimed to elucidate the effect of HO-1 by-products on the generation of VEGF. We, here, demonstrate that HO-1 activity can modulate VEGF synthesis in VSMC. Among the three products of heme degradation by HO activation, i.e. bilirubin, iron and CO, the latter two appear to significantly influence VEGF production in vascular smooth muscle cells.

2. MATERIALS AND METHODS

2.1. Cell Culture and Treatments

Rat thoracic aorta vascular smooth muscle cells (VSMC) were cultured to confluence in 10% FCS/DMEM F-12 medium and then placed in medium containing 0.5% FCS for 24 hours prior to any treatment. Cells were treated in normoxic conditions (air/5% CO_2) with IL-1 (10ng/ml) and TNFα (10ng/ml) for 24h in the presence or absence of tin protoporphyrin (SnPPIX, 1–10M), an inhibitor of HO activity. In other experiments VSMC were treated with cytokines in the presence of zinc (II) deutero-IX-2,4-bisethyleneglycol protoporphyrin (ZnDPPIX) or copper protoporphyrin (CuPPIX) (all at concentration 10µM), the other commonly used inhibitors of HO activity.[10] In another set of experiments, VSMC were pre-treated for 2h with 10M hemin, an inducer of HO-1 expression. The medium was then replaced with fresh medium and cells were incubated for an additional 24–48h with or without SnPPIX (10µM). Cells were also treated with hemin and/or SnPPIX for 6 or 12h, without changing the medium.

2.2. Transfection of Cells with HO-1 Expression Plasmid

Expression plasmids (pcDNA-HO-1) were constructed by cloning rat HO-1 cDNA (kindly provided by Prof. Mahin Maines, Rochester, USA) to pcDNA3 plasmid. Transfection of rat VSMC was performed in 24-well plates as described previously (3–5). Control cells were transfected with pSVβ-gal plasmid (Promega, Madison, USA).

2.3. Effect of HO-1 Expression on Human VEGF Promoter Activity

NIH 3T3 fibroblasts (ATCC, CRL-1658) were cultured in high-glucose DMEM medium supplemented with 5% FCS. Those cells were co-transfected with pcDNA3-

HO-1 or pSVβ-gal plasmids together with VEGF-luciferase plasmid (kindly provided by Dr. Kimura).[11] The VEGF-luciferase plasmid contains the luciferase reporter gene driven by a full sequence of human VEGF promoter. In order to determine the effect of iron on VEGF promoter activity, NIH 3T3 fibroblasts were transfected with VEGF-luc plasmid or HRE-luc plasmid and treated with $FeCl_2$ or deferoxamine. The HRE-luc plasmid (kindly provided by Dr. Kimura) contains the 190-bp fragment of VEGF promoter (−1,014 to −903), including the hypoxia response element. Luciferase activity in the cellular extracts was determined 48 h after transfection according to the vendor's protocol (Luciferase Reporter Gene Assay, High Sensitivity, Roche Biochemicals).

2.4. Effect of Heme Oxygenase by-Products on VEGF Synthesis

Cells were treated for 24 h with either biliverdin or bilirubin (0.5–10 μM), or with ferrous chloride ($FeCl_2$) (30 or 100 μM). VSMC were also treated with 100 or 500 μM of deferoxamine mesylate, an iron chelator. To investigate the role of CO on VEGF synthesis, VSMC were kept for 24 h in 1% CO atmosphere (with the normal level of oxygen) and compared with cells grown under normoxic conditions (air/5% CO_2) or exposed to hypoxia (95% N_2, 5% CO_2; $pO_2 = 2$ mmHg) in a hypoxic chamber (Billups-Rothenberg Inc, Del Marc, CA).

2.5. Statistics

Statistical analyses were performed using ANOVA followed by post hoc Scheffe or Tukey test. Differences at $p < 0.05$ were considered as statistically significant.

3. RESULTS

3.1. Hemin and IL-1β Induce HO-1 Expression and Increase HO Activity in VSMC

Basal expression and activity of HO-1 was detected in non-stimulated rat VSMC (data not shown). Incubation of cells with hemin, substrate and inducer for HO-1, or IL-1β resulted in increased HO-1 protein synthesis and enhancement of enzymatic activity. Basal and stimulated activities of HO-1 were both significantly diminished by SnPPIX, a well-characterized HO inhibitor (data not shown).

3.2. Increased HO Activity is Associated with Augmented VEGF Synthesis

Treatment of VSMC with hemin caused a small (20–30%), but significant increase in VEGF synthesis (Fig. 1A). This effect was reversed by SnPPIX, particularly after longer periods of incubation (Figs. 1, 2). Compared to other protoporphyrins, SnPPIX appeared to be the strongest inhibitor of VEGF synthesis of (Fig. 2). The basal synthesis of VEGF was also significantly diminished by ZnDDPIX

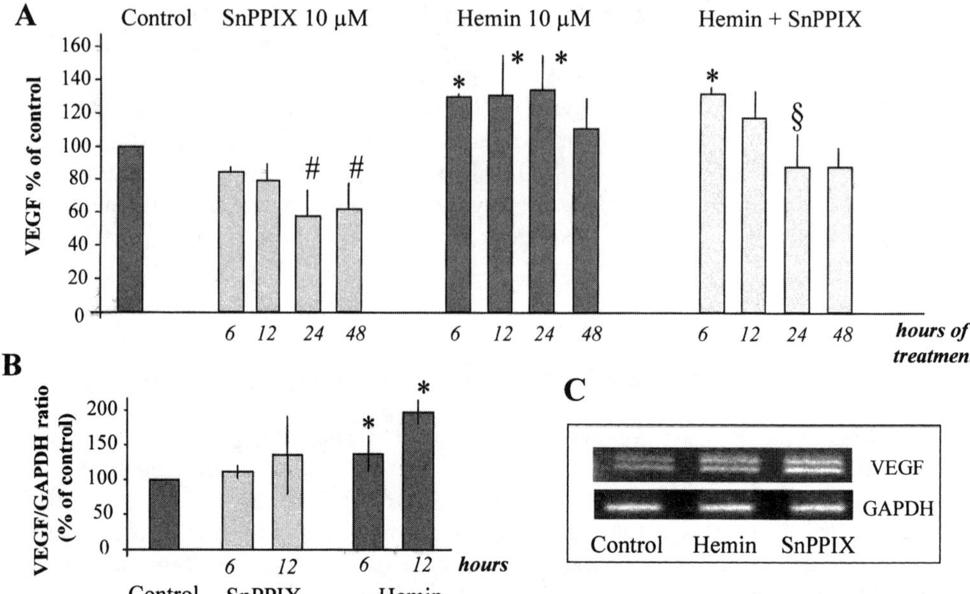

Figure 1. Modulatory effects on VEGF synthesis by the heme oxygenase pathway in VSMC. Rat VSMC were treated for 6–48 h with hemin and/or SnPPIX. VEGF protein was determined by ELISA (A), while gene expression by mRNA ELISA (B) and by RT-PCR (C). VEGF release into the culture media was enhanced by hemin and inhibited by SnPPIX. The enhancement of VEGF synthesis by hemin is partially modulated at the transcriptional level (B,C). Interestingly, SnPPIX exerts stimulatory effect on VEGF transcription (B,C) but an inhibitory effect on VEGF translation (A). Each bar represents mean ± SD of at least two independent experiments, n = 4–8 per group. *p < 0.05 vs. control, # p < 0.01 vs. control, §p < 0.01 vs. hemin treatment at the same time (24 h).

(Fig. 2A). CuPPIX, which is known to have a weak effect on HO activity,[11] did not significantly affect VEGF synthesis (Fig. 2). On the other hand, all tested protoporphyrins potently inhibited cytokine-induced VEGF synthesis, as shown for SnPPIX (Fig. 2B).

Interestingly, and similarly to data reported by others,[12] we found that SnPPIX increased VEGF mRNA expression (Fig 1B and 1C). However, despite an augmented VEGF mRNA expression, VEGF protein synthesis was always inhibited by SnPPIX (Fig. 1A and 2).

3.3. Overexpression of HO-1 Results in Enhanced VEGF Synthesis

Overexpression of the HO-1 gene following transfection to rat VSMC resulted in the enhanced expression of VEGF mRNA and protein synthesis (Fig. 3A). Co-transfection of NIH 3T3 fibroblasts with a pcDNA-HO-1 expression plasmid together with a VEGF-luciferase reporter plasmid resulted in the augmentation of VEGF promoter activity, as determined by increased luciferase production (Fig. 3B).

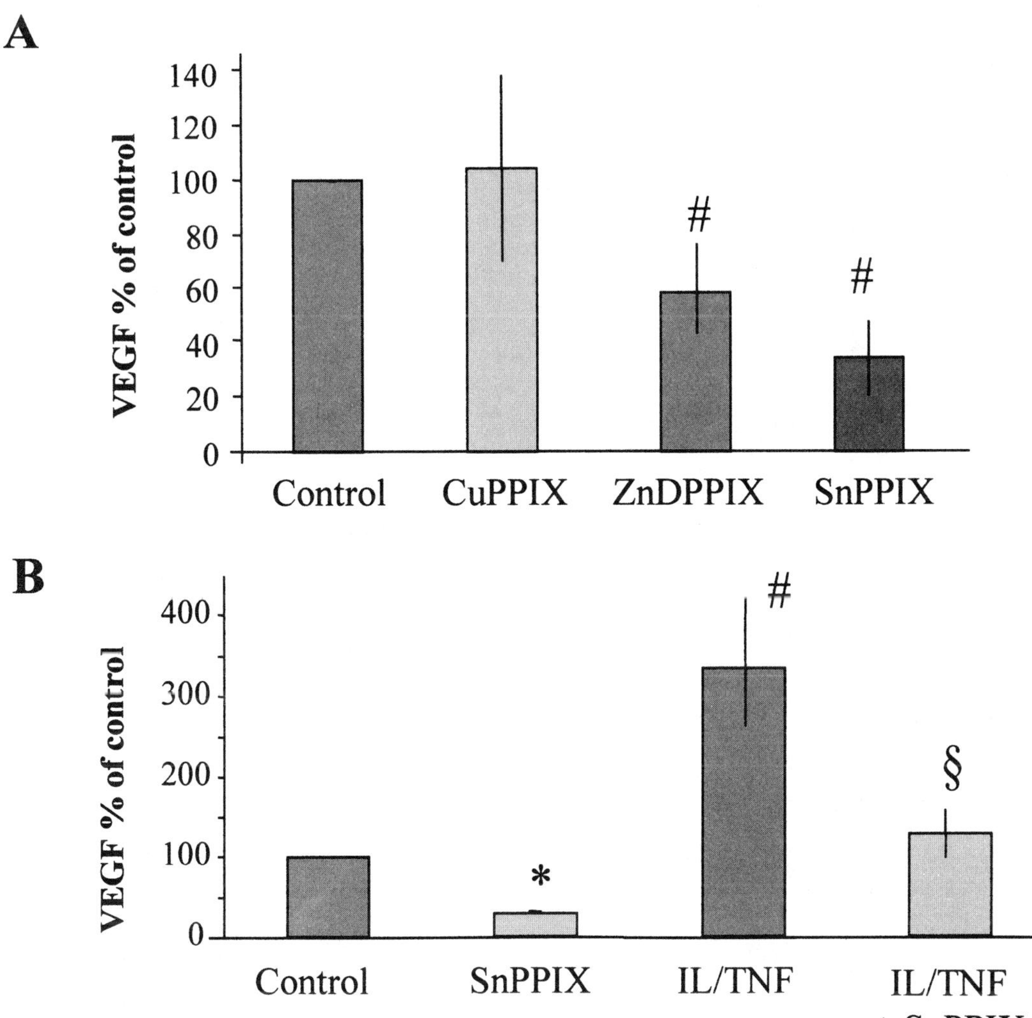

Figure 2. Effect of different protoporphyrins on VEGF synthesis. Among the protoporphyrins tested, the major inhibitory effect on basal VEGF synthesis was exerted by SnPPIX and ZnDDPIX (A). All protoporphyrins inhibited cytokine-induced VEGF synthesis, as shown for SnPPIX (B). Data represent the mean ± SD of at least three independent experiments, n = 7–12 per group. *p < 0.05 vs. control, # p < 0.001 vs. control, § p < 0.001 vs. cytokines.

3.4. Effect of Heme Oxygenase Products on VEGF Synthesis in VSMC

At physiological concentrations (0.5–10 μM), neither biliverdin nor bilirubin have any effect on VEGF synthesis by VSMC (data not shown). Ferrous ions decreased both basal (Fig. 4) and cytokine-induced (data not shown) VEGF synthesis. Interestingly, the iron chelator deferoxamine significantly promoted basal (Fig. 4),

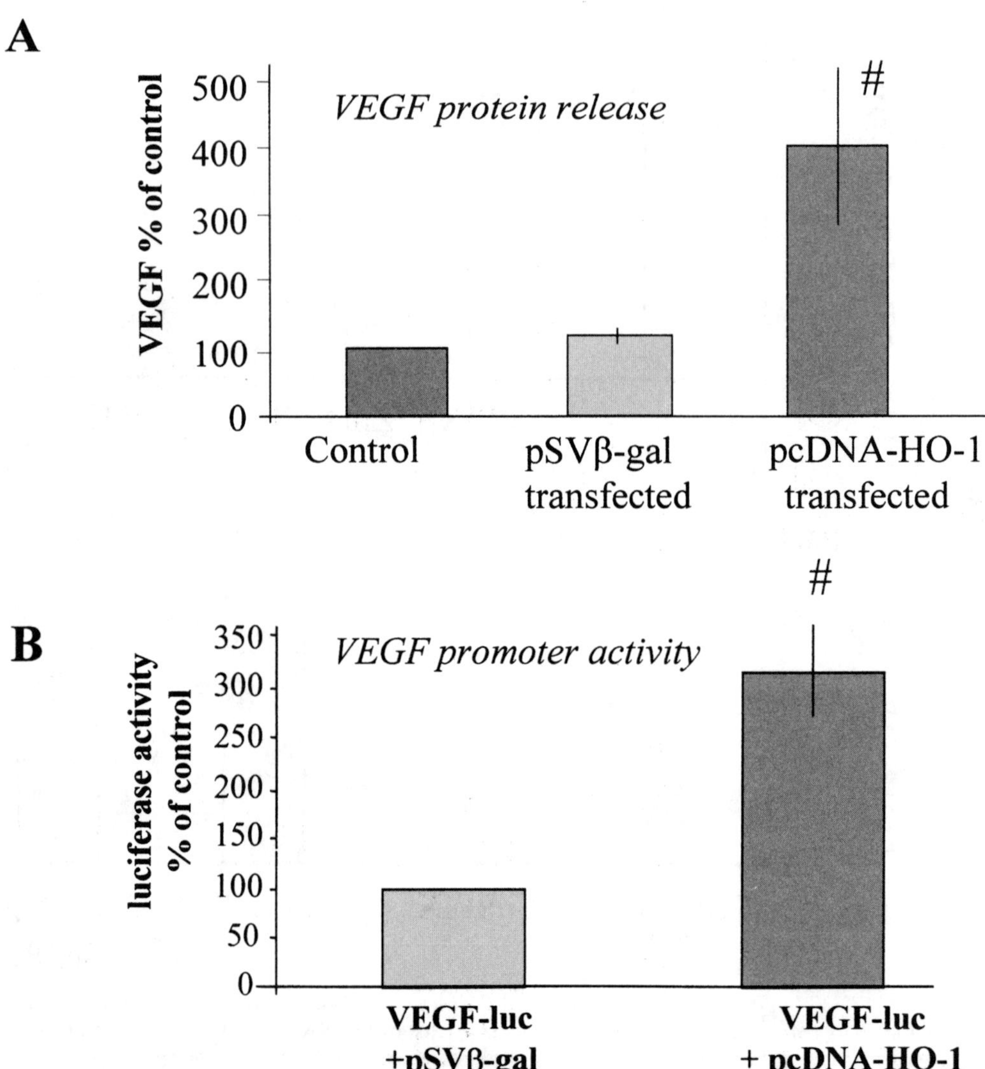

Figure 3. Effect of HO-1 transfection on VEGF expression. Rat VSMC transfected with pcDNA3-HO-1 plasmid cells or cells transfected with control plasmid (A). The effect of HO-1 overexpression is mediated at the transcriptional level, as demonstrated by activation of VEGF promoter (B). To demonstrate this NIH 3T3 were co-transfected with VEGF-luc plasmid containing luciferase gene driven by human VEGF promoter and by pcDNA3-HO-1 or pSVβ-gal control plasmid. Co-transfection with pcDNA-HO-1 resulted in a strong potentiation of VEGF promoter activity. Data represent the mean $\pm$ SD (n = 6–10); similar results were observed in two independent experiments. # p < 0.001 vs. control cells or vs. cells transfected with control plasmid.

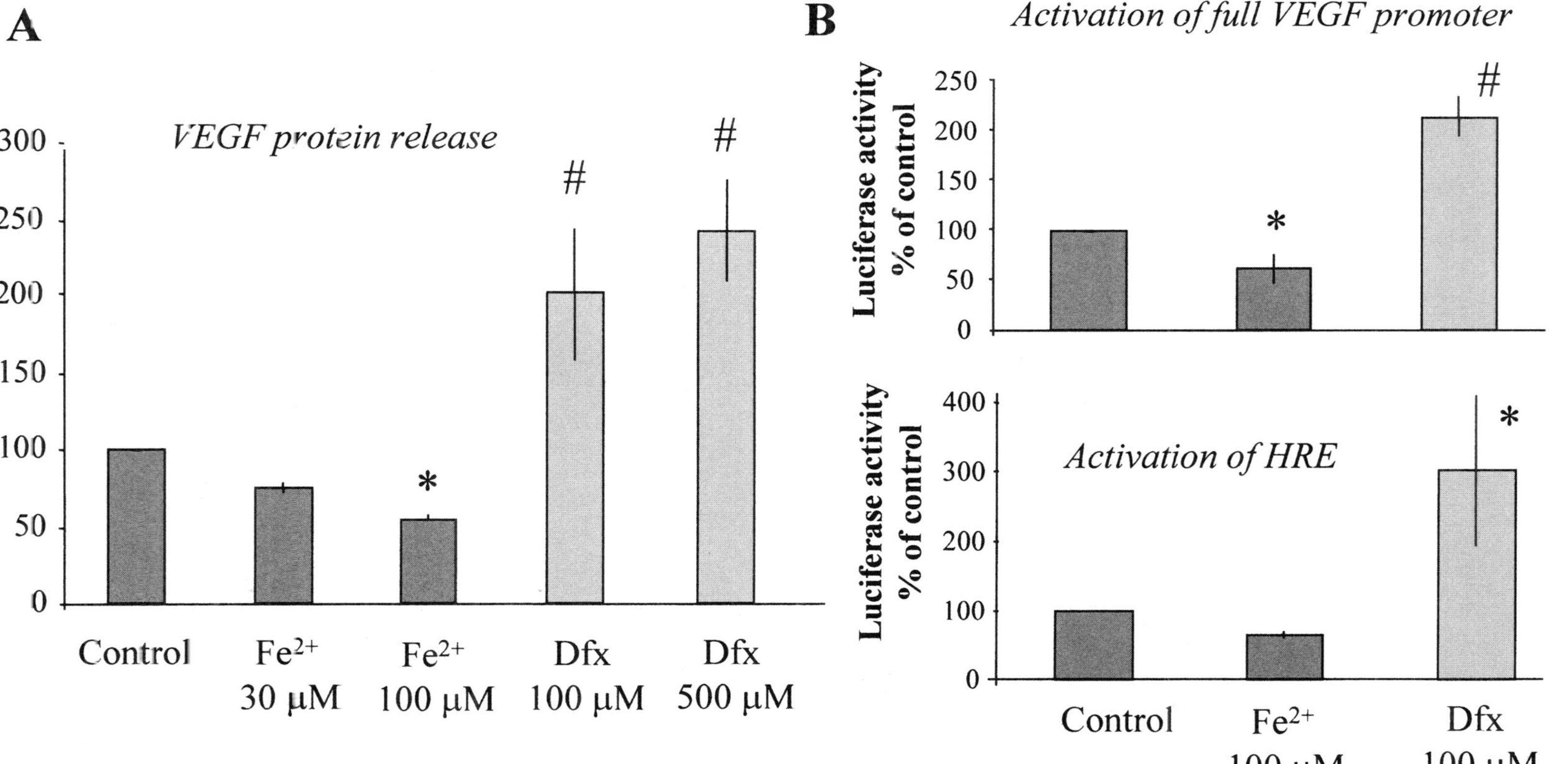

Figure 4. Effect of iron and the iron chelator deferoxamine on VEGF synthesis. Rat VSMC were treated with ferrous chloride or with deferoxamine. Iron ions inhibited VEGF synthesis, deferoxamine enhanced it (A). The effect is mediated by the activation of VEGF- promoter (B), including the HRE-responsive element (C), as demonstrated in NIH 3T3 cells transfected with VEGF-luc or HRE-luc plasmid. Data represent the mean ± SD of two independent experiments (n = 4–5). *p < 0.05 vs. control; # p < 0.001 vs. control.

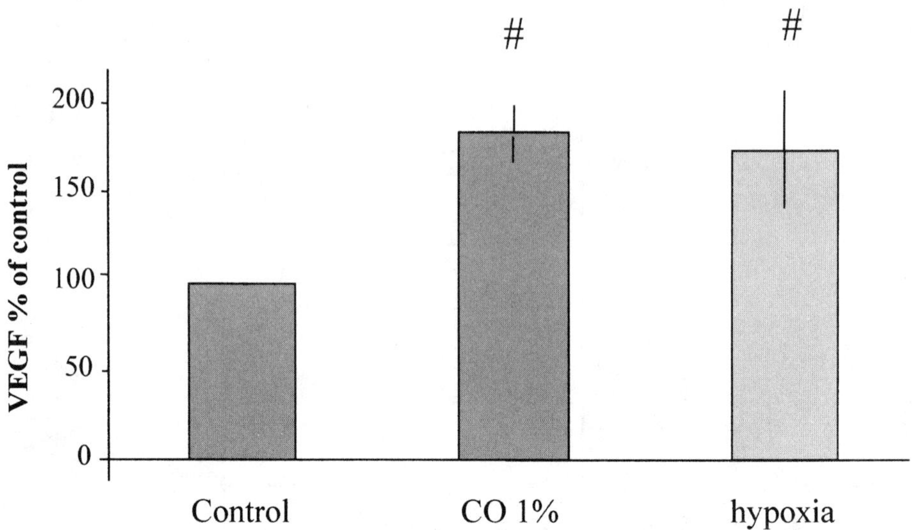

Figure 5. Effect of carbon monoxide (CO) on VEGF synthesis in VSMC. CO (1%) potently enhanced VEGF release by rat VSMC to the level observed in cell kept in hypoxic conditions. Data represent the mean ± SD (n = 6). Similar results were obtained in two independent experiments. *p < 0.0001 vs. control.

cytokine- (not shown) or hemin-induced (not shown) VEGF synthesis. The effect of iron is mediated at the transcriptional level. Deferoxamine strongly enhanced the activity of both full VEGF promoter as well as its HRE-region, while iron inhibited the HRE activation (Fig. 4B).

Therefore, we hypothesized that a third HO by-product, i.e. CO, could be involved in the induction of VEGF synthesis. Indeed, cells exposed for 24 h to 1% CO released significantly more VEGF than cells cultured in normoxic conditions (Fig. 5). The similar enhancement was observed in cells growing in hypoxia (Fig. 5).

4. DISCUSSION

In the present study, we demonstrated that heme oxygenases, the enzymes generating CO, iron and biliverdin from the substrate heme, influence VEGF generation. Increased HO-1 expression, enhanced by chemical treatment or by overexpression of HO-1 cDNA, resulted in augmented VEGF synthesis. Accordingly, inhibition of HO activity led to a decrease in VEGF production.

We investigated the possible role of HO-1-derived catabolites. Neither biliverdin nor bilirubin influenced basal VEGF production whereas iron attenuated the generation of VEGF. This is in agreement with previous reports showing that iron inhibits hypoxia inducible factor-1α (HIF-1α) activity.[13] Furthermore, the iron chelator deferoxamine is a potent activator of VEGF synthesis.[14,15] Deferoxamine can enhance VEGF synthesis by removing iron from the cells. In addition, it may scavenge the hydroxyl radicals. The detailed mechanism of deferoxamine-mediated regulation of VEGF synthesis is unknown and is the subject of our current investigations.

In contrast to effect elicited by iron, CO promoted an increase in VEGF synthesis, as cells exposed for 24 h to an atmosphere containing 1% of CO generated significantly more VEGF than VSMC grown in normoxic conditions. Thus, CO appears to be a crucial regulator of VEGF synthesis. The similar enhancement of VEGF synthesis was observed in cells kept in hypoxic conditions.

Our study is in agreement with recent observations by Marti and Risau,[16] who demonstrated that exposure of mice to 0.1% CO for 6 h resulted in the induction of VEGF expression in numerous organs; the extent of this induction was similar to that obtained in mice kept under 6% oxygen. Thus, it can not be excluded that hypoxia originated from CO exposure enhances VEGF expression. On the other hand, our results appear to be conflicting with other studies[12,17] describing the influence of CO on hypoxia-induced HIF-1α activation[17] and VEGF expression.[12] In a report showing the inhibitory effect of CO on the activity of HIF-1α, HepG2 cells were cultured in hypoxic conditions and were further treated with very high concentration of CO (up to 80%).[17] However, one can speculate that the effect on VEGF synthesis of very high concentrations of CO (10–80%) applied during hypoxia might be different from lower (1%) concentrations of this gas.

The role of HO in regulation of VEGF synthesis is also supported by results of experiments with inhibitors of those enzymes. We observed that SnPPIX consistently, and in a concentration-dependent manner, diminished VEGF synthesis. SnPPIX is a very potent inhibitor of HO activity and it has recently been shown to decrease very strongly CO production in VSMC.[18] Our study shows that SnPPIX at 10 µM concentration inhibited VEGF protein synthesis. The amount of this growth factor was diminished both in culture media and in the cell cytosol. Interestingly, the study of Liu and co-workers[12] reported an induction of VEGF mRNA expression by 100 µM SnPPIX in hypoxia. In our experiments, SnPPIX also enhanced VEGF mRNA expression but the synthesis of VEGF protein was always decreased. As SnPPIX always potently diminished VEGF production, we suggest that SnPPIX influences VEGF synthesis at the post-transcriptional level.

In another previous study, Eyessen-Hernandez and colleagues reported that hemin did not induce VEGF mRNA expression in cardiac myocytes.[19] In view of our recent data indicating a moderate increase in VEGF after hemin treatment, we speculate that iron released from heme might have attenuated the stimulatory effect of HO-derived CO on VEGF synthesis. Accordingly, we found that deferoxamine up-regulated basal VEGF production and enhanced hemin- or IL-1-induced VEGF synthesis. Thus, it is possible that the total amount of VEGF generated due to increased HO activity is dependent on the ratio of CO and free iron.

In conclusion, our data reveal the existence of novel pathways in the modulation of VEGF synthesis. We suggest that HO activity might be an important component of the pathways regulating the production and activity of this potent angiogenic factor.

ACKNOWLEDGMENTS

The research was partially supported by the Polish State Committee for Scientific Research (No. 6 P04B 013 21) and by a fellowship from the Austrian Society of Cardiology (J.D).

REFERENCES

1. P. Carmeliet, Mechanisms of angiogenesis and arteriogenesis. Nat Med **6**:389–395 (2000).
2. M. Hecker, M. Cattaruzza, and A.H. Wagner, Regulation of inducible nitric oxide synthase gene expression in vascular smooth muscle cells. *Gen Pharmacol* **32**:9–16 (1999).
3. J. Dulak, A. Józkowicz, A. Dembinska-Kiec, I. Guevara, A. Zdzienicka, I. Florek, D. Zmudzinska-Grochot, A. Wójtowicz, A. Szuba, and J.P. Cooke, Nitric oxide induces the synthesis of vascular endothelial growth factor by rat vascular smooth muscle cells *Arterioscler Thromb Vasc Biol* **20**:659–666 (2000).
4. A. Jozkowicz, J. Dulak, I. Guevara, A. Zdzienicka, and A. Dembinska-Kiec, Nitric oxide increases the synthesis of vascular endothelial growth factor in vascular smooth muscle cells. *In: The Biology of Nitric Oxide, part 7*, edited by S. Moncada, L. Gustafsson, P. Wiklund and E.A. Higgs, (Portland Press, London, 2000), p. 150.
5. A. Jozkowicz, J.P. Cooke, I. Guevara, I. Huk, P. Funovics, O. Pachinger, F. Weidinger, and J. Dulak, Genetic augmentation of nitric oxide synthase increases the vascular generation of VEGF. *Cardiovasc Res* **51**:773–778 (2001).
6. M.D. Maines, The heme oxygenase system: a regulator of second messenger gases. *Annu Rev Pharmacol Toxicol* **37**:517–554 (1997).
7. R. Foresti and R. Motterlini, The heme oxygenase pathway and its interaction with nitric oxide in the control of cellular homeostasis. *Free Radic Res* **31**:459–475 (1999).
8. B.M. Deramaudt, S. Braunstein, P. Remy, and N.G. Abraham, Gene transfer of human heme oxygenase into coronary endothelial cells potentially promotes angiogenesis. *J Cell Biochem* **68**:121–127 (1998).
9. A. Nishie, M. Ono, T. Shono, J. Fukushi, M. Otsubo, H. Onoue, Y. Ito, T. Inamura, K. Ikezaki, M. Fukui, T. Iwaki, and M. Kuwano, Macrophage infiltration and heme oxygenase-1 expression correlate with angiogenesis in human gliomas. *Clin Cancer Res* **5**:1107–1113 (1999).
10. R.J. Chernick, P. Martasek, R.D. Levere, R. Margreiter, and N.G. Abraham, Sensitivity of human tissue heme oxygenase to a new synthetic metalloporphyrin. *Hepatology* **10**:365–269 (1989).
11. H. Kimura, A. Weisz, Y. Kurashima, K. Hashimoto, T. Ogura, F. D'Acquisto, R. Addeo, M. Makuuchi, and H. Esumi, Hypoxia response element of the human vascular endothelial growth factor gene mediates transcriptional regulation by nitric oxide: control of hypoxia-inducible factor-1 activity by nitric oxide. Blood **95**:189–197 (2000).
12. Y. Liu, H. Christou, T. Morita, E. Laughner, G.L. Semenza, and S. Kourembanas, Carbon monoxide and nitric oxide suppress the hypoxic induction of vascular endothelial growth factor gene via the 5′ enhancer. *J Biol Chem* **273**:15257–15262 (1998).
13. P.H. Maxwell, M.S. Wiesener, G.W. Chang, S.C. Clifford, E.C. Vaux, M.E. Cockman, C.C. Wykoff, C.W. Pugh, E.R. Maher, and P.J. Ratcliffe, The tumour suppressor protein VHL targets hypoxia-inducible factors for oxygen-dependent proteolysis. *Nature* **399**:271–275 (1999).
14. L.V. Beerepot, D.T. Shima, M. Kuroki, K.-T. Yeo, and E.E. Voest, Up-regulation of vascular endothelial growth factor production by iron chelators. *Cancer Res* **56**:3747–3751 (1996).
15. J.M. Gleadle, B.L. Ebert, J.D. Firth, and P.J. Ratcliffe, Regulation of angiogenic growth factor expression by hypoxia, transition metals, and chelating agents. *Am J Physiol* **268**:C1362–1368 (1995).
16. H.H. Marti and W. Risau, Systemic hypoxia changes the organ-specific distribution of vascular endothelial growth factor and its receptors. *Proc Natl Acad Sci USA* **95**:15809–15814 (1998).

17. L.E. Huang, W.G. Willmore, J. Gu, M.A. Goldberg, and H.F. Bunn, Inhibition of hypoxia-inducible factor 1 activation by carbon monoxide and nitric oxide. Implications for oxygen sensing and signaling. *J Biol Chem* **274**:9038–9044 (1999).

18. Y. Morimoto, W. Durante, D.G. Lancaster, J. Klattenhoff, and F.K. Tittel, Real-time measurements of endogenous CO production from vascular cells using an ultransensitive laser sensor. *Am J Physiol Heart Circ Physiol* **280**:H483–H488 (2000).

19. R. Esseyen-Hernandez, A. Ladoux, and C. Frelin, Differential regulation of cardiac heme oxygenase-1 and vascular endothelial growth factor mRNA expression by hemin, heavy metals, heat shock and anoxia. *FEBS* **382**:229–233 (1996).

PHYSIOLOGICAL FUNCTION OF HEME OXYGENASE AND THE CENTRAL NERVOUS SYSTEM

9

CO AND NEONATAL CEREBRAL CIRCULATION

Charles W. Leffler, Jonathan H. Jaggar, and Zheng Fan

Department of Physiology
University of Tennessee Health Science Center
Memphis, TN 38163

INTRODUCTION

The gas, carbon monoxide (CO), is produced physiologically by catabolism of heme to CO, free iron, and biliverdin.[1] This reaction is catalyzed by heme oxygenase with reduction of NADPH. Heme oxygenase (HO) is expressed as three known isoforms that are attached to the endoplasmic reticulum by a hydrophobic sequence near the carboxy terminus of the protein:[1] the easily inducible HO-1, the constituitively expressed poorly inducible HO-2 and a third isoform with much lower heme degrading activity.[2] CO can be a vascular paracrine factor. HO-1 and HO-2 have been identified in vascular endothelial and smooth muscle cells, with HO-2 being constituitively expressed.[1,3,4,5] HO-2 is expressed in highest concentration in the brain[5,92] where it resides in neurons, vascular endothelium and smooth muscle.[1,6] CO can cause increases in cGMP in both autocrine and paracrine fashions[3,7,8,9] and can hyperpolarize vascular smooth muscle via modification of a histidine residue on the external membrane side of the large conductance Ca activated potassium channel (K_{Ca}).[9,10,11] By either of these mechanisms CO could cause endothelial independent dilation of arteries and arterioles.

Our work has concentrated on prenatal and postnatal development and has focused on the cerebral circulation. HO and CO appear to be important in prenatal and postnatal development. HO levels in cerebrum are developmentally regulated with maximal HO expression in the mature fetus, as compared to the immature fetus or adult.[12] CO appears to be important in control of the fetal vasculature with a potential contribution of endogenously produced CO to ductus arteriosus patency.[13] Expressions of both HO-1 and HO-2 are about 15 times higher in the pregnant

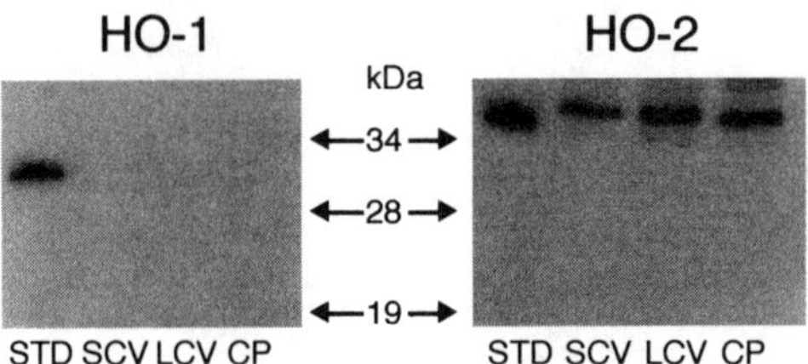

Figure 1. HO in 100,000 × g pellet of small cerebral vessels (SCV), larger cerebral vessels (LCV), and cerebral parenchyma (CP). Copyright 1999 by the American Physiological Society. Reproduced with permission.

compared to nonpregnant myometrium.[14] Sex steroids can induce expression of both HO proteins. These findings further suggest HO may be elevated in perinatal tissues.

We have been pursuing the hypothesis that CO produced from endogenous HO and heme is an important paracrine regulator in the newborn cerebral circulation. This chapter will highlight some of our research to date that address this hypothesis.

HEME OXYGENASE EXPRESSION AND CO PRODUCTION

Newborn pig brain and cerebral vessels express HO that produces CO.[15] HO-2, but not HO-1, was detected on Western blots from larger cerebral vessels, cerebral microvessels, and brain parenchyma freshly isolated from newborn pigs (Fig. 1). HO-2 is clearly expressed in newborn pig cerebral microvascular endothelial cells in primary culture (Fig. 2). The HO-2 expressed in the newborn cerebral microvasculature produces CO at the highest rate of the tissues we examined (Fig. 3) and that rivals production by olfactory neurons.[16]

VASODILATOR INFLUENCE OF CO IN NEWBORN BRAIN

CO is a very potent dilator in the newborn cerebral microcirculation.[15] Our *in vivo* data were obtained using a closed cranial window implanted over the parietal cortex of the anesthetized newborn pig. CO, heme-L-lysinate (HLL), or metal porphyrin inhibitors of HO were placed on the surface arterioles via needle ports on the cranial window. CO dilated both very small and larger pial arterioles (Fig. 4). Similarly, the HO substrate, HLL, caused dilation of pial arterioles that was blocked by

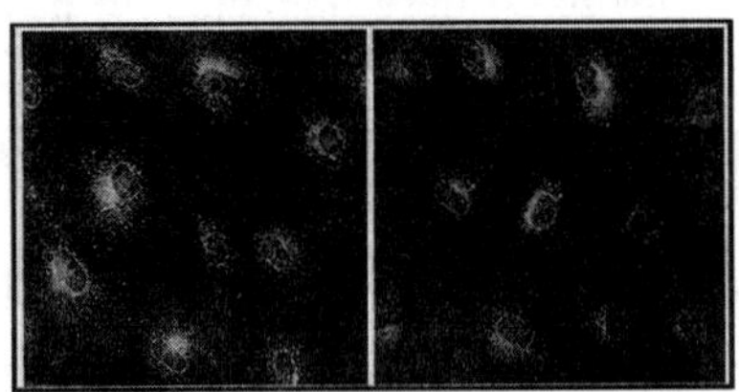

Figure 2. HO-2 in newborn pig cerebral microvascular endothelial cells in primary culture (deconvolution confocal microscopy).

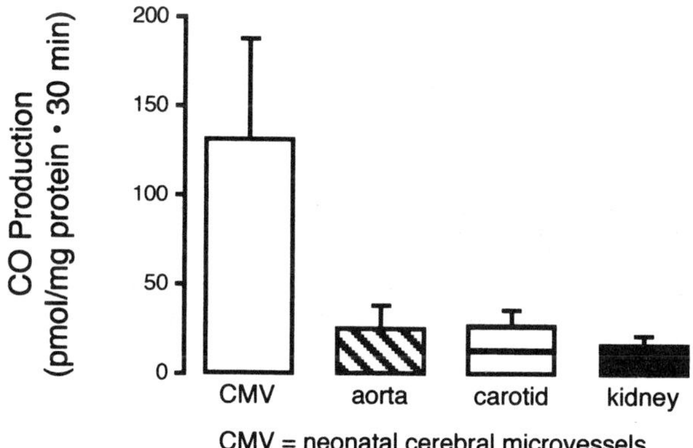

Figure 3. CO production (pmol/mg protein in 30 min). GC-MS of headspace gas.

co-application of the metal porphyrin inhibitor of HO, chromium mesoporphyrin (CrMP) (Fig. 5). These data indicate that endogenous HO can produce sufficient CO to dilate brain arterioles.

Endogenously produced CO appears to contribute physiologically to control of neonatal cerebral circulation. Thus, pial arteriolar dilations to hypoxia and the excitatory amino acid, glutamate, but not to hypercapnia, are inhibited by heme oxygenase inhibitors (Fig. 6). Further, the increase in cerebral blood flow caused by glutaminergic seizures, is markedly reduced following HO inhibition in piglets. These data are consistent with our finding that glutamate increases CO production by piglet cerebral microvascular endothelial cells.

Endogenous CO can also attenuate constrictor responses. For example, pial arterial constriction to mild hypocapnia is markedly enhanced by HO inhibition (Fig. 7). Similar results were obtained when the constrictor was platelet activating factor or autoregulatory vasoconstriction in response to elevated blood pressure.

CO VASODILATOR MECHANISMS

As noted earlier, CO may produce dilation either by activating guanylyl cyclase to increase cGMP or by directly activating K_{Ca} channels, thereby hyperpolarizing the

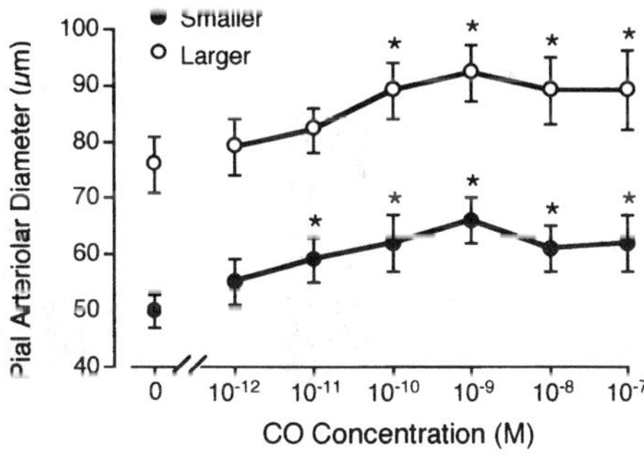

Figure 4. Effect of CO on larger and smaller pial arterioles of newborn pigs. Copyright 1999 by the American Physiological Society. Reproduced with permission.

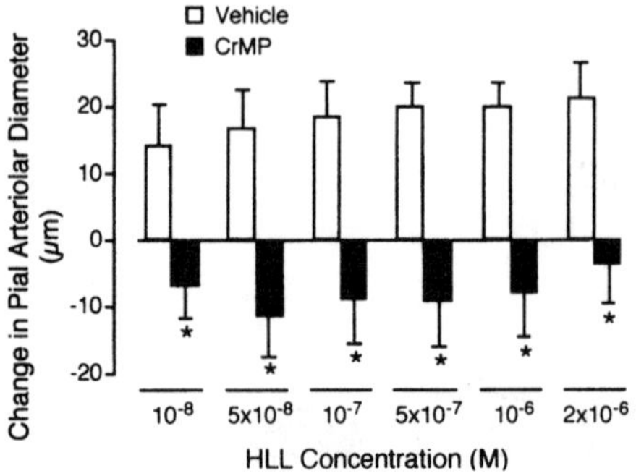

Figure 5. Pial arteriolar dilation to HLL in the absence and in the presence of CrMP. Copyright 1999 by the American Physiological Society. Reproduced with permission.

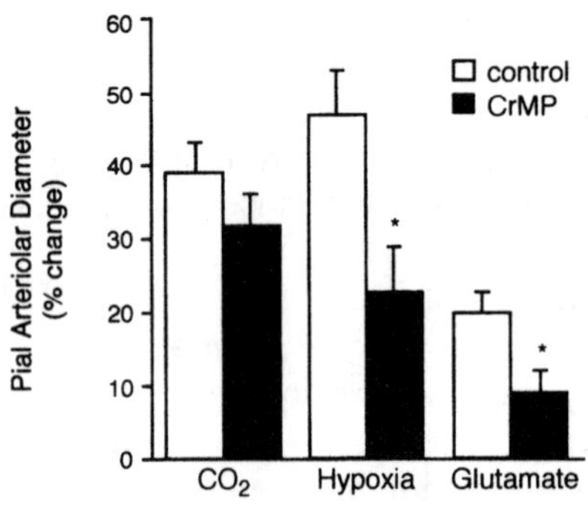

Figure 6. Pial arteriolar dilation in response to systemic hypercapnia, hypoxia, and topical glutamate in newborn pigs before and following topical application of CrMP.

vascular smooth muscle cell. In the newborn cerebral microvasculature activation of K_{Ca} channels appears to account for CO induced dilation. Thus, we could detect no elevation of cerebral cGMP coincident with CO induced dilation while similar dilations attributable to NO markedly increased cGMP.[15] Conversely, the K_{Ca} channel inhibitors, iberiotoxin and TEA, totally blocked dilations of cerebral arterioles to CO or HLL (Fig. 8).

In freshly isolated piglet cerebral microvascular myocytes, CO and HLL activate K_{Ca} channels (Fig. 9).

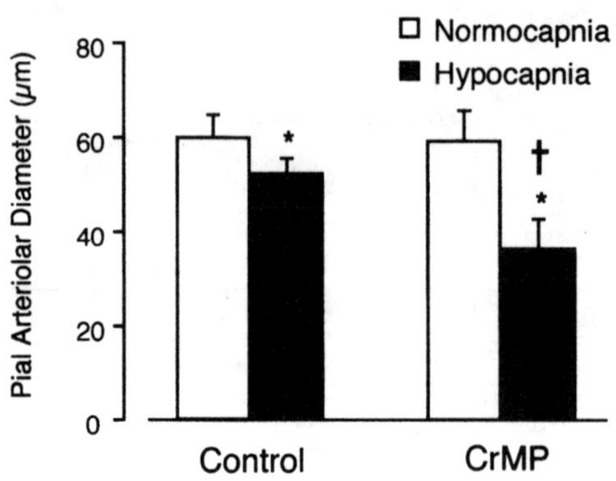

Figure 7. Pial arteriolar constriction in response to systemic hypocapnia before and following topical application of CrMP.

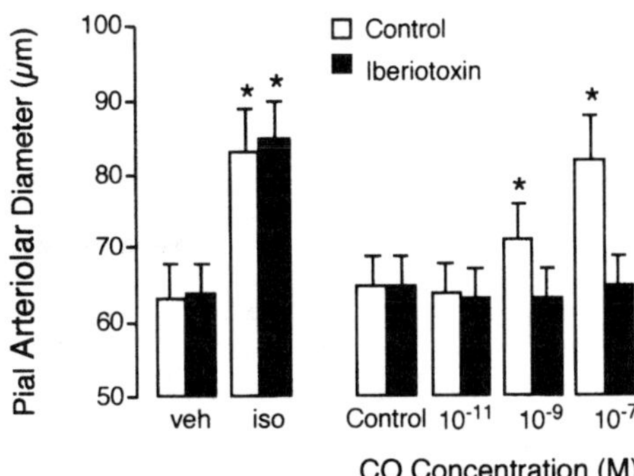

Figure 8. Effect of iberiotoxin on pial arteriolar dilations to isoproterenol (iso) and CO. Copyright 1999 by the American Physiological Society. Reproduced with permission.

Furthermore, piglet cerebral microvascular smooth muscle explants were rapidly and reversibly hyperpolarized by addition of CO to the superfusion media (Fig. 10). HLL also hyperpolarized the isolated microvascular smooth muscle cell (Fig. 11). CrMP blocked HLL induced hyperpolarization but not that caused by authentic CO. These data *in toto* suggest that endogenously generated CO activates microvascular smooth muscle K_{Ca} channels, resulting in hyperpolarization and vasodilation.

CO, NO, AND PROSTACYCLIN

Among the other potential paracrine dilatory influences on the newborn cerebral circulation are prostanoids and NO. The cells that produce CO and can respond to CO can also produce and respond to prostanoids and NO. Thus, the potential for

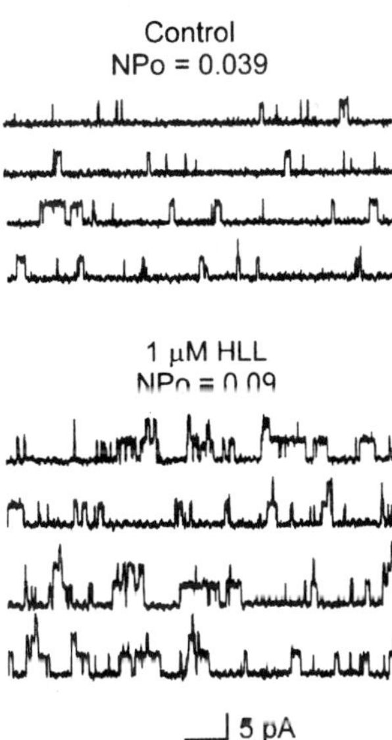

Figure 9. Perforated patch clamp recording from a freshly isolated piglet cerebral microvascular smooth muscle cell before and in the presence of HLL.

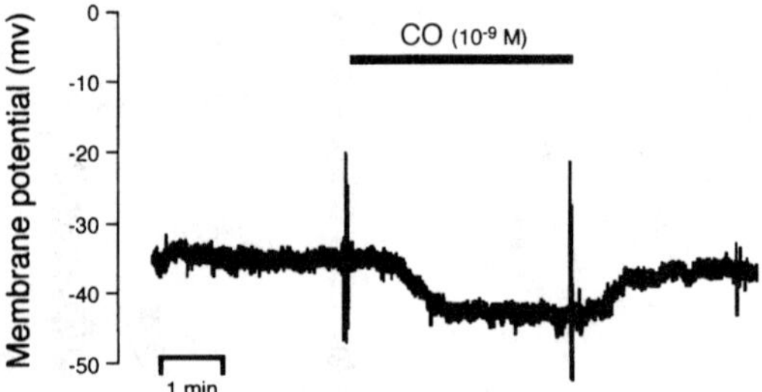

Figure 10. Membrane potential recording of a piglet cerebral microvascular smooth muscle explant super-fused with 10^{-9} M CO where indicated.

interaction among these systems is high and considerable investigation of interactions between the prostanoid and NO systems has been undertaken. We have found that cerebral vasodilation to CO in the newborn brain requires prostacyclin and/or NO.[17] Thus, vasodilation to CO and HLL are blocked by pretreatment with either the cyclooxygenase (COX) inhibitor, indomethacin, or the nitric oxide synthase (NOS) inhibitor, L-NAME (Fig. 12 and 13).

However, the roles of prostacyclin and NO do not appear to be conventional but instead to be permissive. In other words, increasing concentrations of prostacy-clin or NO are not required to allow dose-dependent dilation to CO and HLL to occur. Thus, following treatment with indomethacin that blocks dilation to CO and HLL, addition of a constant and subdilator concentration of the prostacyclin analog, iloprost, totally restores dose-dependent dilation to CO and HLL (Fig. 14).

Similarly, following NOS inhibition that blocks dilation to CO, a constant, approximately dilation threshold concentration of the NO donor, sodium nitroprus-side, allows dose dependent dilation to occur in response to either CO or HLL (Fig. 15). The effects of prostacyclin and NO must be upstream of the CO action on the K_{Ca} channel because neither SNP nor iloprost can restore dilation to CO following inhibition of K_{Ca} channels (Fig. 16).

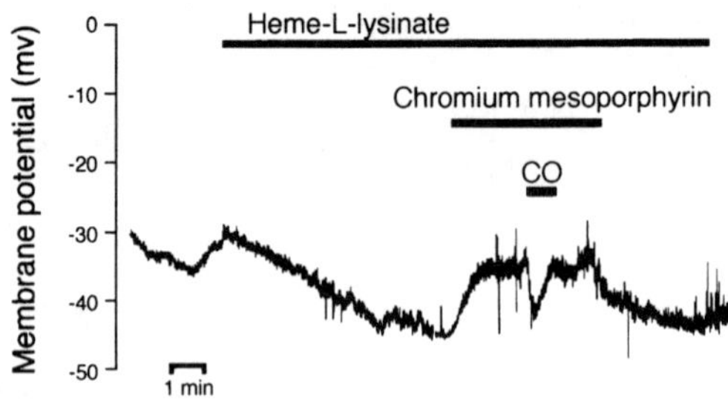

Figure 11. Membrane potential recording of a piglet cerebral microvascular smooth muscle explant super-fused with 10^{-7} M HLL, CrMP, and CO where indicated.

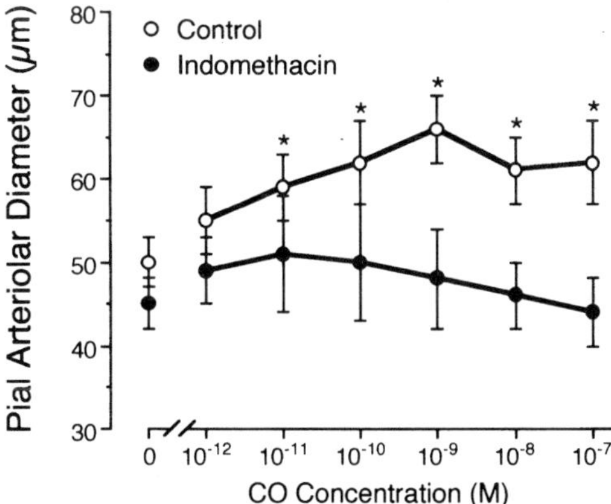

Figure 12. Effect of indomethacin on pial arteriolar dilation in response to CO. Copyright 2001 by the American Physiological Society. Reproduced with permission.

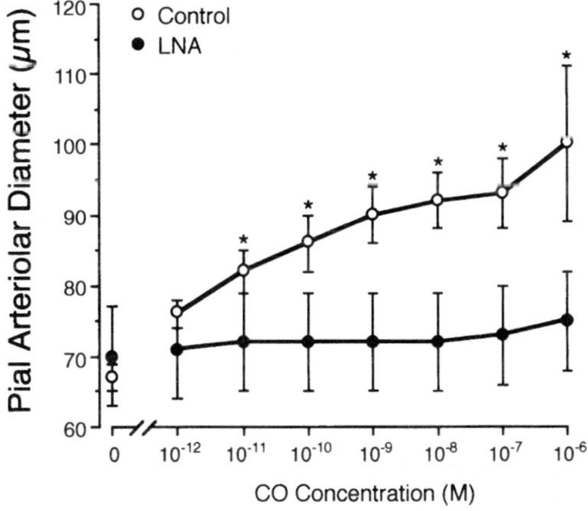

Figure 13. Effect of L-nitro-arginine (LNA) on pial arteriolar dilation in response to CO. Copyright 2001 by the American Physiological Society. Reproduced with permission.

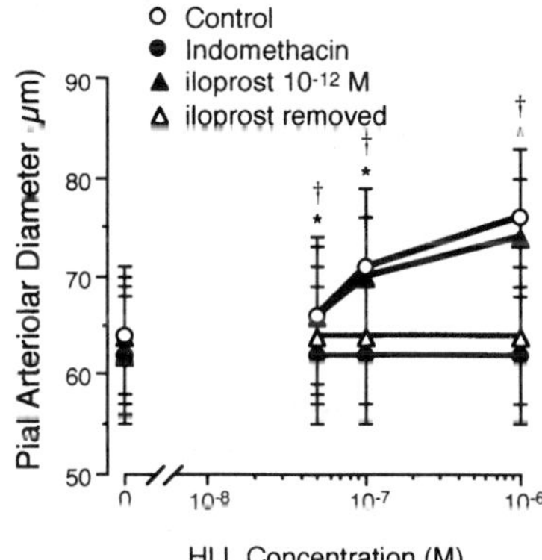

Figure 14. Pial arteriolar dilation to HLL before indomethacin, after indomethacin, and after indomethacin in the presence of iloprost. Copyright 2001 by the American Physiological Society. Reproduced with permission.

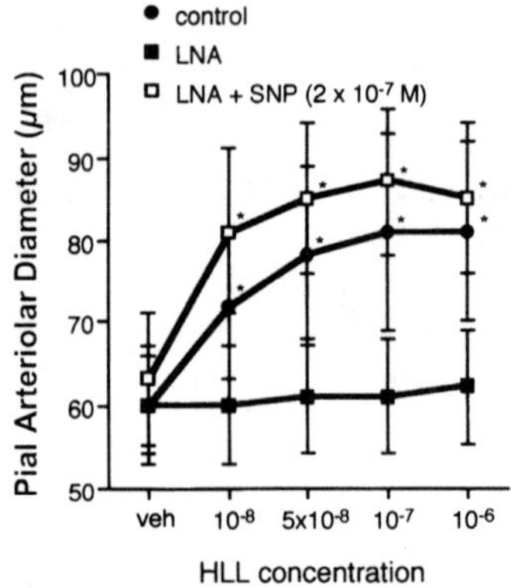

Figure 15. Pial arteriolar dilation to HLL before L-nitro-arginine (LNA), after LNA, and after LNA in the presence of sodium nitroprusside (SNP). Copyright 2001 by the American Physiological Society. Reproduced with permission.

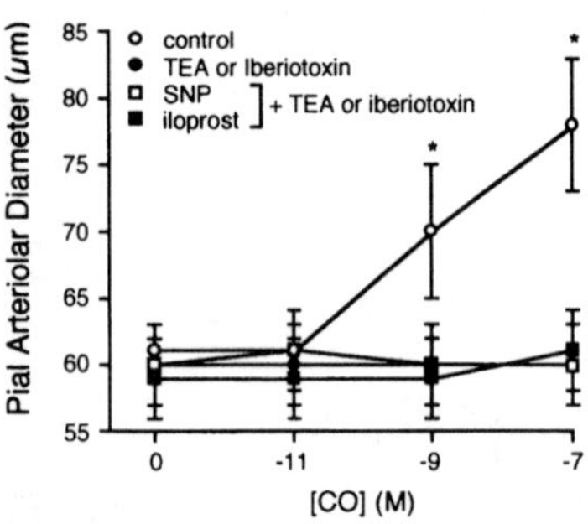

Figure 16. Effect of CO on pial arteriolar diameter in control, following iberiotoxin or TEA, and following iberiotoxin or TEA in the presence of SNP or iloprost. Copyright 2001 by the American Physiological Society. Reproduced with permission.

SUMMARY

Therefore, CO is a potentially important dilator influence in the newborn cerebral circulation under physiological and pathological conditions. The dilator actions of CO involve activation of K_{Ca} channels by CO. Permissive actions of prostacyclin and NO appear to be involved in CO-induced cerebral dilation. Finally, cerebrovascular dilations to hypoxia and glutamate appear to involve CO so that CO could contribute to coupling of cerebral blood flow to neuronal activity. However, research in this area is still in its infancy and much further study is required.

REFERENCES

1. Maines, M.D. The heme oxygenase system: a regulator of second messenger gases. *Annu. Rev. Pharmacol. Toxicol.* 37:517–554, 1997.
2. McCoubrey, W.K., T.J. Huang, and M.D. Maines. Isolation and characterization of a cDNA from the rat brain that encodes hemoprotein heme oxygenase-3. *Eur. J. Biochem.* 247:725–732, 1997.
3. Christodoulides, N., W. Durante, M.H. Kroll, and A.I. Schafer. Vascular smooth muscle cell heme oxygenases generate guanylyl cyclase-stimulatory carbon monoxide. *Circulation* 91:2306–2309, 1995.
4. Zakhary, R., S.P. Gaine, J.L. Dinerman, M. Ruat, N.A. Flavahan, and S.H. Snyder. Heme oxygenase 2: endothelial and neuronal localization and role in endothelium-dependent relaxation. *Proc. Natl. Acad. Sci. U.S.A.* 93:795–798, 1996.

5. Grozdanovic, Z. and R. Gossrau. Expression of heme oxygenase-2 (HO-2)-like immunoreactivity in rat tissues. *Acta Histochem.* 98:203–214, 1996.

6. Maines, M.D. Carbon monoxide: an emerging regulator of cGMP in the brain. *Mol. Cell. Neurosciences* 4:389–397, 1993.

7. Durante, W., N. Christodoulides, K. Cheng, K.J. Peyton, R.K. Sunahara, and A.I. Schafer. cAMP induces heme oxygenase-1 gene expression and carbon monoxide production in vascular smooth muscle. *Am. J. Physiol.* 273:317–323, 1997.

8. Morita, T. and S. Kourembanas. Endothelial cell expression of vasoconstrictors and growth factors is regulated by smooth muscle cell-derived carbon monoxide. *J. Clin. Invest.* 96:2676–2682, 1996.

9. Wang, R., Z. Wang, and L. Wu. Carbon monoxide-induced vasorelaxation and the underlying mechanisms. *Br. J. Pharmacol.* 121:927–934, 1997.

10. Wang, R. and L. Wu. The chemical modification of K_{Ca} channels by carbon monoxide in vascular smooth muscle cells. *J. Biol. Chem.* 272:8222–8226, 1997.

11. Wang, R., L. Wu, and Z. Wang. The direct effect of carbon monoxide on K_{Ca} channels in vascular smooth muscle cells. *Pflugers. Arch.* 434:285–291, 1997

12. Cook, M.N., G.S. Marks, H.J. Vreman, K. Nakatsu, D.K. Stevenson, and J.F. Brien. Ontogeny of heme oxygenase activity in the hippocampus, frontal cerebral cortex, and cerebellum of the guinea pig. *Brain Res. Dev. Brain Res.* 92:18–23, 1996.

13. Coceani, F., L. Kelsey, E. Seidlitz, G.S. Marks, B.E. McLaughlin, H.J. Vreman, D.K. Stevenson, M. Rabinovitch, and C. Ackerley. Carbon monoxide formation in the ductus arteriosus in the lamb: implications for the regulation of muscle tone. *Br. J. Pharmacol.* 120:599–608, 1997.

14. Acevedo, C.H. and A. Ahmed. Heme oxygenase-1 inhibits human myometrial contractility via carbon monoxide and is upregulated by progesterone during pregnancy. *J. Clin. Invest.* 101:949–955, 1998.

15. Leffler, C.W., A. Nasjletti, C. Yu, R.A. Johnson, A.L. Fedinec, and N. Walker. Carbon monoxide and cerebral microvascular tone in newborn pigs. *Am. J. Physiol.* 276:H1641–H1646, 1999.

16. Ingi, T., G. Chiang, and G.V. Ronnett. The regulation of heme turnover and carbon monoxide biosynthesis in cultured primary rat olfactory receptor neurons. *J Neurosci.* 16:5621–5628, 1996.

17. Leffler, C.W., A. Nasjletti, R.A. Johnson, and A.L. Fedinec. Contributions of prostacyclin and nitric oxide to carbon monoxide induced cerebrovascular dilation in newborn pigs. *Am. J. Physiol.* 280: H1490–H1495, 2001.

10

DIFFERENTIAL EXPRESSION OF HEME OXYGENASE-1 IN RAT BRAIN BY ENDOTOXIN (LPS)

G. Scapagnini[a,c], A. M. Giuffrida Stella[a], N. G. Abraham[b], D. Alkon[c], and V. Calabrese[a,c]

[a]Section of Biochemistry & Mol. Biology
Faculty of Medicine, Dept. of Chemistry
University of Catania
[b]Dept. of Pharmacology
NY Medical College
Valhalla, NY, USA
[c]Blanchette Rockefeller Neurosciences Institute
Rockville, MD 20850, USA

SUMMARY

Nitric oxide (NO) has been implicated as a potential contributor to neural cell death in a variety of neurological conditions. Glial cells in the CNS can produce nitric oxide in response to cytokines, and it is known that neurons, which are characterized by low levels of natural antioxidants and free radical scavengers, are more susceptible to oxidative damage. Heme oxygenase-1 (HO-1) is a cellular stress protein (HSP32) which is expressed in several tissues upon stimulation by a large number of potentially noxious stimuli, including oxidative stress. Activation of HO-1, generating the vasoactive molecule carbon monoxide and the potent antioxidant bilirubin, could represent a protective system potentially active against brain oxidative injury.

Address reprint requests to: Prof. Vittorio Calabrese, Section of Biochemistry & Mol. Biology, Department of Chemistry, University of Catania, 95125 Catania—Italy. Phone: 0039-095-738-4067, Fax: 0039-095-336990.

In the nervous system, induction of HO occurs together with the induction of other HSPs during various experimental conditions, including cerebral ischemia, neurodegenerative disorders, epilepsy and trauma. Recent evidence indicates that an increment in either the circulating levels or the local tissue release of cytokines, by activating iNOS in microglial cells and astrocytes to produce toxic amounts of NO, represents the primary pathogenic event contributing to neurodegenerative changes observed during aging. The present study was aimed, therefore, at investigating the expression and activity of HO-1, together with that of iNOS and arginase in different brain regions in response to an acute stress induced by endotoxic treatment, in order to provide a better understanding of the relationship between the NO-dependent signal pathway and HO-1 expression, and their role in neuroprotection.

INTRODUCTION

Oxidative and nitrosative stress have been involved in the pathogenesis of various degenerative disorders. The cellular response to these stresses is a complex adaptive mechanism involving the expression of numerous proteins able to maintain internal homeostasis and cell survival.[1,2] HO-1 (hsp 32) is the inducible isoform of heme oxygenase, a well preserved family of enzymes which catalyze the degradation of heme into gaseous CO, free iron and biliverdin, the latter being quickly converted into bilirubin by biliverdin reductase.[3] Beside its role in hemoglobin heme turnover, HO-1 relevance in cellular stress response has been widely demonstrated in a variety of tissues, including the brain.[4-7] In the CNS, the heme oxygenase pathway has been shown to represent a fundamental defensive mechanism for neurons exposed to an oxidant challenge.[8] Deregulation of the HO system seems to play a crucial role in the pathogenesis of acute and chronic neurodegenerative disorders.[9,10] The mechanisms by which HO-1 can operate as a cytoprotective molecule are not completely understood, but it is conceivable that HO-1 regulates cellular homeostasis through its catalytic by-products: CO, iron and bile pigments. Although all of these are generally considered to be toxic molecules, at low concentrations, such as those produced by HO-1 activity, they mediate fundamental physiological functions. Both biliverdin and bilirubin possess strong direct antioxidant properties, and bilirubin has been shown to be a potent scavenger of peroxyl radicals in the brain.[11] Endogenous CO, similary to nitric oxide (NO), through the activation of guanylyl cyclase plays an important role in regulation of vasomotor tone, and it as been proposed as a putative neuro-transmitter.[12] CO may also exert antiflammatory effects and has been implicated in the modulation of the activity of nitric oxide synthase (NOS).[13,14] Even the release of free iron by HO-1 activity, rather than improve prooxidation through the Fenton reaction, seems to act, at least in some tissues, as a strong enhancer of ferritin expression, rapidly leading to its sequestration and thus showing protective properties.[15]

The HO-1 gene is rapidly induced by its substrate heme as well as by many other non-heme stimulants such as heat shock, cytokines, ultraviolet irradiation, hypoxia, heavy metals, glutathione depletion and oxidants.[16-18] All these different inducers have the common feature of unbalancing the cellular redox state. The role of redox signaling in modulating HO-1 gene expression is also supported by evidence demonstrating that thiols and certain antioxidants suppress its activation.[19] Although different mechanisms of HO-1 induction have been described, the general idea is that a limited number of signal transduction pathways mediate HO-1 gene expression in response to a multitude of cellular perturbations. HO-1 presents several potential consensus regulatory elements in its promoter region, including the NF-κB binding site, the activator protein 1 (AP-1), the heat shock consensus sequence, the metal responsive element and the antioxidant response element (ARE).[20,21] Although AP-1 elements exists in the ARE sequence, recent evidence indicates that trancription factors that activate ARE are quite distinct from AP-1. Two other bZip transcription factors, Nrf1 and Nfr2, are viewed as the primary regulators of ARE.[22] Even if many transcription factors have been proposed as cellular sensors for redox changes, the molecular basis of this regulation has not been elucidated in any eukaryotic DNA binding protein, and also in the case of HO-1 it still remains an open question.[23] Increasing evidence also indicates that NO induces HO-1 expression, as demonstrated in different in vitro models of stress-induced cellular insult.[24,25] Nitric oxide (NO) is a free-radical gas with an unshared electron that is produced by the conversion of L-arginine to L-citrulline catalysed by NO synthase (NOS).[26] It is generally accepted that NO is a major component in signalling transduction pathways controlling smooth muscle tone, platelet aggregation, host response to infection and a wide array of other physiological and pathophysiological processes.[27] NO appears to play several crucial roles in the brain. These include physiological processes such as neuromodulation, neurotransmission and synaptic plasticity. Under conditions of excessive formation, NO is emerging as an important mediator of neurotoxicity in a variety of disorders of the nervous system, such as neurodegeneration and neuroinflammation.[27]

Several biochemical conditions can modulate NOS activity, and many enzymes and trascriptional factors are involved in its functions. Arginine, the major NOS substrate, is a limiting step for the generation of NO. Arginase is the metabolic enzyme that converts arginine into urea plus ornithine, the final step in urea synthesis.[28] The biochemistry and physiology of L-arginine has received recently much consideration in the light of the recent discovery that the amino acid is the only substrate of all isoforms of nitric oxide synthase (NOS). Thus, generation of NO is intertwined with synthesis, catabolism and transport of arginine, which ultimately participates in the regulation of a fine-tuned balance between normal and pathophysiological consequences of NO production. Two different isoforms (I and II) have been isolated[29] and, while type II is present in many tissues, arginase I is expressed exclusively in liver.[30] The complex composition of the brain at the cellular level is reflected in a complex differential distribution of the enzymes of arginine metabolism. Argininosuccinate synthetase (ASS) and argininosuccinate lyase which together can recycle the NOS coproduct L-citrulline to L-arginine are expressed constitutively in neurons, but hardly colocalize with each other or with NOS in the same neuron. Thus, the inducible isoform of arginase (Arginase II), which is expressed in a wide variety of tissues, with highest levels of expression in prostate, brain, and kidney,[31] has been proposed as a crucial element for the regulation of NO synthesis by modulating local arginine concentrations.[32] According to this hypothesis, it has been shown that Arginase II mRNA and NOS mRNA were coinduced by lipopolysaccharide (LPS) in a macrophage-like cell line.[33] LPS, a component of the bacterial wall of gram-negative bacteria, has been recognized as one of the most potent bacterial products in the induction of host inflammatory responses and tissue injury associated with significant alteration of the brain oxidative status.[34] An increment in either the circulating levels or the local tissue release of cytokines (by activating iNOS in microglial cells and astrocytes to produce toxic amounts of NO) is thought to represent the primary pathogenic event contributing to neurodegenerative changes observed during aging.[2,27] The present study was undertaken, therefore, to investigate the expression and activity of HO-1, together with that of iNOS and arginase in different rat brain regions in response to an acute stress induced by endotoxic treatment. Results of such experiments might increase our understanding of the relationship between NO-dependent signal pathway and HO-1 expression and their role in neuroprotection.

MATERIALS AND METHODS

Animals

Male Wistar rats, weighing 140–180 gr., were used for all experiments. The animals were treated with 10 mg/kg LPS endotoxin (Sigma) and sacrificed, by cervical dislocation, at 12 hours. After sacrifice brain was quickly removed and dissected in a cold anatomical chamber following a standardized protocol. In order to exclude the possibility that the small size of nigral or septal samples might affect the results, we analyzed samples of pooled *substantia nigra* or septum areas with a protein content comparable to that of cortical or striatum specimens.

Assay for Heme Oxygenase Activity

Heme oxygenase activity assay was performed with a modified method previously described by Maines et al.[35] Brain tissues were homogenized in 800 µl of a solution containing 100 mM phosphate buffer PBS and 2 mM $MgCl_2$ and sonicated. 400 µl of the homogenate was then added to a reaction mixture containing: NADPH (0.8 mM), glucose 6-phosphate (2 mM), glucose-6-phosphate dehydrogenase (0.2 units), 3 mg of rat liver cytosol prepared from a 105,000 × g supernatant fraction, as a source of biliverdin reductase, potassium phosphate buffer (PBS, 100 mM, pH 7.4), MgCl2 (0.2 mM), and hemin (20 µM). The reaction was conducted at 37 °C in the dark for 1 h and terminated by the addition of 1 ml of chloroform, and the extracted bilirubin was measured calculating the difference in absorbance between 464 and 530 nm ($\varepsilon = 40 \, mM^{-1} \, cm^{-1}$). Heme oxygenase activity was expressed as picomoles of bilirubin/mg of cell protein/h. The total protein content of confluent cells was determined by the BCA assay (Pierce, Rockford, IL) by comparison with a standard curve obtained with bovine serum albumin.

Determination of NOS Activity

NOS activity was determined using the hemoglobin assay as described previously by Hevel and Marletta.[36] The assay is based on the oxidation of oxyhemoglobin with NO and is performed spectrophotometrically by measuring the formation of methemoglobin under initial rate conditions. Brain tissues were homogenized in a solution containing 0.32 M sucrose, 10 mM Tris (pH7.4), 0.5 mM phenylmethylsulfonyl fluoride, and sonicated. The supernatant obtained after centrifugation at 12,000 × g for 30 min was used for NOS activity measurements. Formation of methemoglobin was monitored by the change in absorbance, which occurred over time between 411 and 401 nm using a double beam spectrophotometer (Perkin-Elmer 559).

Arginase Assay

Arginase activity in different brain areas was measured by reading spectrophotometrically at 540 nm the amount of urea formed, according to the method of Corraliza.[37]

Western Blot Analysis for HO-1 and iNOS

Brain tissues were homogenized in 1 ml of lysis buffer (50 mM HEPES, 5 mM EDTA, 50 mM NaCl, 1%Triton X-100, pH 7.5) containing Complete™ protease inhibitor (Boehringer Mannheim). Samples were kept on ice for 1 h and then centrifuged (4 °C) for 30 min at 12,000 × g. After the precipitated unsolubilized fraction was discarded, protein concentration was determined in the supernatant, then an equal amount of proteins (30 µg) for each sample was separated by SDS/polyacrylamide gel electrophoresis, transferred over night to nitrocellulose membranes, and the non-specific binding of antibodies was blocked with 3% non-fat dried milk in PBS. Membranes were than probed with polyclonal rabbit anti HO-1 antibody (Stressgen, Victoria, Canada) (1:1,000 dilution in Tris-buffered saline, pH 7.4). After three washes with PBS containing 0.05% (v/v) Tween 20, blots were visualized using an amplified alkaline phosphatase kit (Extra-3A, Sigma). When probed for NOS protein membranes were incubated for 2 h with anti-iNOS antibody (Santa Cruz Biotechnology) at 1:20,000 dilution in Tris-buffered saline (pH 7.4). After three washes with PBS containing 0.05% (v/v) Tween 20, membranes were incubated with a horseradish peroxidase-conjugated sheep anti-rabbit immonoglobulin G (igG), followed by ECL chemoluminescence (Amersham Pharmacia Biotech). For optical density evaluation, immunoreactive bands were scanned by a laser densitometer.

Detection of S Nitrosothiols

Detection of S-nitrosothiols in the different samples were performed, after cleavage of the S-nitroso bond, by measuring the NO released using the chemiluminescence reaction between NO and the luminol-hydrogen peroxide system.[38]

Analysis of Hydroperoxides

Hydroperoxides were quantitated by the FOX2 method, according to the procedure of Nourooz-Zadeh.[39]

Statistical Analysis

Differences in the data among the groups were analysed by one-way analysis of variance combined with the Bonferroni's test, and all values were expressed as mean ± S.E. The differences between groups were considered to be significant at $p < 0.05$.

RESULTS

In the present study, we evaluated the expression of HO-1 in different rat brain regions in response to endotoxin-induced shock. Figure 1a shows regional distribution of heme oxygenase activity in different regions of brain in control and endotoxin-treated rats. In control rats HO activity exhibited regional variability with highest levels in the substantia nigra (SN) followed by septum, cerebellum, hippocampus and cortex. Although the assay does not distinguish between the two isozymes, HO-2 and HO-1, it is reasonable that in control brain constitutive HO-2 was the principal isoform responsible for the measured basal activity. Treatment with LPS led to a significant increase in HO activity particularly in the SN and hippocampus, followed by cortex, striatum, septum and cerebellum. Western blot analysis revealed a significant increase in brain HO-1 protein expression induced by endotoxic shock in all brain regions studied. As illustrated in Fig. 1b, levels of HO expression varied consistently with the observed changes in HO enzyme activity.

Induction of HO-1 was associated with both increased NOS activity and expression levels. Figure 2a illustrates the effect of endotoxemia on the regional distribution of NOS activity, which was higher in the SN, cortex and striatum compared to the other brain regions examined. A similar pattern was obtained after western blot analysis of iNOS protein expression, which showed the highest levels of induction in the SN compared to the other brain regions examined (Fig. 2b).

We measured also arginase activity in the brain of control and endotoxin-treated rats. Data reported in Fig. 3 demonstrates a significant increase in enzyme activity after LPS treatment in all regions examined.

It has been shown that LPS induces significant alterations in the brain oxidative status and in our study we investigated the effect of endotoxin on the regional distribution of lipid peroxides as an index of oxidative stress. Figure 4a demonstrates that endotoxemia in rat is associated with a significant increase of lipid peroxidation, particularly in the brain regions of SN, cortex and hippocampus and to a lesser extent in the striatum, septum and cerebellum. Figure 4b shows nitrosothiol formation which was higher in the brain of LPS-treated animals compared to untreated controls.

DISCUSSION

In recent years, cellular oxidant/antioxidant balance has become the subject of intense study, particularly by those interested in brain aging and in neurodegenerative mechanisms. Although the etiology and pathogenesis of the major neurodegenerative and neuroinflammatory disorders (Alzheimer's disease, amyotrophic lateral

a

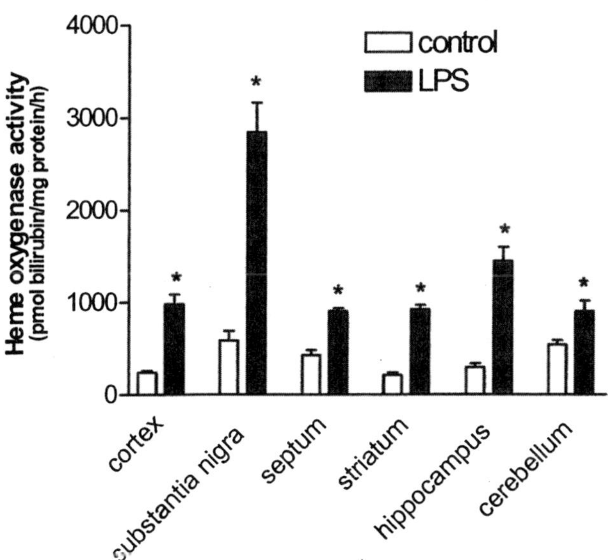

b

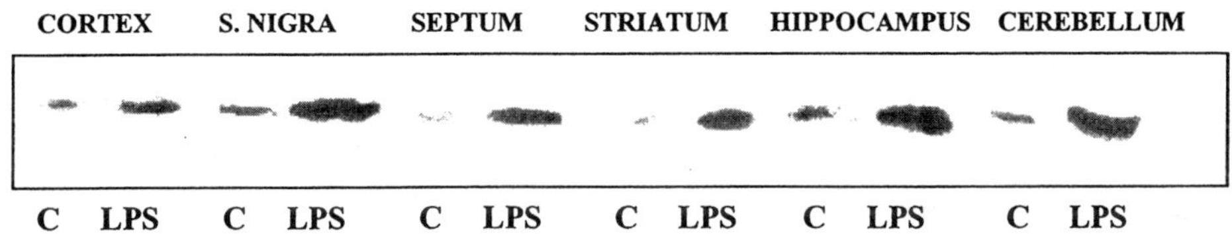

Figure 1. Differential activity and expression of HO-1 in rat brain areas by endotoxin (LPS). (a) HO activity was performed in different regions of rat brain in a control group and in rats treated with LPS 10 mg/kg i.p. (12 h). *$p < 0.05$ versus control. (b) HO-1 protein expression was analyzed by Western immunoblot technique in brain samples treated as above.

sclerosis, Parkinson's disease, Huntington's disease and Multiple sclerosis) are unknown, increasing evidence strongly suggests that reactive oxygen and nitrogen species play a crucial role. The brain has a large potential oxidative capacity due to the high level of tissue oxygen consumption. However, the ability of the central nervous system (CNS) to combact oxidative stress is limited because of the high content of easily oxidizable substrates, such as polyunsaturated fatty acids and catecholamines and relatively low levels of antioxidants. In addition, the CNS contains non-replicating neuronal cells which, once damaged, may be permanently dysfunctional or committed into programmed (apoptotic) and/or passive (necrotic) cell death.[1]

                                                                    G. Scapagnini *et al.*

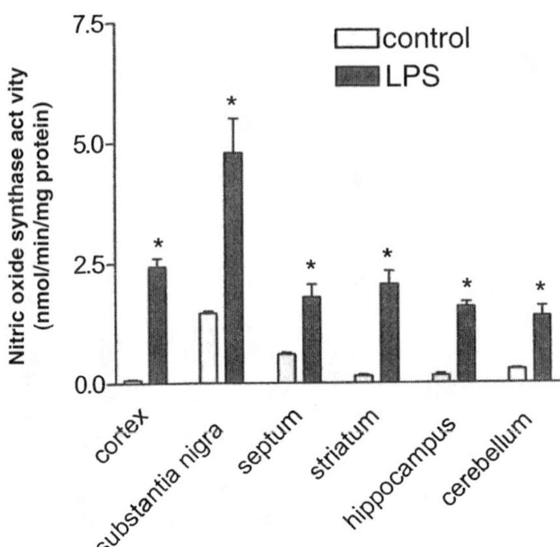

b

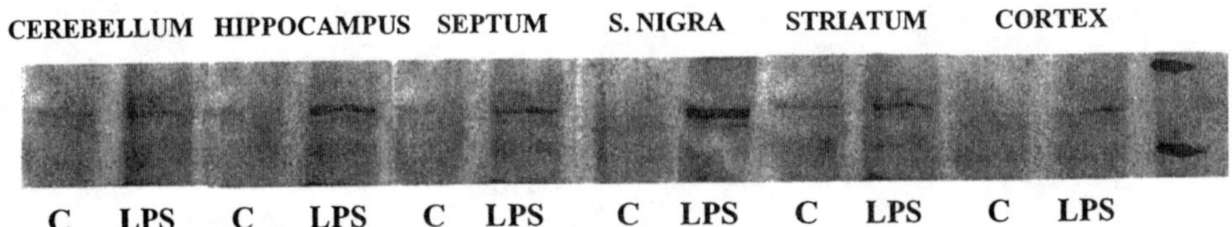

Figure 2. Differential activity and expression of iNOS in rat brain areas by endotoxin (LPS). (a) NOS activity was measured in different regions of rat brain in a control group and in rats treated with LPS 10 mg/kg i.p. after 12 h. $*p < 0.05$ versus control. (b) iNOS protein expression was analyzed by Western immunoblot technique as described in Material and Methods.

Systemic inflammation and sepsis frequently involve neurological abnormalities associated with inhibition of behaviour, motor activity, and level of consciousness, called septic encephalopathy. This medical syndrome is usually associated with a poor outcome, and the precise pathogenic mechanisms of septic encephalopathy have not yet been completely understood.[40] The toxic effects of reactive oxygen and nitrogen species have now been recognized as one of the ultimate mediators of tissue injury during systemic inflammation and sepsis. An excessive level of NO generated by iNOS in states of systemic inflammation is thought to be the primary causative factor underlying the toxic cascades following endotoxemia. Recent evidences demonstrated that transgenic animals with a null mutation of the NOS2 gene are resistant to hypotension and death caused by LPS.[41,42]

In our study we have shown a different expression of iNOS and HO-1 in various brain regions after induction of systemic sepsis. In view of the evidence suggesting that NOS and arginase are two proteins linked to HO-1 in the regulation of cellular redox states, these data suggest a possible interaction between the NO and HO systems, which might occur in the brain following endotoxin treatment. Our *in vivo* data on a co-induction of HO-1 and iNOS induced by LPS treatment in the brain confirm the hypothesis of a fine regulation between these two inducible enzymes. Remarkably, SN and hippocampus, two brain regions critically involved in the pathogenesis of Parkinson's and, respectively, Alzheimer's diseases, are the regions in which we observed the highest induction of both HO and NOS. Hence, a better understanding of the relationship between these two proteins in the various regions of the brain will help to elucidate potentially vital mechanisms involved in the pathogenesis of acute and chronic neurodegenerative diseases.

Taken together, our data suggest the hypothesis that under physiological conditions a balance exists between production of NO, its intracellular association with glutathione, and its targeting by proteins. Under stressful conditions, such as

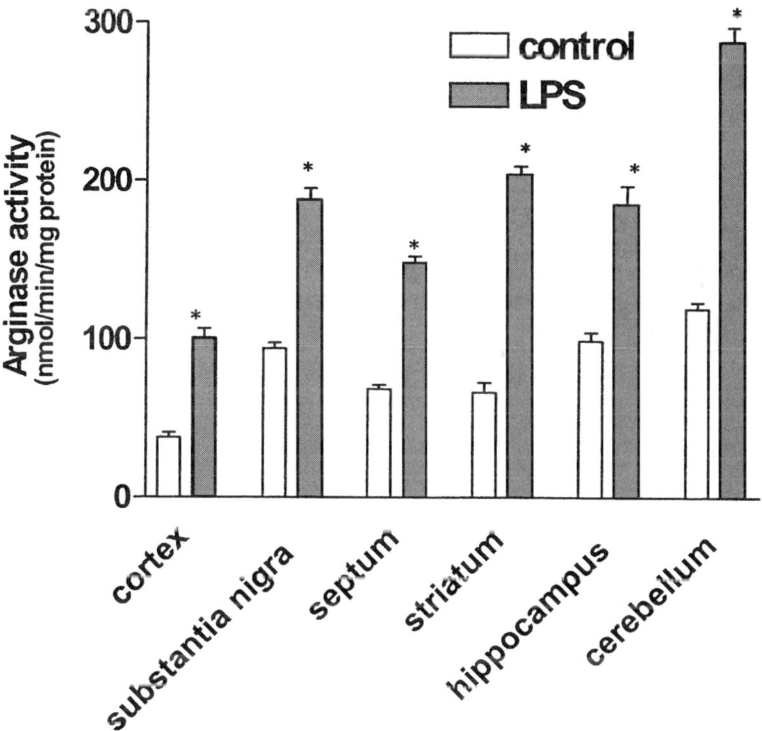

Figure 3. Differential activity of arginase in rat brain areas by endotoxin (LPS). Arginase activity was measured in different regions of rat brain in a control group and in rats treated with LPS 10 mg/kg i.p. after 12 h. *p < 0.05 compared with control.

                                                         G. Scapagnini *et al.*

a

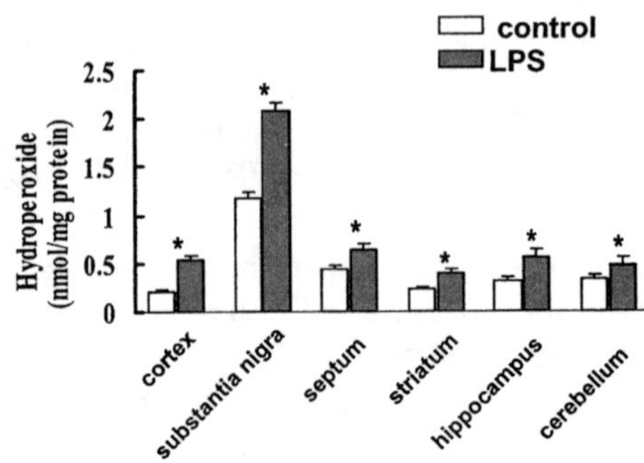

b

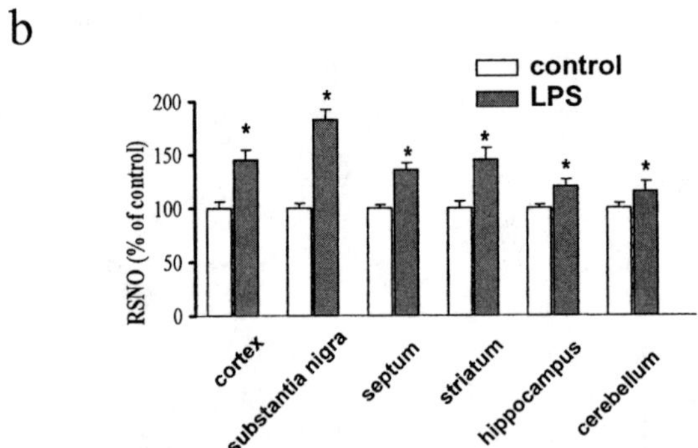

Figure 4. Differential formation of both lipid peroxides and nitrosothiols in rat brain areas by endotoxin (LPS). (a) Evaluation of lipid peroxidation in different regions of rat brain in a control group and in rats treated with LPS 10 mg/kg i.p. after 12 h. *$p < 0.05$ versus control. (b) Nitrosothiols formation in different regions of rat brain in a control group and in rats treated with LPS 10 mg/kg i.p. after 12 h. *$p < 0.05$ versus control.

endotoxic shock, ensuing oxidative stress may alter this balance leading to glutathione depletion, enhanced superoxide formation, and up-regulation of the inducible form of NO synthase. This event leads to increased production of NO and NO reactive species which may act as intracellular signals for the expression of heme oxygenase. The induction of this protein would increase endogenous CO, biliverdin, and bilirubin, all products ultimately implicated in protective mechanisms against oxidative injury. Furthermore, heme activity would provide a sensitive negative feed-back against NO generation by iNOS (Fig. 5).

Until recently, investigations in the field of NO/NO synthase and CO/heme oxygenase have evolved independently from each other, conferring particular emphasis on the role of one pathway or the other in the control of important biological activity. In accordance with other authors[14] our experimental evidence showing that NO stimulates HO-1 gene expression and that the heme oxygenase/CO pathway affects NO synthase activity points to the importance of a relationship between the two systems. This interdependence becomes evident in the CNS where a prominent co-localization of constitutive NOS and HO-2 is observed, and the respective inducible isoforms (HO-1 and iNOS) can be greatly over-expressed following stressful stimuli. Although further investigation is required to clarify the exact role of NO and CO, we suggest that these gaseous molecules might function interdependently, each dynamically influencing the other in regulating crucial signal trasduction mechanisms related to their biological activities. Moreover, the induction of the HO-1 pathway in certain disease states may represent a fine expedient to protect against tissue injury and, above all, to restore cellular homeostasis.

Current research pointing to the possible use of NOS inhibitors in degenerative diseases has produced contradictory results, primarily due to lack of specificity of these agents and also because of their intrinsic toxicity. Acute and chronic neurodegenerative processes which are currently irreversible might be amenable to novel treatment strategies that are based on the modulation of brain NOS. In this context, the substantial body of evidence demonstrating the protective role of HO-1 in different models of pathological oxidative damage strongly favors the view of HO-1 as a potential therapeutical target.

Thus, non toxic modulators of HO-1 induction could be extremely interesting for the treatment of those pathologies involving chronic or acute nitrosative and oxidative stress, such as brain aging and neurodegenerative disorders.

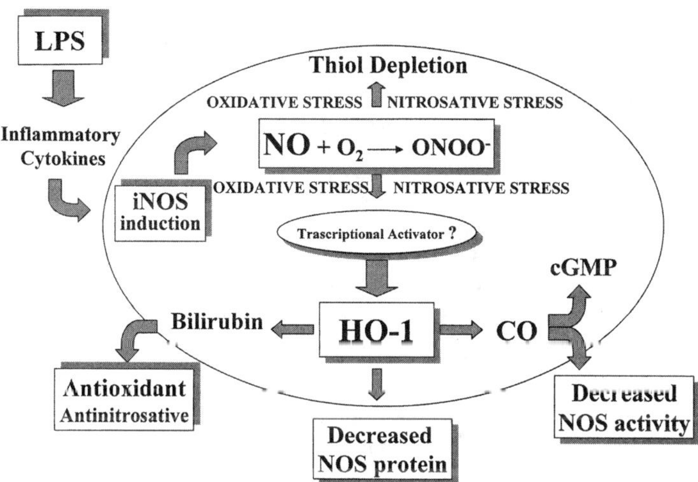

Figure 5. Mutual relationship between HO and NOS systems. Simplified scheme of possible interaction between heme oxygenase and nitric oxide shyntase systems. Increased NO derived from iNOS activity, induced by proinflammatory cytokines, activates cellular stress response including over-expression of highly specialized antioxidant/anti-nitrosative genes, such as HO-1. Increased HO-1 activity, through its byproducts, exerts important feedback mechanisms on NOS activity and NO cytotoxicity.

ACKNOWLEDGMENTS

We thank S. Rosini (Abiogen Pharma) for the precious advice and support to this manuscript, and S. Quattrone for technical assistance and helpful discussions.

REFERENCES

1. R.A. Floyd, Antioxidants, oxidative stress, and degenerative neurological disorders, Proc. Soc. Exp. Biol. Med. 222 (1999) 236–245.
2. V. Calabrese, G. Scapagnini, A.M. Giuffrida Stella, T.E. Bates, and J.B. Clark, Mitochondrial involvement in brain function and dysfunction: relevance to aging, neurodegenerative disorders and longevity, Neurochem. Res. 26 (2001) 739–764.
3. N.G. Abraham, G.S. Drummond, J.D. Lutton, and A. Kappas, The biological significance and physiological role of heme oxygenase, Cell. Physiol. Biochem. 6 (1996) 129–168.
4. K.D. Poss and S. Tonegawa, Reduced stress defense in heme oxygenase 1-deficient cells, Proc. Natl. Acad. Sci. U.S.A. 94 (9-30-1997) 10925–10930.
5. G.F. Vile, S. Basu-Modak, C. Waltner, and R.M. Tyrrell, Heme oxygenase 1 mediates an adaptive response to oxidative stress in human skin fibroblasts, Proc. Natl. Acad. Sci. U.S.A. 91 (3-29-1994) 2607–2610.
6. J.E. Clark, R. Foresti, C.J. Green, and R. Motterlini, Dynamics of haem oxygenase-1 expression and bilirubin production in cellular protection against oxidative stress, Biochem. J. 348 Pt 3 (6-15-2000) 615–619.
7. J.F. Ewing and M.D. Maines, Rapid induction of heme oxygenase 1 mRNA and protein by hyperthermia in rat brain: heme oxygenase 2 is not a heat shock protein, Proc. Natl. Acad. Sci. U.S.A. 88 (6-15-1991) 5364–5368.
8. K. Chen, K. Gunter, and M.D. Maines, Neurons overexpressing heme oxygenase-1 resist oxidative stress-mediated cell death, J. Neurochem. 75 (2000) 304–312.
9. H.M. Schipper, Heme oxygenase-1: role in brain aging and neurodegeneration, Exp. Gerontol. 35 (2000) 821–830.
10. A. Takeda, G. Perry, N.G. Abraham, B.E. Dwyer, R.K. Kutty, J.T. Laitinen, R.B. Petersen, and M.A. Smith, Overexpression of heme oxygenase in neuronal cells, the possible interaction with Tau, J. Biol. Chem. 275 (2-25-2000) 5395–5399.
11. R. Stocker, Y. Yamamoto, A.F. McDonagh, A.N. Glazer, and B.N. Ames, Bilirubin is an antioxidant of possible physiological importance, Science 235 (2-27-1987) 1043–1046.
12. A. Verma, D.J. Hirsch, C.E. Glatt, G.V. Ronnett, and S.H. Snyder, Carbon monoxide: a putative neural messenger, Science 259 (1-15-1993) 381–384.
13. M.D. Maines, The heme oxygenase system: a regulator of second messenger gases, Annu. Rev. Pharmacol. Toxicol. 37 (1997) 517–554.
14. R. Foresti and R. Motterlini, The heme oxygenase pathway and its interaction with nitric oxide in the control of cellular homeostasis, Free Radic. Res. 31 (1999) 459–475.
15. R.S. Eisenstein, D. Garcia-Mayol, W. Pettingell, and H.N. Munro, Regulation of ferritin and heme oxygenase synthesis in rat fibroblasts by different forms of iron, Proc. Natl. Acad. Sci. U.S.A. 88 (2-1-1991) 688–692.
16. D. Lautier, P. Luscher, and R.M. Tyrrell, Endogenous glutathione levels modulate both constitutive and UVA radiation/hydrogen peroxide inducible expression of the human heme oxygenase gene, Carcinogenesis 13 (1992) 227–232.
17. R. Tyrrell, Redox regulation and oxidant activation of heme oxygenase-1, Free Radic. Res. 31 (1999) 335–340.
18. R. Motterlini, R. Foresti, R. Bassi, V. Calabrese, J.E. Clark, and C.J. Green, Endothelial heme oxygenase-1 induction by hypoxia. Modulation by inducible nitric-oxide synthase and S-nitrosothiols, J. Biol. Chem. 275 (5-5-2000) 13613–13620.

19. S.L. Camhi, J. Alam, G.W. Wiegand, B.Y. Chin, and A.M. Choi, Transcriptional activation of the HO-1 gene by lipopolysaccharide is mediated by 5′ distal enhancers: role of reactive oxygen intermediates and AP-1, Am. J. Respir. Cell Mol. Biol. 18 (1998) 226–234.

20. Y. Lavrovsky, M.L. Schwartzman, R.D. Levere, A. Kappas, and N.G. Abraham, Identification of binding sites for transcription factors NF-kappa B and AP-2 in the promoter region of the human heme oxygenase 1 gene, Proc. Natl. Acad. Sci. U.S.A. 91 (6-21-1994) 5987–5991.

21. T. Prestera, P. Talalay, J. Alam, Y.I. Ahn, P.J. Lee, and A.M. Choi, Parallel induction of heme oxygenase-1 and chemoprotective phase 2 enzymes by electrophiles and antioxidants: regulation by upstream antioxidant-responsive elements (ARE), Mol. Med. 1 (1995) 827–837.

22. J. Alam, D. Stewart, C. Touchard, S. Boinapally, A.M. Choi, and J.L. Cook, Nrf2, a Cap′n′Collar transcription factor, regulates induction of the heme oxygenase-1 gene, J. Biol. Chem. 274 (9-10-1999) 26071–26078.

23. H.E. Marshall, K. Merchant, and J.S. Stamler, Nitrosation and oxidation in the regulation of gene expression, FASEB J. 14 (2000) 1889–1900.

24. R. Foresti, J.E. Clark, C.J. Green, and R. Motterlini, Thiol compounds interact with nitric oxide in regulating heme oxygenase-1 induction in endothelial cells. Involvement of superoxide and peroxynitrite anions, J. Biol. Chem. 272 (7-18-1997) 18411–18417.

25. E. Hara, K. Takahashi, T. Tominaga, T. Kumabe, T. Kayama, H. Suzuki, H. Fujita, T. Yoshimoto, K. Shirato, and S. Shibahara, Expression of heme oxygenase and inducible nitric oxide synthase mRNA in human brain tumors, Biochem. Biophys. Res. Commun. 224 (7-5-1996) 153–158.

26. L.J. Ignarro, G.M. Buga, K.S. Wood, R.E. Byrns, and G. Chaudhuri, Endothelium-derived relaxing factor produced and released from artery and vein is nitric oxide, Proc. Natl. Acad. Sci. U.S.A. 84 (1987) 9265–9269.

27. V. Calabrese, T.E. Bates, and A.M. Stella, NO synthase and NO-dependent signal pathways in brain aging and neurodegenerative disorders: the role of oxidant/antioxidant balance, Neurochem. Res. 25 (2000) 1315–1341.

28. H. Wiesinger, Arginine metabolism and the synthesis of nitric oxide in the nervous system, Prog. Neurobiol. 64 (2001) 365–391.

29. E.B. Spector, S.C. Rice, and S.D. Cederbaum, Evidence for two genes encoding human arginase, (Abstract) Am. J. Hum. Genet. 32 (1980) 55.

30. E.B. Spector, S.C. Rice, and S.D. Cederbaum, Immunologic studies of arginase in tissues of normal human adult and arginase-deficient patients, Pediatr. Res. 17 (1983) 941–944.

31. J.G. Vockley, C.P. Jenkinson, H. Shukla, R.M. Kern, W.W. Grody, and S.D. Cederbaum, Cloning and characterization of the human type II arginase gene, Genomics 38 (12-1-1996) 118–123.

32. M. Mori and T. Gotoh, Regulation of nitric oxide production by arginine metabolic enzymes, Biochem. Biophys. Res. Commun. 275 (9-7-2000) 715–719.

33. T. Gotoh, T. Sonoki, A. Nagasaki, K. Terada, M. Takiguchi, and M. Mori, Molecular cloning of cDNA for nonhepatic mitochondrial arginase (arginase II) and comparison of its induction with nitric oxide synthase in a murine macrophage-like cell line, FEBS Lett. 395 (10-21-1996) 119–122.

34. V. Calabrese, A. Copani, D. Testa, A. Ravagna, F. Spadaro, E. Tendi, V.G. Nicoletti, and A.M. Giuffrida Stella, Nitric oxide synthase induction in astroglial cell cultures: effect on heat shock protein 70 synthesis and oxidant/antioxidant balance, J. Neurosci. Res. 60 (6-1-2000) 613–622.

35. M. Maines, Carbon monoxide and nitric oxide homology: differential modulation of heme oxygenases in brain and detection of protein and activity, Methods Enzymol. 268 (1996) 473–488.

36. J.M. Hevel and M.A. Marletta, Nitric-oxide synthase assays, Methods Enzymol. 233 (1994) 250–258.

37. I.M. Corraliza, M.L. Campo, G. Soler, and M. Modolell, Determination of arginase activity in macrophages: a micromethod, J. Immunol. Methods 174 (9-14-1994) 231–235.

38. H. Kojima, K. Kikuchi, M. Hirobe, and T. Nagano, Real-time measurement of nitric oxide production in rat brain by the combination of luminol-H2O2 chemiluminescence and microdialysis, Neurosci. Lett. 233 (9-19-1997) 157–159.

39. J. Nourooz-Zadeh, J. Tajaddini-Sarmadi, and S.P. Wolff, Measurement of plasma hydroperoxide concentrations by the ferrous oxidation-xylenol orange assay in conjunction with triphenylphosphine, Anal. Biochem. 220 (8-1-1994) 403–409.
40. C.F. Bolton, G.B. Young, and D.W. Zochodne, The neurological complications of sepsis, Ann. Neurol. 33 (1993) 94–100.
41. C. Thiemermann, Nitric oxide and septic shock, Gen. Pharmacol. 29 (1997) 159–166.
42. S. Basu and M. Eriksson, Oxidative injury and survival during endotoxemia, FEBS Lett. 438 (1998) 159–160.

11

ROLE OF HEME CATABOLISM IN NEURODEGENERATIVE DISEASES

Atsushi Takeda[a], Yasuto Itoyama[a], Teiko Kimpara[b],
R. Krishnan Kutty[c], Nader G. Abraham[d], Barney E. Dwyer[e],
Robert B. Petersen[f], George Perry[f], and Mark A. Smith[f]

[a]Department of Neurology
Tohoku University School of Medicine
Sendai 980-8574, Japan
[b]Department of Neurology
Nishitaga National Hospital
Sendai 982-8555 Japan
[c]National Eye Institute, National Institutes of Health
LRCMB-NEI, Bethesda, Maryland 20892 USA
[d]Department of Pharmacology
New York Medical College
Valhalla, New York 10595 USA
[e]VA Medical Center, Research Service (151)
White River Junction, Vermont 05009 USA
[f]Institute of Pathology, Case Western Reserve University
Cleveland, Ohio 44106 USA

1. INTRODUCTION

Heme oxygenase (HO), a microsomal enzyme that cleaves heme to produce biliverdin, ferric iron and carbon monoxide, is the rate-limiting step in heme degradation.[1] To date, three HO isoforms (HO-1, HO-2 and HO-3) have been identified that catalyze this reaction. HO-1 is a 32 kDa heat shock protein[2] induced by numerous noxious stimuli,[3] HO-2 is a constitutively synthesized 36 kDa protein which is abundant in brain and testis,[4] and HO-3 has structural homology with HO-2, but its ability to catalyze heme degradation is much less.[5] Within the brain, the majority of HO activity is attributed to the HO-2 isozyme[1] since the expression of HO-1 is normally very low in the brain and restricted to select neuronal and non-neuronal cell populations in the forebrain, diencephalons, cerebellum, and brain stem.[1] However, in the brain, HO-1 increases markedly after heat shock, ischemia or glutathione depletion[4,6,7] and, after heat shock or ischemia, increased HO-1 expression is shown in neuronal and glial cells throughout the brain.[1,7,8] Therefore, HO-1 is known as an oxidative stress-inducible protein and plays a key role in heme catabolism, in which heme, a potential prooxidant, is converted to bilirubin, an antioxidant.[9] However, HO-1 also produces other by-products, such as carbon monoxide, a signal transmitter, and free iron, another prooxidant. Thus, heme catabolism may be followed by a variety of metabolic processes and consequently whether HO-1 acts as an antioxidant or prooxidant seems to be highly dependent on environment.[10]

2. INCREASED EXPRESSION OF HO-1 IN NEURODEGENERATIVE DISEASES

Recent studies show that the oxidative damage associated with neurodegenerative diseases present as lipid peroxidation,[11] nitration,[12] reactive carbonyls[13] and nucleic acid oxidation[14] that are all increased in vulnerable neurons. Therefore, it is perhaps non surprising that HO-1 is also increased in brains affected with certain neurodegenerative diseases such as Alzheimer disease (AD),[15,16] Parkinson disease (PD),[17] Pick disease, progressive supranuclear palsy (PSP) and corticobasal degeneration (CBD).[18] Interestingly, HO-1 expression is coincident with the hallmark pathological structures, such as neurofibrillary tangles, senile plaques and Lewy bodies, suggesting that HO-1 plays a key role in the pathophysiological processes of neurodegeneration (Fig. 1).[1]

To determine whether an increase in HO-1 expression leads to an actual increase in heme catabolism in diseased brains, we investigated the levels of bilirubin and its derivatives in cerebrospinal fluid sampled from cases with a variety of neurodegenerative diseases.[19] We used the monoclonal antibody 24G7, which is highly specific for

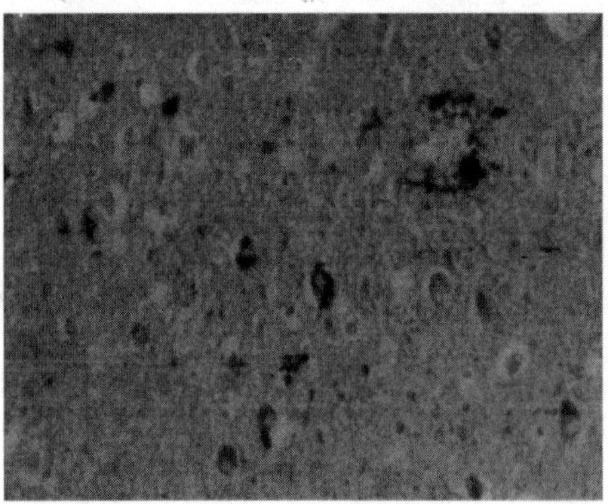

(A)   (B)

Figure 1. In AD brain (A), vulnerable neurons show markedly increased levels of HO-1 in comparison to control brain (B).

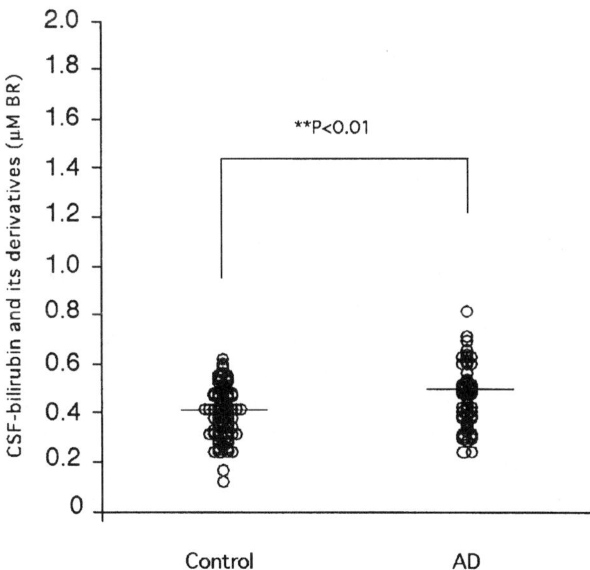

Figure 2. Increased bilirubin and its derivatives in AD CSF. (Modified from *Kimpara T et al., Neurobiology of Aging 21:551, 2000.*)

bilirubin and its degraded derivatives generated by oxidative stress.[19] The ELISA study showed a significant increase of bilirubin and its derivatives in the cerebral spinal fluid from AD cases, suggesting that the induced HO-1 is functional and the heme catabolism does, in fact, increase in the AD brain. We failed to demonstrate such increases in the other neurodegenerative diseases, possibly because the majority of heme catabolism is mediated by HO-2 in the central nervous system.[1]

3. HO-1 INDUCTION IS COINCIDENT WITH CONFORMATIONAL CHANGES OF TAU

In AD, HO-1 induction coincides with the conformational change in tau recognized by the monoclonal antibody, Alz50 (Alz50-epitope),[20] which was originally prepared against an antigen found in the brain of cases of AD.[21,22] The Alz50 epitope is discontinuously located in the N-terminal and C-terminal regions of tau protein and, therefore, recognizes a specific conformation of monomeric tau protein that involves juxtaposition of the N- and C-termini, which is stabilized by its incorporation into structurally-ordered tau filaments.[23] Appearance of the Alz50-epitope in neurons is considered an early pathological change in the tau protein, which is followed by further pathological changes, such as the formation of neurofibrillary tangles that also contain the Alz50-epitope.[24] On the other hand, tau phosphorylation was also found in neurons lacking HO-1, indicating that the phosphorylation of tau probably precedes formation of the Alz50 epitope. Consistent with this interpretation, we found that *in vitro* experiments, treatment of normal tau protein with the reactive lipid peroxidation product, 4-hydroxy-2-nonenal (HNE) or other bifunctional aldehydes or protein modification agents capable of inducing lysine-dependent crosslinking, greatly increased the presence of the Alz50-epitope. Interestingly, bilirubin can bind to proteins through lysine residues. The increased bilirubin may prevent crosslinking of tau by capping lysine-residues and increased expression of HO-1, leading to degradation of bilirubin, would facilitate the formation of the Alz50-epitope. Furthermore, it is intriguing that other neurological conditions where tau accumulation and the Alz50-epitope are found also show increases in HNE adducts, e.g., PSP and frontal temporal dementia (unpublished data).

Together with the observation that tau phosphorylation is accelerated by HNE,[25] our data suggested that oxidative stress and the attendant modification of tau by products of oxidative stress, including HNE as well as other cytotoxic carbonyls, may trigger the cytoskeletal pathology observed in AD. Indeed, while HO-1 plays a pivotal role in tau modification by lipid peroxidation products,[20] only 20–40% of HNE positive neurons also express HO-1 suggesting that the induction of HO-1 is not simply regulated by oxidative stress. Of the known stimulants, heme is one of the most effective inducers of HO-1.[10] In AD, mitochondria, rich in heme molecules, show gross abnormalities.[26] Thus, it is plausible that HO-1 is induced by excess heme release from damaged mitochondria. Furthermore, it is intriguing to consider that HO-1 induction may be a response to the cytochrome c released from mitochondria that is an attempt to protect neurons from apoptosis.[27]

4. POSSIBLE INTERACTION BETWEEN HO-1 AND TAU PROTEIN

To explore possible interactions between HO-1 and tau, we transfected sense- and antisense-human HO-1 cDNA into human neuroblastoma cells.[28] Cells transfected with sense HO-1 cDNA showed a several fold induction of HO-1 transcripts and protein, which was mirrored by a more than 3-fold increase in HO-activity. As would be expected, overexpression of the HO-1 gene considerably enhanced resistance of the cells to oxidative injury produced by hydrogen peroxide, consistent with other reports.[29] Interestingly, HO-1 overexpression specifically led to the suppressed expression of tau protein. This suppression was partially alleviated by treatment with an HO inhibitor, suggesting that tau expression depends on HO activity. While the complete regulatory mechanism for tau gene expression is still unknown, ERKs, which are prominent members of the MAPK family, induce tau promoter activity.[28] In our study,[28] HO-1 overexpression decreased the phosphorylated (activated) forms of ERK-1 and ERK-2, consistent with the observed suppression of tau expression. When the HO inhibitor was added, the phosphorylated form of ERK-1 increased over control levels, suggesting that ERK-1 activation was down regulated by the increased HO activity and that the HO inhibitor suppressed not only the overexpressed HO-1 activity but also the intrinsic HO-2 activity. Although the level of the activated form of ERK-1 was increased and that of ERK-2 was restored to control levels by the addition of HO inhibitor, only partial recovery of tau protein expression was observed.

This suggests that tau gene expression is not directly controlled by ERK cascades. In fact, tau has been classified as a "late gene" and the effect of MAPK cascades on its expression is thought to be indirect and mediated by additional downstream transcription factors.[28] Interestingly, changes in cell growth and morphology were not observed in HO-1 overexpressing cells even though tau protein expression was suppressed. This implies that other molecules can substitute for the tau protein, which has also been suggested in previous reports using either a cell culture model or knockout mice.[30]

5. HEME CATABOLISM AND IRON METABOLISM IN NEURODEGENERATION

The knockout mice[31,32] and a reported human case[33] of a mutant HO-1 gene have a phenotype characterized by severe anemia and iron deposition in certain tissues as well as general vulnerability to oxidative stress, suggesting that heme catabolism plays a key role in iron metabolism including efflux of non-heme iron from the cytoplasm.[34] Since free iron can catalyze Haber-Weiss and Fenton reactions, causing the formation of highly toxic radicals such as peroxynitrite, iron accumulation will increase bulk oxidative stress in tissues. The deposition of non-heme iron within affected neuronal tissues is a common feature in AD, PD and other neurodegenerative diseases,[35] suggesting possible roles for heme catabolism in iron metabolism that has attracted the attention of many researchers. Using *in vitro* models, Shipper et al.[34] showed that HO-1 is responsible not only for extracellular export of iron, but also for intra-mitochondrial sequestration of iron. We demonstrated that overexpression of HO-1 did not affect the level of ferritin expression, suggesting that there were no changes of free iron levels in the cytoplasm following HO-1 overexpression.[28] Moreover, since a recent *in vivo* study failed to demonstrate iron accumulation mediated by HO-1 induction,[36] increased heme catabolism is unlikely to lead to the chronic accumulation of intracellular iron. These data are in good agreement with many previous reports showing that HO-1 overexpression is cytoprotective.[28,37,38]

6. PHYSIOLOGICAL ROLES OF CARBON MONOXIDE

Another product of heme catabolism, carbon monoxide, is a potent neurotransmitter that plays a role in cellular signal transduction. Through guanyl cyclase-cGMP and mitogen-activated protein kinase pathways, carbon monoxide was demonstrated to promote vasodilation,[39] to inhibit inflammatory process[40] and to cause G1/S growth arrest of the cell cycle.[41] In fact, experimental allergic encephalomyelitis (EAE), a model for multiple sclerosis, was shown to be alleviated by the activation of heme catabolism.[42] In AD brain, microglial activation facilitates disease processes while certain anti-inflammatory agents have therapeutic potentials. Moreover, aberrant cell-cycle activation was also described in brains afflicted by neurodegenerative disease.[43] Thus, the heme catabolism, which is the only metabolic pathway to produce carbon monoxide *in vivo*, may attenuate these pathological processes by

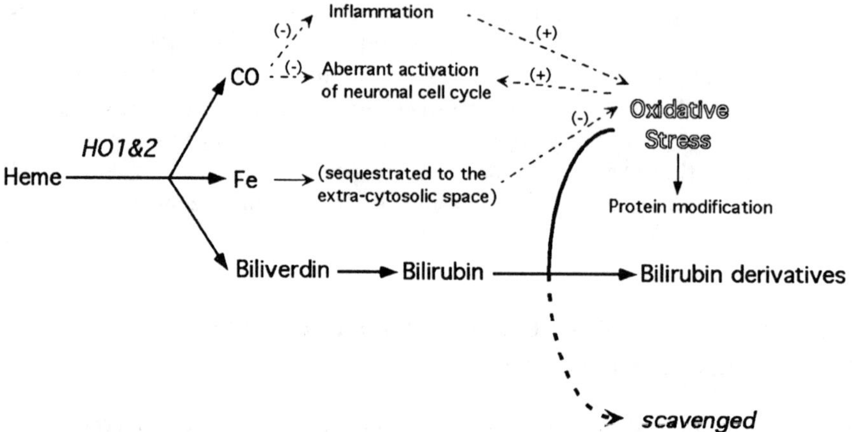

Figure 3. Schematic representation of possible roles of heme catabolism in neurodegeneration.

inhibiting the inflammatory process and arresting the aberrant activation of neuronal cells (Fig. 3).

7. CONCLUSIONS

Taken together the products of heme catabolism seem to be neuroprotective and its decrease may lead to increased vulnerability to oxidative stress. We found that the poly (GT) sequence present in the HO-1 gene promoter is highly polymorphic.[44] Because the longer alleles of this (GT)n repeat were shown to prevent the transcriptional activation of the human HO-1 gene under oxidative stress, we analyzed the allele frequencies in AD and PD. However, we failed to find any association between the longer (GT)n alleles of this polymorphism and neurodegenerative diseases, such as AD and PD.[44] More recently, through yeast two-hybrid screening, the amyloid ß protein precursor (AßPP) was shown to bind the HOs and to inhibit their activity.[45] Furthermore, it was demonstrated that the mutations of AßPP linked to familial AD provided greater inhibition of HO activity as compared to wild type AßPP.[45] These data suggest that although changes in the HO-1 gene may not be directly linked to the pathogenesis of these diseases, heme catabolism plays a pivotal role in neuronal survival and cytoprotection against aging and environmental insults. Further study on the neurodegenerative diseases exploring this ubiquitous but unique metabolic pathway will shed light on other fields and, in turn, mechanisms for modulation of many disease processes.

REFERENCES

1. M.D. Maines, The heme oxygenase system and its functions in the brain, *Cell. Mol. Biol.* **46**(3), 573–585 (2000).

2. S. Shibahara, R.M. Muller, and H. Taguchi, Transcriptional control of rat heme oxygenase by heat shock. *J. Biol. Chem.* **262**(27), 12889–12892 (1987).

3. N.G. Abraham, J.H. Lin, M.L. Schwartzman, R.D. Levere, and S. Shibahara, The physiological significance of heme oxygenase. *Int. J. Biochem.* **20**(6), 543–558 (1988).

4. G.M. Trakshel, R.K. Kutty, and M.D. Maines, Resolution of the rat brain heme oxygenase activity: absence of a detectable amount of the inducible form (HO-1). *Arch. Biochem. Biophys.* **260**(2), 732–739 (1988).

5. W.K. McCoubrey, Jr., T.J. Huang, and M.D. Maines, Isolation and characterization of a cDNA from the rat brain that encodes hemoprotein heme oxygenase-3. *Eur. J. Biochem.* **247**(2), 725–732 (1997).

6. J.F. Ewing and M.D. Maines, Glutathione depletion induces heme oxygenase-1 (HSP32) mRNA and protein in rat brain. *J. Neurochem.* **60**(4), 1512–1519 (1993).

7. A. Takeda, H. Onodera, A. Sugimoto, Y. Itoyama, K. Kogure, and S. Shibahara, Increased expression of heme oxygenase mRNA in rat brain following transient forebrain ischemia. *Brain Res.* **666**(1), 120–124 (1994).

8. A. Takeda, T. Kimpara, H. Onodera, Y. Itoyama, S. Shibahara, and K. Kogure, Regional difference in induction of heme oxygenase-1 protein following rat transient forebrain ischemia. *Neurosci. Lett.* **205**(3), 169–172 (1996).

9. R. Stocker, Y. Yamamoto, A.F. McDonagh, A.N. Glazer, and B.N. Ames, Bilirubin is an antioxidant of possible physiological importance. *Science* **235**(4792), 1043–1046 (1987).

10. S. Okinaga, K. Takahashi, K. Takeda, M. Yoshizawa, H. Fujita, H. Sasaki, and S. Shibahara. Regulation of human heme oxygenase-1 gene expression under thermal stress. *Blood* **87**(12), 5074–5084 (1996).

11. L.M. Sayre, D.A. Zelasko, P.L. R. Harris, G. Perry, R.G. Salomon, and M.A. Smith, 4-Hydroxynonenal-derived advanced lipid peroxidation end products are increased in Alzheimer's disease. *J. Neurochem.* **68**(5), 2092–2097 (1997).

12. M.A. Smith, P.L.R. Harris, L.M. Sayre, J.S. Beckman, and G. Perry, Widespread peroxynitrite-mediated damage in Alzheimer's disease. *J. Neurosci.* **17**(8), 2653–2657 (1997).

13. M.A. Smith, G. Perry, P.L. Richey, L.M. Sayre, V.E. Anderson, M.F. Beal, and N. Kowall, Oxidative damage in Alzheimer's. *Nature* **382**(6587), 120–121 (1996).

14. A. Nunomura, G. Perry, K. Hirai, G. Aliev, A. Takeda, S. Chiba, and M.A. Smith, Neuronal RNA oxidation in Alzheimer's disease and Down's syndrome. *Ann. NY Acad. Sci.* **893**, 362–364 (1999).

15. M.A. Smith, R.K. Kutty, P.L. Richey, S.-D. Yan, D. Stern, G.J. Chader, B. Wiggert, R.B. Petersen, and G. Perry, Heme oxygenase-1 is associated with the neurofibrillary pathology of Alzheimer's disease. *Am. J. Pathol.* **145**(1), 42–47 (1994).

16. H.M. Schipper, S. Cisse, and E.G. Stopa, Expression of heme oxygenase-1 in the senescent and Alzheimer-diseased brain. *Ann. Neurol.* **37**(6), 758–768 (1995).

17. H.M. Schipper, A. Liberman, and E.G. Stopa, Neural heme oxygenase-1 expression in idiopathic Parkinson's disease. *Exp. Neurol.* **150**(1), 60–68.

18. R. Castellani, M.A. Smith, P.L. Richey, R. Kalaria, P. Gambetti, and G. Perry, Evidence for oxidative stress in Pick disease and corticobasal degeneration. *Brain Res.* **696**(1–2), 268–271 (1995).

19. T. Kimpara, A. Takeda, T. Yamaguchi, H. Arai, N. Okita, S. Takase, H. Sasaki, and Y. Itoyama, Increased bilirubins and their derivatives in cerebrospinal fluid in Alzheimer's disease. Neurobiol. Aging **21**(4), 551–554 (2000).

20. A. Takeda, M.A. Smith, J. Avila, A. Nunomura, S.L. Siedlak, X. Zhu, G. Perry, and L.M. Sayre, In Alzheimer's disease, heme oxygenase is coincident with Alz50, an epitope of tau induced by 4-hydroxy-2-nonenal modification. *J. Neurochem.* **75**(3), 1234–1241 (2000).

21. B.L. Wolozin, A. Pruchnicki, D.W. Dickson, and P. Davies, A neuronal antigen in the brains of Alzheimer patients. *Science* **232**(4750), 648–650 (1986).

22. G.A. Jicha, R. Bowser, I.G. Kazam, and P. Davies, Alz-50 and MC-1, a new monoclonal antibody raised to paired helical filaments, recognize conformational epitopes on recombinant tau. *J. Neurosci. Res.* **48**(2), 128–132 (1997).

23. G. Carmel, E.M. Mager, L.I. Binder, and J. Kuret, The structural basis of monoclonal antibody Alz50's selectivity for Alzheimer's disease pathology. *J. Biol. Chem.* **271**(51), 32789–32795 (1996).

24. W.C. Benzing, M.D. Ikonomovic, D.R. Brady, E.J. Mufson, and D.M. Armstrong, Evidence that transmitter-containing dystrophic neurites precede paired helical filament and Alz-50 formation within senile plaques in the amygdala of nondemented elderly and patients with Alzheimer's disease. *J. Comp. Neurol.* **334**(2), 176–191 (1993).

25. M.P. Mattson, N. Robinson, and Q. Guo, Estrogens stabilize mitochondrial function and protect neural cells against the pro-apoptotic action of mutant presenilin-1. *Neuroreport* **8**(17), 3817–3821 (1997).

26. K. Hirai, G. Aliev, A. Nunomura, H. Fujioka, R.L. Russell, C.S. Atwood, A.B. Johnson, Y. Kress, H.V. Vinters, M. Tabaton, S. Shimohama, A.D. Cash, S.L. Siedlak, P.L.R. Harris, P.K. Jones, R.B. Petersen, G. Perry, and M.A. Smith, Mitochondrial abnormalities in Alzheimer's disease. *J. Neurosci.* **21**(9), 3017–3023 (2001).

27. S. Shimizu, M. Narita, and Y. Tsujimoto, Bcl-2 family proteins regulate the release of apoptogenic cytochrome c by the mitochondrial channel VDAC. *Nature* **399**(6735), 483–487 (1999).

28. A. Takeda, G. Perry, N.G. Abraham, B.E. Dwyer, R.K. Kutty, J.T. Laitinen, R.B. Petersen, and M.A. Smith, Overexpression of heme oxygenase in neuronal cells, the possible interaction with tau. *J. Biol. Chem.* **275**(8), 5395–5399 (2000).

29. N.G. Abraham, Y. Lavrovsky, M.L. Schwartzman, R.A. Stoltz, R.D. Levere, M.E. Gerritsen, S. Shibahara, and A. Kappas, Transfection of the human heme oxygenase gene into rabbit coronary microvessel endothelial cells: protective effect against heme and hemoglobin toxicity. *Proc. Natl. Acad. Sci. USA* **92**(15), 6798–6802 (1995).

30. A. Harada, K. Oguchi, S. Okabe, J. Kuno, S. Terada, T. Ohshima, R. Sato-Yoshitake, Y. Takei, T. Noda, and N. Hirokawa, Altered microtubule organization in small-calibre axons of mice lacking tau protein. *Nature* **369**(6480), 488–491 (1994).

31. K.D. Poss and S. Tonegawa, Reduced stress defense in heme oxygenase 1-deficient cells. *Proc. Natl. Acad. Sci. USA* **94**(20), 10925–10930 (1997).

32. K.D. Poss and S. Tonegawa, Heme oxygenase 1 is required for mammalian iron reutilization. *Proc. Natl. Acad. Sci. USA* **94**(20), 10919–10924 (1997).

33. A. Yachie, Y. Niida, T. Wada, N. Igarashi, H. Kaneda, T. Toma, K. Ohta, Y. Kasahara, and S. Koizumi, Oxidative stress causes enhanced endothelial cell injury in human heme oxygenase-1 deficiency. *J. Clin. Invest.* **103**(1), 129–135 (1999).

34. H.M. Schipper, Heme oxygenase-1: role in brain aging and neurodegeneration. *Exp. Gerontol.* **35**(6–7), 821–830 (2000).

35. M.A. Smith, P.L.R. Harris, L.M. Sayre, and G. Perry, Iron accumulation in Alzheimer disease is a source of redox-generated free radicals. *Proc. Natl. Acad. Sci. USA* **94**, 9866–9868 (1997).

36. H.J. Bidmon, B. Emde, E. Oermann, R. Kubitz, O.W. Witte, and K. Zilles, Heme oxygenase-1 (HSP-32) and heme oxygenase-2 induction in neurons and glial cells of cerebral regions and its relation to iron accumulation after focal cortical photothrombosis. *Exp. Neurol.* **168**(1), 1–22 (2001).

37. N. Panahian, M. Yoshiura, and M.D. Maines, Overexpression of heme oxygenase-1 is neuroprotective in a model of permanent middle cerebral artery occlusion in transgenic mice. *J. Neurochem.* **72**(3), 1187–1203 (1999).

38. K. Chen, K. Gunter, and M.D. Maines, Neurons overexpressing heme oxygenase-1 resist oxidative stress-mediated cell death. *J. Neurochem.* **75**(1), 304–313 (2000).

39. A. Verma, D.J. Hirsch, C.E. Glatt, G.V. Ronnett, S.H., et al., Carbon monoxide: a putative neural messenger. *Science* **259**(5093), 381–384 (1993).

40. L.E. Otterbein, F.H. Bach, J. Alam, M. Soares, H. Tao Lu, M. Wysk, R.J. Davis, R.A. Flavell, and A.M. Choi, Carbon monoxide has anti-inflammatory effects involving the mitogen-activated protein kinase pathway. *Nat. Med.* **6**(4), (2000).

41. H.J. Duckers, M. Boehm, A.L. True, S.F. Yet, H. San, J.L. Park, R. Clinton Webb, M.E. Lee, G.J. Nabel, and E.G. Nabel, Heme oxygenase-1 protects against vascular constriction and proliferation. *Nat. Med.* **7**(6), 693–698 (2001).

42. K. Mehindate, D.J. Sahlas, D. Frankel, Y. Mawal, A. Liberman, J. Corcos, S. Dion, and H.M. Schipper, Proinflammatory cytokines promote glial heme oxygenase-1 expression and mitochondrial iron deposition: implications for multiple sclerosis. *J. Neurochem.* **77**(5), 1386–1395 (2001).

43. A.K. Raina, X. Zhu, C.A. Rottkamp, M. Monteiro, A. Takeda, and M.A. Smith, Cyclin' toward dementia: cell cycle abnormalities and abortive oncogenesis in Alzheimer disease. *J. Neurosci. Res.* **61**, 128–133 (2000).
44. T. Kimpara, A. Takeda, K. Watanabe, Y. Itoyama, S. Ikawa, M. Watanabe, H. Arai, H. Sasaki, S. Higuchi, N. Okita, S. Takase, H. Saito, K. Takahashi, and S. Shibahara, Microsatellite polymorphism in the human heme oxygenase-1 gene promoter and its application in association studies with Alzheimer and Parkinson disease. *Human Genet.* **100**(1), 145–147 (1997).
45. M. Takahashi, S. Dore, C.D. Ferris, T. Tomita, A. Sawa, H. Wolosker, D.R. Borchelt, T. Iwatsubo, S.H. Kim, G. Thinakaran, S.S. Sisodia, and S.H. Snyder, Amyloid precursor proteins inhibit heme oxygenase activity and augment neurotoxicity in Alzheimer's disease. *Neuron* **28**(2), 461–473 (2000).

12

HEME OXYGENASE-1 AND ALZHEIMER DISEASE

Yogesh Mawal, Daniel Berlin, Steven Kravitz,
and Hyman M. Schipper

Bloomfield Centre for Research in Aging
Lady Davis Institute for Medical Research
Sir Mortimer B. Davis Jewish General Hospital
McGill University
Montreal, Que., Canada H3T 1E2.3755 Cote St. Catherine Road
Montreal, Quebec, Canada H3T 1E2

1. ALZHEIMER DISEASE

Alzheimer disease (AD) is a dementing illness characterized by progressive neuronal degeneration, gliosis, and the accumulation of intracellular inclusions (neurofibrillary tangles) and extracellular deposits of amyloid (senile plaques) in discrete regions of the basal forebrain, hippocampus, and association cortices (Selkoe, 1991). Sporadic AD is more common than all other aging-related neurodegenerative disorders combined, affecting approximately 5–10% of North Americans aged 65 and as many as 30–50% of those who survive to the end of their ninth decade (Katzman, 1993).

Although the precise mechanisms responsible for neuronal demise in AD remain largely unknown, there exists ample evidence implicating oxidative stress (free radical injury) and mitochondrial insufficiency (ATP depletion) in the pathogenesis of this condition (Reichmann and Riederer, 1994; Beal, 1995; Mattson, 1997). These pathological features may, in turn, be related to the excessive sequestration of redox-active iron, an important generator of damaging reactive oxygen species (ROS), that has been documented in the basal forebrain and association cortices of AD victims (Schipper, 1998a; Sayre et al., 2001). In both the AD and Parkinson brain, regional concentrations of transferrin binding sites remain unchanged or vary inversely with the elevated iron stores. These observations suggest that the transferrin pathway of iron mobilization, important for normal iron delivery to most peripheral tissues, contributes little to the pathological deposition of brain iron in these aging-related neurodegenerative conditions (Schipper, 1999). This realization has led several laboratories to consider alternate mechanisms for aberrant iron compartmentalization in the aging and degenerating CNS. The potential roles of lactoferrin, melanotransferrin (p97) and their receptors in brain iron transport are discussed elsewhere (Jefferies et al., 1996; Schipper, 1998a). In this chapter, we review evidence implicating heme oxygenase-1 (HO-1) in the pathophysiology of AD, with particular emphasis on the role this enzyme may play in pathological iron deposition and free radical-related neural injury.

2. HEME OXYGENASE-1

HO-1 is a 32-kDa member of the stress protein superfamily that mediates the oxidative catabolism of heme to biliverdin in brain and other tissues (Tenhunen et al., 1969; Ewing and Maines, 1991). The HO-1 gene contains heat shock elements and AP1, AP2 and NFκB binding sites in its promoter region rendering it susceptible to rapid upregulation by oxidative stress, metal ions, amino acid analogues, sulfhydryl agents, pro-inflammatory cytokines and hyperthermia (Applegate et al., 1991). In response to oxidative stress, induction of HO-1 may confer cytoprotection by facilitating the degradation of pro-oxidant metalloporphyrins, such as heme, to bile pigments (biliverdin, bilirubin) with free radical scavenging properties (Stocker et al., 1987; Doré et al., 1999). Conversely, free iron and carbon monoxide (CO) released in the course of heme breakdown may, under certain circumstances, amplify intracellular oxidative stress and promote injury to mitochondrial membranes (Zhang and Piantadosi, 1992). Thus, by altering local cellular concentrations of heme, CO, iron and bilirubin, deranged expression of HO-1 in AD-affected brain tissues (see below) could, theoretically, impact disease progression and provide a novel target for therapeutic intervention.

3. HO-1 EXPRESSION IN AD BRAIN

Using Western blot and immunocytochemical techniques, we demonstrated that HO-1 protein is markedly over-expressed in brain specimens derived from subjects with neuropathologically-proven AD relative to age-matched controls without history of neurological illness (Schipper et al., 1995). In the control specimens, the vast majority of neurons exhibited faint or no HO-1 immunoreactivity. Similarly, HO-1 staining appeared as a very weak 32 kDa band or was absent in Western blots of protein extracts derived from control hippocampus, temporal cortex and subcortical white matter. Thus, in the course of normal brain aging, constitutive expression of HO-1 in neurons is low, an observation consistent with earlier findings in adult rats (Ewing et al., 1992). In contrast, many neurons throughout the AD hippocampus (pyramidal and dentate gyrus granule cells) and temporal cortex (layers II-V) exhibited strong, cytoplasmic HO-1 immunoreactivity. In the AD specimens, HO-1 expression was most conspicuous, but not limited to, neurons containing neurofibrillary tangles. In sections double-labeled for HO-1 and tau-2, we noted consistent and robust co-localization of HO-1 to the abundant neurofibrillary tangles observed in the AD specimens as well as to the rare tangles encountered in the control material. HO-1-positive tangles and increased HO-1 mRNA levels were also reported in AD brain by others (Smith et al., 1994; Premkumar et al., 1995). As in the case of neurofibrillary tangles, we observed that senile plaques were replete with immunoreactive HO-1 in AD (and control) tissues stained for both HO-1 and β-amyloid. Furthermore, preparations co-labeled with anti-HO-1 and anti-synaptophysin antisera revealed the presence of HO-1-positive neurites within and surrounding senile plaques (Schipper et al., 1995). Because HO-1 is a very sensitive marker of oxidative stress (Applegate et al., 1991), the latter finding supports the contention that β-amyloid deposits within plaques may subject nearby neuropil constituents to oxidative stress.

Relatively few HO-1-positive astrocytes were encountered in the hippocampus and temporal cortex derived from control specimens dual-labeled for HO-1 and the astrocyte marker, GFAP. For example, only 6.8% of GFAP-positive astrocytes co-expressed immunodetectable HO-1 in control hippocampus whereas 86% of GFAP-positive astrocytes in the AD hippocampus co-expressed HO-1. In the AD substantia nigra, an area relatively unaffected by the disease, glial HO-1 immunoreactivity was minimal and not significantly different from control levels (Schipper et al., 1995).

Augmented expression of HO-1 in AD brain is consistent with the induction of other stress proteins in this condition, including heat shock protein (hsp)27, hsp72, αB-crystallin and ubiquitin (Perez et al., 1991; Wang et al., 1991; Lowe et al., 1992; Renkawek et al., 1993). The genes coding for HO-1 and the latter proteins contain various consensus sequences in their promoter regions which bind, and mediate transcriptional regulation by, redox-sensitive transcription factors (Applegate et al., 1991). Local up-regulation of these proteins is conceivably secondary to the enhanced level of oxidative stress that occurs in AD-affected brain tissues (discussed above). In the following section, we elaborate on downstream effects of neural HO-1 induction which may be germane to the pathogenesis of AD.

4. HO-1 INDUCTION IN AD BRAIN: CYTOPROTECTION OR NEUROENDANGERMENT?

In response to oxidative challenge, up-regulation of HO-1 may protect cells by accelerating the catabolism of pro-oxidant metalloporphyrins, such as heme, to bile pigments (biliverdin, bilirubin) with potent free radical scavenging capabilities (Stocker et al., 1987; Nakagami et al., 1993; Llesuy and Tomaro, 1994; Doré et al., 1999; Barañano et al., 2001). In some situations, however, free iron and CO released during heme degradation have been documented to exacerbate intracellular oxidative stress and predispose the mitochondrial compartment to free radical damage (Zhang and Piantadosi, 1992; Frankel et al., 2000). This Janus-faced behaviour has fostered lively debate as to whether HO-1 induction in the diseased nervous system subserves a cytoprotective function or promotes further neuroendangerment. These conflicting views may not be mutually exclusive; in a given neuropathological condition, the extent and duration of HO-1 induction and the status of the local redox microenvironment may determine whether the antioxidant benefits of a diminished heme: bilirubin ratio or the pro-oxidant potential resulting from intracellular liberation of iron/CO prevail (Galbraith, 1999; Suttner and Dennery, 1999).

(i) Neuroprotective Role of HO-1

Neuroprotective effects of HO-1 have been demonstrated *in vitro* and in intact animals. Primary cultures of cerebellar granule cells derived from transgenic (tg) mice engineered to over-express HO-1 in neurons (Maines et al., 1998) have been shown to be relatively resistant to glutamate- and H_2O_2-mediated oxidative damage (Chen et al., 2000). Similarly, neuroblastoma cell lines transfected with HO-1 cDNA exhibited enhanced resistance to oxidative injury resulting from exposure to H_2O_2 (Le et al., 1999; Takeda et al., 2000) or β-amyloid$_{1-40}$ (Le et al., 1999). In a model of cerebral ischemia, HO-1 tg mice exhibited smaller infarct volumes, decreased tissue staining for lipid peroxidation end-products and enhanced expression of the anti-apoptotic factor, bcl-2 (Panahian et al., 1999). Induction of HO-1 has also been postulated to confer neuroprotection following traumatic brain injury (Beschorner et al., 2000; Fukuda et al., 1996). By promoting smooth muscle relaxation (Verma et al., 1993), heme-derived CO may protect against cerebral vasospasm, a cause of significant morbidity complicating subarachnoid hemorrhage (Matz et al., 1996; Suzuki et al., 1999; Tanaka et al., 2000). It has been argued that the capacity of astrocytes, but not neurons, to mount a robust HO-1 (and other heat shock protein) response may partly account for the relative preservation of the former in the face of oxidative challenge (Dwyer et al., 1995; Manganaro et al., 1995; Snyder et al., 1998).

Snyder and colleagues demonstrated protein-protein interactions between heme oxygenases and the amyloid precursor protein (APP) in a yeast 2-hybrid system and in the intact brain. They determined that cells transfected with wild type and, to a greater extent, mutant forms of APP manifest significant inhibition of HO activity and enhanced vulnerability to the toxic effects of hemin and hydrogen peroxide. The investigators concluded that decreased formation of bilirubin in these cells was likely responsible for their augmented sensitivity to oxidative insults (Takahashi et al., 2000). Smith and co-workers reported down-regulation of tau expression in a neuroblastoma cell line transfected with human HO-1 cDNA (Takeda et al., 2000). The authors proposed that analogous HO-1/tau interactions *in situ* may be beneficial in AD by limiting the substrate (tau) for pathological hyperphosphorylation (a precursor of tangle formation) in this disease. However, tau phosphorylation was not addressed in that study and we are not aware of literature attesting to suppression of tau protein levels in AD-affected brain tissues. Furthermore, down-regulation of tau could theoretically compromise the integrity of the cytoskeleton and thereby predispose to, rather than prevent, cytopathological changes.

(ii) Dystrophic Effects of HO-1 in Neural Tissues

Although low concentrations of heme-derived bilirubin may protect against oxidative stress (see above), excessive exposure to this bile pigment, as occurs in untreated neonatal hyperbilirubinemia, often results in irreversible neurological damage (kernicterus). The latter is preventible in these children by phototherapy (bilirubin degradation) or administration of competitive heme oxygenase inhibitors (metalloporphyrins; Qato et al., 1985). Metalloporphyrin suppression of heme oxygenase activity has been reported to ameliorate traumatic CA1 injury in rat hippocampal slices (Panizzon et al., 1996), reduce tissue necrosis and edema formation following focal cerebral ischemia in intact rats (Kadoya et al., 1995) and provide neuroprotection in an experimental model of intracerebral hemorrhage (Koeppen and Dickson, 1999).

Sustained up-regulation of HO-1 in AD brain may facilitate the development of some of the disease's neuropathological manifestations:

(A) Transfection of human HO-1 cDNA in cultured rat astroglia significantly augments deposition of non-transferrin ($^{55}FeCl_3$–derived) iron within the mitochondrial matrix (Schipper, 1999; Schipper et al., 1999). Antecedent up-regulation of HO-1 is necessary and sufficient for enhanced mitochondrial sequestration of non-transferrin-derived iron observed in astroglia exposed to various pro-oxidants (cysteamine, dopamine, hydrogen peroxide, menadione), pro-inflammatory cytokines (tumour necrosis factor-α, interleukin-1β) or β-Amyloid$_{40/42}$ peptides (Schipper, 1999; Schipper et al., 1999; Ham and Schipper, 2000; Mehindate et al., 2001). Thus, overproduction of HO-1 protein in AD-afflicted neural tissues, possibly in response to excessive amyloid, pro-oxidant and/or cytokine provocation, may contribute to the (transferrin-independent) iron overload and mitochondrial insufficiency observed in this disorder (Reichmann and Riederer, 1994; Beal, 1995). In 1991, we provided evidence that glial mitochondrial iron behaves as a pseudo-peroxidase capable of converting various pro-toxins into potentially neurotoxic semiquinone radicals (Schipper et al., 1991). More recently, using a co-culture paradigm, we demonstrated that sequestration of mitochondrial iron by the glial substratum greatly increases the susceptibility of neuron-like PC12 cells to oxidative injury (Frankel and Schipper, 1999). To the degree that similar glial-neuronal interactions occur *in situ*, it is conceivable that HO-1-related iron redistribution within the glial compartment may endanger nearby neuronal constituents in the brains of AD subjects.

(B) Transient transfection of human HO-1 cDNA into rat astroglia stimulates late, compensatory induction of the manganese superoxide dismutase gene. The latter can be attenuated by antioxidant treatment indicating that HO-1 over-expression, at least in astroglia, may exacerbate intracellular oxidative stress (Frankel et al., 2000). Free iron and CO, products of HO-mediated heme degradation, are likely candidates perpetuating oxidative stress in these cells (Zhang and Piantadosi, 1992) and, by analogy, in AD-affected brain tissues over-expressing HO-1 (Schipper et al., 1995).

(C) HO-1 is a constituent of corpora amylacea (CA), glycoproteinaceous inclusions that accumulate extracellularly and within subpial and periventricular astrocytes of the AD brain and, to a lesser extent, in normal aging neural tissues (Schipper et al., 1995; Cavanagh, 1999). CA can be induced experimentally in rat astroglial cultures (Cissé and Schipper, 1995) and in rat subcortical astrocytes *in situ* (Schipper, 1998b) by long-term exposure to cysteamine, a potent inducer of glial HO-1. We recently determined that cysteamine-induced CA also contain immunoreactive HO-1 and that administration of the HO-1 transcriptional suppressor, dexamethasone attenuates the accumulation of CA in cysteamine-treated glial cultures (Sahlas et al., 2000). Taken together, these data raise the possibility that the accelerated biogenesis of CA in AD astroglia may be contingent upon antecedent overexpression of HO-1 in these cells (Schipper et al., 1995).

(D) Finally, heme-derived CO may facilitate the development of cognitive (Liebson and Albert, 1994), olfactory (Doty, 1991) and neuroendocrine (Sapolsky, 1992) disturbances characteristic of AD by increasing cyclic guanosine monophosphate concentrations in olfactory neurons (Verma et al., 1993), modulating the secretion of corticotropin-releasing hormone (Parks et al., 1994), altering central vasoregulatory mechanisms (Marks et al., 1991) and impacting hippocampal long-term potentiation (Stevens and Wang, 1993; Zhuo et al., 1993).

5. HO-1 EXPRESSION IN AD BLOOD AND CSF

Altered expression of HO-1 may not be restricted to the brain parenchyma in AD subjects. HO-1 protein levels, measured by ELISA, were found to be significantly lower in the plasma of patients with probable early sporadic AD and in the CSF of neuropathologically-definite AD relative to normal elderly control (NEC) values (Schipper et al., 2000). Moreover, initial studies in our laboratory suggested that blood mononuclear cell (MNC) HO-1 mRNA levels, measured by Northern blotting and laser densitometry, were suppressed in early sporadic AD in comparison with NEC and patients with a host of other neurological and medical disorders. Subjects with Mild Cognitive Impairment (MCI), many of whom are in pre-clinical stages of AD, exhibited MNC HO-1 mRNA levels intermediate between NEC and AD patients. On the basis of these findings, we concluded that the decreased HO-1 protein levels in AD plasma/CSF may reflect diminished systemic production of the enzyme rather than accelerated degradation of HO-1 in the circulation. The mechanism responsible for HO-1 suppression in AD peripheral tissues remains unclear. Kimpara and co-workers (1997) could not establish over-representation of distinct HO-1 gene polymorphisms in patients with sporadic AD (relative to NEC and Parkinson subjects) arguing that primary genetic determinants of HO-1 are probably not responsible for reduced levels of this enzyme in AD blood and CSF.

In our initial study (Schipper et al., 2000), MNC pellets from NEC, AD and MCI subjects were stored in cell culture freezing medium-DMSO at −85°C for an average of 15, 12 and 18 months, respectively, prior to RNA extraction. A subsequent analysis using a sensitive and quantitative RT-PCR method (Abraham, 1998) has revealed that absolute HO-1 mRNA copy numbers in MNC derived from AD and MCI subjects may be normal *in vivo* but progressively decline (more rapidly in AD than MCI specimens) as a function of storage time under the above *ex vivo* conditions (Figure; Schipper et al., in press). Similar HO-1 mRNA "instability" was not observed in MNC samples derived from NEC (Fig. 1) and patients with other neurological and medical disorders (Schipper et al., 2000), suggesting that this phenomenon may be AD-specific. Furthermore, in the AD (and other) specimens, mRNA levels for the housekeeping gene, GAPDH remained stable under identical *ex vivo* conditions (Figure and Schipper et al., 2000), indicating that mRNA fragility may

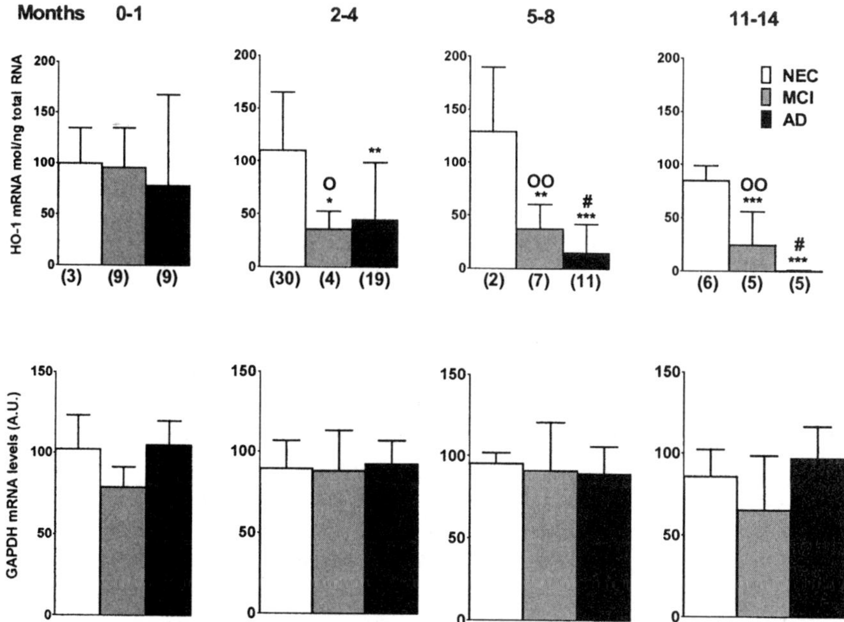

Figure 1. HO-1 and GAPDH mRNA levels (mean ± SD) in blood mononuclear cells derived from NEC, MCI and AD subjects as a function of storage time (months) at −85 °C. ( ) = number of subjects per group, *p < 0.05 vs. NEC, **p < 0.01 vs. NEC, ***p < 0.001 vs. NEC, O p < 0.05 vs. MCI (0–1), OO p < 0.01 vs. MCI (0–1), # p < 0.05 vs. AD (0–1). [From Neurology, in press, with permission.]

be a property unique to certain, but not all, mRNA species expressed in AD MNC.

Redox homeostasis is perturbed in various peripheral AD tissues, and oxidative DNA and RNA damage in AD lymphocytes and other cell types has been documented (Mecocci et al., 1998; Percy et al., 2000; Sayre et al., 2001). We are currently investigating the molecular basis for HO-1 mRNA instability in AD MNC and, specifically, whether oxidative modification of HO-1 mRNA in AD is responsible for a) the accelerated degradation of this nucleic acid *ex vivo* and b) the diminished plasma and CSF HO-1 protein concentrations (impaired translation?) observed in this condition. Because circulating human MNC (especially monocytes) already exhibit a significant degree of HO-1 expression under normal conditions (Schipper, unpublished results), AD-related oxidative stress may promote HO-1 mRNA damage with minimal or no further transcriptional up-regulation of the gene (ceiling effect). Net HO-1 protein levels in such tissues would then remain normal or low depending on whether significant RNA degradation or translational impairment occurs. In contrast, HO-1 protein accumulates in AD-affected brain tissues (Schipper et al., 1995) because robust induction of the HO-1 gene from very low basal (normal) levels of expression in neurons and glia may overide any potential losses of this stress protein accruing from oxidative nucleic acid damage. The reader will appreciate the conjectural nature of this formulation and the need for further experimentation.

6. CONCLUSIONS

Up-regulation of neuronal and glial HO-1 expression in AD-affected brain regions, possibly provoked by β-amyloid peptide or pro-inflammatory cytokines, provides further evidence that these tissues are subjected to increased levels of oxidative stress relative to that which occurs in the course of normal brain aging. There exists considerable controversy, however, concerning the cytophysiological consequences of this HO-1 response in CNS tissues. One school posits that HO-1 induction benefits the ailing brain in so far as the intracellular conversion of pro-oxidant heme molecules to antioxidant bile pigments helps restore a more favourable redox microenvironment. An opposing view maintains that central over-expression of HO-1, with attendant liberation of intracellular free iron and CO, contributes to the pathological iron deposition and mitochondrial insufficiency documented in AD, PD and other aging-associated neurodegenerative conditions. To complicate matters further, plasma and CSF HO-1 protein concentrations appear to be suppressed in blood and CSF of patients with sporadic AD, and mononuclear cell HO-1 mRNA derived from AD and some MCI subjects may be chemically unstable and subject to accelerated denaturation. On the basis of the data reviewed in this chapter, we think it reasonable to surmise that manipulation of brain HO-1 levels or activity, by pharmacological or other means, could significantly impact disease progression in AD subjects. But whether we should be striving to augment or inhibit this enzyme in the brains of these patients remains an open question which merits further exploration.

ACKNOWLEDGMENTS

The skillful secretarial assistance of Mrs. Rhona Rosenzweig is greatly appreciated. The work described herein was supported by grants to HMS from the Canadian Institutes for Health Research, the Fonds de la Recherche en Santé du Québec and the Alzheimer's Association (US).

REFERENCES

Abraham N., 1998. Quantitation of Heme Oxygenase (HO-1) copies in human tissues by competitive RT/PCR. In: Armstrong D., ed. Methods in Molecular Biology, 108: Free Radical and Antioxidant Protocols. Humana Press, Totowa, NJ. pp. 199–209.

Applegate L.A., Luscher P., and Tyrrell R.M., 1991. Induction of heme oxygenase: a general response to oxidant stress in cultured mammalian cells. Cancer Res. 51:974–978.

Barañano D.E., Doré S., Ferris C.D., and Snyder S.H., 2001. Physiological roles for the heme oxygenase products carbon monoxide, bilirubin and iron: links to neuroprotection in stroke and Alzheimer's disease. Clin. Neurosci. Res. 1:46–52.

Beal M.F., 1995. Mitochondrial Dysfunction and Oxidative Damage in Neurodegenerative Diseases. Landes, Austin. pp. 1–128.

Beschorner R., Adjodah D., Schwab J.M., Mattern R., Schluesener H.J., and Meyermann R., 2000. Long-term expression of heme oxygenase-1 (HO-1, HSP-32) following focal cerebral infarctions and traumatic brain injury in humans. Acta Neuropath. 100:377–384.

Cavanaugh J.B., 1999. Corpora amylacea and the family of polyglucosan diseases. Brain Res. Rev. 29:265–295.

Chen K., Gunter K., and Maines M.D., 2000. Neurons overexpressing heme oxygenase-1 resist oxidative stress-mediated cell death. J. Neurochem. 75:304–313.

Cissé S. and Schipper H.M., 1995. Experimental induction of corpora amylacea-like inclusions in rat astroglia. Neuropath. Appl. Neurobiol. 21:423–431.

Doré S., Takahashi M., Ferris C.D., Hester L.D., Guastella D., and Snyder S.H., 1999. Bilirubin, formed by activation of heme oxygenase-2, protects neurons against oxidative stress injury. Proc. Natl Acad. Sci., USA. 96:2445–2450.

Doty R.L., 1991. Olfactory dysfunction in neurodegenerative disorders. In: Getchell T.V., Doty R.L., Bartoshuk L.M., Snow Jr. J.B. (Eds.). Smell and Taste in Health and Disease, Raven, New York. pp. 735–751.

Dwyer B.E., Nishimura R.N., and Lu S.-Y., 1995. Differential expression of heme oxygenase-1 in cultured cortical neurons and astrocytes determined by the aid of a new heme oxygenase antibody. Response to oxidative stress. Mol. Brain Res. 30:37–47.

Ewing J.F. and Maines M.D., 1991. Rapid induction of heme oxygenase 1 mRNA and protein by hyperthermia in rat brain: heme oxygenase 2 is not a heat shock protein. Proc. Natl Acad. Sci., USA. 88:5364–5368.

Ewing J.F., Haber S.N., and Maines M.D., 1992. Normal and heat-induced patterns of expression of heme oxygenase-1 (HSP32) in rat brain: hyperthermia causes rapid induction of mRNA and protein. J. Neurochem. 58:1140–1149.

Frankel D. and Schipper H.M., 1999. Cysteamine pre-treatment of the astroglial substratum (mitochondrial iron sequestration) enhances PC12 cell vulnerability to oxidative injury. Exp. Neurol. 160:376–385.

Frankel D., Mehindate K., and Schipper H.M., 2000. Role of heme oxygenase-1 in the regulation of manganese superoxide dismutase gene expression in oxidatively-challenged astroglia. J. Cell. Physiol. 185:80–86.

Fukuda K., Richmon J.D., Sato M., Sharp F.R., Panter S.S., and Noble L.J., 1996. Induction of heme oxygenase-1 (HO-1) in glia after traumatic brain injury. Brain Res. 736:68–75.

Galbraith R., 1999. Heme oxygenase: who needs it?. PSEBM. 222:299–305.

Ham D. and Schipper H.M., 2000. Heme oxygenase-1 induction and mitochondrial iron sequestration in astroglia exposed to amyloid peptides. Cell. Mol. Biol. 46:587–596.

Jefferies W.A., Food M.R., Gabathuler R., Rothenberger S., Yamada T., Yasuhara O., and McGeer P.L., 1996. Reactive microglia specifically associated with amyloid plaques in Alzheimer's disease brain tissue express melanotransferrin. Brain Res. 712:122–126.

Kadoya C., Domino E.F., Yang G.Y., Stern J.D., and Betz A.L., 1995. Preischemic but not postischemic zinc protoporphyrin treatment reduces infarct size and edema accumulation after temporary focal cerebral ischemia in rats. Stroke. 26:1035–1038.

Katzman R.L., 1993. Education and the prevalence of dementia and Alzheimer's disease. Neurology. 43:13–20.

Kimpara T., Takeda A., Watanabe K., Itoyama Y., Ikawa S., Watanabe M., Arai H., Sasaki H., Higuchi S., Okita N., Takase S., Saito H., Takahashi K., and Shibahara S., 1997. Microsatellite polymorphism in the human heme oxygenase-1 gene promoter and its application in association studies with Alzheimer and Parkinson disease. Hum. Genet. 100:145–147.

Koeppen A.H. and Dickson A.C., 1999. Neuroprotection in intracerebral hemorrhage with tin-protoporphyrin. Ann. Neurol. 46:938.

Le W., Xie W., and Appel S., 1999. Protective role of heme oxygenase-1 in oxidative stress-induced neuronal injury. J. Neurosci. Res. 56:652–658.

Liebson E. and Albert M., 1994. Cognitive changes in dementia of the Alzheimer type. In: Calne D.B. (Ed.). Neurodegenerative Diseases, Saunders, Philadelphia. pp. 615–629.

Llesuy S.F. and Tomaro M.L., 1994. Heme oxygenase and oxidative stress. Evidence of involvement of bilirubin as physiological protector against oxidative damage. Biochem. Biophys. Acta. 1223:9–14.

Lowe J., Errington D.R., Lennox G., Pike I., Spendlove I., Landon M., and Mayer R.J., 1992. Ballooned neurons in several neurodegenerative diseases and stroke contain β-crystallin. Neuropathol. Appl. Neurobiol. 18:341–350.

Maines M., Polevoda B., Coban T., Johnson K., Stoliar S., Huang T.J., Panahian N., Cory-Slechta D.A., and McCoubrey W.K. Jr., 1998. Neuronal overexpression of heme oxygenase-1 correlates with an attenuated exploratory behavior and causes an increase in neuronal NADPH diaphorase staining. J. Neurochem. 70:2057–2069.

Manganaro F., Chopra V.S., Mydlarski M.B., Bernatchez G., and Schipper H.M., 1995. Redox perturbations in cysteamine-stressed astroglia: implications for inclusion formation and gliosis in the aging brain. Free Radic. Biol. Med. 19:823–835.

Marks G., Brien J., Nakatsu K., and McLaughlin B., 1991. Does carbon monoxide have a physiological function?. Trends Pharmacol. Sci. 12:185–188.

Mattson M.P., 1997. Cellular actions of β-amyloid precursor protein and its soluble and fibrillogenic derivatives. Physiol. Rev. 77:1081–1132.

Matz P., Turner C., Weinstein P.R., Massa S.M., Panter S.S., and Sharp F.R., 1996. Heme oxygenase-1 induction in glia throughout rat brain following experimental subarachnoid hemorrhage. Brain Res. 713:211–222.

Mecocci P., Polidori M.C., Ingegni T., Cherubini A., Chionne F., Cecchetti R., and Senin U., 1998. Oxidative damage to DNA in lymphocytes from AD patients. Neurology. 51:1014–1017.

Mehindate K., Sahlas D.J., Frankel D., Mawal Y., Liberman A., Corcos J., Dion S., and Schipper H.M., 2001. Proinflammatory cytokines promote glial heme oxygenase-1 expression and mitochondrial iron deposition: implications for multiple sclerosis. J. Neurochem. 77:1387–1396.

Nakagami T., Toyomura K., Kinoshita T., and Morisawa S., 1993. A beneficial role of bile pigments as an endogenous tissue protector: anti-complement effects of biliverdin and conjugated bilirubin. Biochem. Biophys. Acta. 1158:189–193.

Panahian N., Yoshiura M., and Maines M.D., 1999. Overexpression of heme oxygenase-1 is neuroprotective in a model of permanent middle cerebral artery occlusion in transgenic mice. J. Neurochem. 72:1187–1203.

Panizzon K.L., Dwyer B.E., Nishimura R.N., and Wallis R.A., 1996. Neuroprotection against CA1 injury with metalloporphyrins. NeuroReport. 7:662–666.

Parks D., Kasckow J., and Vale W., 1994. Carbon monoxide modulates secretion of corticotropin-releasing factor from rat hypothalamic cell cultures. Brain Res. 646:315–318.

Percy M., Andrews D.F., and Potter H., 2000. Peripheral markers in Alzheimer's disease. In: Scinto L.F.M., Daffner K.R. (Eds.). Early Diagnosis of Alzheimer's Disease. Humana Press, Totowa, NJ. pp. 191–268.

Perez N., Sugar J., Charya S., Johnson G., Merril C., Bierer L., Perl D., Haroutunian V., and Wallace W., 1991. Increased synthesis and accumulation of heat shock 70 proteins in Alzheimer's disease. Mol. Brain Res. 11:249–254.

Premkumar D.R., Smith M.A., Richey P.L., Petersen R.B., Castellani R., Kutty R.K., Wiggert B., Perry G., and Kalaria R.N., 1995. Induction of heme oxygenase-1 mRNA and protein in neocortex and cerebral vessels in Alzheimer's disease. J. Neurochem. 65:1399–402.

Qato M.K. and Maines M.D., 1985. Prevention of neonatal hyperbilirubinaemia in non-human primates by Zn-protoporphyrin. Biochem. J. 226:51–57.

Reichmann H. and Riederer P., 1994. Mitochondrial disturbances in neurodegeneration. In: Calne D.B. (Ed.). Neurodegenerative Diseases, Saunders, Philadelphia. pp. 195–204.

Renkawek K., Bosman G.C.G.M., and Gaestel M., 1993. Increased expression of heat-shock protein 27 kDa in Alzheimer disease: a preliminary study. NeuroReport. 5:14–16.

Sahlas D.J., Liberman A., and Schipper H.M., 2000. Role of heme oxygenase-1 in the formation of corpora amylacea. Can. J. Neurol. Sci. 27(Suppl 2):S65–S66.

Sapolsky R.M., 1992. Is this relevant to the human? In: Stress, the Aging Brain, and the Mechanisms of Neuron Death. MIT Press, Cambridge, MA. pp. 305–339.

Sayre L.M., Smith M.A., and Perry G., 2001. Chemistry and biochemistry of oxidative stress in neurodegenerative disease. Curr. Med. Chem. 8:721–738.

Schipper H.M., Kotake Y., and Janzen E.G., 1991. Catechol oxidation by peroxidase-positive astrocytes in primary culture: an electron spin resonance study. J. Neurosci. 11:2170–2176.

Schipper H.M., Cissé S., and Stopa E.G., 1995. Expression of heme oxygenase-1 in the senescent and Alzheimer-diseased brain. Ann. Neurol. 37:758–768.

Schipper H.M., 1998a. Glial iron sequestration and neurodegeneration. In: Schipper H.M. (Ed.). Astrocytes in Brain Aging and Neurodegeneration, Landes, Austin. pp. 235–251.

Schipper H.M., 1998b. Experimental induction of corpora amylacea in adult rat brain. Microsc. Res. Tech. 43:43–48.

Schipper H.M., 1999. Glial HO-1 expression, iron deposition and oxidative stress in neurodegenerative diseases. Neurotox. Res. 1:57–70.

Schipper H.M., Bernier L., Mehindate K., and Frankel D., 1999. Mitochondrial iron sequestration in dopamine-challenged astroglia: role of heme oxygenase-1 and the permeability transition pore. J. Neurochem. 72:1802–1811.

Schipper H.M., Chertkow H., Mehindate K., Frankel D., Melmed C., and Bergman H., 2000. Evaluation of heme oxygenase-1 as a systemic biological marker of sporadic Alzheimer disease. Neurology. 54:1297–1304.

Schipper H.M., Mawal, Y., Cherkow H., and Bergman H., (in press). HO-1 mRNA in AD and MCI. Neurology.

Selkoe D.J., 1991. The molecular pathology of Alzheimer's disease. Neuron. 6:487–498.

Smith M.A., Kutty R.K., Richey P.L., Yan S.-D., Stern D., Chader G.J., Wiggert B., Petersen R.B., and Perry G., 1994. Heme oxygenase-1 is associated with the neurofibrillary pathology of Alzheimer's disease. Am. J. Pathol. 145:42–47.

Snyder S.H., Jaffrey S.R., and Zakhary R., 1998. Nitric oxide and carbon monoxide; Parallel roles as neural messengers. Brain Res. Rev. 26:167–175.

Stevens C. and Wang Y., 1993. Reversal of long-term potentiation by inhibitors of heme oxygenase. Nature. 364:147–149.

Stocker R., Yamamoto Y., McDonagh A.F., Glazer A.N., and Ames B.N., 1987. Bilirubin is an antioxidant of possible physiological importance. Science. 235:1043–1046.

Suttner D.M. and Dennery P.A., 1999. Reversal of HO-1 related cytoprotection with increased expression is due to reactive iron. FASEB. J. 13:1800–1809.

Suzuki H., Kanamaru K., Tsunoda H., Inada H., Kuroki M., Sun H., Waga S., and Tanaka T., 1999. Heme oxygenase-1 gene induction as an intrinsic regulation against delayed cerebral vasospasm in rats. J. Clin. Invest. 104:59–66.

Tanaka T., Nishimura Y., Tsunoda H., and Naka M., 2000. Pharmacogenomics and therapeutic target validation in cerebral vasospasm. J. Cardiovasc. Pharma. 36(Suppl 2):S1–S4.

Takahashi M., Doré S., Ferris C.D., Tomita T., Sawa A., Wolosker H., Borchelt D.R., Iwatsubo T., Kim S.-H., Thinakaran G., Sisodia S.S., and Snyder S.H., 2000. Amyloid precursor proteins inhibit heme oxygenase activity and augment neurotoxicity in Alzheimer's disease. Neuron. 28:461–473.

Takeda A., Perry G., Abraham N.G., Dwyer B.E., Kutty R.K., Laitinen J.T., Petersen R.B., and Smith M.A., 2000. Overexpression of heme oxygenase in neuronal cells, the possible interaction with Tau. J. Biol. Chem. 275:5395–5399.

Tenhunen R., Marver H.S., and Schmid R., 1969. Microsomal heme oxygenase: characterization of the enzyme. J. Biol. Chem. 244:6388–6394.

Verma A., Hirsch D.J., Glatt C.E., Ronnett G.B., and Snyder S.H., 1993. Carbon monoxide: a putative neural messenger. Science. 259:381–384.

Wang G.P., Khatoon S., Iqbal K., and Grundke-Iqbal I., 1991. Brain ubiquitin is markedly elevated in Alzheimer disease. Brain Res. 566:146–151.

Zhang J. and Piantadosi C.A., 1992. Mitochondrial oxidative stress after carbon monoxide hypoxia in the rat brain. J. Clin. Invest. 90:1193–1199.

Zhuo M., Small S., Kandel E., and Hawkins R., 1993. Nitric oxide and carbon monoxide produce activity-dependent long-term synaptic enhancement in hippocampus. Science. 260:1946–1950.

13

EARLY MOLECULAR MECHANISMS FOR THE INDUCTION OF NEURONAL MEMBRANE ASYMMETRY AND GENOMIC DNA CLEAVAGE

Kenneth Maiese, Shi-Hua Lin, and Zhao Zhong Chong

Division of Cellular and Molecular Cerebral Ischemia
Departments of Neurology and Anatomy & Cell Biology
Center for Molecular Medicine and Genetics
Center for Molecular and Cellular Toxicology
Wayne State University School of Medicine
Detroit, Michigan, 48201

INTRODUCTION

Although multiple cellular pathways may determine the fate of central nervous system neurons, the opportunity to prevent the progression of neuronal injury once initiated exists within specific time constraints. To this end, it becomes crucial to elucidate the underlying cellular and molecular mechanisms that determine neuronal injury. Generation of the free radical nitric oxide (NO) is considered to be one of the "triggers" for the subsequent induction of neuronal injury. Enhanced expression of the enzyme responsible for NO production, nitric oxide synthase (NOS), has been associated with both chronic neuronal and vascular degeneration (de la Monte et al., 2000). Underlying cellular mechanisms responsible for the detrimental effects of NOS may be related to mitochondrial energy reserves (Almeida and Bolanos, 2001).

Interestingly, each isoenzyme of NOS, such as neuronal NOS (NOS-I), endothelial NOS (NOS-III), and inducible NOS (NOS-II), may differentially modulate cellular survival. During some injury paradigms, the generation of NOS-II in astrocytes (Endoh et al., 1994) or macrophages (Satake et al., 2000) has been suggested to be detrimental to neighboring neurons. As a corollary, it has been demonstrated that absence of NOS-II activity significantly increases neuronal and vascular endothelial cell viability during anoxia (Maiese et al., 1993a; Lin and Maiese, 2001b). Further analyses have illustrated that direct inhibition of NO production in cell culture systems during anoxia is cytoprotective (Maiese et al., 1993a; Maiese et al., 1993b; Grammas et al., 1998; Demerle-Pallardy et al., 2000; Lin and Maiese, 2001b).

Some experimental models have argued for the protective effects of NO in the central nervous system (Buras et al., 2000). Increased production of NO has been shown to decrease infarct size during either neuronal or vascular injury (Dalkara et al., 1994; Guo et al., 1999; Kanno et al., 2000). Although these results have been attributed to improved cerebral perfusion, other studies suggest that improved cerebral perfusion alone to the ischemic zone is insufficient to sustain neuronal survival (Maiese et al., 1992). Direct endothelial production of NO may be an important factor since this has been linked to the preservation of the anti-apoptotic protein Bcl-2 through the down regulation of cytosolic MAP kinase phosphatase-3 (Rossig et al., 2000). It is unclear why certain environmental conditions may predispose NO to function as a protectant rather than a toxin. These divergent observations may involve parameters such as the experimental model, external environmental conditions, duration of the insult, age of the neuronal or vascular system, and the resultant NO species that is generated (Marks et al., 1996; Chiueh, 1999).

TWO INDEPENDENT PATHWAYS FOR PROGRAMMED CELL DEATH

Clinical studies have suggested that programmed cell death (PCD) may lead to neuronal injury in a variety of disease entities such as Parkinson's disease (Hartmann et al., 2001), Alzheimer's disease (Shimohama, 2000), and human immunodeficiency syndrome (Gray et al., 2000). In addition, experimental paradigms have further supported the significance of PCD as a principal contributor to neuronal injury even during free radical exposure. For example, in animal models of epilepsy (Fujikawa et al., 2000), loss of cortical neurons has been attributed to subsequent apoptotic DNA fragmentation. In cell culture studies, rapid induction of PCD occurs with ultraviolet radiation (Kimura et al., 1998), free radical exposure (Maiese, 2001), and beta-amyloid deposition (Kajkowski et al., 2001).

Neuronal PCD proceeds through two distinct pathways that are functionally independent (Maiese and Vincent, 2000b; Maiese, 2001). One pathway involves the degradation of genomic DNA is considered to be a committed event that results in neuronal demise (Okamoto et al., 1993; Tominaga et al., 1993; Vincent and Maiese, 1999b; Vincent et al., 1999a; Vincent et al., 1999b; Love et al., 2000). In contrast, the second component of neuronal PCD consists of the externalization of membrane phosphatidylserine (PS) residues (Chang et al., 2000; Maiese and Vincent, 2000b). Exposure of membrane PS residues is believed to be "upstream" from genomic DNA degradation and serves to "tag" injured cells for phagocytosis (Vincent and Maiese, 1999a; Chang et al., 2000). An additional role of membrane PS externalization in vascular cell systems is the activation of coagulation cascades. The externalization of PS in platelets or endothelial cells can promote the formation of a procoagulant surface (Bombeli et al., 1997).

An important caveat to PCD is that the initial stage of apoptotic death, namely membrane PS residue exposure, may be reversible. Prior techniques used to assess PCD, such as terminal deoxyUTP nick end labelling (TUNEL) or transmission electron microscopy, relied upon fixed tissue (Vincent et al., 1997; Fiorucci et al., 1999). These procedures lacked the ability to assess dynamic or "real-time" changes in PCD in individual cells. Yet, more sophisticated techniques now offer the ability to monitor the induction and change in PCD in individual living cells over a period of time. Such work has supported the premise that neuronal PCD is in fact reversible in nature (Lin et al., 2000; Maiese and Vincent, 2000b; Lin and Maiese, 2001b). The methods employ the reversible labelling of annexin V to exposed membrane PS residues of cells undergoing PCD (Vincent and Maiese, 1999a; Maiese and Vincent, 2000b). By exploiting the dependence of annexin V on extra-cellular calcium to bind to exposed membrane PS residues, the techniques can reversibly label individual cells. During the induction of PCD, progressive externalization of membrane PS residues occurs that is independent of the loss of cellular membrane integrity.

Investigations that examine the efficacy of cytoprotectants during neuronal injury have further garnered support for the premise that neuronal PCD is reversible. For example, the application of trophic factors (Kiprianova et al., 1999), metabotropic glutamate receptor agonists (Maiese et al., 2000; Vincent and Maiese, 2000; Lin and Maiese, 2001b; Lin and Maiese, 2001a), benzothiazole compounds (Maiese and Vincent, 2000b; Maiese and Vincent, 2000a), Bcl-2 expression (Fabisiak et al., 1997), or nicotinamide (Ayoub et al., 1999; Lin et al., 2000; Mokudai et al., 2000; Lin et al., 2001b) has been shown to either prevent or reverse membrane and nuclear changes associated with PCD. It is conceivable that these cytoprotective agents maintain membrane PS asymmetry through the modulation of random "flip-flop" membrane phospholipids (Bratton et al., 1997). Alternatively, they may maintain cellular energy metabolism since exposure of membrane PS residues on the membrane surface is an active process facilitated by an ATP-dependent membrane translocase (Verhoven et al., 1999).

LOSS OF MITOCHONDRIAL MEMBRANE POTENTIAL

Mitochondria play a central role not only in their maintenance of cellular energy metabolism (Schapira, 1996), but also in the determination of whether a cell undergoes PCD (Newmeyer et al., 1994). Maintenance of intact mitochondrial membrane function has been demonstrated to prevent cellular injury during ischemic reperfusion paradigms (Amberger et al., 2001). Initiation of PCD in several cell types can require the depolarization of the mitochondrial membrane to result in the release of pro-apoptotic proteins such as cytochrome c (Liu et al., 1996; Sanchez-Alcazar et al., 2000). Recently, NO exposure has been shown to lead to mitochondrial membrane depolarization in both neurons and endothelial cells (Kang et al., 2001). The generation of NO results in the rapid depolarization of the mitochondrial membrane, the release of cytochrome c, and the subsequent induction of PCD (Lin et al., 2000; Kang et al., 2001; Lin and Maiese, 2001b; Maiese, 2001).

INDUCTION OF CYSTEINE PROTEASE ACTIVITY

Following mitochondrial membrane depolarization and cytochrome c release, activation of cysteine proteases is responsible, at least in part, for the generation of genomic DNA degradation and membrane PS exposure. The cysteine proteases (caspases) are mammalian homologues of the *C. elegans* cell death (CED) genes (Ellis and Horvitz, 1986). Each of the aspartate-specific cysteine proteases is synthesized as a proenzyme that is proteolytically cleaved to subunits that form catalytically active heterodimers during development or injury (Martin and Green, 1995).

Generation of NO in neurons and endothelial cells can elicit cysteine protease activation and directly stimulate caspase 1 and caspase 3-like activities (Brune et al., 1999; Maiese et al., 2000). Investigations with the cysteine proteases caspase 1 and caspase 3 have been implicated to lead to the induction of neuronal PCD (Du et al., 1997; Krohn et al., 1998; Maicse and Vincent, 1999). Caspase 1-like proteases may promote DNA degradation through the activation of other proteins, such as protein kinase C (Emoto et al., 1995) and caspase 3 (Enari et al., 1996). In addition, the caspase 3-like proteases have been directly linked to the development of DNA fragmentation (Enari et al., 1998). Caspase 3-like proteases become active and cleave a constitutive endonuclease inhibitor that can degrade genomic DNA (Enari et al., 1998). Caspase 3-like proteases also cleave poly(ADP-ribose) polymerase (PARP). PARP has been shown to be required for DNA repair and functions by attaching poly(ADP-ribose) at DNA strand breaks (Cristovao and Rueff, 1996). A significant decrease in intact PARP following exposure to NO has been recently observed in neurons (Lin et al., 2000; Maiese et al., 2000).

Induction of cysteine protease activity may also be responsible for the externalization of membrane PS residues. During cytokine mediated injury, the externalization of PS residues has been linked to the activity of caspase 1-like proteases through a mechanism that may involve the cleavage of membrane cytoskeletal proteins such as fodrin (Cryns et al., 1996; Kayalar et al., 1996). In addition, caspase 3-like proteases can cleave fodrin and focal adhesion kinase to induce membrane PS residue exposure during PCD (Levkau et al., 1998). Current studies that employ visualization of PS externalization in neurons and endothelial cells have demonstrated that PS externalization is primarily related to caspase 1-like activity (Lin et al., 2000; Maiese and Vincent, 2000b; Vincent and Maiese, 2000; Lin and Maiese, 2001b).

Another cysteine protease that plays a significant role during NO-induced PCD is caspase 8 (Kang et al., 2001; Lin and Maiese, 2001b). Caspase 8—like activity appears to have a function in mediating the exposure of membrane PS residues in neurons and in endothelial cells (Kang et al., 2001; Lin and Maiese, 2001b). Although increased caspase 8—like activity has been shown to result in genomic DNA degradation through the subsequent activation of caspase 3—like activity (Juo et al., 1998), the mechanism of membrane PS exposure by caspase 8 is less clear. Similar to its role in some non-vascular cellular systems (Takahashi et al., 1999), caspase 8 may invoke the "downstream" activation of caspase 1 which is a direct effector of membrane PS exposure.

THE ROLE OF NEURONAL ENDONUCLEASES

The cleavage of genomic DNA into fragments is considered to be a late event during PCD (Fehsel et al., 1995; Maiese, 1998; Ishikawa et al., 1999). Exclusive of the nervous system, a variety of enzymes responsible for DNA degradation have been differentiated based on their ionic sensitivities to zinc (Torriglia et al., 1997) and magnesium (Sun and Cohen, 1994). Interest also has focused on the calcium/magnesium—dependent endonucleases such as DNase I (Madaio et al., 1996), the acidic cation independent endonuclease (DNase II) (Torriglia et al., 1995), cyclophilins (Montague et al., 1997), and the 97 kDa magnesium—dependent endonuclease (Pandey et al., 1997).

Recent work has demonstrated that endonuclease activity also directly influences cell survival in the nervous system during free radical NO exposure (Vincent and Maiese, 1999b; Vincent et al., 1999b). Through the use of *in vitro* assays of endonuclease activity, three endonucleases have been characterized. Each has specific divalent cation requirements and are physiologically dependent upon the intracellular pH changes induced by NO exposure.

The three independent endonuclease activities present in neurons consist of a constitutive acidic cation independent endonuclease, a constitutive calcium/magnesium-dependent endonuclease, and an inducible magnesium dependent endonuclease (Vincent and Maiese, 1999b) (Fig. 1). The inducible magnesium-dependent endonuclease may be unique for the nervous system (Vincent and Maiese, 1999b). The physiologic characteristics of the magnesium dependent endonuclease, such as a pH range of 7.4–8.0, a dependence on magnesium, and a molecular weight of 95–108 kDa, are consistent with a recently described constitutive 97 kDa endonuclease in non-neuronal tissues, but the endonuclease in the nervous system is believed to be inducible rather than constitutive in nature.

It is interesting to note that free radical exposure alone has been suggested to be sufficient to result in the direct degradation of DNA (Wink et al., 1991). Yet, in some systems, DNA degradation following NO exposure requires a downstream mediator, such as endonuclease activation (Vincent and Maiese, 1999b; Vincent et al., 1999a). As a result, the regulation of either the constitutive or inducible pH dependent endonuclease activity may serve to prevent or reverse neuronal injury. Identifying the role of specific endonucleases during neuronal PCD should offer new avenues for the treatment of neurodegenerative diseases.

Endonuclease	Molecular Weight (kD)	pH	Divalent Cation	Constitutive	Inducible
1	30–40	4.0	None	Yes	No
2	113–130	7.2	Ca^{2+}/Mg^{2+}	Yes	No
3	95–108	8.0	Mg^{2+}	No	Yes

Figure 1. Neuronal endonuclease activities were characterized relative to their molecular weights, dependence on pH, divalent cations, and the nature of their constitutive or inducible expression.

EARLY CELL CYCLE INDUCTION COUPLED TO PROGRAMMED CELL DEATH

Recent investigations have suggested that modulation of a latent cell cycle in post-mitotic neurons may prevent both acute and chronic neuronal degeneration (Arendt et al., 2000; Park et al., 2000). Yet, it is unclear whether induction of a latent cell cycle in post-mitotic neurons is linked to the early induction of PCD and, as a result, can provide an attractive molecular target to prevent the progression of PCD. Prior work has demonstrated that exposure to NO leads to progressive neuronal injury over a twenty-four hour course (Vincent and Maiese, 1999a; Maiese and Vincent, 2000b; Maiese and Vincent, 2000a). This neuronal injury is accompanied by the rapid induction of the initial phase of PCD that involves the exposure of membrane PS residues. This work has now progressed to illustrate that membrane PS residue exposure occurs in concert with the attempted entrance into the cell cycle of post-mitotic neurons (Lin et al., 2001a). This untoward attempt to enter the cell cycle occurs in the same population of neurons that have begun the initial stages of PCD, suggesting that induction of a latent cell cycle in post-mitotic neurons correlates with the early stages of PCD. These studies focus on the initial stages of PCD induction which are potentially responsive to therapeutic modulation. As a result, it becomes critical to identify the subsequent molecular mechanisms that may bridge the cell cycle and membrane PS exposure in order to prevent or reverse PCD (Maiese and Gallant, 1997).

One of the primary modulators of the cell cycle is the retinoblastoma gene product (pRb), a 110 kDa nuclear phosphoprotein that coordinates cellular pathways of growth and differentiation. Hypophosphorylated pRb binds to and inactivates E2F transcription factors (Ludlow et al., 1990). Phosphorylation of pRb deregulates E2F activity and can promote cell proliferation or tumorigenesis (Sherr, 1994). Once neuronal cells have become committed and enter the process of differentiation, loss of pRb activity with subsequent cell cycle induction appears to result in degeneration and PCD in the nervous system. For example, if pRb is de-activated in neuronal ectodermal cells that are destined for differentiation, loss of functional pRb triggers apoptosis (Slack et al., 1995). In addition, the expression of mitotic cyclins and their associated kinases has been reported during periods of neurodegenerative disease (Raina et al., 2000). During injury paradigms, phosphorylated pRb expression appears to promote PCD while the enhancement of hypophosphorylated pRb expression prevents the induction of PCD (Maiese and Gallant, 1997).

SUMMARY

Understanding the molecular mechanisms of neuronal injury has become a requisite for the identification of potential therapeutic modalities that may govern neuronal survival. Recently, neuronal injury has been shown to consist of two independent pathways of PCD that involve the degradation of genomic DNA and the exposure of membrane PS residues. Yet, prior to the initiation of neuronal PCD is the activation of molecular cascades that are considered to be essential for the prevention and reversal of neuronal injury. These include free radical generation, loss of mitochondrial membrane potential, release of cytochrome c, cysteine protease activation, neuronal endonuclease activation, and early cell cycle induction in post-mitotic neurons. Future work that is directed to investigate the underlying cellular and molecular pathways that can contribute to neuronal injury will be crucial for the generation of clinical protocols that favorably influence both clinical plasticity and function.

ACKNOWLEDGMENTS

This work was supported by the following grants to K.M.: American Heart Association (National), Boehringer Ingelheim Training Grant Award, Janssen Neuroscience Award, Johnson and Johnson Focused Investigator Award, and the NIH NIEHS.

REFERENCES

Almeida, A. and Bolanos, J.P., 2001, A transient inhibition of mitochondrial ATP synthesis by nitric oxide synthase activation triggered apoptosis in primary cortical neurons. *J Neurochem* **77(2)**.676–690.

Amberger, A., Weiss, H., Haller, T., Kock, G., Hermann, M., Widschwendter, M., and Margreiter, R., 2001, A subpopulation of mitochondria prevents cytosolic calcium overload in endothelial cells after cold ischemia/reperfusion. *Transplantation* **71(12)**:1821–1827.

Arendt, T., Holzer, M., Stobe, A., Gartner, U., Luth, H.J., Bruckner, M.K., and Ueberham, U., 2000, Activated mitogenic signaling induces a process of dedifferentiation in Alzheimer's disease that eventually results in cell death. *Ann N Y Acad Sci* **920**:249–255.

Ayoub, I.A., Lee, E.J., Ogilvy, C.S., Beal, M.F., and Maynard, K.I., 1999, Nicotinamide reduces infarction up to two hours after the onset of permanent focal cerebral ischemia in Wistar rats. *Neurosci Lett* **259(1)**:21–24.

Bombeli, T., Karsan, A., Tait, J.F., and Harlan, J.M., 1997, Apoptotic vascular endothelial cells become procoagulant. *Blood* **89(7)**:2429–2442.

Bratton, D.L., Fadok, V.A., Richter, D.A., Kailey, J.M., Guthrie, L.A., and Henson, P.M., 1997, Appearance of phosphatidylserine on apoptotic cells requires calcium-mediated nonspecific flip-flop and is enhanced by loss of the aminophospholipid translocase. *J Biol Chem* **272(42)**:26159–26165.

Brune, B., von Knethen, A., and Sandau, K.B., 1999, Nitric oxide (NO): an effector of apoptosis. *Cell Death Differ* **6(10)**:969–975.

Buras, J.A., Stahl, G.L., Svoboda, K.K., and Reenstra, W.R., 2000, Hyperbaric oxygen downregulates ICAM-1 expression induced by hypoxia and hypoglycemia: the role of NOS. *Am J Physiol Cell Physiol* **278(2)**:C292–C302.

Chang, G.H., Barbaro, N.M., and Pieper, R.O., 2000, Phosphatidylserine-dependent phagocytosis of apoptotic glioma cells by normal human microglia, astrocytes, and glioma cells. *Neuro-oncol* **2(3)**:174–183.

Chiueh, C.C., 1999, Neuroprotective properties of nitric oxide. *Ann N Y Acad Sci* **890**:301–311.

Cristovao, L. and Rueff, J., 1996, Effect of a poly(ADP-ribose) polymerase inhibitor on DNA breakage and cytotoxicity induced by hydrogen peroxide and gamma radiation. *Teratog Carcinog Mutagen* **16(4)**:219–227.

Cryns, V.L., Bergeron, L., Zhu, H., Li, H., and Yuan, J., 1996, Specific cleavage of alpha fodrin during Fas- and tumor necrosis factor- induced apoptosis is mediated by an interleukin-1beta-converting enzyme/Ced-3 protease distinct from the poly(ADP-ribose) polymerase protease. *J Biol Chem* **271(49)**:31277–31282.

Dalkara, T., Yoshida, T., Irikura, K., and Moskowitz, M.A., 1994, Dual role of nitric oxide in focal cerebral ischemia. *Neuropharmacol* **33(11)**:1447–1452.

de la Monte, S.M., Lu, B.X., Sohn, Y.K., Etienne, D., Kraft, J., Ganju, N., and Wands, J.R., 2000, Aberrant expression of nitric oxide synthase III in Alzheimer's disease: relevance to cerebral vasculopathy and neurodegeneration. *Neurobiol Aging* **21(2)**:309–319.

Demerle-Pallardy, C., Gillard-Roubert, V., Marin, J.G., Auguet, M., and Chabrier, P.E., 2000, In vitro antioxidant neuroprotective activity of BN 80933, a dual inhibitor of neuronal nitric oxide synthase and lipid peroxidation. *J Neurochem* **74(5)**:2079–2086.

Du, Y., Bales, K.R., Dodel, R.C., Hamilton-Byrd, E., Horn, J.W., Czilli, D.L., Simmons, L.K., Ni, B., and Paul, S.M., 1997, Activation of a caspase 3-related cysteine protease is required for glutamate-mediated apoptosis of cultured cerebellar granule neurons. *Proc Natl Acad Sci USA* **94(21)**: 11657–11662.

Ellis, H.M. and Horvitz, H.R., 1986, Genetic control of programmed cell death in the nematode C. elegans. *Cell* **44(6)**:817–829.

Emoto, Y., Manome, Y., Meinhardt, G., Kisaki, H., Kharbanda, S., Robertson, M., Ghayur, T., Wong, W.W., Kamen, R., and Weichselbaum, R., 1995, Proteolytic activation of protein kinase C delta by an ICE-like protease in apoptotic cells. *Embo J* **14(24)**:6148–6156.

Enari, M., Sakahira, H., Yokoyama, H., Okawa, K., Iwamatsu, A., and Nagata, S., 1998, A caspase-activated DNase that degrades DNA during apoptosis, and its inhibitor ICAD. *Nature* **391(6662)**:43–50.

Enari, M., Talanian, R.V., Wong, W.W., and Nagata, S., 1996, Sequential activation of ICE-like and CPP32-like proteases during Fas- mediated apoptosis. *Nature* **380(6576)**:723–726.

Endoh, M., Maiese, K., and Wagner, J., 1994, Expression of the inducible form of nitric oxide synthase by reactive astrocytes after transient global ischemia. *Brain Res* **651(1–2)**:92–100.

Fabisiak, J.P., Kagan, V.E., Ritov, V.B., Johnson, D.E., and Lazo, J.S., 1997, Bcl-2 inhibits selective oxidation and externalization of phosphatidylserine during paraquat-induced apoptosis. *Am J Physiol* **272(2 Pt 1)**:C675–C684.

Fehsel, K., Kroncke, K.D., Meyer, K.L., Huber, H., Wahn, V., and Kolb-Bachofen, V., 1995, Nitric oxide induces apoptosis in mouse thymocytes. *J Immunol* **155(6)**:2858–2865.

Fiorucci, S., Santucci, L., Federici, B., Antonelli, E., Distrutti, E., Morelli, O., Renzo, G.D., Coata, G., Cirino, G., Soldato, P.D., and Morelli, A., 1999, Nitric oxide-releasing NSAIDs inhibit interleukin-1beta converting enzyme-like cysteine proteases and protect endothelial cells from apoptosis induced by TNFalpha. *Aliment Pharmacol Ther* **13(3)**:421–435.

Fujikawa, D.G., Shinmei, S.S., and Cai, B., 2000, Seizure-induced neuronal necrosis: implications for programmed cell death mechanisms. *Epilepsia* **41(Suppl 6)**:S9–S13.

Grammas, P., Moore, P., Cashman, R.E., and Floyd, R.A., 1998, Anoxic injury of endothelial cells causes divergent changes in protein kinase C and protein kinase A signaling pathways. *Mol Chem Neuropathol* **33(2)**:113–124.

Gray, F., Adle-Biassette, H., Brion, F., Ereau, T., le Maner, I., Levy, V., and Corcket, G., 2000, Neuronal apoptosis in human immunodeficiency virus infection. *J Neurovirol* **6(Suppl 1)**:S38–S43.

Guo, Y., Jones, W.K., Xuan, Y.T., Tang, X.L., Bao, W., Wu, W.J., Han, H., Laubach, V.E., Ping, P., Yang, Z., Qiu, Y., and Bolli, R., 1999, The late phase of ischemic preconditioning is abrogated by targeted disruption of the inducible NO synthase gene. *Proc Natl Acad Sci USA* **96(20)**:11507–11512.

Hartmann, A., Troadec, J.D., Hunot, S., Kikly, K., Faucheux, B.A., Mouatt-Prigent, A., Ruberg, M., Agid, Y., and Hirsch, E.C., 2001, Caspase-8 is an effector in apoptotic death of dopaminergic neurons in Parkinson's disease, but pathway inhibition results in neuronal necrosis. *J Neurosci* **21(7)**: 2247–2255.

Ishikawa, Y., Satoh, T., Enokido, Y., Nishio, C., Ikeuchi, T., and Hatanaka, H., 1999, Generation of reactive oxygen species, release of L-glutamate and activation of caspases are required for oxygen-induced apoptosis of embryonic hippocampal neurons in culture. *Brain Res* **824(1)**:71–80.

Juo, P., Kuo, C.J., Yuan, J., and Blenis, J., 1998, Essential requirement for caspase-8/FLICE in the initiation of the Fas-induced apoptotic cascade. *Curr Biol* **8(18)**:1001–1008.

Kajkowski, E.M., Lo, C.F., Ning, X., Walker, S., Sofia, H.J., Wang, W., Edris, W., Chanda, P., Wagner, E., Vile, S., Ryan, K., McHendry-Rinde, B., Smith, S.C., Wood, A., Rhodes, K.J., Kennedy, J.D., Bard, J., Jacobsen, J.S., and Ozenberger, B.A., 2001, Beta-amyloid peptide-induced apoptosis regulated by a novel protein containing a G protein activation module. *J Biol Chem* **20**:20.

Kang, J., Lin, S.-H., Chong, Z.Z., and Maiese, K., 2001, Nicotinamide prevents endothelial cell injury through the modulation of mitochondrial membrane integrity and cysteine protease activity. *Soc Neurosci Abstr* (in press).

Kanno, S., Lee, P.C., Zhang, Y., Ho, C., Griffith, B.P., Shears, L.L., 2nd, and Billiar, T.R., 2000, Attenuation of myocardial ischemia/reperfusion injury by superinduction of inducible nitric oxide synthase. *Circulation* **101(23)**:2742–2748.

Kayalar, C., Ord, T., Testa, M.P., Zhong, L.T., and Bredesen, D.E., 1996, Cleavage of actin by interleukin 1 beta-converting enzyme to reverse DNase I inhibition. *Proc Natl Acad Sci U S A* **93(5)**:2234–2238.

Kimura, C., Zhao, Q.L., Kondo, T., Amatsu, M., and Fujiwara, Y., 1998, Mechanism of UV-induced apoptosis in human leukemia cells: roles of Ca2+/Mg(2+)-dependent endonuclease, caspase-3, and stress-activated protein kinases. *Exp Cell Res* **239(2)**:411–422.

Kiprianova, I., Freiman, T.M., Desiderato, S., Schwab, S., Galmbacher, R., Gillardon, F., and Spranger, M., 1999, Brain-derived neurotrophic factor prevents neuronal death and glial activation after global ischemia in the rat. *J Neurosci Res* **56(1)**:21–27.

Krohn, A.J., Preis, E., and Prehn, J.H., 1998, Staurosporine-induced apoptosis of cultured rat hippocampal neurons involves caspase-1-like proteases as upstream initiators and increased production of superoxide as a main downstream effector. *J Neurosci* **18(20)**:8186–8197.

Levkau, B., Herren, B., Koyama, H., Ross, R., and Raines, E.W., 1998, Caspase-mediated cleavage of focal adhesion kinase pp125FAK and disassembly of focal adhesions in human endothelial cell apoptosis. *J Exp Med* **187(4)**:579–586.

Lin, S., Chong, Z.Z., and Maiese, K., 2001a, Cell cycle induction in post-mitotic neurons proceeds in concert with the initial phase of programmed cell death in rat. *Neurosci Lett* **310(2–3)**:173–177.

Lin, S.-H., Chong, Z.Z., and Maiese, K., 2001b, Nicotinamide: A nutritional supplement that provides protection against neuronal and vascular injury. *J Med Food* **4(1)**:27 38.

Lin, S.H. and Maiese, K., 2001a, Group I metabotropic glutamate receptors prevent endothelial programmed cell death independent from MAP kinase p38 activation in rat. *Neurosci Lett* **298(3)**:207–211.

Lin, S.H. and Maiese, K., 2001b, The metabotropic glutamate receptor system protects against ischemic free radical programmed cell death in rat brain endothelial cells. *J Cereb Blood Flow Metab* **21(3)**:262–275.

Lin, S.H., Vincent, A., Shaw, T., Maynard, K.I., and Maiese, K., 2000, Prevention of nitric oxide-induced neuronal injury through the modulation of independent pathways of programmed cell death. *J Cereb Blood Flow Metab* **20(9)**:1380–1391.

Liu, X., Kim, C.N., Yang, J., Jemmerson, R., and Wang, X., 1996, Induction of apoptotic program in cell-free extracts: requirement for dATP and cytochrome c. *Cell* **86(1)**:147–157.

Love, S., Barber, R., and Wilcock, G.K., 2000, Neuronal death in brain infarcts in man. *Neuropathol Appl Neurobiol* **26(1)**:55–66.

Ludlow, J.W., Shon, J., Pipas, J.M., Livingston, D.M., and DeCaprio, J.A., 1990, The retinoblastoma susceptibility gene product undergoes cell cycle-dependent dephosphorylation and binding to and release from SV40 large T. *Cell* **60(3)**:387–396.

Madaio, M.P., Fabbi, M., Tiso, M., Daga, A., and Puccetti, A., 1996, Spontaneously produced anti-DNA/DNase I autoantibodies modulate nuclear apoptosis in living cells. *Eur J Immunol* **26(12)**:3035–3041.

Maiese, K., 1998, From the Bench to the Bedside: The Molecular Management of Cerebral Ischemia. *Clinical Neuropharm* **21(1)**:1–7

Maiese, K., 2001, The dynamics of cellular injury: transformation into neuronal and vascular protection. *Histol Histopathol* **16(2)**:633–644.

Maiese, K., Boniece, I., DeMeo, D., and Wagner, J.A., 1993a, Peptide growth factors protect against ischemia in culture by preventing nitric oxide toxicity. *J Neurosci* **13(7)**:3034–3040.

Maiese, K., Boniece, I.R., Skurat, K., and Wagner, J.A., 1993b, Protein kinases modulate the sensitivity of hippocampal neurons to nitric oxide toxicity and anoxia. *J Neurosci Res* **36(1)**:77–87.

Maiese, K. and Gallant, J., 1997, Genetic regulation of neuronal degeneration: The retinoblastoma gene and programmed cell death. *Soc Neurosci Abstr* **23 (Part 1)**.850.

Maiese, K., Pek, L., Berger, S.B., and Reis, D.J., 1992, Reduction in focal cerebral ischemia by agents acting at imidazole receptors. *J Cereb Blood Flow Metab* **12(1)**:53–63.

Maiese, K., Vincent, A., Lin, S.H., and Shaw, T., 2000, Group I and Group III metabotropic glutamate receptor subtypes provide enhanced neuroprotection. *J Neurosci Res* **62(2)**:257–272.

Maiese, K. and Vincent, A.M., 1999, Group I metabotropic receptors down-regulate nitric oxide induced caspase-3 activity in rat hippocampal neurons. *Neurosci Lett* **264(1–3)**:17–20.

Maiese, K. and Vincent, A.M., 2000a, Critical temporal modulation of neuronal programmed cell injury. *Cell Mol Neurobiol* **20(3)**:383–400.

Maiese, K. and Vincent, A.M., 2000b, Membrane asymmetry and DNA degradation: functionally distinct determinants of neuronal programmed cell death. *J Neurosci Res* **59(4)**:568–580.

Marks, K.A., Mallard, C.E., Roberts, I., Williams, C.E., Gluckman, P.D., and Edwards, A.D., 1996, Nitric oxide synthase inhibition attenuates delayed vasodilation and increases injury after cerebral ischemia in fetal sheep. *Pediatr Res* **40(2)**:185–191.

Martin, S.J. and Green, D.R., 1995, Protease activation during apoptosis: death by a thousand cuts? *Cell* **82(3)**:349–352.

Mokudai, T., Ayoub, I.A., Sakakibara, Y., Lee, E.J., Ogilvy, C.S., and Maynard, K.I., 2000, Delayed treatment with nicotinamide (Vitamin B(3)) improves neurological outcome and reduces infarct volume after transient focal cerebral ischemia in Wistar rats. *Stroke* **31(7)**:1679–1685.

Montague, J.W., Hughes, F., Jr., and Cidlowski, J.A., 1997, Native recombinant cyclophilins A, B, and C degrade DNA independently of peptidylprolyl cis-trans-isomerase activity. Potential roles of cyclophilins in apoptosis. *J Biol Chem* **272(10)**:6677–6684.

Newmeyer, D.D., Farschon, D.M., and Reed, J.C., 1994, Cell-free apoptosis in Xenopus egg extracts: inhibition by Bcl-2 and requirement for an organelle fraction enriched in mitochondria. *Cell* **79(2)**:353–364.

Okamoto, M., Matsumoto, M., Ohtsuki, T., Taguchi, A., Mikoshiba, K., Yanagihara, T., and Kamada, T., 1993, Internucleosomal DNA cleavage involved in ischemia-induced neuronal death. *Biochem Biophys Res Commun* **196(3)**:1356–1362.

Pandey, S., Walker, P.R., and Sikorska, M., 1997, Identification of a novel 97 kDa endonuclease capable of internucleosomal DNA cleavage. *Biochemistry* **36(4)**:711–720.

Park, D.S., Obeidat, A., Giovanni, A., and Greene, L.A., 2000, Cell cycle regulators in neuronal death evoked by excitotoxic stress: implications for neurodegeneration and its treatment. *Neurobiol Aging* **21(6)**:771–781.

Raina, A.K., Zhu, X., Rottkamp, C.A., Monteiro, M., Takeda, A., and Smith, M.A., 2000, Cyclin' toward dementia: cell cycle abnormalities and abortive oncogenesis in Alzheimer disease. *J Neurosci Res* **61(2)**:128–133.

Rossig, L., Haendeler, J., Hermann, C., Malchow, P., Urbich, C., Zeiher, A.M., and Dimmeler, S., 2000, Nitric oxide down-regulates MKP-3 mRNA levels: involvement in endothelial cell protection from apoptosis. *J Biol Chem* **275(33)**:25502–25507.

Sanchez-Alcazar, J.A., Ault, J.G., Khodjakov, A., and Schneider, E., 2000, Increased mitochondrial cytochrome c levels and mitochondrial hyperpolarization precede camptothecin-induced apoptosis in Jurkat cells. *Cell Death Differ* **7(11)**:1090–1100.

Satake, K., Matsuyama, Y., Kamiya, M., Kawakami, H., Iwata, H., Adachi, K., and Kiuchi, K., 2000, Nitric oxide via macrophage iNOS induces apoptosis following traumatic spinal cord injury. *Brain Res Mol Brain Res* **85(1–2)**:114–122.

Schapira, A.H., 1996, Oxidative stress and mitochondrial dysfunction in neurodegeneration. *Curr Opin Neurol* **9(4)**:260–264.

Sherr, C., 1994, The ins and outs of Rb: Coupling gene expression to the cell cycle clock. *Trends Cell Biol* **4**:15–18.

Shimohama, S., 2000, Apoptosis in Alzheimer's disease—an update. *Apoptosis* **5(1)**:9–16.

Slack, R.S., Skerjanc, I.S., Lach, B., Craig, J., Jardine, K., and McBurney, M.W., 1995, Cells differentiating into neuroectoderm undergo apoptosis in the absence of functional retinoblastoma family proteins. *J Cell Biol* **129(3)**:779–788.

Sun, X.M. and Cohen, G.M., 1994, Mg(2+)-dependent cleavage of DNA into kilobase pair fragments is responsible for the initial degradation of DNA in apoptosis. *J Biol Chem* **269(21)**:14857–14860.

Takahashi, H., Nakamura, S., Asano, K., Kinouchi, M., Ishida-Yamamoto, A., and Iizuka, H., 1999, Fas antigen modulates ultraviolet B-induced apoptosis of SVHK cells: sequential activation of caspases 8, 3, and 1 in the apoptotic process. *Exp Cell Res* **249(2)**:291–298.

Tominaga, T., Kure, S., Narisawa, K., and Yoshimoto, T., 1993, Endonuclease activation following focal ischemic injury in the rat brain. *Brain Res* **608(1)**:21–26.

Torriglia, A., Chaudun, E., Chany-Fournier, F., Jeanny, J.C., Courtois, Y., and Counis, M.F., 1995, Involvement of DNase II in nuclear degeneration during lens cell differentiation. *J Biol Chem* **270(48)**:28579–28585.

Torriglia, A., Chaudun, E., Courtois, Y., and Counis, M.F., 1997, On the use of Zn2+ to discriminate endonucleases activated during apoptosis. *Biochimie* **79(7)**:435–438.

Verhoven, B., Krahling, S., Schlegel, R.A., and Williamson, P., 1999, Regulation of phosphatidylserine exposure and phagocytosis of apoptotic T lymphocytes. *Cell Death Differ* **6(3)**:262–270.

Vincent, A.M. and Maiese, K., 1999a, Direct temporal analysis of apoptosis induction in living adherent neurons. *J Histochem Cytochem* **47(5)**:661–672.

Vincent, A.M. and Maiese, K., 1999b, Nitric oxide induction of neuronal endonuclease activity in programmed cell death. *Exp Cell Res* **246(2)**:290–300.

Vincent, A.M. and Maiese, K., 2000, The metabotropic glutamate system promotes neuronal survival through distinct pathways of programmed cell death. *Exp Neurol* **166(1)**:65–82.

Vincent, A.M., Mohammad, Y., Ahmad, I., Greenberg, R., and Maiese, K., 1997, Metabotropic glutamate receptors prevent nitric oxide induced programmed cell death. *J Neurosci Res* **50**:549–564.

Vincent, A.M., TenBroeke, M., and Maiese, K., 1999a, Metabotropic glutamate receptors prevent programmed cell death through the modulation of neuronal endonuclease activity and intracellular pH. *Exp Neurol* **155(1)**:79–94.

Vincent, A.M., TenBroeke, M., and Maiese, K., 1999b, Neuronal intracellular pH directly mediates nitric oxide-induced programmed cell death. *J Neurobiol* **40(2)**:171–184.

Wink, D.A., Kasprzak, K.S., Maragos, C.M., Elespuru, R.K., Misra, M., Dunams, T.M., Cebula, T.A., Koch, W.H., Andrews, A.W., Allen, J.S., et al., 1991, DNA deaminating ability and genotoxicity of nitric oxide and its progenitors. *Science* **254(5034)**:1001–1003.

14

HEME OXYGENASE (HO)-1 EXPRESSING MACROPHAGES/MICROGLIAL CELLS ACCUMULATE DURING OLIGODENDROGLIOMA PROGRESSION

Martin H. Deininger, Richard Meyermann,
and Hermann J. Schluesener

Institute of Brain Research
University of Tuebingen
Medical School, Tuebingen
Germany

1. INTRODUCTION

HO-1 (HSP32) is oxidative stress-inducible (Maines, 1997) and catalyzes oxidation of heme to biologically active molecules: iron, a gene regulator, biliverdin, an antioxidant and carbon monoxide. Consecutive downstream mediation is involved in vasodilation, stimulation of guanylate cyclase, and neuronal transmission (Hartsfiled et al., 1997).

In normal brain, HO-1 is present at the limit of immunodetection and is discretely localized in selected neuronal populations. HO-2 is much more widely expressed. It is present in mitral cells in the olfactory bulb, pyramidal cells in the cortex and hippocampus, granule cells in the dentate gyrus, many neurons in the thalamus, hypothalamus, cerebellum and caudal brainstem (Vincent et al., 1994). While the constitutively expressed HO-2 has been reported to be exclusively regulated by glucocorticoids, the inducible HO-1 isozyme is associated with a wide range of

Corresponding author: Martin H. Deininger, Institute of Brain Research, University of Tuebingen Medical School, Calwer Str. 3, D-72076 Tuebingen, Germany. Tel.: +49-7071-2982283 Fax: +49-7071-294846 E-mail. martin.deininger@uni-tuebingen.de pathological conditions in the mammalian brain. Induction of HO-1 expression has been associated with neuroprotection during hyperthermia in glial cells (Ewing et al., 1992) and during hypoxia (Panahian et al., 1999). Consequently, HO-1 expression in neurons, astrocytes and macrophages was observed in a a wide range of experimental of rodent brain such as traumatic injury (Fukuda et al., 1995), ischemia (Nimura et al., 1996) and in human Alzheimer's disease (Schipper et al., 1995). Moreover, HO-1 expression in infiltrating macrophages has been associated with disease severity in atherosclerosis (Wang et al., 1998) and constitutes a marker of oxidative stress in asthma (Horvath et al., 1998). In brain tumors, elevated HO-1 expression was observed, but spatial and cellular expression patterns remain unresolved (Hara et al., 1996; Nishie et al., 1999).

In order to provide a pathological basis for the involvement of HO-1 in oligodendrogliomas, we have analyzed its expression in 69 oligodendroglioma tissue samples, in rat intracranially transplanted C6 gliomas, four rat brains and four neuropathologically unaltered human brains by immunohistochemistry. Twenty-six primary WHO grade II oligodendrogliomas and sixteen primary WHO grade III anaplastic oligodendrogliomas were included. Nineteen grade II tumors progressed, ten were again grade II oligodendrogliomas, and nine had progressed to higher grade lesions. Eight anaplastic oligodendrogliomas progressed, five were again WHO grade III tumors, and three had progressed to glioblastoma multiforme. Double labeling experiments confirmed the nature of HO-1 expressing cells. Reverse transcription-polymerase chain reaction (RT-PCR) was used to demonstrate HO-1 mRNA.

2. HEME OXYGENASE-1 IN NEUROPATHOLOGICALLY UNALTERED BRAINS

Three heme oxygenase isoforms (HO-1, HO-2, and HO-3) have been identified to date. HO-1 is ubiquitous and its mRNA and activity can be increased several-fold by heme, other metalloporphyrins, transition metals, and stimuli that induce cellular stress. In contrast, HO-2 is present chiefly in the brain and testes and is virtually uninducible. HO-3 has very low activity; its physiological function probably involves heme binding (Elbirt et al., 1999). These proteins, which are different gene products, have little in common in primary structure, regulation, or tissue distribution.

To evaluate HO-1 expression in neuropathologically unaltered brains, we analyzed four human autopsy control brains from routine analyses at the Institute of Brain Research in Tuebingen (Table 2). Five male Spraque Dawley rats were sacrificed, perfused with 4% paraformaldehyde and brains were prepared as described (Schluesener et al., 1998). All tissues were fixed in buffered 4% formalin (pH 7.4) and embedded in paraffin by routine methods.

In immunohistochemistry labeling experiments, five μm sections were deparaffinized and rehydrated. For antigen retrieval, the sections were immersed in 0.01 M citrate buffer and irradiated in a microwave oven at 750 W, five cycles of 5 min. Endogenous peroxidase was blocked with 1% H_2O_2 in methanol and the slices were consequently incubated with nonspecific porcine serum. Rabbit polyclonal antibodies directed against recombinant rat HO-1 (StressGen, Victoria, Canada) were diluted

1:200 in 1% BSA (bovine serum albumin) TBS (Tris-balaced salt solution, pH 7.5, containing 0.025 M Tris, 0.15 M NaCl). Secondary antibody biotinylated anti-rabbit IgG (Dako, Hamburg, Germany) was diluted at 1:400 in BSA/TBS and applied to the slices for 30 min. Streptavidin-biotin horseradish peroxidase complex (Dako, Hamburg, Germany) diluted 1:400 was subsequently applied for 30 min. Labeled antigen was visualized with standard diaminobenzidine techniques (Sigma, Deisenhofen, Germany). All sections were counterstained with hematoxylin. Posititve cells were counted in five regions of solid tumor groth in WHO II and WHO III oligodendrogliomas at 400X magnification and compared to the total number of counterstained nuclei in that area. In WHO IV glioblastomas and C6 gliomas of the rat, five regions in the immediate vicinity of focal necrosis were evaluated to demonstrate focal accumulation of HO-1 expressing cells. Then the percentage of HO-1 expressing cells was calculated. Statistical analysis was performed using the Mann-Whitney U-test.

Controls included adjacent sections stained omitting addition of HO-1 antibody. HO-1 immunoreactivity was abolished following overnight incubation of antibody with blocking peptide (Santa Cruz, Santa Cruz, USA) at 4 °C according to the manufacturer's instructions. No cross-reactivity is observed with HO-2.

In rat and human control brains without neuropathological alterations, HO-1 was expressed by singular astrocytes, neurons and macrophages situated in the cortex of the forebrain, diencephalon, cerebellum, and brainstem regions. The number of HO-1 expressing cells was comparably low in rat (Mean = 0.1, SEM = 0.078) and human control brains (Mean = 0.13, SEM = 0.08), (Table 2).

3. HEME OXYGENASE-1 IN OLIGODENDROGLIOMA PATIENTS

All oligodendrogliomas were resected at the Department of Neurosurgery in Tübingen or at the Department of Neurosurgery of the Asklepios Klinik Schildautal in Seesen (Table 1). Resection was documented by the surgeons as incomplete or macroscopically complete. We studied 42 oligodendroglioma tissue samples, 26 primary WHO (World Health Organisation) grade II oligodendrogliomas and 16 primary WHO grade III anaplastic oligodendrogliomas. Nineteen grade II tumors progressed, ten were again grade II oligodendrogliomas and 9 had progressed to higher grade lesions. Eight anaplastic oligodendrogliomas progressed, 5 were again WHO grade III tumors, and 3 had progressed to glioblastoma multiforme. Histological diagnosis was performed by routine neuropathology according to the WHO classification system (Kleihues et al., 1993). WHO grade II oligodendrogliomas were characterized by enlarged rounded cells with a well-defined cell membrane and clear cytoplasm around a central spherical nucleus, uniformly round nuclei, high chromatin density, low mitotic activity, microcalcifications and branching capillary network. WHO grade III oligodendrogliomas were characterized by nuclear atyia, mitotic activity, high proliferation rate, rounded hyperchromatic nuclei, perinuclear swelling, few cellular processes and again microcalcifications and branching capillary network. WHO grade IV glioblastoma multiforme were characterized by a high degree of cellular and nuclear polymorphism with numerous multinucleated cells, high mitotic activity, vascular proliferation and areas of focal necrosis. Seven patients with

Table 1. Oligodendroglioma patients and HO-1 immunoreactivity

Primary tumors						Relapses				
Patient/ gender	Age	Dig	HO-1	Rad	Chem	Dig	HO-1	Surv/ ttp	Local	Res
1 / M	47	O	0.8	–	–			36	LF	I
2 / F	47	O	1.2	–	–			1	LF	C
3 / M	48	O	1.9	+	–			160	LFL	N/A
4 / M	53	O	0.5	N/A	N/A			164	RP	N/A
5 / M	65	O	1.1	–	–			0.1	RF	C
6 / M	36	O	1.3	–	–			168	ROB	I
7 / F	53	O	1.4	+	–			77	RF	N/A
8 / M	50	O	0.8	–	–	O	1.2	12	LFM	N/A
9 / M	27	O	0.6	+	–	GB	33.6	96	LT	N/A
10 / F	36	O	1.3	–	–	N/A	N/A	74	LFMB	N/A
11 / F	32	O	1.2	N/A	N/A	O	2.4	11	LTO	C
12 / F	59	O	1.1	–	–	O	4.3	59	RFP	N/A
13 / M	46	O	0.8	N/A	N/A	O	1.5	100	LP	N/A
14 / F	47	O	0.3	–	–	O	16.4	22	RFP	N/A
15 / M	46	O	N/A	+	–	AO	12.3	142	RFP	N/A
16 / M	39	O	0.8	+	–	O	3.8	27	RT	N/A
17 / F	41	O	N/A	–	–	O	16.4	84	LTPO	I
18 / M	34	O	1.1	–	–	AO	13.7	43	LTM	N/A
19 / M	49	O	0.4	–	–	AO	2.4	22	LT	I
20 / M	56	O	1.3	N/A	N/A	AO	3.1	14	LTP	N/A
21 / F	41	O	6.3	N/A	N/A	O	4.7	52	RF	N/A
22 / M	22	O	N/A	+	–	O	1.3	78	LFB	N/A
23 / F	35	O	0.4	–	–	AO	4.1	36	LFB	I
24 / M	45	O	0.8	–	–	AO	3.8	42	RFP	N/A
25 / M	51	O	0.7	+	–	AO	43.2	8	LFM	I
26 / M	54	O	0.5	–	–	O	17.6	108	LFL	N/A
27 / F	59	AO	1.2	+	–			18	RTB	I
28 / F	46	AO	9.3	+	+			10	LFP	N/A
29 / M	78	AO	0.5	–	–			2	LTB	I
30 / M	53	AO	0.8	+	–			22	LP	N/A
31 / F	26	AO	1.2	+	–			120	RFP	N/A
32 / F	55	AO	0.9	+	–			10	LP	N/A
33 / F	70	AO	1.2	–	–			1	RTM	I
34 / F	43	AO	1.1	+	+			10	RFT	I
35 / F	47	AO	0.9	+	–	AO	15.9	28	LP	N/A
36 / F	33	AO	2.3	+	–	GB	18.3	6	LF	N/A
37 / M	52	AO	1.2	+	–	AO	41.2	42	RTP	N/A
38 / F	43	AO	2.2	+	–	AO	12.4	65	RTPO	N/A
39 / M	55	AO	N/A	–	–	AO	4.5	12	RTPO	C
40 / F	51	AO	1.8	+	+	GB	3.9	12	LPO	I
41 / F	48	AO	1.6	+	–	AO	1.8	24	RF	N/A
42 / M	39	AO	2.2	+	–	GB	12.2	2	RF	N/A

Labeled cells are indicated as per cent of all counterstained nuclei, n/a = not available, Dig = diagnosis, AO = anaplastic oligodendroglioma, O = oligodendroglioma, GB = glioblastoma multiforme, Rad = Radiotherapy, Chem = chemotherapy, ttp = time to progression (months), Surv = survival (months), Local = localization, R = right, L = left, F = frontal, T = temporal, B = basal, P = parietal, O = occipital, M = medial, Res = resection, C = complete; I = incomplete. Labeled cells are indicated as per cent of all counterstained nuclei.

oligodendroglioma received radiotherapy following the resection of the primary tumor, and 14 patients received no post surgical treatment. In 5 patients, no postsurgical therapy was mentioned thus suggesting that no post surgical therapy was applied. Thirteen patients with anaplastic oligodendroglioma received radiotherapy after resection of the primary tumor, 3 radiotherapy and chemotherapy with ACNU and VM26, and 3 received no postsurgical treatment. Involved-field radiotherapy was applied at doses of 36–60 Gy.

In human oligodendrogliomas (Table 1), most prominently, HO-1 expression was observed in macrophages/microglial cells. In both, primary WHO grade II oligodendrogliomas (Mean = 1.157, SEM = 0.2477) and WHO grade III anaplastic oligodendrogliomas (Mean = 1.893, SEM = 0.5476), only singular HO-1 expressing macrophages/microglial cells were observed. However, significantly (p = 0.0292) fewer HO-1 expressing cells were observed in primary WHO grade II oligodendrogliomas (Fig. 1A) than in primary anaplastic oligodendrogliomas. Surprisingly, in areas of infiltrative oligodendroglioma growth, higher numbers of HO-1 expressing cells were observed than in areas of solid tumor growth. Here, HO-1 expressing cells were characterized by morphological characteristics of neurons, astrocytes and macrophages /microglial cells. In oligodendroglioma (Mean = 10.32, SEM = 2.787, P < 0.0001) and anaplastic oligodendroglioma relapses (Mean = 13.78, SEM = 4.449, P = 0.0006), we observed significantly higher numbers of HO-1 expressing macrophages/microglial cells than in the primary tumors (Fig. 1B). The most striking accumulation of HO-1 expressing macrophages/microglial cells was observed adjacent to areas of focal necrosis in glioblastoma relapses (Fig. 1C). Furthermore, in areas of necrosis, single disseminated macrophages/microglial cells expressing HO-1 were readily detected. Prominent infiltration of HO-1 expressing macrophages was frequently observed in the walls of the tumor vasculature of high grade oligodendroglioma relapses and glioblastoma multiforme relapses.

Previous reports have demonstrated that HO-1 overexpression is cytoprotective by attenuating nitric oxide mediated proinflammatory reaction cascades. In this context, it is of note that it has been suggested that HO-1 overexpression constitutes a novel therapeutic approach to disrupt inflammatory tissue deterioration in a wide range of diseases (Wang et al., 1998).

4. HEME OXYGENASE-1 IN RAT C6 GLIOMAS

In order to verify accumulation of HO-1 expressing macrophages/microglial cells adjacent to areas of focal necrosis, we used the rat C6 glioblastoma model (Table 2). C6 glioblastomas are characterized by a high proliferation rate, highly invasive growth and formation of areas of zonal necrosis and are therefore a widely used model for human malignant glioma. At near confluency, C6 glioblastoma cells were harvested using a cell scraper and 5 μl were injected into the basal ganglia region of male Spraque Dawley rats at a concentration of $4 \times 10^5/\mu l$ (Schluesener et al., 1998). After two weeks, rats were sacrificed, perfused with 4% paraformaldehyde and brains were prepared as described.

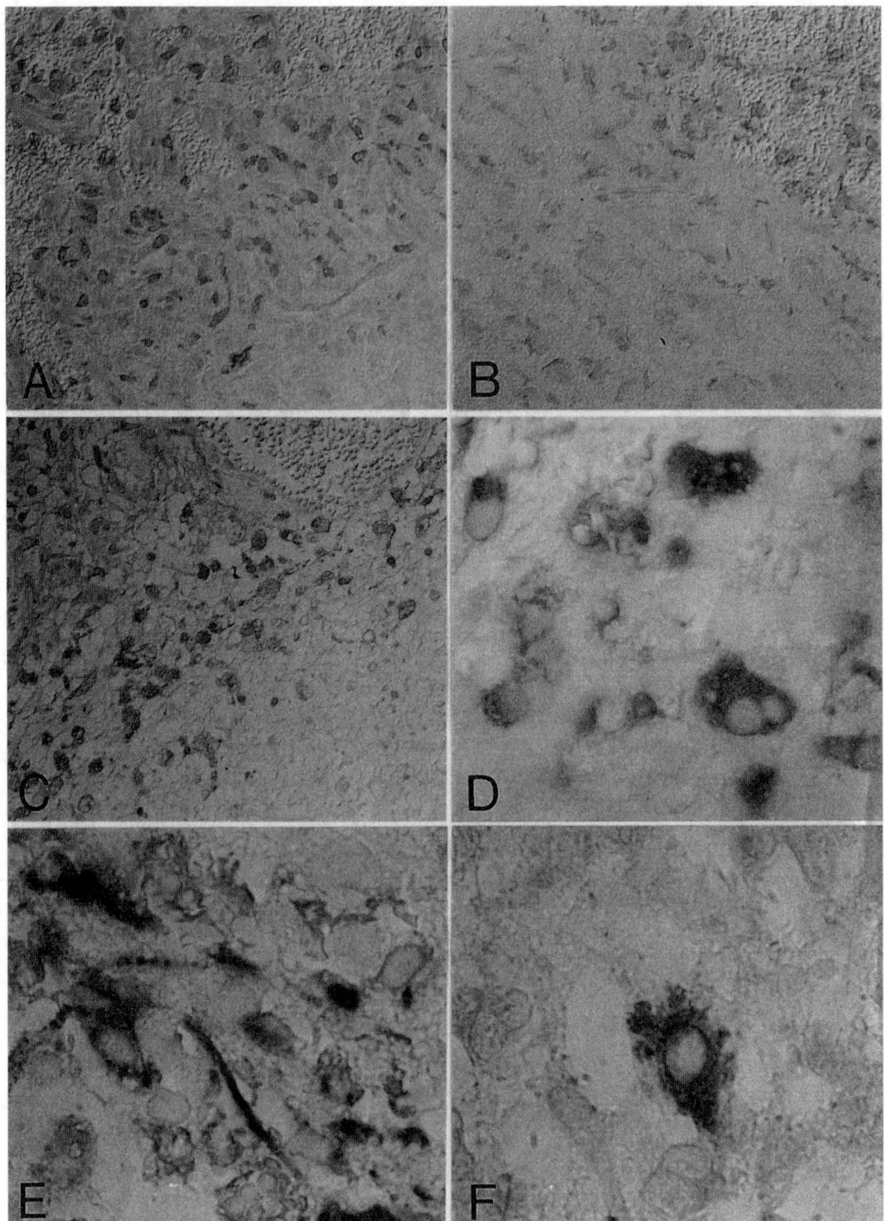

Figure 1. Expression of HO-1 (brown color) in WHO grade II oligodendrogliomas was rarely observed in areas of solid tumor growth (A). In WHO grade III anaplastic oligodendroglioma relapses, disseminated HO-1 expressing cells were observed (B). In human glioblastom multiforme, HO-1 expression was detected in macrophages/microglia cells adjacent to focal necrosis (C). In rat C6 gliomas, HO-1 expressing macrophages/microglial cells accumulated adjacent to focal necrosis (D) and in areas of infiltrative tumor growth (E). Double labeling experiments revealed the coexpression of ED1 (blue) in HO-1 expressing cells (brown) (F), but not of GFAP (blue) in rats.

Table 2. HO-1 immunoreactivity in rat C6 gliomas and in rat and human control brains

ID	Species / Sex / Age	Histopathological diagnosis	HO-1
1	Rat / M / 6 weeks	C6 glioma	43.8
2	Rat / M / 6 weeks	C6 glioma	73.1
3	Rat / M / 6 weeks	C6 glioma	45.6
4	Rat / M / 6 weeks	C6 glioma	42.5
5	Rat / M / 6 weeks	C6 glioma	62.9
6	Rat / M / 6 weeks	Normal brain	0
7	Rat / M / 6 weeks	Normal brain	0
8	Rat / M / 6 weeks	Normal brain	0.1
9	Rat / M / 6 weeks	Normal brain	0
10	Rat / M / 6 weeks	Normal brain	0.4
11	Human / M / 39 years	Normal brain	0
12	Human / M / 26 years	Normal brain	0.3
13	Human / M / 32 years	Normal brain	0.2
14	Human / F / 39 years	Normal brain	0

Labeled cells are indicated as per cent of all counterstained nuclei, M = male, F = female.

Accordingly, we observed accumulation of HO-1 expressing macrophages/microglial cells most prominently in areas of solid tumor growth adjacent to areas of focal necrosis (Fig. 1D). In areas of infiltrative tumor growth, the number of HO-1 expressing macrophages/microglial cells decreased with the distance from the tumor (Fig. 1E). As in human glioblastoma multiforme relapses, we observed significant accumulation of HO-1 expressing macrophages/microglial cells in areas adjacent to focal necrosis (Mean = 53.58, SEM = 6.124, P = 0.0079) compared to normal controls. In the tumor parenchyma, accumulation of HO-1 expressing macrophages/microglial cells was occasionally observed in histopathologically non-necrotic areas. Occasionally, HO-1 expressing macrophages were found in subendothelial localizations of the vasculature.

5. HEME OXYGENASE-1 IS PREDOMINANTLY EXPRESSED BY MACROPHAGES/MICROGLIAL CELLS

In order to determine the cellular origin of HO-1 immunoreactive cells, we used double lableing experiments with antibodies directed against cell-type specific antigens. Slices were pretreated as described above. Then the differentiating mouse monoclonal antibodies were added to the slices all at a dilution of 1:100 in TBS/BSA. We added anti-GFAP (glial fibrillary acidic protein) (Boehringer, Mannheim, Germany) and ED1 (macrophages, Serotec, Oxford, Great Britain) to rat tissues and GFAP, CD68 and HLA-DR, -DP, -DQ (macrophages, Dako, Denmark) to human slices. Visualization was achieved by adding rabbit anti-mouse IgG diluted at 1:20 in TBS for 30 min and then APAAP complex at a dilution of 1:100 in TBS for 30 min. Consecutively, we developed with Fast Blue BB salt (Fluka, Buchs, Switzerland) yielding a blue reaction product. To avoid antibody crossreactivity in double labeling experiments, slices were once more irradiated in a microwave for 20 min in citrate buffer

(Lan et al., 1996). Complete inhibition of alkaline phosphatase function was achieved as previously described (Deininger and Meyermann, 1998). Consecutively, HO-1 was immunolabeled as described above. No counterstain was applied on double-labeled slices.

Double labeling experiments revealed the nature of HO-1 expressing cells. In perinecrotic areas of rat C6 gliomas, the majority of HO-1 expressing macrophages/microglial cells (brown color) coexpressed ED1 (blue color) (Fig. 1F). In contrary, the majority of HO-1 expressing cells did not coexpress GFAP (Fig. 1G). In human oligodendroglioma specimens, predominant colocalization of HO-1 and CD68 (Fig. 1H) and HLA-DR, -DP, -DQ was observed. Occasionally, double labeling of GFAP (blue color) and HO-1 (brown color) was detected (Fig. 1I).

Detailed previous analyses revealed cell-type specific differences of HO-1 bioactivity. In endothelial cells, NO was identified as a determinant in the modulation of the activity of heme oxygenase leading to a major resistance of the endothelium to oxidative stress (Motterlini et al., 1996). In macrophages, HO-1 is a physiological inhibitor of nitrite formation by decreasing haem availability for NOS2 synthesis (Turcanu et al., 1998). Overexpression of HO-1 resulted in the inhibition of several immune effector functions and thus provides an explanation for stress-induced immunosuppression (Woo et al., 1998). In the brain, HO-1 overexpression in macrophages/microglial cells has been suggested to be cytoprotective in the contused spinal cord of the rat (Mautes et al., 1998), following pre- and postganglionic axotomy (Magnusson and Kanje, 1998), hyperosmotic opening of the blood-brain barrier (Richmon et al., 1998), hypoxia-ischemia (Bergeron et al., 1997; Panahian et al., 1999; Takizawa et al., 1998) and trauma (Fukuda et al., 1995). In human gliomas, expression of HO-1 mRNA and protein was correlated with macrophage infiltration and vascular density (Nishie et al., 1999). In this context, however, detrimental HO-1 biological activity has been described. HO-1 is responsible for the physiological breakdown of heme into equimolar amounts of biliverdin, carbon monoxide, and iron and there is convincing evidence that accumulation of HO-1 expressing macrophages in the immediate vicinity of necroses in high grade gliomas is a rather detrimental mechanism that contributes to neoplastic outgrowth and tissue damage. HO-1 expression is induced following multiple pathological stimuli including reactive oxygen species, cytokines, hyperthermia and radiation (Kutty et al., 1995; Maines et al., 1995; Matsuoka et al., 1999; Terry et al., 1999). Therefore, accumulation of HO-1 expressing macrophages/microglial cells in oligodendroglioma relapses may at least in part be induced by irradiation and chemotherapy administered to the patients (Table 1).

6. REVERSE TRANSCRIPTASE POLYMERASE CHAIN REACTION

In order to determine HO-1 mRNA expression in rat C6 glioblastoma and human oligodendroglioma and glioblastoma patients, RT-PCR was performed.

Rat C6 glioblastoma cell lines were obtained from the American Type Culture Collection (ATCC, Manassas, USA) and raised in RPMI 1640 medium with

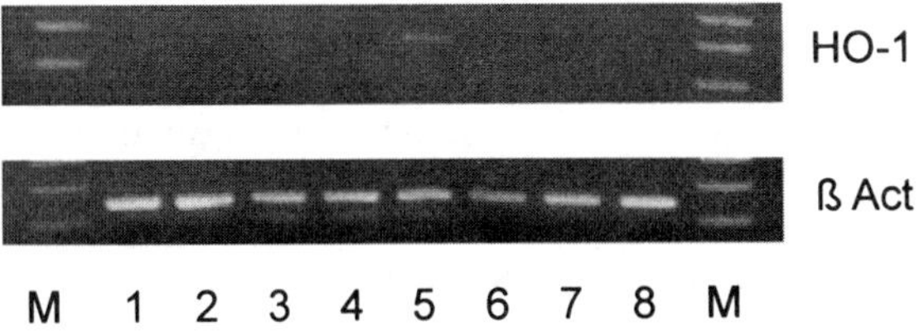

Figure 2. HO-1 mRNA was not observed in WHO grade II oligodendrogliomas (lanes 1–3), WHO grade III anaplastic oligodendroglioma (lanes 4) or WHO grade IV glioblastoma (lanes 6–8). In one glioblastoma multiforme relapse (lane 5), however, HO-1 mRNA was detected.

Glutamax II (Gibco BRL, Paisley, Great Britain) containing 10% fetal calf serum (FCS, Gibco BRL, Paisley, Great Britain) and 1.2% penicillin/ streptomycin (Fluka, Buchs, Switzerland) at 37 °C and 5% CO_2.

Oligodendroglioma tissue of four primary oligodendroglioma and one glioblastoma multiforme relapse patient was collected immediately after resection and frozen in liquid nitrogen. Total RNA extraction was performed using the RNeasy® Mini Kit protocol as suggested by the manufacturer (Qiagen GmbH, Hilden, Germany). The RNA yields of all patients did not differ significantly and were typically in the range of 0.6–1.0 µg/mg tissue. cDNA was prepared using random hexamers, RNAsin® ribonuclease inhibitor and Moloney Murine Leukemia Virus Reverse Transcriptase (Promega, Madison, WI). Samples were incubated 30′ at 37 °C and 5′ at 95 °C. Polymerase chain reaction was then performed using Thermus aquaticus (Taq) DNA Polymerase Mini Kit (Promega) according to the manufacturer's instructions. The forward and reverse primers for HO-1 were: 5′-TGCCCAGCTCCTGGCCCGCCGCTT-3′ and 5′-GTGCATCAACACAGGCGCCTCTTC-3′, respectively, and gave a single band corresponding to a fragment in human HO-1 cDNA (Premkumar et al., 1995). RNA integrity of the examined samples was confirmed using ß-actin forward and reverse primers: 5′-TCACCCTGAAGTACCCCATCGAG-3′ and 5′-TTGGC CTTGGGGGTTCAGGGGGG-3′, respectively, yealding a fragment in human beta-actin cDNA (Nakajima-Iijima et al., 1985). Reaction conditions were: denaturation at 94 °C for 2′ followed by 30 cycles of denaturation at 94 °C for 1′, annealing at 60 °C, 55 °C (HO-1), 55 °C (ß-actin), elongation at 72 °C for 1′. Finally, the reaction was incubated at 72 °C for 10′. The resulting reaction products were analyzed on agarose gels stained with 0.5 mg/ml ethidium bromide.

HO-1 mRNA was not observed in WHO grade II oligodendrogliomas (lanes 1–3), WHO grade III anaplastic oligodendroglioma (lanes 4) or three WHO grade IV glioblastoma (lanes 6–8). In one glioblastoma multiforme relapse (lane 5), however, HO-1 mRNA was detected (Figure 2).

In neoplasia, this is of note, because tumor cells induce reactive oxygen species in monocytes and consequently contribute to spontaneous monocyte cytotoxicity and prolonged tissue damage (Mytar et al., 1999). Moreover, activation of HO-1 is a major stimulus for the synthesis and release of pro-inflammatory cytokines in macrophages (Tamion et al., 1999). It has therefore been suggested that this pathway may play a central role in pathological situations in which local tissue hypoxia/re-oxygenation triggers a systemic inflammatory response. Furthermore, HO-mediated heme degradation is the primary mechanism for cellular CO production, and reactive oxygen intermediates including CO are capable to induce angiogenesis and thus promote neoplastic outgrowth (Kuroki et al., 1996). Iron increases oxidative stress and regulates the expression of proinflammatory mRNAs by affecting the conformation of iron regulatory protein (IRP)-1 and its binding to iron regulatory elements (IREs) in the 5′- or 3′-UTRs of the mRNAs (Elbirt et al., 1999). Accumulation of HO-1 expressing macrophages/microglial cells in the immediate vicinity of necroses during the progression of oligodendrogliomas therfore might indicate a prominent mechanism to confirm neoangiogenesis in hypoxic areas. These data are supported by the finding that HO-1 mRNA was only detectable in one glioblastoma multiforme tissue specimen.

Although we do not have histological evidence, we speculate that HO-1 is only expressed in distinct areas of tumor growth, most likely in perinecrotic regions. Our data provide convincing evidence for the accumulation of HO-1 expressing macrophages/microglial cells in areas of focal necrosis. However, further studies are of need to resolve whether this phenomenon contributes to overall neoplastic outgrowth and cytoprotection or prolonged cytotoxicity.

7. SUMMARY

Heme oxygenases (HO-1, HSP32) catalyze the oxidation of heme to biliverdin and carbon monoxide, a putative neurotransmitter. In the brain, HO-1 expression has been associated with neuroprotection during oxidative stress and hypoxia. However, consecutive downstream mediation is involved in neoangiogenesis and consequent neoplastic outgrowth.

We have analyzed HO-1 expression in 69 oligodendroglioma tissue samples, in rat intracranially transplanted C6 gliomas, and neuropathologically unaltered control brains by immunohistochemistry. Double labeling experiments confirmed the nature of HO-1 expressing cells. Reverse transcription-polymerase chain reaction was used to demonstrate HO-1 gene expression. HO-1 immunoreactivity was predominantly observed in macrophages/microglial cells. The number of HO-1 expressing macrophages/microglial cells was significantly lower in primary oligodendrogliomas than in their matched relapses (P < 0.0001) and lower in primary anaplastic oligodendrogliomas than in their relapses (P = 0.0006). Prominent accumulation of HO-1 expressing macrophages/microglial cells was observed in perinecrotic areas of both experimental rat and human glioblastoma relapses. HO-1 expressing neurons, macrophages/microglial cells and astrocytes were scattered in areas of infiltrative tumor growth. Surprisingly, HO-1 mRNA was detected in only one glioblastoma multiforme relapse.

We conclude from these data that HO-1 expressing macrophages/microglial cells accumulate during oligodendroglioma progression in areas of focal necrosis. However, overall biological function of this phenomenon remains to be determined.

REFERENCES

Bergeron, M., Ferriero, D.M., Vreman, H.J., Stevenson, D.K., and Sharp, F.R., 1997. Hypoxia-ischemia, but not hypoxia alone, induces the expression of heme oxygenase-1 (HSP32) in newborn rat brain. *J Cereb Blood Flow Metab* **17**:647.

Deininger, M.H. and Meyermann, R., 1998. Multiple epitope labeling by the exclusive use of alkaline phosphatase conjugates in immunohistochemistry. *Histochem Cell Biol* **110**:425.

Elbirt, K.K. and Bonkovsky, H.L., 1999. Heme oxygenase: recent advances in understanding its regulation and role. *Proc Assoc Am Physicians* **111**:438.

Ewing, J.F., Haber, S.N., and Maines, M.D., 1992. Normal and heat-induced patterns of expression of heme oxygenase-1 (HSP32) in rat brain: hyperthermia causes rapid induction of mRNA and protein. *J Neurochem* **58**:1140.

Fukuda, K., Panter, S.S., Sharp, F.R., and Noble, L.J., 1995. Induction of heme oxygenase-1 (HO-1) after traumatic brain injury in the rat. *Neurosci Lett* **199**:127.

Hara, E., Takahashi, K., Tominaga, T., Kumabe, T., Kayama, T., Suzuki, H., Fujita, H., Yoshimoto, T., Shirato, K., and Shibahara, S., 1996. Expression of heme oxygenase and inducible nitric oxide synthase mRNA in human brain tumors. *Biochem Biophys Res Commun* **224**:153.

Hartsfield, C.L., Alam, J., Cook, J.L., and Choi, A.M., 1997. Regulation of heme oxygenase-1 gene expression in vascular smooth muscle cells by nitric oxide. *Am J Physiol* **273**:L980.

Horvath, I., Donnelly, L.E., Kiss, A., Paredi, P., Kharitonov, S.A., and Barnes, P.J., 1998. Raised levels of exhaled carbon monoxide are associated with an increased expression of heme oxygenase-1 in airway macrophages in asthma: a new marker of oxidative stress. *Thorax* **53**:668.

Kleihues, P., Burger, P.C., and Scheithauer, B.W., 1993. The new classification of brain tumors. *Brain Pathol* **3**:255.

Kuroki, M., Voest, E.E., Amano, S., Beerepoot, L.V., Takashima, S., Tolentino, M., Kim, R.Y., Rohan, R.M., Colby, K.A., Yeo, K.T., and Adamis, A.P., 1996. Reactive oxygen intermediates increase vascular endothelial growth factor expression in vitro and in vivo. *J Clin Invest* **98**:1667.

Kutty, R.K., Kutty, G., Wiggert, B., Chader, G.J., Darrow, R.M., and Organisciak, D.T., 1995. Induction of heme oxygenase 1 in the retina by intense visible light: suppression by the antioxidant dimethylthiourea. *Proc Natl Acad Sci USA* **92**:1177.

Lan, H.Y., Mu, W., Ng, Y.Y., Nikolic-Paterson, D.J., and Atkins, R.C., 1996. A simple, reliable, and sensitive method for nonradioactive in situ hybridization: use of microwave heating to improve hybridization efficiency and preserve tissue morphology. *J Histochem Cytochem* **44**:281.

Magnusson, S. and Kanje, M., 1998. Differential macrophage responses following pre- and postganglionic axotomy. *Neuroreport* **9**:841.

Maines, M.D., Eke, B.C., Weber, C.M., and Ewing, J.F., 1995. Corticosterone has a permissive effect on expression of heme oxygenase-1 in CA1-CA3 neurons of hippocampus in thermal-stressed rats. *J Neurochem* **64**:1769.

Maines, M.D., 1997. The heme oxygenase system: a regulator of second messenger gases. *Annu Rev Pharmacol Toxicol* **37**:517.

Matsuoka, Y., Kitamura, Y., Kakimura, J., and Taniguchi, T., 1999. Expression of heme oxygenase-1 mediated by non-NMDA and metabotropic receptors in glial cells: possible involvement of reactive oxygen species production and protein kinase C activation. *Neuropharmacology* **38**:825.

Mautes, A.E., Kim, D.H., Sharp, F.R., Panter, S., Sato, M., Maida, N., Bergeron, M., Guenther, K., and Noble, L.J., 1998. Induction of heme oxygenase-1 (HO-1) in the contused spinal cord of the rat. *Brain Res* **795**:17.

Motterlini, R., Foresti, R., Intaglietta, M., and Winslow, R.M., 1996. NO-mediated activation of heme oxygenase: endogenous cytoprotection against oxidative stress to endothelium. *Am J Physiol* **270**:107.

Mytar, B., Siedlar, M., Woloszyn, M., Ruggiero, I., Pryjma, J., and Zembala, M., 1999. Induction of reactive oxygen intermediates in human monocytes by tumour cells and their role in spontaneous monocyte cytotoxicity. *Br J Cancer* **79**:737.

Nakajima-Iijima, S., Hamada, H., Reddy, P., and Kakunaga, T., 1985. Molecular structure of the human cytoplasmic beta-actin gene: interspecies homology of sequences in the introns. *Proc Natl Acad Sci USA* **82**:6133.

Nimura, T., Weinstein, P.R., Massa, S.M., Panter, S., and Sharp, F.R., 1996. Heme oxygenase-1 (HO-1) protein induction in rat brain following focal ischemia. *Brain Res Mol Brain Res* **37**:201.

Nishie, A., Ono, M., Shono, T., Fukushi, J., Otsubo, M., Onoue, H., Ito, Y., Inamura, T., Ikezaki, K., Fukui, M., Iwaki, T., and Kuwano, M., 1999. Macrophage infiltration and heme oxygenase-1 expression correlate with angiogenesis in human gliomas. *Clin Cancer Res* **5**:1107.

Panahian, N., Yoshiura, M., and Maines, M.D., 1999. Overexpression of heme oxygenase-1 is neuroprotective in a model of permanent middle cerebral artery occlusion in transgenic mice. *J Neurochem* **72**:1187.

Premkumar, D.R., Smith, M.A., Richey, P.L., Petersen, R.B., Castellani, R., Kutty, R.K., Wiggert, B., Perry, G., and Kalaria, R.N., 1995. Induction of heme oxygenase-1 mRNA and protein in neocortex and cerebral vessels in Alzheimer's disease. *J Neurochem* **65**:1399.

Richmon, J.D., Fukuda, K., Maida, N., Sato, M., Bergeron, M., Sharp, F.R., Panter, S.S., and Noble, L.J., 1998. Induction of heme oxygenase-1 after hyperosmotic opening of the blood-brain barrier. *Brain Res* **780**:108.

Schipper, H.M., Cisse, S., and Stopa, E.G., 1995. Expression of heme oxygenase-1 in the senescent and Alzheimer-diseased brain. *Ann Neurol* **37**:758.

Schluesener, H.J., Seid, K., Kretzschmar, J., and Meyermann, R., 1998. Allograft-inflammatory factor-1 in rat experimental autoimmune encephalomyelitis, neuritis and uveitis: Expression by activated macrophages and microglial cells. *Glia* **24**:244.

Takizawa, S., Hirabayashi, H., Matsushima, K., Tokuoka, K., and Shinohara, Y., 1998. Induction of heme oxygenase protein protects neurons in cortex and striatum, but not in hippocampus, against transient forebrain ischemia. *J Cereb Blood Flow Metab* **18**:559.

Tamion, F., Richard, V., Lyoumi, S., Hiron, M., Bonmarchand, G., Leroy, J., Daveau, M., Thuillez, C., and Lebreton, J.P., 1999. Induction of haem oxygenase contributes to the synthesis of proinflammatory cytokines in re-oxygenated rat macrophages: role of cGMP. *Cytokine* **11**:326.

Terry, C.M., Clikeman, J.A., Hoidal, J.R., and Callahan, K.S., 1999. TNF-alpha and IL-1alpha induce heme oxygenase-1 via protein kinase C, Ca2+ , and phospholipase A2 in endothelial cells. *Am J Physiol* **276**:H1493.

Turcanu, V., Dhouib, M., and Poindron, P., 1998. Heme oxygenase inhibits nitric oxide synthase by degrading heme: a negative feedback regulation mechanism for nitric oxide production. *Transplant Proc* **30**:4184.

Vincent, S.R., Das, S., and Maines, M.D., 1994. Brain heme oxygenase isoenzymes and nitric oxide synthase are co-localized in select neurons. *Neuroscience* **63**:223.

Wang, L.J., Lee, T.S., Lee, F.Y., Pai, R.C., and Chau, L.Y., 1998. Expression of heme oxygenase-1 in atherosclerotic lesions. *Am J Pathol* **152**:711.

Willis, D., Moore, A.R., Frederick, R., and Willoughby, D.A., 1996. Heme oxygenase: a novel target for the modulation of the inflammatory response. *Nat Med* **2**:87.

Woo, J., Iyer, S., Cornejo, M.C., Mori, N., Gao, L., Sipos, I., Maines, M., and Buelow, R., 1998. Stress protein-induced immunosuppression: inhibition of cellular immune effector functions following overexpression of haem oxygenase (HSP 32). *Transpl Immunol* **6**:84.

CLINICAL IMPLICATIONS OF HEME OXYGENASE SYSTEM IN INFLAMMATION

15

HEME OXYGENASE AND OCULAR SURFACE INFLAMMATION

Michael W. Dunn and Michal Laniado-Schwartzman

Departments of Pharmacology and Ophthalmology
New York Medical College
Valhalla, NY, 10595
supported in part by NIH grants
EY06513 and HL34300

The ocular surface made up of the cornea and conjunctiva is uniquely threatened by processes or agents that produce hypoxic and oxidative tissue injury and inflammation such as exposure to the most energetic wavelengths of sunlight, atmospheric oxygen, airborne irritants, xenobiotics and a host of other agents that generate reactive oxygen species. The ocular surface is also unique among bodily organs in that it routinely experiences prolonged episodes of hypoxia. This occurs with eyelid closure during sleep depriving the cornea of the atmospheric oxygen that this avascular tissue requires. Ocular surface inflammation, corneal swelling and profound changes in the composition of tears are seen in response to this state of hypoxia (Sack et al., 1992) Similar changes are seen during the prolonged wearing of contact lenses, especially those with poor oxygen permeability. A common property of many noxious agents that the ocular surfaces encounters is their ability to enhance cellular hemeprotein prooxidant systems. This protection is denied to the cornea because avasularity is an absolute requirement for corneal transparency. Thus, the cornea surface epithelium is removed from access to circulatory plasma-based antioxidant systems making antioxidant defense far more difficult.

The cornea needs an intrinsic antioxidant system. The detection of heme oxygenase (HO) and NADPH cytochrome P450 (c) reductase in human corneal epithelium was the first description of such a system in the corneal surface epithelium (Abraham et al., 1987) (Table 1).

Expression of the HO-1 isoform was found to be readily induced by agents of oxidative stress and certain non-toxic heavy metals such as stannous chloride

Table 1. Tissue activities of heme oxygenase and NADPH cytochrome P450 (c) reductase

Source	Heme oxygenase (pmol/mg/30/min)*	NADPH cyt. P- 450 (c) reductase (nmol/mg/min)*
Human corneal epithelium	134	16
Bovine corneal epithelium	56	8
Human liver microsomes	645	340

*Results are the mean of three separate measurements. SEM was less than 10%.

(SnCl$_2$)(Davis et al., 1988) Coexisting in the corneal epithelium with HO is the cytochrome P450 dependent pathway that is responsible for the conversion of arachidonic acid into two biologically active metabolites: 12(R)-hydroxy-5,8,10,14-eicosatetraenoic acid (12(R)-HETE) and 12(R)-hydoxy-5,8,14-eicosatrienoic acid (12(R)-HETrE). 12(R)-HETE is an endogenous inhibitor of sodium-potassium activated ATPase while 12(R)-HETrE is a potent vasodilatory, chemotactic and angiogenic factor. Several studies have provided compelling evidence that these two eicosanoids are deeply involved in the elicitation of ocular surface inflammation (Conners et al., 1995b; Davis et al., 1988; Masferrer et al., 1991). It was reported recently that 12-HETrE was detectable in human tear film; notably, greatly increased levels were found in the presence of ocular surface inflammation in response to a wide variety of stimuli, providing further evidence for a role for this agent as an inflammatory mediator (Mieyal et al., 1987). Hypoxia is caused by eyelid closure during sleep which in turn provokes an ocular surface inflammation marked by a massive influx of polymorphonuclear (PMN) cells into the tear film (Sack et al., 1992). Levels of 12-HETrE are increased 3–5-fold in overnight tears, suggesting that the potent chemotactic properties of 12-HETrE may be responsible for the influx of these inflammatory cell in overnight tears (Dunn, 1997). Expression or overexpression of HO, on the other hand, in a number of studies appears to be active in the attenuation of inflammation and in the rescue of cells undergoing oxidative stress. Conners et al. (Conners et al., 1995a) reported that use of the HO-1 inducer, SnCl$_2$, substantially reduced inflammation in a closed eye contact lens rabbit model of hypoxia-generated inflammation. The attenuation of inflammation in this model was found to correlate to the presence and activity of HO-1 (Laniado-Schwartzman et al., 1997). Application of a single drop of SnCl$_2$ (100 µg/nl) before bedtime resulted in a 25% reduction in the overnight influx of PMN cells into the tear film, presumably due to HO-1 induction (Dunn, 1999). Enhanced survival of cells undergoing oxidative stress generated by hemoglobin toxicity was observed in cultured rabbit corneal epithelium and in a human retinal pigment epithelial cell line overexpressing HO-1 (Abraham, 1997). Selective transfection of rat retinal ganglion cells with human HO-1 in an adenovirus vector lead to overexpression of human HO-1 and significantly increased survival of these cell in a model of glaucoma (Hegazy et al., 2000). HO-1 expression in a human corneal epithelial cell line undergoing oxidative stress has been proposed for use as a stress index of ocular irritation since the degree of stress observed correlates to HO-1 expression (Braunstein et al., 1999).

The cytochrome P450 pathway and its two biologically active products are proinflammatory while it is now clear that the HO-1 system is anti-inflammatory and cell protective along with its many other functions. The fact that they coexist in the corneal epithelium and are upregulated by many of the same stimuli indicates a functional relationship. Hypoxia strongly induces both of them and considerable work has been done in this area that demonstrates this interrelationship. This relationship is seen in the rabbit closed eye contact lens model in the midst of hypoxic stress. Hydrophilic contact lenses with relatively poor oxygen transmission qualities were placed in rabbit eyes and the lids were sutured together so that hypoxic conditions exist for the corneal surface epithelium.

DETECTION OF HO-1 mRNA IN A CORNEAL EPITHELIAL CELL LINE

Due to the limited availability of primary cell cultures from rabbit corneal epithelium, we used the rabbit corneal epithelium (RCE) cell line as a model to characterize HO-1 gene expression. Upon reaching confluence, RCE cells were treated with agents known to cause induction of HO-1 mRNA in various cell lines, such as $SnCl_2$, $CoCl_2$, $ZnCl_2$ and CoPP, at concentrations of 10–150 µM. Control experiments consisted of RCE cells treated with the appropriate vehicle. The effects of these agents on HO-1 mRNA levels were assessed, and the results are depicted in Fig. 1.

A basal level of HO-1 mRNA was evident, albeit to a much lesser extent than that of the rabbit liver. Treatment with all agents used resulted in an accumulation of HO-1 mRNA. A quantitative evaluation of the mRNA changes by scanning densitometry relative to GADPH mRNA levels indicated a 50-fold increase in HO-1 mRNA levels in RCE cells treated with $SnCl_2$ (10 µM). CoPP (10 µM) showed similar potency, increasing HO-1 mRNA levels by 30-fold. On the other hand, 150 µM $CoCl_2$ was needed to achieve the same induction; $ZnCl_2$ at 100 µM produced a 10-fold increase of HO-1 mRNA over control. $CoCl_2$ and $ZnCl_2$ at 10 µM did not produce an increase of HO-1 mRNA (data not shown). These results indicate that the RCE

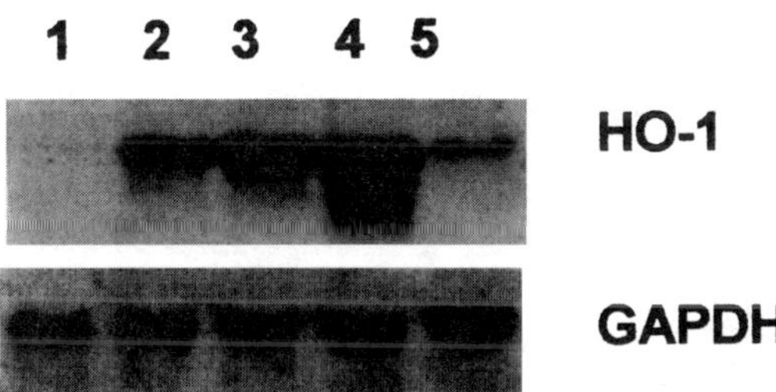

Figure 1. Effects of metals and metalloporphyrins on HO-1 mRNA expression. RCE cells were cultured as previously described. (Laniado-Schwartzman et al., 1997) and treated for 1 hr with the following: Lane 1, control, vehicle-treated cells; Lane 2, $SnCl_2$ (10 µM); Lane 3, $CoCl_2$ (150 µM); Lane 4, CoPP (10 µM); Lane 5, $ZnCl_2$ (100 µM). The lower panel shows that equal amounts of RNA were transferred among samples as indicated by the GADPH probe. Total RNA was extracted and hybridized with ^{32}P-labeled rat HO-1 cDNA as described previously (Laniado-Schwartzman et al., 1997).

cell line responds to HO-1 inducers as do many other tissues. Since the potency of $SnCl_2$ in inducing RCE HO-1 mRNA surpassed that of the other agents studied, we used $SnCl_2$ in the subsequent experiments.

TIME- AND CONCENTRATION-DEPENDENT INDUCTION OF HO-1 mRNA BY $SnCl_2$

RCE cells were grown and maintained in $75\,cm^2$ flasks. Upon reaching confluence, they were treated with $SnCl_2$ and RNA was extracted as described previously (Laniado-Schwartzman et al., 1997). The time-dependent response to $SnCl_2$ is shown in Fig. 2. The vehicle used for solubilization of $SnCl_2$ did not change HO-1 mRNA levels (Lane 2) as compared with control (Lane 1). Optimal induction of HO-1 in RCE was at 16 hr. ($SnCl_2$ 100 μM) (Lane 4), with a subsequent return to control levels at 24 hr (Lane 5). Additional experiments were conducted to demonstrate the long-term effect of $SnCl_2$ in cells grown in a medium containing $SnCl_2$ (100 μM) for up to 6 days. Results showed that, in the continuous presence of $SnCl_2$, cells maintained high levels of HO-1 mRNA while no toxicity (loss of cell viability) was detected (data not shown).

The concentration-response to $SnCl_2$ at 16 hr is shown in Fig. 3. Cells were incubated with $SnCl_2$ at concentrations of 0.1 to 100 μM for 16 hr, after which the RNA was extracted and analyzed. Compared with untreated cells, HO-1 mRNA levels were not increased over controls (Lanes 1 and 2) after the addition of $SnCl_2$ at 0.1 μM (Lane 3). Induction of HO-1 mRNA by $SnCl_2$ (5–100 μM) was concentration-related (Lanes 4–6). Hybridization of the filters with radiolabeled GADPH and ethidium bromide staining of RNA confirmed that similar amounts of total RNA were transferred to the filters in each lane of the paired experiments (data not shown).

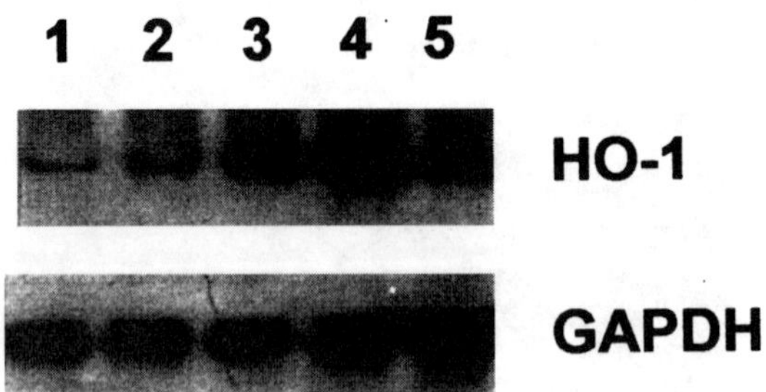

Figure 2. Time-dependent effect of $SnCl_2$ on the induction of HO-1 mRNA in RCE cells. Cells were treated with $SnCl_2$ for various lengths of time. Lane 1, control cells; Lane 2, vehicle-treated cells; Lane 3, $SnCl_2$, 8 hr; Lane 4, $SnCl_2$, 16 hr; Lane 5, $SnCl_2$, 24 hr. The lower panel shows that equal amounts of RNA were transferred among samples as indicated by the GADPH probe. Total RNA was extracted and Northern blot analysis was performed with rat HO-1 cDNA as previously described (Laniado-Schwartzman et al., 1997).

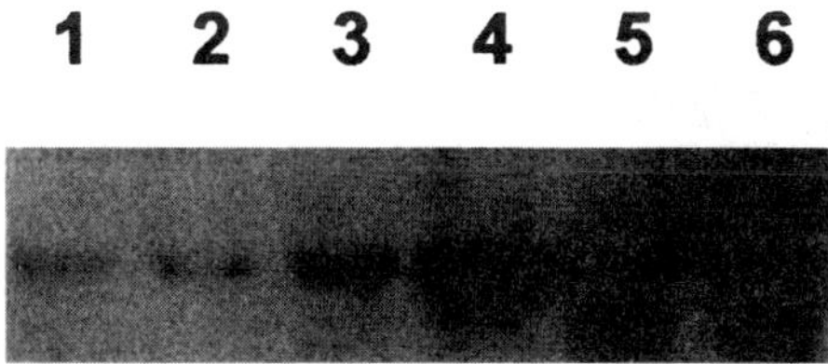

Figure 3. Concentration-dependent effect of $SnCl_2$ on the induction of HO-1 mRNA in RCE cells. Cells were treated with various concentrations of $SnCl_2$ for 16 hr and compared with untreated cells. Lane 1, untreated cells; Lane 2, cells treated with vehicle; Lane 3, $SnCl_2$, 0.1 µM; Lane 4, $SnCl_2$, 5 µM; Lane 5, $SnCl_2$ 10 µM; Lane 6, $SnCl_2$ 100 µM. Total RNA was extracted and Northern blot analysis was performed with the rat HO-1 cDNA probe.

EFFECT OF mRNA-INDUCING AGENTS ON HO ACTIVITY

Translation of HO-1 mRNA into active HO-1 enzyme was verified in the RCE cells. RCE cells grown in $175 \, cm^2$ flasks were treated with either vehicle control or $SnCl_2$ for 24 hr. After combining 6 flasks, microsomes were prepared and HO activity was assessed as previously described (Laniado-Schwartzman et al., 1997). As seen in Fig. 4, HO activity was detectable in untreated (control) cells [16 ± 3 pmol/mg/hr, mean ± SEM (two experiments, triplicate assays, variation within assays <5%)]. With SnCl2 treatment, HO activity significantly increased (6-fold) over the controls (103 ± 21 pmol/mg/hr, mean ± SEM). These results indicate that transcriptional activation of

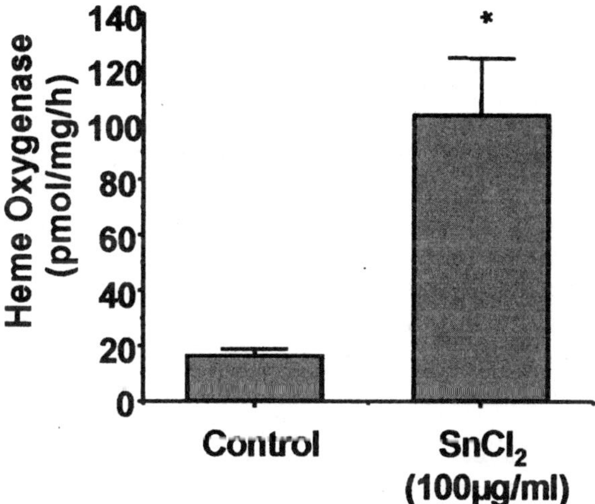

Figure 4. Effect of $SnCl_2$ on HO activity in RCE cells. Confluent cultures were treated with $SnCl_2$ (100 µg/mL; 500 µM) or the vehicle control (0.1 M phosphate buffer, pH 7.4) for 24 hr. Microsomes were prepared and the HO activity was measured as previously described (Laniado Schwartzman et al., 1997) Data are expressed as specific activity in pmol/mg/hr of two separate experiments measured in triplicate. The variation between triplicate assays in each treatment was within 5%. *p < 0.05 versus the vehicle treatment.

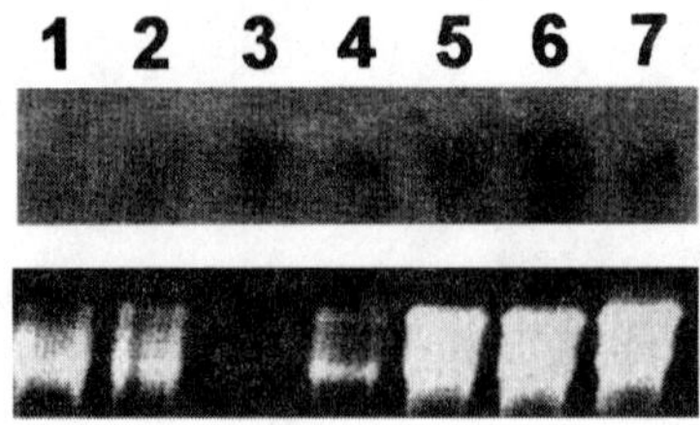

Figure 5. Effect of SnCl$_2$ treatment on HO-1 mRNA in corneal epithelium from 6-day closed eye-hydrogel contact lens-treated eyes. Total epithelial RNA was extracted from control, vehicle, and SnCl$_2$ (100 µg/mL)-treated eyes. RNA was hybridized with the rat HO-1 cDNA probe. Lanes 1–4, untreated eyes; Lanes 5 and 6, SnCl$_2$-treated eyes; Lane 7, vehicle-treated lenses. Lower panel, ethidium bromide staining of the gel (Laniado-Schwartzman et al., 1997).

the HO-1 gene by SnCl$_2$ is followed by translation into functional HO enzyme in the cultures.

EFFECT OF SnCl$_2$ ON CORNEAL EPITHELIAL HO-1 mRNA *IN VIVO*

To determine whether induction of HO-1 would be associated with moderation or suppression of the inflammatory response to contact lens wear, total RNA was extracted from control and closed eye-hydrogel contact lens-treated corneal epithelium ± SnCl$_2$ (500 µM) after 6 days of wear. As shown in Fig. 5, Northern blot analysis of HO mRNA showed detectable levels of HO-1 mRNA for the untreated control eyes (Lanes 1–4). In corneal epithelium from vehicle-treated lenses, HO-1 mRNA was elevated slightly over the control (Lane 7). Corneal epithelium from SnCl$_2$ treated contact lenses, in contrast, demonstrated a marked increase in HO-1 mRNA levels (Lanes 5 and 6). Ethidium bromide staining indicated that similar amounts of RNA were loaded in each lane. The *in vivo* induction of HO mRNA at 6 days was associated with a substantial increase in HO enzyme activity and was further correlated with the severity of the *in situ* inflammatory response (Figs. 6 and 7).

Figure 6 depicts representative slit lamp photographs of the ocular surfaces (n = 3–6) after 6 days of closed eye-hydrogel contact lens wear. In the vehicle-treated eyes, there was a progressive *in situ* inflammatory response becoming prominent at day 3 (Conners et al., 1995a) and more severe at day 6 (Fig. 6a). The inflammatory response consisted of limbal vasodilation, conjunctional swelling, dilation of the iridial vessels, decreased corneal transparency (increased thickness, cloudiness and opacity) and neovascularization. One-time treatment of the hydrogel lenses with SnCl$_2$ (100 µg/mL) resulted in marked attenuation of the inflammatory response (Fig. 6B).

SnCl$_2$ was effective in suppressing both limbal and iridial vasodilation as well as the extent of epithelial defects and neovascularization of the cornea. Quantitative analysis by subjective (blinded) inflammatory scoring showed a significant correlation between the treatment with SnCl$_2$, which attenuated the increase in inflammatory response beginning at day 3 and reaching significance at day 7 (alleviation by approximately 60%). Figure 7 summarizes the effect of SnCl$_2$ treatment on the *in situ* inflammatory response (Fig. 7A). Corneal thickness, indicative of corneal edema was decreased by 60% as compared with eyes with untreated contact lenses (Fig. 7B), while the subjective inflammatory score was reduced by 75% (Fig. 7C).

SnCl$_2$, dose-dependently decreased arachidonic acid metabolism by homogenates of the corneal epithelium to 12 HETE and 12 HETrE as well as corneal trickiness (Fig. 8). Correlations were established between the synthesis of 12-HETE and 12-HETrE, the subjective inflammatory score and the progressive increase in corneal thickness over 9 days (Conners et al., 1995a; Conners et al., 1995b).

These results and their correlations implicate 12-HETE and 12-HETrE as mediators, among others, of the inflammatory response to the hypoxic injury produced by the hydrophilic contact lenses in the closed rabbit eye. Since the cytochrome P450 molecule contains heme, the effect of inducing HO-1 with SnCl$_2$ on this progressive inflammatory response was next studied (Moqattash et al., 1994). Once again, the hydrophilic lens closed eye contact lens wear model provoked an progressive inflammatory response. Coinciding with these events was a time-dependent increase in corneal thickness and 12-HETE and 12-HETrE production rates by corneal epithelial homogenates. Hydration of the lenses with SnCl$_2$ (100 µg/ml) significantly attenuated, by day 7, the inflammatory score (56% decrease), corneal thickness (17% decrease) and 12-HETE and 12-HETrE synthesis (77% and 71% decreases, respectively). This study further substantiates the involvement of cytochrome P450, through the synthesis of 12-HETE and 12-HETrE, in the hypoxia-induced inflammatory response to hydrophilic contact lenses in the closed eye. This dramatic effect was attributed, but not proven to be due, to the induction of HO-1 by both the hypoxia and SnCl$_2$.

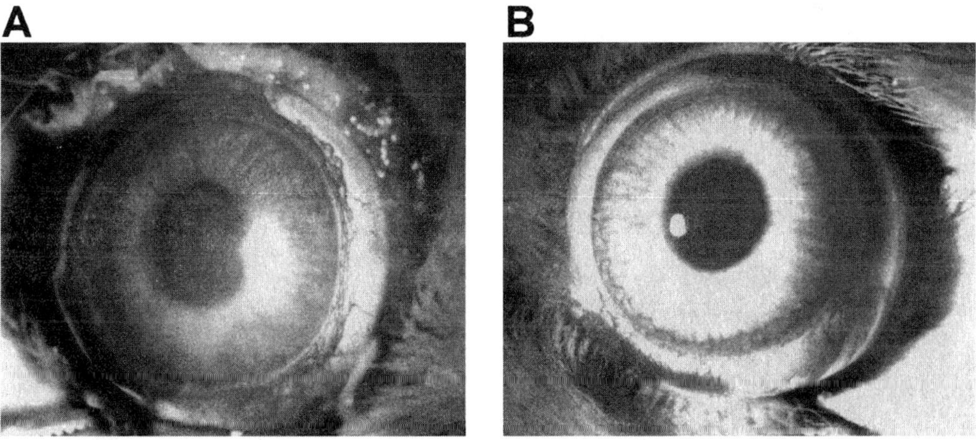

Figure 6. Effect of one-time treatment with SnCl$_2$ on the *in situ* ocular inflammatory response at 6 days of closed eye-hydrogel contact lens wear. Lenses were treated with either vehicle (0.1 M phosphate buffer, pH 7.4) or SnCl$_2$ (100 µg/mL) for 30 min prior to lens placement and tarsorrhaphy. At 6 days, sutures were removed and eyes were examined. Biomicroscopic photographs of the ocular surface were taken with a photographic attachment to the slit lamp. Photographs are representative of n = 5 from each group. (A) Eyes given vehicle-treated lenses (day 6, vehicle); (B) eye given SnCl$_2$-treated lenses. In (A), arrow indicates area of neovascular ingrowth and represents regions of corneal edema, both of which were attenuated significantly in (B).

This increase in HO activity was associated with a dramatic attenuation in inflammation (Hegazy et al., 2000; Laniado-Schwartzman et al., 1997) Its proximate mechanism likely involves a number of factors, including enhanced catabolism of the prooxidant heme to the antioxidant metabolites biliverdin and bilirubin; diminution in the cellular content and activity of hemeprotein species, such as the cytochrome P450 isozymes involved in the production of pro-inflammatory mediators, as shown earlier and, possibly, the local generation of carbon monoxide, which could inactivate other heme-containing proteins that contribute to the inflammatory response.

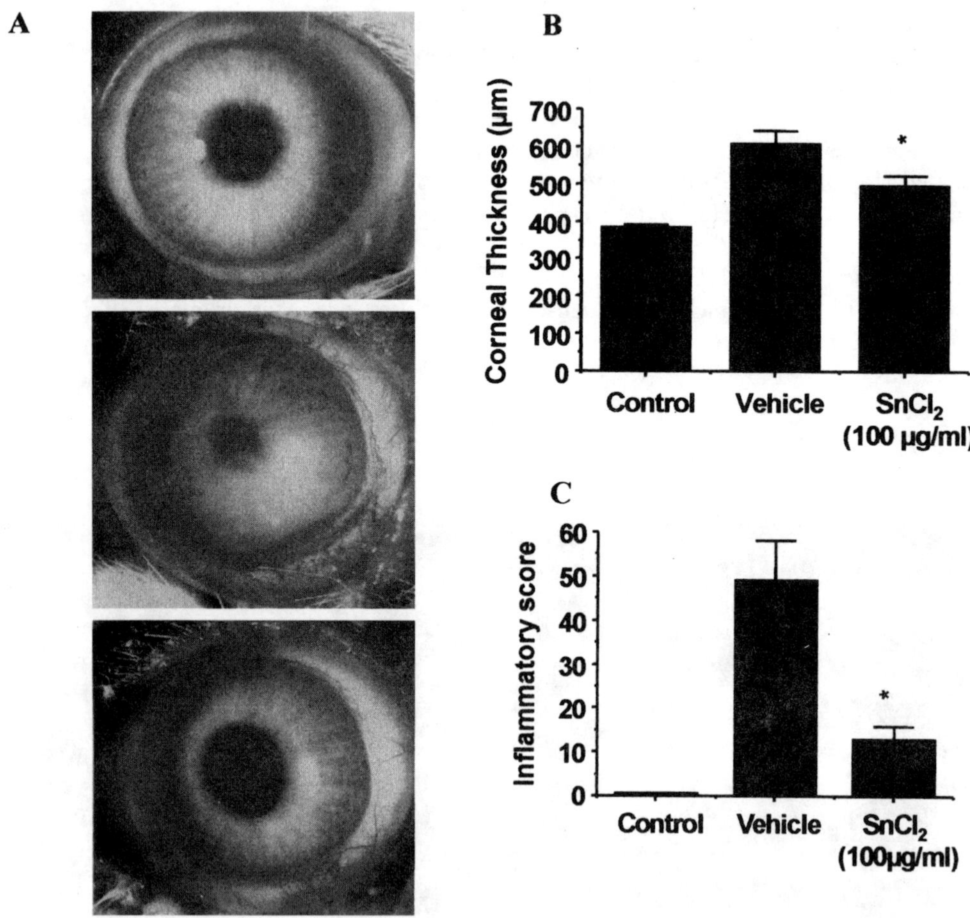

Figure 7. Effect of one-time treatment of SnCl$_2$ on corneal thickness and inflammatory score in the eyes after 6 days of exposure to lenes treated with either vehicle (0.1 M phosphate buffer, pH 7.4) or SnCl$_2$ (100 µg/mL) in phosphate buffer for 30 min prior to lens placement and tarsorrhaphy. At 6 days, sutures were removed and eyes were examined. Corneal thickness was assessed by ultrasonic pachymetry as previously described (Laniado-Schwartzman et al., 1997) Biomicroscopic photographs of the ocular surface were taken with a photographic attachment to the slit lamp (A; in ascending order: control, untreated lens and SnCl2-treated lens). (C) Photographs were subjectively scored as previously described (Laniado-Schwartzman et al., 1997). Data are presented as mean ± SEM (n = 5). *p < 0.05 versus control.

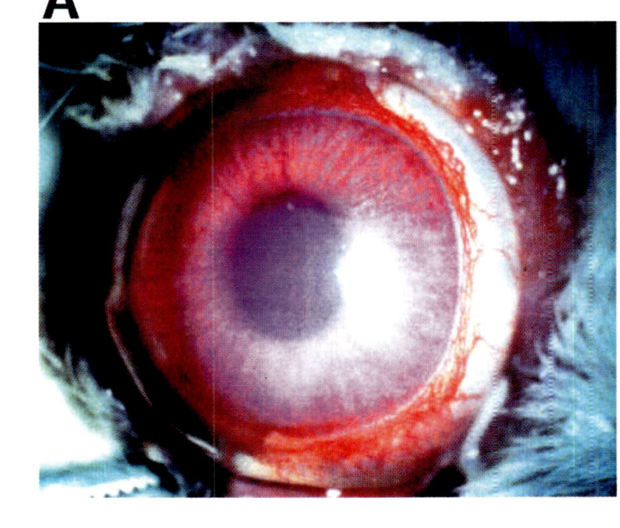

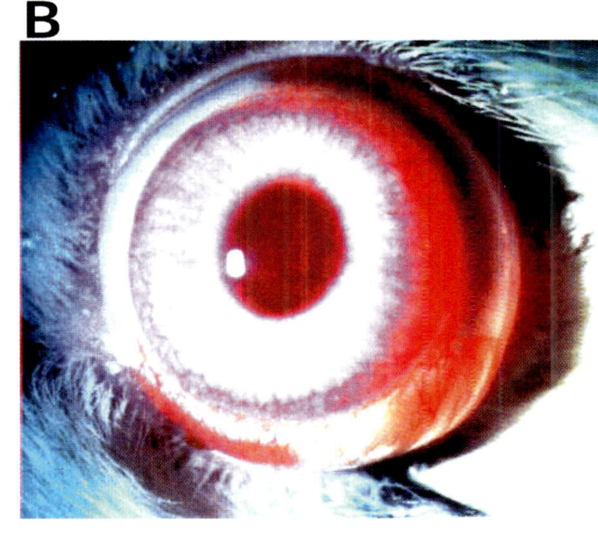

Figure 6. Effect of one-time treatment with SnCl$_2$ on the *in situ* ocular inflammatory response at 6 days of closed eye–hydrogel contact lens wear. Lenses were treated with either vehicle (0.1 M phosphate buffer, pH 7.4) or SnCl$_2$ (100 μg/mL) for 30 min prior to lens placement and tarsorrhaphy. At 6 days, sutures were removed and eyes were examined. Biomicroscopic photographs of the ocular surface were taken with a photographic attachment to the slit lamp. Photographs are representative of n = 5 from each group. (A) Eyes given vehicle-treated lenses (day 6, vehicle); (B) eye given SnCl$_2$-treated lenses. In (A), arrow indicates area of neovascular ingrowth and represents regions of corneal edema, both of which were attenuated sugnificantly in (B).

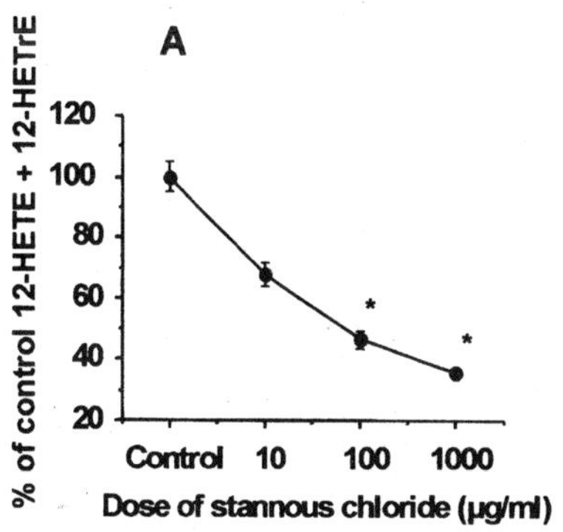
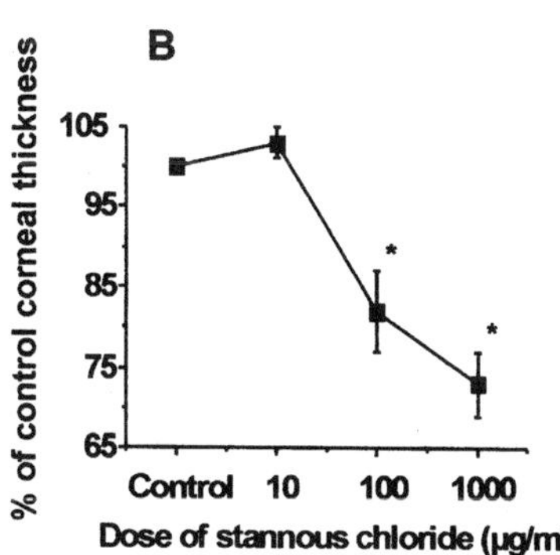

Figure 8. Dose-dependent effect of $SnCl_2$ at 6 days of closed eye hydrogel lens wear on (**A**) 12-HETE + 12-HETrE synthesis by the corneal epithelium and (**B**) the degree of corneal thickness. Data are expressed as a percentage of the control. Control specific activity for 12-HETE + 12-HETrE was 282 ± 50 ng/h mg (n = 4). Control corneal thickness was $607 \pm 48 \mu m$ (n = 4). Significant differences (*$p < 0.05$) were found at $100 \mu g/ml$ and $1,000 \mu g/ml$ doses compared to control. However, there were no significant differences between the $100 \mu g/ml$ and $1,000 \mu g/ml$ doses.

Furthermore, Nakagami et al., have shown that biliverdin and bilirubin are potent anti-complement factors, thus, providing an additional dimension to the anti-inflammatory effects of overexpression of the HO-1 gene. (Nakagami et al., 1993).

The exact mechanism by which $SnCl_2$ attenuates the corneal inflammatory response in the closed eye hydrophilic contact lens model of hypoxic injury remains unknown, but may include some or all of the mechanisms mentioned above. In characterizing this model of hypoxic injury, significant correlations were established between the synthesis of cytochrome P450 eicosanoids, the inflammatory score (r = 0.963) and the progressive increase in corneal thickness (r = 0.971), thus, implicating these eicosanoids as mediators of the inflammatory response. Additionally, the administration of $SnCl_2$ significantly inhibited corneal epithelial synthesis of the cytochrome P450 eicosanoids (82%) (Conners et al., 1995a; Conners et al., 1995b), suggesting that inhibition of cytochrome P450 activity constitutes part of the mechanism of the anti-inflammatory effect of HO-1 induction. Several studies on other organs in *in vivo* animal studies have reported similar results, clearly affirming the role of HO-1 as a tissue-protective response to injury and inflammation (Agarwal et al., 1995; Nath et al., 1992; Willis et al., 1996). The ease and potency of HO-1 induction by $SnCl_2$, associated with a dramatic reduction in inflammation, suggests a potential therapeutic role for this strategy in the future.

REFERENCES

Abraham, N.G., 1997, Retrovirus, and adenovirus mediated interferon gene transfer into stem cells of normal and CML: Ann.Hematol., v. 74(I), p. 20.
Abraham, N.G., J.H. Lin, M.W. Dunn, and M.L. Schwartzman, 1987, Presence of heme oxygenase and NADPH cytochrome P-450 (c) reductase in human corneal epithelium: Invest Ophthalmol.Vis.Sci., v. 28, p. 1464–1472.

Agarwal, A., J. Balla, J. Alam, A.J. Croatt, and K.A. Nath, 1995, Induction of heme oxygenase in toxic renal injury: a protective role in cisplatin nephrotoxicity in the rat: Kidney Int., v. 48, p. 1298–1307.

Braunstein, S., T. Deramaudt, D. Rosenblum, M. Dunn, and N.G. Abraham, 1999, Heme oxygenase-1 gene expression as a stress index to ocular irritation: Curr.Eye Res., v. 19, p. 115–122.

Conners, M.S., R.A. Stoltz, K.L. Davis, M.W. Dunn, N.G. Abraham, and R.D. Levere, 1995a, A closed eye contact lens model of corneal inflammation. Part 2: Inhibition of cytochrome P450 arachidonic acid metabolism alleviates inflammatory sequelae: Invest Ophthalmol.Vis.Sci., v. 36, p. 841–850.

Conners, M.S., R.A. Stoltz, S.C. Webb, J. Rosenberg, M.W. Dunn, and N.G. Abraham, 1995b, A closed eye contact lens model of corneal inflammation. Part 1: Increased synthesis of cytochrome P450 arachidonic acid metabolites: Invest Ophthalmol.Vis.Sci., v. 36, p. 828–840.

Davis, K.L., J.C. McGiff, R.D. Levere, and N.G. Abraham, 1988, Purification and characterization of cytochrome P-450-dependent arachidonic acid epoxygenase from human liver: J.Biol.Chem., v. 263, p. 2536–2542.

Dunn, M.W., 1997, Is 12(R)-HETrE present in overnight tears?: Proc.Ninth Contact Lens Research Congress, Florence Italy, v. Sept.

Dunn, M.W., 1999, Stannous chloride reduces levels of polymorphonuclear cells in overnight tears: Proc.Tenth Int.Contact Lens Research Congress.Thuket, Thailand, v. August.

Hegazy, K.A., M.W. Dunn, and S.C. Sharma, 2000, Functional human heme oxygenase has a neuro-protective effect on adult rat ganglion cells after pressure-induced ischemia: Neuroreport, v. 11, p. 1185–1189.

Laniado-Schwartzman, M., N.G. Abraham, M. Conners, M.W. Dunn, R.D. Levere, and A. Kappas, 1997, Heme oxygenase induction with attenuation of experimentally induced corneal inflammation: Biochem.Pharmacol., v. 53, p. 1069–1075.

Masferrer, J.L., J.A. Rimarachin, M.E. Gerritsen, J.R. Falck, P. Yadagiri, and M.W. Dunn, 1991, 12(R)-hydroxyeicosatrienoic acid, a potent chemotactic and angiogenic factor produced by the cornea: Exp.Eye Res, v. 52, p. 417–424.

Mieyal, P.A., M.W. Dunn, and M. Schwartzman, 1987, Detection of endogenous 12-hydroxyeicosatrienoic acid in human tear film: Invest.Ophthalmol.Vis.Sci., v. 28, p. 1464.

Moqattash, S., N.G. Abraham, and J.D. Lutton, 1994, Modulation of growth supportive and oncogenic properties of murine leukemia cells: Leuk.Res, v. 18, p. 531–539.

Nakagami, T., K. Toyomura, T. Kinoshita, and S. Morisawa, 1993, A beneficial role of bile pigments as an endogenous tissue protector: anti-complement effects of biliverdin and conjugated bilirubin: Biochim.Biophys.Acta, v. 1158, p. 189–193.

Nath, K.A., J. Balla, H.S. Jacob, G.M. Vercellotti, M. Levitt, and M.E. Rosenberg, 1992, Induction of heme oxygenase is a rapid protective response in rhabdomyolysis in the rat: Journal of Clinical Investigation, v. 90, p. 267–270.

Sack, R.A., K.O. Tan, and A. Tan, 1992, Diurnal tear cycle: evidence for a nocturnal inflammatory constitutive tear fluid: Invest Ophthalmol.Vis.Sci., v. 33, p. 626–640.

Willis, D., A.R. Moore, R. Frederick, and D.A. Willoughby, 1996, Heme oxygenase: a novel target for the modulation of the inflammatory response.: Nat Med, v. 2. 2, p. 87–90.

16

TARGETED EXPRESSION OF HEME OXYGENASE-1 AND PULMONARY RESPONSES TO HYPOXIA

Stella Kourembanas[a], Tohru Minamino[a], Helen Christou[a], Chung-Ming Hsieh[b], Yuxiang Liu[a], Vijender Dhawan[a], Laura Atwater[a], Nader G. Abraham[c], Mark A. Perrella[b,d], and S. Alex Mitsialis[a]

[a]Department of Medicine
Division of Newborn Medicine
Children's Hospital and Department of Pediatrics
Harvard Medical School
Boston, Massachusetts 02115 USA
[b]Program of Developmental Cardiovascular Biology
the Cardiovascular Division and
[c]Department of Pharmacology
New York Medical College
Valhalla NY 10595 USA
the [d]Pulmonary and Critical Care Division
Brigham & Women's Hospital Boston
MA 02115 USA

1. INTRODUCTION

Acute hypoxia in the lung causes arteriolar vasoconstriction whereas prolonged hypoxia promotes proliferation and migration of vascular smooth muscle cells (VSMC) and extracellular matrix deposition in the arterial wall, a process known as

Reproduced from *Proc. Natl. Acad. Sci. USA* **98**(15): 8798–8803, 2001. Copyright (2001) National Academy Sciences, USA.
Address correspondence to: Stella Kourembanas, M.D.; Division of Newborn Medicine; Children's Hospital, Enders 9; 300 Longwood Avenue; Boston, MA 02115; Phone: (617) 355–7383, Fax: (617) 355–7677, E-mail: Stella.Kourembanas@tch.harvard.edu

193

vascular remodeling.[1] These abnormalities are characteristic of pulmonary hypertension.[2] Several clinical conditions characterized by lung inflammation have been linked to the development of chronic pulmonary hypertension.[3] Interestingly, perivascular inflammatory cell infiltration as well as increased serum levels of pro-inflammatory cytokines, such as interleukin (IL)-1β and IL-6, have been reported in clinical cases of primary pulmonary hypertension.[4,5] However, little attention has been given up to now to the role of pulmonary inflammation in the pathogenesis of pulmonary hypertension induced by hypoxia.

Heme oxygenase (HO; EC 1.14.99.3) catalyzes the oxidation of heme to carbon monoxide (CO) and biliverdin, which is then converted to bilirubin by biliverdin reductase. Three isoforms of HO have been identified: the inducible, HO-1 and the constitutively-expressed HO-2 and HO-3.[6,7] Our previous *in vitro* data suggest that CO released by HO-1 confers protection against vasoconstriction and vascular remodeling induced by hypoxia.[8-10] More recently, Soares et al. have suggested anti-inflammatory properties of HO-1 in a cardiac transplantation model although the molecular mechanisms have not been fully elucidated.[11] Our recent *in vivo* data using an HO-1 null mouse model suggest that HO-1 plays a central role in protecting the right ventricle from hypoxic pulmonary pressure-induced injury.[12]

In the present study, we established transgenic mice that overexpress HO-1 in the lung and exposed them to hypoxia to investigate the effects of HO-1 activity on the development of pulmonary hypertension. We found that, in wild-type animals, hypoxia caused pulmonary hypertension with vascular remodeling as well as a striking inflammatory response in the lung parenchyma. In contrast, overexpression of HO-1 protected against the development of pulmonary hypertension as well as the hypoxia-induced inflammatory cell infiltration. Significantly, overexpression of HO-1 attenuated the hypoxic induction of pro-inflammatory cytokines and chemokines in the lung. These findings point to novel molecular mechanisms underlying the anti-inflammatory properties of HO-1 that may be crucial for the body's adaptive responses to hypoxia.

2. GENERATION OF HO-1 TRANSGENIC MICE

Founder lines of transgenic mice harboring the human HO-1 gene under the control of the human Surfactant Protein-C (SP-C) promoter were obtained and the inheritance of the transgene was confirmed by Southern blotting of genomic DNA. Expression levels of HO-1 mRNA and protein in the lungs from one line of transgenic mice were 4–6 fold higher than their non-transgenic littermates while no difference in HO-2 expression was detected in the lungs between transgenic and non-transgenic control mice (Fig. 1a and b). Furthermore, 2.5 fold increase in HO activity was observed in the lungs of this line as compared with non-transgenic control mice (163.3 ± 15.1 vs. 60.3 ± 9.4 pmol/hour/mg protein, $p < 0.05$, 6–7 mice for each group, Mann Whitney U-test). A less significant increase in levels of HO-1 protein and enzymatic activity (<2-fold) were detected in the lungs of a second line compared to wildtype animals. Both lines of mice were exposed to hypoxia to determine the development of pulmonary hypertension compared to their non-transgenic littermates, as described below.

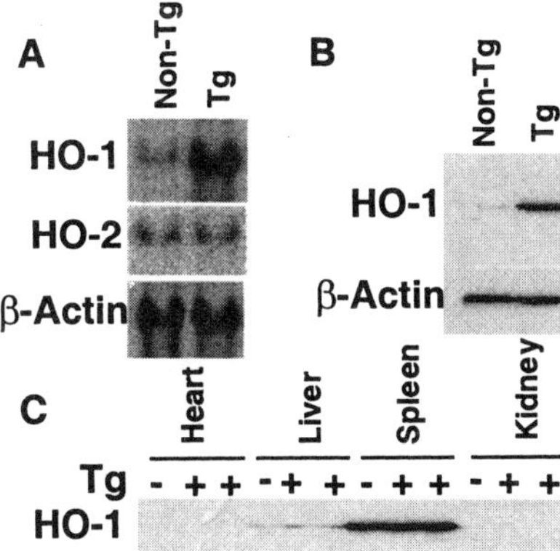

Figure 1. Expression of HO-1 in transgenic mice. (**a**) Nothern blot analysis for HO-1 and HO-2. *HO-1* (upper panel) and *HO-2* mRNA levels (middle panel) were examined in total RNA (30 μg) extracted from the lungs of HO-1 transgenic mice (Tg) or non-transgenic littermates (non-Tg). β-*actin* mRNA levels were examined to confirm equal loading (lower panel). This blot is representative of three independent experiments, six mice per group. (**b**) Western blot analysis for HO-1 in the lung. Lung whole cell lysates (30 μg) from five to six HO-1 Tg and a corresponding number of non-Tg littermates were analyzed for HO-1 protein levels using an anti HO 1 antibody. The same blot was reprobed with an anti β actin antibody to verify equal loading (lower panel). Note that the anti-HO-1 antibody used here is raised against the N-terminal amino acid sequence of the protein, a region conserved between mouse and human HO-1. (**c**) Western blot analysis for HO-1 in non-lung tissues. Tissue distribution of HO-1 was analyzed in whole cell lysates (30 μg) of heart, liver, spleen and kidney from Tg mice (+) or non-Tg littermates (−). This blot represents identical patterns of expression observed in three Tg and three non-Tg animals.

3. LUNG-SPECIFIC EXPRESSION OF HO-1 TRANSGENE

High levels of HO-1 protein were observed in the spleen, moderate levels in the liver and low levels in the heart and kidney by Western analysis (Fig. 1c). In all tissues, HO-1 levels were equal in both transgenic (Tg+) and non-transgenic control mice (Tg−) verifying that overexpression of HO-1 was successfully targeted to the lung. Due to the tissue specificity of the SP-C promoter,[13,14] immunostaining for HO-1 was observed in the alveolar cells of the lungs of transgenic mice, while no HO-1 specific immunostaining was observed in the lungs of non-transgenic control mice (data not shown), consistent with the low levels of endogenous HO-1 expression detected through Western blot analysis (Fig. 1b). No structural abnormalities were noted in the lungs of transgenic mice up to the age of 2 years (data not shown).

4. DEVELOPMENT OF HYPOXIA-INDUCED PULMONARY HYPERTENSION: TIME COURSE AND DOSE-DEPENDENT PROTECTION BY HO-1

Right ventricular hypertrophy (RVH) is a hallmark of pulmonary hypertension resulting from RV pressure overload. The ratio of right ventricle weight to left ventricle plus septum weight (RV/(LV + S)) was measured to assess the development

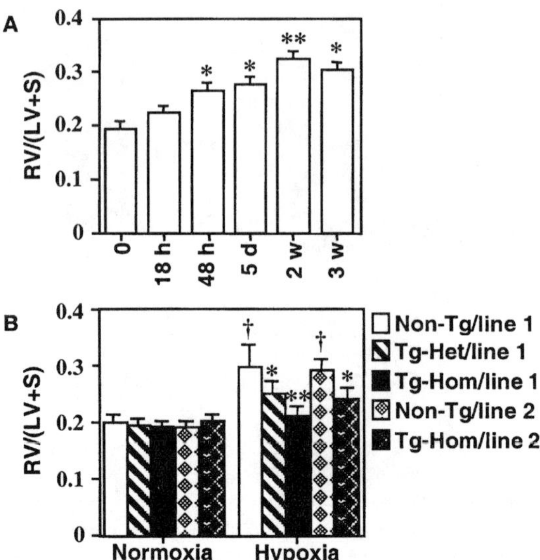

Figure 2. Time course of hypoxic right ventricular hypertrophy: protection by HO-1 overexpression. (**a**) Non-Tg mice were exposed to hypoxia (8–10% O_2) for 18 hours (h), 5 days (d), or 2–3 weeks (w). Development of RVH after hypoxic exposure was determined by the RV/(LV + S) ratio. *p < 0.01, **p < 0.001 vs. time point 0. (n = 6 for each group, one-way ANOVA). (**b**) Tg homozygous, heterozygous and non-Tg control mice from founder 1 and homozygous Tg and non-Tg mice from line 2 were exposed to normoxia or hypoxia for 3 weeks. Under normoxia, line 1 homozygous Tg (n = 19), heterozygous Tg (n = 6) and non-Tg controls (n = 14) as well as line 2 homozygous Tg (n = 5) or non-Tg (n = 5) did not differ in baseline RV/(LV + S) (p > 0.05). After hypoxic exposure, line 1 homozygous Tg (n = 26) and heterozygous Tg mice (n = 8) show significant reduction in RV/(LV + S) as compared with non-Tg controls (n = 22). Similarly, homozygous line 2 Tg mice (n = 5) manifest significantly lower RV/(LV + S) values as compared with their non-Tg controls (n = 5) exposed to hypoxia. (*p < 0.05, **p < 0.005 vs. hypoxic controls; †p < 0.01 vs. normoxic controls, Mann Whitney U-test). Data are mean ± SEM.

of RVH upon exposure of non-transgenic mice to hypoxia (8–10% O_2 environment) for varying periods of time. RV/(LV + S) was significantly increased as early as 48 hours after hypoxia, and RVH was developed in a time-dependent manner peaking at 2 weeks (Fig. 2a). Under these conditions, endogenous HO-1 in non-transgenic mouse lungs was minimally increased after 6 hr of hypoxia and remained significantly lower compared with lungs of transgenic mice for up to two weeks of hypoxic exposure (data not shown). HO-1 protein levels in transgenic mouse lungs were not affected by the hypoxic exposure and remained constitutively elevated (4.6 fold above non-transgenic controls).

To determine whether the overexpression of HO-1 in the lung protects against the development of pulmonary hypertension, transgenic and non-transgenic mice from both the high and low-expressing founder lines were exposed to hypoxia (8–10%) for 3 weeks, and RV/(LV + S) was measured. Under normoxia, there was no difference in RV/(LV + S) between transgenic and non-transgenic control mice (Fig. 2b). After hypoxic exposure, a significant increase in RV/(LV + S) was observed in non-transgenic mice, which was suppressed in transgenic mice. Heterozygous mice from line 1 and homozygous mice from line 2 whose lung HO-1 transgene levels were comparable (≤ 2 fold above controls) demonstrated intermediate protection with mild RVH. RV/(LV + S) was higher than normoxic mice but significantly lower than non-transgenic hypoxic control mice (Fig. 2b). This protection against pulmonary hypertension and secondary RVH persisted in transgenic mice up to 8 weeks of hypoxic exposure (data not shown).

RVSP was measured after 2 and 3 weeks of hypoxia and compared between transgenic and non-transgenic controls. Under normoxia, there was no difference in RVSP measurements, however, after hypoxic exposure a significant increase in RVSP was observed in non-transgenic mice compared with normoxic mice or with transgenic hypoxic mice (Fig. 3a). Although intermediate pressures were noted in transgenic hypoxic mice, they remained significantly elevated above the normoxic controls ($p < 0.01$). Morphometric analysis of the pulmonary vasculature revealed no difference in vessel wall architecture between transgenic and non-transgenic animals under normoxia (23% vs. 26% medial wall thickness, respectively, $p > 0.05$). Two weeks after hypoxic exposure, non-transgenic mice demonstrated a significant increase in the wall thickness of pulmonary arterioles which was completely inhibited in hypoxic transgenic mice (39% vs. 26%, $p < 0.05$) (Fig. 3b). Combined, these findings indicate that overexpression of HO-1 decreases active vasoconstriction and prevents pulmonary vascular remodeling with RVH. It is noted that transgenic and non-transgenic control mice did not differ in hematocrits either at baseline or after hypoxic exposure. Hematocrits were significantly elevated in both groups of mice after hypoxic exposure (data not shown).

5. OVEREXPRESSION OF HO-1 PROTECTS AGAINST PULMONARY INFLAMMATION INDUCED BY HYPOXIA

In addition to vascular remodeling, we particularly noted that hypoxic lungs exhibited hypercellularity compared with normoxic lungs of non-transgenic mice. This difference was more evident at early time points (48 hours) of hypoxic exposure and was significantly diminished after 2 weeks of hypoxia. Interestingly, this cellular infiltration was suppressed in the hypoxic lungs of transgenic mice. To determine whether the protective effects of HO-1 involve anti-inflammatory mechanisms we performed immunohistochemical staining for neutrophils and macrophages in lungs from mice exposed to hypoxia for 48 hours. A prominent infiltration of neutrophils was observed in the hypoxic lungs of non-transgenic mice while the minimal staining detected in hypoxic transgenic mice was comparable to that in normoxic control lungs (data not shown). Although macrophages were observed in the normoxic lung, hypoxia induced further accumulation of macrophages in the lungs of non-transgenic control mice but not in transgenic mice (data not shown). These observations were supported by BAL fluid analysis. The number of neutrophils and macrophages in BAL fluid was significantly increased after hypoxic exposure in non-transgenic mice while this increase was suppressed in transgenic mice under hypoxia (Fig. 4), suggesting that HO-1 overexpression protects against the pulmonary inflammation

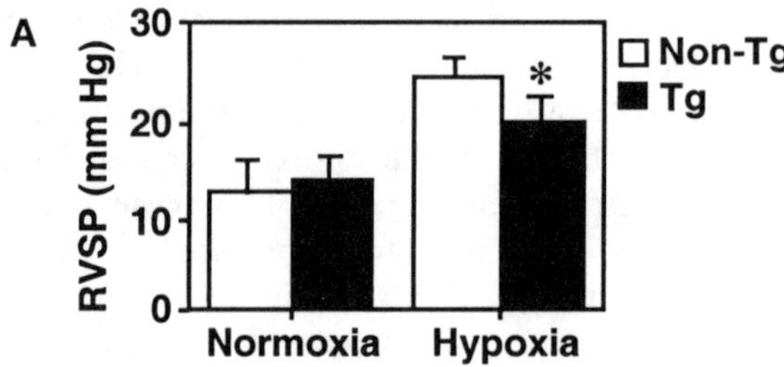

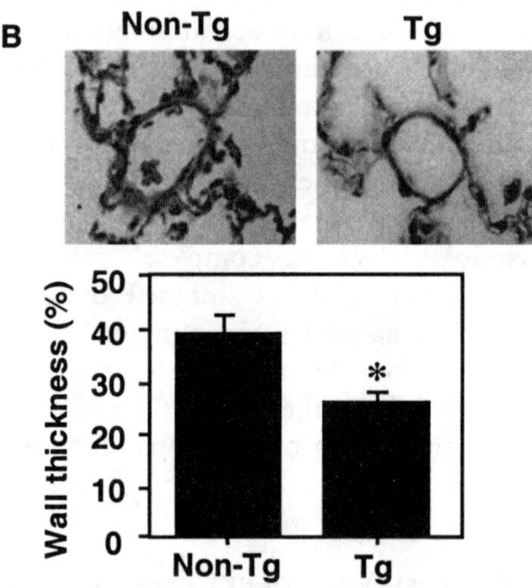

Figure 3. HO-1 overexpressing mice are protected from pulmonary hypertension induced by hypoxia. (a) RVSP measurements were performed in Tg and non-Tg mice exposed to 2–3 weeks of hypoxia. RVSP did not differ between Tg (n = 13) and non-Tg mice (n = 8) at baseline normoxic conditions. Under hypoxia, Tg mice (n = 25) had significantly lower pressure measurements than non-Tg controls (n = 8). (*p < 0.05, Mann Whitney U-Test). Data are mean ± SEM. (b) Hematoxylin and eosin staining of paraffin-embedded lung sections of Tg (right panel) or non-Tg control mice (left panel) exposed to hypoxia for 2–3 weeks. Original magnification: 400×. Percent wall thickness in arterioles of comparable size was estimated based on analysis of area bounded by external (ext) and internal (int) elastic lamina (lower graph). Significant reduction in the wall thickness (wall thickness (%) = $area_{ext} - area_{int}/area_{ext} \times 100$) is observed in Tg mice as compared with non-Tg control. Data are expressed as mean ± SEM (n = 5–7). *p < 0.05 vs. non-Tg control (Mann Whitney U-test).

induced by hypoxia. We have previously reported that mice deficient in HO-1 had a maladaptive response to chronic hypoxia (5–7 weeks exposure to 8–10% O_2) with right ventricular dilatation and infarction.[12] Of interest, the lungs of these mice manifested sustained hypercellularity with inflammatory cell infiltrates even after several weeks of hypoxia compared with normoxic null mice or wild type hypoxic mice (our unpublished observations).

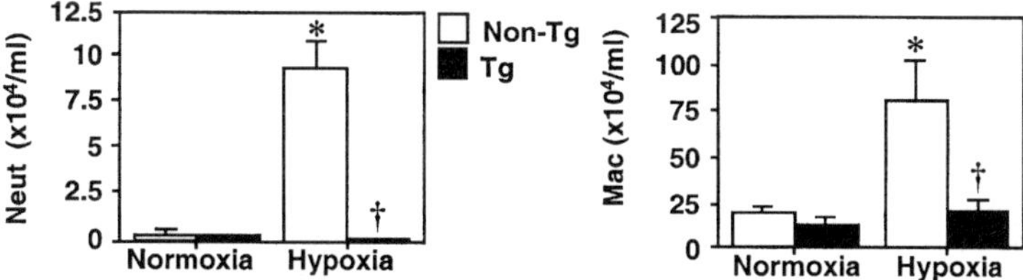

Figure 4. HO-1 overexpressing mice are protected against pulmonary inflammation induced by hypoxia. The number of neutrophils and macrophages in BAL fluid after hypoxia. Differential cell counts for neutrophils (neut, left panel) and macrophages (mac, right panel) were performed in the BAL fluid of Tg (closed bars) and non-Tg control mice (open bars) after 48 hours of hypoxia. Data are expressed as mean $\pm$ SEM (n = 4). *p < 0.05 vs. normoxic non-Tg control. †p < 0.05 vs. hypoxic non-Tg control (Mann Whitney U-test).

6. HO-1 OVEREXPRESSION INHIBITS INDUCTION OF PRO-INFLAMMATORY CYTOKINES AND CHEMOKINES UNDER HYPOXIA

The recruitment of leukocytes is regulated by the action of cytokines and chemokines.[15] These mediators also function as potent mitogens and chemoattractants for VSMC resulting in vascular remodeling.[16,17] To investigate mechanisms underlying the anti-inflammatory properties of HO-1 we examined cytokine expression in the hypoxic lung. We found that the expression of IL-1β, IL-6, monocyte chemotactic protein (MCP)-1 and macrophage inflammatory protein (MIP)-2, a functional murine homologue of human IL-8, were significantly elevated in the lungs of non-transgenic mice in response to hypoxia with a peak at 48 hours after exposure (Fig. 5, non-Tg). In contrast, other pro-inflammatory cytokines, such as tumor necrosis factor (TNF)α, interferon-γ and IL-1α, and chemokines such as

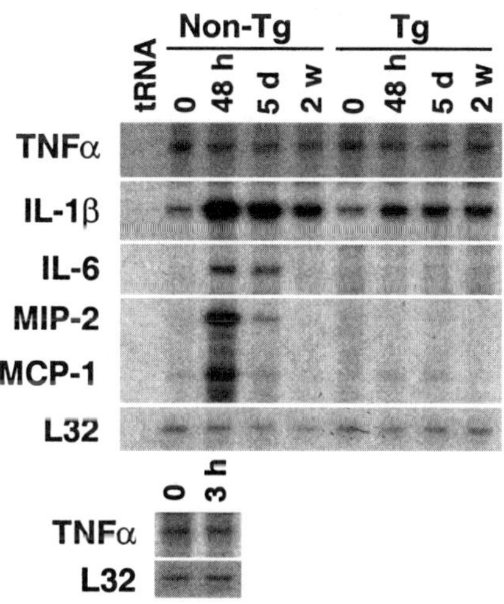

Figure 5. HO-1 overexpression inhibits hypoxic induction of cytokines in the lung. HO-1 Tg and non-Tg control mice were exposed to hypoxia (8–10% O_2) for the times indicated (48 hours (h), 5 days (d), or 2 weeks (w)). After exposure, total lung RNA (10 μg) was examined for cytokine expression by RNase protection assay. Levels of L32, a ribosomal RNA, were unaffected by hypoxia and served as an internal control. Expression of TNFα 3 hours after hypoxia in non-Tg mice is also shown (lower panel). Similar results were observed in four independent experiments.

RANTES were unaffected or only minimally affected (Fig. 5 and data not shown). Anti-inflammatory mediators, such as IL-10 were also unaffected (data not shown). More importantly, the induction of IL-1β, IL-6, MCP-1 and MIP-2 were significantly suppressed in the hypoxic lungs of transgenic mice (Fig. 5, Tg), suggesting that inhibition of these cytokines is involved in the mechanisms mediating the anti-inflammatory effects of HO-1. Along with RNA analysis, we correlated protein levels of MIP-2 in the lungs of these transgenic and non-transgenic hypoxic and normoxic mice. MIP-2 protein was undetectable at baseline in all normoxic transgenic and non-transgenic lungs. After 48 hours and 5 days of hypoxia, an average of 3.75 and 2.9 pg of MIP-2/mg total protein, respectively, were detected in non-transgenic lungs. MIP-2 protein levels remained undetectable in transgenic hypoxic lungs both at 48 hours and 5 days.

7. DISCUSSION

Chronic hypoxia is well-known to cause hypertension and vascular remodeling in the pulmonary vasculature in various animal models of human pathophysiology. Shorter, more pronounced hypoxic exposure (6% O_2 for 4–8 hours), leads to induction of tissue factor and results in fibrin deposition and thrombosis in the pulmonary vasculature of mice.[18] In the present study, we observed that less severe hypoxia (8–10% O_2 for 48 hours to 5 days) lead to a marked induction of pro-inflammatory cytokines and chemokines in association with pulmonary inflammation that preceded the development of pulmonary hypertension. Moreover, we demonstrated that overexpression of HO-1 in the lung inhibited cytokine gene expression and protected against both pulmonary inflammation as well as pulmonary hypertension induced by hypoxia.

In the process of pulmonary inflammation, cytokines are produced and amplified many-fold by target cells, such as monocytes, endothelial cells, and epithelial cells, leading to the recruitment of leukocytes that release a variety of pro-oxidants and proteolytic enzymes capable of causing vasoconstriction and lung injury.[15,19] In addition, pro-inflammatory cytokines and chemokines directly induce changes in vascular permeability and thrombogenicity,[20,21] processes implicated in the pathogenesis of pulmonary hypertension. These cytokines were also reported to promote proliferation and migration of VSMC *in vitro*[16,17] and disruption of the CCR2 chemokine receptor gene decreased vascular remodeling in a model of atherosclerosis *in vivo*.[22] Moreover, it was demonstrated that administration of an IL-1 receptor antagonist or anti-MCP-1 antibody protected against monocrotaline-induced pulmonary hypertension.[23,24] Together with our data of pronounced cytokine and chemokine induction in hypoxic pulmonary hypertension, these findings suggest a crucial role for pro-inflammatory cytokines and chemokines in the development of pulmonary vascular remodeling. Cytokine induction and cellular infiltration peak at 48 hours and decrease by two weeks of hypoxia, yet RVH and vascular remodeling persist for several weeks of continued hypoxia. It is possible that these cytokines initiate a cascade of events leading to VSMC proliferation that is later amplified by the release of vasoactive mediators and growth factors that alter vessel wall structure and function. For example, the initial injury and loss of endothelial barrier function can lead to extravasation of serum and activation of endogenous elastolytic activity from vascular smooth muscle cells.[25,26] Increased serine elastase activity results in proteolytic degradation of extracellular matrix and the release of matrix-bound growth factors amplifying smooth muscle cell proliferation and remodeling of the pulmonary arterioles.[25,27] Indeed, treatment of rats with elastase inhibitors completely reversed malignant pulmonary hypertension induced by monocrotaline.[28] In this report we demonstrate a pronounced neutrophil and macrophage infiltration in the lungs of hypoxic mice that is linked with pulmonary hypertension. It is possible that neutrophil elastase as well as other proteolytic enzymes may trigger similar molecular events in the hypoxic setting. Future studies will determine if inhibition of inflammation can prevent hypoxia-induced pulmonary hypertension.

The mechanism of protection provided by HO-1 may depend on its enzymatic product, CO, to function as a vasodilator against vasoconstriction and bilirubin as an anti-oxidant against the reactive oxygen species that are generated under hypoxic conditions.[29] Since the recruitment of leukocytes was greatly suppressed in the hypoxic lungs of HO-1 transgenic mice, inhibition of cytokine induction may be another important mechanism by which HO-1 activity imparts protection against the development of pulmonary hypertension induced by hypoxia. In two recent studies, HO-1 activity and, in particular, CO, were reported to have potent anti-inflammatory properties in models of vascular injury.[11,30] The anti-inflammatory properties of CO were mediated by the inhibition of pro-inflammatory cytokines, TNFα, IL1-β and MIP-1β in the model of LPS-induced systemic inflammation.[30] HO-1 was also shown to have protective effects in hyperoxic lung injury as well as ischemia-reperfusion in the liver, two well-known models of inflammation-mediated injury.[31,32] In this report, we demonstrate a previously-unrecognized inflammatory response induced by hypoxia. The mechanisms underlying the induction of cytokines by hypoxia are most likely different from other models of systemic or pulmonary inflammation. For example, in endotoxin-induced pulmonary inflammation models, early response cytokines, such as TNFα, are rapidly induced, (within 6 hours) and this early response leads to the activation of cytokine networks including the induction of IL-6, MCP-1 and MIP-2.[19] In contrast, induction of TNFα was not detected in the lungs 3, 6, and 18 hours after hypoxia (Fig. 5, and data not shown) indicating that the hypoxic induction of IL-6, MCP-1 and MIP-2 is not secondary to that of this early response cytokine. Similarly, IL-6 has been suggested to function as an anti-inflammatory cytokine in endotoxin-induced pulmonary inflammation since IL-6 deficiency resulted in an increase in pro-inflammatory cytokines including TNFα and IL-1β.[33] However, despite reduced IL-6 expression in the hypoxic lungs of transgenic mice, we did not observe increased expression of other pro-inflammatory cytokines. In addition, treatment with LPS induced expression of the anti-inflammatory cytokine IL-10, whereas we did not find increased IL-10 expression under hypoxia. Hypoxia, therefore, induces a network of cytokines and chemokines that is distinct from other inflammatory pathways such as those induced by LPS.

In summary, we demonstrated that hypoxia induced an early inflammatory response in the lung prior to the development of pulmonary hypertension. HO-1 overexpression effectively inhibited pulmonary inflammation and prevented pulmonary hypertension induced by hypoxia. HO-1 may be an important therapeutic target to inhibit hypoxia-induced cytokine signaling and prevent the inflammatory pathways that may lead to pulmonary hypertension.

ACKNOWLEDGMENTS

This work is dedicated to the memory of Dr. Arthur Mu-En Lee, an insightful scientist and warm friend. He will remain an inspirational role model to all of us.

Dr. Christou was supported by NIH KO8 HL03917. Dr. Perrella was supported by NIH RO1s HL60788 and GM53249. This work was supported by the American Heart Association and by National Institutes of Health grants RO1 HL55454 and SCOR 1P50 HL56398 to (SK). We also thank Jessica Johnson and Suzanne Thibert for their expert assistance in the preparation of this manuscript.

REFERENCES

1. S. Kourembanas, T. Morita, Y. Liu, and H. Christou, Mechanisms by which oxygen regulates gene expression and cell-cell interaction in the vasculature, *Kidney Intern.* **51**, 438–443 (1997).
2. R.P. Mecham, L.A. Whitehouse, D.S. Wrenn, W.C. Parks, G.L. Griffin, R.M. Senior, E.C. Crouch, K.R. Stenmark, and N.F. Voelkel, Smooth muscle-mediated connective tissue remodeling in pulmonary hypertension, *Science.* **237**, 423–426 (1987).
3. W.M. Zapol and M.T. Snider, Pulmonary hypertension in severe acute respiratory failure, *N. Engl. J. Med.* **3**, 476–480 (1977).
4. R.M. Tuder, B. Groves, D.B. Badesch, and N.F. Voelkel, Exuberant endothelial cell growth and elements of inflammation are present in plexiform lesions of pulmonary hypertension, *Am. J. Pathol.* **144**, 275–285 (1994).
5. M. Humbert, G. Monti, F. Brenot, O. Sitbon, A. Portier, L. Grangeot-Keros, P. Duroux, P. Galanaud, G. Simonneau, and D. Emilie, Increased interleukin-1 and interleukin-6 serum concentrations in severe primary pulmonary hypertension, *Am. J. Respir. Crit. Care Med.* **151**, 1628–1631 (1995).
6. N.G. Abraham, G.S. Drummond, J.D. Lutton and A. Kappas, The biological significance and physiological role of heme oxygenase, *Cell. Physiol. Biochem.* **6**, 129–168 (1996).
7. W.K. McCoubrey, T.J. Huang, and M.D. Maines, Isolation and characterization of a cDNA from the rat brain that encodes hemoprotein heme oxygenase-3, *Eur. J. Biochem.* **247**, 725–732 (1997).
8. T. Morita, M.A. Perrella, M.-E. Lee, and S. Kourembanas, Smooth muscle cell-derived carbon monoxide is a regulator of vascular cGMP, *Proc. Natl. Acad. Sci. USA.* **92**, 1475–1479 (1995).
9. T. Morita and S. Kourembanas, Endothelial cell expression of vasoconstrictors and growth factors is regulated by smooth muscle cell-derived carbon monoxide, *J. Clin. Invest.* **96**, 2676–2682 (1995).
10. T. Morita, S.A. Mitsialis, H. Koike, Y. Liu, and S. Kourembanas, Carbon monoxide controls the proliferation of hypoxic vascular smooth muscle cells, *J. Biol. Chem.* **272**, 32804–32809 (1997).
11. M.P. Soares, Y. Lin, J. Anrather, E. Csizmadia, K. Takigami, K. Sato, S.T. Grey, R.B. Colvin, A.M. Choi, K.D. Poss, and F.H. Bach, Expression of heme oxygenase-1 can determine cardiac xenograft survival, *Nature Med.* **4**, 1073–1077 (1998).
12. S.-F. Yet, M.A. Perrella, M.D. Layne, C.-M. Hsieh, K. Maemura, L. Kobzik, P. Wiesel, H. Christou, S. Kourembanas, and M.-E. Lee, Hypoxia induces severe right ventricular dilatation and infarction in heme oxygenase-1 null mice, *J. Clin. Invest.* **103**, R23–R29 (1999).

13. T.R. Korfhagen, S.W. Glasser, S.E. Wert, M.D. Bruno, C.C. Daugherty, J.D. McNeish, J.L. Stock, S.S. Potter, and J.A. Whitsett, Cis-acting sequences from a human surfactant protein gene confer pulmonary-specific gene expression in transgenic mice, *Proc. Natl. Acad. Sci. USA.* **87**, 6122–6126 (1990).

14. S.W. Glasser, T.R. Korfhagen, S.E. Wert, M.D. Bruno, K.M. McWilliams, D.K. Vorbroker, and J.A. Whitsett, Genetic element from human surfactant protein SP-C gene confers bronchiolar-alveolar cell specificity in transgenic mice, *Am. Phys. Soc.* ••, L349–L356 (1991).

15. A. Xing, M. Jordana, J. Gauldie, and J. Wang, Cytokines and pulmonary inflammatory and immune diseases, *Histol. Histopathol.* **14**, 185–201 (1999).

16. M. Roth, M. Nauck, M. Tamm, A.P. Perruchoud, R. Ziesche, and L.H. Blck, Intracellular interleukin 6 mediates platelet-derived growth factor-induced proliferation of nontransformed cells, *Proc. Natl. Acad. Sci. USA.* **92**, 1312–1316 (1995).

17. T.L. Yue, X. Wang, C.P. Sung, B. Olson, P.J. McKenna, J.L. Gu, and G.Z. Feuerstein, Interleukin-8. A mitogen and chemoattractant for vascular smooth muscle cells, *Circ. Res.* **75**, 1–7 (1994).

18. C.A. Lawson, S.D. Yan, S.F. Yan, H. Liao, Y.S. Zhou, J. Sobel, W. Kisiel, D.M. Stern, and D.J. Pinsky, Monocytes and tissue factor promote thrombosis in a murine model of oxygen deprivation, *J. Clin. Invest.* **99**, 1729–1738 (1997).

19. R.M. Strieter, S.L. Kunkel, M.P. Keane, and T.J. Standford, Chemokines in lung injury, *Chest.* **116**, 103S–110S (1999).

20. M.H. Ali, S.A. Schlidt, N.S. Chandel, K.L. Hynes, P.T. Schumacker, and B.L. Gewertz, Endothelial permeability and IL-6 production during hypoxia: role of ROS in signal transduction, *Am. J. Physiol.* **277**, L1057–L1065 (1999).

21. A.D. Schecter, B.J. Rollins, Y.J. Zhang, I.F. Charo, J.T. Fallon, M. Rossikhina, P.L.A. Giesen, Y. Nemerson, and M.B. Taubman, Tissue factor is induced by monocyte chemoattractant protein-1 in human aortic smooth muscle and THP-1 cells, *J. Biol. Chem.* **272**, 28568–28573 (1997).

22. I. Boring, J. Gosling, M. Cleary, and I.F. Charo, Decreased lesion formation in CCR2$^{-/-}$ mice reveals a role for chemokines in the initiation of atherosclerosis, *Nature.* **394**, 894–897 (1998).

23. N.F. Voelkel, R.M. Tuder, J. Bridges, and W.P. Arend, Interleukin-1 receptor antagonist treatment reduces pulmonary hypertension generated in rats by monocrotaline, *Am J Respir Cell Mol Biol.* **11**, 664–675 (1994).

24. H. Kimura, Y. Kasahara, K. Kurosu, K. Sugito, Y. Takiguchi, M. Terai, A. Mikata, M. Natsume, N. Mukaida, K. Matsushima, and T. Kuriyama, Alleviation of monocrotaline-induced pulmonary hypertension by antibodies to monocyte chemotactic and activating factor/monocyte chemoattractant protein-1, *Lab. Invest.* **78**, 571–581 (1998).

25. K. Thompson and M. Rabinovitch, Exogenous leukocyte and endogenous elastases can mediate mitogenic activity in pulmonary artery smooth muscle cells by release of extracellular matrix-bound basic fibroblast growth factor, *J. Cell. Physiol.* **166**, 495–505 (1996).

26. K. Maruyama, C. Ye, M. Woo, H. Venkatacharya, L.D. Lines, M.M. Silver, and M. Rabinovitch, Chronic hypoxic pulmonary hypertension in rats and increased elastolytic activity, *Am. J. Physiol.* **261**, H1716–H1726 (1991).

27. P.L. Jones, K.N. Cowan, and M. Rabinovitch, Tenascin-C, proliferation and subendothelial accumulation of fibronectin in progressive pulmonary vascular disease, *Am. J. Pathol.* **150**, 1349–1360 (1997).

28. K.N. Cowan, A. Heilbut, T. Humpl, C. Lam, S. Ito, and M. Rabinovitch, Complete reversal of fatal pulmonary hypertension in rats by a serine elastase inhibitor, *Nat. Med.* **6**, 698–702 (2000).

29. N.S. Chandel, E. Maltepe, E. Goldwasser, C.E. Mathieu, M.C. Simon, and P.T. Schumacker, Mitochondrial reactive oxygen species trigger hypoxia-induced transcription, *Proc. Natl. Acad. Sci. USA.* **95**, 11715–11720 (1998).

30. L.E. Otterbein, F.H. Bach, J. Alam, M. Soares, H.T. Lu, M. Wysk, R.J. Davis, R.A. Flavell, and A.M.K. Choi, Carbon monoxide has anti-inflammatory effects involving the mitogen-activated protein kinase pathway, *Nat. Med.* **6**, 422–428 (2000).

31. L.E. Otterbein, J.K. Kolls, L.L. Mantell, J.L. Cook, J. Alam, and A.M.K. Choi, Exogenous administration of heme oxygenase-1 by gene transfer provides protection against hyperoxia-induced lung injury, *J. Clin. Invest.* **103**, 1047–1054 (1999).

32. F. Amersi, R. Beulow, H. Kato, B. Ke, A.J. Coito, X.-D. Shen, D. Zhao, J. Zaky, J. Melinek, C.R. Lassman, J.K. Kolls, J. Alam, T. Ritter, H.-D. Volk, D.G. Farmer, R.M. Ghobrial, R.W. Busuttil, and J.W. Kupiec-Weglinski, Upregulation of heme oxygenase-1 protects genetically fat Zucker rat livers from ischemia/reperfusion injury, *J. Clin. Invest.* **104**, 1631–1639 (1999).
33. Z. Xing, J. Gauldie, G. Cox, H. Baumann, M. Jordana, Z.-F. Lei, and M.K. Achong, IL-6 an anti-inflammatory cytokine required for controlling local or systemic acute inflammatory responses, *J. Clin. Invest.* **101**, 311–320 (1998).

17

HEME OXYGENASE IN SKELETAL MUSCLE

Role in Septic Diaphragmatic Failure

Camille Taillé[a], Roberta Foresti[b], Colin Green[b], Michel Aubier[a],
Roberto Motterlini[b], and Jorge Boczkowski[a]

[a]Institut National poor la Santé et la Recherche Médicale, Unité 408
Faculté de Médecine Xavier Bichat, 75018
Paris, France
[b]Vascular Biology Unit, Surgical Research Department
Northwick Park Institute for Medical Research
Harrow, Middlesex HA13UJ, United Kingdom

1. INTRODUCTION

Sepsis is a common cause of morbidity and mortality, particularly in elderly, immunocompromised, and critically ill patients. Indeed, severe sepsis is at present the most common cause of death in intensive care unit in the United States (Bone et al., 1992).Respiratory failure is a major clinical manifestation of sepsis, greatly contributing to the mortality of this pathologic condition (Montgomery et al., 1985). In this context, respiratory failure has been traditionally related to the development of adult respiratory distress syndrome (Montgomery et al., 1985). However, different experimental studies demonstrated that an impaired contractile function of respiratory muscles, including the diaphragm, is another important mechanism of respiratory failure during sepsis (Boczkowski et al., 1990 and 1996; El Dwairi et al., 1998; Shindoh et al., 1992; Hussain et al., 1985; Van Surell et al., 1992). Diaphragmatic dysfunction during sepsis appears to be mediated mainly by oxidative stress (Shindoh et al., 1992; Supinski et al., 1996; Van Surell et al., 1992). In fact, we and others have recently shown an increased production of oxidants, such as superoxide anion and peroxynitrite, in the diaphragm of endotoxemic rats (El Dwairi et al., 1998; Boczkowski et al., 1999). The expression of the inducible isoform of nitric oxide (NO) synthase (NOS2) is partly responsible for this effect (Boczkowski et al., 1999). Despite experimental evidence showing generation of oxidants during sepsis, little is known about the specific antioxidant systems utilized by the diaphragm to counteract this pathological condition. This is an important point to consider because a better understanding of diaphragmatic antioxidant defenses engaged against sepsis could improve our knowledge on the mechanism of oxidative stress-mediated contractile failure.

The microsomal enzyme heme oxygenase (HO) is widely distributed in mammalian tissues (Maines et al., 1988). HO-1 expression is extremely sensitive to a variety of agents that cause oxidative stress (see reference Foresti et al., 1999 for review), and can be induced in various tissues including skeletal muscle (Essig et al., 1997; Vesely et al., 1998; Pilegaard et al; 2000). In this particular tissue, expression of the HO-1 gene is significantly increased after strenous exercise (Essig et al., 1997; Pilegaard et al., 2000), a condition that can lead tomuscular oxidative stress. HO-2 is expressed in myofibers of the sarcolemma region, endothelial vascular cells and fibroblasts of healthy rat hind limb muscles (Baum et al., 2000). Increasing experimental evidence suggests that, in sepsis and other pathological states, the HO pathway is a powerful protective cellular system. Otterbein and coworkers (Otterbein et al., 1995 and 1997) have previously shown that induction of HO-1 by hemoglobin dramatically decreases mortality and improves biological and hemodynamic parameters of rats challenged with *E. Coli* endotoxin. Intracheal inoculation of an adenovirus encoding HO-1 attenuates LPS-induced acute lung injury (Inoue et al., 2001). This protective effect is, at least for a part, related to its antioxidant properties (Maines et al., 1988). Indeed, elimination of the pro-oxidant free heme and production of biliverdin and bilirubin, two potent free radical scavengers with antioxidant characteristics (Stocker et al., 1987), can account for the defensive role of HO against oxidative stress. Moreover, HO-1 could also be implicated in cellular protection against NO and NO-derived species (Foresti et al., 1999 and 1999).

2. HO-1 IS INDUCED IN DIAPHRAGMATIC MYOCYTES DURING ENDOTOXEMIA

In animals receiving *E. Coli* endotoxin (4 mg/kg IP), a model that has shown to induce marked diaphragmatic failure (Boczkowski et al., 1996), diaphragmatic HO-1 protein levels increases significantly from 24 to 96 hours after LPS injection, while the expression of HO-2 does not change (Taillé et al., 2001) (Fig. 1, a and b). Immunohistochemical analysis shows that HO-1 is exclusively localized in the cytoplasme of diaphragmatic myocytes. Interestingly, HO activity in endotoxemic rats exhibits a biphasic response over time (Fig. 1, c): HO activity transiently decreases during the first 12 h after LPS treatment and this effect is particularly evident at 6 h (33% of control, $P < 0.05$); however, the enzymatic activity augments by 22% ($P < 0.05$) and 17% over control values at 24 h and 48 h, respectively. The increase observed in the late time points directly reflects HO-1 up-regulation, as HO-2 protein levels remains unchanged. As both HO-1 and HO-2 proteins are expressed in the diaphragm of control rats, both isoforms are likely contributing to the detectable basal HO activity. The presence of HO-1 and HO-2 has been already reported in other rat and human skeletal muscles (Vesely et al., 1998 and 1999; Kusner et al., 1999; Baum et al., 2000).

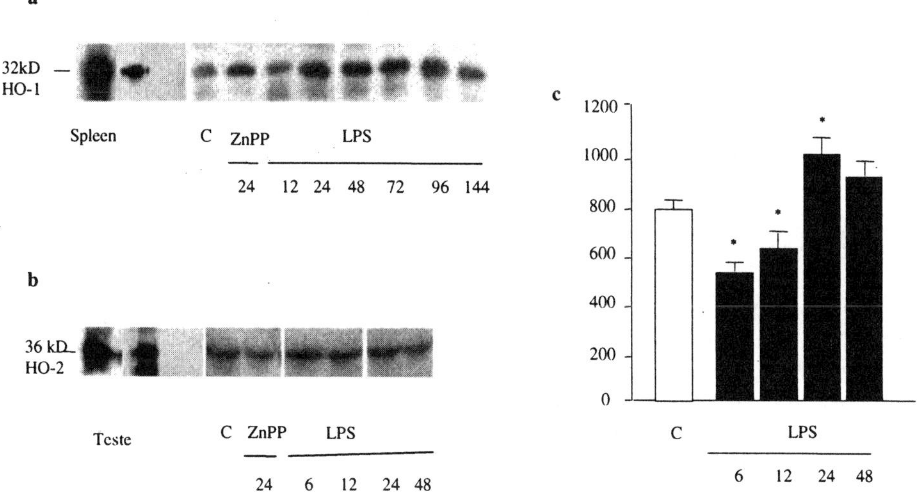

Figure 1. Time course of HO-1 (a) and HO-2 (b) expression and activity (c) in rat diaphragm after LPS inoculation (adapted from Taillé et al., 2001).

In LPS-treated animals, the early (6 and 12h) decrease in diaphragmatic HO activity occurs without changes in the expression of HO protein isoforms, suggesting that this effect is caused by post-translational modifications. Notably, a recent study by Ding and coworkers (Ding et al., 1999) shows that HO-2 enzymatic activity can be inhibited by NO and peroxynitrite *in vitro*; the cysteine residues present in the protein structure appear to be the targets of these two oxidizing molecules. In analogy with our findings, the same authors demonstrated that HO-2 protein expression remained unchanged despite decreased HO-2 activity (Ding et al., 1999). In contrast, HO-1 activity is not affected by NO or peroxynitrite, possibly because this isoform does not contain sulfhydryl groups. Since we have previously reported an early augmented production of NO and peroxynitrite in the diaphragm of endotoxemic rats (Boczkowski et al., 1996 and 1999), the initial decrease in HO activity observed in this study after endotoxin challenge may be the consequence of HO-2 protein inactivation by these nitrogen species. The late (24 and 48h) increase in HO activity detected in response to LPS is likely reflecting up-regulation of HO-1 protein as HO-2 expression in endotoxemic animals did not change over time. However, in view of recent data showing that HO-2 catalytic activity is enhanced by phosphorylation in cultured neuronal cells (Doré et al., 1999), the participation of HO-2 in increasing HO activity can not be excluded *a priori*. Diaphragmatic HO-1 induction during sepsis could be mediated by increased production of NO and reactive oxygen species (Boczkowski et al., 1996 and 1999; Rensing et al., 2001); this hypothesis is plausible since we have demonstrated that NO and NO-derivatives are potent stimulators of HO-1 in skeletal muscle cells and other cell types (Foresti et al., 1999; Vesely et al., 1998). In addition, HO-1 transcript is highly induced in skeletal muscle *in vivo* by strenuous exercise (Essig et al., 1997), a condition known to generate muscular oxidative stress. The time-course of diaphragmatic HO-1 expression observed in this study after endotoxin resembles that described in lung epithelium after tracheal infusion of LPS (Carraway et al., 1998) suggesting that HO-1 induction can be considered a long-lasting response to LPS challenge.

3. EFFECT OF MODULATION OF HO ACTIVITY ON DIAPHRAGMATIC FORCE AND MUSCULAR OXIDATIVE STRESS

Diaphragmatic force is significantly reduced in endotoxemic animals 12 and 24 hours after injection (Taillé et al., 2001). The recovery is complete at 48 h after LPS challenge (Fig. 2). The decrease in diaphragmatic force is markedly worsened by treatment with the heme oxygenase inhibitor ZnPP IX (50 µmol/kg IP) given 1 hour before LPS. Most importantly, ZnPP-IX completely prevents the recovery of diaphragmatic force observed 48 h after treatment with LPS alone (Fig. 2). The same deleterious effect of ZnPP-IX is observed when directly applied *in vitro* on diaphragmatic bundles from septic animals, whereas it had few effects on normal bundles. These experiments avoid systemic effect of the molecule. On the opposite, pre-treatment with hemin (50 mg/kg IP, 24 hours before LPS) completely prevents the decrease in force observed 24 h after LPS injection.

LPS administration induces oxidative stress in rat diaphragm, as provided by increased levels of malondialdehyde (MDA) and oxidized proteins contents (Van Surell et al., 1992; Taillé et al., 2001). ZnPP-IX exacerbates the increase in MDA levels and oxidized proteins observed in the diaphragm 24 h after treatment with LPS alone, while pre treatment with hemin decreases it. Protective effects on diaphragmatic force and oxidative stress afforded by hemin were suppressed by administration of ZnPP-IX.

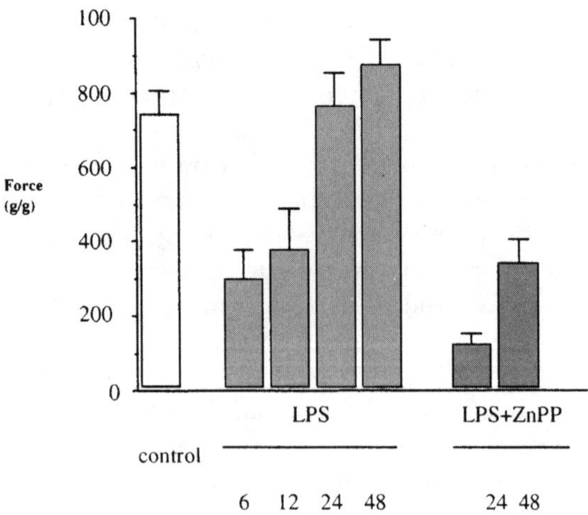

Figure 2. Effect of LPS treatment and HO activity inhibition on diaphragmatic force.

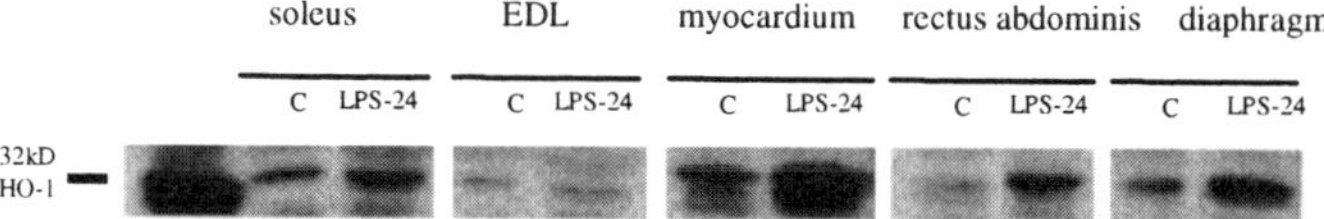

Figure 3. Induction of HO-1 in non-diaphragmatic skeletal muscles after LPS inoculation (reprint from Taillé et al., 2001).

4. INDUCTION OF HO-1 PROTEIN IN NON-DIAPHRAGMATIC SKELETAL MUSCLES OF ENDOTOXEMIC RATS

In control animals, HO-1 is expressed in all muscles examined, with protein levels higher in left ventricular myocardium, soleus and diaphragm than in *extensor digitorium longus* (EDL) or rectus abdominis (Taillé et al., 2001). LPS administration induces a significant increase in HO-1 protein expression in all muscles except EDL, that mean in striated muscles with high oxidative metabolism and high myoglobin content (type I fibers) (Fig. 3). The lack of induction of HO-1 in EDL, a muscle rich in type II fibers (low oxidative metabolism), is in agreement with recent findings by Vesely and coworkers (Vesely et al., 1999) showing a fibre-type specific pattern of HO-1 induction in skeletal muscles of rats treated with hemin. However, the distribution of HO-2 is fiber-type independent (Baum et al., 2000). Collectively, these data suggest that increased generation of oxidant species mediates high HO-1 expression in fibre I type muscles, a response that may lead to protection against oxidative stress. Consistent with this concept, it has been reported that other antioxidant enzymes are greatly expressed in muscles with high oxidative metabolism compared to those characterized by low oxidative metabolism (Ragusa et al., 1996; Vertechy et al., 1989).

5. ROLE OF THE HO PATHWAY IN PROTECTION AGAINST LPS-MEDIATED DIAPHRAGMATIC DYSFUNCTION

Several mechanisms could be responsible for the cytoprotective action of the HO system against diaphragmatic contractile failure induced by LPS. First, systemic induction of the HO-1 gene by LPS could improve diaphragmatic metabolic supply and thus, muscle function (Suplnski et al., 1986), by regulating vessel tone, as demonstrated in the liver during sepsis (Downard et al., 1997). However, in the present study, immunohistochemical analysis revealed that HO-1 is mainly expressed in myocytes and not in the vessels of the diaphragm from endotoxemic rats. This result suggests that the protective effect of HO-1 against LPS-mediated diaphragmatic failure is related to the action of this stress protein to modulate specific muscle functions. This hypothesis is further corroborated by our *in vitro* experiments showing that direct application of ZnPP-IX to diaphragmatic bundles from control and LPS animals reproduces the deleterious effect of systemically administrated ZnPP-IX (Taillé et al., 2001).

Heme oxygenases are important intracellular antioxidant enzymes by virtue of their ability to degrade the pro-oxidant heme and generate biliverdin and bilirubin, two effective free radical scavengers. Bilirubin is probably the most abundant endogenous antioxidant in mammalian tissue (Belanger et al., 1997) and has been shown to efficiently scavenge peroxyl radicals *in vitro* at micromolar concentrations (Stocker et al., 1987). In the present study, endotoxin produces diaphragmatic oxidative stress, as revealed by increased MDA and carbonylated proteins contents, two well established indexes of lipid and protein modifications by oxidative damage (Janero et al., 1990; Levine et al., 1990; Szabo et al., 1997). The fact that diaphragmatic MDA and carbonylated proteins contents are exacerbated by blockade of basal HO activity (likely HO-2 as discussed above), and induction of HO-1 by hemin completely prevented the accumulation of these oxidative products is indicative of both heme oxygenase isozymes serving as important anti-oxidant systems in the diaphragm. This is an interesting concept to consider since the majority of studies on the antioxidant properties of heme oxygenases have focused on HO-1, the isoform highly sensitive to induction by oxidative stimuli. Our data strongly suggest a co-operative action of HO-2 and HO-1 in counteracting oxidant threat in the diaphragm. Indeed, enhanced bilirubin production following activation of HO-2 protects neurons against hydrogen peroxide-mediated (Doré et al., 1999) and HO-1-derived bilirubin ameliorates post-ischemic myocardial dysfunction in isolated heart (Clark et al., 1999). Furthermore bilirubin decreases apoptosis caused by peroxynitrite in endothelial cells (Foresti et al., 1999), and is able to inhibit carbonylation and nitrotyrosine formation in proteins exposed to peroxynitrite donors *in vitro* (Minetti et al., 1998). Since NOS2-derived peroxynitrite significantly contributes to oxidative stress and cellular injury in the diaphragm of endotoxemic animals (Boczkowski et al., 1999), bilirubin could also protect diaphragmatic myocytes from peroxynitrite toxicity. Finally, bilirubin modulates phosphorylation of different proteins (Hansen et al., 1996), including contractile proteins (Samb and Taillé, unpublished data). In addition, recent experimental evidence suggest an alternative protective role for HO-2 and HO-1 isoforms in the modulation of intracellular pro-oxidant iron levels in lung tissue and cultured fibroblasts, respectively (Dennery et al., 1998; Ferris et al., 1999). An additional mechanism of protection by the HO systems could be related to the possible role of HO-1 in down-regulating NOS2 expression by degradation of heme, an essential co-factor for NOS2 protein assembly and activity (Turcanu et al., 1998; Foresti et al., 1999). This hypothesis can be sustained by our findings showing that the increase in NOS2 expression and activity by LPS in the diaphragm were no further detected at the time of induction of HO-1 expression and activity (48 h after LPS) (Boczkowski et al., 1996). Finally, a potential defensive role of CO, the other product of heme oxygenases, can not be excluded *a priori*. Indeed, CO can inhibit NO synthase activity (White et al., 1992) and may influence other intracellular functions such as production of cytokines implicated in the progression of sepsis and other pathophysiological states (Otterbein et al., 1999 and 2000).

In conclusion, the HO pathway is a major cellular protective system against oxidative stress in the diaphragm during sepsis. Specifically, endotoxemia produces a biphasic change in diaphragmatic HO activity which reflects the course of the pathological state: an initial decrease in activity associated with muscle failure followed by a late increase likely involved in the recovery of muscle function. Pharmacological agents that up-regulate HO-1 expression appear therefore as a very promising strategy for preventing sepsis-induced muscular dysfunction, as well as other muscular pathological situations related to oxidative stress, such as muscular fatigue.

REFERENCES

Baum, O., M. Feussner, H. Richter, and R. Gossrau. 2000. Heme oxygenase-2 is present in the sarcolemma region of skeletal muscle fibers and is non-contiuously co-localized with nitric oxide synthase-1. *Acta Histochem.* **102**(3):281–298.

Belanger, S., J. Lavoie, and P. Chessex. 1997. Influence of bilirubin on the antioxidant capacity of plasma in newborn infants. *Biol. Neonate.* **71**:233–238.

Boczkoswski, J., R. Dureuil, Pariente, and M. Aubier. 1990. Preventive effects of indomethacin on diaphragmatic contractile alterations in endotoxemic rats. *Am. Rev. Respir. Dis.* **142**:193–198.

Boczkowski, J., S. Lanone, D. Ungureanu-Longrois, G. Danialou, T. Fournier, and M. Aubier. 1996. Induction of diaphragmatic nitric oxide synthase after endotoxin administration in rats. Role on diaphragmatic contractile dysfunction. *J. Clin. Invest.* **98**:1550–1559.

Boczkowski, J., C. Lisdero, S. Lanone, A. Samb, C. Carreras, A. Boveris, M. Aubier, and J.J. Poderoso. 1999. Endogenous peroxynitrite mediates mitochondrial dysfunction in rat diaphragm during endotoxemia. *FASEB J.* **13**:1637–1646.

Bone, R., R. Balk, F. Cerra, R. Dellinger, A. Fein, W. Knauss, R. Schein, and W. Sibbald. 1992. Definitions for sepsis and organ failure and guidelines for the use of innovative therapies in sepsis. *Chest.* **101**:1644–1655.

Carraway, M., A. Ghio, J. Taylor, and C. Piantadosi. 1998. Induction of ferritin and heme oxygenase-1 by endotoxin in the lung. *Am. J. Physiol.* **275**:L583–L592.

Clark, J., R. Foresti, P. Sarathchandra, H. Kaur, C. Green, and R. Motterlini. 1999. Heme oxygenase-1-derived bilirubin ameliorates postischemic myocardial dysfunction. *Am. J. Physiol.* **278**:H643–H451.

Dennery, P., D. Spitz, G. Yang, A. Tatarov, C. Lee, M. Shegog, and K. Poss. 1998. Oxygen toxicity and iron accumulation in the lungs of mice lacking heme oxygenase-2. *J. Clin. Invest.* **101**:1001–1011.

Ding, Y., W. McCourbey, and M. Maines. 1999. Interaction of heme oxygenase-2 with nitric oxide donors. Is the oxygenase an intracellular "sink" for NO? *Eur. J. Biochem.* **264**:854–861.

Doré, S., M. Takahashi, C. Ferris, L. Hester, D. Guastella, and S. Snyder. 1999. Bilirubin, formed by activation of heme oxygenase-2, protects neurons against oxidative stress injury. *Proc. Natl. Acad. Sci. USA.* **96**:2445–2450.

Downard, P., M. Wilson, D. Spain, P. Matheson, Y. Siow, and R. Garrison. 1997. Heme oxygenase-dependent carbon monoxide production is a hepatic adaptive response to sepsis. *J. Surg. Res.* **71**:7–12.

El Dwairi, Q., A. Comtois, Y. Guo, and S.N. Hussain. 1998. Endotoxin-induced skeletal muscle contractile dysfunction: contribution of nitric oxide synthases. *Am. J. Physiol.* **274**:C770–C779.

Essig, D., D. Borger, and D. Jackson. 1997. Induction of heme oxygenase-1 (HSP32) mRNA in skeletal muscle following contractions. *Am. J. Physiol.* **272**:C59–C67.

Essig, D. and T. Nosek. 1997. Muscle fatigue and induction of stress protein genes: a dual function of reactive oxygen species? *Can. J. Appl. Physiol.* **22**:409–428.

Ferris, C., S. Jaffrey, A. Sawa, M. Takahashi, S. Brady, R. Barrow, S. Tysoe, H. Wolosker, D. Baranano, S. Dore, K. Poss, and S. Snyder. 1999. Heme oxygenase-1 prevents cell death by regulating cellular iron. *Nat. Cell Biol.* **1**:152–157.

Foresti, R. and R. Motterlini. 1999. The heme oxygenase pathway and its interaction with nitric oxide in the control of cellular homeostasis. *Free Radic. Res.* **31**:459–475.

Foresti, R., P. Sarathchandra, J. Clark, C. Green, and R. Motterlini. 1999. Peroxynitrite induces haem oxygenase-1 in vascular endothelial cells: a link to apoptosis. *Biochem. J.* **339**:729–736.

Hansen, T.W.R., S.B.W. Mathiesen, and S.I. Walaas. 1996. Bilirubin has widespread inhibitory effects on protein phosphorylation. *Pediatr Res.* **39**:1072–1077.

Hussain, S.N.A., G. Simkus, and C. Roussos. 1985. Respiratory muscle fatigue: a cause of ventilatory muscle failure in septic shock. *J. Appl. Physiol.* **58**:2033–2040.

Inoue, S., M. Suzuki, Y. Nagashima, S. Suzuki, T. Hashiba, T. Tsuburai, K. Ikehara, T. Matsuse, and Y. Ishigatsubo. 2001. *Hum Gen Therap.* **12**:967–979.

Janero, D. 1990. Malondialdehyde and thiobarbituric acid-reactivity as diagnostic indices of lipid peroxidation and peroxidative tissue injury. *Free Radic. Biol. Med.* **9**:515–540.

Kusner, L., E. Kim, and H. Kaminski. 1999. Heme oxygenase-2 expression at rat neuromuscular junctions. *Neurosci. Letters.* **273**:143–146.

Levine, R., D. Garland, C. Oliver, A. Amici, I. Climent, A. Lenz, B. Ahn, S. Shaltiel, and E. Stadtman. 1990. Determination of carbonyl content in oxidatively modified proteins. *Methods Enzymol.* **186**:464–478.

Maines, M. 1988. Heme oxygenase: function, multiplicity, regulatory mechanisms, and clinical applications. *FASEB J.* **2**:2557–2568.

Maines, M., G. Trakshel, and R. Kutty. 1986. Characterization of two constitutive forms of rat liver microsomal heme oxygenases. Only one molecular species of the enzyme is inductible. *J. Biol. Chem.* **261**:411–419.

Minetti, M., C. Mallozzi, A. Di Stasi, and D. Pietraforte. 1998. Bilirubin is an effective antioxidant of peroxynitrite-mediated protein oxidation in human blood plasma. *Arch. Biochem. Biophys.* **352**:165–174.

Montgomery, A., M. Stager, C. Carrico, and L. Hudson. 1985. Causes of mortality in patients with the adult respiratory distress syndrome. *Am. Rev. Resp. Dis.* **132**:485–489.

Otterbein, L., S. Sylvester, and A. Choi. 1995. Hemoglobin provides protection against lethal endotoxemia in rats: the role of heme oxygenase-1. *Am. J. Respir. Cell Mol. Biol.* **13**:595–601.

Otterbein, L., B. Chin, S. Otterbein, V. Lowe, H. Fessler, and A. Choi. 1997. Mechanism of hemoglobin-induced protection against endotoxemia in rats: a ferritin-independent pathway. *Am. J. Physiol.* **272**:L268–L275.

Otterbein, L., L. Mantell, and A. Choi. 1999. Carbon monoxide provides protection against hyperoxic lung injury. *Am. J. Physiol.* **276**:L688–L694.

Otterbein, L, F.H. Bach, J. Alam, M. Soares, H.T. Lu, M. Wysk, R.J. Davis, R.A. Flavell, and A.M.K. Choi. 2000. Carbon monoxide has anti-inflammatory effects involving the mitogen-activated protein kinase pathway. *Nat Med.* **6**(4):422–428.

Pilegaard H, G.A. Ordway, B. Saltin, and P.D. Neufer. 2000. Transcriptional regulation of gene expression in human skeletal muscle during recovery from exercise. *Am J Physiol Endocrinol Metab.* **279**(4):E806–E814.

Ragusa, R., C. Chow, D.S.T. Clair, and J. Porter. 1996. Extraocular, limb and diaphragm muscle group-specific antioxidant enzyme activity patterns in control and mdx mice. *J. Neurol. Sci.* **139**:180–186.

Rensing, H., H. Jaeschke, I. Bauer, C. Patau, V. Datene, B.H. Pannen, and M. Bauer. 2001. Differential activation pattern of redox-sensitive transcription factors and stress-inductible dilator systems heme oxygenase-1 and inductible nitric oxide synthase in hemorrhagic and endotoxic shock. *Crit Care Med.* **29**(10):1962–1971.

Shindoh, C., A. Dimarco, D. Nethery, and G. Supinski. 1992. Effect of PEG-Superoxide dismutase on the diaphragmatic response to endotoxin. *Am. Rev. Respir. Dis.* **145**:1350–1354.

Stocker, R., Y. Yamamoto, A. McDonagh, A. Glazer, and B. Ames. 1987. Bilirubin is an antioxidant of possible physiological importance. *Science.* **235**:1043–1046.

Supinski, G., D. Nethery, D. Stofan, and A. DiMarco. 1996. Comparison of the effects of endotoxin on limb, repiratory, and cardiac muscles. *J. Appl. Physiol.* **81**:1370–1378.

Supinski, G. 1986. Contol of respiratory muscles blood flow. *Am. Rev. Resp. Dis.* **134**:1078–1079.

Szabo, C., M. O'Connor, and A.L. Salzman. 1997. Endogenously produced peroxynitrite induces the oxidation of mitochondrial and nuclear proteins in immunostimulated macrophages. *FEBS Letters.* **409**:147–150.

Taillé. C., R. Foresti, S. Lanone, C. Zedda, C. Green, M. Aubier, R. Motterlini, and J. Boczkowski. 2001. Protective role of heme oxygenases against endotoxin-induced diaphragmatic dysfunction in rats. *Am J Respir Crit Care Med.* **163**:753–761.

Turcanu, V., M. Dhouib, and P. Poindron. 1998. Nitric oxide synthase inhibition by haem oxygenase decreases macrophage nitric-oxide-dependent cytotoxicity: a negative feedback mechanism for the regulation of nitric oxide production. *Res. Immunol.* **149**:741–744.

Van Surell, C., J. Boczkowski, C. Pasquier, Y. Du, E. Franzini, and M. Aubier. 1992. Effects of N-acetylcysteine on diaphragmatic function and malondialdehyde content in E. coli endotoxemic rats. *Am. Rev. Resp. Dis.* **146**:730–734.

Vertechy, M., M. Cooper, O. Ghirardi, and M. Ramacci. 1989. Antioxidant enzyme activities in heart and skeletal muscle of rats of different ages. *Exp. Gerontol.* **24**:211–218.

Vesely, M., R. Sanders, C. Green, and R. Motterlini. 1999. Fiber type specificity of haem oxygenase-1 induction in rat skeletal muscle. *FEBS Lett.* **458**:257–260.

Vesely, M., D. Exon, J. Clark, R. Foresti, C. Green, and R. Motterlini. 1998. Heme oxygenase-1 induction in skeletal muscle cells: hemin and sodium nitroprusside are regulators in vitro. *Am. J. Physiol.* **275**:C1087–C1094.

White, K., and M. Marletta. 1992. Nitric oxide synthase is a cytochrome P-450 type protein. *J. Biol Chem.* **269**:26390–26395.

18

HEME OXYGENASE-1 (HO-1)

Multiple Effects of a Protective Gene that Prevents Graft Rejection

P. O. Berberat[a], L. Günther[a], S. Brouard[a], M. P. Soares[a,b], and F. H. Bach[a,*]

[a]Immunobiology Research Center
Department of Surgery, Beth Israel Deaconess Medical Center
Harvard Medical School, 99 Brookline Avenue
Boston, MA 02215, USA
[b]Instituto Gulbenkian de Ciência
Apartado 14, 2781-901 Oeiras, Portugal

1. INTRODUCTION

Transplantation is considered one of the most effective ways to overcome terminal dysfunction and failure of some organs. By transplanting organs or tissues between two genetically distinct individuals, an immune mediated response is triggered in the transplant recipient that evokes a potent inflammatory reaction, causing cell injury/death and leading to graft dysfunction and rejection. So far the main therapeutic strategy used to maintain graft function and survival has been to modulate the host immune response directed against the transplanted organ. In the case of an immediately vascularized transplant, the endothelial cell (EC) monolayer lining the blood vessels of the graft acts as the first target of the anti-graft immune response. We have suggested that the survival of grafts may relay on how these cells react to the anti-graft immune response: whether in a rejection-promoting or a "protective"

---

* Fritz H. Bach is a paid consultant by Novarits Pharma. Send correspondence to Pascal O. Berberat, Beth Israel Deaceness Medical Center, Department of Surgery, Immunobiology Research Center, Harvard Medical School, 99 Brookline Avenue, Boston, MA 02215, USA. pascal@berberat.com

manner. Understanding the mechanisms determining the type of reaction may help to develop therapeutic strategies aimed to suppress the rejection of transplanted organs.

In their quiescent state, EC regulate blood flow by fine tuning vessel vaso-constriction/relaxation as well as by inhibiting leukocyte adhesion, coagulation and platelet aggregation.[1] When exposed to pro-oxidant and/or pro-inflammatory conditions associated with rejection of transplanted organs, EC lose their anti-adhesive and anti-thrombotic properties.[1-3] These phenotypic modifications, collectively referred to as EC activation[2,4] are attributed to the expression of a series of pro-inflammatory genes encoding adhesion molecules, i.e. E-selectin, P-selectin, ICAM-1, VCAM-1, cytokines/chemokines, e.g. IL-1α, IL-6, IL-8, MCP-1, MIP-1α, RANTES as well as pro-coagulant molecules, e.g. PAI-1 and tissue factor.[1,3] Overall, the expression of these pro-inflammatory genes is thought to promote host leukocyte activation and transmigration as well as thrombosis leading to graft rejection.[5,6] In addition, the pro-inflammatory environment generated through the expression of these genes can cause EC to undergo cell death through apoptosis. When this occurs the graft endothelial barrier is disrupted and the host coagulation system is triggered to promote thrombosis, vessel occlusion and tissue ischemia leading to loss of graft function and rejection.

We have suggested that EC can counter this pro-inflammatory reaction through the expression of so-called "protective genes" that limit inflammation and apoptosis associated with EC activation. The different protective genes function by varying mechanisms, including the ability of some to suppress signal transduction pathways leading to EC apoptosis. One of these protective genes encodes the heme degrading enzyme heme oxygenase-1 (HO-1). We and others have shown that expression of HO-1 can prevent many of the inflammatory processes involved in the rejection of transplanted organs. We review below several findings of the last few years relating to the expression of this gene, its anti-inflammatory and anti-apoptotic functions and how this translates into suppressing those mechanisms leading to graft rejection.

2. HO-1: A PROTECTIVE GENE

Under the conditions associated with rejection of transplanted organs, HO-1 becomes the rate-limiting enzyme in the catabolism of heme to yield equimolar amounts of biliverdin, CO and free iron.[7-9] Biliverdin is subsequently catabolyzed into bilirubin by biliverdin reductase,[10] whereas free iron leads to rapid induction of the iron sequestering protein ferritin.[11,12] Several stimuli including extracellular heme, cytokines (IL-1, IL-6 or TNF-α), bacterial lipopolysaccharides (LPS) or the gaseous molecule NO can induce the expression of HO-1 in EC.[10,13] Exposure to extracellular heme is particularly important in the context of graft rejection given that oxidation of heme-proteins such as hemoglobin and myoglobin can release pro-oxidant heme that is internalized by EC of a transplanted organ.[11,14] However, upon internalization, heme up-regulates the expression of HO-1, which acts to convert heme into the anti-oxidant molecules bilirubin and CO. Heme derived expression of HO-1 is thought to be regulated primarily at the transcriptional level through the activation of a series of transcription factors including activating protein-1 (AP-1),[15,16] the cap 'n' Collar/basic leucine zipper family member NF-E2-related factor 2 (Nrf2) and the activating transcription factor-4 (ATF4).[17,18]

It has become clear in the past few years that expression of HO-1 has potent anti-inflammatory and anti-apoptotic effects both in vitro[11,19–21] and in vivo.[22–28] These are thought to result from the generation of one or several of the end products of heme catabolism by HO-1, i.e. bilirubin, ferritin, and CO. For example, bilirubin has clear anti-oxidant properties and could, therefore, mediate the anti-inflammatory as well as anti-apoptotic effects of HO-1 by inhibiting the production of reactive oxygen species.[29] Biliverdin as well as conjugated bilirubin have also been shown to block complement activation, which may contribute to the overall anti-inflammatory and cytoprotective effect of HO-1.[30] In this review we will focus on the recently reported anti-inflammatory and anti-apoptotic effects of the two other end products of heme catabolism by HO-1, i.e. ferritin and CO, and how these could contribute to suppress graft rejection.

3. HO-1: MODULATION OF INFLAMMATION AND APOPTOSIS

When a transplanted organ is rejected, EC apoptosis[27] can enhance the inflammatory reaction that leads to graft rejection[31] by disrupting the vascular endothelium and promoting thrombosis.[32,33,34] Expression of HO-1 in cultured EC suppresses apoptosis induced by a variety of pro-apoptotic stimuli including TNF-α.[22] The same phenomenon occurs *in vivo*.[27] The protective effect of HO-1 requires its enzymatic activity, and thus involves the generation of one or more of the end-products of heme catabolism by HO-1.[20] As described above, heme derived Fe^{2+} up-regulates the expression of the iron sequestering protein ferritin.[11,12] Results from our laboratories[20] and from other groups[35] show that the anti-apoptotic effect of HO-1 can be mimicked by the exogenously administered iron chelator deferoxamine[20] and/or by the over-expression of the heavy chain of ferritin (Berberat et al., *submitted for publication*).

In addition to the expression of ferritin[11] or activation of iron pumps,[35] HO-1 generates CO that we have shown to act to suppress EC apoptosis.[20] The anti-apoptotic effect of CO is dependent on at least two signal transduction pathways: the p38 mitogen activated protein kinase (MAPK) and the nuclear factor kappa B (NF-κB) signal transduction pathways. HO-1 in EC activates p38 MAPK very significantly but does not activate the NF-κB signal transduction pathway. However, the specific inhibition of any of these signal transduction pathways suppresses the anti-apoptotic effect of HO-1 derived CO. How the p38 MAPK signal transduction pathway interacts with CO to suppress EC apoptosis is not clear. As for the NF-κB signal transduction pathway, we have shown that in EC HO-1 derived CO requires basal levels of NF-κB activity to suppress apoptosis. Presumably NF-κB activity is required to sustain the expression of a subset of NF-κB dependent anti-apoptotic genes that interact with HO-1/CO to suppress apoptosis. We have shown that this is the case for the anti-apoptotic genes A1 and c-IAP2 (Brouard et al., *submitted for publication*) but not for other NF-κB dependent anti-apoptotic genes. The exact mechanisms by which apoptosis in a graft contributes to graft rejection are not clear. Presumably inhibition of EC apoptosis maintains the vascular endothelial integrity and thus prevents thrombosis, hypoxia and necrosis.[32–34]

4. HO-1 PREVENTS GRAFT ISCHEMIA/REPERFUSION INJURY

Ischemia/reperfusion (I/R) injury such as it occurs during organ transplantation is considered to be a key event in the pathogenesis of graft rejection.[36] I/R injury is triggered through a complex set of events that leads to microvascular dysfunction. These are revealed by impaired vasodilatation, fluid extravasation into the transplanted tissue as well as leukocyte activation and adhesion to the graft endothelium. The events that trigger I/R injury a probably multifactorial but can be summarized as follows. Upon removal from the donor, organs that are to be transplanted under go a period of hypoxia. This is followed by abrupt re-oxygenation at the time of organ re-vascularization. This sequence of events leads to the activation of EC lining the blood vessels of these organs and thus to the expression of a series of pro-inflammatory genes by these cells. As discussed above, the expression of these genes is thought to contribute to the pathogenesis of I/R injury.[37] In addition, once exposed to a cycle of hypoxia/re-oxygenation, EC decrease the production of nitric oxide (NO) and increase that of superoxide. This results in the net generation of reactive oxygen species that trigger pro-inflammatory and pro-apoptotic responses in EC.[38] The decrease in the level of NO production is thought to account in large measure for the inability of blood vessels undergoing I/R to sustain the level of vasodilatation required to maintain blood flow under these pro-oxidant conditions. Presumably this is key in generating the overall pro-thrombotic reaction that characterizes I/R injury.

Expression of HO-1 has been shown by several investigators to be associated with I/R injury.[39] However, the exact role of HO-1 in the pathogenesis of I/R injury has remained elusive. Recently, HO-1 has been shown to have string protective effects during I/R injury.[23,40] This was originally demonstrated using a steatotic rat liver experimental model of *ex vivo* cold I/R injury. In this experimental model, expression of HO-1 significantly improved portal venous blood flow, increased bile production, and decreased hepatocyte injury associated with I/R injury.[23] When these livers were transplanted into syngeneic recipients, the percentage of long-term graft survival was significantly increased from 40% in untreated controls to about 80% in grafts expressing HO-1.[23] A similar protective effect of HO-1 has been reported more recently in a model of renal I/R (Abstract 157, American Society of Transplantation 2001). The mechanism by which HO-1 protects these organs from I/R injury is not clear. A recent report, using a murine lung I/R injury model, suggests that HO-1 derived CO may account in large measure for the protective effect of HO-1.[41] In this model, exogenous CO has been show to protect lungs from warm I/R injury, through a mechanism that relies on the ability of CO to depress the expression of the pro-coagulant molecule plasminogen activator inhibitor-1 (PAI-1) in monocyte/macrophages.[41] This data is in keeping with that showing that CO may also account for the ability of HO-1 to prevent liver I/R injury in rats (Abstract 156, American Society of Transplantation

2001). In this preliminary report, rat livers were stored for at 4 °C in UW solution for a period of 24 hours (ischemia) and then perfused *ex vivo* on a perfusion apparatus with CO supplemented blood. Livers exposed to CO showed significantly less hepatocyte injury and better function in comparison with control livers not exposed to CO (Abtsract 156, American Society of Transplantation 2001). We have also shown that administration of CO can markedly suppress I/R injury to a heart transplanted to a syngeneic rat (Akamatsu et al., *manuscript in preparation*). Taken together, these data point to an important protective effect of HO-1 in several models of I/R injury and suggest that these protective effects of HO-1 can be mediated via the generation of CO.

5. HO-1-PREVENTS ACUTE AND/OR CHRONIC GRAFT REJECTION

Acute rejection of vascularized allogeneic transplants occurs without immunosuppression in the days following transplantation. This type of rejection is driven primarily via the activation of alloreactive T cells that infiltrate the transplanted organ and mediate the cytotoxic effects that cause graft rejection.

In the past decade it has been established that soluble major histocompatibility (MIIC) molecules can have significant immunomodulatory activity that could be used therapeutically to prevent graft rejection.[42,43] It has been suggested that these MHC molecules exert their function at least in part by inducing the expression of HO-1[44] as they bind specifically to HO-1 and up-regulate its level of expression and activity when administered *in vivo*.[44] Among these peptides there is one derived from the conserved heavy chain al helix region of the MHC class I B-2702 molecule (aa 75 to 84) which prolongs the survival of skin and cardiac grafts in rodents[45–47] as well as kidney transplants in humans.[48] This and other related peptides have also been shown to inhibit transplant-associated arteriosclerosis in rodents.[49] The mechanism by which these peptides act to suppress graft rejection is likely to be related to their ability to inhibit the production of pro-inflammatory cytokines[50] as well as to down-regulate the generation of anti-graft antibodies and to inhibit T and NK cell mediated cytotoxicity.[45] Other molecules that specifically up-regulate the expression of HO-1 have similar effects, suggesting that MHC derived peptides can exert their effects by modulating the expression of HO-1.

One of the molecules that up-regulates HO-1 expression specifically, cobalt protoporphyrin IX (CoPPIX), inhibits acute graft rejection by modulating T cell activation and driving T cell activation towards a T helper type 2 response.[51] This is in keeping with the recent observation that HO-1 expression inhibits the proinflammatory phenotype associated with macrophage activation, blocking the generation of pro-inflammatory molecules such as TNF-α while increasing that of antiinflammatory IL 10.[52]

Chronic graft rejection remains the main cause of graft failure in clinical transplantation.[33] This type of rejection is characterized by the development of a transplant related vascular disease comparable to atherosclerosis.[54] In the development of this disease, proliferation of smooth muscle cells in the vessel wall of the transplanted organ is key in causing the pathological features attributed to end stage chronic graft rejection. Cardiac transplants that survive long-term and do not develop chronic graft rejection under a given immunosuppressive protocol express higher levels of several protective genes including HO-1, as compared to cardiac transplants undergoing chronic rejection.[55] In addition, induction of HO-1 expression by CoPPIX can prevent the development of transplant-associated arteriosclerosis that characterizes chronic graft rejection.[55] The protective effect of HO-1 in preventing chronic graft rejection is likely due to its ability to inhibit the initial pro-inflammatory response associated with EC and macrophage activation. In addition, however, HO-1 has been recently shown to act directly on smooth muscle cells to block their proliferation.[56] Presumably this contributes in a critical manner to suppress chronic graft rejection. We have obtained data that also suggests that the ability of HO-1 to suppress smooth muscle cell proliferation and the development of transplant associated arteriosclerosis is mediated via CO (Otterbein et al., *submitted for publication*).

6. HO-1 PREVENTS XENOGRAFT REJECTION AND INDUCES GRAFT ACCOMMODATION

Allogeneic grafts transplanted across the ABO blood group "barrier" are rejected through a process referred to as hyperacute rejection. This type of rejection is mediated through the binding to the graft vascular endothelium of preformed anti-blood group antibodies of the IgM isotype, which expresses antigens of the ABO blood group system.[5] When the recipient's anti-blood group antibodies are depleted prior to transplantation, ABO incompatible grafts are no longer rejected.[57,58] This phenomenon occurs despite the observation that anti-blood group antibodies return to the circulation a few days after transplantation.[57–59] The survival of such grafts has been referred to as "accommodation"[60,61] to reflect the fact that these grafts have developed some kind of resistance to the anti-graft antibodies and complement that usually mediate graft rejection. Accommodation has been shown to occur in several models of xenotransplantation, in which organs are transplanted between different species. For example a mouse or a hamster heart that are transplanted into a rat undergoes acute vascular rejection in 3 to 4 days following transplantation.[62,63,64] Acute vascular rejection is mediated by anti-graft antibodies, which bind to the graft endothelium and activate the classical pathway of complement.[65,66] This type of rejection can be averted when complement activation is suppressed for a few days following transplantation by the administration of cobra venom factor and when T cell activation is suppressed by cyclosporin A.[62,63,67,68] Grafts that survive long term under this immunosuppressive regimen express a series of protective genes in their vascular endothelium.[67] That these grafts are actively protected from undergoing rejection is suggested by the observation that a naïve graft transplanted into a recipient carrying an "accommodating" graft for 10 days undergoes rejection in few minutes whereas the first graft survives long-term.[61,69] We have subsequently demonstrated that the survival of the accommodating mouse heart transplanted to a rat is due to the expression of HO-1. Expression of this protective gene occurs as early as 12 hours after transplantation.

That the expression of HO-1 plays a critical role in establishing accommodation is shown by the observation that HO-1$^{-/-}$ deficient mouse hearts transplanted under the same conditions as those that allow wild type hearts to accommodate, are rejected in 3–7 days.[22] Therefore expression of HO-1 appears to be essential to induce xenograft accommodation. In addition we have shown that CO can substitute for the protective effects of HO-1 in preventing xenograft rejection.[27] Under the same immunosuppressive regimen that allows mouse to rat cardiac transplants to survive long-term, inhibition of HO-1 activity by the specific HO inhibitor tin protoporphyrin IX (SnPPIX) leads to graft rejection. Exogenous CO can restore long-term graft survival despite the inhibition of HO-1 activity by SnPPIX, suggesting that HO-1 prevents xenograft rejection at least in part via the generation of CO.[27]

7. PROTECTIVE EFFECTS OF HO-1 IN ISLET TRANSPLANTATION

The protective effects of HO-1 and its products are not specific to EC. In fact, HO-1 is a ubiquitous stress responsive gene that can be expressed in a large variety of cell types.[8] Given the above, the protective effects of HO-1 could be used therapeutically to suppress the rejection of a variety of transplanted tissues and/or cells such as isolated islets of Langerhans. Islet transplantation is considered as the treatment of choice for autoimmune (type I) diabetes.[70,71] The feasibility of this clinical approach is strongly supported by the recent finding that glucocorticoid-free immunosuppression can sustain the survival of transplanted islets and revert type I diabetes.[71] One major limitation of this approach is the requirement of at least two donor pancreases to obtain enough islets per recipient to be transplanted.[71] This is the direct consequence of the fact that a very significant proportion of transplanted islets are lost in the immediate post-transplantation period through a phenomenon known as "primary non function".[70] Primary non-function results in large measure from a nonspecific inflammatory process that leads to apoptosis of islet β-cells. Islet β-cell apoptosis is thought to result from the action of activated macrophages that infiltrate the transplanted islets and release multiple pro-inflammatory cytokines such as TNF-α, IL-1β, IFN-γ.[72,73] Expression of protective genes, such as the zinc family member A20 or bcl-2 in β-cells provides protection of β-cells from cytokine-mediated apoptosis *in vitro*[74] and allows long-term survival following transplantation.[75] Expression of HO-1 has also been shown to suppress cytokine-mediated β-cell apoptosis[76–78] (Günther et al., *submitted for publication*). This effect of HO-1 probably contributes to the observation that in a marginal mass islet transplantation model where the transplanted islets reverse hyperglycemia with a significant delay, HO 1 expression leads to a significant amelioration of graft function.[78] We have recently obtained data that suggests that this effect of HO-1 can be mediated via CO (Günther et al., *submitted for publication*). Furthermore, we found CO has potent anti-apoptotic effects that can be exerted through brief exposure of β-cells to CO before transplantation (Günther et al., *submitted for publication*). Exposure of β-cells to CO, for a brief period of time before transplantation, may therefore be used to reduce the islet mass required to revert diabetes.

8. CONCLUSION

The ability of the immune system to precipitate the rejection of a graft appears highly dependent on how well these grafts can protect themselves from immune-mediated injury. The ability of a graft to counter this type of injury is mediated via the expression of protective genes. We have shown that among these, HO-1 can play a critical role in dictating graft survival. The action of HO-1 on heme engenders three products (iron/ferritin, biliverdin/bilirubin and CO), which appear to have potent cytoprotective and anti-inflammatory functions. Data from several laboratories, including our own, suggest that CO can account for the ability of HO-1 to prevent graft rejection. Further understanding of the mechanisms of action of CO will help to develop new therapeutic approaches aimed to suppress graft rejection.

REFERENCES

1. D.B. Cines, E.S. Pollak, C.A. Buck, J. Loscalzo, G.A. Zimmerman, R.P. McEver, J.S. Pober, T.M. Wick, B.A. Konkle, B.S. Schwartz, E.S. Barnathan, K.R. McCrae, B.A. Hug, A.M. Schmidt, and D.M. Stern, Endothelial cells in physiology and in the pathophysiology of vascular disorders, *Blood* **91**(10), 3527–3561 (1998).
2. J.S. Pober and R.S. Cotran, The role of endothelial cells in inflammation, *Transplantation* **50**(4), 537–544 (1990).
3. A. Mantovani, F. Bussolino, and M. Introna, Cytokine regulation of endothelial cell function—from molecular level to the bedside, *Immunol Today* **18**(5), 231–240 (1997).
4. F.H. Bach, S.C. Robson, C. Ferran, H. Winkler, M.T. Millan, K.M. Stuhlmeier, B. Vanhove, M.L. Blakely, van, der, Werf, Wj, E. Hofer, and a.l. et, Endothelial cell activation and thromboregulation during xenograft rejection, *Immunological Rev* **141**(5), 5–30 (1994).
5. J.L. Platt, New directions for organ transplantation, *Nature* **392**(6679 Suppl S), 11–17 (1998).
6. F.H. Bach, H. Winkler, C. Ferran, W.W. Hancock, and S.C. Robson, Delayed xenograft rejection, *Immunol Today* **17**(8), 379–384 (1996).
7. R.K. Kutty, R.F. Daniel, D.E. Ryan, W. Levin, and M.D. Maines, Rat liver cytochrome P-450b, P-420b, and P-420c are degraded to biliverdin by heme oxygenase, *Arch Biochem Biophys* **260**(2), 638–644 (1988).
8. M.D. Maines, The heme oxygenase system: a regulator of second messenger gases, *Annu Rev Pharmacol Toxicol* **37**(••), 517–554 (1997).
9. L.E. Otterbein and A.M. Choi, Heme oxygenase: colors of defense against cellular stress, *Am J Physiol Lung Cell Mol Physiol* **279**(6), L1029–1037 (2000).
10. S.W. Ryter and R.M. Tyrrell, The heme synthesis and degradation pathways: role in oxidant sensitivity. Heme oxygenase has both pro- and antioxidant properties, *Free Radic Biol Med* **28**(2), 289–309 (2000).
11. G. Balla, H.S. Jacob, J. Balla, M. Rosenberg, K. Nath, F. Apple, J.W. Eaton, and G.M. Vercellotti, Ferritin: a cytoprotective antioxidant strategem of endothelium, *J Biol Chem* **267**(25), 18148–18153 (1992).
12. R.S. Eisenstein, M.D. Garcia, W. Pettingell, and H.N. Munro, Regulation of ferritin and heme oxygenase synthesis in rat fibroblasts by different forms of iron, *Proc Natl Acad Sci USA* **88**(3), 688–692 (1991).
13. A.M. Choi and J. Alam, Heme oxygenase-1: function, regulation, and implication of a novel stress-inducible protein in oxidant-induced lung injury, *Am J Resp Cell & Mol Biol* **15**(1), 9–19 (1996).
14. J. Balla, H.S. Jacob, G. Balla, K. Nath, J.W. Eaton, and G.M. Vercellotti, Endothelial-cell heme uptake from heme proteins: induction of sensitization and desensitization to oxidant damage, *Proc Natl Acad Sci of the USA* **90**(20), 9285–9289 (1993).

15. C.M. Terry, J.A. Clikeman, J.R. Hoidal, and K.S. Callahan, TNF-alpha and IL-1alpha induce heme oxygenase-1 via protein kinase C, Ca2+, and phospholipase A2 in endothelial cells, *Am J Physiol* **276**(5 Pt 2), H1493–1501 (1999).

16. C.M. Terry, J.A. Clikeman, J.R. Hoidal, and K.S. Callahan, Effect of tumor necrosis factor-alpha and interleukin-1 alpha on heme oxygenase-1 expression in human endothelial cells, *Am J Physiol* **274**(3 Pt 2), H883–891 (1998).

17. J. Alam, D. Stewart, C. Touchard, S. Boinapally, A.M. Choi, and J.L. Cook, Nrf2, a Cap'n'Collar transcription factor, regulates induction of the heme oxygenase-1 gene, *J Biol Chem* **274**(37), 26071–26078 (1999).

18. C.H. He, P. Gong, B. Hu, D. Stewart, M.E. Choi, A.M. Choi, and J. Alam, Identification of activating transcription factor 4 (ATF4) as an Nrf2- interacting protein. Implication for heme oxygenase-1 gene regulation, *J Biol Chem* **276**(24), 20858–20865 (2001).

19. L. Yang, S. Quan, and N.G. Abraham, Retrovirus-mediated HO gene transfer into endothelial cells protects against oxidant-induced injury, *Am J Physiol* **277**(1 Pt 1), L127–133 (1999).

20. S. Brouard, L.E. Otterbein, J. Anrather, E. Tobiasch, F.H. Bach, A.M. Choi, and M.P. Soares, Carbon monoxide generated by heme oxygenase 1 suppresses endothelial cell apoptosis, *J Exp Med* **192**(7), 1015–1026 (2000).

21. N.G. Abraham, Y. Lavrovsky, M.L. Schwartzman, R.A. Stoltz, R.D. Levere, M.E. Gerritsen, S. Shibahara, and A. Kappas, Transfection of the human heme oxygenase gene into rabbit coronary microvessel endothelial cells: protective effect against heme and hemoglobin toxicity, *Proc Natl Acad Sci U S A* **92**(15), 6798–6802 (1995).

22. M.P. Soares, Y. Lin, J. Anrather, E. Csizmadia, K. Takigami, K. Sato, S.T. Grey, R.B. Colvin, A.M. Choi, K.D. Poss, and F.H. Bach, Expression of heme oxygenase-1 (HO-1) can determine cardiac xenograft survival, *Nat Med* **4**(••) 1073–1077 (1998).

23. F. Amersi, R. Buelow, H. Kato, B. Ke, A.J. Coito, X.D. Shen, D. Zhao, J. Zaky, J. Melinek, C.R. Lassman, J.K. Kolls, J. Alam, T. Ritter, H.D. Volk, D.G. Farmer, R.M. Ghobrial, R.W. Busuttil, and J.W. Kupiec-Weglinski, Upregulation of heme oxygenase-1 protects genetically fat Zucker rat livers from ischemia/reperfusion injury, *J Clin Invest* **104**(11), 1631–1639 (1999).

24. L. Otterbein, B.Y. Chin, S.L. Otterbein, V.C. Lowe, H.E. Fessler, and A.M. Choi, Mechanism of hemoglobin-induced protection against endotoxemia in rats: a ferritin-independent pathway, *Am J Physiol* **272**(2 Pt 1), L268–275 (1997).

25. L. Otterbein, S.L. Sylvester, and A.M. Choi, Hemoglobin provides protection against lethal endotoxemia in rats: the role of heme oxygenase-1, *Am J Resp Cell & Mol Biol* **13**(5), 595–601 (1995).

26. K.D. Poss, and S. Tonegawa, Reduced stress defense in heme oxygenase 1-deficient cells *Proc Natl Acad Sci of the USA* **94**(20), 10925–10930 (1997).

27. K. Sato, J. Balla, L. Otterbein, R.N. Smith, S. Brouard, Y. Lin, E. Csizmadia, J. Sevigny, S.C. Robson, G. Vercellotti, A.M. Choi, F.H. Bach, and M.P. Soares, Carbon monoxide generated by heme oxygenase-1 suppresses the rejection of mouse-to-rat cardiac transplants, *J Immunol* **166**(6), 4185–4194 (2001).

28. D. Willis, A.R. Moore, R. Frederick, and D.A. Willoughby, Heme oxygenase: a novel target for the modulation of the inflammatory response, *Nat Med* **2**(1), 87–90 (1996).

29. R. Stocker, Y. Yamamoto, A.F. McDonagh, A.N. Glazer, and B.N. Ames, Bilirubin is an antioxidant of possible physiological importance, *Science* **235**(4792), 1043–1046 (1987).

30. T. Nakagami, K. Toyomura, T. Kinoshita, and S. Morisawa, A beneficial role of bile pigments as an endogenous tissue protector: anti-complement effects of biliverdin and conjugated bilirubin, *Biochim Biophys Acta* **1158**(2), 189–193 (1993).

31. F.A. Haimovitz, C. Cordon-Cardo, S. Bayoumy, M. Garzotto, M. McLoughlin, M. Gallily, C. Edwards III, E.H. Schuchman, Z. Fuks, and R. Kolesnick, Lipopolysaccharide induces disseminated endothelial cell apoptosis requiring ceramide, *J Exp Med* **186**(11), 1831–1841 (1997).

32. T. Bombeli, A. Karsan, J.F. Tait, and J.M. Harlan, Apoptotic vascular endothelial cells become procoagulant, *Blood* **89**(7), 2429–2442 (1997).

33. L.C. Korb and J.M. Ahearn, C1q binds directly and specifically to surface blebs of apoptotic human keratinocytes: complement deficiency and systemic lupus erythematosus revisited, *J Immunol* **158**(10), 4525–4528 (1997).

34. T. Bombeli, B.R. Schwartz, and J.M. Harlan, Endothelial cells undergoing apoptosis become proadhesive for nonactivated platelets, *Blood* **93**(11), 3831–3838 (1999).

35. C. Ferris, S. Jaffrey, A. Sawa, M. Takahashi, S. Brady, R. Barrow, S. Tysoc, H. Wolosker, D. Baranano, S. Dore, K. Poss, and S.H. Snyder, Haem oxygenase-1 prevents cell death by regulating cellular iron, *Nat Cell Biol* **1**(••) 152–157 (1999).

36. N.L. Tilney and R.D. Guttmann, Effects of initial ischemia/reperfusion injury on the transplanted kidney, *Transplantation* **64**(7), 945–947 (1997).

37. P.R. Kvietys and D.N. Granger, Endothelial cell monolayers as a tool for studying microvascular pathophysiology, *Am J Physiol* **273**(6 Pt 1), G1189–1199 (1997).

38. M.B. Grisham, D.N. Granger, and D.J. Lefer, Modulation of leukocyte-endothelial interactions by reactive metabolites of oxygen and nitrogen: relevance to ischemic heart disease, *Free Radic Biol Med* **25**(4–5), 404–433 (1998).

39. N. Maulik, H.S. Sharma, and D.K. Das, Induction of the haem oxygenase gene expression during the reperfusion of ischemic rat myocardium, *J Mol Cell Cardiol* **28**(6), 1261–1270 (1996).

40. H. Kato, F. Amersi, R. Buelow, J. Melinek, A.J. Coito, B. Ke, R.W. Busuttil, and J.W. Kupiec-Weglinski, Heme oxygenase-1 overexpression protects rat livers from ischemia/reperfusion injury with extended cold preservation, *Am J Transpl* **1**(••) 121–128 (2001).

41. T. Fujita, K. Toda, A. Karimova, S.F. Yan, Y. Naka, S.F. Yet, and D.J. Pinsky, Paradoxical rescue from ischemic lung injury by inhaled carbon monoxide driven by derepression of fibrinolysis, *Nat Med* **7**(5), 598–604 (2001).

42. C. Clayberger, P. Parham, J. Rothbard, D.S. Ludwig, G.K. Schoolnik, and A.M. Krensky, HLA-A2 peptides can regulate cytolysis by human allogeneic T lymphocytes, *Nature* **330**(6150), 763–765 (1987).

43. P. Parham, C. Clayberger, S.L. Zorn, D.S. Ludwig, G.K. Schoolnik, and A.M. Krensky, Inhibition of alloreactive cytotoxic T lymphocytes by peptides from the alpha 2 domain of HLA-A2, *Nature* **325**(6105), 625–628 (1987).

44. S. Iyer, J. Woo, M.C. Cornejo, L. Gao, W. McCoubrey, M. Maines, and R. Buelow, Characterization and biological significance of immunosuppressive peptide D2702.75–84(E → V) binding protein. Isolation of heme oxygenase-1, *J Biol Chem* **273**(5), 2692–2697 (1998).

45. M.C. Cuturi, R. Josien, P. Douillard, C. Pannetier, D. Cantarovich, H. Smit, S. Menoret, P. Pouletty, C. Clayberger, and J.P. Soulillou, Prolongation of allogeneic heart graft survival in rats by administration of a peptide (a.a. 75–84) from the alpha 1 helix of the first domain of HLA-B7 01, *Transplantation* **59**(5), 661–669 (1995).

46. R. Buelow, P. Veyron, C. Clayberger, P. Pouletty, and J.L. Touraine, Prolongation of skin allograft survival in mice following administration of ALLOTRAP, *Transplantation* **59**(4), 455–460 (1995).

47. S. Brouard, M.C. Cuturi, P. Pignon, R. Buelow, P. Loth, A. Moreau, and J.P. Soulillou, Prolongation of heart xenograft survival in a hamster-to-rat model after therapy with a rationally designed immunosuppressive peptide, *Transplantation* **67**(12), 1614–1618 (1999).

48. M. Giral, C. Taddei, J.M. Nguyen, J. Dantal, M. Hourmant, D. Cantarovich, G. Blancho, D. Ancelet, and J.P. Soulillou, Single-center analysis of 468 first cadaveric kidney allografts with a uniform ATG-CsA sequential therapy, *Clin Transpl* 257–264 (1996).

49. B. Murphy, K.S. Kim, R. Buelow, M.H. Sayegh, and W.W. Hancock, Synthetic MHC class I peptide prolongs cardiac survival and attenuates transplant arteriosclerosis in the Lewis → Fischer 344 model of chronic allograft rejection, *Transplantation* **64**(1), 14–19 (1997).

50. M.C. Cuturi, F. Christoph, J. Woo, S. Iyer, S. Brouard, J.M. Heslan, P. Pignon, J.P. Soulillou, and R. Buelow, RDP1258, a new rationally designed immunosuppressive peptide, prolongs allograft survival in rats: analysis of its mechanism of action, *Mol Med* **5**(12), 820–832 (1999).

51. J. Woo, S. Iyer, M.C. Cornejo, N. Mori, L. Gao, I. Sipos, M. Maines, and R. Buelow, Stress protein-induced immunosuppression: inhibition of cellular immune effector functions following overexpression of haem oxygenase (HSP 32), *Transpl Immunol* **6**(2), 84–93 (1998).

52. L.E. Otterbein, F.H. Bach, J. Alam, M. Soares, H. Tao Lu, M. Wysk, R.J. Davis, R.A. Flavell, and A.M. Choi, Carbon monoxide has anti-inflammatory effects involving the mitogen-activated protein kinase pathway, *Nat Med* **6**(4), 422–428 (2000).

53. H. Azuma and N.L. Tilney, Chronic graft rejection, *Curr Opin Immunol* **6**(5), 770–776 (1994).

54. P.S. Russell, C.M. Chase, H.J. Winn, and R.B. Colvin, Coronary atherosclerosis in transplanted mouse hearts.I. Time course and immunogenetic and immunopathological considerations, *Am J Pathol* **144**(2), 260–274 (1994).

55. W.W. Hancock, R. Buelow, M.H. Sayegh, and L.A. Turka, Antibody-induced transplant arteriosclerosis is prevented by graft expression of anti-oxidant and anti-apoptotic genes, *Nat Med* **4**(12), 1392–1396 (1998).

56. H.J. Duckers, M. Boehm, A.L. True, S.F. Yet, H. San, J.L. Park, R. Clinton Webb, M.E. Lee, G.J. Nabel, and E.G. Nabel, Heme oxygenase-1 protects against vascular constriction and proliferation, *Nat Med* **7**(6), 693–698 (2001).

57. G.P. Alexandre, J.P. Squifflet, M. De Bruyere, D. Latinne, R. Reding, P. Gianello, M. Carlier, and Y. Pirson, Present experiences in a series of 26 ABO-incompatible living donor renal allografts, *Transpl Proc* **19**(6), 4538–4542 (1987).

58. M. Slapak, R.B. Naik, and H.A. Lee, Renal transplant in a patient with major donor-recipient blood group incompatibility: reversal of acute rejection by the use of modified plasmapheresis, *Transplantation* **31**(1), 4–7 (1981).

59. L.J. West, S.M. Pollock-Barziv, A.I. Dipchand, K.J. Lee, C.J. Cardella, L.N. Benson, I.M. Rebeyka, and J.G. Coles, ABO-incompatible heart transplantation in infants, *N Engl J Med* **344**(11), 793–800 (2001).

60. F.H. Bach, M.A. Turman, G.M. Vercellotti, J.L. Platt, and A.P. Dalmasso, Accommodation: a working paradigm for progressing toward clinical discordant xenografting., *Transpl Proc* **23**(1 Pt 1), 205–207 (1991).

61. M.P. Soares, Y. Lin, K. Sato, K.M. Stuhlmeier, and F.H. Bach, Accommodation, *Immunol Tod* **20**(10), 434–437 (1999).

62. R. Hasan, Van, den, Bogaerde, Jb, J. Wallwork, and D.J. White, Evidence that long-term survival of concordant xenografts is achieved by inhibition of antispecies antibody production, *Transplantation* **54**(3), 408–413 (1992).

63. N. Koyamada, T. Miyatake, D. Candinas, W. Mark, P. Hechenleitner, W.W. Hancock, M.P. Soares, and F.H. Bach, Transient complement inhibition plus T-cell immunosuppression induces long-term survival of mouse-to-rat cardiac xenografts, *Transplantation* **65**(9), 1210–1215 (1998).

64. M.P. Soares, Y. Lin, K. Sato, K. Takigami, J. Anrather, C. Ferran, S.C. Robson, and F.H. Bach, Pathogenesis of and potential therapies for delayed xenograft rejection, *Cur Opin Org Transpl* **4**(••) 80–89 (1999).

65. T. Miyatake, K. Sato, K. Takigami, N. Koyamada, W.W. Hancock, H. Bazin, D. Latinne, F.H. Bach, and M.P. Soares, Complement-fixing elicited antibodies are a major component in the pathogenesis of xenograft rejection, *J Immunol* **160**(8), 4114–4123 (1998).

66. Y. Lin, M.P. Soares, K. Sato, E. Csizmadia, S.C. Robson, N. Smith, and F.H. Bach, Long-term survival of hamster hearts in presensitized rats, *J Immunol* **164**(9), 4883–4892 (2000).

67. F.H. Bach, C. Ferran, P. Hechenleitner, W. Mark, N. Koyamada, T. Miyatake, H. Winkler, A. Badrichani, D. Candinas, and W.W. Hancock, Accommodation of vascularized xenografts: expression of "protective genes" by donor endothelial cells in a host Th2 cytokine environment, *Nat Med* **3**(2), 196–204 (1997).

68. K. Sato, K. Takigami, T. Miyatake, E. Czismadia, D. Latinne, H. Bazin, F.H. Bach, and M.P. Soares, Suppression of delayed xenograft rejection by specific depletion of elicited antibodies of the IgM Isotype, *Transplantation* **68**(6), 844–854 (1999).

69. Y. Lin, M.P. Soares, K. Sato, K. Takigami, E. Csizmadia, N. Smith, and F.H. Bach, Accommodated xenografts survive in the presence of anti-donor antibodies and complement that precipitate rejection of naive xenografts, *J Immunol* **163**(5), 2850–2857 (1999).

70. T. Berney and C. Ricordi, Islet cell transplantation: the future?, *Langenbecks Arch Surg* **385**(6), 373–378 (2000).

71. A.M. Shapiro, J.R. Lakey, E.A. Ryan, G.S. Korbutt, E. Toth, G.L. Warnock, N.M. Kneteman, and R.V. Rajotte, Islet transplantation in seven patients with type 1 diabetes mellitus using a glucocorticoid-free immunosuppressive regimen, *N Engl J Med* **343**(4), 230–238 (2000).

72. C. Benoist and D. Mathis, Cell death mediators in autoimmune diabetes—no shortage of suspects, *Cell* **89**(1), 1–3 (1997).

73. T. Berney, R.D. Molano, P. Cattan, A. Pileggi, C. Vizzardelli, R. Oliver, C. Ricordi, and L. Inverardi, Endotoxin-mediated delayed islet graft function is associated with increased intra-islet cytokine production and islet cell apoptosis, *Transplantation* **71**(1), 125–132 (2001).
74. S.T. Grey, M.B. Arvelo, W. Hasenkamp, F.H. Bach, and C. Ferran, A20 inhibits cytokine-induced apoptosis and nuclear factor kappaB- dependent gene activation in islets, *J Exp Med* **190**(8), 1135–1146 (1999).
75. J.L. Contreras, G. Bilbao, C.A. Smyth, X.L. Jiang, D.E. Eckhoff, S.M. Jenkins, F.T. Thomas, D.T. Curiel, and J.M. Thomas, Cytoprotection of pancreatic islets before and soon after transplantation by gene transfer of the anti-apoptotic Bcl-2 gene, *Transplantation* **71**(8), 1015–1023 (2001).
76. N. Welsh and S. Sandler, Protective action by hemin against interleukin-1 beta induced inhibition of rat pancreatic islet function, *Mol Cell Endocrinol* **103**(1–2), 109–114 (1994).
77. J. Ye and S.G. Laychock, A protective role for heme oxygenase expression in pancreatic islets exposed to interleukin-1beta, *Endocrinology* **139**(10), 4155–4163 (1998).
78. A. Pileggi, R.D. Molano, T. Berney, P. Cattan, C. Vizzardelli, R. Oliver, C. Fraker, C. Ricordi, R.L. Pastori, F.H. Bach, and L. Inverardi, Heme oxygenase-1 induction in islet cells results in protection from apoptosis and improved in vivo function after transplantation, *Diabetes* **50**(9), 1983–1991 (2001).

19

THIOREDOXIN FACILITATES THE INDUCTION OF HEME OXYGENASE-1 IN RESPONSE TO INFLAMMATORY MEDIATORS

Mark A. Perrella[a] and Shaw-Fang Yet

[a]Pulmonary and Critical Care Division
Brigham and Women's Hospital
75 Francis Street, Boston, MA 02115

INTRODUCTION

Heme oxygenase (HO)-1 is a stress response protein that is induced by a variety of stimuli associated with oxidative stress.[1] HO-1 catalyzes the degradation of heme to generate biliverdin, carbon monoxide, and iron.[2] Biliverdin is subsequently converted to bilirubin, a potent endogenous antioxidant.[3] Carbon monoxide shares many similarities with nitric oxide, such as its ability to increase cGMP levels and promote vasodilation, thereby modulating tissue perfusion.[4,5] The induction of HO-1 in response to cellular stress is an important protective mechanism, since mice lacking HO-1 show reduced survival when subjected to endotoxemia,[6] chronic hypoxia,[7] or transplantation of cardiac xenografts.[8]

In macrophages and vascular smooth muscle cells, HO-1 gene transcription is upregulated by inflammatory mediators such as lipopolysaccharide (LPS) and interleukin (IL)-1β.[9,10] Two enhancer regions, located 4 and 10 kilobases (kb) upstream

---

[1] Abbreviations used in this paper: HO, heme oxygenase; LPS, lipopolysaccharide; IL, interleukin; RASMC, rat aortic smooth muscle cells; TRX, thioredoxin; TR, thioredoxin reductase; kb, kilobases; AP, activator protein; ex, extended; NEM, N-ethylmaleimide; DNCB, 1-chloro 2,4 di nitrobenzene; RAW, RAW 264.7A murine macrophages; bp, base pairs; CAT, chloramphenicol acetyltransferase.

This chapter was reproduced from an original article published in The Journal of Biological Chemistry, Vol. 275, No. 32, Issue of August 11, pp. 24840–24846, 2000. Copyright (2000) the American Society of Biochemistry and Molecular Biology, Inc.

from the transcription start site, are important for induction of HO-1 by LPS in macrophages.[11] Both enhancer regions contain multiple activator protein (AP)-1 binding sites, the mutation of which impairs LPS responsiveness in macrophages.[11] Activation of transcription factors such as AP-1 plays a pivotal role in LPS and IL-1β signaling.[12] These transcription factors are responsible for the upregulation of a battery of genes involved in the inflammatory response. These include genes encoding adhesion molecules, matrix metalloproteinases, and inflammatory cytokines.[13,14]

Genes that require AP-1 for transcriptional regulation demonstrate an increase in AP-1 protein-DNA binding, which is often associated with an increase in the level of AP-1 proteins. Post-translational modifications of AP-1 proteins, such as the phosphorylation of critical serine residues in their activation domains, may also enhance the transactivation potential of these factors.[15] However, alterations in redox status play a critical role in the transcriptional activity mediated by AP-1 proteins, and this redox regulation maps to a single conserved cysteine residue in the DNA binding domains of both Fos and Jun.[16] While reduced sulfhydryl groups of this cysteine are required for DNA binding and AP-1 activation, the molecular entity of the inactive, oxidized form of the cysteine is less clear. Abate and colleagues have suggested that oxidation of this cysteine does not involve the formation of intra- or intermolecular disulfide bonds,[16] and recently Klatt and colleagues showed that S-glutathiolation may contribute to the oxidation of c-Jun sulfhydryls and thus interfere with DNA binding.[17] Since AP-1 proteins are important for signaling during oxidative stress, it has been hypothesized that specific endogenous factors may exist to prevent oxidation, and thus activate intracellular AP-1 proteins and allow DNA binding in response to oxidative stress.[16]

An intracellular redox regulator that has been shown to be important for the regulation of transcription factors is thioredoxin (TRX).[18–20] When reduced by TRX reductase (TR) and NADPH, TRX can reactivate oxidatively inactive transcription factors such as Jun and Fos.[21] TRX does not interact directly with AP-1 proteins, but TRX activates AP-1 transcription by direct association with intranuclear redox factor 1 (Ref-1).[19,21] Ref-1 associates transiently with and reduces the conserved cysteine in Fos and Jun, thus enhancing their DNA-binding activity.[23] TRX is usually located in the cytosol,[24] but it translocates into the nucleus in response to various stimuli associated with oxidative stress. Therefore, the nuclear translocation of TRX may be a converging step mediating Ref-1—dependent activation of AP-1 proteins by a variety of stress-inducing agents.

IL-1β INDUCES NUCLEAR PROTEIN BINDING TO AN AP-1 SITE IN THE HO-1 PROMOTER

The binding of AP-1 proteins to cognate *cis*-acting elements in the HO-1 promoter is important for LPS response in macrophages. To determine if binding of nuclear proteins to AP-1 sites in the −4 kb enhancer of the HO-1 promoter is induced by IL-1β in rat aortic smooth muscle cells (RASMC), we performed electromobility gel shift assays (EMSA). Using an oligonucleotide probe encompassing base pairs (bp) 163 to 180 of the −4 kb enhancer, we demonstrated a time-dependent increase in

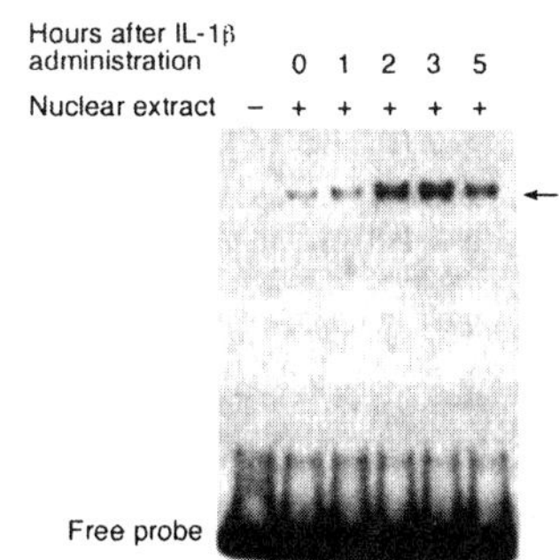

Figure 1. IL-1β induces binding of nuclear proteins to an AP-1 site in the −4 kb enhancer of the HO-1 promoter in RASMC. Nuclear proteins were obtained from RASMC stimulated with IL-1β (10 ng/ml) for the indicated times, and EMSA were performed with a radiolabeled oligonucleotide encoding a 17 bp section of the −4 kb enhancer in the HO-1 promoter region. Arrow indicates specific DNA-protein complex.

protein binding in nuclear extracts obtained from IL-1β-treated RASMC (Fig. 1, arrow). Maximal DNA-protein binding was observed 3 hours after IL-1β stimulation.

This DNA-protein complex was specific because a 100-fold molar excess of unlabeled identical oligonucleotide (I), but not an unrelated oligonucleotide (NI) containing an E2F consensus sequence, competed away binding (Fig. 2A). In addition, an oligonucleotide competitor containing a consensus AP-1 motif was also able to abolish nuclear protein binding. These results indicate that in RASMC, IL 1β induces the binding of nuclear proteins to an AP-1 motif in the −4 kb enhancer of the HO-1 promoter region. To further characterize these proteins, we used specific antibodies against known AP-1 proteins. The addition of an anti-c-Jun antibody produced a slower migrating supershift complex (Fig. 2B, asterisk), suggesting the presence of c-Jun in the nucleoprotein complex. In addition, an antibody against phospho c-Jun (at Ser73) resulted in a similar supershift complex (data not shown). When a polyclonal anti-Fos antibody was added to the incubation mixture, the complex was disrupted. The epitope recognized by this antibody is located within the DNA binding domain

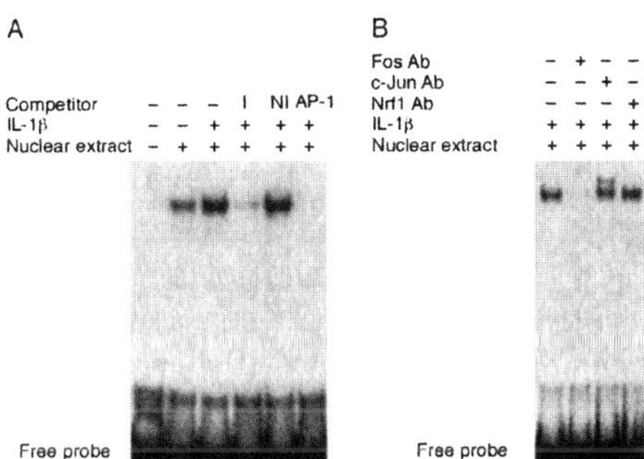

Figure 2. Identification of nuclear proteins in the AP-1 binding complex. A, RASMC were treated with vehicle or IL 1β for 3 h, and EMSA were performed. Where indicated, oligonucleotide competitors were added at a 100-fold molar excess. I, identical competitor; NI, non-identical competitor; AP-1, AP-1 competitor. B, Nuclear proteins from RASMC stimulated with IL-1β for 3 h were incubated with antibodies against Fos, c-Jun, or Nrf1 for 12 h at 4°C. ^{32}P-labeled HO-1/AP-1 probe was added subsequently and the mixture was allowed to incubate for 20 min at room temperature. EMSA were performed. Asterisk indicates supershifted complex.

of Fos. Thus, binding of the antibody to this region prevented Fos, and it's associated complex, from binding to DNA. It has been suggested that proteins belonging to the NF-E2 family of transcription factors may also bind to the AP-1 motifs contained in the HO-1 enhancers.[25] However, antibodies against Nrf1 (Fig. 2B) or Nrf2 (not shown) did not affect DNA-binding. Therefore, our data indicate that AP-1 proteins including c-Jun and Fos were present in the inducible complex after IL-1β stimulation in RASMC.

DNA-PROTEIN BINDING IS DEPENDENT ON THE REDOX STATUS OF NUCLEAR PROTEINS

Both Fos and Jun share a conserved cysteine in their binding domain, which must be reduced to allow DNA-binding. We evaluated the effect of diamide (an agent that catalyzes the oxidation of free sulfhydryl groups) and NEM (N-ethylmaleimide, an agent that alkylates free sulfhydryl groups) on AP-1 binding activity in IL-1β-treated RASMC. Low millimolar concentrations of diamide (Fig. 3, lanes 3–7) and NEM (Fig. 3, lanes 10–14) dose-dependently abolished DNA-binding activity. In RAW cells (RAW 264.7A murine macrophages), LPS treatment for 3 hours also markedly induced the binding of AP-1 proteins to the HO-1 −4kb enhancer (data not shown). Again, both diamide and NEM dose-dependently reduced AP-1-binding activity in cells stimulated with LPS. In both cell types, DTT restored DNA-binding activity in nuclear extracts previously treated with diamide (Fig. 3, lane 8). As expected, DTT had no effect on the irreversible alkylation of free sulfhydryl groups induced by NEM (Fig. 3, lane 15).

BOTH IL-1β AND LPS INDUCE THE NUCLEAR LOCALIZATION OF TRX

TRX has been shown to enhance the DNA-binding activity of AP-1 proteins by interacting with the nuclear protein Ref-1.[19,26] However, TRX is localized in the

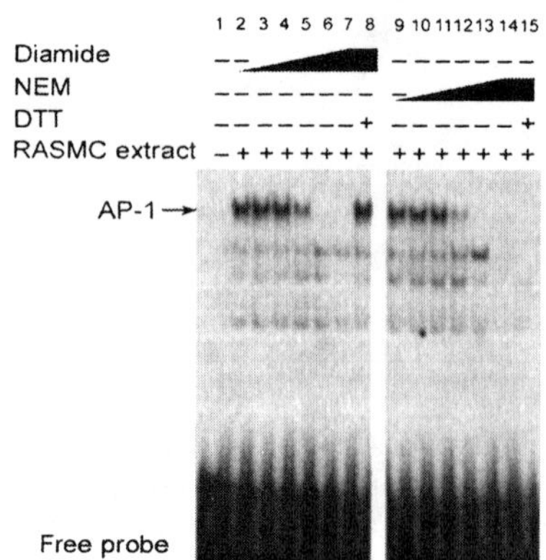

Figure 3. Inhibition of DNA-binding activity by sulfhydryl modifying agents. Nuclear proteins were extracted from RASMC treated with IL-1β for 3 h, and incubated with increasing concentrations of sulfhydryl modifying agents. Diamide concentrations: 0.25, 0.5, 0.75, 1.0, 2.0 and 2.0 mM in lanes 3–8 respectively. NEM concentrations: 0.5, 1.0, 1.5, 2.0, 5.0 and 5.0 mM in lanes 10–15 respectively. Where indicated, diamide or NEM-treated extracts were incubated with 5 mM DTT before addition of the HO-1/AP-1 probe. EMSA were subsequently performed.

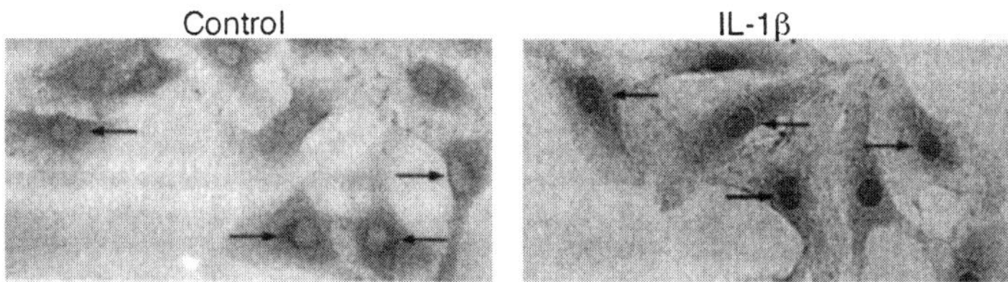

Figure 4. IL-1β induces the nuclear translocation of TRX in RASMC. Cells were grown on cover slips and treated with vehicle (left) or IL-1β (10 ng/ml, right) for 24 h. Immunocytochemical analysis was performed. Arrows indicate nuclei. Histologic sections were processed with an anti-TRX antibody. Arrows indicate nuclei.

cytosol of unstimulated cells[19] and cannot directly reduce AP-1 proteins.[16,23] Therefore, we evaluated whether IL-1β and LPS could induce the translocation of TRX from the cytosol to the nucleus. Nuclear preparations were obtained from RASMC treated with IL-1β, and analyzed by Western blotting with an antibody against TRX. While no signal could be detected in unstimulated cells, nuclear TRX was observed in RASMC as early as 3 hours after IL-1β stimulation, and more prominently after 5 hours (data not shown). To confirm this result with an independent method, we determined the subcellular localization of TRX by immunocytochemistry. In unstimulated RASMC, TRX was predominantly cytosolic, with minimal staining in the nuclear areas (Fig. 4, left panel, arrows). In RASMC treated with IL-1β, intense nuclear staining for TRX was observed (Fig. 4, right panel, arrows). Subcellular localization of TRX was also assessed in aortic tissue *in vivo*. Rats were injected with LPS or saline, and their aortas were harvested 10 hours later and stained with anti-TRX antibody. In untreated rats, TRX was detected in sub-endothelial smooth muscle layers, while little signal was present in the mid and outer parts of the aortic wall or in adventitial tissue. TRX signal was present in the cytoplasm but not the nuclei of the smooth muscle cells (data not shown). In rats treated with LPS for 10 h, TRX was detectable in the entire aortic wall. Moreover, TRX staining was predominantly in a nuclear pattern after LPS.

TRX AND REF-1 COOPERATE TO INDUCE AN HO-1 PROMOTER CONSTRUCT

We next determined whether TRX could induce HO-1 promoter activity. We constructed a heterologous promoter containing the −4 kb enhancer of the HO-1 gene in a luciferase reporter vector. In transient transfection experiments using RAW cells, LPS dose-dependently induced activity of the HO-1 heterologous promoter (Fig. 5A). Maximal induction was obtained with an LPS dose of 0.5 µg/ml. The AP-1 sites in the −4 kb enhancer were then mutated to determine their role in the regulation of HO-1 promoter activity. Mutation of the AP-1 sites abolished LPS-induced promoter activity (Fig. 5B). Taken together with the electromobility gel shift assays, these data

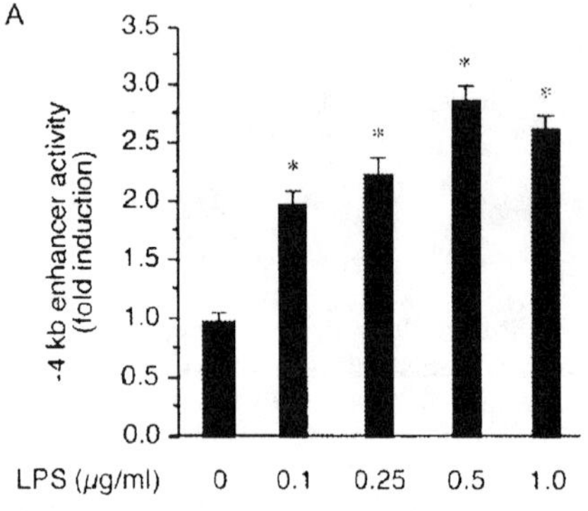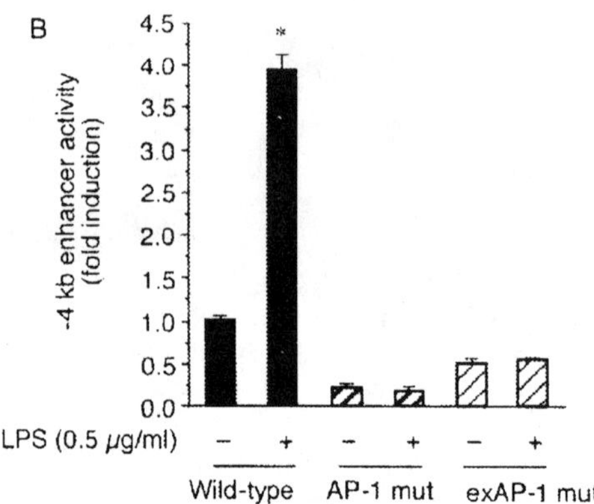

Figure 5. Transactivation of the HO-1 –4 kb enhancer by LPS. A, The pGL3-TK/HO-1-4en reporter construct (1 μg DNA/well) was transiently transfected into RAW cells. pOPRSV1-CAT (1 μg/well) was co-transfected to correct for differences in transfection efficiency. Twenfy-fourth h after transfection. LPS was added to the culture medium at the indicated concentrations. Normalized luciferase activity is presented as fold-induction versus LPS 0 μg/ml. Results represent the mean ± SE of at least three independent experiments (n = 6). B, Reporter constructs containing wild-type (pGL3-TK/HO-1-4en, black bars), AP-1 mutated (pGL3-TK/HO-1-4en/AP-1mut, thick striped bars), or extended AP-1 mutated (pGL3-TK/HO-1-4en/exAP-1mut, thin striped bars) sites were transiently transfected into RAW cells. Twenty-four h after transfection, vehicle (–) or LPS (+) was added to the culture medium. Normalized luciferase activity is presented as fold-induction versus wild-type treated with vehicle (–). Results represent the mean ± SE (n = 6). In A, asterisks indicate significant differences (P < 0.05) in comparison with cells treated with vehicle (0 μg/ml LPS). In B, asterisks indicate significant differences (P < 0.05) in comparison with all other groups.

suggest that AP-1 proteins including c-Jun and Fos are important for LPS induction of HO-1. Interestingly, mutation of the AP-1 sites within the –4 kb enhancer also reduced HO-1 promoter activity in the absence of LPS (Fig. 5B), suggesting that AP-1 proteins may play a role in basal HO-1 expression. Mutation of the 5′ flanking sequence of the AP-1 sites also abrogated transactivation of the HO-1 promoter (Fig. 5B). These extended AP-1 sites (exAP-1) bind other bZIP proteins that are known to heterodimerize with Jun or Fos proteins.[15] Thus, we can not exclude other bZIP family members from contributing to this response.

We then tested the effect of exogenous recombinant TRX on promoter activity (Fig. 6A). Since TRX cooperates with Ref-1 to activate AP-1, we co-transfected a Ref-1 expression plasmid together with the HO-1 heterologous promoter. Twenty hours after transfection, recombinant human TRX in the reduced form was added to the medium. Promoter activity was measured 24 h after adding TRX. Overexpression of Ref-1 alone had no effect on the reporter construct. However, when increasing doses of recombinant reduced TRX were added, promoter activity was induced to a level similar to that achieved by LPS (Fig. 6A). Also, analogous to the LPS response, mutation of the AP-1 sites or the exAP-1 sites in the –4 kb enhancer abolished TRX-induced promoter activity (Fig. 6B).

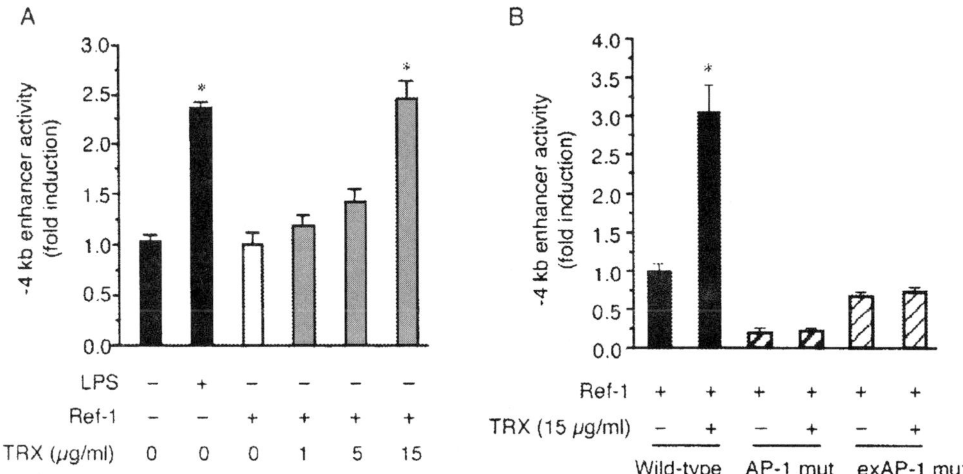

Figure 6. Transactivation of the HO-1 −4 kb enhancer by recombinant TRX. A, The pGL3-TK/HO-1-4en reporter construct (1 µg DNA/well) was transiently transfected into RAW cells. Where indicated, 1 µg hRef-L/pcDNA 3.1 plasmid was co-transfected with pGL3/HO-1-4en in the absence (white bar) or presence (gray bars) of increasing doses of recombinant human TRX. pOPRSV1-CAT (1 µg DNA/well) was co-transfected to correct for differences in transfection efficiency. Normalized luciferase activity is presented as fold-induction versus the respective control. Results represent the mean ± SE of at least three independent experiments (n = 6). B, Reporter constructs containing wild-type (pGL3-TK/HO-1-4en, black bars), AP-1 mutated (pGL3-TK/HO-1-4en/AP-1mut, thick striped bars), or extended AP-1 mutated (pGL3-TK/HO-1-4en/exAP-1mut, thin striped bars) sites were transiently transfected into RAW cells. 1 µg hRef-1/pcDNA3.1 plasmid was co-transfected with constructs in the absence (−) or presence (+) of recombinant human TRX. Normalized luciferase activity is presented as fold-induction versus wild-type treated with no TRX (−), Results represent the mean ± SE (n = 5). In A, Asterisks indicate significant differences (P < 0.05) in comparison with unstimutated cells (—LPS, —Ref-1, and 0 µg/ml TRX). In B, asterisks indicate significant differences (P < 0.05) in comparison with all other groups.

INHIBITION OF TR IMPAIRS IL-1β AND LPS RESPONSE

To determine the biological significance of TRX translocation, we tested the effect of DNCB (1-chloro-2,4-di-nitrobenzene), a specific inhibitor of TR, on HO-1 induction by LPS and IL-1β in macrophages and in RASMC, respectively. DNCB was added to the culture media 30 min before the addition of LPS to RAW cells, and total RNA was extracted 24 h later. By Northern blotting, we demonstrate a 6 fold induction of HO-1 mRNA by LPS (Fig. 7A). With increasing doses of DNCB, we suppressed the induction of HO-1 mRNA. We also hybridized the membranes with a probe against HO-2. HO-2 is the constitutive heme oxygenase isoform, and it is encoded by a different gene. As expected, HO-2 mRNA was not significantly induced by LPS, and DNCB did not decrease the message of HO-2 in control or stimulated cells (Fig. 7A). We then evaluated the effect of DNCB on HO-1 protein by Western blot analysis. As shown in Fig. 7B, LPS treatment for 48 h markedly induced HO-1 protein in RAW cells. Again, pre-treatment with DNCB dose-dependently attenuated

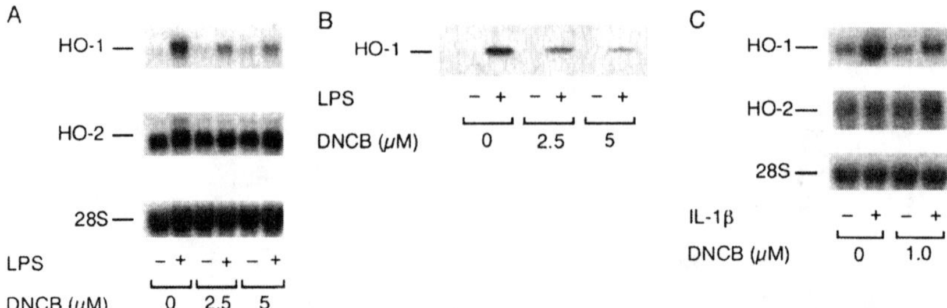

Figure 7. Inhibition of TRX reductase attenuates HO-1 induction by LPS and IL-1β. A, DNCB was added to the RAW cells at the indicated concentration. After thirty min, vehicle or LPS (0.5 µg/ml) was added and total RNA was extracted 24 h later. Northern blot analysis was performed as described and membranes were hybridized to HO-1, HO-2, or 28S probes. B, RAW cells were treated as in A and whole cell protein was used in Western blot analysis. Immunoblots were performed using an anti-HO-1 antibody C. RASMC were stimulated with IL-1β (10 ng/ml) 30 min after the administration of DNCB at the indicated concentration. Twenty-four h later, total RNA was extracted and Northern blot analysis was performed. Blots were hybridized to HO-1, HO-2, or 28S probes, and exposed to film. All experiments were performed at least twice.

the induction of HO-1 protein. A similar effect of TR inhibition on HO-1 mRNA was observed in RASMC stimulated with IL-1β. Consistent with our previous observation,[10] IL-1β (10 ng/ml) induced HO-1 mRNA 3 to 4 fold in RASMC. DNCB at a 0.1 µM concentration reduced the induction of HO-1 mRNA by ~70% (Fig. 7C), but had no effect on HO-2 mRNA.

SUMMARY AND DISCUSSION

TRX is responsible for maintaining many cytosolic proteins in the reduced state. Beyond this homeostatic function, TRX has been shown to play a regulatory role in several cellular processes. TRX potentiates the release of pro-inflammatory cytokines in response to tetradecanoyl phorbol acetate (TPA),[27] and accentuates the transformed phenotype of tumor cells.[28] Also, nuclear TRX may restore the DNA-binding activity of several transcription factors in the presence of reactive oxygen species, including AP-1 proteins and the glucocorticoid receptor.[19,29] However, no data are available on the role of TRX in the regulation of HO-1 in response to inflammatory agents. Our present data suggest that TRX is involved in the AP-1—mediated induction of HO-1 in response to LPS and IL-1β.

We demonstrated previously that LPS and IL-1β induce HO-1 gene transcription in RASMC *in vivo* and *in vitro*.[10] LPS and IL-1β are inflammatory mediators in a number of disease processes including endotoxemia. During endotoxemia, LPS-responsive immune cells (such as macrophages) become activated and release a battery of mediators and proinflammatory cytokines.[30] Similar to what has been described in macrophages stimulated with LPS,[11] we show in RASMC that the proinflammatory cytokine IL-1β increases binding of AP-1 proteins to a cognate motif in the −4 kb enhancer of the HO-1 promoter region (Figs. 1 and 2). In both RASMC and RAW cells, we found that free sulfhydryl groups in nuclear proteins are necessary for DNA-binding activity (Fig. 3). Abate and colleagues[16] had previously demonstrated that the redox status of a conserved cysteine residue in Fos and Jun can alter their ability to interact with DNA, and that exogenous TRX can increase the DNA binding activity of recombinant AP-1 proteins. To do so, TRX needed to be in the reduced form, and required the participation of additional nuclear proteins.[16] One of these nuclear factors cooperating with TRX is Ref-1.[19] TRX may be responsible for the redox cycling of Ref-1, which has been shown to enhance the DNA-binding activity of Fos and Jun.[23] Therefore, the nuclear translocation of reduced TRX has been suggested to represent an important signaling event in the induction of AP-1 driven genes.

Here we show that IL-1β induces the nuclear translocation of TRX in RASMC (Fig. 4) *in vitro*. Previously, the stimuli known to induce the nuclear transloca-tion of TRX *in vitro* included hydrogen peroxide,[31] TPA,[19] UV irradiation[32] and hypoxia/reperfusion.[29,33] We also demonstrate in rats given LPS, a physiological stim-ulus, that nuclear translocation of TRX occurs *in vivo* in vascular smooth muscle cells. Stimuli known to induce the nuclear translocation of TRX cause oxidative stress. We therefore anticipate that reactive oxygen species induced by these stimuli may be involved in the yet undefined mechanisms underlying the nuclear translocation of TRX.

In this chapter, we show that the −4 kb enhancer of the HO-1 gene is induced by LPS (Fig. 5A). Moreover exogenous reduced TRX, in the presence of Ref-1, induces this −4 kb enhancer to a level comparable with that achieved by LPS (Fig. 6A). This effect of TRX on HO-1 promoter activity was abolished by mutation of the AP-1 sites in the −4 kb enhancer (Fig. 6B). We then used an irreversible inhibitor of TR, DNCB, to evaluate the role of reduced TRX in the induction of HO-1 mRNA and protein by IL-1β and LPS. DNCB is an alkylating agent that potently and specif-ically inhibits TR. For example, DNCB reacts with TR about 10,000 times faster than with reduced glutathione.[34] Pretreatment with DNCB dose-dependently suppressed the induction of HO-1 by IL-1β and LPS (Fig. 7). This effect was not due to a general decrease in transcriptional activity, since DNCB did not affect HO-2 mRNA levels (Fig. 7). As expected, DNCB did not decrease basal HO-1 expression, indicating that reduced TRX is not necessary for the binding of AP-1 proteins to their cognate *cis*-acting elements in unstimulated cells. Taken together, our data support a role for TRX in the signaling events promoting the induction of HO-1 by inflammatory mediators.

ACKNOWLEDGMENTS

We dedicate this chapter to the memory of Dr. Arthur Mu-En Lee, who was a constant source of inspiration and support of our work. This chapter is an excerpt from the Arthur Mu-En Lee Memorial Lecture of the 1st International Symposium on Heme Oxygenase-Carbon Monoxide System. We thank Philippe Wiesel, Lauren C. Foster, Andrea Pellacani, Matthew D. Layne, Chung-Ming Hsieh, Gordon S. Huggins, Phyllis Straus, Dorothy Zhang and Bonna Ith for their efforts in these studies.

REFERENCES

1. Keyse S.M. and Tyrrel R.: Heme oxygenase is the major 32-kDa stress protein induced in human skin fibroblasts by UVA irradiation, hydrogen peroxide, and sodium arsenite, Proc. Natl. Acad. Sci. USA., 86:99–103, 1989.

2. Maines M.D.: Heme oxygenase: function, multiplicity, regulatory mechanisms, and clinical applications, FASEB J., 2:2557–2568, 1988.

3. Stocker R., Yamamoto Y., McDonagh A.F., Glazer A.N., and Ames B.N.: Bilirubin is an antioxidant of possible physiological importance, Science., 235:1042–1046, 1987.

4. Morita T., Perrella M.A., Lee M.-E., and Kourembanas S.: Smooth muscle cell-derived carbon monoxide is a regulator of vascular cGMP., Proc. Natl. Acad. Sci. USA., 92:1475–1479, 1995.

5. Christodoulides N., Durante W., Kroll M.H., and Schafer A.I.: Vascular smooth muscle cell heme oxygenases generate guanylyl cyclase-stimulatory carbon monoxide., Circulation., 91:2306–2309, 1995.

6. Poss K.D. and Tonegawa S.: Reduced stress defense in heme oxygenase 1-deficient cells, Proc. Natl. Acad. Sci. U.S.A., 94:10925–10930, 1997.

7. Yet S.-F., Perrella M.A., Layne M.D., Hsieh C.-M., Maemura K., Kobzik L., Wiesel P., Christou H., Kourembanas S., and Lee M.-E.: Hypoxia induces severe right ventricular dilatation and infarction in heme oxygenase-1 null mice, J. Clin. Invest., 103:R23–R29, 1999.

8. Soares M.P., Lin Y.L., Csizmadia E., Takigami K., Sato K., Grey S.T., Colvin R.B., Choi A.M., Poss K.D., and Bach F.H.: Expression of heme oxygenase-1 can determine cardiac xenograft survival., Nat. Med., 4:1073–1077, 1998.

9. Camhi S.L., Alam J., Otterbein L., Sylvester S.L., and Choi A.M.K.: Induction of heme oxygenase-1 gene expression by lipopolysaccharide is mediated by AP-1 activation., Am. J. Respir. Cell Mol. Biol., 13:387–398, 1995.

10. Yet S.-F., Pellacani A., Patterson C., Tan L., Folta S.C., Foster L., Lee W.-S., Hsieh C.-M., and Perrella M.A.: Induction of heme oxygenase-1 expression in vascular smooth muscle cells. A link to endotoxic shock., J. Biol. Chem., 272:4295–4301, 1997.

11. Camhi S., Alam J., Wiegand G., Yoke Chin B., and Choi A.M.K.: Transcriptional activation of the HO-1 gene by lipopolysaccharide is mediated by 5′ distal enhancers: Role of reactive oxygen intermediates and AP-1, Am. J. Respir. Cell Mol. Biol., 18:226–234, 1998.

12. Karin M. and Barnes P.J.: Nuclear factor NF-κB—a pivotal transcription factor in chronic inflammatory disease, N. Engl. J. Med., 336:1066–1071, 1997.

13. Dinarello C.A.: Biological basis for interleukin-1 in disease, Blood., 87:2095–2147, 1996.

14. Friedrichs B., Muller C., and Brigelius-Flohe R: Inhibition of tumor necrosis factor-alpha- and interleukin-1-induced endothelial E-selectin expression by thiol-modifying agents, Arterioscler. Thromb. Vasc. Biol., 18:1829–1837, 1998.

15. Karin M., Liu Z.-g, and Zandi E.: AP-1 function and regulation, Curr. Opin. Cell Biol., 9:240–246, 1997.

16. Abate C., Patel L., Rauscher III F.J., and Curran T.: Redox regulation of Fos and Jun DNA-binding activity in vitro, Science., 249:1157–1161, 1990.

17. Klatt P., Molina E.P., De Lacoba M.G., Padilla C.A., Martinez-Galisteo E., Barcena J.A., and Lamas S.: Redox regulation of c-Jun DNA binding by reversibel S-glutathiolation., FASEB J., 13:1481–1490, 1999.

18. Schenk H., Klein M., Erdebrügger W., Dröge W., and Schulze-Osthoff K.: Distinct effects of thioredoxin and antioxidants on the activation of transcription factors NF-κB and AP-1, Proc. Natl. Acad. Sci. USA., 91:1672–1676, 1994.

19. Hirota K., Matsui M., Iwata S., Nishiyama A., Mori K., and Yodoi J.: AP-1 transcriptional activity is regulated by a direct association between thioredoxin and Ref-1, Proc. Natl. Acad. Sci. USA., 94:3633–3638, 1997.

20. Hirota K., Murata M., Sachi Y., Nakamura H., Takeuchi J., Mori K., and Yodoi J.: Distinct roles of thioredoxin in the cytoplasm and in the nucleus. A two-step mechanism of redox regulation of transcription factor NF-κB., J. Biol. Chem., 274:27891–27897, 1999.

21. Nikitovic D., Holmgren A., and Spyrou G: Inhibition of AP-1 DNA binding by nitric oxide involving conserved cysteine residues in Jun and Fos., Biochem. Biophys. Res. Commun., 242:109–112, 1998.

22. Qin J, M. C.G., Kennedy W.P., Kuszewski J., and Gronenborn A.M.: The solution structure of human thioredoxin complexed with its target from Ref-1 reveals peptide chain reversal, Structure., 4:613–620, 1996.

23. Xanthoudakis S., Miao G., Wang F., Pan Y.-C., and Curran T.: Redox activation of Fos-Jun DNA binding activity is mediated by a DNA repair enzyme, EMBO J., 11:3323–3335, 1992.

24. Holmgren A. and Luthman M.: Tissue distribution and subcellular localization of bovine thioredoxin determined by radioimmunoassay, Biochemistry, 17:4071–4077, 1978.

25. Inamdar N.M., Ahn Y.I., and Alam J.: The heme-responsive element of the mouse heme oxygenase gene is an extended AP-1 binding site that resembles the recognition sequences for MAF and NF-E2 transcription factors, Biochem. Biophys. Res. Com., 221:570–576, 1996.

26. Xanthoudakis S. and Curran T.: Identification and characterization of Ref-1, a nuclear protein that facilitates AP-1 DNA-binding activity, EMBO J., 11:653–665, 1992.

27. Schenk H., Vogt M., Droge W., and Schulze-Osthoff K.: Thioredoxin as a potent costimulus of cytokine expression, J. Immunol., 156:765–771, 1996.

28. Biguet C., Wakasugi N., Mishal Z., Holmgren A., Chouaib S., Tursz T., and Wakasugi H.: Thioredoxin increases the proliferation of human B-cell lines through a proteine kinase C-dependent mechanism, J. Biol. Chem., 269:28865–28870, 1994.

29. Makino Y., Yoshikawa N., Okamoto K., Hirota K., Yodoi J., Makino I., and Tanaka H.: Direct association with thioredoxin allows redox regulation of glucocorticoid receptor function, J. Biol. Chem., 274:3182–3188, 1999.

30. Nathan C.F.: Secretory products of macrophages., J. Clin. Invest., 79:319–326, 1987.

31. Tanaka T., Nishiyama Y., Okada K., Matsui M., Yodoi J., Hiai H., and Toyokuni S.: Induction and nuclear translocation of thioredoxin by oxidative damage in the mouse kidney: independence of tubular necrosis and sulfhydryl depletion, Lab. Invest., 77:145–155, 1997.

32. Masutani H., Hirota K., Sasada T., Ueda-Tanigushu Y., Tanigushi Y., Matsui M., and Yodoi J.: Transactivation of an inducible anti-oxidative stress protein, human thioredoxin by HTLV-I tax., Immunol. Lett., 54:67–71, 1996.

33. Ema M., Hirota K., Mimura J., Abe H., Yodoi J., Sogawa K., Poellinger L., and Fjii-Kuriyama Y.: Molecular mechanisms of transcription activation by HLF and HIF1α in response to hypoxia: their stabilization and redox signal-induced interaction with CBP/p300, EMBO J., 18:1905–1914, 1999.

34. Arnér E.S.J., Björnstedt M., and Holmgren A.: 1-chloro-2,4-dinitrobenzene is an irreversible inhibitor of human thioredoxin reductase, J. Biol. Chem., 270:3473–3482, 1995.

HEME OXYGENASE AND CARDIOVASCULAR SYSTEM

20

INDUCTION OF HEME OXYGENASE-1 AS A PROTECTIVE RESPONSE AGAINST HEME PROTEIN-INDUCED RENAL INJURY

Siobhan T. Pittock[1] and Karl A. Nath[2]

[1]Department of Pediatric and Adolescent Medicine
[2]Division of Nephrology
Mayo Clinic/Foundation, Rochester, MN

INTRODUCTION

The recognition that heme oxygenase (HO) activity is induced when tissues are exposed to heme and other insults raised the question of the functional significance of such induction.[1-4] Evidence attesting to the cytoprotective effects of such induction of HO was first presented in the glycerol model of acute renal failure wherein the kidney is exposed acutely to a large burden of heme proteins originating from damaged skeletal muscle and red blood cells;[5] examination of the induction of HO-1 in this disease model was predicated, in part, on the recognized toxicity of heme when present in inordinate amounts in tissues.[6]

This review summarizes these findings and the line of inquiry pursued over the past decade which examined the induction of HO-1 as a protective response against heme protein-induced renal injury.[5-20] Heme proteins may induce not only acute renal dysfunction and organ failure, but when repetitively administered, heme proteins may incite chronic renal inflammation. In addition to protecting against acute nephrotoxicity, the induction of HO-1 also retards inflammatory responses in the kidney incited by heme proteins,[18] and these latter findings are also discussed.

Address correspondence to: Dr. Karl A. Nath, Mayo Clinic, 200 First St., SW, Guggenheim 542, Rochester, MN 55905. Phone: 507-284-1646; Fax; 507-284-3757; E-mail: *nath.karl@mayo.edu*

TOXICITY OF HEME

In disease states, cells may be exposed to large amounts of heme derived from heme proteins normally present within cells, or from heme proteins incorporated into cells.[6,21,22] Heme proteins are plentiful and ubiquitous within cells, performing numerous indispensable functions as reflected by the diversity of proteins containing the heme prosthetic group—hemoglobin, myoglobin, mitochondrial and microsomal cytochromes, nitric oxide synthase, prostaglandin synthase, guanylate cyclase, catalase, peroxidases, and NADPH oxidase. Injury to cells may destabilize the union between heme and its protein moiety with the attendant release of free heme. Heme proteins may also be incorporated into cells in large amounts as occurs in such diseases as rhabdomyolysis or hemolysis; myoglobin released from injured muscle, or hemoglobin released from lysed erythrocytes, are distributed to organs such as the kidney which incorporate these heme proteins into the intracellular compartment. Following oxidative denaturation and destabilization of these heme proteins, heme is ultimately freed within the intracellular compartment. The hydrophobic nature of heme allows it to permeate plasma and intracellular membranes and to be readily distributed throughout the intracellular compartment via the intracellular canalicular delivery system.[22]

The damaging effects of copious amounts of heme are attested to by a large body of experimental evidence derived from *in vitro* and *in vivo* studies.[6,21] Clinical evidence also indicates the toxicity of heme when present in tissues in large amounts.[23] For example, the administration of relatively small amounts of heme is used to suppress porphyrin synthesis and thereby induce clinical remission in patients with acute intermittent porphyria. However, the administration of excessive amounts of heme to patients with this condition can induce acute renal failure.[23]

Numerous mechanisms contribute to such toxicity of heme.[6,21,22] Heme can peroxidate lipid present in plasma and intracellular membranes, destabilize the attached cytoskeleton, and denature membrane-associated proteins; the structural, transport-related, and other functions of the plasma and cell membranes and membrane-associated proteins are thus impaired. Heme may also impair the activity of cytosolic enzymes and mitochondrial enzymes on the one hand, while on the other, heme may activate cell-damaging proteolytic and lysosomal enzymes. DNA is another target that can be oxidatively denatured by heme *in vitro*. Moreover, the presence of increased, yet nontoxic, quantities of heme may render cells remarkably sensitive to otherwise innocuous amounts of oxidants such as hydrogen peroxide.[24]

The glycerol model of acute renal failure facilitates the study of the toxicity of heme proteins and heme:[5,6,13] the intramuscular injection of hypertonic glycerol induces myolysis and hemolysis, thereby subjecting the kidney to large amounts of myoglobin and hemoglobin. Three main pathways of injury—renal vasoconstriction, nephron cast formation, and direct cytotoxicity—underlie the pathogenesis of renal damage in this disease model,[5,6,13] and in each of these pathways, heme proteins are pathogenetically involved. Heme proteins scavenge nitric oxide, the endogenous vasodilator,[15] and by promoting oxidative stress, heme proteins stimulate the production of such potent vasoconstrictors as 8-isoprostanes.[10] Through their interaction with Tamm-Horsfall proteins, heme proteins also predispose to the formation of obstructing nephron casts. Heme proteins yield large amounts of heme which may injure renal tubular epithelial cells and their contained organelles.[5,10,14,21]

An organelle which is particularly vulnerable to the damaging effects of heme is the mitochondrion: the lipid-enriched mitochondrial membranes are readily permeated by heme, while the normal endogenous generation of peroxides by mitochondria may accentuate the toxicity of heme to mitochondria.[14] In the glycerol model, the content of heme in mitochondria harvested from the kidney is increased 10-fold three hours after the administration of glycerol and is accompanied by evidence of oxidative stress.[14] Derangements in mitochondrial function and structure accompany these changes, and may reflect increased mitochondrial content of heme. For example, when mitochondria, harvested from rats with disease-free kidneys, are exposed to heme concentrations that reproduce the levels observed in mitochondria harvested from the kidney in the glycerol model, the characteristic abnormalities observed in these latter mitochondria can be reproduced. Heme is also a potent stimulus for the cellular generation of hydrogen peroxide by renal epithelial cells.[10] Such enhanced generation of hydrogen peroxide may amplify the damaging effects of heme on mitochondria as well as other cellular targets such as the plasma membrane, cytoskeleton, and the nucleus.

ACUTE HEME PROTEIN-INDUCED NEPHROTOXICITY

To safeguard the vitality of tissues exposed to large amounts of heme, it was hypothesized that induction of HO-1 would confer beneficial effects by facilitating the clearance of heme and mitigating oxidant and other injurious effects of heme.[5] The glycerol model of acute renal injury was employed to test this hypothesis. In these studies, expression of HO-1 mRNA was observed as early as three hours after the administration of glycerol; HO-1 mRNA was markedly upregulated at six hours, and was accompanied by increased HO activity. Other antioxidant enzymes such as catalase and glutathione peroxidase were not concomitantly upregulated, and thus the induction of HO-1 was not part of a more generalized antioxidant response. The biologic significance of increased expression of HO-1 was probed in two ways.[5] The first strategy was to competitively and specifically inhibit HO with tin protoporphyrin. On five sequential days after the administration of glycerol, glycerol-treated rats subjected to tin protoporphyrin exhibited daily serum creatinine measurements which were significantly higher as compared with glycerol-treated rats subjected to vehicle,[5] tin protoporphyrin did not alter serum creatinine in rats with intact, disease-free kidneys. Thus, competitive inhibition of HO exacerbates acute heme protein-mediated nephrotoxicity.[5]

The examination of the functional significance of HO-1 in this model also incorporated a strategy based on the induction of HO-1 prior to the imposition of a large and damaging burden of heme proteins on the kidney; this approach posited that such induction of HO-1 would protect the kidney against heme protein-induced injury.[5] Induction of HO-1 was achieved by administration of small non-toxic doses of hemoglobin 18 hours prior to the intramuscular injection of hypertonic glycerol. In this protocol, the dose of glycerol used caused fulminant acute renal failure and 100%

mortality in vehicle-pretreated rats; hemoglobin-pretreated rats were remarkably protected, evincing less severe acute renal insufficiency and markedly diminished mortality when challenged by glycerol-induced heme protein-mediated injury.[5]

Inducers of HO-1, other than hemoglobin, and even those which may be intrinsically injurious to the kidney, if administered before the induction of the glycerol model, can also protect, quite markedly, against this form of acute renal injury.[9,12] For example, the prior administration of endotoxin protects against glycerol-induced acute renal failure.[9] Such protection conferred by endotoxin is associated with induction of HO-1, and is lost when HO is concomitantly inhibited.[9] Similarly, the administration of immunogenic nephrotoxic serum leads to potent induction of HO-1 in renal tubules, and confers a protective effect; this protection is lost when such HO activity is inhibited with tin protoporphyrin.[12] Thus, the induction of HO-1 by, and even in the presence of, potentially nephrotoxic stimuli (such as endotoxin and nephrotoxic serum) can mitigate acute heme protein-induced renal injury.[9,12]

The role of HO-1 in protecting against heme protein-induced renal injury was also examined in mice in which there is homozygous deletion of HO-1.[17] The administration of hypertonic glycerol induced relatively mild acute reversible renal insufficiency in the wild-type mouse (HO-1 +/+) and did not incur any mortality;[17] in contrast, HO-1 −/− mice, when subjected to the same dose of glycerol, exhibited acute fulminant renal insufficiency and extensive tubular necrosis.[17] Mortality in HO-1 −/− mice subjected to the glycerol model was 100% as compared to 0% in similarly treated HO-1 +/+ mice. Following the administration of glycerol, HO-1 +/+ mice model exhibited striking expression of HO-1 mRNA and protein in the kidney; immunoperoxidase studies localize such expression of HO-1 in renal tubules.[17] In contrast, HO-1 −/− mice failed to express HO-1 either in the basal state or after the administration of glycerol, and this absence of expression of HO-1 in HO-1 −/− mice was accompanied by an eight-fold greater increment in heme content in the kidney after glycerol as compared to HO-1 +/+ mice. These findings thus demonstrate that a buildup in renal heme content attends the exacerbation of acute renal insufficiency in the HO-1 deficiency state.[17]

This sensitivity of HO-1 −/− mice (as reflected by exacerbation of renal injury and increased mortality) to heme proteins, administered in the form of the glycerol model, was also recapitulated when a specific heme protein (hemoglobin) was administered to HO-1 −/− and HO-1 +/+ mice.[17] Studies *in vitro* by Abraham and colleagues also support the importance of HO-1 in protecting cells against the toxic effects of heme and hemoglobin.[25] For example, rabbit coronary microvessel endothelial cells, when transfected with the human HO-1 gene, acquire remarkable resistance to cell injury induced by heme and hemoglobin.[25]

In addition to attenuating the rise in renal heme content that would otherwise occur, induced HO-1 exerts other protective actions that include the elaboration of iron-storage and iron-exporting proteins, the generation of bilirubin, and the production of carbon monoxide. Ferritin is the major intracellular iron-storage protein.[7,26] The exposure of cells to heme induces HO-1 and stimulates the synthesis of ferritin which occurs as a consequence of the release of iron from the heme ring.[7,26] In the glycerol model increased ferritin content is associated with the induction of HO-1.[5] That increased ferritin synthesis contributes to the protective effects of

HO-1 is suggested by the concordant changes in renal ferritin content as expression of HO-1 is manipulated: for example, the prior induction of HO-1 by hemoglobin (which protects against injury in this model) is associated with increased ferritin synthesis whereas the inhibition of HO activity (which exacerbates injury in this model) is associated with diminished ferritin content.[5] Direct evidence that ferritin may contribute to the cytoprotective effects of HO-1 against heme protein-induced toxicity was provided in a cell culture model of heme/hydrogen peroxide-mediated toxicity:[7] in these studies, the prior exposure of these cells to ferritin led to incorporation of ferritin within the intracellular compartment, augmentation in the cellular content of ferritin, and markedly improved cell survival following challenge by heme/hydrogen peroxide. These salutary effects of ferritin likely arise from the chelation of intracellular iron and the capacity of the ferritin H-chain to exhibit ferroxidase activity, the latter abrogating the redox cycling of iron.[7] In addition to increased synthesis of iron-binding proteins such as ferritin, the induction of iron-exporting proteins is also linked to increased expression of HO-1. Iron-exporting proteins facilitate the extracellular transport of iron, and in conjunction with ferritin, restrain the elevation in intracellular iron content that occurs in cells exposed to heme.[27,28]

Bile pigments may also confer cytoprotective effects in the glycerol model.[19] The capacity of biliverdin and bilirubin to protect against oxidative stress has long been recognized,[29-31] and evidence was recently presented demonstrating that low micromolar concentrations of bilirubin confer a dose-dependent protection against acute heme protein induced toxicity in a cell culture model.[19] Moreover, hyperbilirubinemic states as induced by the ligation of the common bile duct, while itself inducing a modest decrement in renal function, are remarkably protective against glycerol-induced acute heme protein-mediated nephrotoxicity. Thus the protective effects of induction of HO-1 in the glycerol model may reflect increased amounts of bile pigments that are generated in the kidney attendant upon the induction of HO-1.

Induction of HO-1 may also confer protection in this model of acute heme protein-mediated renal injury via its vasorelaxant effects. For example, the vasoconstrictive effects of oxyhemoglobin on basilar arteries are markedly attenuated when HO-1 is upregulated in these arteries by adenovirus gene transfer.[32] The attenuation in the vasoconstrictive effects of hemoglobin in these transduced arteries is vitiated by the competitive inhibitor of HO, tin protoporphyrin. The mechanism accounting for this vascular effect may reside in increased generation of the vasorelaxant product of HO-1, namely, carbon monoxide.[33] It is possible such vasorelaxant effects of HO-1 may be germane to the glycerol model of acute heme protein-mediated nephrotoxicity since these actions serve to reduce renal vasoconstriction characteristically observed in this form of acute renal injury.

CHRONIC HEME PROTEIN-ASSOCIATED RENAL INFLAMMATION

Renal inflammation is commonly detected in diseases in which the kidney is intermittently exposed to heme proteins. For example, chronic tubulointerstitial inflammation occurs in hematuric glomerulopathies,[34,35] in sickle cell nephropathy,[36] and in paroxysmal nocturnal hemoglobinuria.[37] That heme proteins *per se* may be instrumental in eliciting such inflammatory responses in the kidney is demonstrated by the appearance of progressive renal interstitial inflammation in the rat following the repetitive administration of heme proteins.[16] Such tubulointerstitial disease is characterized by increased tubular expression of HO-1 and ferritin, and increased deposition of iron; interstitial and basement membrane collagen mRNAs are upregulated along with monocyte chemoattractant protein-1 (MCP-1) mRNA, the latter representing a critical chemokine in the pathogenesis of renal inflammation.[16,18]

Tubular induction of HO-1 also occurs in human nephropathies in which the kidney is repeatedly exposed to heme proteins.[38,39] Such induction occurs in hematuric nephritides wherein the tubules are exposed to erythrocytes which are engulfed and fragmented within tubular epithelial cells.[34] Tubular induction of HO-1 also occurs in sickle cell disease, a condition in which tubules are exposed to sickle hemoglobin which readily undergoes oxidative denaturation within tubular epithelial cells to yield heme.[20] Paroxysmal nocturnal hemoglobinuria is a membrane disorder of the erythrocyte arising from an underlying abnormality in the cytoskeleton which predisposes towards hemolysis. Such hemolysis subjects the kidney, intermittently, to heme proteins, and in a certain subset of patients with this disorder, chronic tubulointerstitial inflammation and progressive renal disease evolve.[37] The kidney in paroxysmal nocturnal hemoglobinuria exhibits marked iron overload, and contains large amounts of ferritin, the major iron-binding protein.[18] Along with these changes, HO-1 is markedly upregulated in the renal tubules where cells are exposed to heme proteins. The induction of HO-1 in the kidney in PNH is due, at least in part, to heme proteins as indicated by the strong induction of HO-1 in renal epithelial cells exposed to urine from such patients (which is heme protein-enriched) but not from heme protein-free urine from healthy controls.[18]

To examine the functional significance of HO-1 induced in the kidney repetitively exposed to heme proteins, HO-1 +/+ and HO-1 −/− mice were injected at weekly intervals to murine hemoglobin, and 7 days after the eighth injection of hemoglobin, the histological appearance of the kidneys was examined.[18] A marked difference in renal inflammation was observed: HO-1 −/− mice exhibited striking interstitial inflammation and tubulointerstitial disease whereas HO-1 +/+ mice, similarly injected with hemoglobin, displayed mild tubulointerstitial inflammation. These findings demonstrate that within the kidney a chronic inflammatory response is influenced by the level of expression of HO-1: specifically, the absence of HO-1 transforms what would be otherwise a mild lesion into an aggressive interstitial cellular infiltrative process. These findings led to the suggestion that the induction of HO-1 in chronic interstitial disease is a countervailing response that restrains the inflammatory changes that would otherwise occur.[18]

This exacerbation of tubulointerstitial disease in kidneys lacking expression of HO-1 raised the possibility that inflammatory mechanisms are amplified when the expected expression of HO-1 fails to occur. In this regard, attention was directed to MCP-1, a chemokine that potently attracts monocytes and T cells, and one widely incriminated in the initiation and perpetuation of chronic tubulointerstitial inflammation.[40–44] Moreover, MCP-1 is upregulated in the kidney in rats subjected to repetitive administration of heme proteins, and heme proteins, such as hemoglobin, directly induce MCP-1.[18] When examined 7 days after the last dose of hemoglobin, fulminant induction of MCP-1 in the kidney occurred in HO-1 –/– mice subjected to repeated exposure to hemoglobin but not in HO-1 +/+ mice subjected to hemoglobin or vehicle, or in HO-1 –/– mice subjected to saline vehicle. Thus, the marked exacerbation of interstitial inflammation in the absence of HO-1 is attended by intense upregulation of a potent chemoattractant for monocytes and T lymphocytes, namely, MCP-1.[18]

The redox-sensitive transcription factor, NF-κB regulates, at least in part, expression of MCP-1, and activation of NF-κB is recognized as a mediator of renal inflammation in certain settings.[43,44] Activation of NF-κB was thus examined in HO-1 +/+ and HO-1 –/– mice subjected to hemoglobin or vehicle.[18] These studies demonstrate that the degree of activation of NF-κB recapitulated the pattern of expression found for MCP-1 and tubulointerstitial disease: activation of NF-κB was much more marked in HO-1 –/– mice subjected to hemoglobin as compared to HO-1 +/+ mice exposed to hemoglobin or vehicle, or HO-1 –/– mice subjected to vehicle.[18] Thus, the deficiency of HO-1 facilitates the activation of NF-κB, and the attendant upregulation of NF-κB-driven cytokines as MCP-1.[18]

That HO-1 restrains activation of NF-κB and the expression of NF-κB-dependent chemokines such as MCP-1 complements recent insights regarding the anti-inflammatory actions of HO-1. Products of heme oxygenase such as carbon monoxide may suppress proinflammatory cytokines such as PDGF and endothelin-1.[45] Carbon monoxide also exerts potent anti-inflammatory effects—the suppression of proinflammatory cytokines (for example, IL-1β and TNF-α) in conjunction with the upregulation of anti-inflammatory cytokines (for example, IL-10) in lipopolysaccharide-stimulated tissues—by activating the p38 MAP kinase pathways.[46] These effects of carbon monoxide along with its anti-apoptotic and other actions also contribute to the protective effects of HO-1.[46,47] Other products of HO-1 such as bilirubin can attenuate adhesion of leukocytes to the venular endothelium,[48] or inhibit the activation of NADPH oxidase.[49] HO-1 reduces the upregulation of ICAM-1 that otherwise occurs in endothelial cells in response to heme.[50] These findings, in conjunction with the recent recognition of the suppressive effect of HO-1 on the expression of MCP-1, thus underscore the diverse anti-inflammatory effects of HO-1.[18]

CONCLUSION

Prior studies in the glycerol model of acute heme protein-mediated nephrotoxicity provided evidence for the cytoprotective effects of HO-1 against tissue injury in general, and against renal injury in particular. Besides this induction in acute heme protein-mediated nephrotoxicity and in chronic heme protein-associated renal inflammation, induction of HO-1 is now recognized in ischemic renal injury,[51] cisplatin and mercury-induced acute renal injury,[52–54] acute transplant rejection,[55] acute glomerular inflammation associated with nephrotoxic serum nephritis,[56,57] chronic administration of angiotensin II,[58] acute obstruction to the urinary tract,[59] and in patients with assorted forms of acute and chronic renal injury.[38,39,60] That findings pertaining to the functional significance of HO-1 derived from experimental models are relevant to human disease is demonstrated by seminal observations derived from a patient with a genetic defect characterized by an inability to express HO-1: cells derived from this patient evince marked sensitivity to heme-induced oxidant stress and injury while the kidney exhibits glomerular and tubular injury accompanied by the presence of oxidative stress.[38,39,60]

REFERENCES

1. Abraham N.G. Lin J.H.-C., Schwartzman M.L. Levere R.D., and Shibahara S.: The physiological significance of heme oxygenase. Int J Biochem 20:543–558, 1988.
2. Maines M.D.: Heme oxygenase: function, multiplicity, regulatory mechanisms, and clinical applications. FASEB J 2:2557–2568, 1988.
3. Keyse S.M. and Tyrrell R.M.: Heme oxygenase is the major 32-kDa stress protein induced in human skin fibroblasts by UVA radiation, hydrogen peroxide and sodium arsenite. Proc Natl Acad Sci, USA 86:99–103, 1989.
4. Stocker R.: Induction of haem oxygenase as a defense against oxidative stress. Free Rad Res Comms 9:101–112, 1990.
5. Nath K.A., Balla G., Vercellotti G.M., Balla J., Jacob H.S., Levitt M.D., and Rosenberg M.E.: Induction of heme oxygenase is a rapid, protective response in rhabdomyolysis in the rat. J Clin Invest 90:267–270, 1992.
6. Nath K.A., Agarwal A., and Vogt B.: Functional consequences of induction of heme oxygenase. In: Contemporary Issues in Nephrology. "Acute renal failure: emerging concepts and therapeutic strategies" Goligorsky MS (ed), Series Editor J Stein, Churchill Livingstone, chapter 5, p 97–118, 1995.
7. Balla G., Jacob H.S., Balla J., Rosenberg M., Nath K., Apple F., Eaton J.W., and Vercellotti G.M.: Ferritin: a cytoprotective antioxidant stratagem of endothelium. J Biol Chem 267:18148–18153, 1992.
8. Balla J., Jacob H.S., Balla G., Nath K.A., Eaton J.W., and Vercellotti G.M.: Endothelial-cell heme uptake from heme proteins: Induction of sensitization and desensitization to oxidant damage. Proc Natl Acad Sci 90:9285–9289, 1993.
9. Vogt B.A., Alam J., Croatt A.J., Vercellotti G.M., and Nath K.A.: Acquired resistance to acute oxidative stress: possible role of heme oxygenase and ferritin. Lab Invest 72:474–483, 1995.
10. Nath K.A., Balla J., Croatt A.J., and Vercellotti G.M.: Heme protein-mediated renal injury: a protective role for 21-aminosteroids in vitro and in vivo. Kidney Int 47:592–602, 1995.
11. Balla J., Nath K.A., Balla G., Juckett M.B., Jacob H.S., and Vercellotti G.M.: Endothelial cell heme oxygenase and ferritin induction in rat lung by hemoglobin in vivo. Am J Physiol 268:L321–L327, 1995.
12. Vogt B.A., Shanley T.P., Croatt A.J., Alam J., Johnson K.J., and Nath K.A.: Glomerular inflammation induces resistance to tubular injury in the rat: a novel form of acquired heme oxygenase-dependent resistance to renal injury. J Clin Invest 98:2139–2145, 1996.
13. Nath K.A.: Adaptation to the nephrotoxicity of heme proteins. Exp Nephrol 4:139–143, 1996.
14. Nath K.A., Grande J.P., Croatt A.J., Likely S., Hebbel R.P., and Enright H.: Intracellular targets in heme protein induced renal injury. Kidney Int: 53:100–111, 1998.
15. Warden D.H., Croatt A.J., Katusic Z.S., and Nath K.A.: Physiologic characterization of acute reversible systemic hypertension in a model of heme protein-induced renal injury. Am J Physiol 277:F58–F65, 1999.
16. Nath K.A., Croatt A.J., Haggard J.J., and Grande J.P.: Renal response to repetitive exposure to heme proteins: Chronic injury induced by an acute insult. Kidney Int 57:2423–2433, 2000.
17. Nath K.A., Haggard J.J., Croatt A.J., Grande J.P., Poss K.D., and Alam J.: The indispensability of heme oxygenase-1 (HO-1) in protecting against heme protein-induced toxicity in vivo. Am J Pathol 156:1527–1535, 2000.
18. Nath K.A., Vercellotti G., Grande J.P., Miyoshi H., Paya C.V., Manivel J.C., Haggard J.J., Croatt A.J., Payne W.D., and Alam J.: Heme protein-induced chronic renal inflammation: Suppressive effect of induced heme oxygenase-1. Kidney Int 59:106–117, 2001.

19. Leung N., Croatt A.J., Haggard J.J., Grande J.P., and Nath K.A.: Acute cholestatic liver disease protects against glycerol-induced acute renal failure in the rat: a novel form of acquired renal resistance to injury. Kidney Int 60:1047–1057, 2001.

20. Nath K.A., Grande J.P., Haggard J.J., Croatt A.J., Katusic Z.S., Solovey A., and Hebbel R.P.: Oxidative stress and induction of heme oxygenase-1 in the kidney in sickle cell disease. Am J Pathol 158:893–903, 2001.

21. Hebbel R.P. and Eaton J.W.: Pathobiology of heme interaction with the erythrocyte membrane. Semin Hematol 26:136–149, 1989.

22. Muller-Eberhard U. and Fraig M.: Bioactivity of heme and its containment. Am J Hematol 42:59–62, 1993.

23. Dhar, G.J., Bossenmaier I., Cardinal R., Petryka Z.J., and Watson C.J.: Transitory renal failure following rapid administration of a relatively large amount of hematin in a patient with acute intermittent porphyria in clinical remission. Acta Med Scan 203:437–443, 1978.

24. Balla G., Vercellotti G.M., Muller-Eberhard U., Eaton J., and Jacob H.S.: Exposure of endothelial cells to free heme potentiates damage mediated by granulocytes and toxic oxygen species. Lab Invest 64:648–655, 1991.

25. Abraham N.G., Lavrovsky Y., Schwartzman M.L., Stoltz R.A., Levere R.D., Gerritsen M.E., Shibahara S., and Kappas A.: Transfection of the human heme oxygenase gene into rabbit coronary microvessel endothelial cells: protective effect against heme and hemoglobin toxicity. Proc Natl Acad Sci, USA 92:6798–6802, 1995.

26. Eisenstein R.S., Garcia-Mayol D., Pettingell W., and Munro H.N.: Regulation of ferritin and heme oxygenase synthesis in rat fibroblasts by different forms of iron. Proc Natl Acad Sci, USA 88:688–692, 1991.

27. Ferris C.D., Jaffrey S.R., Sawa A., Takahashi M., Brady S.D., Barrow R.K., Tysoe S.A., Wolosker H., Baranano D.E., Dore S., Poss K.D., and Snyder S.H.: Haem oxygenase-1 prevents cell death by regulating cellular iron. Nature Cell Biol 1:152–157, 1999.

28. Barañano D.E., Wolosker H., Bae B.I., Barrow R.K., Snyder S.H., and Ferris C.D.: A mammalian iron ATPase induced by iron. J Biol Chem 275:15166–15173, 2000.

29. Stocker R., Yamamoto Y., McDonagh A.F., Glazer A.N., and Ames B.N.: Bilirubin is an antioxidant of possible physiologic significance. Science 235:1043–1046, 1987.

30. Stocker R., Glazer A.N., and Ames B.N.: Antioxidant activity of albumin-bound bilirubin. Proc Natl Acad Sci, USA 84:5918–5922, 1987.

31. Llesuy S.F. and Tomaro M.L.: Heme oxygenase and oxidative stress. Evidence of involvement of bilirubin as physiological protector against oxidative damage. Biochim Biophys Acta 1223:9–14, 1994.

32. Eguchi D., Weiler D., Alam J., Nath K., and Katusic Z.S.: Protective effect of heme oxyenase-1 gene transfer against oxyhemoglobin-induced endothelial dysfunction. J Cereb Blood Flow Metab 21:1215–1222, 2001.

33. Motterlini R., Gonzales A., Foresti R., Clark J.E., Green C.J., and Winslow R.M.: Heme oxygenase-1-derived carbon monoxide contributes to the suppression of acute hypertensive responses *in vivo*. Circ Res 83:568–577, 1998.

34. Hill P.A., Davies D.J., Kincaid-Smith P., and Ryan G.B.: Ultrastructural changes in renal tubules associated with glomerular bleeding. Kidney Int 36:992–997, 1989.

35. Madsen K.M., Applegate C.W., and Tisher C.C.: Phagocytosis of erythrocytes by the proximal tubule of the rat kidney. Cell Tissue Res 226:363–374, 1982.

36. Saborio P. and Scheinman J.I.: Sickle cell nephropathy. J Am Soc Nephrol 10:187–192, 1999.

37. Clark D.A., Butler S.A., Braren V., Hartmann R.C., and Jenkins Jr D.E.: The kidney in paroxysmal nocturnal hemoglobinuria. Blood 57:83–89, 1981.

38. Ohta K., Yachie A., Fujimoto K., Kaneda H., Wada T., Toma T., Seno A., Kasahara Y., Yokoyama H., Seki H., and Koizumi S.: Tubular injury as a cardinal pathologic feature in human heme oxygenase-1 deficiency. Am J Kidney Dis 35:863–870, 2000.

39. Morimoto K., Ohta K., Yachie A., Yang Y., Shimizu M., Goto C., Toma T., Kasahara Y., Yokoyama H., Miyata T., Seki H., and Koizumi S.: Cytoprotective role of heme oxygenase (HO)-1 in human kidney with various renal diseases. Kidney Int 60:1858–1866, 2001.

40. Wenzel U.O. and Abboud H.E.: Chemokines and renal disease. Am J Kid Dis 26:982–994, 1995.

41. Tesch G.H., Schwarting A., Kinoshita K., Lan H.Y., Rollins B.J., and Kelley V.R.: Monocyte chemoattractant protein-1 promotes macrophage-mediated tubular injury, but not glomerular injury, in nephrotoxic serum nephritis. J Clin Invest 103:73–80, 1999.

42. Wada T., Furuichi K., Segawa-Takeda C., Shimizu M., Sakai N., Takeda S.-I., Takasawa K., Kida H., Kobayashi K.-I., Mukaida N., Ohmoto Y., Matsuhima K., and Yokoyama H.: MIP-1α and MCP-1 contribute to crescents and interstitial lesions in human crescentic glomerulonephritis. Kidney Int 56:995–1003, 1999.

43. Ishikawa Y., Sugiyama H., Stylianou E., Kitamura M.: Bioflavonoid quercetin inhibits interleukin-1-induced transcriptional expression of monocyte chemoattractant protein-1 in glomerular cells via suppression of nuclear factor-κB. J Am Soc Nephrol 10:2290–2296, 1999.

44. Gu L., Tseng S.C., and Rollins B.J.: Monocyte hemoattractant protein-1. Chem Immunol 72:7–29, 1999.

45. Morita T. and Kourembanas S.: Endothelial cell expression of vasoconstrictors and growth factors is regulated by smooth muscle cell-derived carbon monoxide. J Clin Invest 96:2676–2682, 1995.

46. Otterbein L.E., Bach F.H., Alam J., Soares M., Tao L.H., Wysk M., Davis R.J., Flavell R.A., and Choi A.M.: Carbon monoxide has anti-inflammatory effects involving the mitogen-activated protein kinase pathway. Nature Med 6:422–428, 2000.

47. Otterbein L.E. and Choi A.M.: Heme oxygenase: color of defense against cellular stress. Am J Physiol 279:L1029–L1037, 2000.

48. Hayashi S., Takamiya R., Yamaguchi T., Matsumoto K., Tojo S.J., Tamatani T., Kitajima M., Makino N., Ishimura Y., and Suematsu M.: Induction of heme oxygenase-1 suppresses venular leukocyte adhesion elicited by oxidative stress. Circ Res 85:663–671, 1999.

49. Kwak J.Y., Takeshige K., Cheung B.S., and Minakami S.: Bilirubin inhibits the activation of superoxide-producing NADPH oxidase in a neutrophil cell-free system. Biochem Biophys Acta 1076:369–373, 1991.

50. Frank A.D., Wagener T.G., Da Silva J.-L., Farley T., De Witte T., Kappas A., and Abraham N.G.: Differential effects of heme oxygenase isoforms on heme mediation of endothelial intracellular adhesion molecule 1 expression. J Pharmacol Exp Therap 291:416–423, 1999.

51. Maines M.D., Mayer R.D., Ewing J.F., and McCoubrey W.K. Jr: Induction of kidney heme oxygenase-1 (HSP32) mRNA and protein by ischemia/reperfusion: Possible role of heme as both promoter of tissue damage and regulator of HSP32. J Pharmacol Exp Therap 264:457–462, 1993.

52. Agarwal A., Balla J., Alam J., Croatt A.J., and Nath K.A.: Induction of heme oxygenase in toxic renal injury: a protective role in cisplatin nephrotoxicity in the rat. Kidney Int 48:1298–1307, 1995.

53. Shiraishi F., Curtis L.M., Truong L., Poss K., Visner G.A., Madsen K., Nick H.S., and Agarwal A.: Heme oxygenase-1 gene ablation or expression modulates cisplatin-induced renal tubular apoptosis. Am J Physiol 278:F726–F36, 2000.

54. Nath K.A., Croatt A.J., Likely S., Behrens T.W., and Warden D.: Renal oxidant injury and oxidant response induced by mercury. Kidney Int 50:1032–1043, 1996.

55. Agarwal A., Kim Y., Matas A.J., Alam J., and Nath K.A.: Gas-generating systems in acute renal allograft rejection in the rat: co-induction of heme oxygenase and nitric oxide synthase. Transplantation 61:93–98, 1996.

56. Mosley K., Wembridge D.E., Cattell V., and Cook H.T.: Heme oxygenase is induced in nephrotoxic nephritis and hemin, a stimulator of heme oxygenase synthesis, ameliorates disease. Kidney Int 53:672–678, 1998.

57. Datta P.K., Koukouritaki S.B., Hopp K.A., and Lianos E.A.: Heme oxygenase-1 induction attenuates inducible nitric oxide synthase expression and proteinuria in glomerulonephritis. J Am Soc Nephrol 10:2540–2550, 1999.

58. Haugen E.N., Croatt A.J., and Nath K.A.: Angiotensin II induces renal oxidant stress *in vivo* and heme oxygenase-1 *in vivo* and *in vitro*. Kidney Int 58:144–152, 2000.

59. Kawada N., Moriyama T., Ando A., Fukunaga M., Miyata T., Kurokawa K., Imai E., and Hori M.: Increased oxidative stress in mouse kidneys with unilateral ureteral obstruction. Kidney Int 56:1004–1013, 1999.

60. Yachie A., Niida Y., Wada T., Igarashi N., Kaneda H., Toma T., Ohta K., Kasahara Y., and Koizumi S.: Oxidative stress causes enhanced endothelial cell injury in human heme oxygenase-1 deficiency. J Clin Invest 103:129–135, 1999.

21

REGULATION AND ROLE OF HEME OXYGENASE-1 IN GLOMERULONEPHRITIS

Prasun K. Datta and Elias A. Lianos

Department of Medicine
Robert Wood Johnson Medical School
University of Medicine & Dentistry of NJ
P. O. Box 19
One Robert Wood Johnson Place
New Brunswick, NJ 08903

INTRODUCTION

Reactive oxygen metabolites have been demonstrated to play a significant pathobiologic role in a number of clinical and experimental models of both immune and non immune glomerular diseases.[1,2] In some forms of these diseases, resident glomerular cells, in addition to recruited inflammatory cells, can be the source of reactive oxygen metabolites. Recent evidence demonstrated a prooxidant role of nitric oxide (NO) as well. Enhanced expression and enzyme activity of the inducible nitric oxide synthase (iNOS) in various forms of glomerulonephritis was described.[3] Activation of iNOS results in a sustained high output production of nitric oxide (NO) which can cause oxidative injury either in its own or by interacting with the superoxide anion to form the relatively stable prooxidant peroxynitrite ($ONOO^-$).[4]

Numerous studies have demonstrated the significant role of the endogenous antioxidant enzymes (such as superoxide dismutase (SOD), glutathione peroxidase and catalase) as an essential defense system against oxidant injury in glomerular diseases.[5] Recent attention has focused on the microsomal enzyme Heme oxygenase (HO)

Address for Correspondance: Prasun K. Datta Ph. D, Department of Medicine, Robert Wood Johnson Medical School, University of Medicine & Dentistry of NJ, P. O. Box 19, One Robert Wood Johnson Place, New Brunswick, NJ 08903. Tel: (732) 235-8472, Fax: (732) 235-8884, e-mail: dattapk@umdnj.edu as an additional antioxidant system. HO controls the initial and rate-limiting step in heme catabolism. Heme oxygenase-1 (HO-1) is the inducible isoform of heme oxygenase. It catalyzes the NADPH, O_2 and cytochrome P450 reductase dependent oxidation of heme to carbon monoxide (CO), iron and biliverdin, which is reduced to bilirubin by biliverdin reductase.[6] Heme is a key constituent of a number of enzymes that are activated in inflammation. Such enzymes include iNOS, cyclooxygenase and cytochrome P450 monooxygenases. Degradation of heme by HO-1 can, therefore, limit heme availability for synthesis or optimal activity of these enzymes. This may reduce production of proinflammatory mediators and prooxidant byproducts i.e. NO from activation of iNOS, prostaglandins and thromboxane from activation of cyclooxygenase, and arachidonate epoxides from activation of cytochrome P450 monooxygenases. In addition, increased HO-1 activity results in generation of bilirubin, that has antioxidant and anti-complement properties.[7–9] Finally, CO has vasodilatory effects mediated via cGMP, and also possess anti-apoptotic and cytoprotective functions.[10–12] An overall anti-inflammatory/antioxidant effect can thus be expected as a result of HO-1 activation. To this extent it was shown that HO-1 induction protects from chemically induced inflammation in the pleural cavity,[13] and protects renal tubules from myoglobin or hemoglobin induced injury.[14] It was further shown that HO-1 induction prolongs survival of cardiac xenografts in mice and protects against chronic rejection.[15–16] Finally, two independent studies demonstrated that HO-1 induction attenuates proteinuria in the rat model of anti-glomerular basement membrane (GBM) nephritis.[17–18]

HEME OXYGENASE-1 SYNTHESIS IN GLOMERULONEPHRITIS

To detect changes in HO-1 gene expression in nephritic glomeruli, we used reverse transcription polymerase chain reaction (RT-PCR) on RNA extracted from glomeruli of rats with anti-GBM nephritis.[18] Glomerulonephritis was induced by injecting subnephritogenic dose of rabbit immune serum raised against rat particulate glomerular basement membarane (GBM) into rats that were preimmunized with rabbit IgG. Glomeruli were isolated by differential seiving and RNA was extracted using Tri-reagent. RNA was then reverse transcribed with Moloney murine leukemia virus reverse transcriptase and PCR was performed using Taq polymerase and primers corresponding to rat HO-1 gene. A PCR product of 284 bp was detectable in glomeruli from nephritic rats (injected with anti-GBM serum) from 2 to 14 days (Fig. 1). The mRNA for HO-1 was also detectable in glomeruli from rats injected with non-immune rabbit serum (controls).

We have also demonstrated increased HO-1 protein expression in nephritic glomeruli by western blot analysis of glomerular protein lysates using a polyclonal rabbit antiserum[18] (Fig. 2A), and observed staining for HO-1 protein in nephritic glomeruli by immunohistochemistry using a monoclonal antibody raised against purified rat native HO-1 (Fig. 2B).

Thus, there is clear evidence that in glomerulonephritis there is induction of HO-1, however, the mechanism of enhanced HO-1 expression in glomerulonephritis is unknown.

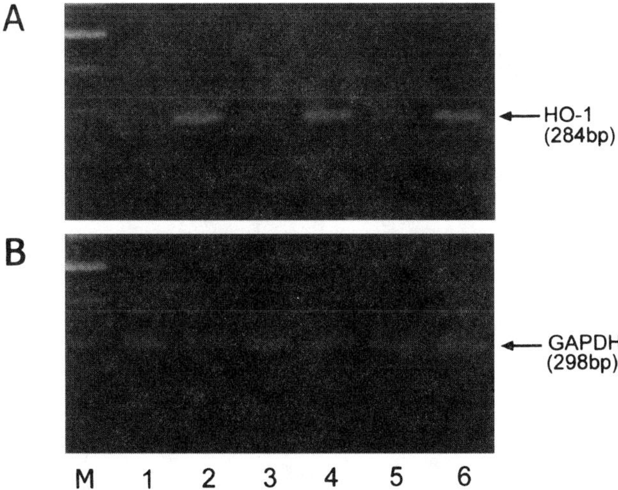

Figure 1. Time course of Heme oxygenase-1 expression in anti-glomerular basement membrane (GBM) nephritis in rats. Glomerular total RNA was isolated from control and nephritic animals and analyzed by RT-PCR for HO-1 (panel A) and GAPDH (panel B). Lane 1: Control day 2; Lane 2: Nephritis day 2; Lane 3: Control day 7; Lane 4: Nephritis day 7; Lane 5: Control day 14; Lane 6: Nephritis day 14; Lane M: 100 bp marker ladder. Reproduced from ref. 18 by permission from Lippincott Williams & Wilkins.

iNOS REGULATES HO-1 EXPRESSION IN GLOMERULONEPHRITIS

In a previous study we reported that NO derived from NO donors or from iNOS activation by cytokines stimulates HO-1 expression and activity in glomerular mesangial cells.[19] This, and the observation that iNOS and HO-1 are coactivated within nephritic glomeruli, prompted us to explore whether in anti-GBM glomerulonephritis iNOS activation stimulates HO-1 expression. The selective iNOS inhibitor L-N^6-(1-iminoethyl) lysine (L-NIL) was used at a dose of 15 mg/kg of body weight to inhibit iNOS activity. L-NIL has an IC_{50} of 3.3 μM for macrophage iNOS compared to an IC_{50} of 92 μM for constitutive NOS.[20] Figure 3 shows a Western blot analysis of HO-1 protein and of the rat macrophage marker ED_1 (as an index of the extent of infiltration of nephritic glomeruli by macrophages) in protein lysates prepared from glomeruli isolated from animals that received anti-GBM serum or from controls that received non-immune serum. In this experiment, two animals received non-immune serum (NIS) (lanes 1 and 2 in the upper and lower panel), two animals received anti-GBM serum (lanes 3, 4, in the upper and lower panel) and two animals received anti-GBM serum and L-NIL treatment (lanes 5, 6 in the upper and lower panel). As expected, ED_1 levels were higher in nephritic glomeruli (upper panel; lanes 3, 4, 5, 6) compared to controls (upper panel: lanes 1 and 2). L-NIL treatment reduced HO-1 levels in nephritic glomeruli that received L-NIL (lanes 5, 6, in lower panel) compared to levels in nephritic glomeruli that did not receive L-NIL (lanes 3, 4, in lower panel).

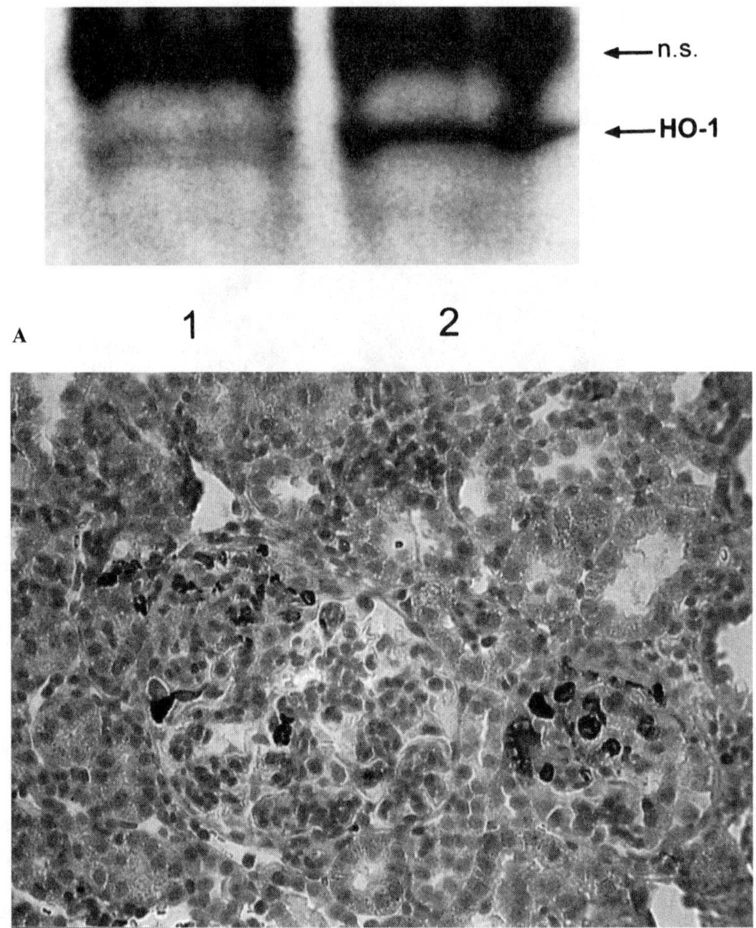

Figure 2. A Changes in glomerular HO-1 expression assessed by Western blot analysis. Lane 1: protein lysates prepared from glomeruli of nephritic animal; Lane 2: protein lysates prepared from glomeruli of nephritic animal pretreated with Hemin. HO-1 denotes heme oxygenase-1; n.s. denotes nonspecific cross-reacting protein band. Reproduced from ref. 18 by permission from Lippincott Williams & Wilkins. **B** Immunohistochemical localization of HO-1 in glomeruli of rats with anti-GBM nephritis.

These observations demonstrate that that iNOS activation upregulates glomerular HO-1 expression in anti-GBM nephritis. Thus, pretreatement with the selective iNOS inhibitor, L-NIL, in doses sufficient to reduce urine nitrite excretion, attenauted glomerular HO-1 protein levels. There was no effect on the extent of glomerular infiltration by macrophages as assessed by glomerular levels of the ED_1. This indicates that the reduction in HO-1 levels observed in L-NIL treated animals was due to inhibition of iNOS rather than to an alteration of the inflammatory repsonse to anti-GBM antibody-induced injury as a result of L-NIL treatment.

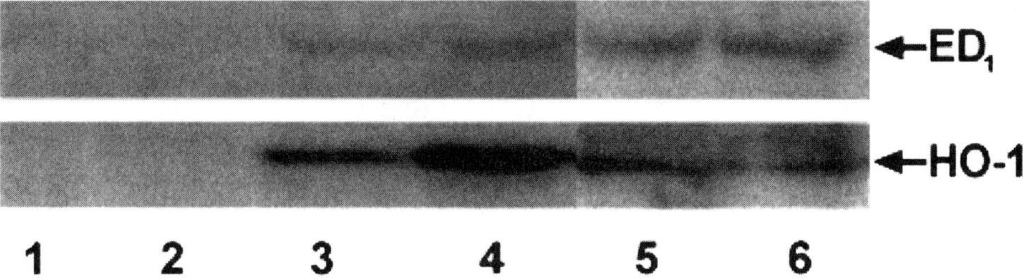

Figure 3. Attenuation of glomerular HO-1 expression in L-NIL treated rats with anti-GBM nephritis.

ROLE OF HO-1 IN GLOMERULONEPHRITIS

In its capacity as the rate-limiting enzyme in heme degradation, HO-1 reduces cellular heme levels thereby limiting availabilty of heme for synthesis or enzyme activity of heme containing enzymes such as iNOS. In these lines we have demonstrated that pretreatment of rats with anti-GBM nephritis with HO-1 inducer, hemin augmented glomerular HO-1 levels (Fig. 2A) and significantly reduced glomerular iNOS enzyme activity and proteinuria.[18] The beneficial effect of HO-1 activation becomes apparent from the effect of Hemin pretreatment of nephritic animals on urine protein excretion (Fig. 4). In these animals proteinuria was reduced without a change in the extent of glomerular infiltration by monocytes/macrophages. The attenuation of proteinuria in Hemin treated nephritic animals confirms a recent study by Mosley and co-workers.[17]

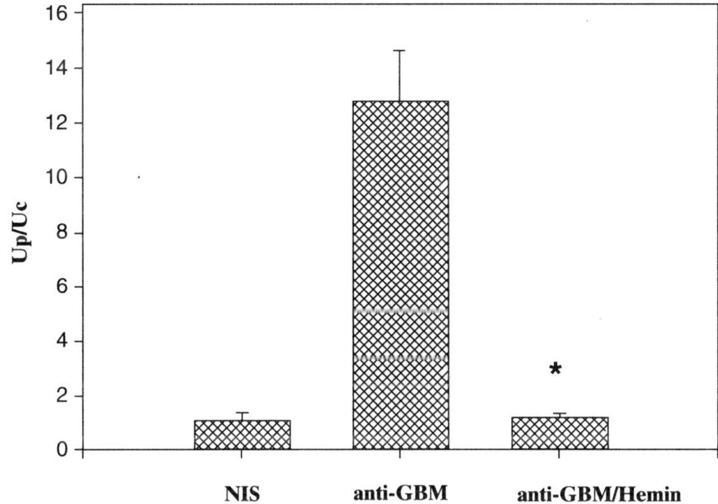

Figure 4. Effect of HO-1 supra-induction on urinary protein excretion in anti-GBM glomerulonephritis. *$P < 0.001$ compared to anti-GBM nephritic animals not treated with Hemin. Reproduced from ref. 18 by permission from Lippincott Williams & Wilkins.

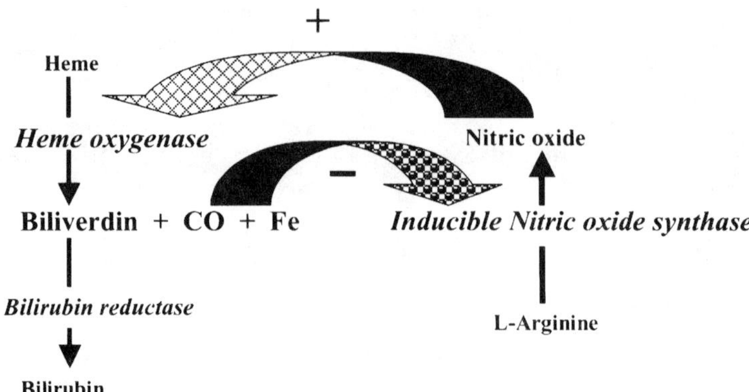

Figure 5. Schematic representation of proposed regulatory interactions between the inducible nitric oxide synthase (iNOS) and Heme oxygenase-1 (HO-1) pathways. Activation of iNOS leads to sustained production of NO from L-arginine. iNOS-derived NO positively (+) modulates the heme metabolizing enzyme HO-1. Activation of HO-1 leads to catabolism of heme to generate biliverdin, carbon monoxide (CO) and iron (Fe). Iron down regulates iNOS transcription while CO inhibits iNOS enzyme activity. Biliverdin and bilirubin are antioxidants.

Whether the reduction of proteinuria in Hemin treated nephritic animals is specifically due to the observed down regulation of iNOS or occurs independently of this effect on iNOS is unknown. Catabolism of heme by HO-1 also generates biliverdin that is converted to bilirubin by biliverdin reductase. Biliverdin and bilirubin can scavenge reactive oxygen species (ROS) and inhibit lipid peroxidation.[7–8] Moreover, biliverdin inhibits complement cascade at the level of C1 activation.[9] Thus, HO-1 activation within nephritic glomeruli may provide these two ROS scavengers that could protect against supra-physiologic levels of ROS generated in the course of inflammatory injury.[21–23] The HO-1 catalyzed production of CO from heme is a stimulator of guanylate cyclase.[10] This could increase intraglomerular levels of cGMP thereby preserving glomerular filtration rate in a manner similar to nitric oxide. Finally, the HO-1 catalyzed release of free iron can induce the cytoprotective protein, ferritin.[24] This protein sequesters free cytosolic iron, the main catalyst of oxygen radical formation. It has been demonstrated that HO-1 induced by mildly oxidized LDL in co-cultures of human aortic endothelial cells and smooth muscle cells may protect these cells through production of the antioxidants biliverdin and bilirubin.[25] Studies in HO-1 knockout mice have also demonstrated a beneficial role of HO-1 against oxidative injury of embryonic fibroblasts by hydrogen peroxide, paraquat or cadmium chloride.[26] Our observations also points to the presence of a negative regulatory interaction between HO-1 and iNOS whereby induction of HO-1 reduces iNOS activity/expression.

SUMMARY

These results support a model of bidirectional regulatory interactions between HO-1 and iNOS systems in glomerulonephritis whereby iNOS-derived NO upregulates HO-1, which negatively modulate iNOS (Fig. 5). HO-1 activation can modulate iNOS by three different mechanisms: a) reduction of cellular heme levels; b) CO-mediated iNOS inactivation;[27] and c) iron mediated ihibition of iNOS transcription.[28]

REFERENCES

1. L. Baud and R. Ardaillou, Reactive oxygen species: production and role in the kidney, *Am. J. Physiol.* **521**, F765–F776 (1986).
2. J.R. Diamond, The role of reactive oxygen species in animal models of glomerular disease. *Am. J. Kidney Dis.* **19**, 292–300 (1992).
3. V. Cattel and H.T. Cook, T. NO, and glomerulonephritis, in Nitric oxide and Kidney, edited by M.S. Goligorsky and S.S. Gross (Chapman and Hall, New York, 1997), pp. 307–331.
4. J.S. Beckman, T.W. Beckman, J. Chen, P.A. Marshall, and B.A. Freeman, Apparent hydroxyl radical production by peroxynitrite: Implications for endothelial cell injury from nitric oxide. *Proc. Natl. Acad. Sci. USA.* **87**, 1620–1624 (1987).
5. I. Ichikawa, S. Kiyama, and T. Yoshioka, Renal antioxidant enzymes: Their regulation and function. *Kidney Int.* **45**, 1–9 (1994).
6. M.D. Maines, The heme oxygenase system: A regulator of second messenger gases. *Annu. Rev. Pharmacol. Toxicol.* **37**, 517–554 (1997).
7. R. Stocker, Y. Yamamoto, A.F. McDonagh, A.N. Glazer, and B.N. Ames, Bilirubin is an important antioxidant of possible physiological importance. *Science* **235**, 1043–1046 (1987).
8. R. Stocker, A.N. Glazer, and B.N. Ames, Antioxidant activity of albumin bound bilirubin. *Proc. Natl. Acad. Sci. USA* **84**, 5918–5922 (1987).
9. T., Nakagami, K. Toyomura, T. Kinoshita, and S. Morisawa, A beneficial role of bile pigments as an endogenous tissue protector: Anti complement effects of biliverdin and conjugated bilirubin. *Biochim. Biophys. Acta.* **158**, 189–193 (1993).
10. T. Morita, M.A. Perrella, M.E. Lee, and S. Kourembanas, Smooth muscle cell-derived carbon monoxide is a regulator of vascular cGMP. *Proc. Natl. Acad. Sci. USA.* **92**, 1475–1479 (1995).
11. S.R. Them, D. Fisher, Y.A. Xu, K. Notarfrancesco, and H. Ischiropoulos, Adaptive reponses and apoptosis in endothelial cells exposed to carbon monoxide. *Proc. Natl. Acad. Sci. USA.* **97**, 1305–1310 (2000).
12. L.E. Otterbein, L.L. Mantell, and A.M.K. Choi, Carbon monoxide provides protection against hyperoxic lung injury. *Am. J. Physiol..* **276**, L688–L694 (1999).
13. D. Willis, A.R. Moore, R. Frederick and D.A. Willoughby, Heme oxygenase: A novel target for the modulation of the inflammatory response. *Nature Med.* **2**, 87–90 (1996).
14. K.A. Nath, G. Balla, G.M. Vercellotti, J. Balla, H.S. Jacob, M.D. Levitt, and M.E. Rosenberg, Induction of heme oxygenase is a rapid, protective response in rhabdomyolysis in the rat. *J. Clin. Invest.* **90**, 267–270 (1992).
15. M. Soares, Y. Lin, J. Anrather, E. Csizmadia, K. Takigami, K. Sato, S.T. Gray, R.B. Colvin, A.M. Choi, K.D. Poss, and F.H. Bach, Expression of heme oxygenase-1 can determine cardiac xenograft survival. *Nature Med.* **4**, 87–90 (1998).
16. W.W. Hancock, R. Buelow, M.H. Saygeh, and L.A. Turka, Antibody-induced transplant arteriosclerosis is prevented by graft expression of antioxidant and anti-apoptotic genes. *Nature Med.* **4**, 1392–1396 (1998).
17. K. Mosley, D.E. Wembridge, V. Cattell, and H.T. Cook, Heme oxygenase is induced in nephrotoxic nephritis and hemin, a stimulator of heme oxygenase synthesis, ameliorates disease. *Kidney Int.* **53**, 672–678 (1998).
18. P.K. Datta, S.B. Koukouritaki, K.A. hopp, and E.A. Lianos, Heme oxygenase-1 induction attenuates inducible nitric oxide synthase expression and proteinuria in glomerulonephritis. *J. Am. Soc. Nephrol.* **10**, 2540–2550 (1999).
19. P.K. Datta and E.A. Lianos, Nitric oxide induces Heme oxygenase-1 in mesangial cells. *Kidney Intl.* **55**, 1734–1739 (1999).
20. W.M. Moore, R.K. Weber, G.M. Jerome, F.S. Tjoeng, M.P. Misko, and M.G. Currie, L-N^6-(1-Iminoethyl) lysine: A selective inhibitor of inducible nitric oxide synthase. *J. Med. Chem.* **37**, 3886–3888 (1994).

21. N.W. Boyce and S.R. Holdsworth, Hydroxyl radical mediation of immune renal injury by desferrixamine. *Kidney Int.* **30**, 813–817 (1986).
22. S.V. Shah, The role of reactive oxygen metabolites in glomerular disease. Annu. Rev. Physiol. **57**, 245–262 (1995).
23. K.A. Nath, Reactive oxygen species in renal injury, in International Year book of Nephrology, edited by V.E. Andreucci and L.J. Fine (Academic Press, Orlando, 1991), pp 47–69.
24. M. Linder-Horowitz, R.T. Ruettinger, and H.N. Munro, Iron induction of electrophoretically different ferritins in rat liver, heart and kidney. *Biochim Biophys. Acta* **31**, 442–448 (1970).
25. K. Ishikawa, M. Navab, N. Leitinger, A.M. Fogelman, and A.J. Lusis, Induction of heme oxygenase-1 inhibits the transmigration induced by mildly oxidized LDL. *J. Clin. Invest.* **100**, 1209–1216 (1997).
26. K.D. Poss and S. Tonegawa, Reduced stress defense in heme oxygenase l-deficient cells. *Proc. Natl. Acad. Sci. USA* **94**, 10925–10930 (1997).
27. K. McMillan, D.S. Brendt, D.J. Hirsch, S.H. Synder, J.E. Clark, and B.S. Master, Cloned, expressed rat cerebellar nitric oxide synthase contains stoichiometric amounts of heme, which binds CO. *Proc. Natl. Acad. Sci. USA* **89**, 11141–11145 (1992).
28. G. Weiss, G. Werner-Felmayer, E.R. Werner, K. Grunewald, H. Wachter, and M.W. Henze, Iron regulates NO synthase activity by controlling nuclear transcription. *J. Exp. Med.* **180**, 969–976 (1994).

22

HEME OXYGENASE-1 MEDIATES ATRIAL NATRIURETIC PEPTIDE INDUCED PROTECTION OF RENAL CELLS FROM CYCLOSPORIN TOXICITY

Tobias Polte[a], Anke Hemmerle[b], Nina Grosser[b], Aida Abate[b], and Henning Schröder[b]

[a]University of Bath
Bath, BA2 7AT, UK
[b]School of Pharmacy, Martin Luther University
06099 Halle (Saale), Germany

INTRODUCTION

Cyclosporin A (CsA) is an immunosuppressive drug which has played an important role in the improvement of solid-organ transplant patients and graft survival but also in the treatment of auto-immune diseases such as psoriasis or rheumatoid arthritis. However, its use is complicated by the development of hypertension or nephropathy in 40–90% of patients.[1,2] A variety of mechanisms is held responsible for CsA-induced nephrotoxicity, with oxidant stress, according to recent studies, playing a central role as pathogenetic factor. By altering intrarenal mediators such as renin, nitric oxide or endothelin-1, CsA leads to vasoconstriction, which in turn results in hypoxia-reoxygenation and free radical injury.[3,4] In addition, direct cytotoxic effects due to activation of pro-oxidant pathways are thought to contribute to the deleterious action of CsA in the kidney. CsA enhances generation of hydrogen peroxide in vitro and stimulates lipid peroxidation in vitro and in vivo.[3,5,6] Furthermore, antioxidants have been shown to be protective in cyclosporin A-dependent renal toxicity indicating functional relevance and a causative role for oxygen radicals in this drug-related disorder.[6,7]

Atrial natriuretic peptide has beneficial therapeutic effects in CsA-induced renal toxicity. In CsA-treated transplant patients ANP improves renal function and lowers systolic blood pressure.[8,9] Whereas effects of ANP on renal hemodynamics and electrolytes are well recognized as important in preserving renal function and normotension under CsA therapy, it is not known whether ANP may also antagonize CsA toxicity directly at the cellular level. This question has gained even more relevance in the light of recent findings demonstrating that ANP acts as an antioxidant and cytoprotective agent in different models of ischemia-reperfusion injury. Thus, ANP was reported to reduce oxidant cell damage in a rat liver model as well as in canine cardiac Tissue.[10–12] Although the second messenger cGMP appeared to be involved in those tissue protective actions of ANP, further downstream targets with potential antioxidant functions were not identified. Recently, we found that in endothelial cells, the cytoprotective and antioxidant protein heme oxygenase(HO)-1 is cGMP-regulated and sensitive to induction by ANP.[13] Increased HO-1 expression leads to degradation of heme and accumulation of iron, bilirubin, and carbon monoxide followed by reduced sensitivity of tissues to oxidant damage.[14–17] Of these metabolites, bilirubin acts as a direct antioxidant,[18] whereas carbon monoxide may exert tissue protective actions primarily through its vasodilator and anti-platelet effects.[14,19] The HO-1 signaling pathway was previously shown to be crucial in confering resistance to inflammatory and oxidant injury in the kidney.[20–23] Using LLC-PK1 renal tubular cells, the present study investigates the effect of ANP on CsA-induced cytotoxicity and explores a protective role of the HO metabolite bilirubin under these conditions.

MATERIALS AND METHODS

Materials

LLC-PK1 cells (ATCC CL 101) were obtained from the American Type Culture Collection, Rockville, MD, USA. Fetale bovine serum, Ham's F-12 medium, Dulbecco's modified Eagle medium and penicillin-streptomycin were purchased from Gibco. Tin protoporphyrin-IX (SnPP) and CsA were purchased from Alexis. Stock solutions of CsA were made in ethanol. The Chemiluminescence Western Blotting Kit was from Boehringer Mannheim. 8-Bromo cGMP, human atrial natriuretic peptide (ANP) and all other chemicals were bought from Sigma.

Cell Culture

LLC-PK1 cells were maintained and subcultured in Ham's F12 medium supplemented with 20% Dulbecco's modified Eagle medium, 15% fetal bovine serum, 100 U/ml penicillin and 100 μg/ml streptomycin.[24] The cells were grown in a humidified incubator at 37 °C and 5% CO_2.

Cell Viability Analysis

LLC-PK1 cells were seeded at 2×10^4 cells/well in 96-well microtiter plates in 100 μl of media containing 15% fetal bovine serum. After a 24-hour incubation at 37 °C, cells reached confluence and were incubated for 8 hours in the presence of ANP, bilirubin, or 8-bromo cGMP. SnPP was added 10 minutes prior to ANP. Then, CsA was given to the cells without washing out the previously added agents. Incubation at 37 °C was continued for 48 hours, followed by a cytotoxicity assay. Cell viability was measured by staining with crystal violet as previously described.[13,25] This colorimetric test allows assessment of the remaining viable cells after the incubation procedure. Cells were washed with phosphate-buffered saline, fixed with methanol for 5 minutes and then stained for 10 minutes with a 0.1% crystal violet solution. Following 3 washes with tap water, the dye was eluted with 0.1 M trisodium citrate in 50% ethanol for 10 minutes. Optical density at 630 nm was measured using a microtiter plate reader (Biotek EL 311s).

Western Blot Analysis

LLC-PK1 were cultured in 150-mm dishes as described above. After a 16-hour incubation with control media, ANP, or 8-bromo cGMP, cells were washed and extracted as described previously.[13] Protein (100 µg) was applied to sodium dodecyl sulfate polyacrylamide gel electrophoresis. After electrophoresis, protein was transferred to a nitrocellulose membrane, and a polyclonal antibody to rat HO-1 (Stressgene) was used to identify HO-1 protein content. Antigen antibody complexes were visualized with the horseradish peroxidase chemiluminescence system according to the manufacturer's instructions (Boehringer, Mannheim). Quantitation of HO-1 protein content was performed using computer-assisted videodensitometry (Eagle Eye II-system, Stratagene).

HO Activity

Confluent LLC-PK1 cells in 150-mm culture dishes were incubated for 16 hours in the presence of control media, ANP, or 8-bromo cGMP. The method used for the determination of HO activity follows the protocol published by Motterlini and co-workers.[26] Briefly, after the incubation, cells were washed twice with phosphate-buffered saline, gently scraped off the dish and centrifuged (1,000 × g, 10 min, 4 °C). The cell pellet was suspended in $MgCl_2$ (2 mM) phosphate (100 mM) buffer (pH 7.4), frozen at minus 70 °C, thawed 3 times, and finally sonicated on ice before centrifugation at 18,000 × g for 10 minutes at 4 °C. The supernatant (400 µl) was added to a NADPH-generating system containing 0.8 mM NADPH, 2 mM glucose-6-phosphate, 0.2 U glucose-6-phosphate-1-dehydrogenase, and 2 mg protein of rat liver cytosol prepared from the 105,000 × g supernatant fraction as a source of biliverdin reductase, potassium phosphate buffer (100 mM, pH 7.4), and hemin (10 µM) in a final volume of 200 µl. The reaction was conducted for 1 hour at 37 °C in the dark and terminated by addition of 1 ml chloroform. The extracted bilirubin was calculated by the difference in absorption between 464 and 530 nm using a quartz cuvette (extinction coefficient, $40\,mM^{-1} \times cm^{-1}$ for bilirubin). HO activity was measured as picomoles of bilirubin formed per milligram of endothelial cell protein per hour. Basal HO activity was in a range between 200 and 600 pmol bilirubin/mg protein/h.

RESULTS

Protective Effects of ANP and cGMP

Incubation with CsA (48 hours) resulted in a marked cytotoxicity and reduction of cell viability. Preincubation with ANP at 1–100 nM (8 hours) diminished CsA toxicity in a concentration-dependent fashion and increased the surviving cell fraction by up to 70% (Fig. 1A). Cytoprotection by ANP was less pronounced after shorter preincubation times and not detectable when CsA and ANP were added simultaneously to the cells (not shown). A similar cytoprotective effect was observed with the membrane permeable cGMP analogue 8-bromo cGMP (1–100 μM). 8-Bromo cGMP at 1–100 μM increased resistance of LLC-PK1 cells to CsA toxicity concentration-dependently and augmented cell viability by up to 65% (Fig. 1B). ANP and 8-bromo cGMP alone had no significant effect on cell viability under these conditions (not shown).

Role of Bilirubin in ANP Protection

Preincubation with exogenously added bilirubin produced a substantial reduction of CsA toxicity (Fig. 2A) and at the highest bilirubin concentration used (10 μM) cells became almost completely resistant to CsA toxicity (surviving fraction: 92%). The protective effect of bilirubin was detected at concentrations that are within the range of normal bilirubin plasma levels.[27] In order to explore a potential involvement of HO products such as bilirubin in the observed protection by ANP, the HO inhibitor SnPP was used. At a concentration (20 μM) that was previously shown to specifically block HO-1-dependent effects,[13,28] SnPP abrogated cellular protection afforded by ANP against CsA toxicity (Fig. 2B). Bilirubin, ANP and SnPP alone had no significant effect on cell viability under these conditions (not shown).

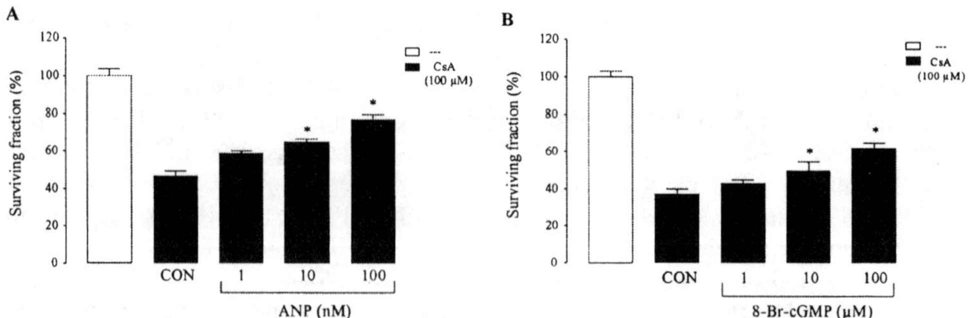

Figure 1. Effect of ANP (A) and 8-bromo cGMP (8-Br-cGMP, B) on CsA-induced cytotoxicity in LLC-Pk1 cells. Incubations and viability determinations were carried out as described under Materials and Methods. *): $P < 0.05$, treatment vs. control (CON), one-way ANOVA and Bonferroni's multiple comparison test. All data shown are mean ± SEM of n = 6 independent observations in separate cell culture wells.

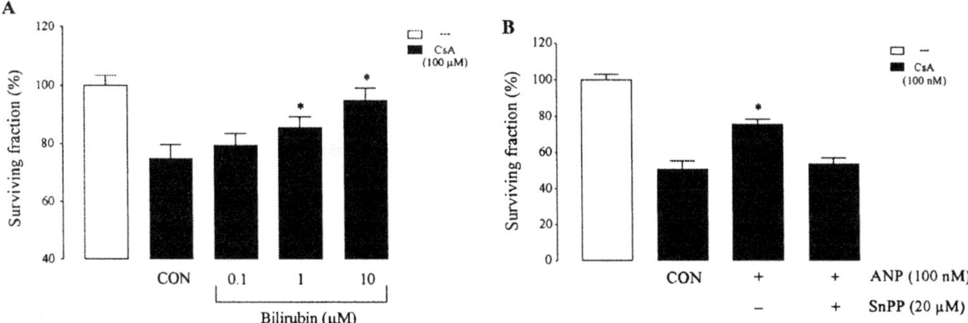

Figure 2. Effect of bilirubin on CsA-induced cytotoxicity in LLC-Pk1 cells (A). Effect of SnPP on ANP-induced cytoprotection from CsA-induced cytotoxicity in LLC-PK1 cells (B). Incubations and viability determinations were carried out as described under Materials and Methods. *): $P < 0.05$, treatment vs. control (CON), one-way ANOVA and Bonferroni's multiple comparison test. All data shown are mean ± SEM of n = 6 independent observations in separate cell culture wells.

ANP and cGMP Increase HO Activity and Expression

Cells were exposed to ANP or 8-bromo cGMP for 16 hours. HO activity was assessed in the cell lysate by measuring formation of the HO metabolite bilirubin. ANP (0.1–1 µM) produced a concentration-dependent increase in HO activity up to 2.9-fold over basal levels (Fig. 3A). Similar results were obtained when cells were incubated with 8-bromo cGMP (0.1–100 µM). A concentration-dependent stimulation of HO activity with a maximal 2.4-fold increase was detected in the presence of 8-bromo cGMP (Fig. 3B). Stimulations of HO activity corresponded to increases in HO-1 protein synthesis (Fig. 4A and B). ANP (1–100 nM) and 8-bromo cGMP (1–100 µM) induced the synthesis of HO-1 protein in a concentration-dependent fashion and

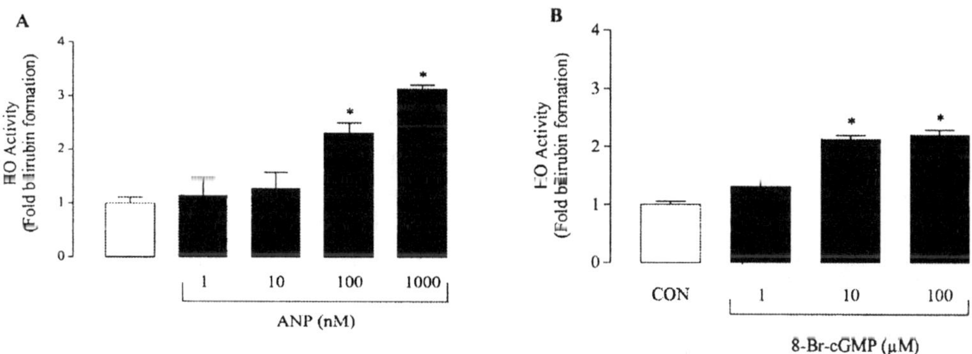

Figure 3. Effect of ANP (A) and 8-bromo cGMP (8-Br-cGMP, B) on HO activity in endothelial cells. HO activity was determined via bilirubin formation as described under Materials and Methods. *): $P < 0.05$, treatment vs. control (CON), one-way ANOVA and Bonferroni's multiple comparison test. All data shown are mean ± SEM of n = 3 independent observations in separate cell culture wells.

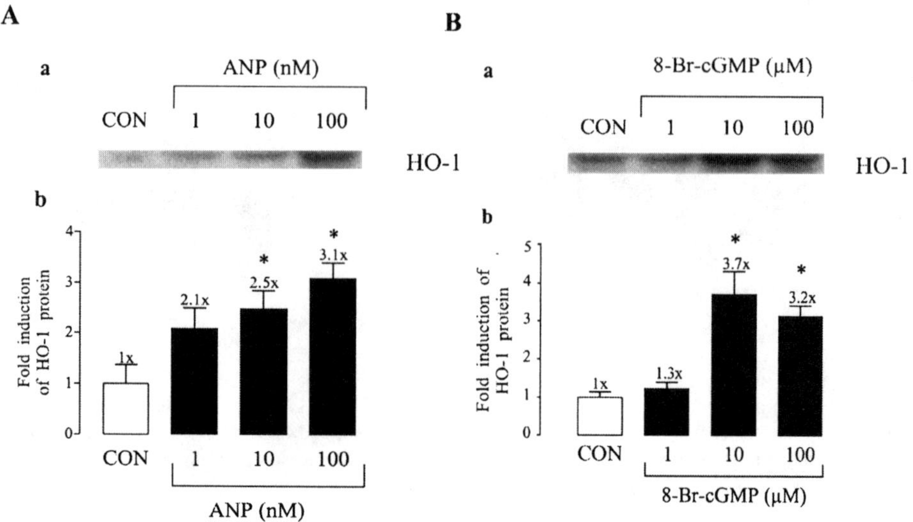

Figure 4. Effect of ANP (A) and 8-bromo cGMP (8-Br-cGMP, B) on HO-1 protein expression in LLC-PK1 cells. Incubations, protein isolation and Western blot analysis were performed as described under Materials and Methods. The densitometric data (b) are mean ± SEM of n = 3 separate experiments. *): P < 0.05, treatment vs. control (CON), one-way ANOVA and Bonferroni's multiple comparison test. A representative Western blot analysis is shown in (a).

produced maximal 3.1-fold and 3.7-fold elevations of basal HO-1 protein expression, respectively (Fig. 4A and B).

DISCUSSION

The present investigation demonstrates, for the first time, that ANP is capable of antagonizing renal toxicity of CsA at the cellular level and that the antioxidant HO metabolite bilirubin is a crucial downstream mediator in renal cell protection afforded by the ANP/cGMP signaling pathway.

The experiments were carried out in LLC-PK1 renal tubular epithelial cells which were previously introduced as a model system to study mechanisms of CsA-induced nephrotoxicity and its prevention.[29–31] Our incubation protocol follows procedures outlined in those earlier publications.

CsA has been shown to reduce cellular viability predominantly through pro-oxidant pathways such as the increased oxygen radical flux resulting from cytochrome P-450-dependent metabolism.[32,33] That oxidative stress is likewise a major cause of CsA toxicity in renal tissue was demonstrated in several investigations including those using LLC-PK1 cells.[6,7,29,31]

Possible therapeutic strategies to prevent oxidant damage to the kidney comprise administration of antioxidant drugs, e.g. vitamin E,[6] but also novel, genomic approaches such as induction or overexpression of cytoprotective genes. It has been known for some time that HO-1 is involved in the adaptive response of the kidney to oxidant stress.[15,22,23,34] Moreover, recent work shows that HO-1 has a protective role in drug-induced nephrotoxicity suggesting that targeted gene expression of HO-1 may be used as a therapeutic and preventive modality in high-risk settings of acute renal failure.[20,35] According to our results, ANP-induced renal protection, to some extent, represents such an approach since ANP attenuates CsA toxicity through increased expression and activity of HO-1.

We observed renal cell protection from CsA-induced toxicity at nanomolar ANP concentrations. This clearly indicates that in addition to counteracting CsA-dependent vasoconstriction and sodium retention, ANP protects from CsA toxicity at the cellular level. ANP increased HO-1 protein expression and activity at concentrations that were also effective in cytoprotection. Moreover, renal cell protection by ANP was abrogated in the presence of the HO inhibitor SnPP[14,36] suggesting that HO-1 and its enzymatic products are of functional relevance to the observed protective action. In agreement with this, the HO metabolite bilirubin, when added exogenously to the cells, profoundly increased cellular resistance to CsA toxicity with the surviving cell fraction nearing that of untreated cells (92%). This effect was seen at low micromolar concentrations of bilirubin which are in the upper range of the reference interval for plasma levels. High-normal serum levels of bilirubin were reported to be inversely related to atherogenic risk and to provide protection against endothelial damage.[27,37] Our findings lend support to the concept of bilirubin as a biologically important antioxidant[18] and corroborate the role of oxygen radical formation in CsA-evoked renal toxicity. The HO metabolite carbon monoxide has long been considered as being tissue protective solely by its vasodilatory and anti-platelet effects.[19] However, recent evidence points to direct anti-inflammatory properties of carbon monoxide[38] which may complement and support the cytoprotective actions of bilirubin.

The effects of ANP reported in this investigation were mimicked by cGMP. Renal cell protection as well as increased HO-1 expression or bilirubin formation showed cGMP-dependence, suggesting that the ANP/cGMP signaling pathway is crucial for the observed effects. This is consistent with previous findings showing that LLC-PK1 cells only express ANP-A and, to a much smaller extent, ANP-B receptors which are both coupled to particulate guanylyl cyclase.[24,39] Interestingly, ANP was reported to reduce oxidant cell damage in a rat liver model as well as in canine cardiac tissue via cGMP-dependent pathways.[10–12] It is conceivable and would explain those observations that induction of HO-1 and bilirubin formation contributes to antioxidant protection by ANP/cGMP not only in renal tissue.

Other tissues in which HO-1 regulation is sensitive to cGMP include hepatocytes[40] and endothelial cells.[13] The pathway, through which cGMP induces HO-1 and as a consequence cellular resistance to cytotoxic stress, could be a direct one, i.e., via cGMP sensitive transcription factors such as activator protein-1.[41] Alternatively, cGMP may act through secondary increases in cAMP which we reported earlier to occur in endothelial cells, possibly due to cGMP-elicited inhibiton of cAMP breakdown.[42] Since AP-1 as well as cAMP response elements have been identified in the promotor region of HO-1,[10,41,43,44] HO-1 induction by ANP/cGMP may be regulated through different mechanisms depending on species and tissue.

Beyond a generally improved defense against oxidant injury, induction of an anti-inflammatory protein by activators of particulate guanylyl cyclase may have additional clinical implications since HO-1 is considered to protect against the chronic rejection of transplants.[15,36,45] Through increased HO activity, ANP and related compounds may protect transplant patients not only from drug-dependent nephrotoxicity but likewise from early transplant rejection and deterioration of graft function. It is noteworthy in this regard, that both ANP and HO-1 inducers were independently reported to reduce cold-storage injury to transplant organs and to improve allograft survival.[46-49]

We have shown, for the first time, that ANP attenuates CsA nephrotoxicity at the cellular level via a cGMP-dependent pathway. According to our data, induction of HO-1 and increased formation of the antioxidant HO metabolite bilirubin are responsible for renal cell protection afforded by the ANP/cGMP system. Thus, HO-1 is an important cellular target of ANP, with clinical implications for the therapy or prevention of drug-induced nephrotoxicity and other inflammatory disorders.

ACKNOWLEDGMENT

This work was supported by the Deutsche Forschungsgemeinschaft (Schr 298/8-3 and PO 731/1-1)

REFERENCES

1. A.J. Olyaei, A.M. De Mattos, and W.M. Bennett, Immunosuppressant-induced nephropathy: pathophysiology, incidence and management, *Drug Saf.* **21**, 471–488 (1999).
2. H. Zachariae, Renal toxicity of long-term cyclosporin, *Scand. J. Rheumatol.* **28**, 65–68 (1999).
3. R. Baliga, N. Ueda, P.D. Walker, and S.V. Shah, Oxidant mechanisms in toxic acute renal failure, *Drug. Metab. Rev.* **31**, 971–997 (1999).
4. Z. Zhong, G.E. Arteel, H.D. Connor, M. Yin, M.V. Frankenberg, R.F. Stachlewitz, J.A. Raleigh, R.P. and Mason, R.G. Thurman, Cyclosporin A increases hypoxia and free radical production in rat kidneys: prevention by dietary glycine, *Am. J. Physiol.* **275**, F595–F604 (1998).
5. Y. Chancerelle, M. De Lorgeril, R. Viret, B. Chiron, G. Dureau, S. Renaud, and J.F. Kergonou, Increased lipid peroxidation in cyclosporin-treated heart transplant recipients, *Am. J. Cardiol.* **68**, 813–816 (1991).
6. C. Wang and A.K. Salahudeen, Lipid peroxidation accompanies cyclosporine nephrotoxicity: effects of vitamin E, *Kidney Int.* **47**, 927–934 (1995).
7. K.V. Kumar, M.U. Naidu, A.A. Shifow, A. Prayag, and K.S. Ratnakar, Melatonin: an antioxidant protects against cyclosporine-induced nephrotoxicity, *Transplantation* **67**, 1065–1068 (1999).
8. G. Capasso, C. Rosati, F. Ciani, D.R. Giordano, F. Russo, and N.G. De Santo, The beneficial effect of atrial natriuretic peptide on cyclosporine nephrotoxicity, *Am. J. Hypertens.* **3**, 204–210 (1990).
9. C.C. Lang, I.S. Henderson, R. Mactier, W.K. Stewart, and A.D. Struthers, Atrial natriuretic factor improves renal function and lowers systolic blood pressure in renal allograft recipients treated with cyclosporin, *A. J. Hypertens.* **10**, 483–488 (1992).
10. M. Bilzer, H. Jaeschke, A.M. Vollmar, G. Paumgartner, and A.L. Gerbes, Prevention of Kupffer cell-induced oxidant injury in rat liver by atrial natriuretic peptide, *Am. J. Physiol.* **276**, G1137–G1144 (1999).
11. M. Bilzer, R. Witthaut, G. Paumgartner, and A.L. Gerbes, Prevention of ischemia/reperfusion injury in the rat liver by atrial natriuretic peptide, *Gastroenterology* **106**, 143–151 (1994).
12. Y. Takata, Y. Hirayama, S. Kiyomi, T. Ogawa, K. Iga, T. Ishii, Y. Nagai, and C. Ibukiyama, The beneficial effects of atrial natriuretic peptide on arrhythmias and myocardial high-energy phosphates after reperfusion, *Cardiovasc. Res.* **32**, 286–293 (1996).

13. T. Polte, A. Abate, P.A. Dennery, and H. Schröder, Heme oxygenase-1 is a cyclic GMP-inducible endothelial protein and mediates the cytoprotective action of nitric oxide, *Arterioscler. Thromb. Vasc. Biol.* **20**, 1209–1215 (2000).

14. M.D. Maines, The heme oxygenase system: A regulator of second messenger gases, *Annu. Rev. Pharmacol. Toxicol.* **37**, 517–554 (1997).

15. J.L. Platt and K.A. Nath, Heme oxygenase: Protective gene or Trojan horse, *Nature Med.* **4**, 1364–1365 (1998).

16. D.M. Suttner, K. Sridhar, C.S. Lee, T. Tomura, T.N. Hansen, and P.A. Dennery, Protective effects of transient HO-1 overexpression on susceptibility to oxygen toxicity in lung cells, *Am. J. Physiol.* **276**, L443–L451 (1999).

17. L. Yang, S. Quan, and N.G. Abraham, Retrovirus-mediated HO gene transfer into endothelial cells protects against oxidant-induced injury, *Am. J. Physiol.* **277**, L127–L133 (1999).

18. R. Stocker, Y. Yamamoto, A.F. McDonagh, A.N. Glazer, and B.N. Ames, Bilirubin is an antioxidant of possible physiological importance, *Science* **235**, 1043–1046 (1987).

19. F. Coceani, Carbon monoxide in vasoregulation, *Circ. Res.* **86**, 1184–1186 (2000).

20. A. Agarwal, J. Balla, J. Alam, A.J. Croatt, and K.A. Nath, Induction of heme oxygenase in toxic renal injury: a protective role in cisplatin nephrotoxicity in the rat, *Kidney Int.* **48**, 1298–1307 (1995).

21. K. Mosley, D.E. Wembridge, V. Cattell, and H.T. Cook, Heme oxygenase is induced in nephrotoxic nephritis and hemin, a stimulator of heme oxygenase synthesis, ameliorates disease, *Kidney Int.* **53**, 672–678 (1998).

22. K.A. Nath, G. Balla, G.M. Vercellotti, J. Balla, H.S. Jacob, M.D. Levitt, and M.E. Rosenberg, Induction of heme oxygenase is a rapid, protective response in rhabdomyolysis in the rat, *J. Clin. Invest.* **90**, 267–270 (1992).

23. B.A. Vogt, T.P. Shanley, A. Croatt, J. Alam, K.J. Johnson, and K.A. Nath, Glomerular inflammation induces resistance to tubular injury in the rat. A novel form of acquired, heme oxygenase-dependent resistance to renal injury, *J. Clin. Invest.* **98**, 2139–2145 (1996).

24. D.C. Leitman, V.L. Agnost, R.M. Catalano, H. Schröder, S.A. Waldman, B.M. Bennett, J. Tuan, and F. Murad, Atrial natriuretic peptide, oxytocin and vasopressin increase cyclic GMP in LLC-PK$_1$ kidney epithelial cells, *Endocrinology* **122**, 1478–1485 (1988).

25. D.A. Flick and G.E. Gifford, Comparison of in vitro cell cytotoxic assays for tumor necrosis factor, *J. Immunol. Methods* **68**, 167–175 (1984).

26. R. Motterlini, R. Foresti, M. Intaglietta, and R.M. Winslow, NO-mediated activation of heme oxygenase: Endogenous cytoprotection against oxidative stress to endothelium *Am. J. Physiol.* **270**, H107–H114 (1996).

27. P.N. Hopkins, L.L. Wu, S.C. Hunt, B.C. James, G.M. Vincent, and R.R. Williams, Higher serum bilirubin is associated with decreased risk for early familial coronary artery disease, *Arterioscler. Thromb. Vasc. Biol.* **16**, 250–255 (1996).

28. S.D. Appleton, M.L. Chretien, B.E. McLaughlin, H.J. Vreman, D.K. Stevenson, J.F. Brien, K. Nakatsu, D.H. Maurice, and G.S. Marks, Selective inhibition of heme oxygenase, without inhibition of nitric oxide synthase or soluble guanylyl cyclase, by metalloporphyrins at low concentrations, *Drug. Metab. Dispos.* **27**, 1214–1219 (1999).

29. F. Massicot, C. Martin, H. Dutertre-Catella, S. Ellouk-Achard, C. Pham-Huy, M. Thevenin, P. Rucay, J.M. Warnet, and J.R. Claude, Modulation of energy status and cytotoxicity induced by FK506 and cyclosporin A in a renal epithelial cell line, *Arch. Toxicol.* **71**, 529–531 (1997).

30. R.J. Walker, V.A. Lazzaro, G.G. Duggin, J.S. Horvath, and D.J. Tiller, Structure-activity relationships of cyclosporines. Toxicity in cultured renal tubular epithelial cells, *Transplantation* **48**, 321–327 (1989).

31. J.J. Yang, W.F. Finn, Effect of oxypurinol on cyclosporine toxicity in cultured EA, LLC-PK1 and MDCK cells, *Ren. Fail.* **20**, 85–101 (1998).

32. S.S. Ahmed, H.W. Strobel, K.L. Napoli, and J. Grevel, Adrenochrome reaction implicates oxygen radicals in metabolism of cyclosporine A and FK-506 in rat and human liver microsomes. *J. Pharmacol. Exp. Ther.* **265**, 1047–1054 (1993).

33. F. Serino, J. Grevel, K.L. Napoli, B.D. Kahan, and H.W. Strobel, Oxygen radical formation by the cytochrome P450 system as a cellular mechanism for cyclosporine toxicity, *Transplant. Proc.* **26**, 2916–2917 (1994).

34. M.D. Maines, R.D. Mayer, J.F. Ewing, and W.K. McCoubrey, Induction of kidney heme oxygenase-1 (HSP32) mRNA and protein by ischemia/reperfusion: possible role of heme as both promotor of tissue damage and regulator of HSP32, *J. Pharmacol. Exp. Ther.* **264**, 457–462 (1993).

35. F. Shiraishi, L.M. Curtis, L. Truong, K. Poss, G.A. Visner, K. Madsen, H.S. Nick, and A. Agarwal, Heme oxygenase-1 gene ablation or expression modulates cisplatin-induced renal tubular apoptosis, *Am. J. Physiol. Renal Physiol.* **278**, F726–F736 (2000).

36. D. Willis, A.R. Moore, R. Frederick, and D.A. Willoughby, Heme oxygenase: A novel target for the modulation of the anti-inflammatory response, *Nature Med.* **2**, 87–90 (1996).

37. A. Yachie, Y. Niida, T. Wada, N. Igarashi, H. Kaneda, T. Toma, K. Ohta, Y. Kasahara, and S. Koizumi, Oxidative stress causes enhanced endothelial cell injury in human heme oxygenase-1 deficiency, *J. Clin. Invest.* **103**, 129–135 (1999).

38. L.E. Otterbein, F.H. Bach, J. Alam, M. Soares, H.T. Lu, M. Wysk, R.J. Davis, R.A. Flavell, and A.M.K. Choi, Carbon monoxide has anti-inflammatory effects involving the mitogen-activated protein kinase pathway, *Nature Med.* **6**, 422–428 (2000).

39. S. Suga, K. Nakao, M. Mukoyama, H. Arai, K. Hosoda, Y. Ogawa, and H. Imura, Characterization of natriuretic peptide receptors in cultured cells, *Hypertension* **19**, 762–765 (1992).

40. S. Immenschuh, V. Hinke, A. Ohlmann, S. Gifhorn-Katz, N. Katz, K. Jungermann, and T. Kietzmann, Transcriptional activation of the haem oxygenase-1 gene by cGMP via a cAMP response element activator protein-1 element in primary cultures of rat hepatocytes, *Biochem. J.* **334**, 141–146 (1998).

41. R.B. Pilz, M. Suhasini, S. Idriss, J.L. Meinkoth, and G.R. Boss, Nitric oxide and cGMP analogs activate transcription from AP-1-responsive promoters in mammalian cells, *FASEB J.* **9**, 552–558 (1995).

42. T. Polte and H. Schröder, Cyclic AMP mediates endothelial protection by nitric oxide, *Biochem. Biophys. Res. Commun.* **251**, 460–465 (1998).

43. J. Alam and Z. Den, Distal AP-1 binding sites mediate basal level enhancement and TPA induction of the mouse heme oxygenase-1 gene, *J. Biol. Chem.* **267**, 21894–21900 (1992).

44. W. Durante, N. Christodoulides, K. Cheng, K.J. Peyton, R.K. Sunahara, and A.I. Schafer, cAMP induces heme oxygenase 1 gene expression and carbon monoxide production in vascular smooth muscle, *Am. J. Physiol.* **273**, H317–H323 (1997).

45. W.W. Hancock, R. Buelow, M.H. Sayegh, and L.A. Turka, Antibody-induced transplant arteriosclerosis is prevented by graft expression of anti-oxidant and anti-apoptotic genes, *Nature Med.* **4**, 1392–1396 (1998).

46. L.A. DeBruyne, J.C. Magee, R. Buelow, and J.S. Bromberg, Gene transfer of immunomodulatory peptides correlates with heme oxygenase-1 induction and enhanced allograft survival, *Transplantation* **69**, 120–128 (2000).

47. Am. Salahudeen, H. Huang, K. Ndebele, J. Jenkins, and A.B. Salahudeen, Heme oxygenase (HO-1) reduces cell death during cold storage: studies utilizing hemin induction and HO-1 gene transfer, *Acta Haematol.* **103**(Suppl. 1), 85 abstr. (2000).

48. F. Marumo, Y. Masaki, T. Ida, K. Sato, and K. Ando, Prolongation of the kidney preservation period by simple cold storage up to 72 hours by human atrial natriuretic peptide, *Transplantation* **51**, 982–986 (1991).

49. R. Lewis, R. Janney, R. Osgood, J. McAndrew, R. Verani, and T. Fried, Effect of exogenous ANP on initial renal function following 24-hour cold preservation, *Kidney Int.* **36**, 562–569 (1989).

23

HEME OXYGENASE AND ATHEROSCLEROSIS

Jesus Araujo[a], Kazunobu Ishikawa[b], and Aldons J. Lusis[a]

[a]Department of Medicine
University of California-Los Angeles
Los Angeles, CA 90095
[b]The First Department of Internal Medicine
Fukushima Medical University
Fukushima, Japan 960-1295

I. ATHEROSCLEROSIS: AN INFLAMMATORY DISEASE

Atherosclerosis is characterized by the accumulation of lipids and fibrous elements in large and medium-sized elastic and muscular arteries. It may result in the obstruction of blood flow which can result in ischemia of the heart, brain, or extremities.[1,2] For many decades, it was considered a degenerative disease; currently it is better understood as a chronic inflammtory condition that is converted to an acute clinical event by the induction of plaque rupture, which in turn leads to thrombosis.[3] In westernized societies, atherosclerosis is the underlying cause of about 50% of total mortality.

Stary has described the microscopical evolution of atherosclerosis.[4] The earliest stage (Stary I lesion) which is not apparent macroscopically, consists of subendothelial accumulations of lipid-loaded macrophages, called "foam cells", and is found in 45% of infants up to eight months of age. By puberty, it is common to see the classical "fatty streak" (Stary II lesions), visualized macroscopically on Sudan IV stain as a flat or slightly raised streak, consisting of lipid-loaded macrophages, smooth-muscle cells also containing lipid droplets and minimal, scattered, extracellular lipid. More advanced lesions include those with multiple extracellular lipid cores (Stary III lesion) which appear as a raised fatty streak or an atheroma (Stary IV lesion), characterized by a single confluent extracellular lipid core. By the third decade of life, the fibrous lesions (Stary V lesion) appear when Stary III or Stary IV lesions are surrounded and/or cap by smooth-muscle cells and collagen. These plaques can harbor calcification and areas of microthrombi in different stages of fibrotic organization. The great majority of acute coronary ischemic events occur as a result of plaque fissuring/rupture with exposure of the subendothelial matrix and the subsequent development of an occluding thrombus on the surface of the plaque.[5]

Artery wall lipid deposition is indeed a hallmark of the pathology of atherosclerosis High plasma concentrations of cholesterol and in particular those of low-density lipoprotein (LDL) cholesterol are the principal risk factors for atherosclerosis. Infiltrating lipids come from LDL particles that travel into the artery wall in a concentration-dependent process that does not require receptor-mediated endocytosis.[6,7] Once in the artery wall, the lipoproteins may be trapped in the three-dimensional cagework of fibers and fibrils present in the subendothelial space,[8] establishing in time associations with the extracellular matrix of the subendothelial space.[9] The build-up of LDL is determined by the rate of retention as well as the rate of influx. LDL may be modified by oxidation, glycation (in diabetes), aggregation, association with proteoglycans, or incorporation into immune complexes.[10–13] It is internalized by macrophages at least in part by means of the scavenger receptors on the surfaces of these cells to give rise to foam cells.[10,14] Oxidatively-modified LDL increases the oxidative stress in the artery wall and may be cytotoxic.[15,16] Modified LDL can also up-regulate the expression of genes for a variety of inflammatory proteins, including macrophage colony-stimulating factor[17,18] and monocyte chemotactic protein[19] in endothelial cells, expanding the inflammatory response by stimulating the entry of new monocytes into lesions and the differenciation into macrophages within the vessel.[1,2] A vicious circle of inflammation ensues as inflammatory conditions promote the retention and modification of LDL and more and more modified LDL stimulates inflammation (Fig. 1).

II. HEME OXYGENASE: OXIDATION AND INFLAMMATION

Accumulating evidence suggests that oxidative stress plays a central role in many pathogenic processes. Free radicals have been implicated in over 100 diseases, from malaria and haemorrhagic shock to AIDS,[21] and many of the biological consequences of excess radiation exposure may result from free radical damage to proteins, DNA and lipids. In a number of human diseases, the increased production of free radicals, although secondary, may make an important contribution to disease pathology. This may be the case in rheumatoid arthritis, inflammatory bowel diseases, pulmonary diseases such as adult respiratory distress syndrome, emphysema, asthma, bronchopulmonary dysplasia and interstitial pulmonary fibrosis, the sequalae of ischaemic or traumatic injury to the brain, myocardial stunning and coronary artery disease.[22]

Oxidative stress can elicit the induction of "classic" antioxidant enzymes such as superoxide dismutase, catalase, glutathione peroxidase[23] and also a variety of stress-response proteins such as heat shock proteins, metallothionein, and heme oxygenase-1.[22] Heme oxygenase (HO) appears to play a key role in the response of cells and organisms to oxidative and other stresses (the so-called stress response), and the products of the reaction, namely, carbon monoxide, biliverdin, and iron, have important functions of their own.[24]

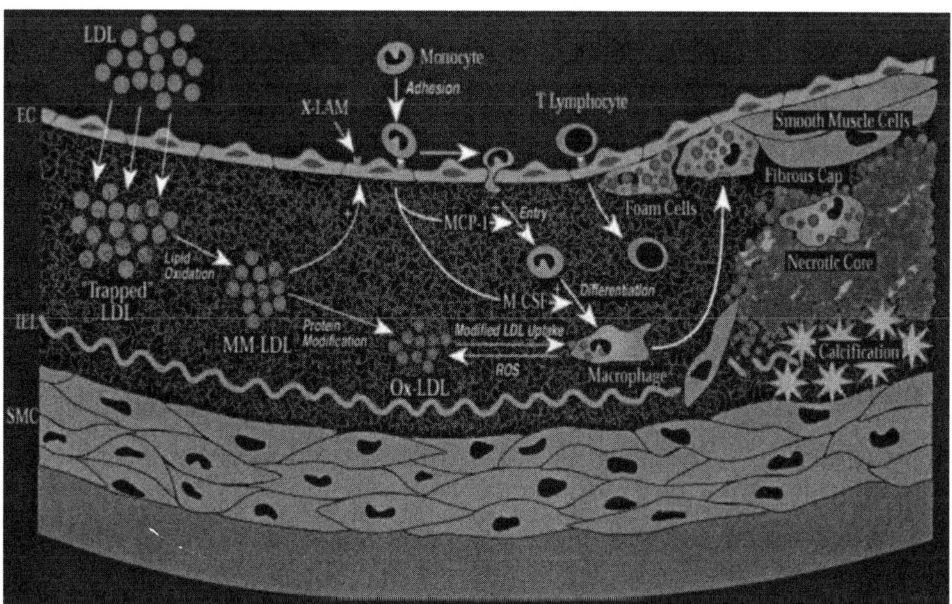

Figure 1. Atherosclerosis model. Low density lipoproteins (LDL) or other atherogenic lipoproteins enter and accumulate in the subendothelial space. Trapped LDL particles become oxidized probably as a result of interactions with reactive oxygen species (ROS) produced by vascular cells. Such minimally modified LDL (MM-LDL) induce endothelial cells to express adhesion molecules for monocytes (X-LAM), monocyte chemotactic protein (MCP-1), and macrophage-colony stimulating factor (M-CSF), resulting in more recruitment of blood monocytes and subsequent differenciation into macrophages. With time, LDL particles become highly oxidized (Ox-LDL), then recognized by scavenger receptors on macrophages which endocytose them, giving rise to cholesterol-engorged foam cells. Such foam cells, the hallmark of fatty streaks, contribute to the development of advanced lesions by the production of various cytokines and growth factors.[20]

HO plays an important biological role in the degradation of heme, which has a central part in eukaryotic metabolic pathways, as the prosthetic moiety of hemoproteins involved in cell respiration and oxidative biotransformations such as hemoglobin, myoglobin, cytochrome P450, nitric oxide synthase and soluble guanylyl cyclase.[25] HO regulates hemoprotein levels and protects cells from the deleterious effects of free heme.[25] In association with NADPH ferrihemoprotein reductase (NADPH cytochrome P-450 reductase) and in the presence of oxygen and NADPH or NADH, HO catalyzes the initial reaction in heme catabolism to yield equimolar quantities of biliverdin IXa, carbon monoxide, and iron.[26] Biliverdin is subsequently converted to bilirubin by the action of biliverdin reductase.[27] Three isoforms of HO have been described in mammals, HO-1, HO-2 and HO-3, which are the products of separate genes. HO-1 seems to be very prominent since it is widely distributed in tissues and the only isoform highly inducible in virtually all cells. In terms of inflammatory disorders such as atherosclerosis, HO-1 is likely to be particularly important since it is up-regulated during inflammation and has important antioxidant properties.

III. HO-1 IN ATHEROSCLEROSIS

The propensity for LDL oxidation in the subendothelial space will depend on the balance between pro-oxidant and antioxidant elements present in this environment. HO-1 induction may play a key role in the cellular antioxidant defences in atherogenesis. HO may act as an antioxidant system by the production of bilirrubin, which acts as a natural antioxidant in several in vitro systems.[28,29] It may protect LDL from oxidation,[30] possibly acting with vitamin E as a co-antioxidant.[31] Epidemiological studies have shown that low serum concentrations of bilirubin could be an independent risk factor for coronary artery disease.[32–34] We have observed that induction of HO-1 mRNA expression and activity in co-cultures of human aortic endothelial and smooth muscle cells by mildly oxidized LDL, oxidized lipid metabolites or hemin inhibits monocyte chemotaxis; this was attenuated by pretreatment with these agents and reversed by inhibition of HO activity by Sn-protoporhyrin IX.[35]

In addition to heme, multiple chemical and environmental stimuli are known to induce HO-1 in different cells. We have found that HO-1 is expressed in atherosclerotic lesions of C57BL/6J mice fed a high fat diet, of LDL-receptor knockout mice fed a high fat diet and of spontaneously hyperlipidemic apoE knockout mice. Using immunohistochemistry, HO-1 was localized to the vascular endothelium, macrophages and smooth muscle cells in the lesions, most abundantly expressed in macrophages but not in neighbooring normal arterial wall (Fig. 2).

HO-1 expression in atherosclerotic lesions has also been observed by Wang et al.[37] who have shown that expression of HO-1 was prominent in endothelial cells and foam cells/macrophages of thickened intima in lesions from both humans and apo-E deficient C57BL/6J mice and also in medial smooth muscle cells of advanced lesions. Induction of HO-1 in endothelial and smooth muscle cells by pro-atherogenic agents, such as heme, oxidized LDL, lipid metabolites, peroxynitrite, hypoxia, heavy metals, pro-inflammatory cytokines and angiotensin II could provide a secondary line of antioxidant defence and maintain vessel patency through the generation of CO that could act as a vasodilator.[25]

The presence of elevated heme levels in atherosclerotic plaques due to the release of hemoproteins from necrotic and ruptured cells present within lesions may lead to further LDL oxidation.[38] Heme is an iron-containing compound that will spontaneously asociate with hydrophobic structures such as cell membranes or lipoproteins, making them extraordinarily susceptible to the oxidant damage caused by activated inflammatory cells or soluble H_2O_2. Induction of ferritin chelates free iron generated by increased HO activity, preventing further oxidative damate.[39] We have observed that within the vascular cells of atherosclerotic lesions, there is concomitant induction of ferritin and HO-1.

In support of the protective role of HO-1, we have shown that HO-1 up-regulation achieved by 6-week administration of hemin or hemin + deferoxamine or HO inhibition by SnPPIX of 10 week-old-LDL-r KO mice resulted in lower or greater atherosclerotic lesion development, respectively, than non-treated control animals, on both a "western" diet and a high cholesterol diet (Fig. 3). On both type of diets, HO-1 up-regulation in the hemin-treated groups correlated with lower plasma lipoperoxides and higher levels of plasma nitrite/nitrate. Also, HO inhibition in the Sn-

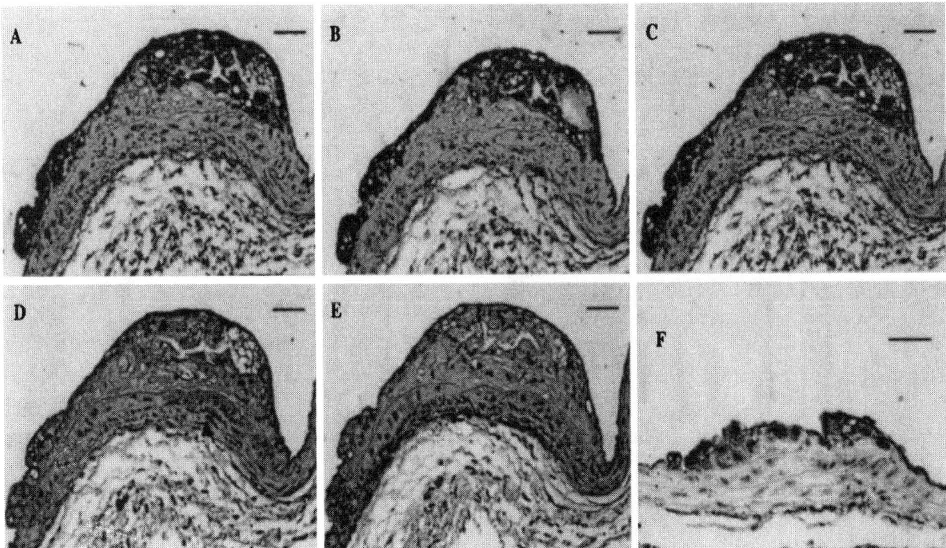

Figure 2. Immunohistochemical staining of atherosclerotic lesions of LDL-receptor knocknout mice after feeding a western diet for 6 weeks. A darker staining is seen at the top of these aortic sections, specific for HO-1 (A), macrophages (B), oxidized phospholipids (C) where specific antibodies were used and absent in the control section where noimmune sera was employed (D). E, stained with anti-vascular smooth muscle actin, shows a darker staining in the aortic media. F shows an early atherosclerotic lesion stained with anti-HO-1 antibody. All sections were immunostained using the immunoperoxidase technique and then counterstained with hematoxylin. Magnification bars = 50 µm. (Ishikawa et al., 36.)

PPIX-treated group correlated with higher levels of plasma lipoperoxides and lower levels of plasma nitrite/nitrate. We hypothesize that the protective effect of HO-1 may be explained, in part, by the suppression of lipid peroxidation and by modulation of the NO pathway.[36]

There is also evidence that adenovirus-mediated HO-1 gene transfer results in lesser atherosclerotic lesions in apoE-KO mice and that this effect could be partly mediated by a decrease in the iron deposition as well as iron content in the aortic wall.[40] Interestingly, the effect was observed only following intraventricular administration of the adenovirus resulting in expression in liver and aorta, but not with intravenous administration resulting in expression in liver but not in the aortic tissue. If thus, it appears that HO-1 must be expressed in the local tissue in order to render its protective anti-atherogenic effect.

To clearly established if up-regulation of HO-1 results in inhibition of atherosclerosis whithout having to consider any other possible secondary effects derived from the pharmacological manipulations mentioned above, we generated rat HO-1 transgenic mice. We have used a 52 kb rat HO-1 insert containing ~27.7 kb of the 5′ upstream region, 8.3 kb of the HO-1 gene and ~16 kb of the 3′ downstream region for the development of our rHO-1 transgenic mice. Cloning was achieved by screening of a rat genomic P1 library with two sets of PCR primers corresponding to the first and last exon (1 and 5) of the rHO-1 gene. Three clones were positive for both

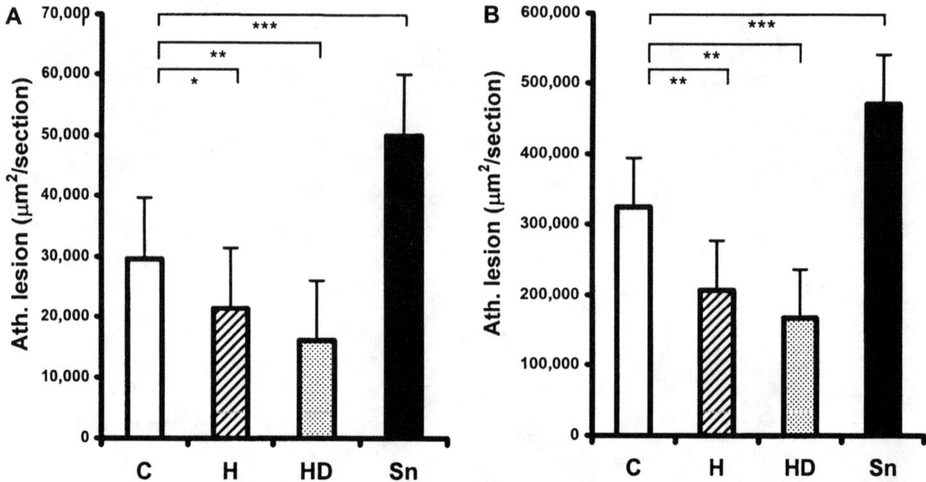

Figure 3. Effect of pharmacological HO modulation on the aortic atherosclerotic lesion formation on LDL-receptor KO mice on a "western" diet (A) or high-cholesterol diet (B). Mice were treated by intraperitoneal injections of saline (C), hemin (H), hemin + deferoxamine (HD), or Sn-protoporphyrin IX (Sn) during 6-week feeding of the hypercholesterolemic diets. Atherosclerotic lesion areas were determined using serial sections of the first 400 μm of the ascending aorta. Data represents mean ± SD. *p < 0.05, **p < 0.01, ***p < 0.001. (Ishikawa et al., 36).

sets of primers, and restriction mapping revealed that two of them contained the same insert. The SfiI fragment of one of the two clones was microinjected into C57BL/6J eggs for the transgenic mice generation. Four founders were initially obtained but one died early. Three transgenic lines are currently breeding.[41] The transgene has been transferred onto the apoE-KO background and we are currently conducting a study to determine if the transgene provides a protective effect.

IV. HO-1: A MARKER FOR GENETIC SUSCEPTIBILITY?

Surveys of strains of mice have revealed striking differences in susceptibility to atherosclerosis when fed cholesterol-enriched diets. Some strains such as C3H/HeJ, BALB/cJ, A/J, SWR/J, and NZB/J are quite resistant, whereas other strains such as C57BL/6J, DBA/2J, AKR/J, and 129/J are susceptible.[20] This difference is also present in a mouse model of spontaneous hyperlipidemia and atherosclerosis such as the apoE-KO mice. We have constructed congenic apoE-KO mice on a C3H genetic background by repeatedly backcrossing B6.apoE-KO mice to C3H mice with selection for the apoE-null allele resulting in apoE-KO mice with background of ~97% C3H. C3H.apoE-KO mice were highly resistant to atherosclerosis, developing much smaller lesions than B6.apoE-KO mice on either chow or a "western" diet.[42]

Differences in between C3H/HeJ (C3H) and C57BL/6J (B6) are not due to differences in blood-derived cells since reciprocal bone marrow transplantation between the strains, with congenics carrying the same H-2 haplotype, showed little effect on lesion development.[42] There are however, substantial differences in the endothelial cells (ECs). Cultured ECs from B6 are much more responsive to minimally oxidized LDL, as judged by the induction of MCP-1, M-CSF, and HO-1, than ECs from C3H.[43] Interestingly, the baseline level of HO-1 mRNA was significanty higher in C3H than in B6 mice raising the possibility that this could be one of the factors making C3H more resistant than B6. On the other hand, induction of HO-1 by minimally oxidized LDL was ~8–9 fold that of control, much higher than the 2–3 fold induction observed in C3H ECs.[43] Presumably, the greater induction in C57BL/6J mice is due to the fact that C3H ECs better resist oxidative stress resulting from exposure to minimally oxidized LDL.

In a set of recombinant inbred (RI) strains designated as BXH derived from the wild-type B6 and C3H strains, ECs responses to oxidized LDL were seen to cosegregate with atherosclerosis susceptibility (Fig. 4). This suggests that the response of ECs to minimally oxidized LDL is a primary determinant of atherosclerosis susceptibility between the strains.

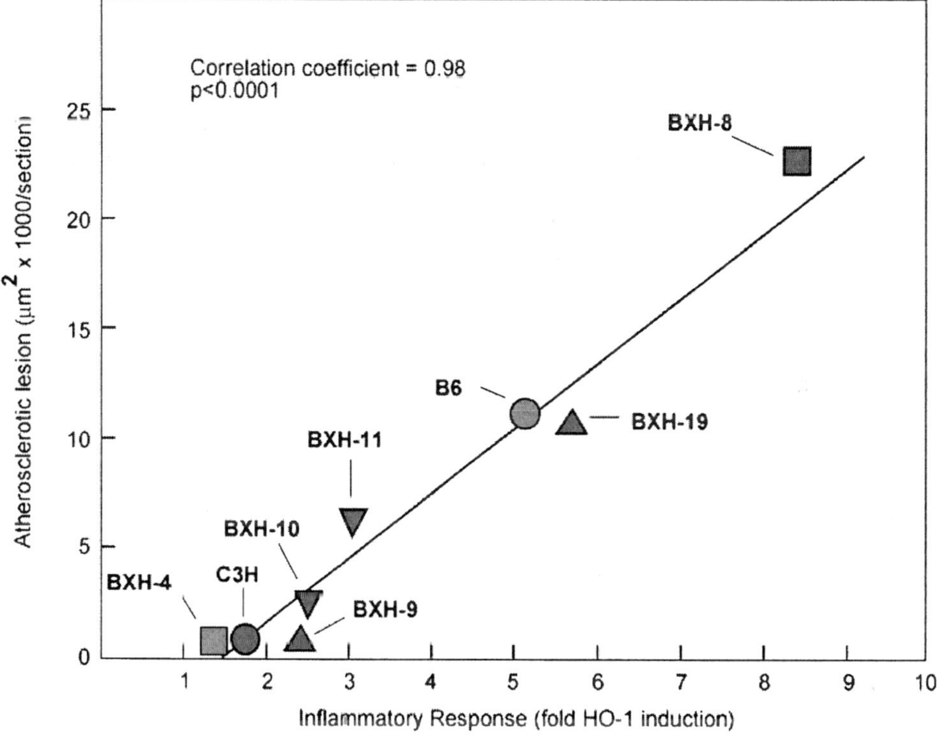

Figure 4. Cosegregation between the HO-1 fold induction of endothelial cells in response to minimally oxidized LDL and atherosclerotic lesion formation in a set of recombinant inbred strains from parental strains C57BL/6J (B6) and C3H/HeJ (C3H). Strains which displayed a higher HO-1 mRNA induction had larger lesions.[20]

V. HO-1 IN HUMAN POPULATION STUDIES

Whether naturally occurring variations affecting HO expression contribute to susceptibility to inflammatory diseases is unknown, but there is some suggestive evidence. A (GT)n repeat microsatellite polymorphism has been described in the promoter region of human HO-1[44,45] in association with chronic pulmonary emphysema.[45] The shorter number of repeats results in higher transcriptional activity and the longer number of repeats in lower activity.

It has been reported in a study of Japanese diabetic patients that longer number of repeats (>27 repeats) correlate with significant coronary artery disease defined angiographically.[46] It will be important to evaluate this polymorphism in different populations and determine if there is association with cardiovascular disease.

VI. SUMMARY

Atherosclerosis is a chronic inflammatory condition of large and medium-sized elastic and muscular arteries characterized by accumulation of lipids and various fibrous and cellular elements. Lipids derive mostly from circulating low-density lipoproteins that get trapped in the artery wall and are modified in the subendothelial space mainly by oxidation. Oxidized LDL is endocytosed by macrophages leading to the generation of foam cells. Lipid oxidation is of paramount importance both in early and advanced atherogenesis and vascular anti-oxidative strategies are fundamental in the prevention of the process. HO-1 is an antioxidant enzyme that is highly expressed by endothelial cells, macrophages and smooth muscle cells of atherosclerotic lesions but essentially absent in normal artery wall. HO-1 induction may be one of the protective pathways that vascular cells employ. Consistent with this, pharmacological induction of HO in mice resulted in lower atherosclerotic lesion formation and HO inhibition resulted in greater lesion formation. We have recently generated rat HO-1 transgenic mice and the transgene has been transferred onto the apoE-KO background to evaluate whether overexpression of HO-1 would result in lesser lesions as hypothesized.

HO-1 also seems to be important as a marker of genetic susceptibility to atherosclerosis. HO induction is part of the inflammatory response observed in endothelial cells in response to minimally oxidized LDL. In a genetic study in mice, HO-1 inducibility correlated with atherosclerotic lesion formation.

REFERENCES

1. Lusis A. Atherosclerosis. Nature 2000;407:233–241.
2. Ross R. Atherosclerosis—An inflammatory Disease. N Engl J Med 1999;340:115–126.
3. Berliner J., Navab M., Fogelman A.M., Frank J.S., Demer L.L., Edwards P.A., Watson A.D., and Lusis A.J. Atherosclerosis: Basic Mechanisms. Circulation 1995;91:2488–2496.
4. Stary H.C. Evolution and progression of atherosclerotic lesions in coronary arteries of children and young adults. Arteriosclerosis 1989;99:Suppl I:19–32.

5. Fuster V., Badimon L., Badimon J.J., and Chesebro J.H. The pathogenesis of coronary artery disease and the acute coronary syndromes. N Engl J Med 1992;326:242.

6. Steinberg D., Parthasarathy S., Carew T.E., Khoo J.C., and Witztum J.L.. Beyond cholesterol: modifications of low density lipoprotein that increase its atherogenicity. N Engl J Med 1989;320:915–924.

7. Young S.G. and Parthasarathy S. Why are low density lipoproteins atherogenic?. West J Med 1994; 160:153–164.

8. Frank J.S. and Fogelman A.M. Ultrastructure of the intima in WHHL and cholesterol-fed rabbit aortas prepared by ultra-rapid freezing and freeze-etching. J Lipid Res 1989;30:967–978.

9. Camejo G., Olafsson F., Lopez P., Carisson P., and Bondjers G. Identification of apoB-100 segments mediating the interaction of low-density lipoproteins with arterial proteoglycans. Arteriosclerosis 1988;8:368–377.

10. Steinberg D. Low density lipoprotein oxidation and its pathobiological significance. J Biol Chem 1997;272:20963–20966.

11. Khoo J.C., Miller E., McLoughlin P., and Steinberg D. Enhanced macrophage uptake of low density lipoprotein after self-aggregation. Arteriosclerosis 1988;8:348–358.

12. Khoo J.C., Miller E., Pio F., Steinberg D., and Witztum J.L. Monoclonal antibodies against LDL further enhance macrophage uptake of LDL aggregates. Arterioscler Thromb 1992;12:1258–1266.

13. Navab M., Berliner J.A., Watson A.D., Hama S.Y., Territo M.C., Lusis A.J., Shih D.M., Van Lenten B.J., Frank J.S., Demer L.L., Edwards P.A., and Fogelman A.M. The Yin and Yang of oxidation in the development of the fatty streak: a review based on the 1994 George Lyman Duff Memorial Lecture. Arterioscler Thromb Vasc Biol 1996;16:381–342.

14. Han J., Hajjar D.P., Febbraio M., and Nicholson A.C. Native and modified low density lipoproteins increase the functional expression of the macrophage class B scavenger receptor, CD 36. J Biol Chem 1997;272:21654–21659.

15. Morel D.W., Hessler J.R., and Chisholm G.M. Low density lipoprotein cytotoxicity induced by free radical peroxidation of lipid. J Lipid Res 1983;24:1070–1076.

16. Griendling K.K. and Alexander R.W. Oxidative stress and cardiovascular disease. Circulation 1997;96:3264–3265.

17. Quinn M.T., Parthasarathy S., Fong L.G., and Steinberg D. Oxidatively modified low density lipoproteins: a potential role in recruitment and retention of monocyte/macrophages during atherogenesis. Proc Natl Acad Sci USA 1987;84:2995–2998.

18. Rajavashisth T.B., Andalibi A., Territo M.C., Berliner J.A., Navab M., Fogelman A.M., and Lusis A.J. Induction of endothelial cell expression of granulocyte and macrophage colony-stimulating factors by modified low-density lipoproteins. Nature 1990;344:254–257.

19. Leonard E.J. and Yoshimura T. Human monocyte chemoatractant protein-1 (MCP-1). Immunol Today 1990;11:97–101.

20. Villa-Colinayo V., Shi W., Araujo J., and Lusis A.J. Genetics of Atherosclerosis: The search for genes acting at the level of the vessel wall. Curr Ath Rep 2000;2:380–389.

21. Halliwell B. The role of oxygen radicals in human disease, with particular reference to the vascular system. Haemostasis 1993;23(suppl 1):118–126.

22. Choi A.M.K. and Alam, J. Heme Oxygenase-1: Function, Regulation and implication of a novel stress-inducible protein inoxidant-induced lung injury. Am J Respir Cell Mol Biol 1996;15:9–19.

23. Camhi S., Lee P., and Choi A.M.K. The oxidative stress response. New Horizons 1995;3:170–182.

24. Elbirt K.K. and Bonkovsky H.L. Heme Oxygenase: Recent advances in understanding its regulation and role. Procc Assoc Amer Phys 1999;5:438–447.

25. Maines M.D. The heme oxygenase system: a regulator of second messenger gases. Annu Rev Pharmacol Toxicol 1997;37:517 554

26. Tenhunen R., Marver, H.S., and Schmid R. The enzymatic conversion of heme to bilirubin by microsomal heme oxygenase. Proc Natl Acad Sci USA 1968;61:748–755.

27. Maines M.D. Heme oxygenase: function, multiplicity, regulatory mechanisms, and clinical applications. FASEB J 1988;2:2557–2568.

28. Stocker R., McDonagh A.F., Glazer A.N., and Ames B.N. Antioxidant activities of bile pigments: biliverdin and bilirubin. Methods Enzymol 1990;186:301–309.

29. Neuzil J. and Stocker R. Bilirubin attenuates radical-mediated damage to serum albumin. FEBS Lett 1993;331:281–284.

30. Wu T.W., Fung K.P., and Yang C.C. Unconjugated bilirubin inhibits the oxidation of human low density lipoprotein better than Trolox. Life Sci 1994;54:P477–P481.

31. Neuzil J. and Stocker R. Free and albumin-bound bilirubin are efficient co-antioxidants for alpha-tocopherol, inhibiting plasma and low density lipoprotein lipid peroxidation. J Biol Chem 1994; 269:16712–16719.

32. Madhavan M., Wattigney W.A., Srinivasan S.R., and Berenson G.S. Serum bilirubin distribution and its relation to cardiovascular risk in children and young adults. Atherosclerosis 1997; 141:107–113.

33. Schwertner H.A., Jackson W.G., and Tolan G. Association of low serum concentration of bilirubin with increased risk of coronary artery disease. Clin Chem 1994;40:18–23.

34. Hopkins P.N., Wu L.L., Hunt S.C., James B.C., Vincent G.M., and Williams R.R. Higher serum bilirubin is associated with decreased risk for early familial coronary artery disease. Arterioscler Thromb Vasc Biol 1996;16:250–255.

35. Ishikawa K., Navab M., Leitinger N., Fogelman A.M., and Lusis A.J. Induction of heme oxygenase-1 inhibits the monocyte transmigration induced by mildly oxidized LDL. J Clin Invest 1997;100:1209–1216.

36. Ishikawa K., Sugawara D., Wang X., Suzuki K., Itabe H., Maruyama Y., and Lusis A.J. Heme oxygenase-1 inhibitis atherosclerotic lesion formation in ldl-receptor knockout mice. Circ Res 2001; 88:506–512.

37. Wang L.J., Lee T.S., Lee F.Y., Pai R.C., and Chau L.Y. Expression of heme oxygenase-1 in atherosclerotic lesions. Am J Pahol 1998;152:711–720.

38. Paganga G., Riceevans C., Rule R., and Leake D. The interaction between ruptured erythrocytes and low-density lipoproteins. FEBS Lett 1992;303:154–158.

39. Balla G., Jacob H.S., Balla J., Rosenberg M., Nath K., Apple F., Eaton J.W., and Vercellotti, G.M. Ferritin: a cytoprotective antioxidant strategem of endothelium. J Biol Chem 1992;267:18148–18153.

40. Juan S.H., Lee T.S., Tseng K.W., Liou J.Y., Shyue S.K., Wu K.K., and Chau L.Y. Adenovirus-mediated heme oxygenase-1 gene transfer inhibits the development of atherosclerosis in apolipoprotein E-deficient mice. Circulation 2001;104:1519–1525.

41. Araujo J., Tward A., Ishikawa K., Shih D., and Lusis A.J. Development of transgenic C57BL/6J mice for the rat heme oxygenase-1 gene. Acta Haematol 2000;103(Suppl 1):66.

42. Shi W., Wang N.J., Shih D.M., Sun V.Z., Wang X., and Lusis A.J. Determinants of atherosclerosis susceptibility in the C3H and C57BL/6 mouse model. Circ Res 2000;86:1078–1084.

43. Shi W., Haberland M., Jien M.L., Shih D.M., and Lusis A.J. Endothelial responses to oxidized lipoproteins determine genetic susceptibility to atherosclerosis in mice. Circulation 2000;102:75–81.

44. Kimpara S., Takeda A., Watanabe K., Itoyama Y., Ikawa S., Watanabe M., Arai K., Sasaki H., Higuchi S., Okita N., Takase S., Saito H., Takahashi K., and Shibahara S. Microsatellite polymorphism in the human heme oxygenase-1 gene promoter and its application on association studies with Alzheimer and Parkinson disease. Human Genet 1997;100:145–147.

45. Yamada N., Yamaya M., Okinaga S., Nakayama K., Sekizawa K., Shibahara S., and Sasaki H. Microsatellite polymorphism in the heme oxygenase-1 gene promoter is associated with susceptibility to emphysema. Am J Hum Genet 2000;66:187–195.

46. Kaneda H., Taguchi J., Hashimoto H., Aizawa T., and Ohno M. Microsatellite polymorphism in the heme oxygenase-1 gene promoter is associated with coronary artery disease susceptibility in japanese diabetic patients. Circulation 2000;102(Suppl II):757.

24

ATHEROGENICITY OF HYPERCHOLESTEROLEMIA IN THE PRECENSE OF HEMOLYSIS IN SPITE OF HEME OXYGENASE-1 INDUCTION

A Possible Consequence of Its Interaction with Plasma Lipoproteins

Germán Camejo[a], Ana Z. Fernandez[b], Flor López[b],
Alfonso Tablante[b], Egidio Romano[b], Eva Hurt-Camejo[a],
and Rafael Apitz-Castro[1]

[a]Wallenberg Laboratory for Cardiovascular Research
University of Gothenburg
Sahlgrenska Hospital
Gothenburg, S-41345, Sweden, and
AstraZeneca, Mölndal, S-431 83, Sweden
[b]Laboratorio de Trombosis Experimental
Centro de Biofísica y Bioquímica
Instituto Venezolano de Investigaciones Científicas (IVIC)
Caracas-Venezuela

1. INTRODUCTION

Atherosclerosis appears to be caused mainly by a focal and complex tissue response to apoB-containing lipoproteins (LDL, IDL, VLDL) deposition in the arterial intima (Camejo et al., 1998b; Williams and Tabas, 1995). The retention of apoB-lipoproteins by proteoglycans of the arterial intima seems to provide the opportunity for enzymatic and non-enzymatic alterations of their lipid and protein moieties. This appears to cause formation of substances, mainly bioactive lipids that acting locally on arterial cells could initiate an inflammatory atherogenic response (Camejo et al., 1998b; Hurt-Camejo et al., 2001). Among the most studied potentially atherogenic modifications of lipoproteins are those caused by free radical-mediated reactions, probably taking place in the intima (Steinberg and Witztum, 1999). However, the origin of reactive oxygen species (ROS) that affect lipoproteins and cells in arteries still remains uncertain. The major cause for this is that agents, like transition metals, used as a model of pro-oxidant conditions *in vitro* do not oxidize lipoproteins in the presence of other plasma components that are known to exist in circulation and in the intima (Steinberg and Witztum, 1999). Heme is an iron-containing catabolic product of hemoglobin that has high affinity for human low density lipoprotein (LDL) in plasma and that is capable of oxidizing this particle in diluted plasma (Camejo et al., 1998a; Tribble et al., 1995) Heme interacts with the lipid surface monolayer and with specific positively charged segments of apoB-100 in the LDL particle (Camejo et al., 1998a). Heme transported by apoB-lipoproteins can accumulate in the arterial intima and become an *in situ* generator of ROS.

Heme release from erythrocytes by hemodynamic stress appears to occur constantly in what is called "trivial hemolysis" (Balla et al., 1991). Free heme in plasma is associated with lipoproteins and appears to accumulate in progressing plaques at sites of lipoprotein deposition (Scannapieco et al., 1988). In conditions where intravascular hemolysis is augmented, heme could be rapidly bound to circulating lipoproteins and be transported into the intima at sites where hydrogen peroxide produced by mononuclear cells could increase the pro-oxidant actions of heme-iron on the lipoproteins (Balla et al., 1991). Additionally, oxidative degradation of heme could contribute to iron accumulation in phagocytes, a phenomenon that characterizes lesions caused by hypercholesterolemia in rabbits and men (Araujo et al., 1995; Lee et al., 1999). Heme is a potent pro-inflammatory agent that increases vascular permeability, adhesion molecule expression and leukocyte recruitment; all these actions are considered atherogenic factors (Ross, 1999; Wagener et al., 2001). Most tissues have protective mechanisms to control excess generation of ROS by iron-containing heme. Heme oxygenase-1 (EC 1.4.99.3; HO-1) is a stress protein that initiates the decomposition of heme into carbon monoxide, iron and biliverdin that is further transformed into bilirubin. HO-1 has been suggested to participate in defense mechanism against oxidative injury mediated by hemoglobin and heme (Wagener et al., 2001; Willis et al., 1996). *In vivo*, heme formation has been shown to induce the gene and enzyme activity of HO-1. This suggests that if heme becomes available chronically to cells it could induce HO-1 and other heme oxidases and this may control the potential atherogenic effect of the iron-carrying protoporphyrin (Hayashi et al., 1999; Ishikawa et al., 2001).

Recent research has explored the possible anti-atherogenic action of HO-1 using different animal models. Paul et al. (Paul et al., 1999) showed that apoE-deficient mice develop less atherosclerosis when intravascular hemolysis induced by phenylhydrazine causes upregulation of the HO-1 gene and the enzyme. Juan et al. (Juan et al., 2001), in addition, showed that adenovirus mediated over-expression of HO-1 also causes reduction of atherosclerosis in apoE-knockout mice. Finally, Ishikawa et al. (Ishikawa et al., 2001), showed that induction of HO-1 by intraperitoneal injection of hemin caused an increase in HO-1 activity and a decrease in atherosclerosis of LDL-receptor defective mice in a western diet. In this article we will discuss the nature and effects of the interaction of hemin with lipoproteins and some of the reasons why this association may be atherogenic. These concepts will be extended to the interpretation of recent results about the potential protective role that HO-1 may have in atherogenesis in different animal models, including data from cholesterol-fed rabbits with increased intravascular hemolysis and induction of heme oxygenase. We found that in these animals hemolysis worsened atherosclerosis in spite of increased HO-1 expression. The results suggest that in such model heme overproduction and its possible action on plasma lipoproteins overrides the anti-atherosclerotic action of increased heme-oxygenase (Fernandez et al., 2001).

2. ASSOCIATION OF HEMIN WITH PLASMA LIPOPROTEINS

Free radical mediated oxidation of lipids and a protein of lipoproteins entrapped in the arterial intima occurs in atherosclerotic lesions (Steinberg and Witztum, 1999). Several free radical generating systems that may oxidize lipoproteins in the intima microenvironment have been described and some of them appear to be heme-mediated. Hemoglobin from ruptured erythrocytes oxidizes LDL *in vitro* (Miller et al., 1997), probably through a pathway that involves a ferric/ferryl cycle of the heme iron and hydroperoxides (Miller et al., 1997). Blood lipoproteins transport porphyrins, including heme and hemin, in plasma and may be the vehicles that take them to atherosclerotic lesions and tumors (Jori and Reddi, 1993; Kongshaug, 1992). Atherosclerotic lesions concentrate porphyrins and, although not specifically documented, probably heme. We found that the Fe^{3+}-hemin in diluted human serum binds to LDL and HDL with high affinity and that it remains associated with the lipoproteins even after ultracentrifugal fractionation under a density gradient (Fig. 1).

Also, high affinity with lipoproteins has been reported for porphyrins containing no iron, thus indicating that the association depends on the porphyrin ring system (Beltramini et al., 1987; Kongshaug, 1992). The binding affinity measured for the hemin-lipoproteins associations (Kd) was 49 ± 5 nmol/L for human LDL and 291 nmol/L for human HDL and at saturation each molecule of LDL could carry 45–65 molecules of hemin and HDL 4–6 molecules (Camejo et al., 1998a). The Kd for albumin was 1000 ± 220 nmol/L. These affinities and total binding capacities (Bt) explain why when added to serum or plasma hemin or protoporphyrin IX are found associated with lipoproteins, indicating that lipoproteins can compete effectively with hemopexin and albumin for the binding of the porphyrins (Kongshaug, 1992; Camejo, 1998).

LDL is a spherical particle with a surface monolayer made of polar charged segments of the apoB-100 protein and a mosaic of polar phospholipids and free cholesterol (Segrest et al., 2001). Binding experiments, thermal titration of fluorescent quantum yield and polarization fluorescence of LDL complex with hemin and protoporphyrin IX indicate that the porphyrin ring is inserted into the lipoprotein surface monolayer. Probably, the planar non-polar region of the ring is in contact with non-polar chains of the fatty acids of surface phospholipids. The fact that protoporphyrin IX and hemin increase the negative charge of LDL indicates that the negative charges of the carboxylic acids of the porphyrin ring are in contact with lysine and arginine positive charged aminoacids of apoB-100 segments in the LDL surface and interact

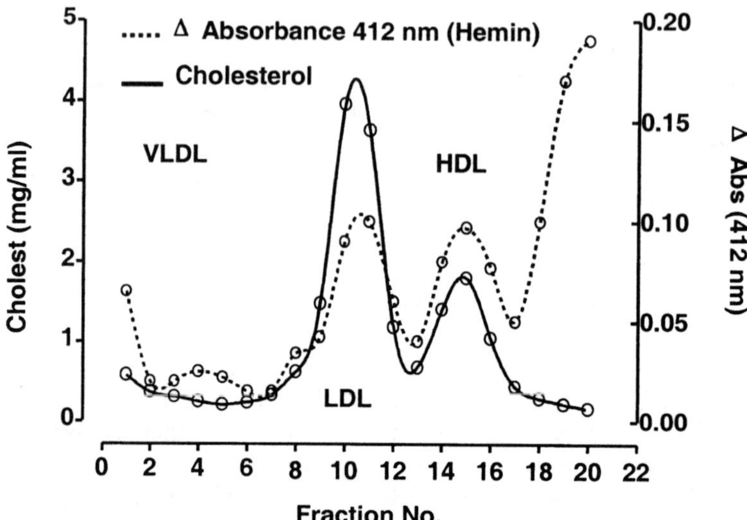

Figure 1. Distribution of 125 μmol/L hemin equilibrated with serum diluted 5 fold with buffer. To avoid oxidation of hemin the solutions contained 10 μmol/L butylated hydroxytoluene and EDTA. The hemin-serum solution was fractionated by ultracentrifugation in an exponential KBr gradient and hemin and cholesterol were measured after collection into 0.4 ml fractions (reproduced with permission from Camejo et al. *J Lipid Res* **39**:755).

with the external solvent (Camejo et al., 1998a). Figure 2 shows a scheme of the proposed location of hemin and protoporphyrin IX in the LDL surface. This high affinity of heme or hemin for LDL and HDL may facilitate oxidative processes in arterial tissue if electron-sequestering agents oxidize the Fe^{3+} of heme. Hydrogen peroxide, a product of macrophages and other immunocompetent cells, may be the biological relevant oxidizing agent. Hemin oxidizes polyunsaturated fatty acids of the lipoprotein more efficiently than ferric ions and this is potentiated by H_2O_2. These alterations caused an increase uptake of LDL by macrophages in culture (Balla et al., 1991; Tribble et al., 1995; Tribble et al., 1994).

3. OXIDATION OF LIPOPROTEINS BY HEMIN IN DILUTED SERUM AND INTERACTIONS OF LDL WITH MACROPHAGES

The extracellular arterial intima is bathed by a dilute ultrafiltrate of blood plasma with albumin concentrations around 100 μM and other plasma proteins similarly diluted (Smith and Staples, 1980). ApoB-lipoproteins, in the other hand reside, in progressing lesions at concentrations above those of the circulating plasma (Collins and Carew, 1997). Free radical mediated oxidation of the polyunsaturated fatty acids of phospholipids and cholesterol esters of LDL is thought to contribute to generation of cytotoxic molecules that contribute to the inflammatory tissue response of

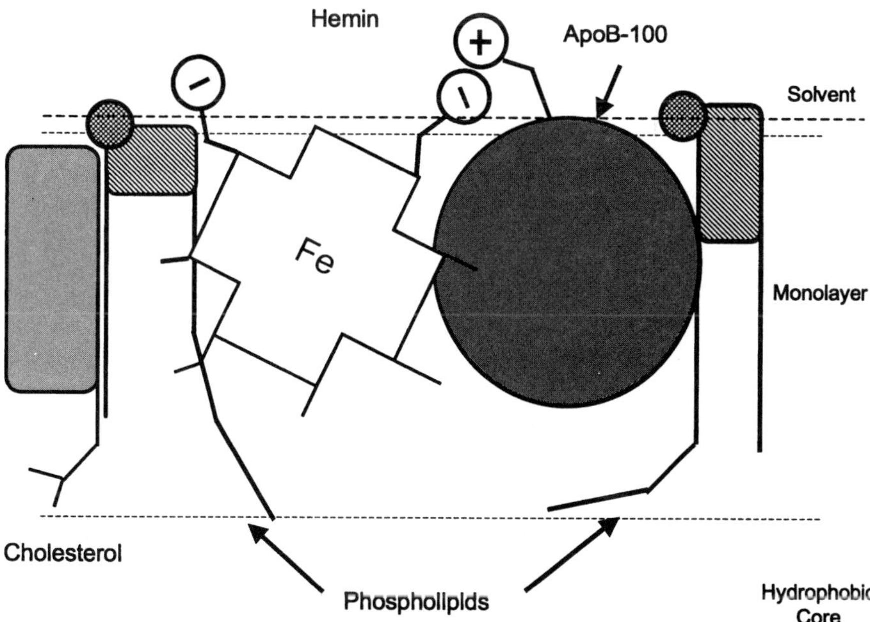

Figure 2. Scheme of the hypothetical location of heme at the surface monolayer of apoB-100 lipoproteins. Binding data and electrophoresis suggest that the carboxylic groups of heme interact with positively charged segments of apoB-100 in contact with the solvent inducing an increase in the negative charge of the particle. Fluorescence quantum yield studies and spectral alterations indicate that the heme hydrophobic ring and side chains interact with the non-polar regions of the surface monolayer in contact with the cholesteryl ester core of LDL (reproduced with permission from Camejo et al. *J Lipid Res* **39**:755).

atherogenesis. Much of the *in vitro* evidence supporting this hypothesis was gathered using transition metal (Cu^{2+} and Fe^{3+}) mediated oxidation of purified lipoproteins. However the biological significance of many of the results is diminished because it is highly unlikely that in the intima in the presence of plasma components enough free Cu^{2+}, or Fe^{3+} is available to act directly as catalysts of free radical-mediated oxidations. We found that the polyunsaturated fatty acids of LDL and HDL can be oxidized in the presence of 20% diluted human serum by 10–100 µM hemin and 250 µM H_2O_2, as judged by the production of conjugated dienes (Fig. 3).

In addition we found F_2-isoprostanes in LDL and HDL isolated from diluted serum treated with hemin even in the absence of hydrogen peroxide Fig. 4 (Camejo et al., 1998a). F_2-isosprotane production is a sensitive measurement of polyunsaturated fatty acid oxidation and these products have been found in human atherosclerotic lesions (Praticò et al., 1997). EDTA did not inhibit the oxidation, indicating that free transition metals were not involved (Fig. 5).

Increased uptake of oxidative modified LDL is believed to be a key pro-atherogenic action contributing to foam cell formation (Steinberg and Witztum, 1999). LDL isolated from diluted serum after treatment with hemin-H_2O_2 was more readily taken up and degraded by human monocyte-derived macrophages than LDL isolated from non-treated serum. Addition of probucol to the lipoproteins in diluted

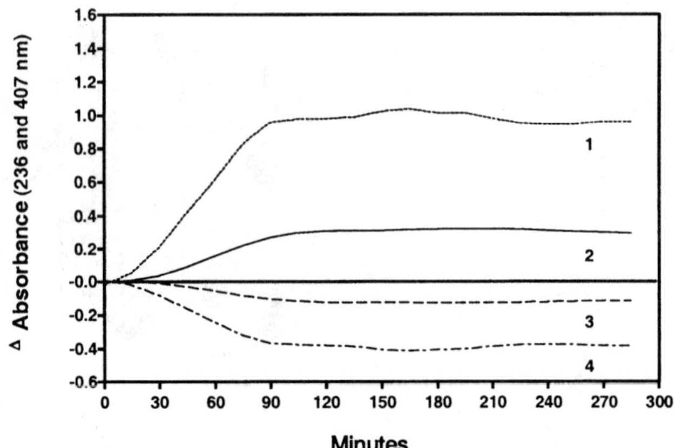

Figure 3. Conjugated diene formation in serum. Increased in dienes, absorbance at 236 nm (curves 1 and 2), and disappearance of hemin, decrease in absorbance at 407 nm (curves 3 and 4) were monitored for 300 minutes at 37 °C, VLDL-depleted serum diluted 5 fold with PBS and incubated with 10 µmol/L (curves 2 and 3) and 100 µmol/L (curves 1 and 4) and 250 µmol/L H_2O_2. All samples contained 0.1 mg EDTA/ml (reproduced with permission from Camejo et al. *J Lipid Res* **39**:755).

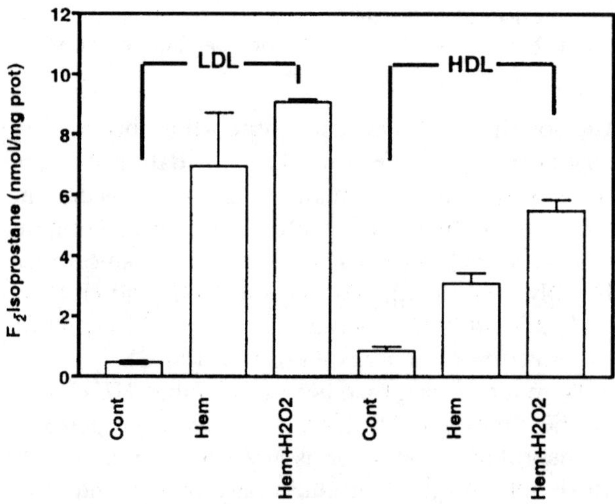

Figure 4. Content of 8-epi-PFG$_{2\alpha}$ in lipids of LDL and HDL isolated from VLDL-depleted serum after incubation with 150 µmol/L hemin and 150 µmol/L hemin plus 750 µmol/L H_2O_2 for 16 hours at 37 °C. The samples contained 0.1 mg/ml EDTA. The F_2-isoprostane was measured in the lipids extracted from the isolated lipoproteins by gas chromatography/mass spectrometry (reproduced with permission from Camejo et al. *J Lipid Res* **39**:755).

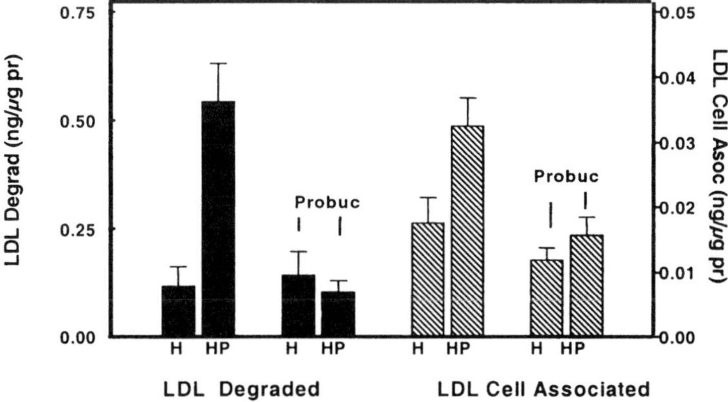

Figure 5. Uptake and degradation by human monocyte-derived macrophages of LDL isolated from serum treated with hemin (H) or with hemin and H_2O_2 (HP) in the absence or in presence of 10 µmol/L probucol (reproduced with permission from Camejo et al. *J Lipid Res* **39**:755).

serum before treatment with hemin -hydrogen peroxide decreased oxidation and consequently reduced macrophage uptake and degradation. EDTA had no effect in.

The results discussed show that hemin, even in the absence of hydrogen peroxide, can produce oxidation products of arachidonate, F_2-isoprostanes, in LDL and HDL in diluted serum. With added hydrogen peroxides, hemin in diluted serum oxidized the lipoproteins to the extent of conjugated dienes and thiobarbituric acid reacting substances (TBARS) format ion. If such processes take place in the intima, it is possible that such modified LDL retained in the subendothelial matrix may contribute to the atherogenic response. In addition, oxidative alterations of HDL may compromise its anti-atherogenic properties as part of the "reverse cholesterol transport" from vessels to liver. The alterations caused by hemin on the LDL particle led to an increase in its uptake and degradation by macrophages. Together with lipid oxidation products, LDL may be in addition a vehicle for internalization of Fe $^{3+}$ into vascular cells; a situation that could contribute to the cytotoxicity of hemin-oxidized lipoproteins, as suggested by Yuan et al. (Yuan et al., 1995). The experiments described thus support the hypothesis that at sites of LDL retention in the extracellular intima associated heme may be a pro-oxidant of pathological significance, even in the presence of other plasma components.

4. ATHEROGENICITY OF INTRAVASCULAR HEMOLYSIS

The described results suggest the possibility that increased intravascular heme generation, like in intravascular hemolysis, could contribute to atherogenesis if the augmented heme production is not compensated by increased activity of heme-oxidases. We explored this possibility in an *in vivo* model of atherogenesis, the cholesterol fed rabbit in which intravascular hemolysis was induced by phenylhydrazine (PHZ) intraperitoneal injections.

Animals and Experimental Procedures

Male New Zealand white rabbits (age 2–3 months, 2 to 4 kg) were randomly sorted into four groups of 10 rabbits/group: a control group (C), a phenylhydrazine-induced intravascular hemolysis group (IVH), a moderate hypercholesterolemia group (MHC), and a group with intravascular hemolysis and moderate hypercholesterolemia (MHC-IVH). Rabbits were kept in individual cages, at 25 °C, with water *ad libitum*. Commercial rabbit chow (Protein: 17%, animal fat: 2%, fiber: 15%; Alimentos Protinal, Caracas) that was found to contain 0.17% cholesterol (w/w) was given to the control and IVH groups. The MHC and MHC-IVH groups received a similar chow supplemented with 0.30% cholesterol (w/w) (Sigma Chemical, St. Louis, MO, USA) for the first 4 weeks and 0.25% w/w in subsequent weeks up to ten weeks. The cholesterol-enriched diet was prepared by dissolving cholesterol in dehydrated ethanol at 60 °C and then impregnating the chow pellets with it. Ethanol was eliminated by evaporation (Araujo et al., 1995). Each animal in each group had daily access to 120 g of chow. PHZ was freshly prepared in 0.85% NaCl and filtered through 0.22 μm filter at the time of administration (Redondo et al., 1995). Blood specimens (anticoagulated in 0.2 g/ml EDTA), from rabbits fasted for 14 to 16 hours, were obtained every two weeks until week 10, when animals were sacrificed by i.v. injection of pentothal 50 mg/kg (Abbot Laboratories, Caracas, Venezuela). Liver, spleen and kidney were removed after perfusion of the animal with ice-cold PBS, pH 7.4 for immediate processing. Aortas were excised from the aortic arch to the iliac bifurcation and immersed in ice-cold preservative solution (5 mM Tris, 0.15 mM NaCl, 0.5 mM EDTA, 0.1% thimerosal, 0.2 mM PMSF, 10 mM EACA, 10 mM Cu_2SO_4, pH 7.4) for macroscopic evaluation of the damaged areas. For western blotting analysis, segments of intima media were homogenized in a buffer containing 10 mM Tris, 10 mM NaCl, 0.1 mM EDTA, 0.5% Triton X-100, 0.02% sodium azide and 2 mM PMSF with pH 7.5 (Bauer et al., 1998). All the animal protocols were approved by the institute's bioethical committee.

Total Hemoglobin was determined as cyanometahemoglobin (Drabkin and Austin, 1935). Free hemoglobin in plasma was measured by its absorbance at 415 nm using purified hemoglobin as standard. Cholesterol and triglycerides were determined by colorimetric methods (Biggs et al., 1975; Bowman and Wolf, 1951). The plasma total cholesterol exposure was calculated for each group (Bocan et al., 1993). Plasma was used to obtain lipoproteins by sequential ultracentrifugation (Havel et al., 1955). Anionic electrophoretic mobility was determined after non-denaturing electrophoresis on agarose gels (submarine mode) and staining with Oil Red O (Sigma Chemical, St. Louis, MO, USA) as previously described (Hallberg et al., 1994). Cholesterol content of lipoproteins fractionated by agarose electrophoresis was evaluated as described (Camejo et al., 1981).

For evaluation of atherosclerosis, aortas were cut open longitudinally for macroscopic documentation of intimal lesions. After removal of the adventitia, aortas were immersed in 0.1% Oil Red-O in 60% isopropanol for 30 min. The aortas were washed with 60% isopropanol three times for 5-min. Aortas were placed on plastic templates and luminal surfaces were photographed. Lesion, stained red, and no lesion areas were drawn on the photographic prints then cut and weighed. Lesion area was estimated as percentage of the intimal surface area affected by atherosclerotic lesion.

Expression of heme oxygenase in homogenates of thoracic aortic intima-media segments was evaluated by western blotting after SDS polyacrylamide gel electrophoresis essentially as described by Bauer et al. (Bauer et al., 1998).

Alterations Induced in Total and Free Hemoglobin

The groups treated with PHZ showed, as expected, a decrease in total hemoglobin of 10 to 20%. Interestingly, also the hypercholesterolemic group (MHC) had lower levels of total hemoglobin than the control group. Figure 6 shows that administration of PHZ, increased free hemoglobin in the PHZ-treated animals (IVH and MHC-IVH groups). The MHC group also showed a significant higher level of free Hb than the controls between weeks 2 and 6 of the study. In this regard, it is worthy to note during this same period, the MHC-IVH group showed a similar, although higher, peak of free hemoglobin in plasma. This suggests that hypercholesterolemia by itself increases the susceptibility to hemoglobin leakage from erythrocytes. No differences in feeding or drinking patterns were observed between the groups at any time along the protocol. Rabbits treated with PHZ alone did not show changes in their behavior or any other visible physical change.

Hypercholesterolemia

Control and IVH groups maintained cholesterol level without significant variation from the basal range. However, groups fed the 0.30% w/w cholesterol-

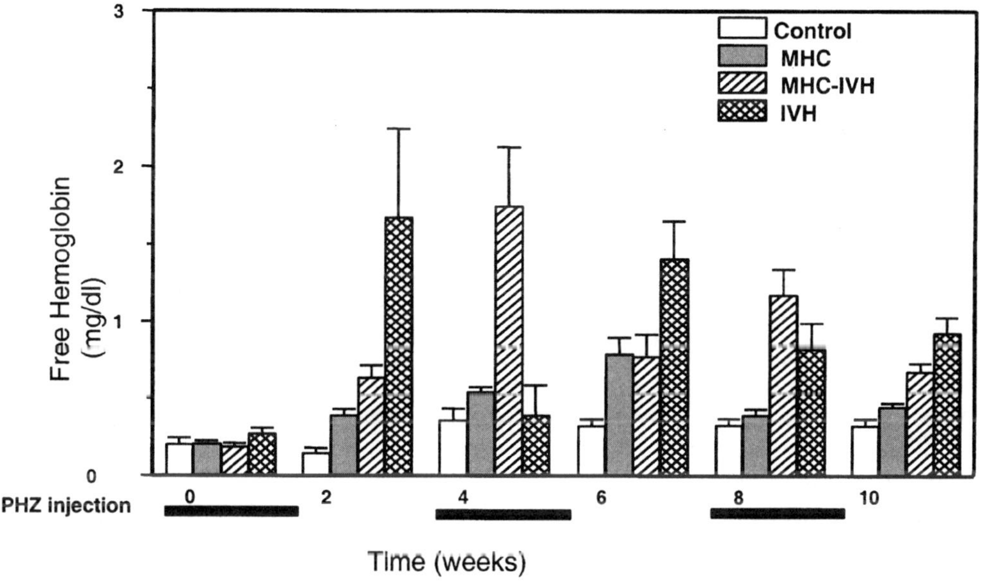

Figure 6. Free hemoglobin in plasma of the control, IVH, MHC and MHC-IVH groups (N = 10/group; mean ± s.d). Horizontal black bars indicate the time of the PHZ injection of the IVH and the MHC-IVH groups (reproduced with permission from Fernandez et al. *Atherosclerosis* **58**:103).

Table 1. Cholesterol levels in fractionated lipoproteins from agarose electrophoresis from control rabbits (C), rabbits with moderate hypercholesterolemia (MHC), rabbits with intravascular hemolysis (IVH) and with combined moderate hypercholesterolemia and intravascular hemolysis (MHC-IVH) (reproduced with permission from Fernandez et al. *Atherosclerosis* **58**:103)

	CONTROL	IVH	HC	HC-IVH
VLDL	12.56 ± 1.78	10.94 ± 2.29	793.43[a] ± 101.67	868.45[a] ± 86.06
LDL	15.04 ± 2.67	23.52 ± 4.65	Non-detected	Non-detected
HDL	25.84 ± 4.41	21.02 ± 1.44	82.20 ± 5.89	67.87 ± 6.96

The results are expressed as mean ± SEM.
[a]VLDL-cholesterol in the HC and HC-IVH groups corresponds to a β-VLDL, a lipoprotein class observed in plasma from cholesterol-fed rabbits.

supplemented diet showed a similar rapid increase in plasma cholesterol, reaching up to 600–800 mg/dl at four weeks. After the fourth week, cholesterol supplementation of the diet was changed to 0.25% w/w until the end of the experimental period, in order to maintain the moderate level of plasma cholesterol reached in the Previous 4 weeks.

Table 1 shows the cholesterol content of the main lipoprotein fractions separated by agarose electrophoresis. VLDL-cholesterol in the MHC and MHC-IVH groups corresponds to a β-VLDL, a lipoprotein class observed in plasma from cholesterol-fed rabbits (Camejo et al., 1973). VLDL- and HDL-cholesterol in the MHC and MHC-IVH groups showed no significant difference. Also no significant difference was found between the groups in their levels of plasma triglycerides.

Lesion Induction

The aortas of control and only IVH rabbits showed less than 1% of their area covered with lesions. However, those from animals with moderate hypercholesterolemia (MHC) induced with diet supplemented with 0.30–0.25% w/w cholesterol showed typical atherosclerotic lesions after ten weeks that covered 11 ± 7% of the luminal area. When intravascular hemolysis and moderate hypercholesterolemia were combined (MHC-IVH), a significant larger area of the aortic inner surface was involved than in the MHC, or the IVH group (27 ± 8%). These differences were significant at a p < 0.05 using ANOVA. Bocan et al. (Bocan et al., 1993) made a detailed study in cholesterol-fed rabbits of the correlation of lesion extent with plasma lipoproteins and found that a threshold for total cholesterol exposure of 31,800 mg × day/dl is needed for lesion development. These researchers also found that the percentage of thoracic aorta surface covered by lesions was better correlated to the levels of VLDL-cholesterol plus LDL-cholesterol (r = 0.9) than with total cholesterol exposure (r = 0.74). In our experiments, the animals in the MHC group and the MHC + IVH all reached exposure values between 32,500 and 40,000 mg × day/dl and the content of cholesterol in the beta lipoproteins (VLDL + LDL) was not different between the two groups (Table 1). Therefore we attributed the difference in lesion to factors associated with the intravascular hemolysis. The results presented seem to confirm the

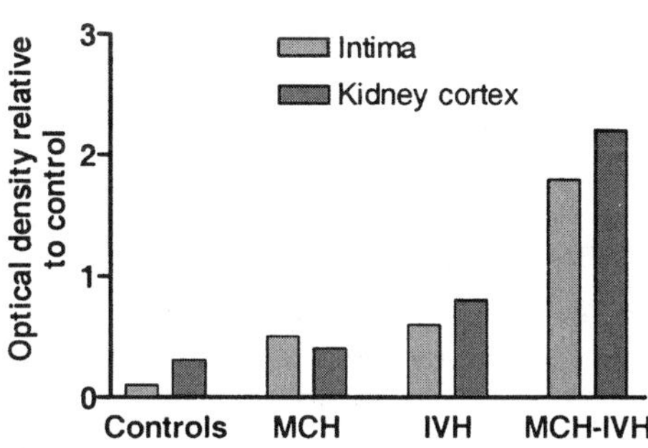

Figure 7. Densitometric evaluation of western blots used to measure Heme Oxigenase-1 in homogenates of aortic intima and kidney cortex from control rabbits (C), rabbits with moderate ercholesterolemia (MHC), rabbits with intravascular hemolysis (IVH) and with combined moderate hypercholesterolemia and intravascular hemolysis (MHC-IVH) (reproduced with permission from Fernandez et al. *Atherosclerosis* **58**:103).

hypothesis that combined intravascular hemolysis IVH) and moderate hypercholesterolemia (MHC) can cause more extensive lesions than MHC or IVH alone.

Induction of Heme Oxygenase I

In order to assess the capacity of the animals to respond to the oxidative stress induced by intravascular hemolysis we determined the level of expression of HO-1 protein in aortic intima media homogenates and in kidney cortex homogenates (Fig. 7).

Expression level was almost two-fold higher in the MHC-IVH group than in the IVH group in kidney cortex. In aortic intima-media, minimal expression of HO-1 protein was detected in the control group, while the MHC and IVH groups expressed a significant higher level of the HO-1 protein (Fig. 7).

Balla et al. have shown that heme as well as heme-transporting proteins (hemoglobin) can be incorporated into endothelial cells and induce oxidative cellular damage. At the same time, depending on the time of exposure, heme and heme-carrying proteins may induce a resistance to oxidative damage that has been defined as "desensitization" by the authors and that is believed to be caused by induction of heme oxygenase (Balla et al., 1991; Balla et al., 1993; Balla et al., 1995; Hayashi et al., 1999). Our results suggest that the degree of intravascular hemolysis achieved is sufficient to provoke a significant upregulation of heme oxygenase-1 protein in the intima and other organs. However this endogenous inductions seems not to be sufficient to eliminate the atherogenic stimuli that intravascular hemolysis add to moderate hypercholesterolemia.

5. CONCLUSIONS

Recent studies in apoE and LDL-receptor defective mice indicate that marked induction of heme oxygenase expression can reduce the atherosclerosis provoked in these models by extreme dyslipidemia (Ishikawa et al., 2001; Nakayama et al., 2001; Paul et al., 1999). The authors suggest that this may be caused by a decreased oxidation of apoB-lipoproteins induced by the augmented heme oxygenase-1 that degrades the oxidizing heme thus reducing foam cell formation caused by oxidized LDL. In the rabbit model, of phenylhydrazine-induced hemolysis, increased release of hemoglobin and probably heme was not sufficient to induce lesions. However when combined with moderate hypercholesterolemia it potentiated its atherogenicity in spite of provoking a marked upregulation of heme oxygenase-1. We have no explanation for the differences between the mice and rabbit models; especially since the dissimilar lipoprotein profiles and metabolism caused by the lack of apoE or the LDL receptor in mice allow no direct comparison with the rabbit model. The more extensive lesions observed in rabbits with hypercholesterolemia plus hemolysis may be a combination of increased apoB-lipoprotein deposition with augmented oxidation of the particles by heme, this last one an effect that has been found to stimulate LDL uptake by macrophages *in vitro* (Camejo et al., 1998a). We believe that the observation that cholesterol feeding by itself induced substantial hemoglobin release, probably contributing to the pro-oxidant status in plasma and in arteries, is worth exploring further.

Heme oxygenase-1 action on heme results in release of iron and carbon monoxide with opening of the tetrapyrrolic ring and production of biliverdin that is finally transformed in bilirubin. These two last compounds are potent antioxidants *in vitro* and they may add to the anti-atherogenic action of the enzyme (Balla et al., 1995; Siow et al., 1999; Stocker et al., 1987). Released heme iron is rapidly taken by ferritin, which is also induced in endothelial cells by heme. This suggests that induction of HO-1 in different cells of the vascular wall is an important pathway for protection against oxidative stress injury. However contrary to what has been described in apoE and LDL receptor defective mice, in the rabbit model described endogenous upregulation of HO-1 appears not to be sufficient to counteract the atherogenicity of the oxidative condition induced by intravascular hemolysis associated with moderate hypercholesterolemia. There are possible several causes for the discrepancy of the results from gene manipulated mice and cholesterol-fed rabbits. The most obvious one is that the susceptibility to heme-mediated modifications of lipoproteins from apoE and LDL receptor KO mice is lower than that of cholesterol fed rabbits and that increased expression of heme oxygenases is sufficient to prevent such alterations. Another possibility is that in rabbit's lesion development depends more on lipoprotein deposition than in lipoprotein oxidation its uptake by macrophages and foam cell generation. These unresolved issues indicate that further studies are required about the levels and structure of lipoproteins susceptible to heme-mediated modification and about the activities of heme oxygenase-1 required to counteract this oxidation in different animal models of the disease. Such studies are important because heme'oxygenase-1 can be up regulated by administration of small molecular weight substances that can diminish the pro-inflammatory vascular actions of heme (Ishikawa et al., 2001; Wagener et al., 2001). Development of substances with these characteristics and compatible with long term use may open new ways for anti-atherosclerotic therapy.

REFERENCES

Araujo, J., Romano, E., Brito, B., Parthe, V., Romano, M., Bracho, M., Montano, R., and Cardier, J., 1995, Iron overload augments the development of atherosclerotic lesions in rabbits, *Arterocler, Thromb, Vasc Biol* **15**:1172.

Balla, G., Jacob, H., Eaton, J., Belcher, J., and Vercellotti, G., 1991, Hemin: a possible physiological mediator of low density lipoprotein oxidation and endothelial injury, *Arterioscler Thromb* **11**:1700.

Balla, J., Jacob, H., Balla, G., Nath, K., Eaton, J., and Vercellotti, G., 1993, Endothelial-cell heme uptake from heme proteins: induction of sensitization and desensitization to oxidant damage, *Proc NaT Acad Sci USA* **90**:9285–9289.

Balla, J., K., N., G., B., Juckett, M., Jacob, H., and G., V., 1995, Endothelial cell heme oxygenase and ferritin induction in rat lung by hemoglobin in vivo, *Am J Physiol* **268**:L321.

Bauer, I., Wanner, G., Rensing, H., Alte, C., Miescher, E., Wolf, B., and Pannen, B., 1998, Expression patern of heme oxygenase isoenzymes 1 and 2 in normal and stress-exposed rat liver, *Hepathology* **27**:829.

Beltramini, M., Firey, P., Ricchelli, F., Rodgers, M., and Jori, G., 1987, Steady-state and time-resolved spectroscopic studies on the hematoporphyrin-lipoprotein complex, *Biochemistry* **26**:6852.

Biggs, H., Erikson, J., and Moorehead, W., 1975, A manual colorimetric assay of triglycerides in serum, *Clin Chem* **21**:437.

Bocan, T., Mueller, S., Mazur, M., Uhlendorf, P., and Brown, E.K., K., 1993, The relationship between the degree of dietary-induced hypercholesterolemia in the rabbit and atherosclerotic lesion formation, *Atherosclerosis* **102**:9.

Bowman, R. and Wolf, R., 1951, A rapid and specific ultramicro method for total serum cholesterol, *Clin Chem* **8**:302.

Camejo, G., Bosch, V., Arreaza, C., and Mendez, H., 1973, Early changes in plasma lipoprotein structure and biosynthesis in cholesterol-fed rabbits, *J Lipid Res* **14**:61.

Camejo, G., Cortez, M.M., López, C., and Mosquera, B., 1981, Photometric measurement of lipoprotein-cholesterol after agarose electrophoresis: comparison with single spin ultracentrifugal analysis, *Clin Chim Acta* **111**:239.

Camejo, G., Halberg, C., Manschick-Lundin, A., Camejo-Hurt, E., Rosengren, B., Olsson, H., Hansson, G.I., Forsberg, G.-B., and Ylhen, B., 1998a, Hemin binding and oxidation of lipoproteins in serum:mechanisms and effect on the interaction of LDL with human macrophages, *J Lipid Res* **39**:755.

Camejo, G., Hurt-Camejo, E., Wiklund, O., and Bondjers, G., 1998b, Association of lipoproteins with arterial proteoglycans:Pathological significance, *Atherosclerosis* **139**:205.

Collins, E. and Carew, T., 1997, Residence time of low-density lipoprotein in the normal and atherosclerothic rabbit aorta, *Circ Res* **80**:208.

Drabkin, D. and Austin, J., 1935, Spectrophotometric studies. III. Metahemoglobin, *J Biol Chem* **112**:67.

Fernandez, A., Lopez, F., Tablante, A., Romano, E., Hurt-Camejo, E., Camejo, G., and Apitz-Castro, R., 2001, Intravascular hemolysis increases atherogenicity of diet-induced hypercholesterolemia in rabbits in spite of heme oxygenase-1 gene and protein induction, *Atherosclerosis* **158**:103.

Hallberg, C., Håden, M., Bergstrom, M., Hansson, G., Pettersson, K., Westerlund, C., Bondkjers, G., Lindqvist-Östlund, A.-M., and Camejo, G., 1994, Lipoprotein fractionation in deuterium oxide gradients: a procedure for evaluation of antoioxidant binding susceptibility to oxidation, *J Lipid Res* **35**:1.

Havel, R., Eder, H., and Bragdon, J., 1955, Distribution chemical composition of ultracentrifugally separated lipoproteins in human serum, *J Clin Invest* **34**:1345.

Hayashi, S., Takamiya, R., Yamaguchi, T., Matsumoto, K., Tojo, S., Tamatani, T., Kitajima, M., Makino, N., Ishimura, Y., and Suematsu, M., 1999, Induction of heme oxygenase-1 supresses venular leukocyte adhesion elicited by oxidative stress: Role of bilirubin generated by the enzyme, *Circ Res* **85**:663.

Hurt-Camejo, E., Camejo, G., Peilot, H., Öörni, K., and Kovanen, P., 2001, Phospholipase A2 in vascular disease, *Circ Res* **89**:298.

Ishikawa, K., Sugawara, D., Wang, X.-P., Suzuki, K., Itabe, H., Maruyama, Y., and Lusis, A.J., 2001, Heme oxygenase-1 inhibits atherosclerotic lesion formation in LDL-receptor knockout mice, *Circ Res* **88**(5):506.

Jori, G. and Reddi, E., 1993, The role of lipoproteins in the delivery of tumor targeting photosensitizers, *Int f Biochem Cell Biol* **25**:1369.

Juan, S.-H., Lee, T.-S., Tseng, K.-W., Liou, J.-Y., Shyue, S.-K., Wu, K.K., and Chau, L.-Y., 2001, Adenovirus-mediated heme oxygenase-1 gene transfer Inhibits the development of atherosclerosis in apolipoprotein E-deficient mice, *Circulation* **104**(13):1519.

Kongshaug, M., 1992, Distribution of tetrapyrrole photosensitizers among human plasma proteins, *Intl J Biochem Cell Biol* **24**:1239.

Lee, T., Lee, F., Pang, J., and Chau, L., 1999, Erythrophagocytosis and iron deposition in atherosclerotic lesions, *Chin J Physiol* **42**:17.

Miller, Y., Altamentova, S., and Shaklai, N., 1997, Oxidation of low density lipoprotein by hemoglobin stems from a heme-initiatiated globin radical: antioxidant role of haptoglobin, *Biochemistry* **36**:12189.

Nakayama, M., Takahashi, K., Komaru, T., Fukuchi, M., Shioiri, H., Sato, K.-I., Kitamuro, T., Shirato, K., Yamaguchi, T., Suematsu, M., and Shibahara, S., 2001, Increased Expression of heme oxygenase-1 and bilirubin accumulation in foam cells of rabbit atherosclerotic lesions, *Arterioscler, Thromb Vasc Biol* **21**(8):1373.

Paul, A., Calleja, L., Villela, E., Martinez, R., Osada, J., and Joven, J., 1999, Reduced progression of atherosclerosis in apolipoprotein E-defficient mice with phenylhydrazine-induced anemia, *Atherosclerosis* **147**:61.

Praticò, D., Iuliano, L., Mauriello, A., Spagnoli, L., Lawson, J., MacLouf, J., Violi, F., and Fitzgerald, G., 1997, Localization of distint F_2.isoprostanes in human atherosclerotic lesions, *J Clin Invest* **100**:2028.

Redondo, P., Alvarez, A., Diez, C., Fernandez-Rojo, F., and Prieto, J., 1995, Physiological response to experimental induced anemia in rats: a comparative study, *Lab Animal Sc* **45**:578.

Ross, R., 1999, Atherosclerosis: an inflammatory disease, *New Engl J Med* **340**:115.

Scannapieco, G., Pauletto, P., Pagnan, A., Mattiello, A., Jori, G., and Dal Palu, C., 1988, retention of haematoporphyrin in the aorta of hypertensive rats: in vivo and in vitro studies, *Europ J Clin Invest* **18**:614.

Segrest, J.P., Jones, M.K., De Loof, H., and Dashti, N., 2001, Structure of apolipoprotein B-100 in low density lipoproteins, *J Lipid Res* **42**(9):1346.

Siow, R., Sato, H., and Mann, G., *Cardiovac. Res.* 1999, 1999, Hemeoxygenase-carbon monoxide signalling pathway in atherosclerosis: anti-atherogenic actions of bilirubin and carbon monoxide?, *Cardiovasc Res* **41**:385.

Smith, E.B. and Staples, E.M., 1980, Distribution of plasma proteins across the human aortic wall: Barrier functions and endothelium and internal elastic lamina, *Atherosclerosis* **37**:579.

Steinberg, D. and Witztum, J., 1999, Lipoproteins, lipoprotein oxidation and atherogenesis, in: *Molecular Basis of Cardiovascular Disease* (Breslow, J., Leiden, J., Rosenberg, R., Seidman, D., eds.), W. B. Sauders Co., Philadelphia, p. 458.

Stocker, R., Yamamoto, Y., McDonagh, A., Glazer, A., and Ames, B., 1987, Bilirubin is an antioxidant of possible physiological importance, *Science* **235**:1043.

Tribble, D., Krauss, R.M., Langsberg, M., Thiel, P.M., and van den Berg, J., 1995, Greater oxidative susceptibility of the surface monolayer in small dense LDL may contribute to differences in copper-induced oxidation among LDL subfractions, *J Lipid Res* **36**:662.

Tribble, D., van den Berg, J.M., Motchnik, P.A., Ames, B.N., Lewis, D.M., Chait, A., and Krauss, R.M., 1994, Oxidative susceptibility of low density lipoprotein subfractions is related to their ubiquinol-10 and α-tocopherol content, *Proc Natl Acad Sci USA* **91**:1183.

Wagener, F., Eggert, A., Boerman, O., Oyen, W., Verhofstad, A., Abraham, N., Adema, G., van Kooyk, Y., and Figdor, C., 2001, Heme is a potent inducer of inflammation in mice and is counteracted by heme oxygenase, *Blood* **98**:1802.

Williams, K.J. and Tabas, I., 1995, The response-to-retention hypothesis of early atherogenesis, *Arterioscler, Thromb, Vasc Biol* **15**:551.

Willis, D., Moore, A., Frederick, R., and Willoughby, D., 1996, Heme oxygenase: a novel target for the modulation of the inflammatory response, *Nature Med* **2**:87.

Yuan, X.M., Brunk, U.T., and Olsson, A.G., 1995, Effects of iron- and hemoglobin-loaded human monocyte-derived macrophages on oxidation and uptake of LDL, *Arterioscler, Thromb, Vasc Biol* **15**(9):1345.

25

ANTI-ATHEROGENIC PROPERTIES OF HEME OXYGENASE

Kazunobu Ishikawa

The First Department of Internal Medicine
Fukushima Medical University
1 Hikarigaoka, Fukushima-city, Fukushima
960-1295, Japan

1. INTRODUCTION

Recent advances in molecular biology and transgenic techniques have led us to better understand the mechanisms of atherosclerosis (Ross, 1999; Lusis, 2000). Oxidative modification of lipoproteins in vascular wall has been considered to trigger the initiation and development of atherosclerosis (Halliwell, 1994; Thomas, 2000; Tomasian et al., 2000). To inhibit these oxidative processes, exogenous antioxidants such as vitamin E and vitamin C or endogenous enzymes such as superoxide dismutase, catalase and glutathione peroxidases have been challenged in various experimental studies and mega-clinical trials. However, since clinical benefits elicited from these antioxidants have been inconsistent (Steinberg, 1995; Yusuf et al., 2000; Stephens et al., 1996), more potent and reliable antioxidants which reduce vascular events, are desired at present.

Heme oxygenase (HO) is a rate limiting enzyme of heme degradation and three isoforms have been identified (Tenhunen et al., 1969; Yoshida et al., 1978; Maines, 1997; Ishikawa et al., 1995; McCoubrey et al., 1997). Among these isoform, an inducible form of HO (HO-1) is transcriptionally induced by a variety of stresses including oxidized LDL, cytokines and growth factors in artery wall and inflammatory cells. Since these stresses have been suggested to involve in various cardiovascular disorders, potential roles of HO-1 in atherogenesis have recently been investigated. Our recent observations have suggested that HO-1 response functions as an intrinsic anti-atherogenic pathway through inhibiting lipid peroxidation and through affecting nitric oxide metabolism (Ishikawa et al., 2001A; Ishikawa et al., 2001B). HO-1 modulation in vascular wall will be a novel therapeutic target for cardiovascular disorders.

2. HEME OXYGENASE-1 INDUCTION BY OXIDIZED LIPOPROTEINS

A number of studies have shown that oxidized LDL takes a central role for the development of atherosclerosis (Ross, 1999; Lusis, 2000). We have shown that HO-1 was remarkably induced by oxidized LDL whereas HO-1 was hardly present in vascular endothelial cells or smooth muscle cells at physiological condition or when exposed to native LDL (Ishikawa et al., 1997). HO-1 induction in macrophages and renal tubular epithelial cells by oxidized LDL has also reported (Wang et al., 1998; Yamaguchi et al., 1993; Agarwal et al., 1996). The responsible components of oxidized LDL to induce HO-1 are oxidized arachidonic acid-containing phospholipid such as 1-palmitoyl-2-isoprostanoyl-sn-glycero-3-phosphoryl-choline (Ishikawa et al., 1997) and linoleyl hydroperoxide (Agarwal et al., 1998). These oxidized phospholipids are also potent to activate endothelial cells to bind monocytes. The most potent inducer of HO-1 seems to be 1-palmitoyl-2-isoprostanoyl-sn-glycero-3-phosphoryl-choline, which posseses an isoprostanoyl residue at the sn-2 position (Ishikawa et al., 1997). In contrast, HO-1 was not induced by lysophosphatidyl-choline (Wang et al., 1998).

In vivo HO-1 expression has demonstrated in the atherosclerotic lesions of autopsied patients, Apo E knockout mice and LDL-receptor knockout mice by immunohistochemistry and in situ hybridization (Wang et al., 1998; Ishikawa et al., 2001A). HO-1 was expressed in atherosclerotic lesions from early fatty streak formation to advanced complex lesions (Fig. 1). HO-1 was expressed in vascular endothelial cells and macrophages in early atherosclerotic lesion. In the advanced

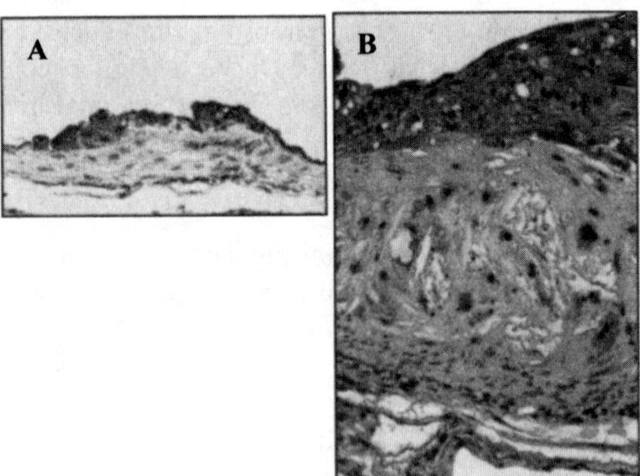

Figure 1. HO-1 expression in the atherosclerotic lesions of LDL-receptor knockout mice. (×200) (A) Early atherosclerotic lesion. (B) Advanced atherosclerotic lesion. HO-1 expressions are shown as brown color by immunoperoxidase technique.

atherosclerotic lesions, HO-1 was predominantly expressed in macrophages/foam cells, which reside in the necrotic core, and vascular smooth muscle cells. Recently, we also observed HO-1 expression in the atherosclerotic lesions of human coronary atherectomy specimens (K. Ishikawa, unpublished data) and in Watanabe heritable hyperlipidemic (WHHL) rabbits (Ishikawa et al., 2001B). Interestingly, HO-1 was co-localized with oxidized phospholipids as judged by specific antibodies against these molecules (Ishikawa et al., 2001A; Ishikawa et al., 2001B) in all of these atherosclerotic lesions. These observations may strongly support the idea that HO-1 was induced by oxidized phospholipids in vivo. Accumulation of bilirubin with HO-1 in atherosclerotic lesions of high fat-fed rabbits have been also reported (Nakayama et al., 2001). Thus, HO-1 induction seems to be general responses in atherosclerotic lesion developments.

3. POTENTIAL ANTI-ATHEROGENIC PROPERTIES OF HO-1

Since HO-1 appears to be expressed entire period of atherosclerotic lesion formation (Ishikawa et al., 2001A), HO-1 may be involved in a variety of atherosclerotic processes. We have first examined the effect of HO-1 expression on monocyte trans migration into artery wall, which is an essential event for fatty streak formation (Ishikawa et al., 1997). HO-1 induction significantly inhibited monocyte transmigration induced by oxidized LDL whereas HO inhibition had adverse effect. Inhibitory effect of HO-1 on monocyte transmigration seems to be conducted through the action of biliverdin and bilirubin because direct addition of these compounds had inhibitory effects on monocyte transmigration. As these compounds have been shown to scavenge reactive oxygen spieces and to inhibit lipid peroxidation in vitro (Stocker et al., 1987; Neuzil et al., 1994; Wu et al., 1994; Hulea et al., 1995), these properties might have reduced the inflammatory responses of vascular wall cells against oxidized LDL. Indeed, inhibitory effects of HO-1 through the action of bilirubin on leukocyte adhesion to activated vascular endothelium by inflammatory stresses have also reported (Hayashi et al., 1999).

Next, we examined whether HO affects the development of atherosclerotic lesion in vivo using several experimental atherosclerotic models. Since we observed that HO-1 was abundantly expressed in the atherosclerotic lesions of LDL-receptor knockout mice fed high fat diets, we modulated HO activities in these animals during high fat diet trials (Ishikawa et al., 2001A). HO-1 induction and inhibition were achieved by hemin and Sn-protoporphyrin IX treatment, respectively. HO overexpression resulted in the reduction of atherosclerotic lesion formation while HO inhibition had adverse effect. In addition, we have recently observed that HO-1 also exhibits anti-atherogenic properties in WHHL rabbits model (Ishikawa et al., 2001B). Although probucol, a potent antioxidant, exhibited different effects on atherogenesis in murine and rabbit atherosclerotic models (Kita et al., 1987; Zhang et al., 1997), HO inhibition resulted in the promotion of atherosclerotic formation in both species.

One of the most possible mechanisms of HO against atherogenesis seems to be inhibitory effects on lipid peroxidation. Plasma lipid hydroperoxide level was significantly lower in the LDL-receptor knockout mice and WHHL rabbits treated by a HO-1 inducer, while opposing effects were observed by a HO inhibitor (Ishikawa et al., 2001A; Ishikawa et al., 2001B). In WHHL rabbits, HO inhibition resulted in the accumulation of lipid peroxidation in liver and aorta. Biliverdin and bilirubin produced from HO reaction may be strong candidates for inhibiting lipid peroxidation in these models, however, additional studies must be performed to support this idea. Another mechanisms of HO against atherogenesis seem to be the effects on nitric oxide (NO) pathway. Impaired NO synthesis and availability have been suggested in the atherosclerotic arteries under hypercholesteremic conditions (Stroes et al., 1995). We examined the effect of HO modulation on NO pathway by measuring plasma NOx (nitrite and nitrate) concentration after high fat diet feeding in LDL-receptor knockout mice. HO modulation was inversely correlated with plasma NOx levels, which were significantly decreased after high fat diet feeding (Ishikawa et al., 2001A). At present, the precise mechanism how HO affect NO pathway, however, these results indicate that anti-atherogenic properties of HO under hypercholesteremia may be conducted through the influences on NO pathway. Other than these two mechanisms of HO-1 against atherogenesis, inhibitory actions of CO, an another product of HO reaction, on cell cycle genes in vascular smooth muscle cells (Duckers et al., 2001) and on platelet aggregations (Wagner et al., 1997) have been reported. These effects may also influence on the development of atherosclerotic lesions.

4. THE FUNCTIONS OF HO REACTION PRODUCTS IN ATHEROGENESIS

Bilirubin

Although biliverdin and bilirubin have been considered to be waste products that form bile pigments. Stocker and his colleagues reported that both free and albumin-bound forms of bilirubin could prevent oxidation of linoleic acid by peroxy radicals in vitro (Stocker et al., 1987). In terms of atherosclerosis, bilirubin has shown to inhibit oxidation of LDL (Neuzil et al., 1994; Wu et al., 1994; Hulea et al., 1995). In addition, bilirubin has also shown to have inhibitory effect on protein kinase C activity, which leads to the inactivation of various proatherogenic factors (Amit et al., 1993). A recent population study reported that higher plasma bilirubin concentration reduces the risk of coronary heart disease (Schwertner et al., 1994; Hopkins et al., 1996). The molecular mechanism how bilirubin reacts with oxygen radical species has not completely understood although the presence of its metabolites biotripyrroles has been reported (Yamaguchi et al., 1995).

Carbon Monoxide

Carbon monoxide (CO), a gaseous molecule derived from HO reaction in vivo, has suggested exhibiting properties similar to NO through enhancing intracellular cGMP level (Stone et al., 1994; Sharma et al., 1999) though the potency of vasodilatory action is still controversial. This argument is based on the differences between

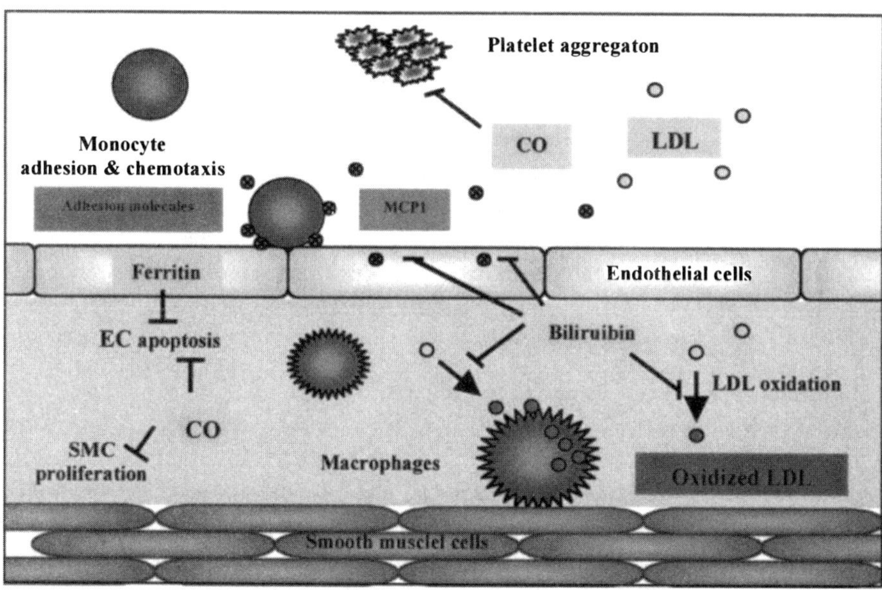

Figure 2. Proposed anti-atherogenic properties of HO pathway. CO, carbon monoxide; LDL, low density lopoprotein; SMC, smooth muscle cells; MCP1, monocyte chemotactic protein 1.

two gaseous molecules in the potency producing cGMP through the conformational change of soluble guanylate cyclase (Wang et al., 1997). NO is potent to increase cGMP production about 130-fold whereas CO only about 4.4-fold. Nevertheless, since CO is much chemically stable than NO and there apprears to be no enzymatic pathway catalyzing CO in vivo, the biological availabilities of NO and CO may be different. The biological effects of CO on cGMP pathway need to be further examined.

Another direct action of CO on Ca-dependent potassium channels in vascular smooth muscle cells, which also results in vascular relaxation through Ca desensitization, has also reported (Wang et al., 1997). In atherosclerotic arteries, hypoxic conditions under insufficient blood circulation of vasa vasorum may arise. Morita and his colleagues have shown that HO-1 is induced in vascular smooth muscle cells in such hypoxic condition and that resultant CO increase inhibits unfavorable proliferation of vascular smooth muscle cells (Morita et al., 1997). Zakhary and his colleagues have reported that CO from constitutive form of HO in porcine pulmonary artery can function as a vasodilator similar to NO (Zakhary et al., 1997). Suematsu and his colleagues have reported that CO is a predominant vasodilator in hepatic microvascular circulation (Suematsu et al., 2000). In addition, several observations have reported that CO is potent to relieve acute hypertensive state in rats (Motterlini et al., 1998; Seki et al., 1997). Other than blood vessels, effects on sphincter muscles (Battish et al., 2000) and carotid body (Prabhakar, 1998) have reported. Thus, CO produced by HO seems to be involved in various pathophysiological conditions. Recently, involvement of CO on mitogen activated protein kinase pathway has been reported (Otterbein et al., 2000).

Iron-Ferritin

Iron produced from HO reaction may promote Fenton reaction, which produces highly reactive oxygen species (Tyrrell, 1999). Although HO reaction produces such deteriorative free iron within cells, HO have been reported to be cytoprotective under various pathophysiological conditions. This disagreement may be partly explained by ferritin induction, which scavenges intracellular free iron. Enhancement of intracellular ferritin level through HO-1 induction in endothelial cells has bben reported to reduce cytotoxic effects against hydrogen peroxide and hemoglobin (Balla et al., 1992; Oberle et al., 1999). As ferritin is expressed in human atherosclerotic lesion (Pang et al., 1996), its role in atherogenesis should be explored.

5. SUMMARY

HO-1 expression during atherosclerotic lesion formation seems to be adaptive response or defense rather than preventive event. This idea may be supported by the findings that HO-1 was highly induced in proatherogenic C57BL/6J mice whereas HO-1 was scarcely induced in athero-resistant C3H/HeJ mice (Liao et al., 1993). A recent study using bone marrow transplantation showed the importance of the difference in inflammatory responses of endothelial cells such as HO-1 induction to oxidized LDL (Shi et al., 2000). As homozygous HO-1 knockout mice with proatherogenic backgrounds are poorly bred, in vivo studies using HO-1 knockout mice examining the effect on atherosclerosis seems to be difficult at present and new animal models for HO research should be established.

HO-1 seems to contribute as an intrinsic defense system against pro-oxidative stresses in vascular wall. In fact, HO-1 gene responds to most of the established atherogenic risk factors such as oxidized lipoproteins (Ishikawa et al., 1997; Wang et al., 1998; Agarwal et al., 1996), hypertension (Motterlini et al., 1998; Seki et al., 1997), hyperglycemia (Yan et al., 1994) and hypoxia (Morita et al., 1995; Motterlini et al., 2000). Since our recent studies have shown that HO-1 responses significantly reduce the promotion of atherosclerotic formation, HO modulation may lead to a novel target for gene therapies preventing atherosclerotic diseases. However, the development of therapeutic strategies modulating HO pathway in vascular wall should be considered for cellular heme metabolism because heme, the substrate of HO, also functions as a prosthetic group in various other enzymes such as cytochrome P450 and NO synthase,

REFERENCES

Agarwal, A., Balla, J., Balla, G., Croatt, A.J., Vercellotti, G.M., and Nath, K.A., 1996. Renal tubular epithelial cells mimic endothelial cells upon exposure to oxidized LDL. *Am J Physiol* **271** (4 Pt 2), F814.

Agarwal, A., Shiraishi, F., Visner, G.A., and Nick, H.S., 1998. Linoleyl hydroperoxide transcriptionally upregulates heme oxygenase-1 gene expression in human renal epithelial and aortic endothelial cells. *J Am Soc Nephrol* **9** (11), 1990.

Amit, Y. and Boneh, A., 1993. Bilirubin inhibits protein kinase C activity and protein kinase C-mediated phosphorylation of endogenous substrates in human skin fibroblasts. *Clin Chim Acta* **223** (1–2), 103.

Balla, G., Jacob, H.S., Balla, J., Rosenberg, M., Nath, K., Apple, F., Eaton, J.W., and Vercellotti, G.M., 1992. Ferritin: A cytoprotective antioxidant stratagem of endothelium. *J Biol Chem* **267** (25), 18148.

Battish, R., Cao, G.Y., Lynn, R.B., Chakder, S., and Rattan, S., 2000. Heme oxygenase-2 distribution in anorectum: colocalization with neuronal nitric oxide synthase. *Am J Physiol* **278** (1), G148.

Duckers, H.J., Boehm, M., True, A.L., Yet, S.F., San, H., Park, J.L., Webb, R.C., Lee, M.E., Nabel, G.J., and Nabel, E.G., 2001. Heme oxygenase-1 protects against vascular constriction and proliferation. *Nat Med* **7** (6), 693.

Halliwell, B., 1994. Free radicals, antioxidants, and human disease: curiosity, cause, or consequence? *Lancet* **344** (8924), 721.

Hayashi, S., Takamiya, R., Yamaguchi, T., Matsumoto, K., Tojo, S.J., Tamatani, T., Kitajima, M., Makino. N., Ishimura, Y., and Suematsu, M., 1999. Induction of heme oxygenase-1 suppresses venular leukocyte adhesion elicited by oxidative stress: role of bilirubin generated by the enzyme. *Circ Res* **85** (8), 663.

Hopkins, P.N., Wu, L.L., Hunt, S.C., James, B.C., Vincent, G.M., and Williams, R.R., 1996. Higher serum bilirubin is associated with decreased risk for early familial coronary artery disease. *Arterioscler Thromb Vasc Biol* **16** (2), 250.

Hulea, S.A., Wasowicz, E., and Kummerow, F.A., 1995. Inhibition of metal-catalyzed oxidation of low-density lipoprotein by free and albumin-bound bilirubin. *Biochim Biophys Acta* **1259** (1), 29.

Ishikawa, K., Takeuchi, N., Takahashi, S., Matera, K.M., Sato, M., Shibahara, S., Rousseau, D.L., Ikeda-Saito, M., and Yoshida, T., 1995. Heme oxygenase-2: Properties of the heme complex of the purified tryptic fragment of recombinant human heme oxygenase-2. *J Biol Chem* **270** (11), 6345.

Ishikawa, K., Navab, M., Leitinger, N., Fogelman, A.M., and Lusis, A.J., 1997. Induction of heme oxygenase-1 inhibits the monocyte transmigration induced by mildly oxidized LDL. *J Clin Invest* **100** (5) 1209.

Ishikawa, K., Sugawara, D., Wang, W.-P., Suzuki, K., Itabe, H., Maruyama, Y., and Lusis, A.J., 2001. Heme oxygenase-1 inhibits atherosclerotic lesion formation in LDL receptor knockout mice. *Circ Res* **88** (5), 506.

Ishikawa, K., Sugawara, D., Goto, J., Watanabe, Y., Kawamura, K., Shiomi, M., Itabe, H., and Maruyama, Y., 2001. Heme oxygenase-1 inhibits atherogenesis in Watanabe heritable hyperlipidemic rabbits. *Circulation* **104** (15), 1831.

Kita, T., Nagano, Y., Yokode, M., Ishii, K., Kume, N., Ooshima, A., Yoshida, H., and Kawai, C., 1987. Probucol prevents the progression of atherosclerosis in Watanabe heritable hyperlipidemic rabbit, an animal model for familial hypercholesterolemia. *Proc Natl Acad Sci USA* **84** (16), 5928.

Liao, F., Andalibi, A., deBeer, F.C., Fogelman, A.M., and Lusis, A.J., 1993. Genetic control of inflammatory gene induction and NF-kappa B-like transcription factor activation in response to an atherogenic diet in mice. *J Clin Invest* **91** (6), 2572.

Lusis, A.J., 2000. Atherosclerosis. *Nature* **407** (6801), 233.

Maines, M.D., 1997. The heme oxygenase system: a regulator of second messenger gases. *Annu Rev Pharmacol Toxicol* **37**, 517.

McCoubrey, W.K., Huang, T.J., and Maines, M.D., 1997. Isolation and characterization of a cDNA from the rat brain that encodes hemoprotein heme oxygenase-3. *Eur J Biochem* **247** (2), 725.

Morita, T., Perrella, M.A., Lee, M., and Kourembanas, S., 1995. Smooth muscle cell-derived carbon monoxide is a regulator of vascular cGMP. *Proc Natl Acad Sci USA* **92** (5), 1475.

Morita, T., Mitsialis, S.A., Koike, H., Liu, Y., and Kourembanas, S., 1997. Carbon monoxide controls the proliferation of hypoxic vascular smooth muscle cells. *J Biol Chem* **272** (52), 32804.

Motterlini, R., Gonzales, A., Foresti, R., Clark, J.E., Green, C.J., and Winslow, R.M., 1998. Heme oxygenase-1-derived carbon monoxide contributes to the suppression of acute hypertensive responses in vivo. *Circ Res* **83** (5), 568.

Motterlini, R., Foresti, R., Bassi, R., Calabrese, V., Clark, J.E., and Green, C.J., 2000. Endothelial heme oxygenase-1 induction by hypoxia. Modulation by inducible nitric-oxide synthase and S-nitrosothiols. *J Biol Chem* **275** (18), 13613.

Nakayama, M., Takahashi, K., Komaru, T., Fukuchi, M., Shioiri, H., Sato, Ki., Kitamuro, T., Shirato, K., Yamaguchi, T., Suematsu, M., and Shibahara, S., 2001. Increased expression of heme oxygenase-1 and bilirubin accumulation in foam cells of rabbit atherosclerotic lesions. *Arterioscler Thromb Vasc Biol* **21** (8), 1373.

Neuzil, J. and Stocker, R., 1994. Free and albumin-bound bilirubin are efficient co-antioxidants for alpha-tocopherol, inhibiting plasma and low density lipoprotein lipid peroxidation. J Biol Chem **269** (24), 16712.

Oberle, S., Schwartz, P., Abate, A., and Schroder, H., 1999. The antioxidant defense protein ferritin is a novel and specific target for pentaerithrityl tetranitrate in endothelial cells. *Biochem Biophys Res Commun* **261** (1), 28.

Otterbein, L.E., Bach, F.H., Alam, J., Soares, M., Tao Lu, H., Wysk, M., Davis, R.J., Flavell, R.A., and Choi, A.M.K., 2000. Carbon monoxide has anti-inflammatory effects involving the mitogen-activated protein kinase pathway. *Nat Med* **6** (4), 422.

Pang, J.H., Jiang, M.J., Chen, Y.L., Wang, F.W., Wang, D.L., Chu, S.H., and Chau, L.Y., 1996. Increased ferritin gene expression in atherosclerotic lesions. *J Clin Invest* **97** (10), 2204.

Prabhakar, N.R., 1998. Endogenous carbon monoxide in control of respiration. *Respir Physiol* **114** (1), 57.

Ross, R., 1999. Atherosclerosis—an inflammatory disease. *N Engl J Med* **340** (2), 115.

Schwertner, H.A., Jackson, W.G., and Tolan, G., 1994. Association of low serum concentration of bilirubin with increased risk of coronary artery disease. *Clin Chem* **40** (1), 18.

Seki, T., Naruse, M., Naruse, K., Yoshimoto, T., Tanabe, A., Tsuchiya, K., Hirose, S., Imaki, T., Nihei, H., and Demura, H., 1997. Roles of heme oxygenase/carbon monoxide system in genetically hypertensive rats. *Biochem Biophys Res Commun* **241** (2), 574.

Sharma, V.S. and Magde, D., 1999. Activation of soluble guanylate cyclase by carbon monoxide and nitric oxide: a mechanistic model. *Methods* **19** (4), 494.

Shi, W., Wang, N.J., Shih, D.M., Sun, V.Z., Wang, X., and Lusis, A.J., 2000. Determinants of atherosclerosis susceptibility in the C3H and C57BL/6 mouse model: evidence for involvement of endothelial cells but not blood cells or cholesterol metabolism. *Circ Res* **86** (10), 1078.

Steinberg, D., 1995. Clinical trials of antioxidants in atherosclerosis: are we doing the right thing? *Lancet* **346** (8966), 36.

Stephens, N.G., Parsons, A., Schofield, P.M., Kelly, F., Cheeseman, K., and Mitchinson, M.J., 1996. Randomised controlled trial of vitamin E in patients with coronary disease. *Lancet* **347** (9004), 781.

Stocker, R., Yamamoto, Y., McDonagh, A.F., Glazer, A.N., and Ames, B.N., 1987. Bilirubin is an antioxidant of possible physiological importance. *Science* **235** (4792), 1043.

Stone, J.R. and Marletta, M.A., 1994. Soluble guanylate cyclase from bovine lung: Activation of nitric oxide and carbon monoxide and spectral characterization of the ferrous and ferric states. *Biochemistry* **33** (18), 5636.

Stroes, E.S., Koomans, H.A., de Bruin, T.W., and Rabelink, T.J., 1995. Vascular function in the forearm of hypercholesterolaemic patients off and on lipid-lowering medication. *Lancet* **346** (8973), 467.

Suematsu, M. and Ishimura, Y., 2000. The heme oxygenase-carbon monoxide system: a regulator of hepatobiliary function. *Hepatology* **31** (1), 3.

Tenhunen, R., Marver, H.S., and Schmid, R., 1969. Microsomal heme oxygenase. Characterization of the enzyme. *J Biol Chem* **244** (23), 6388.

Thomas, M.J., 2000. Physiological aspects of low-density lipoprotein oxidation. Current Opin Lipid **11** (3), 297.

Tomasian, D., Keaney, J.F., and Vita, J.A., 2000. Antioxidants and the bioactivity of endothelium-derived nitric oxide. *Cardiovasc Res* 47(3) 426.

Tyrrell, R., 1999. Redox regulation and oxidant activation of heme oxygenase-1. *Free Rad Res* **31** (4), 335.

Wagner, C.T., Durante, W., Christodoulides, N., Hellums, J.D., and Schafer, A.I., 1997. Hemodynamic forces induce the expression of heme oxygenase in cultured vascular smooth muscle cells. *J Clin Invest* **100** (3), 589.

Wang, L.J., Lee, T.S., Lee, F.Y., Pai, R.C., and Chau, L.Y., 1998. Expression of heme oxygenase-1 in atherosclerotic lesions. *Am J Pathol* **152** (3), 711.

Wang, R. and Wu, L., 1997. The chemical modification of KCa channels by carbon monoxide in vascular smooth muscle cells. *J Biol Chem* **272** (13), 8222.

Wu, T.W., Fung, K.P., and Yang, C.C., 1994. Unconjugated bilirubin inhibits the oxidation of human low density lipoprotein better than trolox. *Life Sciences* **54** (25), 477.

Yamaguchi, M., Sato, H., and Bannai, S., 1993. Induction of stress proteins in mouse peritoneal macrophages by oxidized low-density lipoprotein. *Biochem Biophys Res Commun* **193** (3), 1198.

Yamaguchi, T., Horio, F., Hashizume, T., Tanaka, M., Ikeda, S., Kakinuma, A., and Nakajima, H., 1995. Bilirubin is oxidized in rats treated with endotoxin and acts as a physiological antioxidant synergistically with ascorbic acid in vivo. *Biochem Biophys Res Commun* **214** (1), 11.

Yan, S.D., Schmidt, A.M., Anderson, G.M., Zhang, J., Brett, J., Zou, Y.S., Pinsky, D., and Stern, D., 1994. Enhanced cellular oxidant stress by the interaction of advanced glycation end products with their receptors/binding proteins. *J Biol Chem* **269** (13), 9889.

Yoshida, T. and Kikuchi, G., 1978. Purification and properties of heme oxygenase from pig spleen microsomes. *J Biol Chem* **253** (12), 4224.

Yusuf, S., Dagenais, G., Pogue, J., Bosch, J., and Sleight, P., 2000. Vitamin E supplementation and cardiovascular events in high-risk patients. *New Eng J Med* **342** (3), 154.

Zakhary, R., Poss, K.D., Jaffrey, S.R., Ferris, C.D., Tonegawa, S., and Snyder, S.H., 1997. Targeted gene deletion of heme oxygenase 2 reveals neural role for carbon monoxide. *Proc Natl Acad Sci USA* **94** (26), 14848.

Zhang, S.H., Reddick, R.L., Addievich, E., Surles, L.K., Jones, R.G., Reynolds, J.B., Quarfordt, S.H., and Maeda, N., 1997. Paradoxical enhancement of atherosclerosis by probucol treatment in apolipoprotein E-deficient mice. *J Clin Invest* **99** (12), 2858.

26

HEME OXYGENASE AND THE NOVEL TUMOUR-SPECIFIC ANTI-VASCULAR COMPOUND COMBRETASTATIN A4-PHOSPHATE

Amel F. Khelifi[a], Vivien E. Prise[a], Roberta Foresti[b], James E. Clark[b], Chryso Kanthou[a], Roberto Motterlini[b], and Gillian M. Tozer[a]

[a]Gray Cancer Institute, Mount Vernon Hospital
Northwood, Middelesex
HA6 2JR, UK
[b]Department of Surgical Research
Northwick Park Institute for Medical Research
Harrow, Middlesex HA1 3UJ, UK

1. INTRODUCTION

In recent years there has been increasing interest in exploiting the tumour vasculature as a target for cancer therapy. It is now well established that the growth and expansion of malignant tumours is critically dependent on the establishment of a vascular network assuring nutritive blood flow (Folkman, 1990). However, a number of solid tumours are now known to be characterised by an inadequate vascularisation consisting of a temporally and spatially heterogeneous blood supply (Vaupel et al., 1989). Vascular targeting has received increasing interest in the past few years. This strategy aims at targeting the established tumour vasculature causing direct shut down in tumour blood flow, leading to secondary tumour cell death (Chaplin et al., 1998). Combretastatin A4-phosphate (CA-4-P) is the lead compound of a number of tubulin binding agents which cause selective vascular shut down in experimental tumours. It entered clinical trial in 1998.

Currently, there is still a need to understand both the mechanisms by which CA-4-P causes selective tumour vascular injury as well as the processes which lead from the initial blood flow shut down to tumour cell death. With this regard, we are investigating the role of the heme oxygenase (HO) system. The potential cytoprotective roles of this enzyme system in vascular injury have been addressed and reviewed in a number of recent papers (Aizawa et al., 1999; Togane et al., 2000; Duckers et al., 2001; Durante and Schafer, 1998; Schwartz, 2001). In our laboratory, we have examined the role of HO in the vasculature of tumours with regard to both the role of this enzyme in the control of tumour blood flow and its potential protective effects against the vascular injury mediated by the novel anti-vascular agent, CA-4-P.

In this chapter, we will report data from our laboratory on the effects of CA-4-P on HO in both a tumour model and normal tissues. We will review the potential roles of HO in tumour pathophysiology and cancer therapy and the potential protective roles of this enzyme in vascular injury and inflammation. Finally, both the anti-vascular effects attributed to CA-4-P as well as the effects of this compound on the HO enzyme system will be described.

2. HEME OXYGENASE IN TUMOURS

In recent years, a number of studies have addressed the question of what might be the role of HO expression/activity in tumours. In human brain tumour specimens, HO-1 mRNA was found to be expressed at higher levels than in normal brain tissue (Hara et al., 1996). Furthermore, increased HO-1 immunoreactive staining was observed in benign prostatic hyperplasia and malignant prostate tissue (Maines and Abrahamsson, 1996). With regard to animal tumour models, there is evidence for HO-1 expression from an experimental solid tumour model (AH136B hepatoma) in rats (Doi et al., 1999). We also found high HO activity in subcutaneous transplants of the P22 rat carcinosarcoma, which was of the same order as that found in normal rat liver (Tozer et al., 1998). *In vitro*, studies using tumour spheroids of the A431 squamous carcinoma cell line have shown that HO expression and activity is enhanced in this three dimensional tumour model as compared to monolayer cultures (Murphy et al., 1993).

In the light of these findings other studies have made an attempt to examine the actual role of HO in cancer progression. Two recent clinical studies have linked HO-1 expression with angiogenesis in human gliomas and human vertical growth melanomas (Nishie et al., 1999; Torisu-Itakura et al., 2000). Moreover, direct evidence showing the relationship between HO-1 expression and new blood vessel formation can be found from *in vitro* studies investigating the effects of gene transfer of human HO into endothelial cells (Deramaudt et al., 1998). This report provides direct evidence that over-expression of the HO-1 gene in endothelial cells enhances angiogenesis i.e. induces an increase in the formation of both branches and anastomosing capillary-like cords in an *in vitro* model of angiogenesis. At present the mechanism involving HO-1 in angiogenesis remains unclear. However, studies have shown that the HO-1 gene is up regulated in response to mediators (e.g. tumour necrosis factor-α and interleukin 1-α) which have also been shown to up-regulate known angiogenic stimulating factors, such as vascular endothelial growth factor (VEGF) (Terry et al., 1998; Torisu et al., 2000). Moreover, binding sites for several transcription factors (e.g. AP-1, AP-2 and NF-κB) on the VEGF gene promoter can also be found on the HO-

1 gene promoter (Tischer et al., 1991; Gille et al., 1997; Lavrovsky et al., 1994; Alam and Den, 1992). This therefore suggests possible cross talk between these different proteins in promoting/mediating angiogenesis. Further evidence for a putative role of HO-1 in tumour progression relates to its possible involvement in promoting cell proliferation and tumour cell growth. In this regard, zinc protoporphyrin IX, a HO inhibitor, was shown to significantly suppress tumour growth when injected intra-arterially to a solid tumour (Doi et al., 1999). To further support this idea, a recent study has demonstrated the presence of regulatory sequences for the transcription factors ETS-1, FLI-1 or ERG in the promoter region of the human HO-1 gene (Dera-maudt et al., 1999). These transcription factors belong to the Ets-family of proteins, which have been shown to be widely expressed during development and cell prolifer-ation. Moreover, the evidence for the role of this family of transcription factors in increasing angiogenesis (Remy et al., 1996; Deramaudt et al., 1999) further supports a possible involvement of HO-1 in mediating/promoting this phenomenon. In addi-tion, it has been proposed that the iron released as a result of heme degradation by HO-1 could play an important role in cellular proliferation (Voest et al., 1993).

More recently, it has been suggested that HO-1 could be used as a predictive factor in cancer therapy. A recent clinicopathological and immunohistochemical study has examined the role of HO-1 in predicting the response to radiotherapy in human oesophageal squamous cell carcinomas (Yokoyama et al., 2001). Although the study was carried out on only a small number of specimens (from 13 oesophageal squa-mous cell cancer patients), a strong correlation was found between the presence of HO-1 in cancer tissues and increased radiation response. Both the former and the latter were independent from tumour size, stage or histologic grade. It, however, remains unclear as to why tumours with high HO-1 expression are better responders to radiotherapy than tumours with a lower HO-1 expression. It is possible that tumours expressing HO-1 have higher levels of carbon monoxide (CO). Since CO has been shown to possess vasodilatory properties (Suematsu et al., 1995; Wang et al., 1997; Wakabayashi et al, 1999) it is possible that these tumours are better oxygenated than tumours with lower HO activity and are therefore better responders to radio-therapy (Gray et al., 1953).

Collectively these studies point to the importance of the HO system both in cancer progression and therapy and clearly further studies are warranted in order to fully elucidate the role of this enzyme system in cancer.

3. ANTI-VASCULAR EFFECTS OF COMBRETASTATIN A4-PHOSPHATE

CA-4-P belongs to the family of combretastatins, which are compounds that have been isolated from the South African tree *Combretum caffrum* (Pettit et al., 1987). The active drug CA-4 is structurally similar to the tubulin binding agent colchicine (Fig. 1a, b). It has also been shown to bind to tubulin at or near the colchicine binding site causing disruption of microtubule function leading to destabilisation of the cell cytoskeleton (Pettit et al., 1989; Woods et al., 1995). CA-4-P pro-drug (Fig. 1c) is a

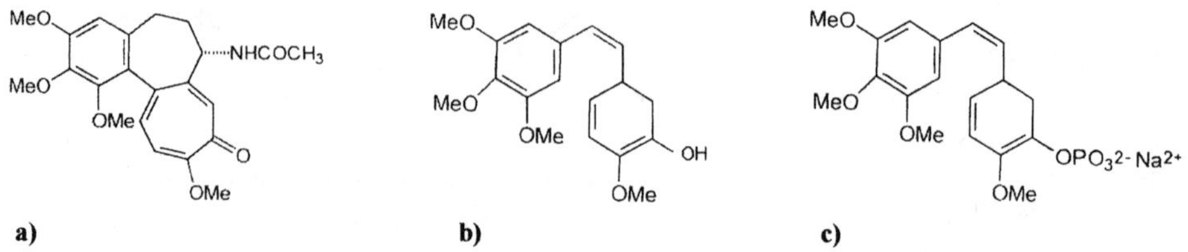

Figure 1. Structures of colchicine (a), CA-4 (b) and CA-4-P (c).

more soluble derivative, which is cleaved *in vivo* to the active form CA-4 by endogenous non-specific phosphatases.

CA-4-P has been shown to cause vascular collapse in a range of transplanted and spontaneous murine tumours as well as xenografted human tumours at relatively non-toxic doses (Dark et al., 1997; Horsman et al., 1998; Chaplin et al., 1999). Its selectivity for decreasing blood flow in the tumour as compared to a range of normal tissues was demonstrated in a rat tumour model (Tozer et al., 1999). Histological studies have shown extensive haemorrhagic necrosis resulting from vascular damage with only a small rim of viable tumour remaining in the periphery (Dark et al., 1997, Lingyun et al., 1998).

Initial *in vitro* studies have demonstrated that short-term drug exposure can lead to profound long-term antiproliferative/cytotoxic and apoptotic effects against proliferating but not quiescent endothelial cells (Dark et al., 1997; Iyer et al., 1998). However, these effects cannot explain the rapid tumour blood flow shut down that occurs in tumours *in vivo* (Tozer et al., 2001). CA-4-P causes a rapid increase in permeability of an endothelial cell monolayer to high molecular weight dextrans (Kanthou and Tozer, 2000) and an increase in vascular permeability of tumour blood vessels *in vivo*, within minutes of drug exposure (Tozer et al., 2001). This increase in vascular permeability is a likely trigger for vascular shut down via development of tissue oedema, followed by an increase in blood viscosity and/or an increase in interstitial fluid pressure. These effects are probably enhanced by direct effects on tumour vascular resistance arising from changes in endothelial cell shape. These effects have been reported for endothelial cells *in vitro* (Galbraith et al., 2001). These vascular effects of CA-4-P *in vivo* closely resemble that of an acute inflammatory-type of reaction (Tozer et al., 2001). At later times, significant recruitment of neutrophils into tumours is observed (Parkins et al., 2000), suggesting their involvement in the late phase events, namely tumour cell kill via the production of oxygen free radicals. We are examining the role of the HO enzyme system and its possible involvement in cytoprotection during these late phase events following the vascular injury mediated by CA-4-P.

4. HEME OXYGENASE AND PROTECTION AGAINST VASCULAR INJURY AND INFLAMMATION

Durante and Schafer have reviewed the cytoprotective role of HO-1 induction during blood vessel injury (Durante and Schafer, 1998). In this review they suggest that in pathological states where vascular injury has occurred, HO-1 could be induced in vascular smooth muscle cells because of their exposure to both inflammatory cytokines and fluid shear stress. The consequent release of CO would stimulate cyclic guanosine monophosphate (cGMP) production leading to blood vessel relaxation, inhibition of smooth muscle cell proliferation and platelet activation thereby promoting vascular homeostasis. Furthermore, the release of biliverdin/bilirubin would also be important in scavenging any free radicals generated as a result of tissue injury and inflammation. Indeed, HO has been reported by a number of studies to mediate cytoprotection during vascular injury. Togane and colleagues have demonstrated that HO-1 is induced in vascular smooth muscle cells following arterial balloon injury (Togane et al., 2000). This induction led to inhibition of neointimal formation, an effect mediated by CO, and modulated by HO inhibitors and inducers. In addition, Duckers et al. have further demonstrated the potential protective role of HO-1 in vascular wound repair (Duckers et al., 2001). These authors report the beneficial effects of the HO-1 pathway in moderating the severity of vascular injury by both stimulating vascular relaxation and reducing vascular smooth muscle cell growth and proliferation by a mechanism involving cGMP and independent from nitric oxide (NO). In an inflammatory context HO has also been implicated in the modulation of the inflammatory response and HO-1 was shown to be induced by various inflammatory mediators (Terry et al., 1998). In a model of acute inflammation, HO was shown to dramatically increase as the inflammation progressed being highest as the inflammatory response was resolving thereby providing strong evidence for a role of this enzyme in the resolution of inflammation (Willis et al., 1996; Willis et al., 2000).

In the light of these findings, we hypothesised that the HO enzyme system could play a protective role during/following the vascular injury and inflammatory-type of reaction mediated by the tubulin binding agent CA-4-P *in vivo*. We therefore investigated the effects of CA-4-P on the HO enzyme system *in vivo*.

5. POTENTIAL PROTECTIVE ROLE OF HO-1 AGAINST CA-4-P VASCULAR INJURY

We have examined the effects of CA-4-P administration on the activity and protein levels of HO in the liver, kidney and a transplanted tumour. Tumour pieces of the P22 rat carcinosarcoma were transplanted subcutaneously into the left flank of 8–9 weeks old BDIX rats. The animals were injected with either hemin (10 mg/kg; i.p.) or CA-4-P (30 mg/kg; i.p.) and the tumour, liver and kidney were excised 6 or 24 hours following treatment. In the microsomal fractions of tissues, HO activity was measured using a spectrophotometric assay of bilirubin production while HO-1 protein levels were assessed by western blotting. Analysis of the tumour samples confirmed that the level of HO activity in the P22 tumour is high and comparable to normal rat liver (Tozer et al., 1998) (Fig. 2). However, we also found that the level of HO-1 protein in the P22 tumour is higher than in the liver (Fig. 3), thereby suggesting that most of the activity in the tumour is probably due to HO-1 rather than the constitutive HO-2 isozyme, and the opposite holds for the liver.

Clearly these results give rise to two major questions. First, why does the tumour tissue contain higher levels of HO-1 than the normal tissues examined and second,

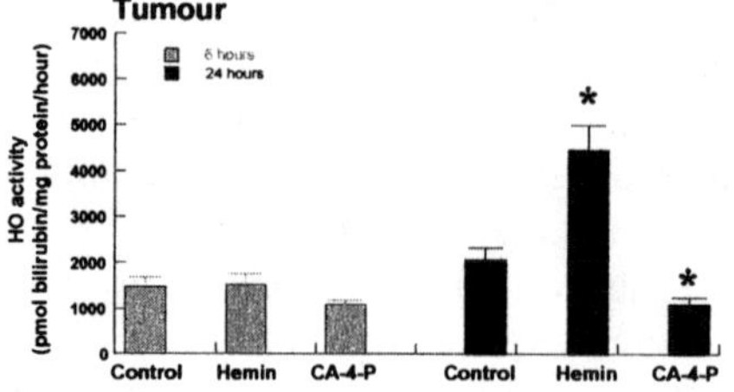
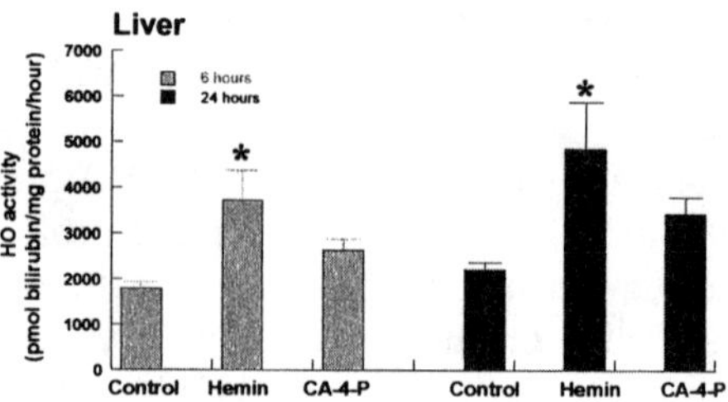
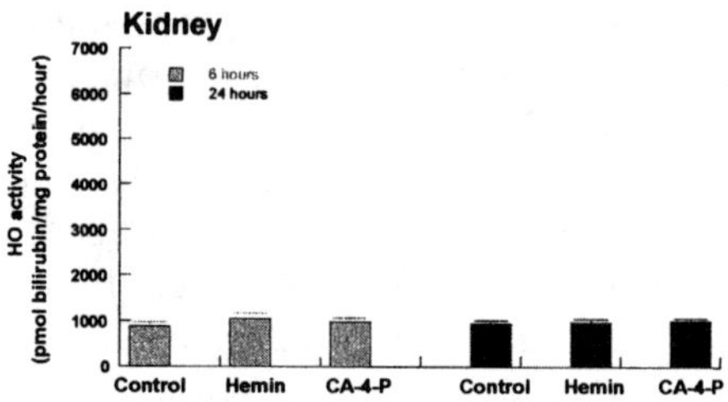

Figure 2. Effects of hemin (10 mg/kg) and C-4-P (30 mg/kg) on HO enzyme activity in the P22 tumour, liver and kidney 6 and 24 hours post-treatment. The data are means of 6 animals per group. Error bars are ±1 standard error. *represents a significant difference from controls (p < 0.05).

what could be the putative roles of this enzyme in this tumour model and possibly others? The particular characteristics of the tumour microenvironment may contribute to the high resting levels of HO-1 in tumours. Hypoxia for example, which is a tumour specific condition, has been shown to induce HO-1 (Motterlini et al., 2000; Panchenko et al., 2000). Furthermore, as described earlier, the vascular network of tumours is rather inefficient due to a spatial and temporal heterogeneity in blood supply. Therefore, it is possible that transient fluctuations in blood flow could cause the generation of oxygen free radicals due to ischaemia-reperfusion injury. Again the latter has been demonstrated to induce HO-1 in other systems (Maines et al., 1993; Maulik et al., 1996).

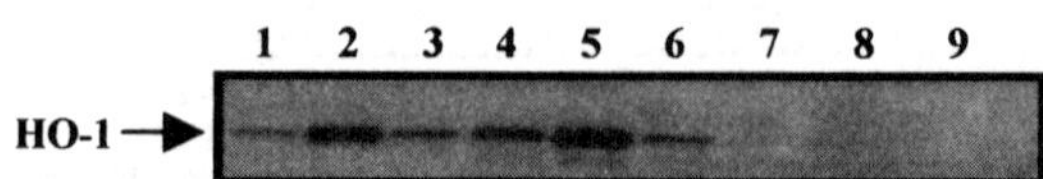

Figure 3. Effects of hemin (10 mg/kg) and CA-4-P (30 mg/kg) on HO-1 protein levels in the P22 tumour, liver and kidney 24 hours post-treatment. The blot is one of 6 independent experiments with similar results. Liver control (1), liver hemin (2), liver CA-4-P (3), tumour control (4), tumour hemin (5), tumour CA-4-P (6), kidney control (7), kidney hemin (8), kidney CA-4-P (9).

With regard to the putative roles of HO-1, we can suggest possible involvement of the products of heme degradation, namely bilirubin and CO, in cytoprotection against radical damage and maintenance of vascular homeostasis, respectively. In addition to that, a role of HO-1 in tumour progression via mediating/promoting cell proliferation and angiogenesis cannot be excluded and these have been discussed in section 2.

Our studies also established that hemin treatment, a known inducer of the enzyme, led to further HO-1 induction in the P22 tumour (Figs. 2, 3). Furthermore, CA-4-P was shown to decrease HO-1 protein levels and HO activity in the P22 tumour to around 53% of control at 24 hours post-treatment. On the other hand, a tendency for HO activity and HO-1 protein levels to increase to around 154% of control in the liver at 24 hours following CA-4-P treatment was measured whereas no detectable changes were found in the kidney (Figs. 2, 3). Collectively these results indicate that there could be a differential effect of CA-4-P on HO-1 induction between tumour and normal tissues. The decrease in HO activity and HO-1 protein in the P22 tumour could be a non-specific effect due to cell loss as a result of the extensive vascular damage which is known to occur in this tumour model (Tozer et al., 1999). To elucidate further the effects of CA-4-P on the HO enzyme system, later time points *in vivo* need to be examined and *in vitro* studies are also warranted. On the other hand, the increase in HO-1 protein levels and HO activity in the liver following CA-4-P administration could indicate a possible role of HO-1 in cytoprotection in this tissue. The liver is the main metabolising organ in the body and is involved in the metabolism of CA-4-P. It is possible that HO-1 induction in this organ is involved in a protective/detoxifying pathway against CA-4-P, which is not present in the tumour. Indeed, there is evidence from the literature that stilbene derivatives, of which CA-4-P is one, are able to induce HO-1 in the liver of rats as a result of oxidative stress caused by glutathione depletion (Oguro et al., 1996). Investigation of this pathway would help elucidate the specificity of CA-4-P for the tumour tissue. It is possible that normal tissues possess efficient pathways such as HO-1 and possibly other stress proteins to overcome the detrimental effects mediated by CA-4-P following the initial blood flow shut down.

6. SUMMARY

This review has outlined the emerging importance of the HO enzyme system in both tumour progression and cancer therapy. Furthermore, our studies on the novel vascular targeting agent CA-4-P have shown that at 24 hours after a single dose treatment, HO-1 levels were increased in normal liver but decreased in tumour tissue with the levels in the kidney remaining unaffected. These results suggest that HO-1 may play a cytoprotective/detoxifying role in some normal tissues such as the liver following treatment. Although the mechanism involved remains unclear at present, this differential effect of CA-4-P on HO-1 induction in normal versus tumour tissue may be possible to exploit for improving future therapy with this agent.

REFERENCES

Aizawa, T., Ishizaka, N., Taguchi, J., Kimura, S., Kurokawa, K., and Ohno, M., 1999. Balloon injury does not induce heme oxygenase-1 expression, but administration of hemin inhibits neointimal formation in balloon-injured rat carotid artery. *Biochem Biophys Res Commun* **261**:302.

Alam, J. and Den, Z., 1992. Distal AP-1 binding sites mediate basal level enhancement and TPA induction of the mouse heme oxygenase-1 gene. *J Biol Chem* **267** (30):21894.

Chaplin, D.J., Hill, S.A., Bell, K.M., and Tozer, G.M., 1998. Modification of tumour blood flow: Current status and future directions. *Semin Radiat Oncol* **8** (3):151.

Chaplin, D.J., Pettit, G.R., and Hill, S.A., 1999. Anti-vascular approaches to solid tumour therapy: evaluation of combretastatin A4-Phosphate. *Anticancer Res* **19**:189.

Dark, G.G., Hill, S.A., Prise, V.E., Tozer, G.M., and Pettit, G.R., Chaplin, D.J., 1997. Combretastatin A-4, an agent that displays potent and selective toxicity towards tumour vasculature. *Cancer Res* **57**:1829.

Deramaudt, B.M., Braunstein, S., Remy, P., and Abraham, N.G, 1998. Gene transfer of human heme oxygenase into coronary endothelial cells potentially promotes angiogenesis. *J Cell Biochem* **68**:121.

Deramaudt, B.M., Remy, P., and Abraham, N.G., 1999. Upregulation of human heme oxygenase gene expression by Ets-family proteins. *J Cell Biochem* **72**:311.

Doi, K., Akaike, T., Fujii, S., Tanaka, S., Ikebe, N, Beppu, T., Shibahara, S., Ogawa, M., and Maeda, H., 1999. Induction of haem oxygenase-1 by nitric oxide and ischaemia in experimental solid tumours and implications for tumour growth. *Br J Cancer* **80** (12):1945.

Duckers, H.J., Boehm, M., True, A.L., Yet, S.F., San, H., Park, J.L., Webb, R.C., Lee, M.E., Nabel, G.J., and Nabel, E.G., 2001. Heme oxygenase-1 protects against vascular constriction and proliferation. *Nat Med* **7** (6):693.

Durante, W. and Schafer A.I., 1998. Carbon monoxide and vascular cell function (Review). *Int J Mol Med* **2**:255.

Folkman, J., 1990. What is the evidence that tumours are angiogenesis dependent? *J Natl Cancer Inst* **82** (1):4.

Galbraith, S.M., Chaplin, D.J., Lee, F., Stratford, M.R.L., Locke, R.J., Vojnovic, B., and Tozer, G.M., 2001. Effects of combretastatin A4 phosphate on endothelial cell morphology in vitro and relationship to tumour vascular targeting activity in vivo. *Anticancer Res* **21**:93.

Gille, J., Swerlick, R.A., and Caughman, S.W., 1997. Transforming growth factor-alpha-induced transcriptional activation of the vascular permeability factor (VPF/VEGF) gene requires AP-2-dependent DNA binding and transactivation. *EMBO J* **16** (4):750.

Gray, L.H., Conger, A.D., Ebert, M., Hornsey, S., and Scott, O.C., 1953. Concentration of oxygen dissolved in tissues at the time of irradiation as a factor in radiotherapy. *Br J Radiol* **26**:638.

Hara, E., Takahashi, K., Tominaga, T., Kumabe, T., Kayama, T., Suzuki, H., Fujita, H., Toshimoto, T., Shirato, K., and Shibahara, S. (1996). Expression of heme oxygenase and inducible nitric oxide synthase mRNA in human brain tumours. *Biochem Biophys Res Commun* **24**:153.

Horsman, M.R., Ehrnrooth, E., Ladekarl, M., and Overgaard, J., 1998. The effects of combretastatin A-4 disodium phosphate in a C3H mouse mammary carcinoma and a variety of murine spontaneous tumours. *Int J Radiat Oncol Biol Phys* **42** (4):895.

Iyer, S., Chaplin, D.J., Rosentha D.S., Boulares, A.H., Li, L.-Y., and Smulson, M.E., 1998. Induction of apoptosis in proliferating human endothelial cells by the tumour-specific anti-angiogenesis agent combretastatin A-4. *Cancer Res*, **58**:4510.

Kanthou, C. and Tozer, G.M., 2000. Studies on the mechanism of action of the vascular targeting agent combretastatin A-4 phopsphate. *Br J Cancer* **83** (Suppl. 1):12.

Lavrovsky, Y., Schwartzman, M.L., Levere, R.D., Kappas, A., and Abraham, N.G., 1994. Identification of binding sites for transcription factors NF-kappa B and AP-2 in the promoter region of the human heme oxygenase 1 gene. *Proc Natl Acad Sci U S A* **91** (13):5987.

Lingyun, L., Rojiani, A., and Siemann, W., 1998. Targeting the tumour vasculature with combretastatin A-4 disodium phosphate: effects on radiation therapy. *Int J Radiat Oncol Biol Phys*, **42** (4):899.

Maines, M.D., Mayer, R.D., Ewing, J.F., and McCoubrey, W.K. Jr., 1993. Induction of kidney heme oxygenase-1 (HSP32) mRNA and protein by ischemia/reperfusion: possible role of heme as both promotor of tissue damage and regulator of HSP32. *J Pharmacol Exp Ther* **264** (1):457.

Maines, M.D. and Abrahamsson, P.A., 1996. Expression of heme oxygenase-1 (HSP32) in human prostate: normal, hyperplastic, and tumour tissue distribution. *Urology* **47**:727.

Maulik, N., Sharma, H.S., and Das, D.K., 1996. Induction of the haem oxygenase gene expression during the reperfusion of ischemic rat myocardium. *J Mol Cell Cardiol* **28** (6):1261.

Motterlini, R., Foresti, R., Bassi, R., Calabrese, V., Clark, J.E., and Green, C.J., 2000. Endothelial heme oxygenase-1 induction by hypoxia. Modulation by inducible nitric-oxide synthase and S-nitrosothiols. *J Biol Chem* **275** (18):13613.

Murphy B.J., Laderoute K.R., Vreman H.J., Grant T.D., Gill N.S., Stevenson D.K., and Sutherland R.M., 1993. Enhancement of heme oxygenase expression and activity in A431 squamous carcinoma multicellular tumor spheroids. *Cancer Res* **53** (12):2700.

Nishie, A., Ono, M., Shono, T., Fukushi, J., Otsubo, M., Onoue, H., Ito, Y., Inamura, T., Ikezaki, K., Fukui, M., Iwaki, T., and Kuwano, M., 1999. Macrophage infiltration and heme oxygenase-1 expression correlate with angiogenesis in human gliomas. *Clin Cancer Res* **5**:1107.

Oguro, T., Kaneko, E., Numazawa, S., Imaoka, S., Funae, Y., and Yoshida, T., 1996. Induction of hepatic heme oxygenase and changes in cytochrome P-450s in response to oxidative stress produced by stilbenes and stilbene oxides in rats. *J Pharmacol Exp Ther* **280** (3):1455.

Parkins, C.S., Holder, A.L., Hill, S.A., Chaplin, D.J., and Tozer, G.M., 2000. Determinants of anti-vascular action by combretastatin A-4 phosphate: role of nitric oxide. *Br J Cancer* **83** (6):811.

Panchenko, M.V., Farber, H.W., and Korn, J.H., 2000. Induction of heme oxygenase-1 by hypoxia and free radicals in human dermal fibroblasts. *Am J Physiol Cell Physiol* **278** (1):C92.

Pettit, G.R., Cragg, G.M., and Singh, S.B., 1987. Antineoplastic agents, 122. Constituents of of Combretum cattrum. *J Nat Prod* **50**:386.

Pettit, G.R., Singh, S.B., Hamel, E., Lin, C.M., Alberts, D.S., and Garia-Kendall, D., 1989. Isolation and structure of the strong cell growth and tubulin inhibitor combretastatin A4. *Experientia* **45**:205.

Remy P., Senan F., Meyer D., Mager A.M., and Hindelang C., 1996. Overexpression of the Xenopus Xl-fli gene during early embryogenesis leads to anomalies in head and heart development and erythroid differentiation. *Int J Dev Biol* **40** (3):577.

Schwartz, S.M., 2001. A protective player in the vascular response to injury. *Nat Med* **7** (6):656.

Suematsu, M., Goda, N., Sano, T., Kashiwagi, S., Egawa, T., Shinoda, Y., and Ishimura, Y., 1995. Carbon monoxide: an endogenous modulator of sinusoidal tone in the perfused rat liver. *J Clin Invest* **96** (5):2431.

Terry, C.M., Clikeman, J.A., Hoidal, J.R., and Callahan, K.S., 1998. Effect of tumour necrosis factor-α and interleukin-1α on heme oxygenase-1 expression in human endothelial cells. *Am J. Physiol.* **274** (*Heart Circ. Physiol.* 43):H883.

Tischer, E., Mitchell, R., Hartman, T., Silva, M., Gospodarowicz, D., Fiddes, J.C., and Abraham, J.A. 1991. The human gene for vascular endothelial growth factor. Multiple protein forms are encoded through alternative exon splicing. *J Biol Chem* **266** (18):11947.

Togane, Y., Morita, T., Suematsu, M., Ishimura, Y., Yamazaki, J.I., and Katayama, S., 2000. Protective roles of endogenous carbon monoxide in neointimal development elicited by arterial injury. *Am J Physiol Heart Circ Physiol* **278**:H623.

Torisu, H., Ono, M., Kiryu, H., Furue, M., Ohmoto, Y., Nakayama, J., Nishioka, Y., Sono, S., and Kuwano, M., 2000. Macrophage infiltration correlates with tumor stage and angiogenesis in human malignant melanoma: possible involvement of TNFalpha and IL-1alpha. *Int J Cancer* **85** (2):182.

Torisu-Itakura, H., Furue, M., Kuwano, M., and Ono, M., 2000. Co-expression of thymidine phosphorylase and heme-oxygenase-1 in macrophages in human malignant vertical growth melanomas. *Jpn J Cancer Res* **91**:906.

Tozer, G.M., Prise, V.E., Motterlini, R., Poole, D.J., and Chaplin, D.J., 1998. The comparative effects of the NOS inhibitor, N^ω-nitro-L-arginine, and the haemoxygenase inhibitor, zinc protoporphyrin IX, on tumour blood flow. *Int J Radiat Oncol Biol Phys* **42** (4):849.

Tozer, G.M., Prise, V.E., Wilson, J., Locke, R.J., Vojnovic, B., Stratford, M.R.L., Dennis, M.F., and Chaplin, D.J., 1999. Combretastatin A4-Phosphate as a tumour vascular targeting agent: early effects in tumours and normal tissues. *Cancer Res* **59**:1626.

Tozer, G.M., Prise, V.E., Wilson, J., Cemazar, M., Shan, S., Dewhirst, M.W., Vojnovic, B., and Chaplin, D.J., 2001. Mechanisms associated with tumour vascular shut-down induced by combretastatin A-4 phosphate: intravital microscopy and measurement of vascular permeability. *Cancer Res* **61** (17):6413.

Vaupel, P., Kalliniwski, F., and Okunieff, P., 1989. Blood flow, oxygen and nutrient supply, and metabolic microenvironment of human tumours: A review. *Cancer Res* **49**:6449.

Voest, E.E., Rooth, H., Neijt, J.P., van Asbeck, B.S., and Marx, J.J., 1993. The in vitro response of human tumour cells to desferrioxamine is growth medium dependent. *Cell Prolif* **26** (1):77.

Wakabayashi, Y., Takamiya, R., Mizuki, A., Kyokane, T., Goda, N., Yamaguchi, T., Takeoka, S., Tsuchida, E., Suematsu, M., and Ishimura, Y., 1999. Carbon monoxide overproduced by heme oxygenase-1 causes a reduction of vascular resistance in perfused rat liver. *Am J Physiol* **277** (5 Pt 1):G1088.

Wang, R., Wang, Z., and Wu, L., 1997. Carbon monoxide-induced vasorelaxation and the underlying mechanisms. *Br J Pharmacol* **121** (5):927.

Willis D., Moore A.R., Frederick R., and Willoughby D.A., 1996. Heme oxygenase: a novel target for the modulation of the inflammatory response. *Nat Med* **2** (1):87.

Willis D., Moore A.R., and Willoughby D.A., 2000. Heme oxygenase isoform expression in cellular and antibody-mediated models of acute inflammation in the rat. *J Pathol* **190** (5):627.

Woods, J.A., Hadfield, J.A., Pettit, G.R., Fox, B.W., and McGown, A.T., 1995. The interaction with tubulin of a series of stilbens based on combretastatin A-4. *Br J Cancer* **71**:705.

Yokoyama, S., Mita, S., Okabe, A., Abe, M., and Ogawa, M., 2001. Prediction of radiosensitity in human oesophageal squamous cell carcinomas with heme oxygenase-1: a clinicopathological and immuno-histochemical study. *Oncol Rep* **8**:355.

27

THE HEME OXYGENASE/CARBON MONOXIDE SYSTEM IN HEPATOBILIARY PATHOPHYSIOLOGY

David Sacerdoti and Angelo Gatta

Department of Clinical and Experimental Medicine
University and Azienda Ospedaliera of Padova
Via Giustiniani 2, 35100 Padova, Italy
Tel.: 39-0498212300
Fax: 39-0498754179
E-mail: david.sacerdoti@unipd.it

1. CHARACTERIZATION OF HEME OXYGENASE

Heme-oxygenase (HO), the rate-limiting enzyme in heme catabolism, was first described in 1968 by Tenhunen et al.[1] This enzyme catalyzes the degradation of heme to biliverdin and iron, with the concurrent release of carbon monoxide (CO). In mammals, biliverdin is then converted to bilirubin by the cytosolic enzyme biliverdin reductase; bilirubin is subsequently conjugated with sugars (mainly glucuronic acid) by UDP-glucuronyl transferase and then excreted into the bile. HO is also responsible for the recycling of iron from senescent red blood cells and extra hematopoietic cells such as liver. Some 80–85% of the bilirubin formed in vivo is derived from hemoglobin released from aging or damaged erythrocytes; this accounts for the high basal activity of HO within those tissues rich in reticuloendothelial cells, such as the spleen and bone marrow. HO mRNA levels are high in fetal rat liver during prenatal maturation (9 days before birth) and reach a maximum 24 hrs after birth[2] when levels decline but remain above adult levels for at least a month. This correlates with a greater capacity of the liver for bilirubin production in fetuses compared with adults and such circumstances could render the fetus more susceptible to drug injuries because of a depressed heme-cytochrome P450 system. However, Dennery et al.[3]

found that serum bilirubin protects against serum oxidative damage in the first days
of life in neonatal Gunn rats exposed to hypoxia.

1.1. Heme Oxygenase Isozymes

HO was first purified to homogeneity from rat liver and pig and bovine spleen
and shown to have a molecular weight of 32,000 Daltons.[4] Three HO isoforms (HO-
1, HO-2, HO-3) have been identified.[5,6] HO-1 is a 32 kDa heat shock protein which is
inducible by numerous noxious stimuli. HO-2, a 36 kDa protein, is constitutively
expressed and is abundantly found in brain, testis, liver and endothelium. HO-3,
although related to HO-2, is a product of a different gene with far lesser ability than
HO-2 to catalyze heme degradation. HO-1 activity is increased in whole animal tissues
following treatment with its natural substrate heme, as well as various metals, xenobiotics, endocrine factors and synthetic metalloporphyrins. Many cells in culture,
including hemopoietic, hepatic, epithelial, endothelial, and retinal pigment epithelial
cells, also respond to these agents with a marked increase in HO-1 activity.[5] HO-1 is
a heat-shock protein, and also a stress protein induced, for example, by oxidative
damage. Induction of HO-1 may be an essential event for some types of acute reactions and for cellular protection following injury. This hypothesis implies that the
induction of HO-1 removes the potentially toxic molecule, heme, a lipid soluble, trans-
missible form of iron, and generates bilirubin and biliverdin, metabolites with antioxidant properties.[7] Another condition in which there is induction of HO is liver
regeneration following 2/3 hepatectomy.[8]

In HO-1 deficient mice, there is an increased susceptibility of the liver to endotoxin and interruption of a major pathway of iron recycling.[9] Most importantly, the
phenotype of the first human case of the HO-1 deficiency includes endothelial cell
damage, iron accumulation in the liver and kidney, and an increasing cell susceptibility to heme overloading in-vitro.[10] Conversely, overexpression of HO-1 through
gene transfer or induction attenuates inflammation and endothelial damage related
to oxidative stress.[11] Further studies are needed to establish whether these events are
caused solely by actions of CO or also reflect changes such as a reduced capacity for
oxidant generation via heme and/or antioxidant effects of bilirubin.

The role of HO-2 in cells is not as well understood. However, it is becoming
apparent that HO-2 has an important role in epidermal cells, germ cell development,
signal transduction in neural tissues, NO-independent endothelial-derived relaxation.
In rat liver[12] HO-1 is prominent in Kupffer cells, whereas HO-2 is most abundant in
hepatocytes. Sinusoidal endothelial cells and/or hepatic stellate cells in culture appear
to express HO-2.[13] These findings suggest cooperative roles of these enzymes in the
catabolism of hemoglobin heme in different cellular compartments.

1.2. Substrate Specificity

The presence of a central iron atom for the binding of molecular oxygen is a
prerequisite for a metalloporphyrin to serve as a substrate for HO capable of being
enzymatically degraded to bile pigments. Metalloporphyrins in which the central iron
atom is replaced by either Sn, Co, Zn or Mn act as potent competitive inhibitors of

the enzyme; metalloporphyrins such as Mg, Ni and Cu protoporphyrins which can also bind tot he catalytic side of HO have a much lesser inhibitory effect on heme degradation.[1] Hemeproteins to which heme is bound relatively loosely, such as methemoglobin, metalbumin, the a and b chains of hemoglobin, heme-hemopexin and hemoglobin-haptoglobin complexes, are all substrates for HO.[1] Hemeproteins to which heme is firmly bound, such as myoglobin, oxyhemoglobin and carboxyhemoglobin, are not substrates unless first denatured. Porphyrins having the vinyl groups at positions 2 and 4 replaced by methyl groups are also excellent substrates. Type c heme is also a substrate for HO after its proteolytic cleavage from cytochrome c.

2. CARBON MONOXIDE

The major source of CO in animals is the degradation of heme by HO; CO produced by HO may serve as an important cellular signal in the microenvironment. Nitric oxide synthase (NOS) is an important heme-containing enzyme which generates NO; certain of the effects produced by NO can be duplicated by CO. In particular, the action of some neurotransmitters and muscle relaxants may be regulated by both molecules. In studies on vascular smooth muscle, data indicates that the regulation of vascular tone and platelet thrombus formation is due, in part, to HO-mediated CO production.[14] It is considered unlikely that CO and NO represent redundant messenger molecules, even though both are active in the vessel wall. CO may be a major physiological regulator of cGMP levels in the brain. An important role for HO-2 in the brain may be to produce CO in the microenvironment which could then function as a type of messenger. The function of CO produced by HO in tissues other than the brain is unclear, although it appears that CO can act as a smooth muscle relaxant. HO-2 has been localized by immunohistochemistry to endothelial cells and adventitial nerves of blood vessels. Inhibition of HO activity by tin protoporphyrin IX (SnPP) reverses the component of endothelial-derived relaxation of porcine distal pulmunary arteries,[15] suggesting coordinated physiologic roles for NOS and HO. It is well known that CO is toxic to cells, but less is known about pico-levels of CO in the cellular microenvironment.

2.1. Control of Liver circulation by CO

In 1994 Suematsu et al.[16] have shown that CO is present at submicromolar levels in the liver effluent and that inhibition of HO with ZnPP increases perfusion pressure in the isolated perfused rat liver, an effect that can be reversed by adding CO or a cGMP analogue to the perfusate. CO serves as an endogenous factor that reduces sinusoidal tone involving, at least in part, hepatic stellate cells.[17] We have shown (Fig. 1) that overproduction of CO by induction of HO with $CoCl_2$ reduces the response to the vasoconstrictor endothelin-1, but not to phelylephrine, in the isolated perfused rat liver.[19] The importance of CO as an endogenous modulator of vascular portal perfusion has been confirmed by Pannen et al.[18] who demonstrated how NO serves as a potent vasodilator in the hepatic arterial circulation, but exerts only a

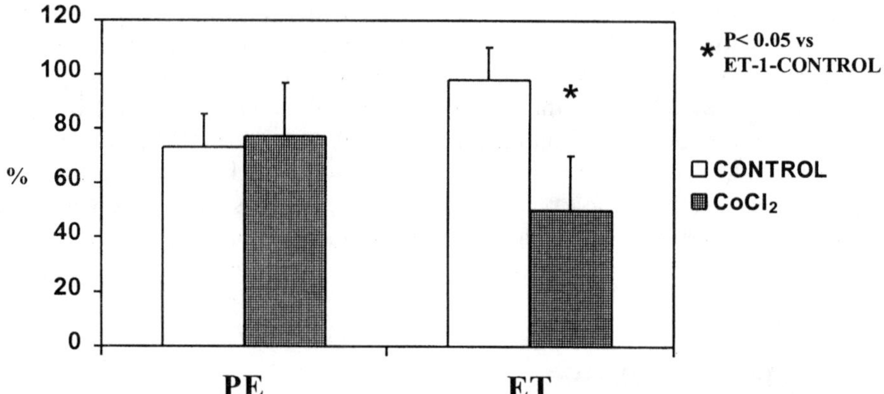

Figure 1. Effects of induction of heme oxygenase by $CoCl_2$ (24 mg/Kg i.p. for 2 days) on phenylephrine (PE) (5 µg) and endothelin-1 (ET-1) (40 ng) induced increase in portal perfusion pressure in isolated perfused rat liver (n = 7).

minor vasodilatory effect in the portal venous vascular bed, while CO does not regulate hepatic artery tone, but acts to maintain portal vascular tone in a relaxed state.

2.2. Effects of CO on Bile Formation

Elimination of constitutive CO generation through administration of ZnPP not only increases sinusoidal tone but also stimulates bile acid-dependent bile flow.[20] The choleretic action coincides with an increase in microvascular tone and oxygen consumption and may thus reflect a prolonged duration of bile acid uptake by hepatocytes. Exogenous CO at micromolar concentrations completely reverse these changes. The effect of ZnPP is mimicked by administration of methilene blue, an sGC inhibitor, but not fully reversed by membrane-permeable 8Br-cGMP, suggesting some involvement of cGMP-independent mechanisms. CO generated in hepatocytes may also affect bile excretion through altering the contractility of the bile canaliculus (BC).[20] Inhibition of CO with ZnPP shortens intervals of contraction of BCs and increases intracellular Ca^{2+}, an effect that is reversed by CO at a micromolar level without increasing cGMP.[21] As to mechanisms of the CO effect, several lines of experimental evidence suggest that CO modulates BC functions through its action on cytochrome P450-mediated calcium mobilization.

3. ROLE OF HEME OXYGENASE IN PATHOPHYSIOLOGY

As we define the molecular mechanisms that regulate the HO gene and identify the cascade of intracellular changes that accompany a change in HO levels, are we now able to appreciate the role of HO in pathophysiology. Induction of HO-1 may minimize the amount of hemoglobin (heme) to which tissues are exposed and lessens toxicity and injury to these tissues. The induced levels of HO-1, brought about by the increased heme levels, cause an increase in free iron, CO, biliverdin, and bilirubin, important mediators in modulating inflammation. Induction of HO by $SnCl_2$ is able to decrease blood pressure in young SHRs[22] and to prevent the development of hypertension.[23]

Bilirubin and biliverdin are both antioxidant agents in vitro and in vivo;[7] increased local concentrations of antioxidative agents following HO-1 induction protect cells from oxidative injury. CO may mimic NO and serve as an important modulator of endothelial cell functions following hemorrhagic injury. CO is a powerful vasodilator as is NO; together they counteract the vasoconstrictor properties of hemoglobin and heme.

3.1. Hypoxia and Ischemia

Reperfusion injury has been defined as the conversion of reversibly injured cells (myocardial, endothelial, etc.) to irreversibly injured cells, and is mediated by a burst of free radical generation as the previously hypoxic cells are flooded with oxygen. HO mRNA increases within 4 hrs of reperfusion of non-necrogenic ischemic rat liver.[24] The role of HO induction in these events, whether protective or cytotoxic, is still under debate. However, overexpression of the human HO gene by gene transfer in rabbit coronary endothelial cells[25] considerably enhances resistance to oxidative injury produced by concomitant exposure to free heme or hemoglobin.

Hemorrhagic shock (HS) causes hepatic dysfunction or failure related to decreased hepatic microcirculatory flow, and results in enhanced hepatic expression of HO-1.[13] Furthermore, the increase in portal resistance, upon blockade of the HO-CO pathway, is much more pronounced after HS compared with sham controls. After HS endogenously generated CO was shown to act to preserve sinusoidal perfusion, mitochondrial redox state, and secretory function in the isolated perfused rat liver.[26] This protective role of CO was mediated via a relaxing mechanism in part involving Ito cells. Similar results have been obtained by Kyokane et al.[27] in endotoxemic rats that have stimulation of both i-NOS and HO-1. In this condition, inhibition of CO, but not of NO, causes a marked vasoconstriction and cholestasis. Thus, CO may exert a protective function against hepatobiliary dysfunction after HS and endotoxemia.

3.2. Jaundice

Bilirubin production is 2-3-fold greater in newborns than in an adult. This increase in plasma bilirubin levels is due in large part to the combination of the rapid degradation of fetal hemoglobin in the first few days of life and the immaturity of the hepatic bilirubin conjugating system thus leading to an increase in unconjugated bilirubin. If the levels of unconjugated bilirubin become too high, the bilirubin may cross the blood-brain barrier resulting in bilirubin encephalopathy or kernicterus. Phototherapy is the method of choice to lower serum bilirubin levels, but its safety and efficiency have been called into question. The clinical use of HO inhibitors is an alternative therapy. Sn-PP causes a significant decrease in the levels of serum (mean decrease, 38%) and biliary bilirubin (mean decrease: 47%) in normal subjects. The decrease in these parameters lasts for a minimum of 4 days after administration of the metalloporphyrin.[28] The tin porphyrins have been used on newborns with ABO incompatibility,[29] on patients with hereditary porphyria,[30] liver disease[31] or Crigler-Najjar type I syndrome.[32] Results[33] indicate that the use of SnMP within 24 hours of birth in premature newborns substantially moderates the development of hyperbilirubinemia and reduces the requirement for phototherapy markedly (»75%) in inhibitor treated infants compared with control subjects. When administered at the appropriate time to near-term and term newborns with hyperbilirubinemia it can entirely eliminate the need for phototherapy to control this problem. In patients with biliary cirrhosis and hemochromatosis, SnPP was able to reduce bilirubin levels for about four days. Biliary bilirubin concentrations decreased (mean decrease, 49%) in the haemochromatosis patients after SnPP administration. No decrease in biliary bilirubin concentrations could be detected in the primary biliary cirrhosis patients under the same conditions.[31]

3.3. Cirrhotic and Pre-Hepatic Portal Hypertension

We have recently evaluated the role of CO in the pathophysiology of hemodynamic alterations related to portal hypertension in experimental CCl4-induced cirrhosis in the rat. In isolated perfused cirrhotic livers, the vasoconstrictive response to HO inhibition with SnMP was completely absent (Fig. 2A), while in control livers portal perfusion pressure could be increased by about 35%.[37] While the response to different vasoconstrictors—phenylephrine (PE) (Fig. 3), KCl, and endothelin-1 (ET-1)—was enhanced by SnMP in normal animals, in cirrhotic rats it was not affected. Furthermore, HO-2 expression in the portal vessels was decreased in

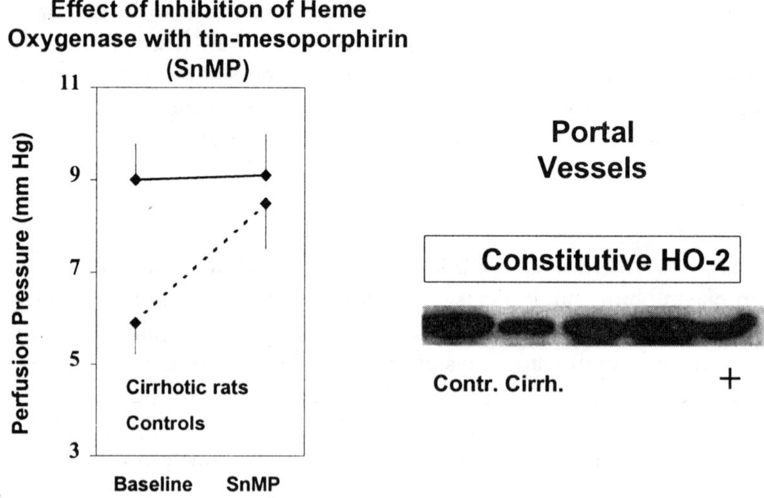

Figure 2. (A) Perfusion pressure in the isolated perfused liver from normal and cirrhotic rats before and after inhibition of heme oxygenase with SnMP. B) Western blot analysis of constitutive heme oxygenase-2 in portal vessels from normal and cirrhotic animals.

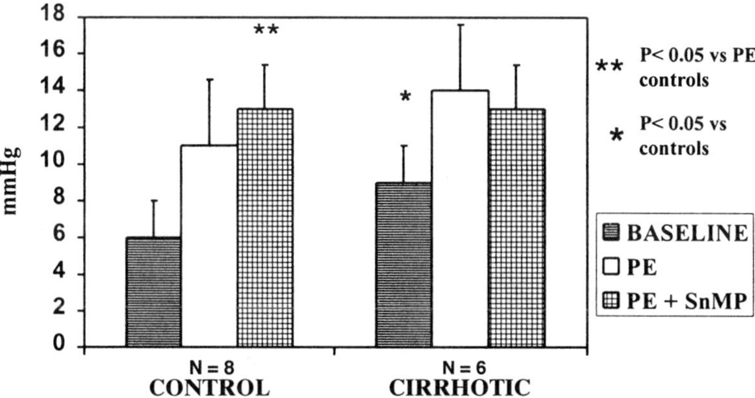

Figure 3. Effects of inhibition of heme-oxygenase with SnMP (10 µM) on phenylephrine (PE) induced increase in portal perfusion pressure in isolated perfused rat liver.

cirrhotic animals (Fig. 2B), suggesting that decreased CO in the porto-sinusoidal circulation may participate in the pathophysiology of increased resistance to portal flow in cirrhosis.

As increased portal inflow through vasodilatation of the mesenteric vasculature (SMV) is a major component of the increase in portal pressure of cirrhosis, we have evaluated the role of CO in SMV vasodilatation and hyporesponsiveness to vasoconstrictors. HO inhibition with SnMP increased SMV perfusion pressure only in cirrhotic animals, reaching values similar to normal rats. It also reversed almost completely the decreased response to KCl, PE, and ET-1. In SMV from cirrhotic animals expression of HO-2, but not of HO-1, was increased as was HO activity.[37] To further confirm the role of increased CO in the alterations of SMV of cirrhosis, we attempted to reproduce the hemodynamic alterations in normal animals by increasing HO. This was done by administering an adenovirus containing human HO-1 gene. A week after transfection, human HO-1 could be demonstrated by RT-PCR in the SMV (Fig. 4A) and HO activity was increased by about 300%. Treated animals had similar, or even more evident, alterations of SMV as cirrhotic rats, i.e. decreased perfusion pressure and reactivity to vasopressors (Fig. 4B).[37]

In experimental pre-hepatic portal hypertension, obtained by partial portal vein ligation in rats, Fernandez et al.[34] have shown that HO activity was increased in the liver. HO-1 expression was present in hepatocytes and Kupffer cell of portal hypertensive rats but not of normal rats, while HO-2 was similarly expressed in all liver cell types of normal and portal-vein ligated rats. In patients with post-hepatitic cirrhosis, Makino et al.[35] have shown that HO-1 is increased in the liver, being mainly distributed in Kupffer cells and hepatocytes. By contrast, in livers in which portal hypertension was idiopathic and due to increase in presinusoidal resistance, there was a decreased expression of HO-1 in Kupffer cells and absence in hepatocytes. Fernandez et al.[38] have also evaluated the role of CO in hyporeactivity of the mesenteric vascular beds of prehepatic portal hypertension in rat. In this model, inhibition

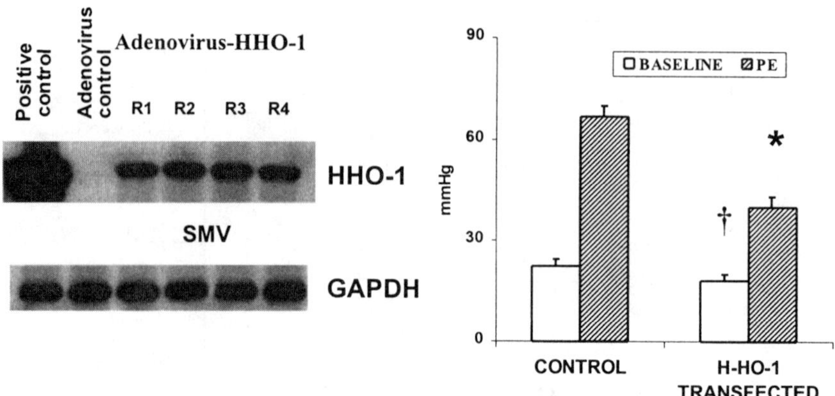

Figure 4. Human HO-1 (HHO-1) mRNA expression in superior mesenteric vasculature (SMV) from rats injected with recombinant adenovirus expressing human HO-1 gene (And-HHO-1). Different rats (R1–R4) were injected with And-HHO-1 viral particles (10^{12} PFU/300 g body weight). PCR analysis of HHO-1 specific product was done 7 days after injection and P^{32} dCTP-incorporated products were subjected to gel electrophoresis and exposed to X ray film. The figure shows equal amplification of GAPDH transcripts. B) Effect of H-HO-1 transfection on SMV perfusion pressure and response to phenylephrine (PE).

of HO with ZnMP did not modify the hyporeactivity to KCl, that was partially attenuated by NOS inhibition and completely corrected by simultaneous inhibition of HO and NOS. Also the hyporeactivity to methoxamine was not affected by ZnMP, but it was completely overcome by L-NAME, without any increase in response after combined inhibition of NOS and HO.

4. CONCLUSIONS

The HO/CO system appears to be deeply involved in hepatobiliary physiology and pathophysiology. It regulates xenobiotic metabolism, bile acid synthesis and bile flow, macro- and micro-circulation, and modulates oxidative stress. The study of its role in pathophysiology is giving everyday new information and clearly, the more we define and understand the role in cellular physiology and pathology, the more the possibility arises of a useful potential in the regulation of this system for clinical purposes.

REFERENCES

1. Tenhunen R., Marver H.S., and Schmid R. The enzymatic conversion of heme to bilirubin by microsomal heme oxygenase. *Proc Natl Acad Sci USA* 61:748–755, 1968.
2. Lin J.H.-C., Villalon P., Nelson J., and Abraham N.G. Expression of rat liver heme oxygenase gene during development. *Arch Biochem Biophys* 270:623–629, 1989.

3. Dennery P.A., McDonagh A.F., Spitz D.R., Rodgers P.A., and Stevenson O.K. Hyperbilirubinemia results in reduced oxidative injury in neonatal Gunn rats exposed to hypoxia. *Free Rad Biol Med* 9:395–404, 1995.

4. Maines M.D., Abraham N.G., and Kappas A. Solubilization and partial purification of heme oxygenase from rat liver. *J Biol Chem* 252:5900–5903, 1977.

5. Abraham N.G., Lin J.H.-C., Schwartzman M.L., Levere R.D., and Shibahara S. The physiological significance of heme oxygenase. *Int J Biochem* 20:543–558, 1988.

6. McCoubrey Jr. W.K., Huang T.J., and Maines M.D. Isolation and characterization of a cDNA from the rat brain that encodes hemoprotein heme oxygenase-3. *Eur J Biochem* 247:725–732, 1997.

7. Stocker R., Yamamoto Y., McDonagh A.F., Glazer A.N., and Ames B.N. Bilirubin is an antioxidant of possible physiological significance. *Science* 235:1043–1046, 1987.

8. Solangi K., Sacerdoti D., Goodman A., Schwartzman M.L., Abraham N.G., and Levere R.D. Differential effects of partial hepatectomy on hepatic and renal heme and cytochrome P450 metabolism. *Am J Med Sci* 296:387–391, 1988.

9. Poss K.D. and Tonegawa S. Heme oxygenase-1 is required for mammalian iron reutilization. *Proc Natl Acad Sci USA* 94:10919–10924, 1997.

10. Yachie A., Niida Y., Wada T., Igarashi N., Kaneda H., Toma T., and Ohta K. Oxydative stress causs enhanced vascular endothelial injury in human heme oxygenase-1 deficiency. *J Clin Invest* 103:129–135, 1999.

11. Abraham N.G., Lavrosky Y., Schwartzman M.L., Stoltz R.A., Levere R.D., Gerritsen M.E., and Shibahara S. Transfection of the human heme oxygenase gene into rabbit coronary microvessel endothelial cells: protective effect against heme and hemoglobin toxicity. *Proc Natl Acad Sci USA* 92:6798–6802, 1995.

12. Goda N., Suzuki K., Naito M., Takeoka S., Tsuchida E., Ishimura Y., Tamatani T., and Suematsu M. Distribution of heme oxygenase isoforms in rat liver. Topographic basis for carbon monoxide-mediated microvascular relaxation. *J Clin Invest* 101:604–612, 1998.

13. Bauer I., Wanner G.A., Rensing H., Alte C., Miescher E.A., Wolf B., and Pannen B.H. Expression pattern of heme oxygenase isoenzyme 1 and 2 in normal and stress-exposed rat liver. *Hepatology* 27:829–838, 1998.

14. Christodoulides N., Durante W., Kroll M.H., and Schafer A.I. Vascular smooth muscle heme oxygenases generate guanyl cyclase-stimulatory carbon monoxide. *Circulation* 91:2306–2309, 1995.

15. Zakhary R., Gaine S.P., Dinerman J.L., Ruat M., Flavahan N.A., and Snyder S. Heme oxygenase 2: endothelial and neuronal localization and role in endothelium-dependent relaxation. *Proc Natl Acad Sci USA* 93:795–798, 1996.

16. Suematsu M., Kashiwagi S., Sano T., Goda N., Shinoda Y., and Ishimura Y. Carbon monoxide as an endogenous modulator of hepatic vascular perfusion. *Biochem Biol Res Commun* 205: 1333–1337,1994.

17. Suematsu M., Goda N., Sano T., Kashiwagi S., Egawa T., Shinoda Y., and Ishimura Y. Carbon monoxide: an endogenous modulator of sinusoidal tone in the perfused rat liver. *J Clin Invest* 96(5):2431–2437, 1995.

18. Pannen B.H. and Bauer M. Differential regulation of hepatic arterial and portal venous vascular resistance by nitric oxide and carbon monoxide in rats. *Life Sci* 62:2025–2033, 1998.

19. Sacerdoti D., McGiff J.C., Oyekan A.O., Yang L., Gatta A., and Abraham N.G. Increase in portohepatic resistance in cirrhosis: the role of the carbon monoxide/heme oxygenase system. Acta Haematologica abstract 268, p. 67.

20. Suematsu M. and Yshimura Y. The heme oxygenase-carbon monoxide system: a regulator of hepatobiliry function. *Hepatology* 31:3 6, 2000.

21. Dufour J.-F.J., Turner T.J., and Arias I.R. Nitric oxide blocks bile canalicular contraction by inhibiting inositol triphosphate-dependent calcium mobilization. *Gastroenterology* 108:841–849, 1995.

22. Sacerdoti D., Escalante B., Abraham N.G., McGiff J.C., Levere R.D., and Schwartzman M.L. Treatment with tin prevents the development of hypertension in spontaneously hypertensive rats. *Science* 243:388 390, 1989.

23. Escalante B., Sacerdoti D., Davidian M.M., Schwartzman M.L., and McGiff. Chronic treatment with tin normalizes blood pressure in spontaneously hypertensive rats. *Hypertension* 17:776–779, 1991.

24. Tacchini L., Schiaffonati L., Pappalardo C., Gatti S., and Bernelli-Zazzera A. Expression of HSP70, intermediate-early response and heme oxygenase genes in ischemic-reperfused rat liver. *Lab Invest* 68:465–471, 1993.

25. Abraham N.G., Lavrovsky Y., Schwartzman M.L., Stoltz R.A., Levere R.D., Gerritsen M.E., Shibahara S., and Kappas A. Transfection of the human heme oxygenase gene into rabbit coronary microvessel endothelial cells: Protective effect against heme and hemoglobin toxicity. *Proc Natl Acad Sci USA* 92:6798–6802, 1995.

26. Pannen B.H.J., Kohler N., Hole B., Bauer M., Clemens M.G., and Geiger K.K. Protective role of endogenous carbon monoxide in hepatic microcirculatory dysfunction after hemorrhagic shock in rats. *J Clin Invest* 102:1220–1228, 1998.

27. Kyokane T., Norimizu S., Taniai H., Yamaguchi T., Takeoka S., Tsuchida E., Naito M., Nimura Y., Ishimura Y., and Suematsu M. Carbon monoxide from heme catabolism protects against hepato-biliary dysfunction in endotoxin-treated rat liver. *Gastroenterology* 120:1227–1240, 2001.

28. Berglund L., Angelin B., Blomstrand R., Drummond G.S., and Kappas A. Sn-protoporphyrin lowers serum bilirubin levels, decreases biliary bilirubin output, enhances biliary heme excretion and potently inhibits hepatic heme oxygenase activity in normal human subjects. *Hepatology* 8:625–631, 1988.

29. Kappas A., Drummond G.S., Manola T., Petmezaki S., and Valaes T. Sn-protoporphyrin use in the management of hyperbilirubinemia in term newborns with direct Coombs-positive ABO incompatibility. *Pediatrics* 81:485–497, 1988.

30. Galbraith R.A. and Kappas A. Pharmokinetics of tin-mesoporphyrin in man and the effects of tin-chelated porphyrins on hyperexcretion of heme pathway precursors in patients with acute inducible porphyria. *Hepatology* 9:882–888, 1989.

31. Berglund L., Angelin B., Hultcrantz K., et al. Studies with the haem oxygenase inhibitor Sn-protoporphyrin in patients with primary biliary cirrhosis and idiopathic haemochromatosis. *Gut* 31:899–904, 1990.

32. Galbraith R.A., Drummond G.S., and Kappas A. Suppression of bilirubin production in the Crigler-Najjar type I syndrome: studies with the heme oxygenase inhibitor tin-mesoporphyrin. *Pediatrics* 89:175–182, 1992.

33. Valaes T., Petmezaki S., Henschke C., Drummond G.S., and Kappas A. Control of jaundice in preterm newborns by an inhibitor of bilirubin production: studies with tin-mesoporphyrin. *Pediatrics* 93:1–11, 1994.

34. Fernandez M. and Bonkovsky H.L. Increased heme oxygenase-1 gene expression in liver cells and splanchnic organs from portal hypertensive rats. *Hepatology* 29:1672–1679, 1999.

35. Makino N., Suematsu M., Sugiura Y., Morikawa H., Shiomi S., Goda N., Sano T., Nimura Y., Sugimachi K., and Ishimura Y. Altered expression of heme oxygenase-1 in the livers of patients with portal hypertensive disease. *Hepatology* 33:32–42, 2001.

36. Sacerdoti D., Oyekan A., Jiang S., Gatta A., McGiff J.C., and Abraham N.G. The role of carbon monoxide abd heme-oxygenase in the alterations of mesenteric and porto-hepatic circulation of cirrhotic rats. *Hepatology* 30:237A, abstract 305, 1999.

37. Sacerdoti D., McGiff J.C., Oyekan A.O., Yang L., Gatta A., and Abraham N.G. Transfection of rats with human heme-oxygenase-1 gene reproduces abnormalities of mesenteric circulation of cirrhosis. *J Hepatol* 34; suppl.1: abstract 43, 2001.

38. Fernandez M., Lambrecht R., and Bonkovsky H.L. Increased heme oxygenase activity in splanchnic organs from portal hypertensive rats: role in modulating mesenteric vascular reactivity. *J Hepatol* 34:812–817, 2001.

Section V

Heme Oxygenase System and Oxidative Stress Response

28

HUMAN HEME OXYGENASE (HO)-1 DEFICIENCY AND THE OXIDATIVE INJURY OF VASCULAR ENDOTHELIAL CELLS

A. Yachie[a], T. Toma[b], S. Shimura[b], L. Yue[b], K. Morimoto[b],
K. Maruhashi[b], Y. Niida[b], K. Ohta[b], Y. Kasahara[b], Y. Saikawa[b],
and S. Koizumi[b]

[a]Dept. Lab. Sci.
[b]Dept. Pediatr., Kanazawa Univ.
Kanazawa, 920-8641 Japan

1. INTRODUCTION

Heme oxygenase-1 (HO-1) constitutes one of the three isozymes of HO, which catalyze the degradation of heme into biliverdin, carbon monoxide (CO) and free iron.[1-5] Recent experimental data indicate that one of the heme degradation products, CO acts on cellular metabolism to protect cells from oxidative stress and regulate production of inflammatory molecules.[6-8] As a result, CO directly controls the inflammatory state of a given tissue and at the same time, regulates the level of microcirculation within the target organs acting as a gaseous vasodilator.[9-11] Among three isozymes with similar enzyme activities, HO-1 is the only protein which is rapidly induced upon stimulation with various oxidative stresses.[12-14] Therefore, it is suggested that any defect in its function will lead to uncontrollable inflammatory responses upon certain exogenous insults, such as infection and hemolysis.

2. THE FIRST CASE OF HUMAN HEME OXYGENASE-1 DEFICIENCY

We have recently experienced the first case of human heme oxygenase (HO)-1 deficiency.[15 17] The patient was 2 years old when he first exhibited relapsing fever, erythematous rash and joint pain. Hepatomegaly was marked but the spleen was absent.

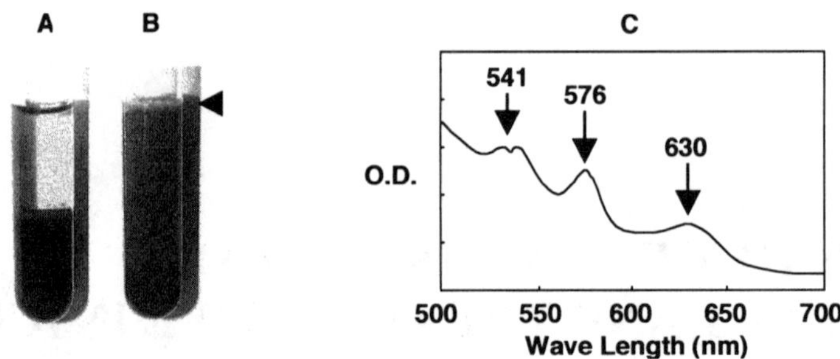

Figure 1. Heme accumulation in the HO-1 deficient serum. Note turbid, dark brown appearance of the serum and thick layer of fat (arrow head) above the patient serum (**B**), as compared with clear, control serum (**A**). Absorption spectrum of the patient serum (**C**) detected peaks representing both oxyhemoglobin (541 nm and 576 nm) and methemoglobin (630 nm).

Thrombocytopenia and leukocytosis were predominant. Peripheral blood smear showed numerous fragmented erythrocytes and erythroblasts. Both gross appearance of the serum (Fig. 1A, B) and absorption spectrum analysis (Fig. 1C) indicated that heme in the form of hemoglobin and methemoglobin was markedly increased in the patient serum. Haptoglobin concentration was extremely elevated and large amount of haptoglobin-hemoglobin complex was detected in the patient urine sample. Repeated measurement of serum bilirubin concentration was always low. Low bilirubin level in the presence of significant intravascular hemolysis suggested that a certain abnormality in heme degradation pathway existed. In particular, in vivo and in vitro HO-1 productions were examined because it is the enzyme actively induced upon oxidative stress. Immunohistochemical analysis of the liver biopsy specimen showed that Kupffer cells did not produce HO-1 in the patient liver (Fig. 2A, B). Futhermore, immunoblotting of Epstein-Barr virus-transformed lympho-blastoid cell lines (LCLs) revealed that the patient lacked HO-1 protein (Fig. 2C, D).

Endothelial cells were detached from basement membrane of glomerular capillary within the patient kidney. Diffuse subendothelial deposition of unidentifiable

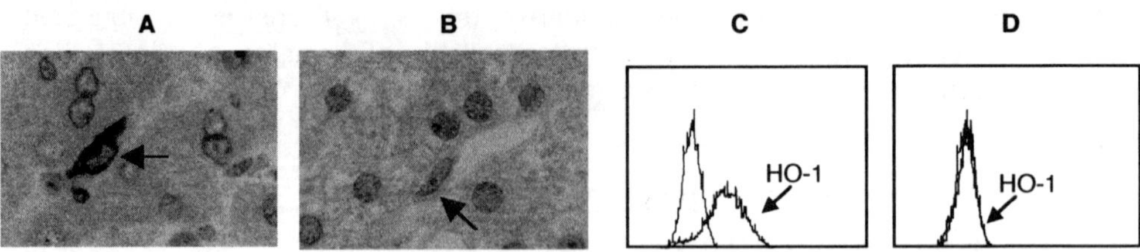

Figure 2. HO-1 production in vivo and in vitro. Biopsy specimens of the liver from control (**A**) and the patient (**B**) were stained for HO-1. Kupffer cells are indicated by arrows. LCLs were established from normal control (**C**) and the patient (**D**) and stimulated with hemin for 24 hrs. HO-1 expression was evaluated by flow cytometry.

A   **B**

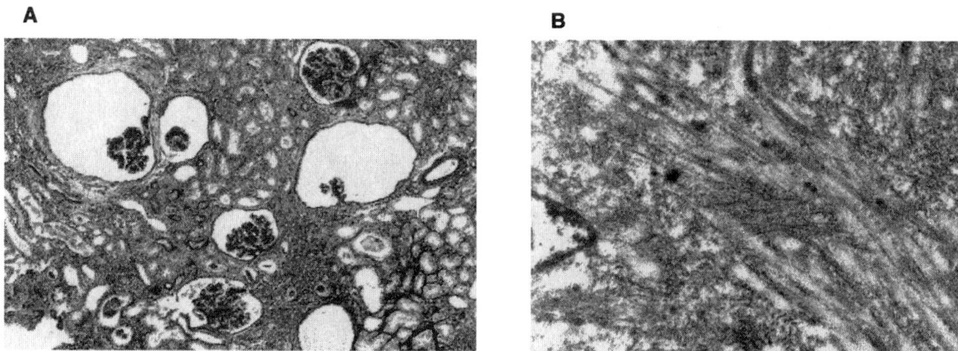

Figure 3. Oxidative injury of the kidney and the liver. **A**: The kidney at autopsy, Azan staining. Atrophic tubular epithelial cells, tubular narrowing and marked dilatation of Bowman's capsules are seen. **B**: Electron microscopic view of the liver. Massive deposition of amyloid fibrils were identified within the liver parenchyma.

material was ovserved.[16] Degeneration of renal tubular epithelium was prominent, with progressive narrowing of the tubular canals (Fig. 3A). Secondary enlargement of Bowman's capsules were found. Massive amyloid deposition within the liver was the characteristic finding observed at the time of autopsy (Fig. 3B). Peripheral blood monocytes showed morphological changes, together with significant reduction in CD36 and HLA-DR expressions. Accumulation of excess heme and lipids, together with fragmented erythrocytes and non-functioning platelets suggested that phagocytic scavanger functions were markedly impaired. Endothelial injury, tubular injury, liver parenchymal damage and monocyte dysfunction, all of which seemed to be the consequences of exposure to concentrated heme, comprised the major sequela of the first case of human HO-1 deficiency.

3. HO-1 MRNA AND HO-1 GENE MUTATION ANALYSIS

The human HO-1 gene is about 14 kb long and organized into five exons.[18] The exon 1 contains the start codon and the exons 2 through 5 consist of 121 bp, 492 bp, 100 bp, and 734 bp long sequences, respectively. At least two of the functionally active regulatory elements have been identified in this gene. The proximal elements within 121 bp of the sequence upstream of the mRNA cap site respond to various agents such as sodium arsenite, hydrogen peroxide, hemin, and cadmium chloride.[19] A 10-bp *cis*-acting element about 4 kb upstream from the transcription initiation site is responsible for cadmium-mediated induction as a distal enhancer.[20]

A complete defect of HO-1 induction after stimulation with several stress-inducing factors was documented in the patient. LCLs from the patient, family members, and normal controls were established under informed consent. The LCLs were preincubated with 10 μM of cadmium chloride for induction of HO-1 gene transcripts and subjected to the molecular analysis of HO-1 mRNA. Reverse transcriptase (RT)-

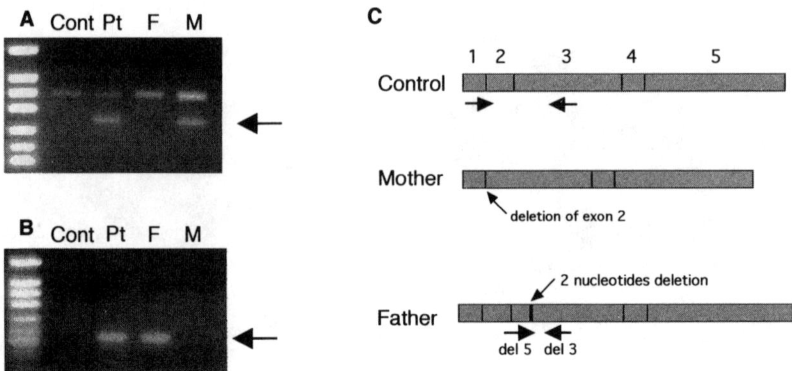

Figure 4. Genomic structures of HO-1 gene. **A**: RT-PCR of HO-1 mRNA. The patient and the mother express HO-1 mRNAs of two different sizes. **B**: Mutation-specific PCR to detect two nucleotides deletion within exon 3 of the paternal allele. Only the patient and the father are positive for mutation-specific sequences. **C**: Genomic structures of control, the maternal and the paternal alleles. Cont; control, Pt; patient, F; father and M; mother.

PCR was performed to amplify the entire open reading. Using the primer pairs to amplify the fragments containing exon 2, RT-PCR products from the patient and his mother contained an extra, shorter fragment (278 bp) (Fig. 4A, arrow) in addition to a normal size of HO-1 mRNA fragment (399 bp). A single size of HO-1 mRNA fragment (399 bp) was amplified in the father, sister, and brother. Direct sequencing of the shorter fragment observed in the patient and the mother showed a complete loss of exon 2 (Fig. 4C, middle panel). Loss of exon 2 led to a frame shift and created a premature termination codon at the 10th residue. Analysis of the paternal PCR product revealed a two-nucleotide (AC) deletion within exon 3 (Fig. 4C, lower panel). Genomic amplification by mutation-specific PCR using synthesized mutation-specific primers confirmed the mutation in the patient and the father as well as in his sister (Fig. 4B, arrow). The AC deletion led a frame shift resulting in a premature termination codon at the 131st residue. Dinucleotide tandem repeats (ACACAC) were observed at the deletion site of exon 3. This structural feature suggests the slipped strand mispairing for the mechanism of AC deletion in exon 3.[21]

In summary, the patient was a compound heterozygote for a complete deletion of exon 2 from the maternal allele and a two-nucleotide deletion within exon 3 from the paternal allele. Both the maternal and the paternal mutations led to frameshifts and created premature termination codons into the HO-1mRNA. The parents and the sister carrying the paternal mutation were shown to be heterozygotes, providing evidence for autosomal recessive inheritance.

There could be at least two possibilities for the molecular mechanisms of exon 2-deletion mutation in HO-1mRNA from the maternal allele, exon skipping due to alternative splicing or genomic deletion. We examined the nucleotide sequence flanking splice junctions for exons 1, 2 and 3 to investigate the former possibility. The approximately 200 bp-genomic regions surrounding each of the exons were amplified from the genomic DNA of the family members. Sequence analysis of the PCR products revealed no substitution at the splice-acceptor or donor sites of indicated exons in the patient and the mother (data not shown). We next considered a large deletion or genomic rearrangements involving exon 2 rather than exon-skipping mutation. Genomic fragments spanning exons 1 through 3 of the human HO-1 gene were amplified. The genomic PCR products from the patient and his mother contained an extra 4.3 kb-fragment in addition to a normal 6.0 kb-fragment from the intact allele. The sequences of this shorter fragment from the patient and the mother were identical and demonstrated the deletion of 1730 bp-long sequences which contained complete sequences of exon 2 (121 bp) and flanking intron sequences (1609 bp).

To further confirm this genomic deletion in the maternal allele, southern blot analysis was performed using PCR-generated genomic probes. The deletion junction contained the sequences homologous to a consensus *Alu* element. Characteristically, *Alu* elements are composed of a tandem repeat of two highly homologous sequences separated by a short A-rich region and are 300-bp long.[22–25] The flanking sequences of exon 2 contained a tandem repeat of two *Alu-Sx* (*Sx*-1 and *Sx*-2) elements in the intron 1 and an *Alu-Sq* element in the intron 2 (Fig. 5A). Analysis of the deletion junction sequence revealed fusion of a 133-bp 5′ of *Sx*-1 element with a 169-bp 3′ of *Sq* element (Fig. 5B). In addition, the sequences with high homology (22/26) to the 26-bp *Alu* "core" sequence (CCTGTAATCCCAGCACTTTGGGAGGC), considered to be a recombinogenic hotspot,[28] were present in the deletion joint.

Alu elements have been evolutionarily conserved in primates and are known to be involved in gene arrangements, including deletions, insertions and partial tandem duplications in several different genes.[22] Several mechanisms for *Alu*-mediated gene deletion have been proposed based on homologous recombination that occurs imperfectly or is associated with unequal crossing-over in meiotic cells.[21] These structural features of the HO-1 gene with respect to *Alu* elements and the deletion joint sequences led us to support the hypothesis that recombination event between homologous *Alu* sequences encompassing exon 2 could be the mechanism responsible for genetic deletion in the case with this defect.

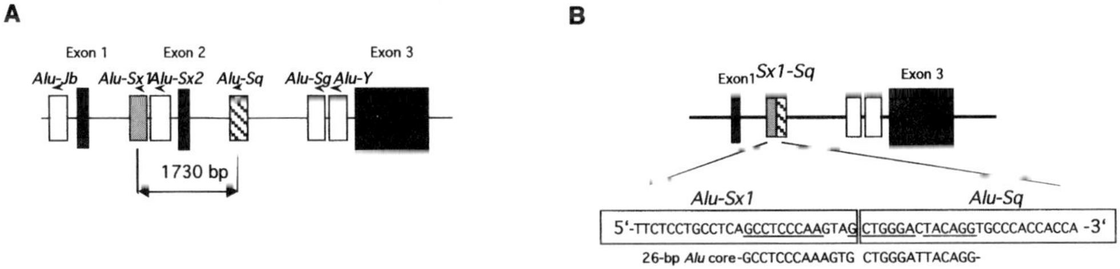

Figure 5. **A** Diagrammatic representation of *Alu* elements present in the HO 1 gene (exons 1 through 3 only). The location of *Alu*-family sequences is shown by open boxes. The arrow heads above the open boxes indicate the 5′ to 3′ direction of each of the *Alu* sequences. **B.** Alignment of *Alu-Sx1* and *Alu-Sq* sequences results in a 1730-bp genomic deletion encompassing exon 2. The sequences of the deletion junction are shown. The junction contains the sequences (underlined) with homology to 26-bp *Alu* "core" sequences.

4. HO-1 DEFICIENCY AND OXIDATIVE CELL INJURY

4.1. Functional Significance of HO-1 Deficiency

As it has been summarized in the previous sections, direct consequences of HO-1 deficiency are either the extensive cell injury by the lack of the enzyme and the induction of cell dysfunction, in particular, scavenger functions of macrophages. Although it is extremely difficult to prove that these are the general phenomenon observed in every HO-1 deficient patient, we tried to analyze the functional significances of HO-1 deficiency.

LCL derived from the HO-1 deficient patient was extremely sensitive to hemin-induced cell injury. When the cell injury was determined by propidium iodide uptake and annexin-V surface binding, significant cell injury was induced in patient LCL after 24 h. In accord with the sensitivity to hemin-induced cell injury, patient LCL did not produce detectable level of HO-1 even at the highest concentration of hemin (Fig. 6). In contrast, low level of HO-1 production was already detectable in control LCL without stimulation, and the level increased in a dose-dependent fashion with added hemin. HO-2 was produced constitutively by both control and patient LCLs.

Because the patient serum contained more than $400\,\mu M$ of heme consistently, it is plausible to assume that the cells exposed to this high heme concentration were invariably vulnerable to hemin-induced cell injury, unless heme-degrading enzyme HO-1 is produced appropriately. Extreme sensitivity of HO-1 deficient LCL to hemin-stimulation was not reversed by addition of apoferritin or bilirubin to the hemin-stimulated cultures. Furthermore, ferritin production by HO-1 deficient LCL was comparable with normal control LCL with or without added hemin. Although ferritin and bilirubin may act as antioxidant in certain situations, they do not contribute much to protect cells from hemin-induced cell injury, at least at high hemin concentration. Rather, direct degradation of heme by HO-1 was thought to have a significant role in protecting cells from high concentration of hemin.

4.2. Hemin-Induced Cell Injury of LCL Transfected with HO-1 Gene

We next tried to explain the direct role of HO-1 in the protection of cells from hemin-induced cell injury by transfecting HO-1 gene into patient LCL using a

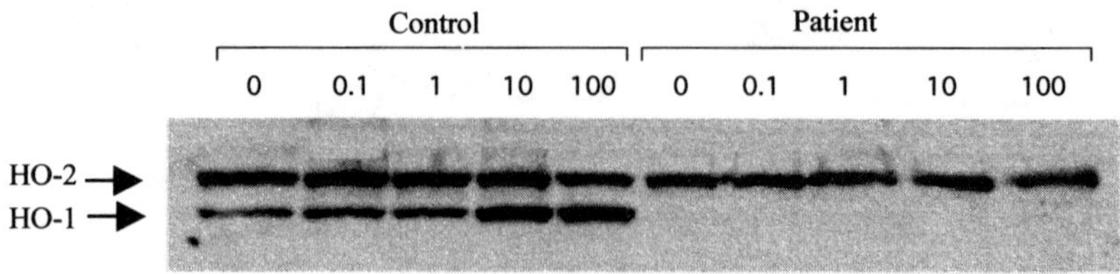

Figure 6. Hemin-induced HO productions. LCLs derived from normal control and the patient were cultured for 24 hrs in the presence of different concentrations of hemin (0 to $100\,\mu M$). Immunoblotting analysis for both HO-1 and HO-2 was performed after the culture.

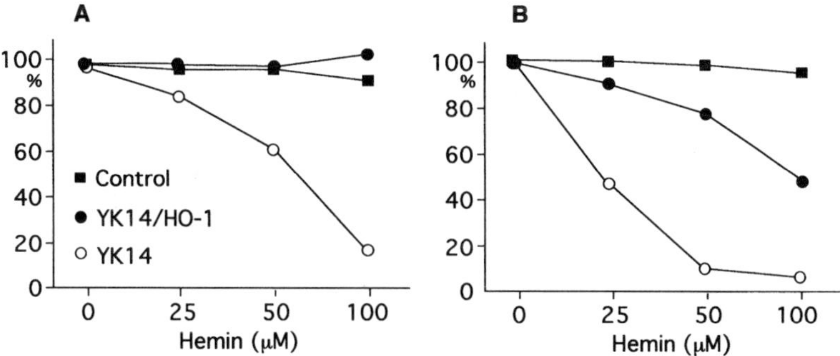

Figure 7. Hemin-induced oxidative cell injury and the effect of HO-1 gene transfection. LCLs derived from normal control and the patient (YK14) were cultured for 24 hrs (**A**) and 48 hrs (**B**) in the presence of different concentrations of hemin (0 to 100 μM). LCLs transfected with HO-1 gene (YK14/HO-1) were similarly cultured and cell injury was compared.

retrovirus vector. The human HO-1 cDNA was cloned into pGCsamEN so that the HO-1 translational start site was positioned precisely where the env translational initiation site in the wild type virus.

LCL successfully transfected with pGCsamEN were selected by G418.[27] HO-1 expression was confirmed in patient LCL transfected with HO-1 gene by a flow cytometry and by immunohistochemistry. HO-1 transfected LCL expressed HO-1 constitutively, without any stimulation. The level of HO-1 protein did not increase even after stimulation. Patient LCL transfected with *neo* control vector as a control was not producing HO-1. The HO-1 protein expression was further confirmed by immunoblotting assay. When LCL transfected with pGCsamENHO-1 were cultured with different concentrations of hemin, significant inhibition of cell injury was observed. LCL transfected with pGCsamEN without HO-1 gene were injured dose-dependently, as the patient original LCL (Fig. 7). These results supported the notion that degradation of heme by HO-1 is directly responsible for the reversal of hemin-induced cell injury.

5. PROTECTION OF ENDOTHELIAL CELL INJURY BY HO-1

Injury of vascular endothelial cells plays pivotal roles in pathological phenomenon induced by several inflammatory factors and oxidative stresses. HO-1 is stress-induced enzyme that acts as antioixidant and has potent anti-inflammatory properties.[28,29] The significance of protective role of HO-1 in cell damage induced by inflammatory stress is suggested by the fact that a HO-1 deficient human case showed severe endothelial cell injury and systemic inflammation.[15] To clarify the role of HO-1, cell death induced by hydrogen peroxide and heme with death signals through TNF families were examined in human endothelial cell line ECV304 and HO-1 transfected cell line. Cell death were induced by treatment with H_2O_2 in a dose-dependent manner, but hemin, anti-Fas antibody or soluble form of TNF-related apoptosis inducing ligand (sTRAIL) alone did not result in cell death. Hemin treatment in combination

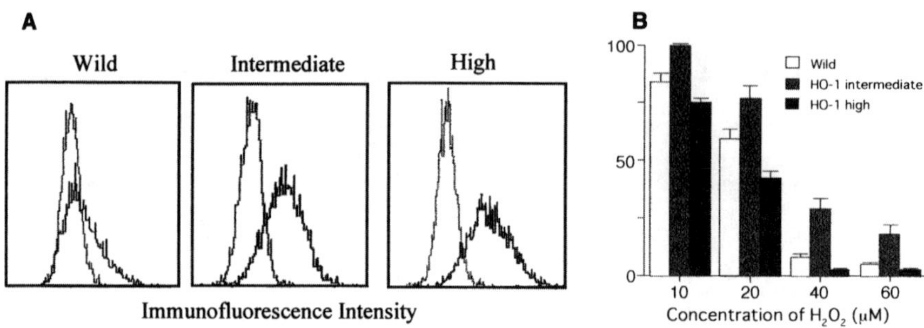

Figure 8. **A**: HO-1 gene transfection into a endothelial cell line ECV304. **B**: Sensitivities to oxidative injury were significantly different among wild type cells and transfectant with intermediate or high HO-1 expression.

with anti-Fas mAb (0.1–1 mg/ml) or sTRAIL (0.1–1 mg/ml) induced apoptosis of wild type ECV304. HO-1 was not expressed in wild type of ECV304 and the HO-1 expression was induced by treatment with hemin in a dose-dependent manner.

We obtained two ECV304 subclones which expressed stable and constitutive HO-1 protein expression at intermediate and high levels by transfection of human HO-1 gene using a retrovirus vector. ECV304 clone with intermediate HO-1 expression showed apparent decrease of cell death induced by H_2O_2, but sensitivity to Fas-induced apoptosis in combination with hemin was not changed. Unexpectedly, the clone with high HO-1 expression exhibited enhanced cell death induced by H_2O_2 or by combination of hemin and anti-Fas mAb. The increase in cell death was associated with decreased rate of cell growth. These results indicate that HO-1 acts as repressor of cell death induced by oxidative stress, but not as inhibitor of apoptosis through Fas death signal in vascular endothelial cells. Furthermore, prolonged expression of HO-1 at high level seems to be detrimental, resulting in accelerated cell death by increased oxidative stress (Figure 8).

6. SUMMARY

The in vitro experimental models presented here have been performed with extremely high concentration of hemin as a stimulant. Furthermore, it is obviously dangerous and probably too early to draw any conclusion based on the results obtained with a single HO-1 deficienct patient. However, our experience of the first case of HO-1 deficient patient revealed that the inducible heme oxygenase is indispensible in the protection of the vulnerable cells from various forms of oxidative injury and maintaining vascular functions, and most likely preventing progressive dysfunction of mononuclear phagocytes. Simple gene transfer may not work for the treatment of the illness. As indicated by the in vitro study using an endothelial cell line and the transfectant, constitutive expression of HO-1 may have unexpected, detrimental effect. Modulation of regulatory elements of HO-1 gene may become more feasible approach to control the level of the gene expression. Further studies are mandatory for the better description of the functional roles of this critical, albeit little understood enzyme of heme degradation.

REFERENCES

1. N.G. Abraham, G.S. Drummond, J.D. Lutton, and A. Kappas, The biological significance and physiological role of heme oxygenase, *Cell. Physiol. Biochem.*, **6**, 129–168 (1996).
2. P.A. Dennery, Regulation and role of heme oxygenase in oxidative injury, *Cur. Top. Cell. Regul.*, **36**, 181–199 (2000).
3. L.E. Otterbein and A.M.K. Choi, Heme oxygenase: colors of defence against cellular stress, *Am. J. Physiol. Lung. Cell. Mol. Physiol.*, **279**, L1029–1037 (2000).
4. M.D. Maines, Heme oxygenase: function multiplicity, regulatory mechanisms, and clinical applications, *FASEB J.*, **2**, 2557–2568 (1988).
5. W.K. McCoubrey Jr., T.J. Huang, and M.D. Maines, Isolation and characterization of a cDNA from the rat brain that encodes hemoprotein heme oxygenase-3, *Eur. J. Biochem.*, **247**, 725–732 (1997).
6. A.M.K. Choi and J. Alam., Heme oxygenase-1: function, regulation and implication of a novel stress-inducible protein in oxidant-induced lung injury, *Am. J. Respir. Cell. Mol. Biol.*, **15**, 9–19 (1996).
7. D. Willis, A.R. Moore, R. Frederick, and D.A. Willoughby, Heme oxygenase: a novel target for the modulation of the inflammatory response *Nature Med.*, **2**, 87–90 (1996).
8. K.D. Poss and S. Tonegawa, Reduced stress defence in heme oxygenase1-deficient cells, *Proc. Natl. Acad. Sci. USA*, **94**, 10925–10930 (1997).
9. M. Suematsu and Y. Ishimura, The heme oxygenase-carbon monoxide system: a regulator of hepatobiliary function, *Hepatology*, **31**, 3–6 (2000).
10. H. Suzuki, K. Kanamaru, H. Tsunoda, H. Inada, M. Kuroki H. Sun, S. Waga, and T. Tanaka, Heme oxygenase-1 gene induction as an intrinsic regulation against delayed cerebral vasospasms in rats, *J. Clin. Invest.*, **104**, 59–66 (1999).
11. N. Goda, K. Suzuki, M. Naito, S. Takeoka, E. Tsuchida, Y. Ishimura, T. Tamatani, and M. Suematsu, Distribution of heme oxygenase isoforms in rat liver, *J. Clin. Invest.*, **101**, 604–612 (1998).
12. K.A. Nath, G. Balla, G.M. Vercellotti, J. Balla, H.S. Jacob, M.D. Levitt, and M.E. Rosenberg, Induction of heme oxygenase is a rapid, protective response in rhabdomyolysis in the rat, *J. Clin. Invest.*, **90**, 267–270 (1992).
13. J. Balla, H.S. Jacob, G. Balla, K. Nath, and G.M. Vercellotti, Endothelial cell heme oxygenase and ferritin induction by heme proteins: a possible mechanism limiting shock damage, *Trans. Assoc. Am. Physicians*, **105**, 1 (1992).
14. F. Amersi, Upregulation of heme oxygenase-1 protects genetically fat Zucker rat livers from ischemia/reperfusion injury, *J. Clin. Invest.*, **104**, 1631–1639 (1999).
15. A. Yachie, Y. Niida, T. Wada, N. Igarashi, H. Kaneda, T. Toma, K. Ohta, Y. Kasahara, and S. Koizumi, Oxidative stress causes enhanced endothelial cell injury in human heme oxygenase-1 deficiency, *J. Clin. Invest.*, **103**, 129 (1999).
16. K. Ohta, A. Yachie, K. Fujimoto, H. Kaneda, T. Wada, T. Toma, A. Seno, Y Kasahara, H. Yokoyama, H. Seki, and S. Koizumi, Tubular injury as a cardinal pathological feature in human heme oxygenase-1 deficiency, *Am. J. Kidney Dis.*, **35**, 863–870 (2000).
17. Y. Saikawa, H. Kaneda, L. Yue, S. Shimura, T. Toma, Y. Kasahara, A. Yachie, and S. Koizumi, Structural evidence of genomic exon-deletion mediated by *Alu-Alu* recombination in a human case with heme oxygenase-1 deficiency, *Human Mut.*, **16**, 178–179 (2000).
18. S. Shibahara, M Sato, R.M. Muller, and T. Yoshida, Structural organization of the human heme oxygenase gene and the function of its promoter, *Eur. J. Biochem.*, **171**, 557–563 (1988).
19. R.M. Tyrrell, L.A. Applegate, and Y. Tromvoukis, The proximal promoter region of the human heme oxygenase gene contains elements involved in stimulation of transcriptional activity by a variety of agents including oxidants *Carcinogenesis* **14**, 761–765 (1993).

20. K. Takeda, S. Ishizawa, M. Sato, T. Yoshida and S. Shibahara, Identification of a cis-acting element that is responsible for cadmium-mediated induction of the human heme oxygenase gene, *J. Biol. Chem.*, **269**, 22858–22867 (1994).

21. D.N. Cooper, M. Krawczak, and S.E. Antonarakis, The nature and mechanisms of human gene mutation, in: *The Metabolic and Molecular Biology*, edited by C.R Scriver, et al. (McGraw-Hill, New York, 1995), pp. 259–291.

22. Y. Miki, Retrotransposal integration of mobile genetic elements in human disease, *J. Hum. Genet.*, **43**, 77–84 (1998).

23. C.W. Schmid and W.R. Jelinek, The *Alu* family of dispersed repetitive sequences, *Science*, **216**, 1065–1070 (1982).

24. J. Jurka and T. Smith, A fundamental division in the *Alu* family of repeated sequence, *Proc. Natl. Acad. Sci. USA*, **85**, 4775 (1998).

25. J. Jurka and A. Milosavljevic, Reconstruction and analysis of human *Alu* genes, *J. Mol. Evol.*, **32**, 105–121 (1991).

26. N.S. Rudiger, N. Gregersen, and M.C. Kielland-Brandt, One short well conserved region of *Alu*-sequences is involved in human gene rearrangements and has homology with prokaryotic chi, *Nucleic Acids Res.*, 23, 256–260 (1995).

27. M. Onodera, A. Yachie, D.M. Nelson, H. Welchlin, R.A. Morgan, and R.M. Blaese, A simple and reliable method for screening retroviral producer clones without selectable markers, *Hum Gene Ther*, **8**, 1189–1194 (1997).

28. L.J. Wang, T.S. Lee, F.Y. Lee, R.C. Pai, and L.Y. Chau, Expression of heme oxygenase-1 in atherosclerotic lesions, *Am. J. Pathol.* **152**, 711 (1998).

29. A. Agarwal, H. de Leon, J.B. Laursen, J.N. Wilcox, G. De Keulenaer K.K. Griendling, and R.W. Alexander, Linoleyl hydroperoxide transcriptionally upregulates heme oxygenase-1 gene expression in human renal epithelial cells, *J. Am. Soc. Nephrol.*, **9**, 1990–1997 (1998).

29

HEME OXYGENASE INDUCTION IN LIVER CELLS BY HEPATOCYTE GROWTH FACTOR AND OXIDATIVE STRESS

Maria Alfonsina Desiderio and Lorenza Tacchini

Istituto di Patologia Generale-Università degli Studi di Milano e Centro di Studio sulla Patologia Cellulare del C.N.R.-via L. Mangiagalli 31-20133 Milano-Italy.

1. INTRODUCTION

Hepatocyte growth factor (HGF), initially identified as "scatter factor", is produced by mesenchimal cells and exerts multiple biological activities such as morphogenesis, motogenesis, mitogenesis and pro- or anti-apoptotic effects on a variety of normal and neoplastic epithelial cells. The receptor capable of signal transduction of HGF is the transmembrane tyrosine kinase encoded by the proto-oncogene c-*met*. Met receptor is activated by autophosphorylation, and the bulk of receptor signalling activity is funneled through a single multisubstrate docking site. Different signal transduction pathways mediate cell proliferation (Grb2/Sos/Ras/Raf/MAPK), survival (PI3K/Akt/PAK1), and cell dissociation (PI3K/Rac) (Zarnegar and Michalopoulos, 1995; Maggiora et al., 1997).

While normal HGF/Met signalling is involved in various kinds of normal growth processes, such as embryogenesis, liver regeneration and angiogenesis, abnormal HGF/Met signalling has been implicated in both tumor development and progression. Germline missense mutations in the tyrosine kinase domain of Met were detected in a small number of genetically inherited tumors of the kidney papilla, while somatic mutations have been found in a small proportion of sporadic tumors, such as papillary renal carcinoma and childhood hepatocellular carcinoma. Met is overexpressed at high frequency in sporadic cancers including primary liver carcinomas, hepatoblastoma and intrahepatic cholangiocarcinoma, and is amplified during progression from primary tumors to metastatic cells. Recently a more direct link between paracrine effects of HGF and motility/invasion at the secondary site of tumor development has been reported. Met activating mutations might confer the ability to establish metastatic colonies to a subpopulation of head and neck squamous cell carcimomas (Di Renzo et al., 2000).

We have previously observed that the intracellular signal transduction cascade(s) triggered by HGF/Met binding in human hepatoma cells HepG2 activate(s) downstream transcription factors involved in growth and apoptosis, such as AP-1 and c-Myc/Max (Tacchini et al., 2000). Consistent with these data, in the same experimental model HGF causes an initial stimulation of DNA synthesis before the occurrence of a cytostatic and apoptotic effect (Matteucci et al., 2001). Taking also into account the rather surprising activation of the transcription factor NF-kB (Tacchini et al., 2000), known to regulate transcription of genes under stress, injury and inflammatory conditions, we undertook further studies to clarify specific transduction pathways ultimately leading to transcription factor activation and possibly implicated in the invasive phenotype induced by HGF during tumor progression.

2. HEME OXYGENASE INDUCTION BY HGF

Among the transcription factors activated by HGF, we identified hypoxia-inducible factor-1 (HIF-1). Searching for putative target genes, i.e. genes containing the corresponding hypoxia responsive element (HRE) consensus sequence in their promoters, we demonstrated a time-dependent induction of heme-oxygenase (HO) mRNA with the maximum (3.4-fold) 4 hours after HGF treatment (Tacchini et al., 2001). The contemporaneous activation by HGF of a transcription factor, which was first described as a DNA-binding activity essential for hypoxia-inducible genes, and the induction of a gene, which was considered typical of oxidative stress, seem to be counterintuitive. But a series of facts put this observation under a more logical perspective. First of all, we observed that the activation of HIF-1 by HGF was strongly decreased by N-acetylcysteine (NAC), a free radical scavenger, suggesting a role of reactive oxygen species (ROS) in the mechanism of action of HGF: the idea was reinforced by the fact that NAC reduced the stimulatory effect of HGF on the activity of c-Jun-N-terminal kinase and p38 mitogen activated protein kinase, two well known stress-induced protein kinases. Secondly, the concept was beginning to emerge that the cell treatment with HGF was associated with the generation of ROS (Arakaki et al., 1999). This was in analogy with the observations done with Platelet derived growth factor, thrombin and serum (Rao et al., 1999), with Epidermal growth factor (Bae et al., 1997) and with basic Fibroblast growth factor (Tannickal et al., 2000). Third, data had been reported that activation of HIF-1 by hypoxia could be mediated by small amounts of ROS, generated by mitochondria during hypoxia (Richard et al., 1999; Semenza et al., 2000). These facts could explain: (1) the similarity of the responses to HGF and hypoxia, and (2) the HGF-dependent concomitant activation of HIF-1 and induction of HO, which are respectively considered responsive to hypoxia and oxidative stress.

3. OXIDATIVE STRESS AND OTHER KINDS OF STRESSES

Data from this laboratory have shown the induction of HO and the concurrent activation of HIF-1, during oxidative stress induced by different chemical agents, capable of changing the redox state of the liver cells by different mechanisms (Figure 1). Nitrofurantoin, which causes the formation of superoxide anion by redox cycling (Rossi et al., 1988; Hoener et al., 1989) and Phorone, which strongly depletes the glutathione (GSH) pool through the action of glutathione-S-transferase (Sunahara and Chiesa,1992), gave a prompt and persisting increase of HO mRNA which is

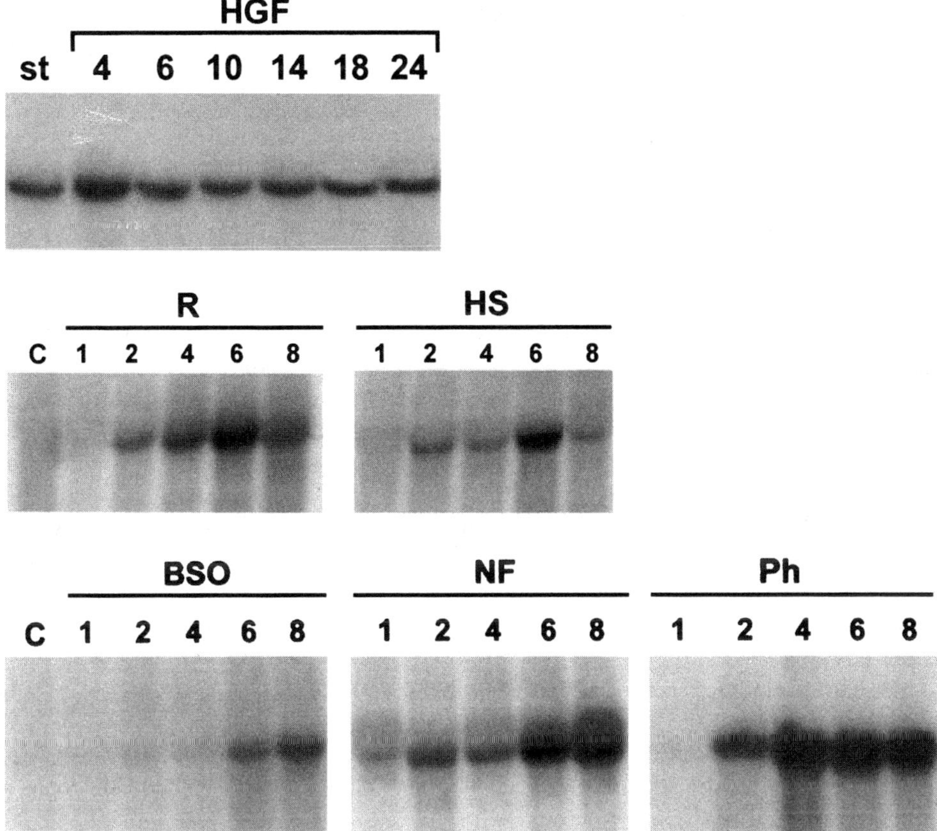

Figure 1. Induction of HO mRNA in HepG2 hepatoma cells by HGF, and in rat liver after post-ischemic reperfusion (R), heat shock (HS), Buthionine sulphoximine (BSO), Nitrofurantoin (NF) and Phorone (Ph). st, starved cells; C, control liver. The numbers on the top of the lanes represent hours after treatment. From: Tacchini et al. *Carcinogenesis* **22**, 1363 (2001); *J Cell Physiol* **180**, 255 (1999); *Biochem Pharmacol* (2002, in press) by permission of Oxford University Press, Wiley-Liss Inc., a subsidiary of John Wiley & Sons, Inc., and Elsevier Science B.V., respectively.

particularly marked in the case of Phorone (Tacchini et al., 2002). Buthionine sulphox-imine, which inhibits GSH synthesis by inhibiting γ-glutamylcysteine synthetase (Martensson et al., 1991) and decreases GSH concentration more slowly than the other two compounds, acted less rapidly and gave a less pronounced response. In run-on experiments with Phorone-treated rat liver nuclear preparations, we were able to demonstrate that the response has a strong transcriptional component (Tacchini et al., 1995).

We were also able to show induction of HO under different conditions of cellular stress. During the re-establishment of blood flow that follows previous ischemia there are changes, generally known as reperfusion injury, that are not only the expression of a harm superimposed on those caused by ischemia, but may represent an active response involved in the protection against the injury or in the repair of the damage previously inflicted on the tissue. Post-ischemic reperfusion of the liver is characterised by progressively increased levels of HO mRNA, which remained stable for a long time (Tacchini et al., 1999). Heat shock is the best known kind of cellular stress, and has been the subject of a vast literature (Nover, 1991). Heat shock had substantially the same effects as post-ischemic reperfusion on HO mRNA (Tacchini et al., 1999), which is understandable considering the nature of HO described for many years as Hsp32kD.

4. TRANSCRITPION FACTORS

The enhanced messenger expression of HO under many different conditions addresses the question of the mechanism(s) of induction and the transcription factors possibly involved. The promoter of HO contains two HIF-1 consensus sequences. We have shown that HIF-1 was clearly activated by HGF stimulation (Tacchini et al., 2001), slightly activated by ischemia, but disappearing at reperfusion (Tacchini et al., 1999), strongly but transiently activated by heat shock (Tacchini et al., 1999) and also by chemically induced oxidative stress (Tacchini et al., 1995). However HO also contains consensus sequences for HSF, AP-2 and NFkB (Lavrovsky et al., 1994) which are subjected to a complex regulation and are differently activated during the above mentioned types of stresses. The respective role of these transcription factors can vary in importance under different conditions, and the degree to which a promoter responds to a factor can also vary depending on the presence of other site-specific transcription factors. Regulatory sequences of eukaryotic genes commonly contain binding sites for multiple transcription factors, and provide a basis for combinational interactions for different factors. These regions are usually referred to as composite response elements, that mediate the integration of multiple regulatory signals (Stephanou et al., 1999).

Complex and sometimes seemingly contradictory data can be better understood in the frame of the recently expressed concept of context-dependent transcriptional regulation (Fry and Farnham, 1999). The concept refers to instances in which the transcriptional properties of a particular factor are influenced either by its position relative to other factors bound to a given promoter, or by the abundance of transcription cofactors in a given cell type. Transcriptional activity can also be influenced by the cellular environment in which the assays are performed.

5. PERMISSIVE CONDITIONS

A case in point seems to be represented by the effects of polyamine imbalance on the levels of HO mRNA in FAO hepatoma cells exposed to heat shock. The study addressed primarily the expression of hsp 70, but HO was also investigated (Desiderio et al., 1996). Depletion of polyamines was obtained in short term experiments (24–48 hours) by the use of α-difluoromethylornithine (DFMO), a classical inhibitor of ornithine decarboxylase, or of the combination of two newly available inhibitors of ornithine decarboxylase and S-adenosylmethionine decarboxylase, known respectively as MAP and AbeAdo. Under our experimental conditions, imbalance of the polyamine pool was realized without appreciable growth arrest and impaired expression of growth-related genes. The level of HO mRNA was only slightly increased immediately after the exposure to high temperature (1 hour at 42 °C), but became much higher during the period (30–60 min) of recovery after heat shock. Decreases of putrescine and spermidine content caused by DFMO treatment reduced the accumulation of HO mRNA, while the depletion of spermidine and spermine obtained with the treatment with MAP plus AbeAdo practically abolished the response of HO. Therefore, the integrity of the polyamine pool seems to be a necessary permissive condition for an effective induction of HO.

The induction of HO under different conditions of the liver cell poses the problem of the significance of this phenomenon. One common feature of these inducers is their capacity to increase the level of ROS, either due to their increased generation (Nitrofurantoin, HGF and post-ischemic reperfusion) or decreased concentration of GSH (Phorone and Buthionine sulphoximine). This observations not only demonstrate that HO can be induced by agents causing oxidative stress, but hint to the possibility that HO can act as a protective molecule against ROS-induced cell injury, in particular because of its antiapoptotic effects (Otterbein and Choi, 2000). On the other hand, the increasingly accepted idea that ROS can have a function as signalling molecules (Thannickal and Fanburg, 2000) has its counterpart in the recognition that the heme oxygenase system can also have a role as a regulator of second messenger gases (Maines, 1997).

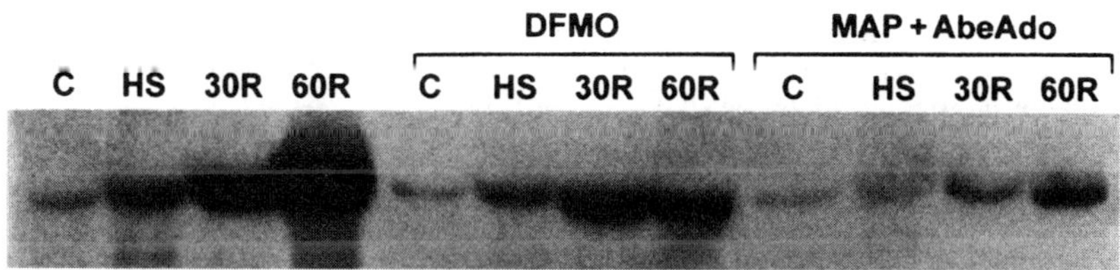

Figure 2. Induction of HO in FAO hepatoma cells exposed to heat shock, and effects of polyamine depletion. C, control; HS, at the end of heat shock (42 °C for 60 min); 30 R, after 30 min of recovery at 37 °C; 60 R, after 60 min of recovery at 37 °C; DFMO and MAP + AbeAdo, cells pretreated with these polyamine-depleting agents. From: Desiderio et al. *Hepatology* **24**, 150 (1996) by permission of W.B. Saunders Co. a Harcourt Health Sciences Co.

6. SUMMARY

Treatment of hepatoma cells with HGF induces HO and activates the transcription factor HIF-1. The concomitant induction of a gene commonly connected with oxidative stress and of a hypoxia-associated transcription factor is explained by a series of facts and is reinforced by the observation that various kinds of oxidative stressful conditions lead to the same result. Reperfusion stress and heat shock are also good inducers of HO, which seems also to depend on permissive cellular conditions. Imbalance of the polyamine pool in hepatoma cells, caused by different polyamine depleting agents, decreases or suppresses the induction of HO. The role of various transcription factors in the induction of HO is discussed in the frame of the recently expressed concept of "context-dependent transcriptional regulation".

ACKNOWLEDGMENTS

We thank Prof. Aldo Bernelli-Zazzera for helpful advice and criticism.

REFERENCES

Arakaki, N., Kajihara, T., Arakaki, R., Ohnishi, T., Kazi, J.A., Nakashima, H., and Daikuhara, Y., 1999. Involvement of oxidative stress in tumor cytotoxic activity of hepatocyte growth factor/scatter factor. *J Biol Chem* **274**:13541.

Bae, Y.S., Kang, S.W., Seo, M.S., Baines, I.C., Tekle, E., Chock, P.B., and Rhee, S.G., 1997. Epidermal growth factor (EGF)—induced generation of hydrogen peroxide. Role of EGF receptor mediated tyrosine phosphrylation. *J Biol Chem* **272**:217.

Desiderio, M.A., Tacchini, L., Anzon, E., Pogliaghi, G., Radice, L., and Bernelli-Zazzera, A., 1996. Effects of polyamine imbalance on the induction of stress genes in hepatocarcinoma cells exposed to heat shock. *Hepatology* **24**:150.

Di Renzo, M.F., Olivero, M., Mrtone, T., Maffe, A., Maggiora, P., De Stefani, A., Valente, G., Giordano, S., Cortesina, G., and Comoglio, P.M., 2000. Somatic mutations of the *Met* oncogene are selected during metastatic spread of human HNSC carcinomas. *Oncogene* **19**:1547.

Fry, C.J. and Farnham, P.J., 1999. Context-dependent transcriptional regulation. *J Biol Chem* **274**:29583.

Hoener, B., Noach, A., Andrup, M., and Yen, T.S., 1989. Nitrofurantoin produces oxidative stress and loss of glutathione and protein thiols in the isolated perfused rat liver. *Pharmacology* **38**:363.

Lavrovsky, Y., Schwartzman, M.L., Levere, R.D., Kappas, A., and Abraham, N.G. 1994. Identification of binding sites for transcription factors NF-kappa B and AP-2 in the promoter region of the human heme oxygenase 1 gene. *Proc Natl Acad Sci USA* **91**:5987.

Maggiora, P., Gambarotta, G., Olivero, M., Giordano, S., Di Renzo, M.F., and Comoglio M.P., 1997. Control of invasive growth by the HGF-receptor family. *J Cell Physiol* **173**:183.

Martensson, J., Jain, A., Stole, E., Frayer, W., Auld, P.A., and Meister, A., 1991. Inhibition of glutathione synthesis in the newborn rat: a model for endogenously produced oxidative stress. *Proc Natl Acad Sci USA* **88**:9360.

Matteucci, E., Castoldi, R., and Desiderio, M.A., 2001. Hepatocyte growth factor induces pro-apoptotic genes in HepG2 hepatoma but not in B16-F1 melanoma cells. *J Cell Physiol* **186**:387.

Maines, M.D., 1997. The heme oxygenase system: a regulator of second messenger gases. *Annu Rev Pharmacol Toxicol* **37**:517.

Nover, L. (Ed), 1991. Heat shock response. Boca Raton; CRC Press.

Otterbain, L.E. and Choi, A.M.K., 2000. Heme oxygenase: colors of the defense against cellular stress. *Am J Physiol Lung Cell Mol Physiol* **279**:L1029.

Rao, G.N., Katki, K.A., Madamanchi, N.R., Wu, Y., and Birrer, M.J. 1999. JunB forms the majority of the AP-1 complex and is a target for redox regulation by receptor tyrosine kinase and G protein-coupled receptor agonists in smooth muscle cells. *J Biol Chem* **274**:6003.

Richard, D.E., Berra, E., and Pouyssegur, J. 1999. Angiogenesis: how a tumor adapts to hypoxia. *Biochem Biophys Res Commun* **266**:718.

Rossi, L., Silva, J.M., McGirr, L.G., and O'Brien, P.J., 1988. Nitrofurantoin-mediated oxidative stress cytotoxicity in isolated rat hepatocytes. *Biochem Pharmacol* **37**:3109.

Semenza, G.L., 2000. Expression of hypoxia-inducible factor 1: mechanisms and consequences. *Biochem Pharmacol* **59**:47.

Stephanou, A., Isenberg, D.A., Nakajima, K., and Latchman, D.S., 1999. Signal transducer and activator of transcription-1 and heat shock factor-1 interact and activate the transcription of the Hsp-70 and Hsp-90beta gene promoters. *J Biol Chem* **274**:1723.

Sunahara, G.I. and Chiesa, A., 1992. Phorone (diisopropylidene acetone), a glutathione depletor, decreases rat glucocorticoid receptor binding in vivo. *Carcinogenesis* **13**:1083.

Tacchini, L., Dansi, P., Matteucci, E., and Desiderio, M.A., 2000. Hepatocyte growth factor signal coupling to various transcription factors depends on triggering of Met receptor and protein kinase transducers in human hepatoma cells HepG2. *Exp Cell Res* **256**:272.

Tacchini, L., Dansi, P., Matteucci, E., and Desiderio, M.A., 2001. Hepatocyte growth factor signalling stimulates hypoxia inducible factor-1 (HIF-1) activity in HepG2 hepatoma cells. *Carcinogenesis* **22**:1363.

Tacchini, L., Fusar-Poli, D., and Bernelli-Zazzera, A., 2002. Activation of transcription factors by drugs inducing oxidative stress in rat liver. *Biochem Pharmacol* in press.

Tacchini, L., Pogliaghi, G., Radice, L., Anzon, E., and Bernelli-Zazzera, A., 1995. Differential activation of heat-shock and oxidation-specific stress genes in chemically induced oxidative stress. *Biochem J* **309**:453.

Tacchini, L., Radice, L., and Bernelli-Zazzera A., 1999. Differential activation of some transcription factors during rat liver ischemia, reperfusion, and heat shock. *J Cell Physiol* **180**:255.

Thannickal, V.J., Day, R.N., Kling, S.J., Bastien, M.C., Larios, J.M., and Fanburg, B.L., 2000. Ras-dependent and -independent regulation of reactive oxygen species by mitogenic growth factors and TGF-b1. *FASEB J* **14**:1741.

Thannickal, V.J. and Fanburg, B.L., 2000. Reactive oxygen species in cell signaling. *Am J Physiol Lung Cell Mol Physiol* **279**:L1005.

Zarnegar, R. and Michalopoulos, G.K., 1995. The many faces of hepatocyte growth factor: from hepatopoiesis to hematopoiesis. *J Cell Biol* **129**:1177.

30

RESPONSE OF HEME OXYGENASE AND TELOMERASE ENZYMES TO OXIDATIVE STRESS IN HUMAN HEPATOMA CELL LINE (HepG2)

Maivel H. Ghattas[a], Linus T. Chuang[c], and Nader G. Abraham[a,b]

Department of Pharmacology
New York Medical College
Valhalla, NY[a] and The Rockefeller University
NY[b]
Department of Obstetric and Gynecology,
Westchester Medical Center, Valhalla, NY[c]

INTRODUCTION

Telomerase, a specialized reverse transcriptase enzyme, maintains the telomere length by adding new telomeric repeats onto chromosome ends using part of a constituent RNA molecule as a template for synthesizing a telomeric repeat.[1] The enzymatic core of human telomerase minimally consists of telomerase RNA (hTR) and the catalytic subunit homology protein (hTERT).[2,3] Other protein components of the telomerase enzyme, such as hTEP1[4] are not required for telomerase action in vitro, but telomerase-associated proteins may serve key roles in allowing telomerase to act at the telomeric complex in vivo.[5] Telomerase is expressed widely during human embryogenesis but, in adults, expression is limited primary to germ cells, early progenitor/stem cells, and stimulated cycling lymphocytes.[6] Telomerase activity has been reported in many types of malignant tumors, including neuroblastomas,[7] lung carcinomas,[8] gastric and colon carcinomas,[9] bladder carcinoma[10] and brain tumors.[11] Ogami et al.[12] detected high level of telomerase activity in most hepatocellular carcinoma tissues; however, very weak telomerase activity was detected in approximately half of nontumorous chronic liver disease tissues.

The regulation of telomerase is executed primarily through the transcriptional control of the gene encoding the catalytic subunit TERT. Expression of hTERT is observed at high levels in cancer cell lines, but not in telomerase-negative cell lines.

The life span of a normal telomerase negative human fibroblast is significantly extended if these cells are transfected with the human telomerase reverse transcriptase (hTERT) gene.[13] On the other hand, it is well established that telomerase activity strongly correlates with cell immortality.[14] Based on this fact, telomerase becomes a potential target for anticancer applications.[15]

Oxygen free radicals are generated under both normal and pathological circumstances and have been implicated in the pathogenesis of diseases such as atherosclerosis and cancer, as well as in aging and in some inflammatory disorders.[16,17] The involvement of free radicals in the carcinogenic process suggests that oxidative stress might have an effect on the intrinsic signal transduction cascades leading to cell division. At the same time, reactive oxygen species (ROS) generation has been found to be an important event in apoptotic tumor cell death induced by various anti-cancer agents such as camptothecin, vinblastine, inostamycin and adrimycin.[18] In human fibroblasts, the rate of telomere shortening was increased by mild hyperoxia and hydrogen peroxide.[19] It is thus important to study the interrelation between oxidative stress and telomerase in human tumor cells.

Heme oxygenase (HO) is the enzyme that controls the initial and rate-limiting step in heme catabolism, catalyzing the cleavage of the heme ring to form ferrous iron, carbon monoxide (CO) and biliverdin. Three isoforms of HO have been cloned, HO-1, HO-2 and HO-3.[20] HO-1 is a 32-KDa-heat shock protein[21] that is inducible by numerous noxious stimuli.[22] HO-2 is a constitutively synthesized 36-KDa protein that is abundant in brain and testis.[23] HO-3 is related to HO-2, but is the product of a different gene, and its ability to catalyze heme degradation is lower than that of HO-1.[23] A growing body of data reveals that HO and its metabolized product, CO, play an important role in numerous biological processes, especially antioxidative reactions.[24–26] Human HO-1 gene-transduced endothelial cells acquired substantial resistance to toxicity produced by exposure to heme and H_2O_2 compared with that in nontransduced cells.[27] Moreover, overexpression of HO-1 gene in endothelial cells enhances angiogenesis.[28]

In contrast to the antioxidant effects, a relationship between malignant behavior and alteration of HO may exist; HO activity was reported as increased in several tumors and malignant cells.[29] Maines and Abrahamsson[30] detected increased expression of HO-1 in benign prostate hyperplesia and prostate cancer. Goodman et al.[31] demonstrated elevation of HO activity in renal adenocarcinoma compared with juxtatumor or normal renal tissues and this elevation may be attributed solely to the HO-1 gene expression. The authors suggest that the differential elevation of HO-1 gene may be a contributing factor for the undetectable levels of the heme dependent cytochrome P450 arachidonic acid metabolite, 20-HETE; a potent mitogen in the adenocarcinoma. This alteration of HO may be the result of local or circulating factors released from malignant cells, as well as oxidative stress occurring through uncontrolled and uncoordinated cell growth.[31] The same elevation of HO-1 was observed in the acute hepatitis stage and hepatoma stage in a rat model and was probably due to the oxidative stress caused by the accumulation of free copper, free iron and free heme levels, as well as by the inflammatory cytokines produced by the surrounding tissues at the hepatoma stage.[32] In this chapter, we review the response of HO and telomerase enzyme to oxidative stress and the preventive role of up-regulation of HO-1 in protecting HepG2 against oxidative stress.

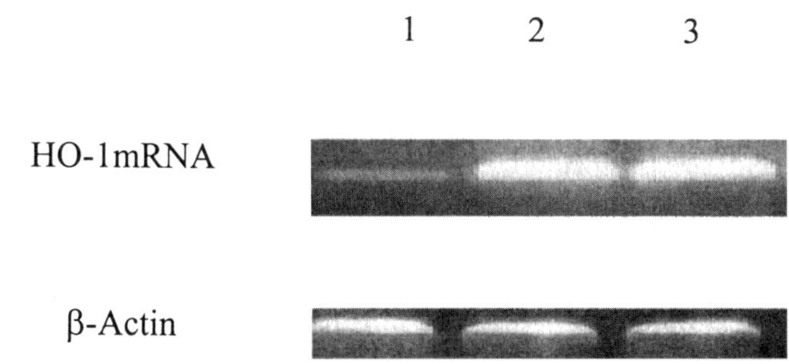

Figure 1. Induction of HO-1 mRNA in HepG2 cells. HepG2 cells were grown in $75\,cm^2$ Falcon tissue culture flasks until they reached 80% confluence and then were treated with $10\,\mu M$ heme or $100\,\mu M$ H_2O_2, for 4 h. Total RNA from the cells were subjected to RT-PCR to evaluate HO-1 mRNA expression. Induction of HO-1 was noticed in cells treated with heme, H_2O_2 (Lanes 2, 3 respectively) in comparison to non-treated cells (Lane 1). β-Actin mRNA expression for each sample was used as internal control.

HUMAN HEPATOMA CELL LINE (HepG2)

Human hepatoma cell line (HepG2) was grown in Minimum Essential Medium Eagle (MEM) supplemented with 10% (vol/vol) heat-inactivated fetal bovine serum (FBS), 100 U/ml of penicillin, 100 μg/ml of streptomycin, (pH 7.4) at 37 °C in a 5% CO_2/95% air chamber.

HO-1 A STRESS RESPONSIVE GENE

Many agents that elicit oxidative stress are potent inducers of HO-1 gene; stimulation of HO-1 provides protection against cellular oxidant stress both in vivo and in vitro.[25,33–35] Our study has demonstrated the potential use of HO-1, a major cellular stress gene in mammalian tissues, as a stress index in hepatoma cell line (HepG2); these cells when treated with oxidants, (heme 10 μM and H_2O_2 100 μM), respond by elevation of HO-1 mRNA (Fig. 1) and HO-1 protein (Fig. 2) in comparison to untreated cells.

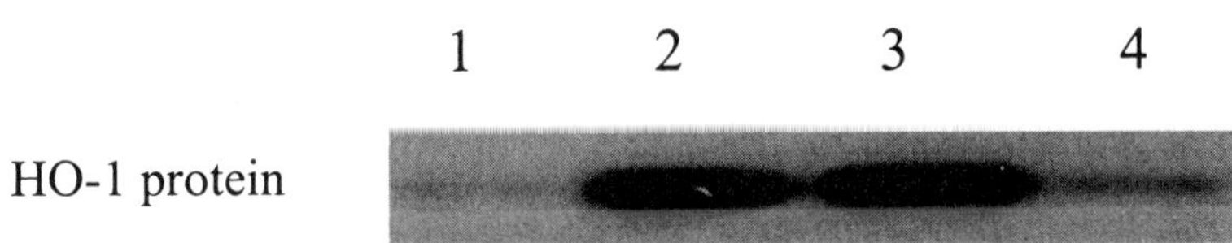

Figure 2. Western blot analysis of HO-1 and HO-2 in HepG2 cell homogenate. Cells were treated with 10 μM heme, $SnCl_2$ or 100 μM H_2O_2 for, 24 h. Cells were lysed and equal amount of cytosolic protein were subjected to immunoblotting to detect HO-1 protein expression. Nontreated cells were used as a control. Control cells containing low basal levels of HO-1 protein (Lane 1). Induction by (heme, $SnCl_2$ and H_2O_2) was seen in lanes 2, 3 and 4 respectively.

DOSE DEPENDENT EFFECT OF OXIDANTS ON HepG2 CELL VIABILITY

Hydrogen peroxide is a major component of ROS produced intracellularly during many physiological and pathological processes, and causes oxidative damage.[27] We tested the effect of different concentration of heme and H_2O_2 on cell viability. HepG2 cells were cultured and exposed to heme and H_2O_2 at a concentration of 50, 100 and 200 μM for 24h. (Fig. 3). The cytotoxic effect of heme and H_2O_2 on HepG2 cells was demonstrated by the strong decrease of cell viability in a dose dependent manner.

PROTECTIVE ROLE OF HO-1 AGAINST OXIDATIVE INJURY

Previous studies have hypothesized that HO may play a role in protecting cells from oxidative injury by elevating levels of bilirubin and biliverdin. Poss and Tonegawa[36] analyzed the response of cells, from mice lacking functional HO-1 gene expression, to oxidative challenges. Their results showed that cultural HO-1 deficient embryonic fibroblasts could produce higher oxygen free radicals when exposed to heme, H_2O_2, paraquat or cadmium chloride, and that they were hypersensitive to cytotoxicity caused by heme and H_2O_2. Furthermore, retroviral-mediated overexpression

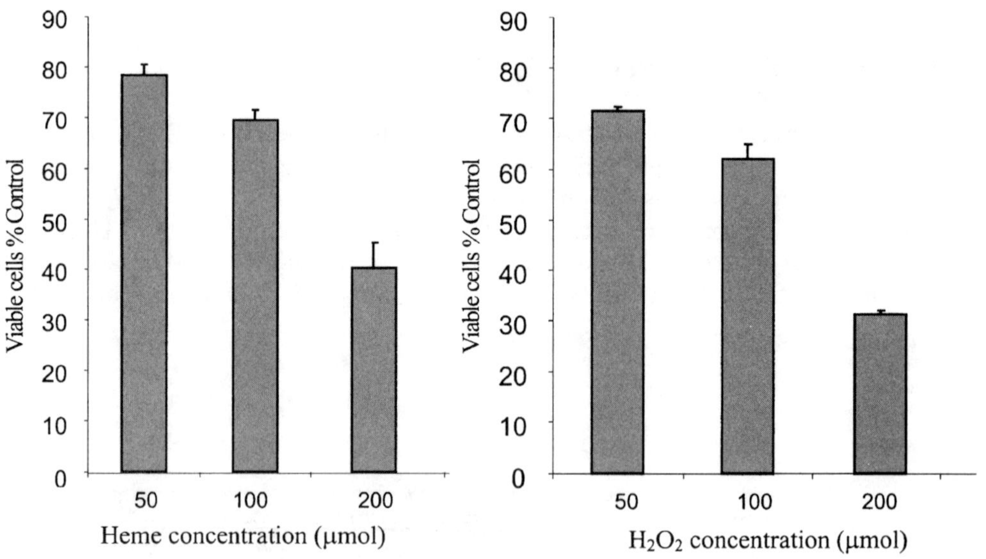

Figure 3. Dose dependent effect of oxidants on HepG2 cell viability. HepG2 cells were cultured in 96-well plates (5000 cells/well in 100 μl of complete MEM), incubated with heme (50–200 μM) and H_2O_2 (50–200 μM), and were allowed to grow for 24h. Cell viability was assessed by Aqueous cell proliferation ELISA kit assay.

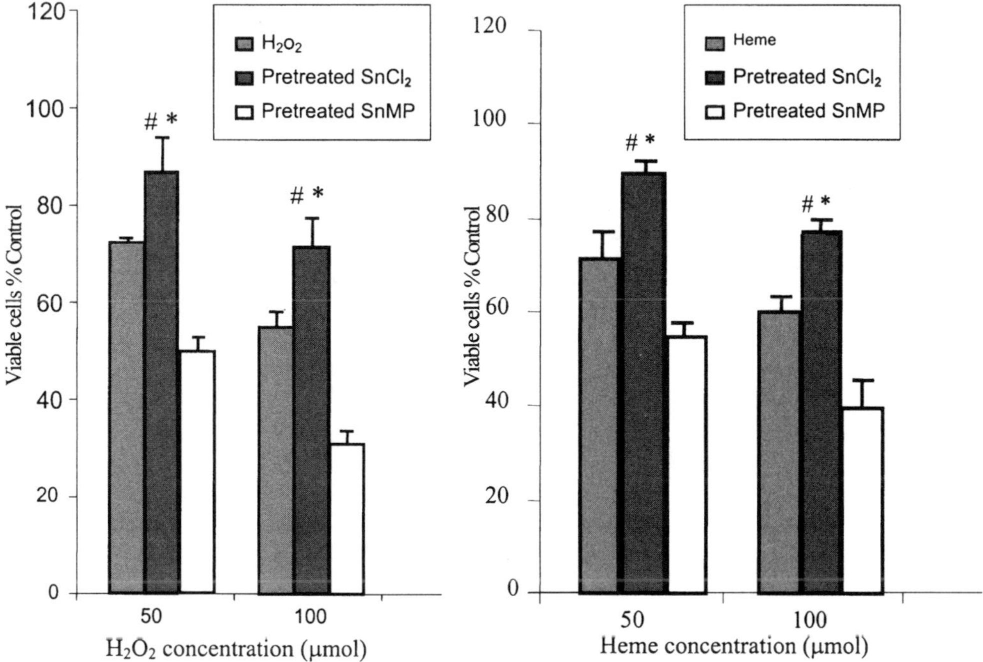

Figure 4. Protective effect of HO-1 against oxidative stress HepG$_2$ cells were cultured in 96-well plates (5000 cells/well in 100 μl of complete MEM) then incubated with SnCl$_2$ (10 μM), SnMP (10 μM) for 24 h. Heme and H$_2$O$_2$, at the indicated concentrations, were added for an additional 24 h. There was a significant difference between the inducer, SnCl$_2$, *p $<$ 0.01, and inhibitor, SnMP (# p $<$ 0.05), of HO-1 in comparison to cells treated with H$_2$O$_2$ or heme alone at concentrations of 50 and 100 μM.

of HO-1 was found to protect cells against heme-induced cytotoxicity.[37] We demonstrate that the ability of H$_2$O$_2$, as well as heme, to decrease cell viability is significantly reduced by the upregulation of HO-1 gene by SnCl$_2$, whereas downregulation of HO-1 by SnMP significantly increases this effect (Fig. 4). This defense function of HO eliminates heme[33] and produces bilirubin, which functions as a neutral antioxidant.[38]

RESPONSE OF TELOMERASE ENZYME TO OXIDATIVE STRESS

Most human cancer cells have mechanisms that compensate for telomere shortening, mainly through the activation of telomerase. Thus, the activation of telomerase is an important step in human carcinogenesis, and the repression of this enzyme activity acts as a tumor suppressive mechanism.[39] We found that oxidative stress induced by heme and H$_2$O$_2$ did not affect the activity of the telomerase enzyme (Fig. 5), suggesting that oxidative stress may have direct effect on telomere itself.

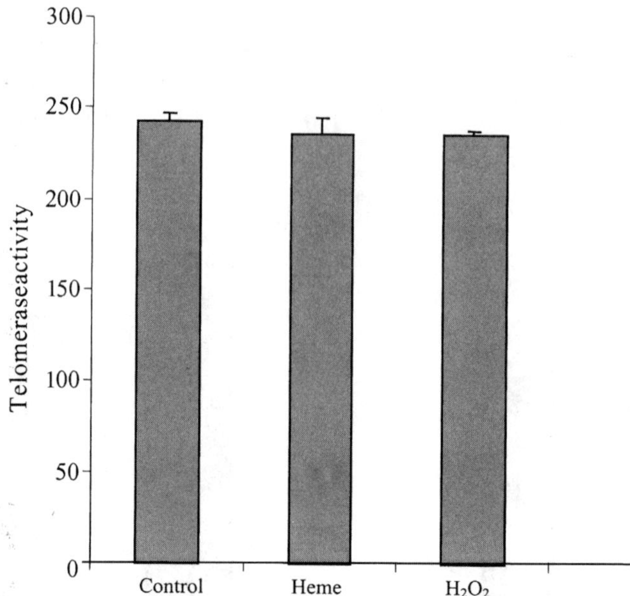

Figure 5. Effect of oxidative stress on telomerase activity in HepG2 cells. HepG2 cells (2×10^6 cells) were treated for 24 h with heme (50 µM) and H_2O_2 (100 µM). Untreated HepG2 cells were used as a control. Telomerase activity was assessed using PCR-based TRAP ELISA kit. Telomerase activity was found to be positive in all HepG2 cells.

TELOMERASE ACTIVITY AND TELOMERASE CATALYTIC SUBUNIT hTERT

Expression of hTERT is one of the rate-limiting determinants of the enzymatic activity of human telomerase. Upregulation of hTERT mRNA appears to play a major role in telomerase activation and may be a critical step in carcinogenesis.[40] To examine the expression of the catalytic subunit hTERT in HepG2 exposed or not exposed to oxidants and the relationship between the expression of hTERT and telomerase activity, total RNA was extracted from HepG2 cells treated or non-treated with different concentration of heme and H_2O_2. hTERT-mRNA was assessed by RT-PCR. The catalytic subunit of telomerase enzyme hTERT was expressed in all telomerase positive HepG2 cells. Thus, this results raises the possibility of using hTERT as a tumor marker in the diagnosis of hepatocellular carcinoma. Interestingly, the level of hTERT-mRNA was not affected by oxidants (Fig. 6).

SUMMARY

In summary, we demonstrate that upregulation of HO-1, a stress response gene, protects HepG2 cells against the effects of oxidants, while downregulation of HO-1

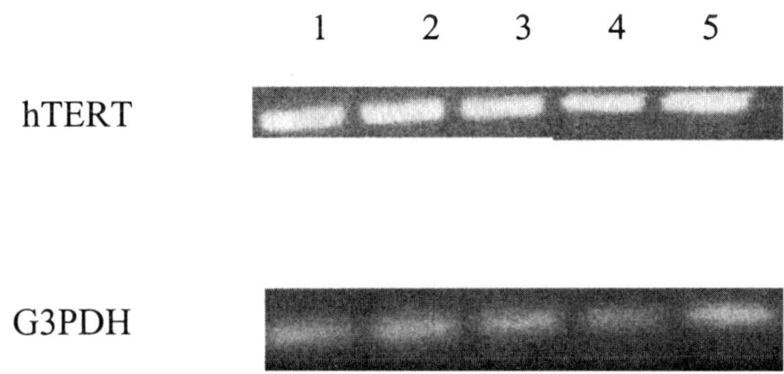

Figure 6. Expression of hTERT mRNA in HepG2 cells. hTERT expression in human HepG2 cells was determined by RT-PCR. Amplification of G3PDH was used to demonstrate the integrity of cDNA templates. Telomerase activity, determined by TRAP assay, was positive in all samples. hTERT was expressed in treated and nontreated HepG2 cells and this expression was correlate with telomerase activity. Lane 1 (control); Lanes 2 and 3 (heme 10 and 50 μM); Lanes 4 and 5 (H_2O_2 50 and 100 μM).

gene increases susceptibility to oxidative damage. These effects of the HO-1 on cell viability were independent from telomerase activity. The catalytic subunit of telomerase enzyme (hTERT) is expressed in all telomerase positive HepG2 cells and its expression was not affected by oxidative stress. Further analysis of the effect of oxidative stress on telomere length will be needed to clarify the roles oxidative stress and heme oxygenase gene play in cancer. Understanding this mechanism will provide critical insights into the molecular bases of human carcinogenesis.

REFERENCES

1. D. Shippen-Lentz and E.H. Blackburn, Functional evidence for an RNA template in telomerase, Science 247 (1990) 546–552.
2. T.M. Nakamura, G.B. Morin, K.B. Chapman, S.L. Weinrich, W.H. Andrews, J. Lingner, C.B. Harley, and T.R. Cech, Telomerase catalytic subunit homologs from fission yeast and human, Science 277 (1997) 955–959.
3. S. Kyo, T. Kanaya, M. Takakura, M. Tanaka, and M. Inoue, Human telomerase reverse transcriptase as a critical determinant of telomerase activity in normal and malignant endometrial tissues, Int. J. Cancer 80 (1999) 60–63.
4. L. Harrington, T. McPhail, V. Mar, W. Zhou, R. Oulton, M.B. Bass, I. Arruda, and M.O. Robinson, A mammalian telomerase-associated protein, Science 275 (1997) 973–977.
5. M.J. McEachern, A. Krauskopf, and E.H. Blackburn, Telomeres and their control, Annu. Rev. Genet. 34 (2000) 331–358.
6. S.E. Artandi and R.A. DePinho, A critical role for telomeres in suppressing and facilitating carcinogenesis, Curr. Opin. Genet. Dev. 10 (2000) 39–46.
7. E. Hiyama, K. Hiyama, T. Yokoyama, Y. Matsuura, M.A. Piatyszek, and J.W. Shay, Correlating telomerase activity levels with human neuroblastoma outcomes, Nat. Med. 1 (1995) 249–255.
8. J. Albanell, F. Lonardo, V. Rusch, M. Engelhardt, J. Langenfeld, W. Han, D. Klimstra, E. Venkatraman, M.A. Moore, and F. Dmitrovsky, High telomerase activity in primary lung cancer: association with increased cell proliferation rates and advanced pathologic stage, J. Natl. Cancer Inst. 89 (1997) 1609–1615.

9. H. Tahara, H. Kuniyasu, H. Yokozaki, W. Yasui, J.W. Shay, T. Ide, and E. Tahara, Telomerase activity in preneoplastic and neoplastic gastric and colorectal lesions, Clin. Cancer Res. 1 (1995) 1245–1251.

10. K. Yoshida, T. Sugino, H. Tahara, A. Woodman, J. Bolodeoku, V. Nargund, G. Fellows, S. Goodison, E. Tahara, and D. Tarin, Telomerase activity in bladder carcinoma and its implication for noninvasive diagnosis by detection of exfoliated cancer cells in urine, Cancer 79 (1997) 362–369.

11. L.A. Langford, M.A. Piatyszek, R. Xu, S.C. Schold, Jr., and J.W. Shay, Telomerase activity in human brain tumours, Lancet 346 (1995) 1267–1268.

12. M. Ogami, Y. Ikura, S. Nishiguchi, T. Kuroki, M. Ueda, and M. Sakurai, Quantitative analysis and in situ localization of human telomerase RNA in chronic liver disease and hepatocellular carcinoma, Lab Invest 79 (1999) 15–26.

13. A.G. Bodnar, M. Ouellette, M. Frolkis, S.E. Holt, C.P. Chiu, G.B. Morin, C.B. Harley, J.W. Shay, S. Lichtsteiner, and W.E. Wright, Extension of life-span by introduction of telomerase into normal human cells, Science 279 (1998) 349–352.

14. M. Meyerson, C.M. Counter, E.N. Eaton, L.W. Ellisen, P. Steiner, S.D. Caddle, L. Ziaugra, R.L. Beijersbergen, M.J. Davidoff, Q. Liu, S. Bacchetti, D.A. Haber, and R.A. Weinberg, hEST2, the putative human telomerase catalytic subunit gene, is up-regulated in tumor cells and during immortalization, Cell 90 (1997) 785–795.

15. A. Hoos, H.H. Hepp, S. Kaul, T. Ahlert, G. Bastert, and D. Wallwiener, Telomerase activity correlates with tumor aggressiveness and reflects therapy effect in breast cancer, Int. J. Cancer 79 (1998) 8–12.

16. Y.M. Janssen, B. Van Houten, P.J. Borm, and B.T. Mossman, Cell and tissue responses to oxidative damage, Lab Invest 69 (1993) 261–274.

17. J.L. Witztum and D. Steinberg, Role of oxidized low density lipoprotein in atherogenesis, J. Clin. Invest 88 (1991) 1785–1792.

18. S. Simizu, M. Takada, K. Umezawa, and M. Imoto, Requirement of caspase-3(-like) protease-mediated hydrogen peroxide production for apoptosis induced by various anticancer drugs, J. Biol. Chem. 273 (1998) 26900–26907.

19. T. von Zglinicki, R. Pilger, and N. Sitte, Accumulation of single-strand breaks is the major cause of telomere shortening in human fibroblasts, Free Radic. Biol. Med. 28 (2000) 64–74.

20. N.G. Abraham, G.S. Drummond, J.D. Lutton, and A. Kappas, The biological significance and physiological role of heme oxygenase, Cell. Physiol. Biochem. 6 (1996) 129–168.

21. S. Shibahara, M. Muller, and H. Taguchi, Transcriptional control of rat heme oxygenase by heat shock, J. Biol. Chem. 262 (1987) 12889–12892.

22. B.E. Dwyer, R.N. Nishimura, S.Y. Lu, and A. Alcaraz, Transient induction of heme oxygenase after cortical stab wound injury, Brain Res Mol. Brain Res 38 (1996) 251–259.

23. W.K. McCoubrey, Jr., T.J. Huang, and M.D. Maines, Isolation and characterization of a cDNA from the rat brain that encodes hemoprotein heme oxygenase-3, Eur. J. Biochem. 247 (1997) 725–732.

24. T. Ingi and G.V. Ronnett, Direct demonstration of a physiological role for carbon monoxide in olfactory receptor neurons, J. Neurosci. 15 (1995) 8214–8222.

25. L.E. Otterbein and A.M. Choi, Heme oxygenase: colors of defense against cellular stress, Am. J. Physiol Lung Cell Mol. Physiol 279 (2000) L1029–L1037.

26. R. Zakhary, K.D. Poss, S.R. Jaffrey, C.D. Ferris, S. Tonegawa, and S.H. Snyder, Targeted gene deletion of heme oxygenase 2 reveals neural role for carbon monoxide, Proc. Natl. Acad. Sci. USA 94 (1997) 14848–14853.

27. C.F. Yang, H.M. Shen, and C.N. Ong, Protective effect of ebselen against hydrogen peroxide-induced cytotoxicity and DNA damage in HepG2 cells, Biochem. Pharmacol. 57 (1999) 273–279.

28. B.M. Deramaudt, S. Braunstein, P. Remy, and N.G. Abraham, Gene transfer of human heme oxygenase into coronary endothelial cells potentially promotes angiogenesis, J. Cell Biochem. 68 (1998) 121–127.

29. B.A. Schacter and P. Kurz, Alterations in microsomal drug metabolism and heme oxygenase activity in isolated hepatic parenchymal and sinusoidal cells in Murphy-Sturm lymphosarcoma-bearing rats, Clin. Invest Med. 9 (1986) 150–155.

30. M.D. Maines and P.A. Abrahamsson, Expression of heme oxygenase-1 (HSP32) in human prostate: normal, hyperplastic, and tumor tissue distribution, Urology 47 (1996) 727–733.
31. A.I. Goodman, M. Choudhury, J.L. da-Silva, M.L. Schwartzman, and N.G. Abraham, Overexpression of the heme oxygenase gene in renal cell carcinoma, Proc. Soc. Exp Biol Med. 214 (1997) 54–61.
32. A. Matsumoto, R. Hanayama, M. Nakamura, K. Suzuki, J. Fujii, H. Tatsumi, and N. Taniguchi, A high expression of heme oxygenase-1 in the liver of LEC rats at the stage of hepatoma: the possible implication of induction in uninvolved tissue, Free Radic. Res 28 (1998) 383–391.
33. N.G. Abraham, Y. Lavrovsky, M.L. Schwartzman, R.A. Stoltz, R.D. Levere, M.E. Gerritsen, S. Shibahara, and A. Kappas, Transfection of the human heme oxygenase gene into rabbit coronary microvessel endothelial cells: protective effect against heme and hemoglobin toxicity, Proc. Natl. Acad. Sci. USA 92 (1995) 6798–6802.
34. K. Chen, K. Gunter, and M.D. Maines, Neurons overexpressing heme oxygenase-1 resist oxidative stress-mediated cell death, J. Neurochem. 75 (2000) 304–313.
35. I. Petrache, L.E. Otterbein, J. Alam, G.W. Wiegand, and A.M. Choi, Heme oxygenase-1 inhibits TNF-alpha-induced apoptosis in cultured fibroblasts, Am. J Physiol Lung Cell Mol. Physiol 278 (2000) L312–L319.
36. K.D. Poss and S. Tonegawa, Heme oxygenase 1 is required for mammalian iron reutilization, Proc. Natl. Acad. Sci. USA 94 (1997) 10919–10924.
37. L. Yang, S. Quan, and N.G. Abraham, Retrovirus-mediated HO gene transfer into endothelial cells protects against oxidant-induced injury, Am. J. Physiol. 277 (1999) L127–L133.
38. R. Stocker, Y. Yamamoto, A.F. McDonagh, A.N. Glazer, and B.N. Ames, Bilirubin is an antioxidant of possible physiological importance, Science (Washington DC) 235 (1987) 1043–1047.
39. T. Sumida and H. Hamakawa, Telomerase and oral cancer, Oral Oncol. 37 (2001) 333–340.
40. G.A. Ulaner, J.F. Hu, T.H. Vu, H. Oruganti, L.C. Giudice, and A.R. Hoffman, Regulation of telomerase by alternate splicing of human telomerase reverse transcriptase (hTERT) in normal and neoplastic ovary, endometrium and myometrium, Int. J. Cancer 85 (2000) 330–335.

31

THE INFLUENCE OF GLUCURONIDATION ON IN VITRO ASSESSMENT OF BILIRUBIN PRODUCTION AS MEASURE OF HO ACTIVITY

Studies in Microsomal and Cellular Systems

Gerben J. Schaaf[a], Roel F. M. Maas[a], Els M. de Groene[b], and Johanna Fink-Gremmels[a],*

[a]Department of Veterinary Pharmacology
Pharmacy and Toxicology (VFFT)
Faculty of Veterinary Medicine
Utrecht University
P.O. Box 80152 NL 3508 TD
Utrecht, The Netherlands
[b]Unilever Health Institute
P.O. Box 114, 3130 AC Vlaardingen
The Netherlands

ABSTRACT

Heme oxygenase (HO) catalyzes the rate-limiting step in the degradation of heme to bilirubin. Glucuronidation of the cytotoxic bilirubin is a critical step in heme catabolism. HO and UDP-glucuronyl transferase (UGT) enzymes are co-localized in the membranes of the endoplasmic reticulum allowing efficient glucuronidation of bilirubin. Besides its physiological significance, this co-localization has practical consequences as microsomal protein is used to assess HO activity in biological samples. In the present study the influence of sample specific glucuronidation capacity on the production of chloroform-extractable bilirubin was investigated. Addition of

* Corresponding author: Tel.: +31 30 2535453; Fax: +31 30 2534125 Email address: j.fink@vfft.vet.uu.nl

exogenous UDP-glucuronic acid significantly decreased the levels of measurable bilirubin. In contrast, dialysis of the tissue-cytosol fraction, which is routinely added to the mixture as source of biliverdin reductase, completely eliminated glucuronidation activity of the HO mixture and significantly increased the concentration of chloroform-extractable bilirubin. Thus, varying levels of UGT-activity in the individual samples may greatly influence the assessment of microsomal HO activity.

In renal tubular cells, basal levels of HO are low. Exposure of these cells to hemin resulted in an approximately 10-fold increase in microsomal HO activity and HO-1 protein expression. However, induction of HO did not result in a comparable increase in the production of chloroform-extractable bilirubin by heme-exposed intact cells as was anticipated. The present data suggest that bilirubin is conjugated progressively in HO-induced cells, resulting in a relatively lower percentage of chloroform-extractable bilirubin.

INTRODUCTION

Heme oxygenase (HO) catalyzes the rate-limiting step in the degradation of heme (ferriprotoporphyrin IX) to biliverdin IXa, water, CO and iron. Biliverdin is further reduced to bilirubin by NAD(P)H: biliverdin reductase.

There is increasing interest in the role of HO in diverse physiological and pathological cellular processes including the regulation of cell growth and differentiation (reviewed by Abraham et al.[1]). HO-1, the inducible isoform of heme oxygenase, is quickly and strongly induced by many stressors including heme, heavy metals and X-ray irradiation.[2] The induction of HO by oxidative stress is considered to comprise a cytoprotective response. However, as a result of HO activity iron is released into the cytosol, which can result in additional oxidative stress. Increased iron levels are normally neutralized by ferritin, an iron-chelating protein, which expression is often co-induced with HO-1.[3]

The HO activity of biological samples is determined by measuring the bilirubin formation in a specified reaction mixture. A liver cytosolic fraction, as source of biliverdin reductase (BVR), is commonly added to ensure complete conversion of heme to bilirubin. Bilirubin formed in this mixture is extracted with chloroform and subsequently quantified as described previously by Kutty and Maines.[4] Addition of BVR is essential as this method relies on complete conversion of heme and its primary product biliverdin, both water-soluble compounds, to water-insoluble bilirubin.

Glucuronidation of bilirubin, by hepatic UDP-glucuronyltransferases (UGTs), is responsible for the elimination of bilirubin from the body.[5] HO and the bilirubin-UGT are co-localized in the endoplasmic reticulum,[2,6,7] ensuring efficient and rapid excretion of the cytotoxic bilirubin. Because of the co-localization of HO and UGT in the cell, both enzymes are present in the reaction mixture for measuring HO activity. This potentially allows glucuronidation of bilirubin that will result in an increase of water-soluble bilirubin-glucuronides, which are not extracted by chloroform and may lead to underestimation of HO activity of the sample. Furthermore, bilirubin-UGT (UGT1A1) is, in vivo and in vitro, induced by several compounds including ethanol, dexamethasone and by bilirubin itself[5,8,9] and may serve to protect against unbalanced increases in the levels of the potentially toxic bilirubin. Varying levels of UGT-activity in the individual samples may thus further influence the determination of HO activity.

The liver is quantitatively the most important site for glucuronidation.[10] Much less, however, is known about the glucuronidation activity of the kidney. The kidney may, due to its role as excretion organ, frequently be exposed to toxicants causing strong variations in both HO-1 and bilirubin levels. Regulation of bilirubin UGT activity, similarly as mentioned above for the liver, may protect the kidney from toxic variations in intracellular bilirubin levels.

In the present study the possible interference of hepatic UGT activity with the measurement of microsomal HO activity was determined. Furthermore, to study the possible interference of glucuronidation on bilirubin formation in cultured kidney cells, the production of chloroform-extractable bilirubin in control and HO-stimulated cultured tubular cells upon exposure to exogenous heme was assessed. In order to characterize bilirubin-handling in intact cells, bilirubin production by control and HO-induced cells was compared to that in cell homogenates with basal and elevated HO activity.

MATERIALS AND METHODS

Materials

Hemin was obtained from Porphyrin Products Laboratories (Logan, UT, USA). The polyclonal antibody raised against rat HO-1 antibody was obtained from Sanbio b.v. (Uden, the Netherlands, Stressgen catalog no. SPA 895). The cell culture solutions and chemicals (M199, FCS, glutamine, penicillin and streptomycin) were obtained from Life Technologies (Breda, the Netherlands). All other chemicals were of analytical grade and obtained from Sigma (St. Louis, MA, USA).

Cell Culture

LLC-PK$_1$ cells, obtained from ATCC (number CL-101) were routinely seeded at 5×10^4 cells/flask and were subcultured every 6–7 days. Two days prior to experiments LLC-PK$_1$ cells were plated at a density of 1×10^6 cells/25 cm^2 culture area in medium 199 supplemented with 5% FCS, 2 mM L-glutamine, 100 U/ml penicillin, 100 μg/ml streptomycin. Cells were grown in a 5% CO_2 humidified atmosphere at 37 °C.

Determination of Glucuronyl Transferase Activity

Microsomal glucuronyl transferase activity was determined by addition of 2 mM 1-naphthol to the standard mixture for assessing HO activity (see below). Glucuronyl transferase activity in cultured cells was determined by incubating the cells for 0.5 h in medium containing 1.25 mM 1-naphthol. Medium was collected and stored at

–20 °C until analysis. Samples were centrifuged (5 min. 12,000 x g) and the formation of 1-naphthyl-glucuronides was determined by HPLC analysis as described previously.[11] A standard curve was prepared using authentic standards (α-naphthyl-β-D-glucuronide, obtained from Sigma, St. Louis, USA) and was used to quantify formation of metabolites.

Biliverdin Reductase Activity

Activity of biliverdin reductase in samples containing pig liver cytosol was determined as described previously by Nutter et al.[12] with minor modifications. Briefly, the reaction mixture was composed of 10 μM biliverdin, 100 μM NADPH and 20 μl liver cytosol (0.8–1.0 mg protein) in 1 ml 0.1 M phosphate buffer (pH 7.4). After 15 min. at 37 °C bilirubin formation was determined by measuring the spectral difference on a Shimadzu split-beam spectrophotometer (UV 2101PC, UV VIS scanning, Shimadzu Benelux, 's Hertogenbosch, the Netherlands) between 464 and 530 nm using the molecular extinction coefficient of 40 mM^{-1}.[13] Chloroform-extractable bilirubin production was expressed as nmol bilirubin/h.mg protein.

Microsomal Heme Oxygenase Activity

The activity of microsomal heme oxygenase was assayed as described by Kutty and Maines with minor modifications.[4] The reaction mixture (1.5 ml) contained 3–4 mg protein of the cell homogenate, 25 μM hemin, 2 mM NADP, 5 mM glucose-6-phosphate, 1.5 units glucose-6-phosphate dehydrogenase, 1.5 mM $MgCl_2$ and 3 mg 100,000 x g pig liver supernatant (cytosol), in 0.05 M phosphate buffer (pH 7.4). The incubation was carried out at 37 °C for 60 min. The reaction was terminated by the addition of 7.5 ml chloroform. After centrifugation (4,300 x g, 5 °C, 10 min.) the organic phase was transferred to a clean tube and concentrated to 1 ml by evaporation under a stream of N_2. All steps were performed under subdued light conditions. The production of bilirubin was determined as described for the biliverdin reductase assay.

Measurement of Unconjugated-Bilirubin Production by Cultured Cells

LLC-PK$_1$ cells were seeded on 100 mm cell culture dishes (Costar, USA) at a density of 2.6×10^4/cm^2 and allowed to grow for 2 days. Cells were exposed to 0 (solvent only; controls) and 100 μM hemin for 1.5 h, then hemin was removed and cells were incubated in hemin-free medium for 6 h. Control and hemin-pretreated cells were then either directly harvested for determination of microsomal HO-activity or were subsequently exposed for 2 h to 100 μM hemin in fresh culture medium. The incubation was terminated by collecting the medium, the cells were scraped in PBS and stored as cell pellets at −20 °C for determination of microsomal HO activity, Western blot analysis or determination of total cellular protein. All steps were performed under subdued light conditions. The production of chloroform-extractable bilirubin in the incubation medium was determined as described for the biliverdin reductase assay.

Western Blot Analysis

HO-1 protein expression in homogenates from LLC-PK$_1$ cells was determined by Western Blot analysis as described previously.[11]

Statistical Analysis

Data was subjected to one-way ANOVA to determine homogeneity of results followed by pairwise comparison of the data using the Bonferroni test. $P < 0.05$ was considered statistically different.

RESULTS

HO Activity and Bilirubin-Glucuronidation in Liver Microsomes

Figure 1A shows that addition of 1-naphthol to the standard mixture used to measure HO activity, resulted in the formation of 1-naphtyl-glucuronides, indicating functional UGT activity. In these experiments liver microsomes were used as they contain the highest levels of bilirubin-UGT. Furthermore, the presence of exogenous glucuronic acid (UDPGA) in the mixture significantly stimulated glucuronidation of 1-naphthol (Fig. 1A). However, when dialyzed cytosol was used as source of BVR the glucuronidation activity towards 1-naphthol as substrate was completely eliminated from the standard HO mixture (Fig. 1B).

A   **B**

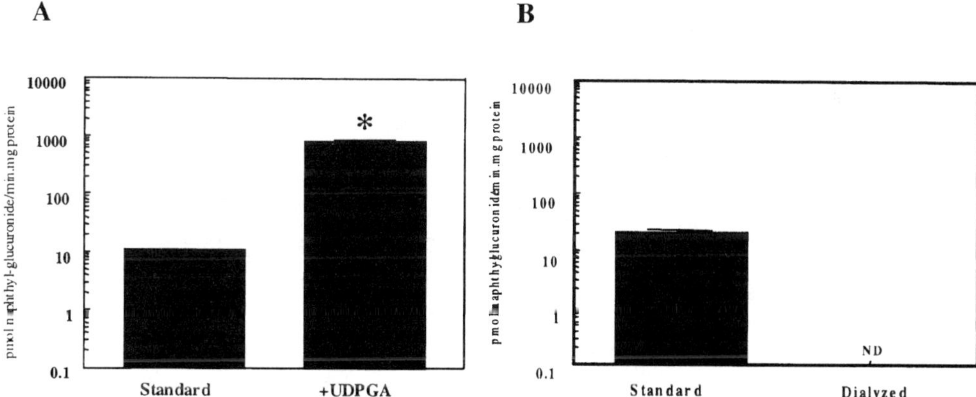

Figure 1. The effect of exogenous UDPGA and dialysis of the liver cytosol fraction on the glucuronidation of 1-naphthol in the standard HO mixture. (A) A standard mixture to measure HO activity (see method section) was prepared and the glucuronidation of 1-naphthol in the HO mixture with or without exogenous UDPGA was determined. (B) The effect of dialysis of pig liver cytosol, used as source of biliverdin reductase, on the 1-naphthol glucuronidation activity in the HO mixture. The data represent mean ± SD from at least 2 independent experiments. *Significantly different from value in standard HO mixture ($P < 0.05$). ND not detectable.

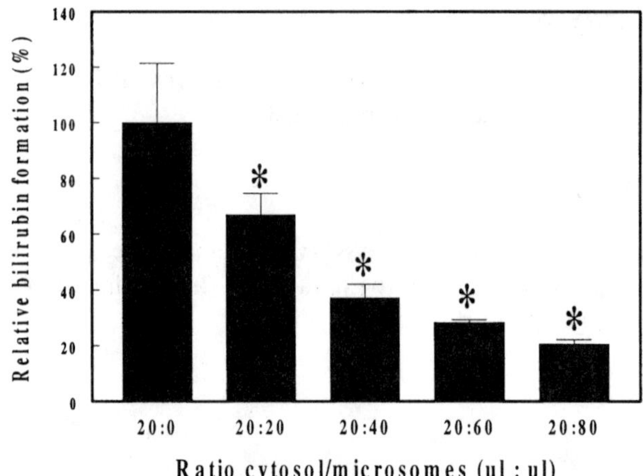

Figure 2. The effect of increasing amounts of microsomal protein on the formation of chloroform-extractable bilirubin by biliverdin reductase in liver cytosol. A standard mixture to measure biliverdin reductase activity in liver cytosol was composed. Increasing amounts of microsomal protein was added and the production of chloroform-extractable chloroform was determined. The data represent the mean ± SD of triplicate measurements. 100% equals ± 600 nmol/min.mg protein. * Significant difference (P < 0.05).

UDPGA in Liver Cytosol

The biliverdin reductase (BVR) assay showed that the pig liver cytosol used in the HO activity mixture contained significant amounts of biliverdin reductase activity (Fig. 2; ratio 20:0). Introduction of increasing amounts of microsomal protein to the BVR mixture, which is basically composed of hemin and cytosol, significantly reduced the formation of unconjugated bilirubin. The ratio of cytosol:microsomes of 20:40 (µl:µl) is routinely used in the HO mixture.

Furthermore, the presence of UDPGA in the incubation mixture resulted in a significant decrease in the production of chloroform-extractable bilirubin (Fig. 3A). Dialysis of the cytosolic fraction prior to the assay resulted in a significant increase (almost 1.6-fold) in the production of chloroform-extractable bilirubin (Fig. 3B). In the experiments as depicted in Figs. 1B and 3B the biliverdin reductase activity in control and dialyzed cytosol was equal (approximately 600 nmol/min.mg protein).

HO-1 Induction and Bilirubin Production in Renal Tubular Cells

HO-1 is rapidly and strongly induced after short-term hemin treatment in LLC-PK$_1$ cells (Fig. 4A). As soon as 2h after withdrawal of hemin a significant increase in HO-1 protein was observed, reaching an apparent maximum 6h after withdrawal of hemin. This increase in protein expression after a 6h recovery period in hemin-free medium was accompanied by a strong increase in microsomal HO activity in LLC-PK$_1$ cells (Fig. 4B; "direct" treatment).

The standard mixture to measure HO contains substrate (hemin), HO (in the microsomal fraction), NADPH-CYP450 reductase and biliverdin reductase (liver cytosol). An intact cell contains all the requirements for producing bilirubin, provided that ample substrate is provided. To investigate if in cultured cells bilirubin production correlates with (microsomal) HO activity, control and HO-induced LLC-PK$_1$ cells were exposed to exogenous hemin and the production of chloroform-extractable bilirubin was determined. To obtain maximal HO-1 induction LLC-PK$_1$ cells were exposed to hemin for 1.5 h (hemin-pretreatment) followed by 6 h incubation in hemin-free medium, a treatment corresponding to the results presented in Fig. 4A. Following the pretreatment period contral and HO-induced cells were exposed to exogenous hemin, as it is known that during routine culture intracellular heme levels are too low to allow quantification of bilirubin. Exposure of control (solvent treated) and HO-induced (hemin-pretreated) tubular cells to 100 µM of hemin for 2 h did result in an increase in chloroform-extractable bilirubin production into medium (~2.5-fold; Fig. 4B "Medium"). However, this increase was not as strong as would be expected from the observed increase in microsomal HO activity (~10-fold) ("direct" treatment; Fig. 4B). To investigate if HO activity was significantly affected by the 2 h hemin-exposure in control and HO-induced cells, microsomal HO activity and HO-1 protein expression following the 2 h substrate (second hemin) period was determined. The second hemin treatment decreased the induction ratio (HO activity of [HO-induced/control]) to approximately 7-fold (Fig. 4B; "2nd hemin" treatment). Figure 4C shows that during this 2 h substrate-exposure period ("2nd hemin") the HO-1 expression especially in control cells wes induced. However, the difference in microsomal HO induction and increase in bilirubin levels in medium were still statistically significant, suggesting that

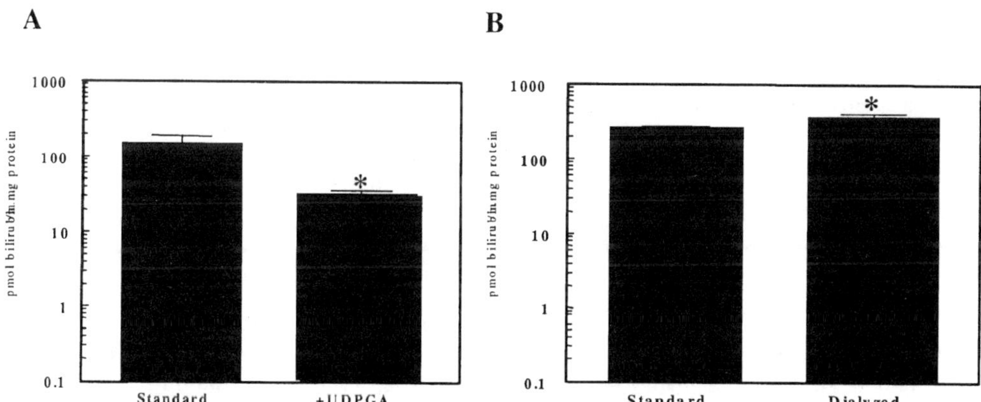

Figure 3. The effect of exogenous UDPGA and dialysis of the liver cytosolic fraction on the bilirubin formation by rat liver microsomes. (A) The formation of chloroform-extractable bilirubin by rat liver microsomes was determined in the presence and absence of exogenous UDPGA in the standard HO mixture. (B) The effect of dialysis of pig liver cytosol, used as source of biliverdin reductase, on the production of bilirubin in the standard HO mixture. The data represent the mean ± SD of triplicate measurements. * Significant difference (P < 0.05).

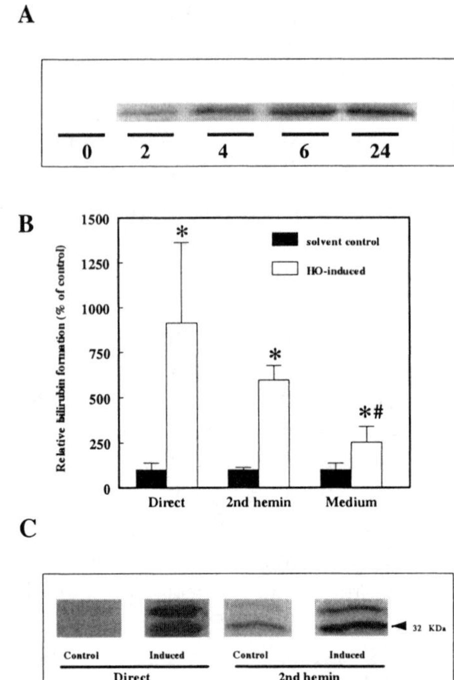

Figure 4. Effect of hemin on HO-1 protein expression (A and C) and on the production of chloroform-extractable bilirubin by cell homogenates or in heme-exposed cultured cells (B). (A) LLC-PK$_1$ cells were exposed to 100 µM hemin for 1.5 h and subsequently incubated in hemin-free medium. Cell homogenates were prepared 0, 2, 4, 6 and 24 h after removal of hemin. HO-1 protein expression in homogenates was determined by Western analysis. (B) LLC-PK$_1$ cells were cultured for 1.5 h in medium containing 0.1% DMSO (solvent control; black bars) or 100 µM hemin (HO-induced; white bars). Subsequently all cells were maintained for 6 h in hemin-free medium. Cells were then either directly harvested for determination of microsomal HO activity ("direct") or again exposed for 2 h to 100 µM hemin. Following this second hemin incubation cells were harvested for determination of microsomal HO activity ("2nd hemin") or the production of chloroform-extractable hemin in medium was determined ("medium"). (C) Alterations in HO-1 protein expression in homogenates harvested "direct" or after a "second hemin" treatment. * Significantly different from solvent-treated control. # Significantly different from HO activity in HO-induced "direct" and "2nd hemin" preparation.

in HO-induced renal tubular cells the production of chloroform-extractable bilirubin is decreased through altered bilirubin-handling. LLC-PK$_1$ cells were able to glucuronidate 1-naphthol, but 1-napthol glucuronidation activity did not differ significantly in cells treated with hemin for 24 h as compared to the cells treated for 2 h (data not shown).

DISCUSSION

Heme oxygenase and UDP-glucuronyl transferase are not only expressed in close proximity of one another in the ER, but are also functionally linked. Co-induction of UGT-activity during periods of increased HO activity may limit imminent bilirubin toxicity. Increased HO activity regulates glucuronidation of its potentially toxic endproduct, as bilirubin was shown to upregulate UGT-1A1 activity.[5]

Rapid glucuronidation of bilirubin has an obvious physiological importance, but also practical implications for assessment of HO activity. The widely used assay to determine HO activity in microsomes or cell homogenates is debated, among other reasons, for spectral interference.[14] Indeed, using the modified method of Kutty et al.,[4] extracting bilirubin from a mixture containing microsomal and cytosolic protein leads to spectral interference under certain conditions, especially in the 400–420 nm range (data not shown).

Furthermore, we show in the present study that measurement of production of chloroform-extractable bilirubin is influenced by intrinsic UGT activity of the sample. The conjugated bilirubin is not extractable by chloroform and consequently escapes quantification. This is demonstrated by an almost complete absence of bilirubin from the chloroform-extract after addition of glucuronic acid to the reaction mixture. These suggestions were confirmed by the observation that the standard mixture composition is capable of glucuronidating a marker substrate, 1-naphthol, for glucuronidation activity. Dialysis of pig liver cytosol, the source for biliverdin reductase, eliminates glucuronidation activity. Thus, we conclude that non-dialysed liver cytosol introduces the cofactor UDPGA into the reaction mixture. Therefore, it can be concluded that intrinsic UGT activity of the sample leads to underestimation of HO activity. In addition, UGT activity of differentially treated samples, such as for instance ethanol and dexamethasone, may vary considerably. Differences in UGT activities between samples can further influence the determination of HO activity. Alternative methods to accurately quantify HO activity have been described. Ryter et al. described a chromatographic method for determining HO activity in cultured cells using HPLC analysis.[14,15] The method of Ryter detects biliverdin isomers, bilirubin isomers and bilirubin conjugates and, thus, allows calculation of HO activity independent of BVR activity in the reaction mixture.

Our data show that renal cells are able to produce bilirubin in culture, although addition of exogenous heme was necessary to obtain quantifiable levels of bilirubin. This was also observed by Nutter and colleagues[12] who reported that HO induction in MCF-7 cells did not result in increased (or even detectable) intracellular or extracellular bilirubin levels[12] and concluded that free intracellular heme levels are rate-limiting in bilirubin production. In contrast to LLC-PK$_1$ cells, incubation of primary rat proximal tubular cells (PT-cells) with hemin did not result in detection of chloroform-extractable bilirubin (data not shown). Rat primary PT-cells express considerably higher glucuronidation activity towards 1-naphthol as compared to LLC-PK$_1$ cells (approximately 50–100 times, calculated from[11] and unpublished observation). It is therefore likely to assume that the primary cells also glucuronidate bilirubin to a significantly higher level, resulting in low or non-detectable levels of unconjugated bilirubin in the culture media, an assumption in line with the data presented here.

A-priori induction of HO did not result in a strong induction of production of chloroform-extractable bilirubin by hemin-exposed tubular cells as would be expected by the approximately 10-fold increase in microsomal HO-activity and HO-1 protein expression. These preliminary data suggest that tubular cells react to increased HO activity and the inherent increase in bilirubin levels by decreasing the formation of unconjugated-bilirubin, most likely through induction of bilirubin-glucuronidation. It might be argued that the extremely rapid induction of HO-1 expression (significant in 2h hemin exposure) may have decreased the difference in HO activity between control and HO-induced cells during the 2h period of exposure to substrate. However, the difference microsomal HO activity and HO-1 protein expression between control- and hemin-pretreated cells remained approximately 6–7 fold after the second 2h hemin incubation. The latter observation further supports our assumption of increased bilirubin-glucuronidation in HO-induced renal tubular cells.

In summary, the co-localization of HO and UGT in the microsomal membranes allows for the interference of bilirubin-glucuronidation on the measurement of bilirubin formation as measure for HO activity, especially when unpurified liver cytosol as source for biliverdin reductase is used. Intrinsic bilirubin-glucuronidation activity of the sample can be excluded by dialysis of the liver cytosol fraction prior to use in the assay for assessing HO activity. Differences in bilirubin production in the microsomal-based assay and the cell-based measurement of bilirubin production may be partly explained by the glucuronidation activity of the cells. Therefore, it seems unlikely that the approach of using hemin-exposed cultured cells can be used for reliably determining HO activity in cell cultures.

REFERENCES

1. Abraham N.G., Drummond G.S., Lutton J.D., and Kappas A.: The biological significance and physiological role of heme oxygenase. Cell Physiol Biochem 6:129–168, 1996.
2. Maines M.D.: Heme oxygenase: function, multiplicity, regulatory mechanisms, and clinical applications. Faseb J 2:2557–2568, 1988.
3. Dennery P.A.: Regulation and role of heme oxygenase in oxidative injury [In Process Citation]. Curr Top Cell Regul 36:181–199, 2000.
4. Kutty R.K. and Maines M.D.: Oxidation of heme c derivatives by purified heme oxygenase. Evidence for the presence of one molecular species of heme oxygenase in the rat liver. J Biol Chem 257:9944–9952, 1982.
5. Li Y.Q., Prentice D.A., Howard M.L., Mashford M.L., and Desmond P.V.: Bilirubin and bile acids may modulate their own metabolism via regulating uridine diphosphate-glucuronosyltransferase expression in the rat. J Gastroenterol Hepatol 15:865–870, 2000.
6. Lohr J.W., Willsky G.R., and Acara M.A.: Renal drug metabolism. Pharmacol Rev 50:107–141, 1998.
7. Mulder G.J.: Glucuronidation and its role in regulation of biological activity of drugs. Annu Rev Pharmacol Toxicol 32:25–49, 1992.
8. Jemnitz K., Veres Z., Monostory K., and Vereczkey L.: Glucuronidation of thyroxine in primary monolayer cultures of rat hepatocytes: in vitro induction of UDP-glucuronosyltranferases by methylcholanthrene, clofibrate, and dexamethasone alone and in combination. Drug Metab Dispos 28:34–37, 2000.
9. Ritter J.K., Kessler F.K., Thompson M.T., Grove A.D., Auyeung D.J., and Fisher R.A.: Expression and inducibility of the human bilirubin UDP-glucuronosyltransferase UGT1A1 in liver and cultured primary hepatocytes: evidence for both genetic and environmental influences. Hepatology 30:476–484, 1999.
10. McGurk K.A., Brierley C.H., and Burchell B.: Drug glucuronidation by human renal UDP-glucuronosyltransferases. Biochem Pharmacol 55:1005–1012, 1998.

11. Schaaf G.J., de Groene E.M., Maas R.F., Commandeur J.N., and Fink-Gremmels J.: Characterization of biotransformation enzyme activities in primary rat proximal tubular cells. Chem Biol Interact 134:167–190, 2001.

12. Nutter L.M., Sierra E.E., and Ngo E.O.: Heme oxygenase does not protect human cells against oxidant stress [see comments]. J Lab Clin Med 123:506–514, 1994.

13. Agarwal A., Balla J., Alam J., Croatt A.J., and Nath K.A.: Induction of heme oxygenase in toxic renal injury: a protective role in cisplatin nephrotoxicity in the rat. Kidney Int 48:1298–1307, 1995.

14. Ryter S., Kvam E., and Tyrrell R.M.: Heme oxygenase activity determination by high-performance liquid chromatography. Methods Enzymol 300:322–336, 1999.

15. Ryter S., Kvam E., Richman L., Hartmann F., and Tyrrell R.M.: A chromatographic assay for heme oxygenase activity in cultured human cells: application to artificial heme oxygenase overexpression. Free Radic Biol Med 24:959–971, 1998.

32

MOLECULAR MECHANISM OF HEME OXYGENASE-1 GENE INDUCTION BY ACTIVATION OF THE PROTEIN KINASE A-DEPENDENT SIGNALING PATHWAY

Stephan Immenschuh[a],* and Thomas Kietzmann[b]

[a]Institute of Clinical Chemistry and Pathobiochemistry
Justus-Liebig-University, 35392 Giessen
Gemany
[b]Institute of Biochemistry
Georg-August-University, 37073 Göttingen
Germany

1. ABSTRACT

Heme oxygenase-1 (HO-1) is the inducible isoform of the rate-limiting enzyme of heme degradation and is induced by a host of oxidative stress stimuli. We have previously demonstrated that HO-1 mRNA expression is up-regulated by the second messenger Bt_2cAMP in primary rat hepatocytes cultures. The HO-1 induction by Bt_2cAMP occurred transcriptionally via activation of the protein kinase A-dependent signaling pathway. To examine the *cis*-acting regulatory elements (REs) responsive to Bt_2cAMP and glucagon, luciferase constructs of the rat HO-1 gene were tested in transiently transfected primary rat hepatocyte cultures. A functional RE was

* Correspondence: Stephan Immenschuh, MD; Institute of Clinical Chemistry and Pathobiochemistry, Justus-Liebig-University, Gaffkystr. 11; 35392 Giessen, Germany. Phone: -49-641-99-41578. Fax: -49-641-99-41559. e-mail: Stephan.Immenschuh@klinchemie.med.uni-giessen.de
Supported by grants from the Deutsche Forschungsgemeinschaft (Bonn, Germany) SFB 402 A8 (SI) and SFB 402 A1 (TK).

identified in the HO-1 5′-flanking promoter region between position −664 and −657 from the cap side which was previously termed HO-1 cAMP RE (CRE)/activating protein-1 (AP-1) element. In electrophoretic mobility shift assays the HO-1 CRE/AP-1 element specifically bound to the recombinant transcription factor CRE binding protein (CREB) and formed a specific DNA-protein complex with nuclear extracts of rat hepatocyte cultures which was unchanged by the treatment with Bt_2cAMP or glucagon. Since the HO-1 CRE/AP-1 element is also involved in HO-1 gene activation by the protein kinase G-dependent signaling pathway this regulatory site may be a crucial nuclear target for potential pharmacological interventions.

2. INTRODUCTION

Heme oxygenase-1 is the inducible isoform of the rate limiting enzymatic step of heme degradation (Maines, 1997; Foresti and Motterlini, 1999). HO-1 is induced by a host of oxidative stress stimuli such as heme, heavy metals, tumor necrosis factor-α (TNFα) or lipopolysaccharide (Choi and Alam, 1998). Although the exact functional role of the HO-1 gene induction after stress stimuli is not completely understood it has been demonstrated that HO-1 has beneficial effects against oxidative stress and is considered an adaptive autoprotective cellular response. Targeted induction of HO-1 via signaling pathways that are independent of oxidative stress stimuli may ultimately provide a basis for future therapeutic interventions. Therefore, we were interested in HO-1 gene regulation by physiological second messengers such as cAMP and cGMP. We have previously shown that both cAMP induce HO-1 gene expression in primary rat hepatocyte cultures (Immenschuh et al., 1998A). To further delineate the molecular mechanism of the cAMP-dependent induction of HO-1 gene expression we investigated the 5′-flanking region of the rat HO-1 gene promoter for potential *cis*-acting REs responsive to Bt_2cAMP and glucagon. For this purpose, transient transfection of primary rat hepatocytes with luciferase reporter gene constructs and electrophoretic mobility shift assays (EMSA) were performed. The data show that the transcriptional cAMP-dependent induction of HO-1 gene expression is mediated by a HO-1 CRE/AP-1 element (−668 to −657) which has previously been shown to be also involved in cGMP-dependent induction of HO-1 gene expression. This element is the nuclear target for the transcription factor (TF) CREB.

3. EXPERIMENTAL PROCEDURES

Cell isolation and culture—Hepatocytes were isolated from male Wistar rats by circulating perfusion with collagenase under sterile conditions as described previously (Immenschuh et al., 1998A). The cells were cultured under air/CO_2 (19/1) in medium 199 with Earle's salts containing bovine serum albumin (2 g/l), $NaHCO_3$ (20 mM), HEPES (10 mM), streptomycin sulfate (117 mg/l), penicillin (60 mg/l), insulin (1 nM) and dexamethasone (10 nM). 5% fetal calf serum was present during the plating phase up to 4h, and cell cultures were incubated in serum-free medium for another 18h before treatment.

RNA isolation, Northern blot analysis and hybridization—Total RNA isolation and Northern blot analysis from hepatocytes was performed as described (Immenschuh et al., 1998A). Probes were the cDNAs of HO-1 and glyceraldehyde-3-phosphate dehydrogenase (GAPDH) of rat. Labeling of cDNAs was performed by the oligomer method with $\alpha[^{32}P]$-dCTP using the multiprime DNA labeling kit according to the manufacturer's instructions.

Isolation of nuclei from rat hepatocyte cultures and nuclear run off-transcription assay—Isolation of nuclei and nuclear run-off assay were performed as described previously (Immenschuh et al., 1998A).

Plasmid constructs—All basic recombination techniques referred to were standard protocols. A PCR fragment of the rat HO-1 promoter 5'-flanking region from −754 to +71 was amplified from genomic rat DNA and was ligated into the *KpnI/BglII* site in front of the luciferase gene of pGl3basic (Promega) after subcloning into pUC18 and was designated pHO-754 (construct 1; Fig. 3). To create pHO-754del (construct 2; Fig. 3) the −714/−549 fragment of construct 1 was deleted by partial digestion with *HindIII*. A deletion within the HO-1 CRE-like element (pHO-754ΔCRE/AP-1) was introduced by primer directed mutagenesis of 9bp between −67 to −659 (construct 3; Fig. 3). The correctness of all constructs was confirmed by sequencing in both directions.

Cell transfection, luciferase assay—Rat hepatocyte cultures (about 1×10^6 cells per dish) were transiently transfected with 2.5μg plasmid DNA containing 500ng of pRL-SV40 (Promega) to control transfection efficiency and 2μg of the HO-1 promoter luciferase construct (Immenschuh et al., 1998B). Luciferase activity was determined as described previously (Immenschuh et al., 1998B).

Preparation of nuclear extracts and EMSA—Nuclear extracts were prepared from hepatocyte cultures (6×10^6 cells) according to Christ et al. (1991). The following oligonucleotides were synthetized by Nucleic Acid Product Supply (Göttingen, Germany) and were used as probes for EMSA experiments: rat HO-1 CRE-like element (−684/−653), 5'-TGTGTCAGAGCCATGTGTCCTGACTTCAGTCT-3', rat phosphoenolpyruvate carboxykinase (PCK) CRE-1 element (−97/−78), 5'-GGCCCCTTACGTCAGAGGCG-3', rat PCK-CRE-2 element (−152/−132), 5'-TGTGTTAGGTCAGTTCCAAAC-3', and the rat PCK-NF1/CTF element (−118/−100) 5'-TGGCTATGATCCAAAGGC-3'. The oligonucleotides were labeled with $\gamma[^{32}P]$-ATP using the 5'-end labeling kit and were purified with the nucleotide removal kit. The binding reaction was carried out with 3μg nuclear extract and 40fmol of labeled DNA for 10 minutes at room temperature in a final volume of 20μl of the following buffer: 15mM Hepes pH 7.9, 15% glycerol, 120mM KCl, 5mM MgCl2, 0.5 mM EDTA, 0.75mM dithioerythritol, 4mM spermine, 2μg BSA, 2μg poly-dIdC. A 10-fold molar excess of unlabeled probes was used as cold competitors in competition assays. The truncated recombinant CREB-1 protein, corresponding to the amino acids 254–327 (DNA binding and dimerization domains), was from Santa Cruz Biotechnology. The DNA-protein complexes were resolved on a 5% native polyacrylamide gel in $1 \times$ Tris-borate buffer (pH 8.0) at 4°C. The gels were dried and autoradiographed using a phosphorimager (Molecular Dynamics).

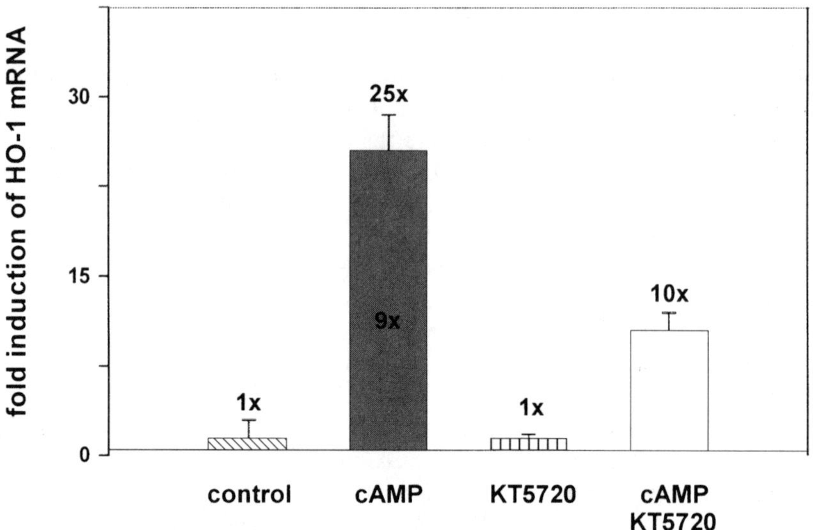

Figure 1. Inhibition of Bt₂cAMP-dependent HO-1 mRNA induction by KT5720 in rat hepatocyte cultures—
Primary rat hepatocytes were isolated and cultured as described in Experimental Procedures. Hepatocytes
were treated for 18 h with serum-free medium before cell culture was continued in the presence of Bt₂cAMP
(250 μM) and KT5720 (1 μM) alone or a combination of Bt₂cAMP plus KT5720 for 6 h. Autoradiograms
from Northern blots were quantitated and values represent the fold induction rate of HO-1 mRNA normalized to GAPDH from at least three independent experiments (mean ± standard error).

4. RESULTS

4.1. Bt₂cAMP-Dependent Induction of HO-1 mRNA Expression in Primary Rat Hepatocyte Cultures

Bt₂cAMP (250 μM) induced HO-1 mRNA steady state levels 25-fold in primary
rat hepatocyte cultures as determined by Northern blot analysis (Fig. 1). The up-
regulation of HO-1 mRNA was time-dependent with a maximum after 6 h and
returned to basal expression levels after 24 h. The increase in HO-1 mRNA expression
during Bt₂cAMP treatment was dose-dependent reaching a peak level of induction at
a concentration of 250 μM Bt₂cAMP (data not shown). HO-1 mRNA expression was
also induced by glucagon, a hormone which stimulates the adenylate cyclase via a
receptor-mediated mechanism and which, in turn, produces increased levels of cellular cAMP (data not shown).

To find out whether Bt₂cAMP-dependent induction of HO-1 mRNA was regulated by the protein kinase A-dependent (PKA) signaling pathway, hepatocyte cultures were preincubated for 30 min with the specific PKA inhibitor KT5720 at a
concentration of 1 μM before treatment with Bt₂cAMP. HO-1 mRNA induction by
Bt₂cAMP was inhibited by more than 50% by KT5720 (Fig. 2B).

4.2. Transcriptional Induction of HO-1 Gene Expression by Bt₂cAMP

To probe into the mechanism of the cAMP-dependent HO-1 mRNA induction hepatocytes were treated with actinomycin D (1 µg/ml) and cycloheximide (1 µg/ml) before addition of Bt₂cAMP. Induction of HO-1 mRNA by Bt₂cAMP was inhibited by actinomycin D and also to a lesser degree by cycloheximide (Fig. 2A). The transcriptional mode of HO-1 gene induction by Bt₂cAMP was confirmed by nuclear run-off assay (Fig. 2B).

4.3. Regulation of Transfected Rat HO-1 Gene Sequences by Bt₂cAMP and Glucagon in Rat Hepatocyte Hultures

The regulation of the 5′-flanking region and the promoter linked to a reporter gene by *cis*-acting regulatory elements (REs) of the rat HO-1 gene was determined in transiently transfected primary rat hepatocyte cultures. A construct with the proximal 754 bp of the HO-1 5′-flanking region inserted in front of the luciferase reporter

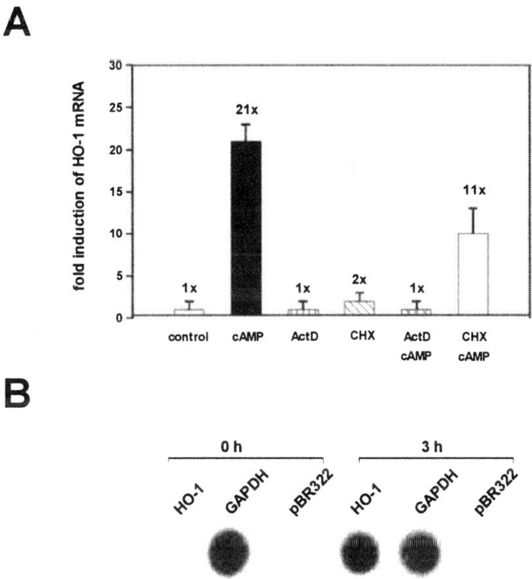

Figure 2. Transcriptional regulation of the Bt₂cAMP-dependent gene activation—(A) Hepatocytes were pretreated for 30 min with actinomycin D (ActD; 1 µg/ml) or cycloheximide (CHX; 1 µg/ml) and Bt₂cAMP (250 µM) was added for another 6 h, after which total RNA was isolated and subjected to Northern blot analysis. The blots were probed sequentially with the ³²P-labeled cDNAs of HO-1 and GAPDH. Values given represent the fold induction rate of HO-1 mRNA normalized to GAPDH levels from two or three independent experiments (mean ± standard error). (B) After 18 h in serum-free medium, hepatocyte culture was continued in the presence of Bt₂cAMP (250 µM). At 0 and 3 h, nuclei were prepared and subjected to nuclear run-off transcription assay as described in Experimental Procedures.

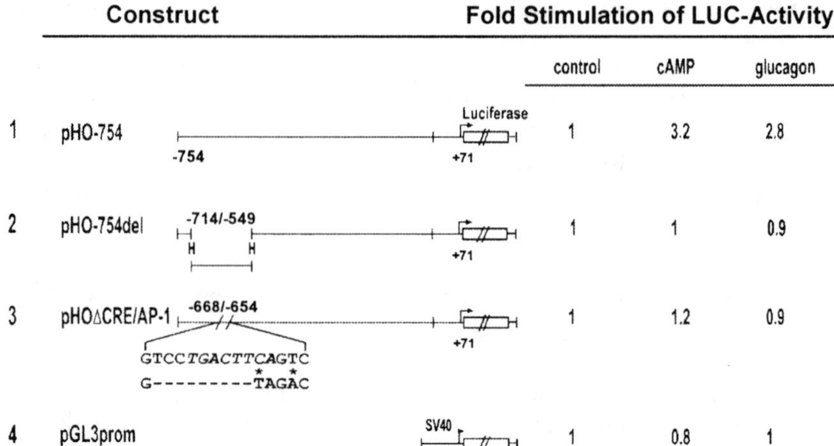

Figure 3. Bt₂cAMP-dependent regulation of rat HO-1 gene sequences in transiently transfected rat hepato-cyte cultures—The indicated rat HO-1 gene sequences were cloned into pGL3Basic (constructs 1–3) as described in Experimental Procedures. The reporter constructs were transiently transfected into primary rat hepatocyte cultures and after 24 h the transfected cells were treated for 12 h with Bt₂cAMP (250 μM). The rate of induction in each experiment relative to the control was determined. Regulation of luciferase activity of pGL3prom is shown as a control. The values are from at least three independent experiments. In construct 3, H is for *Hind*III restriction site.

gene (pHO-754) was induced about 3-fold by treatment with Bt₂cAMP and glucagon, respectively (construct 1; Fig. 3). By contrast, no induction of luciferase activity by Bt₂cAMP was observed for a control construct with the minimal SV40 promoter (construct 4; Fig. 3). When hepatocyte cultures were transfected with a deletion construct in which the sequence between −714 and −549 was eliminated (pHO-754del; construct 2; Fig. 3) luciferase activity was not affected by either treatment with Bt₂cAMP or glucagon. The deleted promoter fragment contained the previously characterized HO-1 CRE/AP-1 element (Immenschuh et al., 1998B, Immenschuh et al., 2000). To prove the functional significance of the HO-1 CRE/AP-1 element the sequence between −667 and −659 was deleted (pHO-754ΔCRE/AP-1; construct 3; Fig. 3). No Bt₂cAMP- or glucagon-dependent respon-siveness was observed when this construct was transiently transfected into hepatocytes. To investigate the functional relevance of this element independent from the rat HO-1 promoter a synthetic oligonucleotide with three copies of the HO-1 CRE/AP-1 sequence was cloned into the pLUCSV40 reporter plasmid. A 2-to 3-fold responsiveness to Bt₂cAMP and glucagon was observed for this reporter construct (data not shown).

The data indicate that the rat HO-1 CRE/AP-1 element mediates the Bt₂cAMP-and glucagon-dependent transcriptional induction of the rat HO-1 gene in rat hepa-tocyte cultures. This element does not, however, confer the full Bt₂cAMP response of the endogenous HO-1 gene in primary rat hepatocytes to luciferase reporter constructs.

4.4. The HO-1 CRE/AP-1 Element as a Binding Site for the Nuclear Factor CREB

To determine whether the identified functional HO-1 CRE/AP-1 element is a binding site for hepatic nuclear proteins, a radiolabeled synthetic oligonucleotide containing the native rat HO-1 sequence between −684 and −653 was used as a probe for EMSA studies. Nuclear extracts of untreated primary rat hepatocyte cultures produced a DNA-protein complex which was unchanged with nuclear extracts of hepatocytes treated for 1 h with glucagon (Fig. 4; left panel). In competition EMSA experiments it could be shown that the formation of the DNA-protein complex was inhibited by addition of a 10-fold molar excess of the non-labeled HO-1 CRE-like oligonucleotide indicating the sequence specificity of this DNA-protein complex (Fig. 4). In addition, competititon EMSAs were carried out with non-labeled oligonucleotides containing sequences of the promoter 5′-flanking region of the rat PCK gene which have been demonstrated to mediate the cAMP-dependent PCK transcription

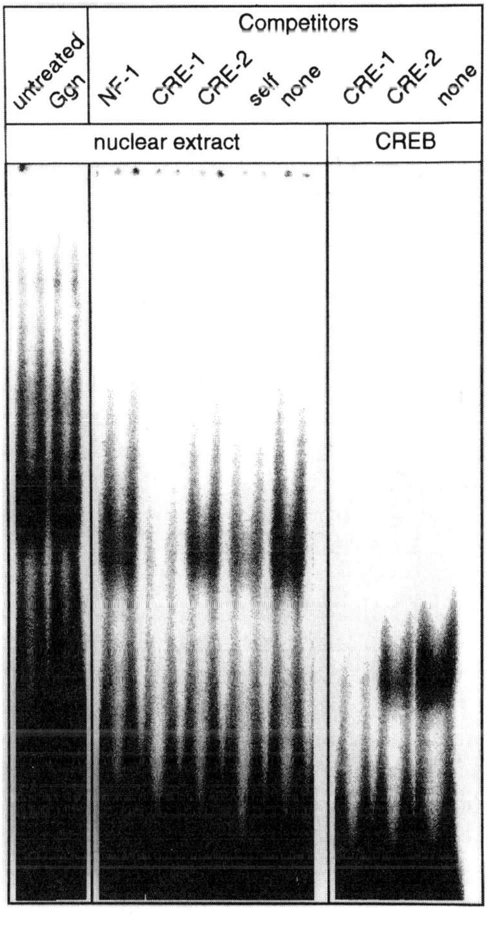

Figure 4. EMSA and competititon EMSA of the rat HO-1 CRE/AP-1 element with nuclear extracts of primary rat hepatocyte cultures and recombinant CREB—A radiolabeled synthetic oligonucleotide (−684 to −653) containing the rat HO-1 CRE/AP-1 element was incubated with nuclear extracts of untreated primary rat hepatocyte cultures (3 μg; left panel, left lane) or nuclear extracts of hepatocytes treated for 1 h with 0.05 μM glucagon (Ggn; left panel, right lane). Competition EMSA with the −684/−653 oligonucleotide was performed with either 3 μg of nuclear extracts of hepatocyte cultures (middle panel) or recombinant CREB (right panel). Incubation of the oligonucleotide was performed either without or after preincubation for 10 min with a 10-fold molar excess of oligonucleotides containing the sequences of the NF-1/CTF, the CRE-1 or the CRE-2 sites of the rat PCK gene, as indicated.

regulation (Roesler et al., 1995). Whereas the addition of a 10-fold molar excess of the non-labeled rat PCK CRE-1 oligonucleotide prevented the formation of the DNA-protein complex, neither the cold oligonucleotides with the CRE-2 nor the NF-1/CTF sites of the rat PCK gene significantly inhibited the DNA-protein band formation.

Incubation of the HO-1 CRE-like sequence with the recombinant CREB produced a specific DNA-protein complex in EMSAs (Fig. 4; right panel). Similar to the data with the nuclear extracts of hepatocyte cultures the formation of a DNA-protein complex with recombinant CREB was prevented by preincubation with a 10-fold molar excess of the cold CRE-1. Complex formation was only slightly affected by preincubation with an excess of the CRE-2 of the rat PCK gene. The obvious difference in the extent of migration of the DNA-protein complexes between the HO-1 CRE-like site, incubated with the hepatocyte nuclear extracts (Fig. 4; left panel), and the recombinant CREB (Fig. 4; right panel) can be explained by the different molecular weights of native and recombinant CREB. Native CREB has a molecular weight of 43 kDa (Montminy and Bilezikjian, 1987), and the recombinant CREB, which represents the DNA-binding domain of the nuclear factor, has a molecular weight of 14 kDa. The results suggest that the HO-1 CRE-like sequence of the HO-1 5′-flanking region serves as a specific binding site for the nuclear factor CREB.

5. DISCUSSION

5.1. Transcriptional cAMP-dependent HO-1 Gene Induction via the HO-1 CRE/AP-1 Element

We have previously shown that cAMP induces HO-1 gene induction in primary rat hepatocyte cultures (Immenschuh et al., 1998A). Initially we assumed that the cAMP-dependent induction of HO-1 could be a hepatocyte-specific regulatory pathway. Other groups, however, have also observed *in vivo* (Bakken et al., 1972) or in vascular smooth muscle cells (Durante et al., 1997) a cAMP-dependent induction of HO-1 gene expression suggesting that this regulatory pathway may not be limited to hepatocytes. To extend our initial study on cAMP-dependent induction of HO-1 and to further identify the molecular mechanism of the cAMP-dependent HO-1 gene induction we perfomed transient transfection studies with HO-1 gene promoter constructs. As demonstrated in Fig. 3 the previously identified HO-1 CRE/AP-1 element (Immenschuh et al., 1998B; Immenschuh et al., 2000) confers cAMP-dependent regulation to luciferase reporter constructs in transfected rat hepatocyte cultures. The magnitude of the transcriptional regulation of the endogenous HO-1 gene by PKA stimulatory agents (about 20-fold), however, was not reflected by the level of Bt$_2$cAMP- and glucagon-induced stimulation elicited when HO-1 promoter constructs were transfected into hepatocytes. This observation could be due to a lack of (an) additional element(s) in the tested HO-1 gene sequence that is (are) necessary to confer the full cAMP response of HO-1 induction. Alternatively, it has to be considered that the HO-1 promoter in the luciferase reporter plasmid is in a structurally distinct context from that of the chromosomal HO-1 gene.

5.2. Role of the TF CREB for HO-1 Gene Induction via the HO-1 CRE/AP-1 Element

The finding that the HO-1 CRE/AP-1 element is a specific binding site for the TF CREB (Fig. 4) strongly suggests that CREB may be involved in the transcriptional induction of HO-1 gene expression by Bt_2cAMP. This hypothesis is also supported by the finding that a combination of Bt_2cAMP and the serine/threonine protein phosphatase inhibitor okadaic acid caused a synergistic induction of HO-1 mRNA expression in rat hepatocytes (Immenschuh et al., 2000). Activation of gene transcription by CREB is mediated by the phosphorylation of the regulatory subunit serine 133 of CREB (Hagiwara et al., 1992) the dephosphorylation of which can be inhibited by okadaic acid (Wadzinski et al., 1993). For a final conclusion of the regulatory role of CREB for HO-1 gene regulation, however, cotransfection studies will have to be performed. Another class of REs responsive to cAMP is represented by the AP-2 binding site (Imagawa et al., 1987) as demonstrated for the acetyl carboxylase gene (Park and Kim, 1993). An AP-2 binding site was not identified by computer search in the HO-1 promoter 5'-flanking region within the first 1,300 bp of the 5'-flanking region of the rat HO-1 gene. It is also conceivable that other REs with no sequence homology to the CRE or the AP-2 consensus sequences could mediate the PKA-dependent induction of the HO-1 gene. Moreover, it is likely that the maximal effect of cAMP on the transcriptional activation of the HO-1 gene is mediated by a synergism of more than one *cis*-acting element and transcription factor, as has been suggested for the rat PCK gene (Roesler et al., 1995).

5.3. Physiological and Clinical Implications of cAMP-dependent HO-1 Gene Induction

The HO-1 gene has essential physiological functions in various organs such as liver, heart and kidney. This has been demonstrated in HO-1 deficient mouse models (Poss and Tonegawa, 1997; Yet et al., 1999) as well as in genetic human HO-1 deficiency (Yachie et al., 1999). Induction or overexpression of the HO-1 gene not only protects against the toxic effects of heme or hemoproteins (Abraham et al., 1995; Nath et al., 1992), but, in addition, HO-1 has also been demonstrated to have beneficial effects during inflammation (Willis et al., 1996) and after organ transplantation (Soares et al., 1999). Therefore, the targeted up-regulation of HO-1 by non-toxic cyclic nucleotides such as cAMP or cGMP may provide novel pharmacological approaches for treatment of such clinically relevant conditions. Similar to the targeted induction of the TF NFκB which has been demonstrated to be an antiinflammatory target (Barnes and Karin, 1997) HO-1 induction by cAMP and/or cGMP may be an even more specific therapeutic target (Immenschuh and Ramadori, 2000). In accordance with this concept, the group of Schröder has previously shown that TNFα-dependent toxicity in bovine endothelial cell cultures is prevented by pretereatment with cGMP (Polte et al., 2000).

REFERENCES

Abraham, N.G., Lavrovsky, Y., Schwartzman, M.L., Stoltz, R.A., Levere, R.D., Gerritsen, M.E., Shibahara, S., and Kappas, A., 1995, Transfection of the human heme oxygenase gene into rabbit coronary microvessel endothelial cells: Protective effect against heme and hemoglobin toxicity, *Proc. Natl. Acad. Sci. USA* **92**:6798.

Bakken, A.F., Thaler, M.M., and Schmid, R., 1972, Metabolic regulation of heme catabolism and bilirubin production. Hormonal control of hepatic heme oxygenase activity, *J. Clin. Invest.* **51**:530.

Barnes, P.J. and Karin, M., 1997, Nuclear factor-κB-a pivotal transcription factor in chronic inflammatory disease, *N. Engl. J. Med.* **336**:1066.

Choi, A.M.K. and Alam, J., 1996, Heme oxygenase-1: Function, regulation, and implication of a novel stress-inducible protein in oxidant-induced lung injury, *Am. J. Respir. Cell. Mol. Biol.* **15**:9.

Christ, B., Nath, A., and Jungermann, K., 1991, Interactions of nuclear protein from cultured rat hepatocytes with the cyclic nucleotide reponsive elements and NF1-CTF site in the promoter of the phosphoenolpyruvate carboxykinase gene, *Biochem. Biophys. Res. Comm.* **181**:367.

Foresti, R. and Motterlini, R., 1999, The heme oxygenase pathway and its interaction with nitric oxide in the control of cellular homeostasis, *Free Rad. Res.* **31**:459.

Durante, W., Christodoulides, N., Cheng, K., Peyton, K.J., Sunahara, R.K., and Schafer A.I., 1997, cAMP induces heme oxygenase-1 gene expression and carbon monoxide production in vascular smooth muscle, *Am. J. Physiol.* **273**:H317.

Hagiwara, M., Alberts, A., Brindle, P., Meinkoth, J., Feramisco, J., Deng, T., Karin, M., Shenolikar, S., and Montminy, M., 1992, Transcriptional attenuation following cAMP induction requires PP-1-mediated dephosphorylation of CREB, *Cell* **70**:105.

Imagawa, M., Chiu, R., and Karin, M., 1987, Transcription factor AP-2 mediates induction by two different signal transduction pathways: protein kinase C and cAMP, *Cell* **51**:251.

Immenschuh, S., Kietzmann, T., Hinke, V., Wiederhold, M., Katz, N., and Muller-Eberhard, U., 1998A, The rat heme oxygenase-1 gene is transcriptionally induced via the protein kinase A signaling pathway in rat hepatocyte cultures, *Mol. Pharmacol.* **53**:483.

Immenschuh, S., Hinke, V., Ohlmann, A., Gifhorn-Katz, S., Katz, N., Jungermann, K., and Kietzmann, T., 1998B, Transcriptional activation of the haem oxygenase-1 gene by cGMP via a cAMP response element/activator protein-1 element in primary cultures of rat hepatocytes, *Biochem. J.* **334**:141.

Immenschuh, S., Hinke, V., Katz, N., and Kietzmann, T., 2000, Transcriptional induction of the heme oxygenase-1 gene expression by okadaic acid in primary rat hepatocyte cultures, *Mol. Pharmacol.* **57**:610.

Immenschuh, S. and Ramadori, G., 2000, Gene regulation of heme oxygenase-1 as a therapeutic target, *Biochem. Pharmacol.* **60**:1121.

Maines, M.D., 1997, The heme oxygenase system: a regulator of second messenger gases, *Annu. Rev. Pharmacol. Toxicol.* **37**:517.

Montminy, M.R. and Bilezikjian, L.M., 1987, Binding of a nuclear protein to the cyclic-AMP response element of the somatostatin gene, *Nature* **328**:175.

Nath, K.A., Balla, G., Vercellotti, G.M., Balla, J., Jacob, H.S., Levitt, M.D., and Rosenberg, M.E, 1992, Induction of heme oxygenase is a rapid, protective response in rhabdomyolysis in the rat, *J. Clin. Invest.* **90**:267.

Park, K. and Kim, K.H., 1993, The site of cAMP action in the insulin induction of gene expression of acetyl-CoA carboxylase is AP-2, *J. Biol. Chem.* **268**:17811.

Polte, T., Abate, A., Dennery, P.A., and Schroder, H., 2000, Heme oxygenase-1 is a cyclic GMP-inducible endothelial protein and mediates the cytoprotective action of nitric oxide, *Arterioscler. Thromb. Vasc. Biol.* **20**:1209.

Poss, K.D. and Tonegawa, S., 1997, Reduced stress defense in heme oxygenase 1-deficient cells, *Proc. Natl. Acad. Sci. USA* **94**:10925.

Roesler, W.J., Graham, J.G., Kolen, R., Klemm, D.J., and McFie, P.J., 1995, The cAMP response element binding protein synergizes with other transcription factors to mediate cAMP responsiveness, *J. Biol. Chem.* **270**:8225.

Soares, M.P., Lin, Y., Anrather, J., Csizmadia, E., Takigami, K., Sato, K., Grey, S.T., Colvin, R.B., Choi, A.M., Poss, K.D., and Bach, F.H., 1998, Expression of heme oxygenase-1 can determine cardiac xenograft survival, *Nat. Med.* **4**:1073.

Wadzinski, B.E., Wheat, W.H., Jaspers, S., Peruski, L.F., Lickteig, R.L., Johnson, G.L., and Klemm, D.J., 1993, Nuclear protein phosphatase 2A dephosphorylates protein kinase A-phosphorylated CREB and regulates CREB transcriptional stimulation, *Mol. Cell. Biol.* **13**:2822.

Willis, D., Moore, A.R., Frederick, R., and Willoughby, D.A., 1996, Heme oxygenase: A novel target for the modulation of the inflammatory response, *Nat. Med.* **2**:87.

Yachie, A., Niida, Y., Wada, T., Igarashi, N., Kaneda, H., Toma, T., Ohta, K., Kasahara, Y., and Koizumi, S., 1999, Oxidative stress causes enhanced endothelial cell injury in human heme oxygenase-deficiency, *J. Clin. Invest.* **103**:129.

Yet, S.-F., Perrella, M.A., Layne, M.D., Hsieh, C.M., Maemura, K., Kobzik, L., Wiesel, P., Christou, H., Kourembanas, S., and Lee, W.-S., 1999, Hypoxia induces severe right ventricular dilatation and infarction in heme oxygenase-1 null mice, *J. Clin. Invest.* **103**:R23.

33

REGULATION OF HEME OXYGENASE-1 GENE TRANSCRIPTION VIA THE STRESS-RESPONSE ELEMENT

Daniel Stewart, Julia L. Cook, and Jawed Alam[a]

[a]Department of Molecular Genetics
Alton Ochsner Medical Foundation, New Orleans
Louisiana 70121

1. INTRODUCTION

Heme oxygenase (HO) enzymes regulate cellular heme and iron levels by catalyzing the rate-limiting reaction in heme catabolism—the oxidative cleavage of b-type heme molecules to yield equimolar quantities of biliverdin IXα, carbon monoxide (CO) and iron. Biliverdin is subsequently converted to bilirubin by the action of biliverdin reductase. Two enzymatically active isoforms of HO, HO-1 and HO-2, encoded by distinct genes, have been identified. While expression of HO-2 is generally constitutive, expression of HO-1 is greatly enhanced by the substrate heme and a diverse array of stimuli. HO-1 expression is also stimulated in a variety of pathologies including atherosclerosis, AIDS, diabetes mellitus and Parkinson's and Alzheimer's diseases. From the initial discovery of HO-1 activity in the mid- to late-1960's until quite recently, the biological "purpose"of this enzyme has been relegated primarily to the essential, but unassuming, task of heme metabolism. Studies in the past several years, however, have convincingly demonstrated that, largely as a consequence of its inducibility and the catalytic products generated, HO-1 manifests both antioxidant and anti-inflammatory activities and participates in the more generalized processes of cell signaling and homeostasis in response to injury. These roles are deduced from the convergence of several observations: 1) in addition to the substrate heme, a variety of stress conditions and agents including, but not limited to, ultra violet irradiation, hyperthermia, ischemia-reperfusion, heavy metals, hydrogen peroxide, endotoxin and inflammatory cytokines strongly stimulate HO-1 expression, 2) these and other stimuli share in common the ability to stimulate cellular oxidative stress; 3) in dismantling heme, HO-1 eliminates a cellular pro-oxidant and simultaneously produces molecules (biliverdin and bilirubin) that are potent antioxidants; 4) even the released heme-iron, which can catalyze the production of reactive oxygen species, may ultimately be protective by virtue of its ability to stimulate, in certain settings, the synthesis of the iron-sequestering protein apo-ferritin; 5) CO, a diffusable gas, functions as a neural messenger and, similar to nitric oxide (NO), exhibits vasoactive properties. While the protective function of HO-1 has been documented in various cellular and animal models of oxidant-mediated injury, the importance of HO-1, and presumably its inducibility, is most dramatically demonstrated by the physiological abnormalities, including growth retardation, hepatomegaly, leukocytosis, anemia, and tissue iron deposition, observed in cases of HO-1 deficiency.[1–3] The biological activities of HO-1 and the reaction products of heme catabolism are discussed in more detail in various review articles.[4–7]

2. TRANSCRIPTIONAL REGULATION AND *CIS*-ELEMENTS

Nuclear run-on assays have provided direct evidence for transcriptional induction of the *ho-1* gene by several stimuli including heme, cadmium, UV-irradiation, 12-*O*-tetradecanoylphorbol 13-acetate (TPA), lipopolysaccharide (LPS) and NO. While all inducers have not been tested in this manner, it is generally accepted that stimulation of HO-1 expression by practically all agents is regulated primarily at the level of gene transcription. The basic mechanism of transcription is essentially identical for all genes and involves the initiation and procession of RNA synthesis by RNA polymerase and the basic transcription machinery. Temporal, spatial, tissue-specific and inducer-dependent variations in gene expression result primarily from the binding of specific transcription factors to their target DNA sequences (*cis*-elements) located within the body of the gene, downstream of the gene or, more commonly, upstream of the gene in the proximal and/or distal promoter regions. These sequence-specific DNA-binding proteins, in turn, modulate the rate of transcription initiation by interacting directly, or indirectly via co-activators, with the basic transcription machinery. To date, understanding the mechanism of *ho-1* gene regulation has largely been a quest to identify inducer-responsive *cis*-elements and their cognate binding proteins. All inducible *cis*-elements of the *ho-1* genes thus far identified are located within three segments of the 5′ flanking region (Fig. 1). These elements and other characteristics are summarized in Table 1.

2.1. The Stress-Response Element

Many of the *cis*-elements thus far identified are typically present as single copy sequences and are responsive to a single, or only a few, agents. Our studies on the mouse *ho-1* gene, however, have identified a dominant *cis*-element, termed the stress-response element (StRE), that is present in multiple copies within the distal enhancer regions, DE1 and DE2, and mediates gene activation by multiple agents. Furthermore, as is apparent from the information in Table 1, different elements and, consequently, different transcription factors have been implicated in *ho-1* gene activation by the same inducer. The reason for these discrepancies is unclear but may reflect cell- and

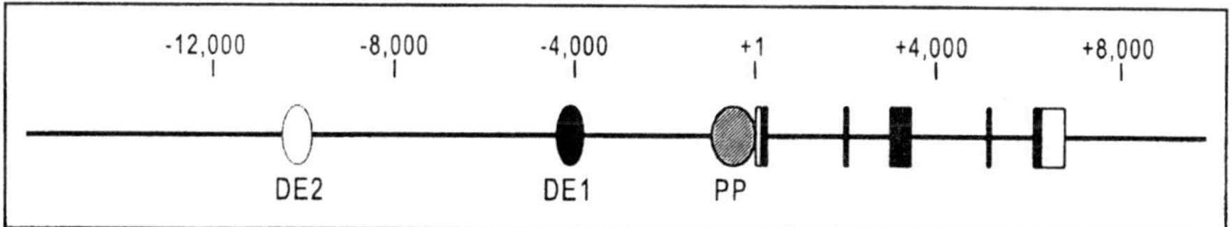

Figure 1. Cis-elements have been localized within the promoter proximal (PP) and distal enhancer (DE) regions. Exons are designated by rectangles; open segments indicate 5′ and 3′ untranslated regions. +1, transcription start site. Numbers are in base pairs.

Table 1. Inducer-Responsive Elements of *ho-1* Genes

Gene[a]	*Cis*-Element	Loc[b]	Putative *Trans*-Factor	Inducing Agent[c]	Ref
C	TGGCTGAGTCAGGCT	PP	AP-1	ARS, $CoCl_2$	8, 9
C	TGCTCACGTGG	PP	???	ARS	8, 9
C	−4.1 to −3.6 kbp	DE1	???	Heme, CoPP	10
R	GTTCTGGAACCTTCC	PP	HSF	Hyperthermia	11
R	GCCATGTGTCCTGAG	PP	???	PGJ_2	12
R	CTGACTTCAGTC	PP	AP-1/CREB	cGMP, OA	13, 14
H	GTCATATGAC	PP	bHLH	TPA	15
H	AAGGGTCAT	PP	???	IL-6	16
H	TGCTAGATTT	DE1	???	Cd	17
H	AGGCGGATTTTGCTAGATTT TGCTGAGTCACC	DE1	???	SIN-1	18
H	AGGCGGATTTTGCTAGATTT	DE1	???	Tobacco smoke, H_2O_2, Cd	19
H	Region −120 to −1	PP	???	ARS, Hemin, H_2O_2, TPA, Cd	20
H	CCTGGCCCAC	PP	USF	UV-Irradiation	21
H	GCCCGGGGC	PP	AP-2	Heme	22, 23
H	TTTCCCAAGGG	PP	NF-κB	Heme	22, 23
H	−4.5 kbp to +80 bp		???	Heme	24
M	Not Defined		NF-κB	TPA, LPS	25, 26
M	TGCTG(T/A)GTCA	DE1	AP-1	LPS, IL-1β	27
M	GAGGGGTCG	DE2	Egr-1	ZnPP	28
M	(T/C)GCTGAGTCA	DE1 DE2	AP-1 ATF/CREB MAF CNC	Heme, Metals, TPA, NO, LPS, IL-6, H_2O_2, ARS, DEP, quinones, PGJ_2 hyperoxia, electrophiles	29–38
M	TTCCGGGAA, TTCTGGAAA	PP	STAT	Hyperoxia	39
M	GACGTGC(T/C)	DE2	HIF-1	Hypoxia	40

[a]C, chicken; R, rat; H, human; M, mouse.
[b]Location (see Fig. 2).
[c]ARS, arsenite; CoPP/ZnPP, cobalt/zinc protoporphyrin; PGJ_2, prostaglandin J_2; OA, okadaic acid; TPA, 12-*O*-tetradecanoylphorbol 13-acetate; Cd, cadmium; SIN-1, 3-morphilinosydnonimine; LPS, lipopolysaccharide; NO, nitric oxide; DEP, deisel exhaust particles.

species-specific differences in the induction mechanism. It is noteworthy, however, that many of the earlier studies cited in Table 1 utilized 1,500 bp or less of the *ho-1* gene 5′ flanking regions; these analyses obviously would not have included the distal enhancers and may have led to incomplete results. Furthermore, in several cases, the role of a particular *cis*-element in *ho-1* gene regulation has not been rigorously established using both functional and protein-binding criteria. Finally, it should be noted that some of the elements identified are not conserved across species[17] whereas all the StREs of the mouse *ho-1* gene are conserved within the human gene, both spatially and with regard to primary structure. (The latter genes and their flanking regions have been sequenced in their entirety as part of the Human Genome Project; Accession # AC005290 and Z82244).

3. STRE-BINDING PROTEINS (StRE-BPs)

Aside from the the multiple copies of the StREs, elements common to both mouse DE1 and DE2 include binding sites for CCAAT/enhancer binding protein (C/EBP) and specificity protein 1 (Sp1). However, the StREs, but not the other motifs, are sufficient and necessary for gene activation.[31,33,34,41] The minimal, consensus StRE, derived from sequence similarity and functional analysis, is shown in Table 2. This sequence resembles the consensus binding sites for several families of DNA-binding proteins.

3.1. AP-1 Factors

AP-1 factors are comprised of Jun family homo- and hetero-dimers and Jun/Fos heterodimers. *fos* and *jun* were initially identified as the oncogenes carried by the Finkel-Biskis-Jinkins murine sarcoma virus and the avian sarcoma virus 17, respectively. Their cellular counterparts, *c-fos* and *c-jun*, and other family members have been implicated in various cellular processes including proliferation, differentiation

Table 2. Transcription Elements Similar to the StRE

Element	Consensus Sequence[a]	Canonical Binding Proteins	Ref
StRE	TGCTGA GTCA C		
NF-E2 binding site	TGCTCA GTCAC C C T	CNC-bZIP/Maf heterodimers	46
Maf binding sites	TGCTGA GTCAGCA TGCTGACGTCAGCA	Maf	47
ARE: antioxidant response element	GCNNNGTCACC T		48
TRE: TPA response element	TGA GTCA C	AP-1 (Fos/Jun)	42
CRE: cAMP response element	TGACGTCA	CREB/ATF	43

[a]Sequences are spaced to give maximal alignment.

and apoptosis. These proteins are widely distributed, being found in many tissues and cell types. Additionally, the expression and/or DNA-binding activity of certain members, such as c-Fos and c-Jun, are potently stimulated in many cell types by various pro-oxidants including TPA, heavy metals, H_2O_2 and UV-irradiation.[42] Given these characteristics and the prominent role these proteins play in directing the genetic program under pro-oxidant states, we initially proposed that AP-1 proteins mediate induction of the *ho-1* gene.[31,33]

3.2. ATF/CREB Factors

The activating transcription factor (ATF) and the cAMP response element binding proteins (CREB) recognize a common palindromic sequence, termed the cAMP response element (CRE). The ATF/CREB family contains more than 10 members, not all of which are responsive to cAMP. Some ATF/CREB proteins are known to form cross-family hetero-dimers with AP-1 proteins and the activity/expression of certain members, such as ATF2 and ATF3, is stimulated by stress.[43]

3.3. Maf Factors

The *v-maf* oncogene is a constituent of the avian musculoaponeurotic fibrosarcoma virus AS42 and it's product, v-Maf, is a nuclear protein that recognizes two types of palindromic consensus sequences—an extended AP-1 binding site and an extended CRE (Table 2). Like other oncogenes, the cellular *c-maf* proto-oncogene is a member of a family of related genes of which six have been identified to date and most members are expressed in a wide range of tissues.[44] The Maf family consists of "big" Maf and "small" Maf proteins; the latter lack apparent transcription activation domains. Maf family members are analogous to Jun proteins in that they can form intra-family homo- and hetero-dimers. Unlike Jun (and Fos) proteins, Maf transcription factors are not thought to be involved in cell growth regulation. Cross-family heterodimerization between Maf and AP-1 proteins has been observed.

3.4. Nuclear Factor-Erythroid 2 (NF-E2) and Related Factors

NF-E2, a heterodimer of an erythroid-specific 45 kDa subunit (p45) and p18, a "small" Maf protein, was initially identified as a protein that binds to the locus control regions of the β-globin gene. The p45 subunit contains a Cap'n'Collar (CNC) domain homologous to a region within the fruit fly homeotic selector protein encoded by the *cap'n'collar* gene. Other mammalian CNC proteins include Nrf (*N*F-E2 *r*elated *f*actors) 1, 2, and 3 and Bach proteins. In contrast to p45, the Nrf proteins are more widely expressed. CNC proteins do not form intra-family dimers and heterodimerize most prominently with "small" Maf proteins.[45]

3.5. The Antioxidant Response Element Binding Proteins

The StRE also conforms to the antioxidant/electrophile response element (ARE/EpRE), an element that regulates the induction of several genes in response to xenobiotics, oxidants or peroxides. These genes encode proteins, such as NAD(P)H:quinone reductase, glutathione *S*-transferase, and γ-glutamylcysteine synthetase, that function in xenobiotic detoxification and, in this capacity, exhibit antioxidant activity. Many transcription factors, including members of the AP-1, CNC-bZIP and Maf families are known to bind to the ARE.[49]

All of the proteins described above belong to the basic-leucine zipper (bZIP) superfamily of transcription factors that are characterized by two conserved structural motifs: 1) a heptad repeat with leucine residues at every 7th position (the "leucine zipper") that is necessary for polypeptide dimerization; and 2) an adjacent DNA-binding domain enriched in basic amino acids. bZIP proteins function as obligate homo-dimers or hetero-dimers. Because of intra-family and cross-family dimerization (not only between the protein families listed in Table 2 but also between some of these proteins and other bZIP (*e.g.*, C/EBP) and non-bZIP factors), the number of potential StRE-binding, dimeric species is quite extensive (>100). Indeed, one or more members of each of the sub-families listed above binds to the StRE.[29,31,50,51] One consequence of the multiplicity of StRE binding proteins is that it permits inducer-dependent variation in the induction mechanism (*i.e.*, inducer-dependent utilization of different StRE-BPs) even when such induction is mediated by a single type of element.

4. NRF2—THE MASTER REGULATOR?

We have initiated studies to identify the StRE-BP(s) that mediate inducer-dependent *ho-1* gene activation by utilizing two complementary strategies: examining 1) the activation of StRE-dependent reporter gene activation by overexpression of putative StRE-BPs and 2) the inhibition of StRE activity and *ho-1* gene induction by dominant-negative mutants of individual StRE-BPs. These initial studies point to Nrf2 as a dominant regulator of *ho-1* gene activation in response to several inducers including heme, cadmium, cobalt and arsenite in in various cell types including fibroblast, hepatoma and human mammary epithelial cells.[50–52] This conclusion is further strengthened by the recent demonstration that macrophages derived from *nrf2*[-/-] mice exhibit reduced *ho-1* gene activation by several stimuli.[53] Additionally, Nrf2 appears to be the primary regulator of ARE-dependent genes including those encoding NAD(P)H:quinone reductase, glutathione *S*-transferase, and γ-glutamyl-cysteine synthetase. These observations suggest that Nrf2 is an important regulator of the genetic program of cells during the adaptive response to oxidative stress.[54] Consistent with this idea, Nrf2-deficient mice are exquisitely sensitive to butylated hydroxytoluene and succumb from acute-respiratory distress syndrome[55] and macrophages derived from such mice exhibit reduced resistance to toxic electrophiles.[53]

The primary structure within the leucine zipper region of Nrf2 precludes self-dimerization and, thus, Nrf2 functions as an obligate heterodimer. Based on the NF-E2 paradigm, Nrf2 dimerizes most prominently with small Maf proteins but whether such dimers mediate positive or negative regulation of ARE/StRE dependent genes remains unclear.[50,52,56,57] Jun:Nrf2 complexes, either as heterodimers or in other configurations, are also proposed to positively regulate ARE-mediated gene expression.[49] Given the propensity of bZIP proteins to form cross-family dimers, it is likely that other Nrf2:bZIP heterodimers also exist.

In order to identify such dimeric species we have screened for Nrf2 interacting proteins using the yeast two-hybrid system. Surprisingly, of the various interacting proteins identified, only one was a bZIP protein—ATF4, a member of the ATF/CREB subfamily. Interestingly, expression of ATF4 is up-regulated in response to specific stimuli such as anoxia and TPA, two stressors that also enhance *ho-1* gene expression. Interaction between Nrf2 and ATF4 in mammalian cells was confirmed by co-immunoprecipitation experiments and mammalian two hybrid analysis. Furthermore, Nrf2:ATF4 dimers specifically bind to the StRE and a dominant-negative mutant of ATF4 inhibits basal and cadmium-induced expression of an StRE-regulated reporter gene in Hepa cells but effects only basal activity in MCF-7 cells.[51] These results imply that ATF4, possibly in association with Nrf2, may modulate inducer-dependent *ho-1* gene activation in a cell-specific manner.

5. FUTURE PROSPECTS

Establishing the role of ATF4 in *ho-1* gene regulation will require further analysis, including examination of ATF-deficient mice or cells. Generalized information regarding the cellular genetic response to oxidative stress, however, is more likely to be achieved by investigating the various characteristics of Nrf2 function or activity. Unlike AP-1 and NF-κB, two well-characterized oxidative stress-responsive mammalian transcription factor families, the mechanisms of action of Nrf2 (and other CNC-bZIP factors) are only poorly understood. For Nrf2, much remains to be investigated including identification of other Nrf2 target genes, identification of other dimerization partners, types and consequence of post-translation modification and characterization of signal transduction pathways that modulate various Nrf2 activities (*e.g.*, cytoplasmic retention/nuclear import, DNA binding, transcription activation and co-activator interaction). One cannot dismiss the possibility that some of these properties will exhibit inducer-specific or inducer-selective variations. Given that Nrf2 has been implicated in the regulation of multiple genes encoding anti-oxidant proteins, the substantial interest generated in this transcription factor is likely to lead to a clearer understanding of the mechanisms of action in the not too distant future.

ACKNOWLEDGMENTS

Work by the authors was supported by United States Public Health Service Grants DK-43135.

REFERENCES

1   K. D. Poss and S. Tonegawa, Heme oxygenase 1 is required for mammalian iron reutilization, *Proc. Natl. Acad. Sci. U.S.A.* **94**, 10919–10924 (1997).
2. K.D. Poss and S. Tonegawa, Reduced stress defense in heme oxygenase 1-deficient cells, *Proc. Natl. Acad. Sci. U.S.A.* **94**, 10925–10930 (1997).

3. A. Yachie, Y. Niida, T. Wada, N. Igarashi, H. Kaneda, T. Toma, K. Ohta, Y. Kasahara, and Ss. Koizumi, Oxidative stress causes enhanced endothelial cell injury in human heme oxygenase-1 deficiency, *J. Clin. Invest.* **103**, 129–135 (1999).

4. A.M.K. Choi and J. Alam, Heme oxygenase-1: function, regulation, and implication of a novel stress-inducible protein in oxidant-induced lung injury, *Am. J. Respir. Cell Mol. Biol.* **15**, 9–19 (1996).

5. M.D. Maines, The heme oxygenase system: a regulator of second messenger gases, *Ann. Rev. Pharmacol. Toxicol.* **37**, 517–554 (1997).

6. A. Agarwal and H.S. Nick, Renal response to tissue injury: lessons from heme oxygenase-1 gene ablation and expression, *J. Am. Soc. Nephrol.* **11**, 965–973 (2000).

7. L.E. Otterbein and A.M. Choi, Heme oxygenase: colors of defense against cellular stress, *Am. J. Physiol. Lung Cell. Mol. Physiol.* **279**, L1029–1037 (2000).

8. K.K. Elbirt, A.J. Whitmarsh, R.J. Davis, and H.L. Bonkovsky, Mechanism of sodium arsenite-mediated induction of heme oxygenase-1 in hepatoma cells, *J. Biol. Chem.* **273**, 8922–8931 (1998).

9. T.H. Lu, Y. Shan, J. Pepe, R.W. Lambrecht, and H.L. Bonkovsky, Upstream regulatory elements in chick heme oxygenase-1 promoter: a study in primary cultures of chick embryo liver cells, *Mol and Cell. Biochem.* **209**, 17–27 (2000).

10. Y. Shan, J. Pepe, T.H. Lu, K.K. Elbirt, R.W. Lambrecht, and H.L. Bonkovsky, Induction of heme oxygenase-1 gene by metalloporphyrins, *Arch. Biochem. Biophys.* **380**, 219–227 (2000).

11. S. Okinaga and S. Shibahara, Identification of a nuclear protein that constitutively recognizes the sequence containing a heat-shock element, *Eur. J. Biochem.* **212**, 167–175 (1993).

12. T. Koizumi, N. Odani, T. Okuyama, A. Ichikawa, and M. Negishi, Identification of a cis-regulatory element for Δ^{12}-prostaglandin J_2-induced expression of the rat heme oxygenase gene, *J. Biol. Chem.* **270**, 21779–21784 (1995).

13. S. Immenschuh, V. Hinke, A. Ohlmann, S. Gifhorn-Katz, N. Katz, K. Jungermann, and T. Kietzmann, Transcriptional activation of the heme oxygenase-1 gene by cGMP via a cAMP response element/activator protein-1 element in primary cultures of rat hepatocytes, *Biochem. J.* **334**, 141–146 (1998).

14. S. Immenschuh, V. Hinke, N. Katz, and T. Kietzmann, Transcriptional induction of heme oxygenase-1 gene expression by okadaic acid in primary rat hepatocyte cultures, *Mol. Pharmacol.* **57**, 610–618 (2000).

15. Y. Muraosa and S. Shibahara, Identification of a cis-regulatory element and putative trans-acting factors responsible for 12-O-Tetradecanoylphorbol-13-Acetate (TPA)-mediated induction of heme oxygenase expression in myelomonocytic cell lines, *Mol. Cell. Biol.* **13**, 7881–7891 (1993).

16. K. Mitani, H. Fujita, A. Kappas, and S. Sassa, Heme Oxygenase is a positive acute-phase reactant in human Hep3B hepatoma cells, *Blood* **5**, 1255–1259 (1992).

17. K.S. Takeda, M. Ishizawa, T. Sato, T. Yoshida, and S. Shibahara, Identification of a cis-acting element that is responsible for cadmium-mediated induction of the human heme oxygenase gene, *J. Biol. Chem.* **269**, 22858–22867 (1994).

18. E. Hara, K. Takahashi, K. Takeda, M. Nakayama, M. Yoshizawa, H. Fujita, K. Shirato, and S. Shibahara, Induction of heme oxygenase-1 as a response in sensing the signals evoked by distinct nitric oxide donors, *Biochem. Pharmacol.* **58**, 227–236 (1999).

19. F. Favatier and B.S. Polla, Tobacco-smoke inducible human haem oxygenase-1 gene expression: role of distinct transcription factors and reactive oxygen intermediates, *Biochem. J.* **353**, 475–482 (2001).

20. R.M. Tyrrell, L.A. Applegate, and Y. Tromvoukis, The proximal promoter region of the human heme oxygenase gene contains elements involved in stimulation of transcriptional activity by a variety of agents including oxidants, *Carcinogenesis* **14**, 761–765 (1993).

21. A.L. Nascimento, P. Luscher, and R.M. Tyrrell, Ultraviolet A (320–380 nm) radiation causes an alteration in the binding of a specific protein/protein complex to a short region of the promoter of the human heme oxygenase 1 gene, *Nucl. Acids. Res.* **21**, 1103–1109 (1993).

22. Y. Lavrovsky, M.L. Schwartzman, and N.G. Abraham, Novel regulatory sites of the human heme oxygenase-1 promoter region, *Biochem. Biophys. Res. Commun.* **196**, 336–341 (1993).

23. Y. Lavrovsky, M.L. Schwartzman, R.D. Levere, A. Kappas, and N.G. Abraham, Identification of binding sites for transcription factors NF-κB and AP-2 in the promoter region of the human heme oxygenase 1-gene., *Proc. Natl. Acad. Sci. U.S.A.* **91**, 5987–5991 (1994).

24. A. Agarwal, F. Shiraishi, G.A. Visner, and H.S. Nick, Linoleyl hydroperoxide transcriptionally upregulates heme oxygenase-1 gene expression in human renal epithelial and aortic endothelial cells, *J. Am. Soc. Nephrol.* **9**, 1990–1997 (1998).

25. S.-I. Kurata, M. Matsumoto, Y. Tsuji, and H. Nakajima, Lipopolysaccharide activates transcription of the heme oxygenase gene in mouse M1 cells through oxidative activation of nuclear factor κB, *Eur. J. Biochem.* **239**, 566–571 (1996).

26. S.-I. Kurata, M. Matsumoto, and H. Nakajima, Transcriptional control of the heme oxygenase gene in mouse M1 cells during their TPA-induced differentiation into machophages, *J. Cell. Biochem.* **62**, 314–324 (1996).

27. P.L.C. Wiesel, A. Foster, M.D. Pellacani, C.M. Layne, G.S. Hsieh, G.S. Huggins, P. Strauss, S.F. Yet, and M.A. Perrella, Thioredoxin facilitates the induction of heme oxygenase-1 in response to inflammatory mediators, *J. Biol. Chem.* **275**, 24840–24846 (2000).

28. G. Yang, X. Nguyen, J. Ou, P. Rekulapelli, D.K. Stevenson, and P.A. Dennery, Unique effects of zinc protoporphyrin on HO-1 induction and apoptosis, *Blood* **97**, 1306–1313 (2001).

29. J. Alam and D. Zhining, Distal AP-1 binding sites mediate basal level enhancement and TPA induction of the mouse heme oxygenase-1 gene, *Biol. Chem.* **267**, 21894–21900 (1992).

30. J. Alam, J. Cai, and A. Smith, Isolation and characterization of the mouse heme oxygenase-1 gene: distal 5′ sequences are required for induction by cadmium or heme, *Biol. Chem.* **269**, 1001–1009 (1994).

31. J. Alam, Multiple elements within the 5′ distal enhancer of the mouse heme oxygenase-1 gene mediate induction by heavy metals, *J. Biol. Chem.* **269**, 25049–25056 (1994).

32. S.L. Camhi, J. Alam, L. Otterbein, and A.M.K. Choi, Activation of the heme oxygenase gene by lipopolysaccharide is AP-1 dependent: transcriptional regulation and signal transduction, *Am. J. Resp. Cell Mol. Biol.* **3**, 387–398 (1995).

33. J. Alam, S. Camhi, and A.M.K. Choi, Identification of a second region of the mouse heme oxygenase-1 gene that functions as a basal level and inducer-dependent transcription enhancer, *J. Biol. Chem.* **270**, 11977–11984 (1995).

34. T. Prestera, P. Talalay, J. Alam, Y.I. Ahn, P.J. Lee, and A.M.K. Choi, Parallel induction of heme oxygenase-1 and chemoprotective phase 2 enzymes by electrophiles and antioxidants: regulation by upstream antioxidant responsive elements (ARE), *Mol. Med.* **1**, 827–837 (1995).

35. P. Lee, J. Alam, N. Inamdar, S.L. Sylvester, L. Otterbein, and A.M.K. Choi, Regulation of heme oxygenase-1 gene expression *in vivo* and *in vitro* in hyperoxic lung injury, *Am. J. Resp. Cell Mol. Biol.* **14**, 556–568 (1996).

36. C.L. Hartsfield, J. Alam, and A.M.K. Choi, Transcriptional regulation of the heme oxygenase-1 gene by pyrrolidine dithiocarbamate, *FASEB J.* **12**, 1675–1682 (1998).

37. N. Li, R. Kaplan, S. Gujuluva, J. Alam, and A. Nel, Induction of heme oxygenase-1 in macrophages by diesel exhaust particle and quinones is mediated through antioxidant responsive elements, *J. Immunol.* **165**, 3393–3401 (2001).

38. P. Gong, B. Hu, D. Stewart, C. Wicks, M. Ellerbe, Y. Gutierrez, B. Beckman, and J. Alam, Cobalt induces heme oxygenase-1 expression by a hypoxia inducible factor-independent mechanism in chinese hamster ovary cells. *J. Biol. Chem.* **276**, 27018–27025 (2001).

39. P.J. Lee, S.L. Camhi, B.Y. Chin, J. Alam, and A.M. Choi, AP-1 and STAT mediate hyperoxia-induced gene transcription of heme oxygenase-1, *Am. J. Physiol.* **279**, L175–182 (2000).

40. P.J. Lee, B.-H. Jiang, B.Y. Chin, G.L. Semenza, J. Alam, and A.M.K. Choi, Activation of the mouse heme oxygenase-1 gene by hypoxia is mediated by hypoxia inducible factor-1 response elements, *J. Biol. Chem.* **272**, 5375 5381 (1997).

41. N.M. Inamdar, Y.I. Ahn, and J. Alam, The heme-responsive element of the mouse heme oxygenase-1 gene is an extended AP-1 binding site that resembles the recognition sequences for Maf and Nf-E2 transcription factors, *Biochem. Biophys. Res. Comm.* **221**, 570–576 (1996).

42. M. Karin, Z.-g. Liu, and E. Zandi, AP-1 function and regulation, *Curr. Opin. Cell Bio.* **9**, 240–246 (1997).

43. T. Hai, C.D. Wolfgang, D.K. Marsee, A.E. Allen, and U. Sivaprasad, ATF3 and stress responses, *Gene Express* **7**, 321–335 (1999).

44. H. Motohashi, J.A. Shavit, K. Igarashi, M. Yamamoto, and J.D. Engel, The world according to Maf. *Nucl. Acids Res.* **25**, 2953–2959 (1997).

45. N.C. Andrews, Molecules in focus. The NF-E2 Transcription factor, *Int. J. of Biochem. Cell Biol.* **30**, 429–432 (1998).

46. N. Andrews, H. Erdjument-Bromage, M.B. Davidson, P. Tempst, and S.H. Orkin, Erythroid transcription factor NF-E2 is a haematopoietic-specific basic-leucine zipper protein. *Nature*, **362**, 722–728 (1993).

47. K. Kataoka, M. Noda, and M. Nishizawa, Maf nuclear oncoprotein recognizes sequences related to an AP-1 site and forms heterodimers with both Fos and Jun, *Mol. Cell Biol.* **14**, 700–712 (1994).

48. T.H. Rushmore, M.R. Morton, and C.B. Pickett, The antioxidant responsive element, *J. Biol. Chem.* **266**, 11632–11639 (1991).

49. S. Dhakshinamoorthy, D.J. Long 2nd, and A.K. Jaiswal, Antioxidant regulation of genes encoding enzymes that detoxify xenobiotics and carcinogens, *Curr. Top. Cell Regul.* **36**, 201–216 (2000).

50. J. Alam, C. Wicks, D. Stewart, P. Gong, C. Touchard, S. Otterbein, A.M.K. Choi, M.E. Burow, and J.s. Tou, Mechanism of heme oxygenase-1 gene activation by cadmium in MCF-7 mammary epithelial cells. Role of p38 kinase and Nrf2 transcription factor, *J. Biol. Chem.* **275**, 27694–27702 (2000).

51. C. He, P. Gong, B. Hu, D. Stewart, M.E. Choi, A.M. Choi, and J. Alam J, Identification of activating transcription factor 4 (ATF4) as a Nrf2 interacting protein: implication for heme oxygenase-1 gene regulation. *J. Biol. Chem.* **276**, 20858–20865 (2001).

52. J. Alam, D. Stewart, C. Touchard, S. Boinapally, A.M.K. Choi, and J.L. Cook, Nrf2, a Cap "N" Collar transcription factor regulates induction of the heme oxygenase-1 gene, *J. Biol. Chem.* **274**, 26071–26078 (1999).

53. T. Ishii, K. Itoh, S. Takahashi, H. Sato, T. Yanagawa, Y. Katoh, S. Bannai, and M. Yamamoto, Transcription factor Nrf2 coordinately regulates a group of oxidative stress-inducible genes in macrophages, *J. Biol. Chem.* **275**, 16023–16029 (2000).

54. K. Itoh, N. Wakabayashi, Y. Katoh, T. Ishii, K. Igarashi, J.D. Engel, and M. Yamamoto, Keap1 represses nuclear activation of antioxidant responsive elements by Nrf2 through binding to the amino-terminal Neh2 domain, *Genes Dev.* **13**, 76–86 (1999).

55. K. Chan and Y.W. Kan, Nrf2 is essential for protection against acute pulmonary injury in mice, *Proc. Natl. Acad. Sci. USA* **96**, 12731–12736 (1999).

56. A.C. Wild, H.R. Moinova, and R.T. Mulcahy, Regulation of gamma-glutamylcysteine synthetase subunit gene expression by the transcription factor Nrf2, *J. Biol. Chem.* **274**, 33627–33636 (1999).

57. B.S. Dhakshinamoorthy and A.K. Jaiswal, Small Maf (MafG and MafK) proteins negatively regulate ARE-mediated expression and antioxidant induction of the NAD(P)H:quinone oxidoreductase1 gene, *J. Biol. Chem.* **275**, 40134–40141 (2000).

34

HEME OXYGENASE-1

A Major Player in the Defense Against the Oxidative Tissue Injury

Toru Takahashi[a], Reiko Akagi[b], Hiroko Shimizu[a],
Masahisa Hirakawa[a], and Shigeru Sassa[c]

[a]Department of Anesthesiology and Resuscitology
Okayama University Medical School
2-5-1 Shikata-cho, Okayama
700-8558, Japan
[b]Department of Nutritional Science
Faculty of Health and Welfare Science
Okayama Prefectural University
111 Kuboki, Soja, Okayama
719-1197, Japan
[c]Laboratory of Biochemical Hematology
The Rockefeller University
Nes York, N.Y., 10021

1. INTRODUCTION

Oxidative stresses such as oxidant stimuli, inflammation, exposure to xenobiotics, and ionizing irradiation elicit various tissue injuries and provoke cellular responses, principally involving transcriptional activation of genes encoding proteins which participate in the defense reactions (Camhi et al., 1995). One of them is microsomal heme oxygenase-1 (HO-1), the rate-limiting enzyme in heme degradation, as well as the 32-kDa heat shock protein. In an oxidative tissue injury, HO-1 induction confers protection, while its abrogation accelerates cellular injuries (Otterbein and Choi, 2000). In this context, HO-1 plays a major protective role against oxidant stimuli. In this article, we summarized recent evidence from our laboratory as well as from others on the role of HO-1 in the reperfusion injury, and the oxidative tissue injury by volatile inhaled gases.

2. DIFFERENTIAL EFFECTS OF HALOTHANE AND ISOFLURANE ON HEAT SHOCK PROTEINS

Oxidative tissue injuries can occur as the result of inhalation of volatile anesthetics. It has been largely studied with respect to hepatic heat shock proteins (Van Dyke et al., 1992; Odaka et al., 2000; Yamasaki et al., 2001), but our recent studies demonstrated that *ho-1* gene activation plays also an important role in the protection of hepatic injury induced by volatile anesthetics such as halothane.

Induction of HSP70 and HO-1 by Halothane

Halothane anesthesia is known to cause hepatic injury, such as severe hepatitis (Kenna and Jones, 1995). Halothane is known to be metabolized via two main pathways, both of which are catalyzed by microsomal cytochrome P450 (Kenna and Jones, 1995). Under a normal oxygen concentration, halothane is metabolized oxidatively to a trifluoroacetyl chloride, an unstable and reactive intermediate that induces covalent trifluoroacetylation of several proteins of the endoplasmic reticulum (Kenna and Jones, 1995). The trifluoroacetylated proteins are then involved in the immune response leading to severe hepatitis (Kenna and Jones, 1995). In contrast, under a hypoxic condition, halothane is metabolized by a reductive pathway to yield a free radical intermediate(s) which initiates lipid peroxidation (Awad et al., 1996) and hepatic injury (Kenna and Jones, 1995). In an experimental rat model of halothane-hypoxia toxicity, concomitant induction of hepatic cytochrome P450 by phenobarbital (PB) treatment is necessary in order to provoke a hepatic injury (Odaka et al., 2000; Yamasaki et al., 2001). Induction of HSP70 following halothane-hypoxia exposure occurs in perivenular zones both at transcript and protein levels (Yamasaki et al., 2001). Importantly, we found that halothane-hypoxia exposure resulted in a rapid increase in hepatic intracellular free heme in PB-pretreated rats, which is presumably derived from PB-induced cytochrome P450, and resulted in the induction of HO-1 mRNA and its enzyme activity in hepatocytes around the central vein (Odaka et al., 2000). Pretreatment with hemin, a chemically available form of heme and a potent inducer of HO-1 (Shibahara et al., 1979), increased hepatic HO-1 and resulted in the abrogation of the halothane-induced hepatic injury (Odaka et al., 2000). HO-1 induction thus plays an important role in the protection of the hepatic injury due to oxidative damages in halothane hepatotoxicity.

Induction of HSP70 and HO-1 by Isoflurane

Isoflurane preserves a better hepatic blood flow than halothane (Gelman et al., 1984) and is considered to be a much safer inhalation anesthetic (Kenna and Jones, 1995). Isoflurane is thus a preferred anesthetic for major hepatic surgeries such as liver transplantation, in which ischemic insult to the liver can frequently be anticipated (Chapin et al., 1989). We examined the effect of isoflurane on the levels of hepatic HSP70 and HO-1, and compared them with those induced by halothane under a hypoxic condition in PB-treated rats. Our findings indicate that isoflurane-hypoxia exposure caused much less induction of HSP70 mRNA and its protein than

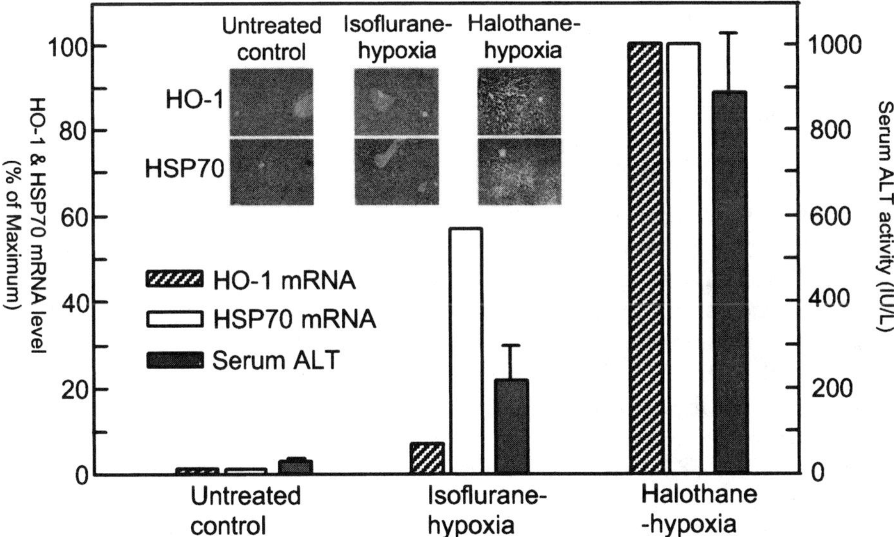

Figure 1. Differential effects of isoflurane and halothane on the induction of HO-1 and HSP70 in the liver, and on serum ALT activity. Phenobarbital (PB)-pretreated rats were exposed for 2 h to isoflurane or halothane under a hypoxic condition. Twenty-four hours after gas treatment, whole blood was collected for determination of serum ALT activity. Hepatic HSP70 and HO-1 gene expression was analyzed by Northern blot analysis 6 h after gas treatment. Shown in the inset are immunohistochemical detections of hepatic HSP70 and HO-1 from rats 12 h after gas treatment, using mouse monoclonal anti-human HSP70 and rabbit polyclonal anti-rat HO-1 as a primary antibody, respectively (Original magnification; ×100). Levels of HSP70 mRNA, HO-1 mRNA and serum ALT activity are shown as the bar graph. Serum ALT level represents the mean ± S.D. of a group of 6 animals.

halothane-hypoxia exposure in PB-pretreated rats (Fig. 1) (Yamasaki et al., 2001). In contrast to halothane-hypoxia, induction of HO-1 was also insignificant in rats treated with isoflurane-hypoxia, suggesting that much less free heme is formed by isoflurane-hypoxia exposure (Fig. 1) (Yamasaki et al., 2001). Our findings thus indicate differential effects of isoflurane and halothane on the expression of two HSPs, i.e., HSP 70 and HO-1, under a hypoxic condition. Consistent with these findings, isoflurane caused a significantly lesser hepatic injury than halothane as shown by lower serum alanine aminotransferase activity (~1/4 of the halothane treatment) and insignificant centrilobular necrosis (Fig. 1). Our findings therefore substantiate the fact that isoflurane is a much safer volatile anesthetic than halothane, and provide a biochemical basis that isoflurane induces a lesser oxidative stress than halothane.

3. HO-1 INDUCTION IN THE EXPERIMENTAL MODEL OF ACUTE RENAL FAILURE (ARF)

ARF is one of the best models in which the role of reactive oxygen species (ROS) and HO-1 has been definitively studied. There are several experimental models of

ARF, each with a unique feature. Each model has been used for assessing the role of HO-1 in the protection of the kidney from oxidative tissue injuries.

Glycerol-Induced ARF

The glycerol-induced ARF in rats is the most commonly used model and is prepared by subcutaneous or intramuscular injection of hypertonic glycerol to animals (Nath et al., 1992). In this model, there are skeletal muscle injuries, termed "rhabdomyolysis", resulting in the release of myoglobin into plasma. Approximately one third of the patients with rhabdomyolysis are known to develop ARF (Guglielminotti and Guidet, 1999) and rhabdomyolysis accounts for approximately 10% of all cases of ARF (Guglielminotti and Guidet, 1999). A large amount of heme released from myoglobin may be directly responsible for attendant lipid peroxidation which is associated with rhabdomyolysis (Guglielminotti and Guidet, 1999).

In the kidney of rats treated with glycerol for 6 h, HO-1 mRNA was found to increase more than 50-fold, compared with untreated animals (Nath et al., 1992). The blockade of increased HO activity by tin protoporphyrin (Sn-PP), a competitive inhibitor of HO activity, significantly aggravated the renal injury in this model. In contrast, induction of HO-1 by pretreatment of animals with a hemoglobin solution resulted in the protection against ARF development (Nath et al., 1992). Thus the exposure of the kidney to an inordinate amount of hemoglobin elicited an adaptive cellular response that facilitates the clearance of cytotoxic free heme. These findings indicate that induction of HO-1, which itself is a free heme-mediated process, serves to clear an excess amount of free heme, ultimately resulting in a beneficial adaptive response.

Cisplatin-Induced toxic renal injury

Cisplatin is a commonly used anticancer drug, but its use is often curtailed by its well-known nephrotoxicity, particularly that on proximal tubules (Ries and Klastersky, 1986). However, by virtue of this unique effect, cisplatin has been utilized to prepare an experimental model of ARF. In contrast to the glycerol-induced ARF, iron derived from renal cytochrome P450 appears to be important in ROS formation in the cisplatin-induced ARF (Baliga et al., 1998). HO-1 has been shown to be induced in a time and dose-dependent fashion in the kidney following cisplatin administration (Agarwal et al., 1995). Administration of Sn-PP aggravated the renal injury (Agarwal et al., 1995), while pre-induction of HO-1 by hemin treatment, or an overexpression of the *ho-1* gene by gene transfer resulted in a significant amelioration of the cisplatin-induced renal injury (Shiraishi et al., 2000).

Ischemic ARF (IARF)

IARF, the reperfusion injury of the kidney, is the major form of ARF and accompanies an acute tubular epithelial cell injury (Liano et al., 1998). The IARF injury is thought to be due to ROS generated by reperfusion (Paller et al., 1984), which has been suggested to be a result of the rapid release of heme from microsomal cytochrome P450 (Paller and Jacob, 1994). The reversibility of renal function in IARF depends on the length of the ischemic pretreatment prior to reperfusion, e.g., longer than 60 min ischemia resulting in an irreversible renal damage (Finn and Chevalier, 1979). Rats with a unilateral nephrectomy and the ligation of a contralateral renal artery or with bilateral ligation followed by reperfusion, have been used as experimental model systems of IARF (Shimizu et al., 2000; Maines et al., 1993). We found that both HO-1 mRNA and its enzyme activity were significantly increased in the reversible IARF model (Shimizu et al., 2000). Inhibition of HO activity by Sn-mesoporphyrin (Sn-MP), a specific competitive inhibitor of HO, resulted both in a marked increase in intracellular heme content, and in the aggravation of renal function (Fig. 2). HO-1 induction thus plays an important role in the protection of renal dysfunction due to oxidative damages (Shimizu et al., 2000).

4. AMELIORATION OF ARF BY TIN CHLORIDE TREATMENT

Tin chloride ($SnCl_2$) treatment is known to potently induce HO-1 in the kidney (Kappas and Maines, 1976). We examined the effect of $SnCl_2$ treatment prior to ischemia/reperfusion, and found that it improved renal dysfunction, as shown by a marked decrease in serum creatinine concentration in the control IARF animals (Fig. 2). While there were significant damages in proximal tubular cells in IARF control animals, these cells were hardly affected in $SnCl_2$-pretreated animals (Fig. 2) (Toda et al., in press). Following $SnCl_2$ treatment, a marked elevation of renal HO-1 mRNA was also observed, followed by increases in HO-1 protein expression and HO activity (Toda et al., in press). HO-1 protein accumulated also specifically in the renal tubular epithelial cells, following $SnCl_2$ treatment (Fig. 2). In contrast, inhibition of HO activity by the administration of Sn-MP abolished the beneficial effect of $SnCl_2$ pretreatment on IARF, indicating the fundamental role of HO-1 in the protection of renal epithelial cell injuries in IARF (Toda et al., in press).

5. HO-1 INDUCTION AS AN ADAPTIVE RESPONSE TO OXIDATIVE STIMULI

HO-1 is highly inducible by a vast array of stimuli including an oxidative stress, heat shock, UV radiation, ischemia-reperfusion, heavy metals, bacterial lipopolysaccharide, cytokines such as IL-6, and nitric oxide, and its own substrate heme (Otterbein and Choi, 2000). Accumulating evidence indicates overwhelmingly that induction of HO-1 provides cytoprotective effects in various *in vitro* and *in vivo* models of the oxidative cellular injury (Otterbein and Choi, 2000). The importance of HO-1 in the protection from oxidant stresses is further substantiated in mice and humans which are deficient in HO-1 (Poss and Tonegawa, 1997; Yachie et al., 1999). For example, the absence of HO-1 in a patient with hereditary HO-1 deficiency accompanied an abnormally elevated serum heme concentrations (~0.5 mM), and various oxidative as well as inflammatory complications (Yachie et al., 1999).

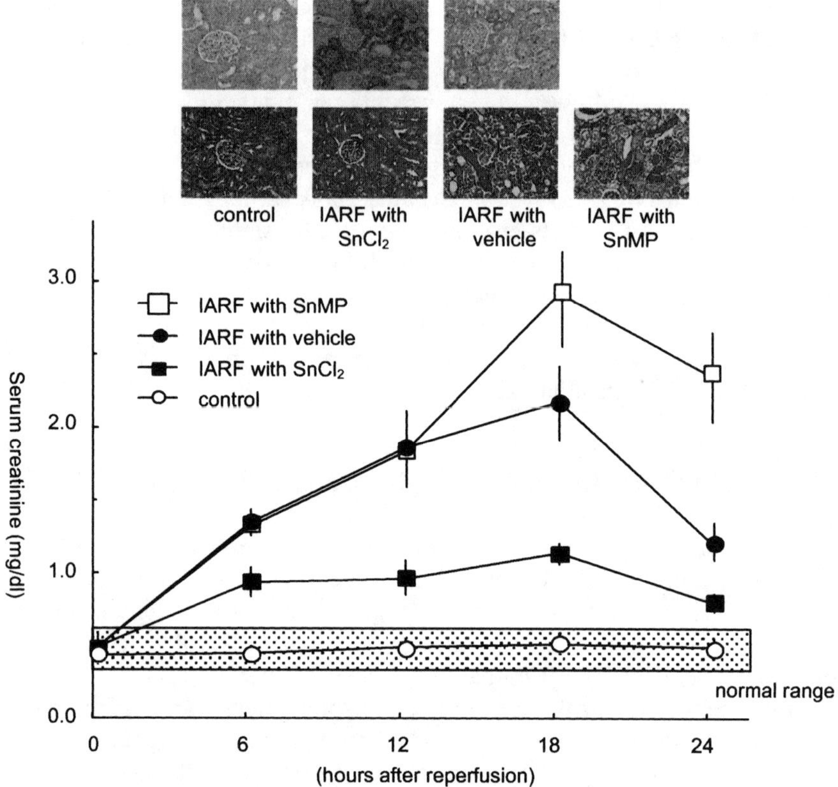

Figure 2. Effect of $SnCl_2$ and Sn-MP administration on serum creatinine and renal histopathology in rats with IARF. Rats were uninephrectomized and subjected to unilateral ischemia for 40 min to produce IARF. $SnCl_2$ (10 mg/100 g body weight) was administered subcutaneously, and Sn-MP (1 μmol/kg body weight) was administered intravenously 24 hrs prior to the uninephrectomy. After the initiation of reperfusion, whole blood was collected for the determination of serum creatinine concentrations at 0, 6, 12, 18 and 24 hrs. They are shown as the mean ± SEM (n = 6). Shown in the inset are histochemical changes in the renal cortex of rats following various treatments. Top, Kidney sections stained immunohistochemically, using anti-rat HO-1 as a primary antibody. After IARF, HO-1 protein was induced in tubular epithelial cells, while the induction became obvious in IARF with $SnCl_2$ pretreatment. Bottom, Kidney sections stained with hematoxylin and eosin. Extensive necrosis with cast formation in the proximal tubular cells was observed in IARF with vehicle treatment. It was aggravated in IARF with Sn-MP, in contrast to the relief from the injury in IARF with $SnCl_2$ pretreatment (Original magnification; ×200).

Free heme, which can be liberated from hemeproteins under oxidative conditions, is highly lipophilic and likely intercalates into the lipid bilayer of adjacent cells. Exposure of cells to heme is known to stimulate the expression of adhesion molecules ICAM-1, VCAM-1, and E-selectin on endothelial cells *in vitro*, probably through heme-mediated generation of ROS, which underscores various reactive inflammatory changes (Wagener et al., 2001). HO-1 catalyzes the decomposition of the pro-oxidant heme into three elements, i.e., iron, biliverdin IXα and carbon monoxide (CO). Iron, which itself is an oxidant, but is directly sequestered and rapidly inactivated by co-induced ferritin (Vogt et al., 1995). Biliverdin IXα is rapidly converted by biliverdin reductase to bilirubin IX*a, which is an anti-oxidant (Stocker et al., 1987). CO produced from heme by HO can suppress apoptosis of endothelial cells via the activation of p38 MAPK (Brouard et al., 2000). Thus, in addition to the removal of the pro-oxidant heme, HO produces a series of metabolites from heme which all act as a member of the protective response, and contribute to the suppression of oxidative tissue injuries. The strong adaptive response of HO-1 to various oxidative stimuli suggests an entirely new paradigm for HO-1, and HO-1 should now be recognized as a major player in the protection of inflammatory processes and oxidative tissue injuries.

6. MECHANISM OF *ho*-1 GENE ACTIVATION IN THE OXIDATIVE TISSUE INJURY

Although the precise mechanism(s) of renal injury caused by ischemia/reperfusion and hepatic injury caused by halothane-hypoxia treatment remains elusive, ROS produced by reperfusion following ischemia, or halothane-hypoxia exposure, appears to be significantly involved in such tissue injuries (Awad et al., 1996; Paller et al., 1984). For example, there is a rapid increase in microsomal heme concentration and/or hepatic free heme concentration, which is thought to be liberated from microsomal cytochrome P450 (Paller and Jacob, 1994), immediately after reperfusion in the kidney (Shimizu et al., 2000), and after halothane-hypoxia exposure in the liver (Odaka et al, 2000). Free heme is not associated with apoproteins, since it acts as a potent pro-oxidant, leading to the generation of ROS. ROS also induces the breakdown of hemeproteins which further aggravates oxidative tissue damages by liberating more heme. In these conditions, *ho-1* gene expression is also transcriptionally increased in various cell types (Otterbein and Choi, 2000). In contrast, HO-1 deficient embryonic fibroblasts are exquisitely sensitive to the cytotoxicity of both hemin and hydrogen peroxide (Poss and Tonegawa, 1997).

Recent evidence points to the fact that a group of oxidative stress inducible genes is under the immediate transcriptional control of Nrf2-small Maf heterodimer-regulatory protein (Fig. 3). Nrf2 forms a heterodimer with a small Maf protein, interacts with an anti-oxidant responsive element (ARE), and induces transcription of a set of genes that encode anti-oxidant functions. Nrf2 thus regulates stress inducible protein genes via ARE (Fig. 3). For example, Nrf2-deficient cells have been shown to be hypersensitive to oxidative stresses (Itoh et al., 1997). Various ROS-inducing agents increase the DNA binding activity of Nrf2 in the nucleus without influencing its mRNA level (Ishii et al., 2000). In this manner, Nrf2 regulates a wide-range of metabolic responses to oxidative stresses, which include, among others, HO-1 (Ishii et al., 2000). The ARE cognate sequence shares a high degree of homology to the consensus MARE sequence, permitting AREs competitively bound by a number of bZip transcription factors in the Maf, Jun, Fos and Cap'n'Collar families (Ishii et al., 2000) (Fig. 3). Because of their responsiveness to a wide variety of stress agents, the MAREs were also named stress-responsive elements (StREs) (Alam et al, 1995). Multiple StREs are also found in HO-1 enhancers, and have been shown to play an important

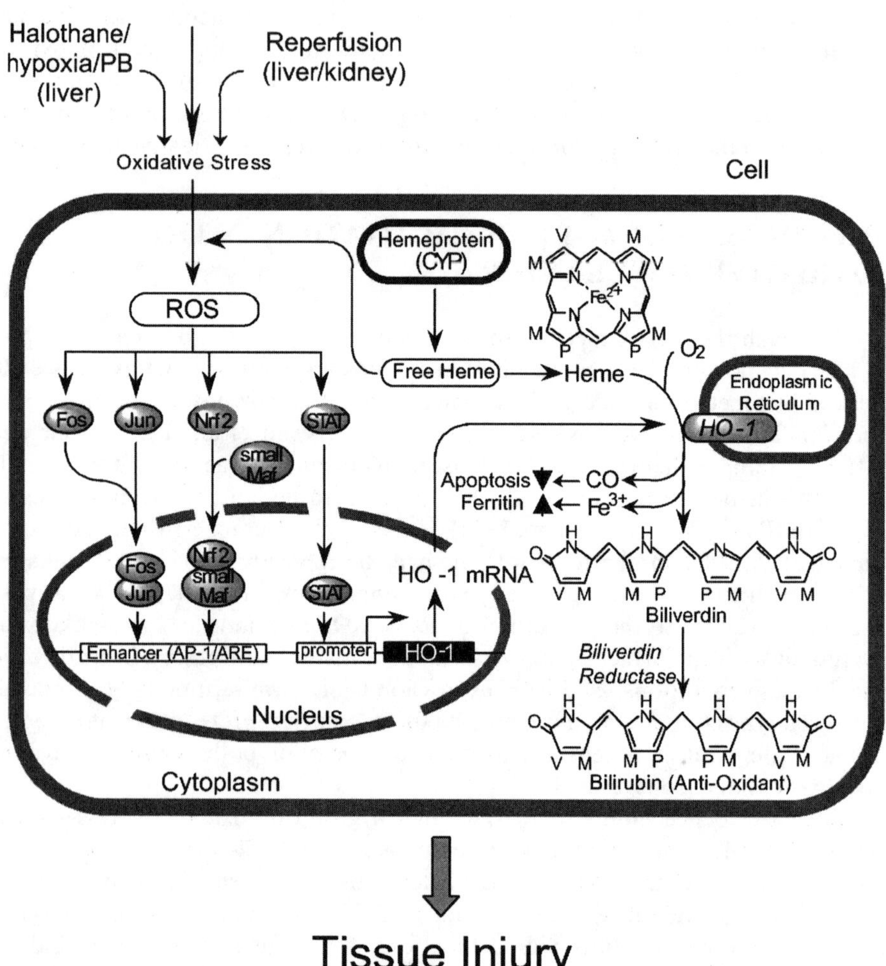

Figure 3. Hypothetical mechanism of HO-1 induction in oxidative tissue injuries. Oxidative stresses liberate Nrf2 from the cytoskeleton into the nucleus (Itoh et al., 1999). Nrf2 then forms a heterodimer with a small Maf protein, interacts with ARE, and induces transcription of a set of genes that encode antioxidant activities (Ishii et al., 2000), one of which is ho-1. Cytosolic free heme contributes to the generation of ROS, which in turn facilitates the formation of a Nrf2/small Maf heterodimer. Induction of HO-1 protein results in the removal of the pro-oxidant free heme, in the production of the anti-oxidant bile pigments, and in the formation of the anti-apoptotic CO. When the synthesis of ROS exceeds the level of detoxifying capacity of the cells, the oxidative tissue injury ensues.

role in the regulation of *ho-1* gene expression by various oxidative stresses (Ishii et al., 2000). For example, the DNA binding activity of transcription factor AP-1 was inhibited when AP-1 was exposed to an oxidative stimulus *in vitro* (Abate et al., 1990), which in turn may allow the competitive binding of other activating transcription factors, such as an Nrf2-small Maf heterodimer, to ARE and hence induce genes with protective properties against oxidative stresses, e.g., HO-1 (Ishii et al., 2000). While exposure of human MCF-7 cells to $CdCl_2$, a potent inducer of HO-1, stimulates phosphorylation of ERK, JNK and p38 mitogen-activated kinases, an inhibitor of p38, or co-expression of a dominant-negative mutant of p38α, but not of ERK1, ERK2, JNK1, or JNK2, decreased HO-1 mRNA levels significantly (Alam et al., 2000). A dominant negative mutant of Nrf2 (a CNC-bZip member), but not of c-Jun or C/EBPβ, inhibited $CdCl_2$-mediated transactivation (Alam et al., 2000). These findings, together with others, suggest that an oxidative stress induces *ho-1* gene expression via sequential activation of the p38 kinase pathway and Nrf2 (Alam et al., 2000).

7. SUMMARY

In this chapter, recent evidence indicating the protective role of HO-1 in oxidative tissue injuries is summarized. Both inhibition of HO activity, and suppression of *ho-1* gene expression, led to aggravation of oxidative tissue injuries. In contrast, induction of HO-1 and exogenous administration of HO-1 by gene transfer conferred significant protection in a rat model of inflammation, or a hemoglobin-heme toxicity model (Yang et al., 1999). These findings suggest that HO-1 may be a major player in the defense against the oxidative tissue injury, and that tissue-specific HO-1 expression, either by HO-1 induction by chemicals, or by an overexpression of the *ho-1* gene by gene transfer, may hold therapeutic usefulness. Thus, $SnCl_2$, which has been simply thought to be toxic, may offer a new mode of treatment of IARF, because of its highly specific HO-1 inducing property in the kidney.

ACKNOWLEDGMENTS

This work was in part supported by grants from USPHS DK32890, and from Grant-in-Aid for Scientific Research (09671564 & 13671582) from the Ministry of Education, Science, and Culture of Japan.

REFERENCES

Abate, C., Patel, L., Rauscher, F.J., and Curran, T., 1990. Redox regulation of fos and jun DNA-binding activity in vitro. *Science* **249**:1157.

Agarwal, A., Balla, J., Alam, J., Croatt, A.J., and Nath, K.A., 1995. Induction of heme oxygenase in toxic renal injury: a protective role in cisplatin nephrotoxicity in the rat. *Kidney Int* **48**:1298.

Alam, J., Camhi, S., and Choi, A.M., 1995. Identification of a second region upstream of the mouse heme oxygenase-1 gene that functions as a basal level and inducer-dependent transcription enhancer. *J Biol Chem* **270**:11977.

Alam, J., Wicks, C., Stewart, D., Gong, P., Touchard, C., Otterbein, S., Choi, A.M., Burow, M.E., and Tou, J., 2000. Mechanism of heme oxygenase-1 gene activation by cadmium in MCF-7 mammary epithelial cells. Role of p38 kinase and Nrf2 transcription factor. *J Biol Chem* **275**:27694.

Awad, J.A., Horn, J.L., Roberts, L.J. II., and Franks, J.J., 1996. Demonstration of halothane-induced hepatic lipid peroxidation in rats by quantification of F2-isoprostanes. *Anesthesiology* **84**:910.

Baliga, R., Zhang, Z., Baliga, M., Ueda, N., and Shah, S.V., 1998. Role of cytochrome P-450 as a source of catalytic iron in cisplatin-induced nephrotoxicity. *Kid Int* **54**:1562.

Brouard, S., Otterbein, L.E., Anrather, J., Tobiasch, E., Bach, F.H., Choi, A.M., and Soares, M.P., 2000. Carbon monoxide generated by heme oxygenase 1 suppresses endothelial cell apoptosis. *J Exp Med* **192**:1015.

Camhi, S.L., Lee, P., and Choi, A.M., 1995. The oxidative stress response (Review). *New Horiz* **3**:170.

Chapin, J.W., Newland, M.C., and Hurlbert, B.J., 1989. Anesthesia for liver transplantation (Review). *Semin Liver Dis* **9**:195.

Finn, W.F. and Chevalier, R.L., 1979. Recovery from postischemic acute renal failure in the rat. *Kidney Int* **16**:113.

Gelman, S., Fowler, K.C., and Smith, L.R., 1984. Liver circulation and function during isoflurane and halothane anesthesia. *Anesthesiology* **61**:726.

Guglielminotti, J. and Guidet, B., 1999. Acute renal failure in rhabdomyolysis (Review). *Minerva Anestesiol* 1999 **65**:250.

Ishii, T., Itoh, K., Takahashi, S., Sato, H., Yanagawa, T., Katoh, Y., Bannai, S., and Yamamoto, M., 2000. Transcription factor Nrf2 coordinately regulates a group of oxidative stress-inducible genes in macrophages. *J Biol Chem* **275**:16023.

Itoh, K., Chiba, T., Takahashi, S., Ishii, T., Igarashi, K., Katoh, Y., Oyake, T., Hayashi, N., Satoh, K., Hatayama, I., Yamamoto, M., and Nabeshima, Y., 1997. An Nrf2/small Maf heterodimer mediates the induction of phase II detoxifying enzyme genes through antioxidant response elements. *Biochem Biophys Res Commun* **236**:313.

Itoh, K., Wakabayashi, N., Katoh, Y., Ishii, T., Igarashi, K., Engel, J.D., and Yamamoto M., 1999. Keap1 represses nuclear activation of antioxidant responsive elements by Nrf2 through binding to the amino-terminal Neh2 domain. *Genes & Develop* **13**:76.

Kappas, A. and Maines, M.D., 1976. Tin: A potent inducer of heme oxygenase in kidney. *Science* **192**:60.

Kenna, J.G. and Jones, R.M., 1995. The organ toxicity of inhaled anesthetics (Review). *Anesth Analg* **81**:S51.

Liano, F., Junco, E., Pascual, J., Madero, R., and Verde, E., the Madrid Acute Renal Failure Study Group, 1998. The spectrum of acute renal failure in the intensive care unit compared with that seen in other settings. *Kidney Int* **53**:S16.

Maines, M.D., Mayer, R.D., Ewing, J.F., and McCoubrey, W.K. Jr., 1993. Induction of kidney heme oxygenase-1 (HSP32) mRNA and protein by ischemia/reperfusion: possible role of heme as both promotor of tissue damage and regulator of HSP32. *J Pharmacol Exp Ther* **264**:457.

Nath, K.A., Balla, G., Vercellotti, G.M., Balla, J., Jacob, H.S., Levitt, M.D., and Rosenberg, M.E., 1992. Induction of heme oxygenase is a rapid, protective response in rhabdomyolysis in the rat. *J Clin Invest* **90**:267.

Odaka, Y., Takahashi, T., Yamasaki, A., Suzuki, T., Fujiwara, T., Yamada, T., Hirakawa, M., Fujita, H., Ohmori, E., and Akgaki, R., 2000. Prevention of halothane-induced hepetotoxicity by hemin pretreatment: The protective role of heme oxygenase-1 induction. *Biochem Pharmacol* **59**:871.

Otterbein, L.E. and Choi, A.M., 2000. Heme oxygenase: colors of defense against cellular stress (Review). *Am J Physiol Lung Cell Mol Physiol* **279**:L1029.

Paller, M.S., Hoidal, J.R., and Ferris, T.F., 1984. Oxygen free radicals in ischemic acute renal failure in the rat. *J Clin Invest* **74**:1156.

Paller, M.S. and Jacob, H.S., 1994. Cytochrome P-450 mediates tissue-damaging hydroxyl radical formation during reoxygenation of the kidney. *Proc Natl Acad Sci USA* **91**:7002.

Poss, K.D. and Tonegawa, S., 1997. Reduced stress defense in heme oxygenase 1-deficient cells. *Proc Natl Acad Sci USA* **94**:10925.

Ries, F. and Klastersky, J., 1986. Nephrotoxicity induced by cancer chemotherapy with special emphasis on cisplatin toxicity (Review). *Am J Kid Dis* **8**:368.

Shibahara, S., Yoshida, T., and Kikuchi, G., 1979. Mechanism of increase of heme oxygenase activity induced by hemin in cultured pig alveolar macrophages. *Arch Biochem Biophys* **197**:607.

Shimizu, H., Takahashi, T., Suzuki, T., Yamasaki, A., Fujiwara, T., Odaka, Y., Hirakawa, M., Fujita, H., and Akagi, R., 2000. Protective effect of heme oxygenase induction in ischemic acute renal failure. *Crit Care Med* **28**:809.

Shiraishi, F., Curtis, L.M., Truong, L., Poss, K., Visner, G.A., Madsen, K., Nick, H.S., and Agarwal, A., 2000. Heme oxygenase-1 gene ablation or expression modulates cisplatin-induced renal tubular apoptosis. *Am J Physiol Renal Physiol* **278**:F726.

Stocker, R., Yamamoto, Y., McDonagh, A.F., Glazer, A.N., and Ames, B.N., 1987. Bilirubin is an antioxidant of possible physiological significance. *Science* **235**:1043.

Toda, N., Takahashi, T., Mizobuchi, S., Fujii, H., Nakahira, K., Takahashi, S., Yamashita, M., Morita, K., Hirakawa, M., and Akagi, R., Tin chloride pretreatment prevents the renal injury in rats with ischemic acute renal failure. *Crit Care Med* (In Press).

Van Dyke, R.A., Mostafapour, S., Marsh, H.M., Li, Y., and Chopp, M., 1992. Immunocytochemical detection of the 72-kDa heat shock protein in halothane-induced hepatotoxicity in rats. *Life Sci* **50**:PL41.

Vogt, B.A., Alam, J., Croatt, A.J., Vercellotti, G.M., and Nath, K.A., 1995. Acquired resistance to acute oxidative stress. Possible role of heme oxygenase and ferritin. *Lab Invest* **72**:474.

Wagener, F.A.D.T.G., Eggert, A., Boeman, O.C., Oyen, W.J.G., Verhofstadt, A., Abraham, N.G., Adema, G., van Kooyk, Y., de Witte, T., and Figdor, C.G., 2001. Heme is a potent inducer of inflammation in mice and is counteracted by heme oxygenase. *Blood* **98**:1802.

Yachie, A., Niida, Y., Wada, T., Igarashi, N., Kaneda, H., Toma, T., Ohta, K., Kasahara, Y., and Koizumi, S., 1999. Oxidative stress causes enhanced endothelial cell injury in human heme oxygenase-1 deficiency. *J Clin Invest* **103**:129.

Yamasaki, A., Takahashi, T., Suzuki, T., Fujiwara, T., Hirakawa, M., Ohmori, E., and Akagi, R., 2001. Differential effects of isoflurane and halothane on the induction of heat shock proteins. *Biochem Pharmacol* **62**:375.

Yang, L., Quan, S., and Abraham, N.G., 1999. Retrovirus-mediated HO gene transfer into endothelial cells protects against oxidant-induced injury. *Am J Physiol* **277**:L127.

35

HEME OXYGENASE, GINKGO BILOBA EXTRACT AND ITS TERPENOIDS PROTECT MYOCYTES AGAINST OXIDATIVE INJURY

Jian-Xiong Chen[a], Heng Zeng[a], Xiu Chen[b], Ching-Yuan Su[c], and Chen-Ching Lai[c]

[a]Department of Pathology
Center for Lung Research
Vanderbilt University Medical Center
Nashville, TN, USA
[b]Hunan Medical University
China
[c]Southern Illinois University
School of Medicine

INTRODUCTION

A number of pharmacological approaches to protect against myocardial damage during ischemia reperfusion events (e.g. β-blockers, calcium antagonists, nitrates, and free radical scavengers) have met with limited success. For example, administration of exogenous antioxidants during and after myocardial ischemial-reperfusion were not sufficient to protect intracellular targets against reactive oxidant species, because exogenous antioxidants are membrane impermeable and cannot gain access to intracellular sites of free radical production and reaction. Development of other novel approaches to enhance endogenous cardioprotective mechanisms to minimize myocardial damage during ischemial-reperfusion is currently a major area of investigation. Since its first description by Murry et al. manipulation of myocardial protection via "ischemic preconditioning (I/P)" has been revealed to trigger endogenous protective mechanisms that increase cardiomyocytes resistance to oxidative stress (Murry et al., 1986; Parratt, 1994). Two classes of cellular protectants, heat shock proteins (HSPs) and endogenous anti-oxidant enzymes (Yellon and Baxter, 1995), have been postulated to participate in this I/P induced cardioprotection. Accumulation of

HSP70 in myocytes was associated with an enhancement structural stabilization in cardiac tissue. It was shown that transfecting cells with HSP70 gene conferred cyto-protection against hydrogen peroxide, hypoxia/reoxygenation injury (Chong et al., 1998; Plumier et al., 1995). Overexpression of HSP70 dramatically reduced infarct size in hearts of transgenic HSP70 mouse model (Marber et al., 1995). HSPs can be categorized into several major families with distinct molecular weights: HSP110, HSP90s, HSP60s, HSP47, and other small HSPs including ubiquitin, HSP32, HSP27. Some proteins function as antioxidant enzymes have been received increasing inter-est in recent years. One of these, HSP32 (heme oxygenase-1, HO-1) has been exten-sively investigated both *in vivo* and *in vitro* models of oxidant-mediated cellular and tissue injury. Many laboratories have shown that HO-1 could work as an important molecule in the host's defense against oxidative stress.

HEME OXYGENASE AND CYTOPROTECTION

Heme oxygenase (HO) is the rate-limiting enzyme in heme catabolism catalyz-ing the oxidative cleavage of heme to biliverdin, carbon monoxide and iron. Biliverdin is subsequently converted to bilirubin by biliverdin reductase (Maines, 1988). Two distinct isoforms of HO, an inducible form (HO-1) and a constitutive form (HO-2), have been identified and cloned (Maines et al., 1986, 1989). These iso-enzymes are products of different genes and differ markedly in their tissue expression as well as their molecular properties. The HO-2 isoform was classified as constitutive and is present in high concentration in the brain and testes (Maines, 1988). In contrast, HO-1 is an inducible form that is transcriptionally up-regulated by a variety of chemical or physical stress-inducing factors such as heavy metals, certain organic compounds including its substrate, heme, cytokines, (Taketani et al., 1989; Yoshida et al., 1988), heat shock (Shibahara et al., 1987), ultraviolet (UV) irradiation (Keyse and Tyrell, 1989). A large number of inducing agents that directly or indirectly provoke HO have been shown to stimulate HO-1; none of them has been reported to affect HO-2 (Abraham et al., 1988). Such a pattern of regulation in response to a various condi-tions of stress suggested that HO-1 may posse potential protective effects against these stresses.

Indeed, many studies have demonstrated that increased HO-1 production pro-vided cellular protection against oxidative-mediated injury (Otterbein and Choi, 2000; Maines, 1997). For example, transfection of HO-1cDNA and increased synthesis of HO has been shown to protect pulmonary epithelial cells against oxidant stress *in vitro* (Lee et al., 1996). Similarly, overexpression OH-1 protein in coronary vessel endothe-lial cells also conferred protection against hemoglobin-induced injury (Abraham et al., 1995). On the other hand, treatment of cells with HO-1 specific antisense oligonu-cletides have shown that reduced HO-1 protein levels were correlated with facilitated UVA radiation and hydrogen peroxide induced damage (Lautier et al., 1992; Keyse and Tyrell, 1989). *In vivo* studies, Nath et al. (1992) showed that induction of HO-1 by infusion of hemoglobin prevented kidney failure and reduced mortality in the rha-bodomyolysis rat model. It has been reported that administrated inducers of HO-1, such as heme arginate or heme lysinate, caused a marked decrease in blood pressure in spontaneously hypertension rats and dramatically increased coronary blood flow in isolated perfusion hearts (Levere et al., 1990; Johnson et al., 1996). Most recently it has been reported that increased cardiac HO-1 expression and heme oxygenase activity by pretreatment with heme restored myocardial function and reduced infarct size in ischemial/reperfusion rat heart model. HO inhibitors completely abolished the protective effect of heme mediated HO-1 induction (Clark et al., 2000). The cytoprotection of HO-1 is further supported by Poss et al. (1997), who demonstrated that HO-1 knock out mice exhibited increased susceptibility to injury when cells were exposed to endotoxin. In contrast, by using isolated perfused heart preparation and an in vivo myocardial infarction model, overexpression of HO-1 in cardiac specific transgenic mice protected the heart against ischemia/reperfusion tissue injury and improved cardiac function (Yet et al., 2001). The mechanisms underlying these phenomena are not clearly understood, however, these experimental data do suggest that moderate increases in HO activity by pharmacological intervention or gene therapy will be beneficial to cells in providing resistance to oxidative injury and playing a cytoprotective role in cardiovascular disease.

How does HO-1 confer cytoprotection against cellular and oxidative stress? It has been hypothesized that HO may serve a role in protection against oxidative damage through its end products, biliverdin and bilirubin. For many years, bilirubin was considered as a toxic waste product formed during catabolism. However, it has been demonstrated bilirubin is an effective free radicals scavenger (Stocker et al., 1987) and is able to prevent lipid peroxidation (Neuzil and Stocker, 1994). It has been reported that bilirubin reduced free radical-mediated damage to serum albumin and attenuated hydrogen peroixde induced injury in vascular endothelial cells (Neuzil and Stocker, 1993; Liesuy and Tomaro, 1994). Bilirubin has been shown to protect against peroxidation of human plasma and low-density lipoproteins and inhibit ox-LDL-induced monocyte chemotaxis *in vitro* (Wu et al., 1994; Ishikawa et al., 1997). More recently, Foresti et al. (1999) have demonstrated that HO-1 derived bilirubin protects against endothelial apoptosis mediated by peroxynitrite. Other intracellular effects of bilirubin relating to antioxidant effect include inhibition of protein kinase C activity and suppression of superoxide production by activated NADPH oxidase (Sano et al., 1985; Kwak et al., 1991). Despite these promising features, few studies have been done to elucidate the potential role and mechanisms of biliverdin and bilirubin in the cardiovascular system, particularly in cardiomycytes in which HO-1 gene expression and HO activity have been found to be up-regulated during ischemia-reperfusion/ reoxygenation (Maulik et al., 1996).

Another beneficial effect of HO-1 is ascribed to its vasoactive gas product carbon monoxide (CO). CO has been identified as an endogenous biological messenger in the brain, and recent studies suggest an important role for CO in the cardiovascular system. CO could affect cardiovascular function via activation of soluble guanylate cyclase and increase in intracellular cGMP levels. CO has potent cardioprotective effects. These involved in the regulation of vasomotor tone, suppression of platelet aggregation and inhibition of vascular smooth muscle cell proliferation (Durante et al., 1998; Siow et al., 1999; Johnson et al., 1999).

In this chapter, we will discuss our data concerning the induction of heme oxygenase-1 by ginkgo biloba extract (EGB), and its terpenoids protection against

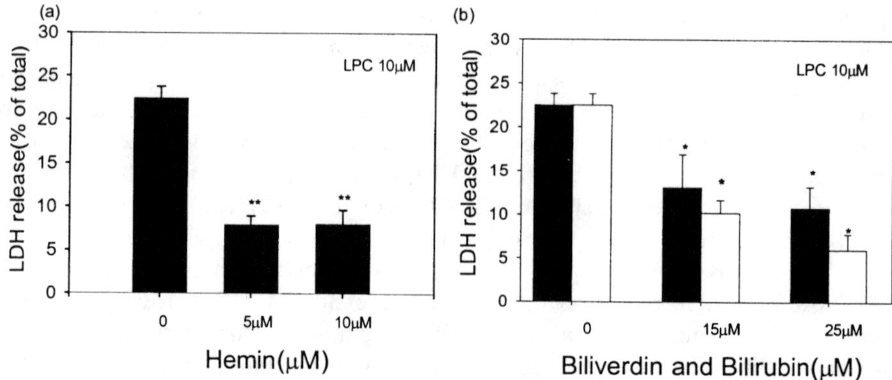

Figure 1. Pretreatment with hemin (a), biliverdin and bilirubin (b) protected mycytes against LPC induced injury (Mean ± S.D., n = 5, *P < 0.05, **P < 0.01 compared with LPC. Biliverdin (black box) and Bilirubin (open box).

lysophosphatidylcholine (LPC)-induced myocytes damage. We also will discuss the potential roles of biliverdin, bilirubin and hemin.

POTENTIAL PROTECITVE ROLE OF BILIVERDIN, BILIRUBIN AND HEMIN ON LPC-INDUCED INJURY

Lysophosphatidylcholine (LPC) is an amphiphilic metabolite derived from membrane phospholipids by activated of phospholipase A_2 (PLA_2). It accumulates in the heart during ischemia and reperfusion. In our previous studies, myocytes exposure to LPC (10–30 μmol/L) for 30 min caused a significant increase of LDH release into the medium in a concentration dependent manner. Morphological studies also showed that exposure to LPC leads to cell apoptosis and DNA damage. HO activity is induced by its substrate heme or heme analogues, and by a variety of chemical agents and physical types of stress. We have investigated the effect of hemin on HO-1 gene expression and HO activity in myocytes. Pretreatment of myocytes with 5 μM hemin for 24 h caused a significant increase in HO-1 gene expression and elevation of HO activity (control = 126.7 ± 30.1 pmol bilirubin/mg protein/h, hemin-treated = 680.0 ± 183.0 pmol bilirubin/mg protein/h, n = 5, P < 0.05). Myocytes pretreated with 5 μM and 10 μM hemin for 24 h significantly decreased LDH release from the LPC treated cells (Fig. 1A). Therefore, the protective effect of hemin against LPC-induced damage is likely to be the result of increase in HO activity.

We further examined the potential role of biliverdin and bilirubin in cardiomycytes against oxidative stress. Biliverdin and bilirubin when accumulated at abnormally high concentrations in the tissue were considered as toxic catabolites. The micro-molar concentrations used in our experiments were within physiological range (Stocker et al., 1987) and did not have any toxicity to myocytes. Pretreatment of myocytes with biliverdin or bilirubin at 15 μM and 25 μM for 2 h, then exposure to

LPC 10 µmol/L for 30 min in the present of biliverdin and bilirubin. Biliverdin and bilirubin pretreatment significantly attenuated LPC-induced damage (Fig. 1B). Our data showed pretreatment with biliverdin and bilirubin protected myocytes against LPC-induced damage. Although we did not directly measure the intracellular biliverdin or bilirubin concentrations, our data revealed that bilirubin and biliverdin either produced from the induction of HO-1 or administered exogenously were most likely correlated with their protective effect against oxidative stress.

EGB AND ITS TERPENOIDS PRETREATMENT INDUCED OXIDATIVE RESISTANCE IN CARDIOMYOCYTES

EGB is a complex mixture extracted from the leaves of Ginkgo Biloba. The standardized extract contains 6% terpenoids and 24% flavonoids. Previous studies have demonstrated that EGB were effective against myocardial ischemia-reperfusion injury in isolated heart preparation or whole animal (Tosaki et al., 1993; Shen et al., 1995). We have also shown that pretreatment with EGB exerted protective effects against oxidative injury in cultured endothelial cells (Chen et al., 1998). The terpenoid constituents of EGB, ginkgolides and bilobalide (Fig. 2), which occur uniquely in Ginkgo biloba, are bioavailable after oral ingestion. Treatment with terpenoids significantly improved heart functions and protected cardiomyocytes from ischemia-reperfusion injury (Pietri et al., 1997). We examined the effects of EGB and its terpenoids on endogenous antioxidant enzymes and stress gene expression, to see whether such changes would provide cardiomyocytes more resistance against oxidative stress.

As shown in Table 1, pretreatment myocytes with EGB for 24 h significantly elevated glutathione peroxidase (GPx), superoxide dismutase (SOD) and catalase(Cat) activities in a concentration dependent manner. Pretreatment with ginkgolide B and bilobalide for 24 h resulted in a significant increase in SOD activity. However, the GPx and catalase activities were increased only in ginkgolide B treated cells. Ginkgolide A pretreatment had little effect on the antioxidant enzyme's activities. The mRNA levels of two isoforms of heat shock protein 70(HSP70c for constitutive form and HSP70i for inducible form) were examined by RT-PCR analysis. As shown in Fig. 3A,

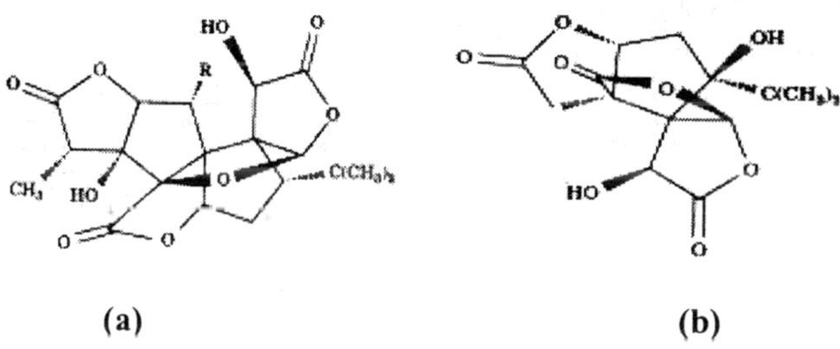

(a)   (b)

Figure 2. Structures of terpenoids: (a) ginkgolide A, ginkgolide B, (b) bilobalide.

Table 1. Effect of EGB and its terpenoids on antioxidant enzyme's activities and on LPC induced LDH release in myocytes (GA = Ginkgolide A, GB = Ginkgolide B, BL = Bilobalide, Mean ± S.D, n = 6, *P < 0.05 compared with LPC or control)

	µg/ml	Cat(U/mg pro)	SOD(U/mg pro)	GPx(U/mg pro)	LDH(%)
CON	0	1.6 ± 0.4	55.1 ± 12.1	12.7 ± 1.5	16.4 ± 3.6
EGB	100	2.3 ± 0.2*	95.7 ± 18.7*	34.3 ± 1.5*	7.9 ± 2.4*
EGB	200	2.5 ± 0.1*	108.8 ± 16.5*	36.1 ± 2.8*	2.4 ± 0.9*
EGB	300	2.7 ± 0.4*	133.6 ± 19.7*	37.2 ± 2.8*	2.3 ± 1.2*
EGB	500	2.8 ± 0.5*	158.0 ± 22.3*	43.0 ± 2.7*	2.6 ± 0.5*
GA	30	1.6 ± 0.4	53.9 ± 7.1	12.7 ± 2.3	16.1 ± 3.7
GB	30	2.2 ± 0.4*	75.8 ± 12.7*	21.0 ± 2.4*	6.4 ± 2.1*
BL	30	1.6 ± 0.1	79.5 ± 12.3*	14.7 ± 1.5	6.7 ± 1.4*

pretreatment with EGB for 24 h caused a dose-dependent increase in HSP70c mRNA expression (top panel). HSP70i gene expression was undetectable in untreated cells and pretreatment with EGB led to induction of HSP70i gene expression in a dose-dependent manner (bottom panel). As shown in Fig. 3B, pretreatment of myocytes with Ginkgolide B or Bilobalide increased HSP70c mRNA expression. The induction of HSP70i was not detected in control and terpenoids-treated cells. Pretreatment of myocytes with EGB, Ginkgolide B and bilobalide for 24 h resulted in a dose-dependent decline of LDH release from LPC-treated cells, while ginkgolide A had little effect (Table 1). Our data revealed that preconditioning myocytes with EGB and its terpenoids increased endogenous antioxidant enzymatic activities and HSP70, which further enhanced myocytes resistance to oxidative stress.

These results strongly suggested that protective effects of the EGB and its terpenoids against oxidative stress are clearly linked to its ability to activate endogenous protective proteins and intracellular antioxidant enzymatic pathways. In light of these findings, we examined whether EGB and its terpenoids constituents protect against oxidative stress through actions on HO-1 gene expression and HO activity.

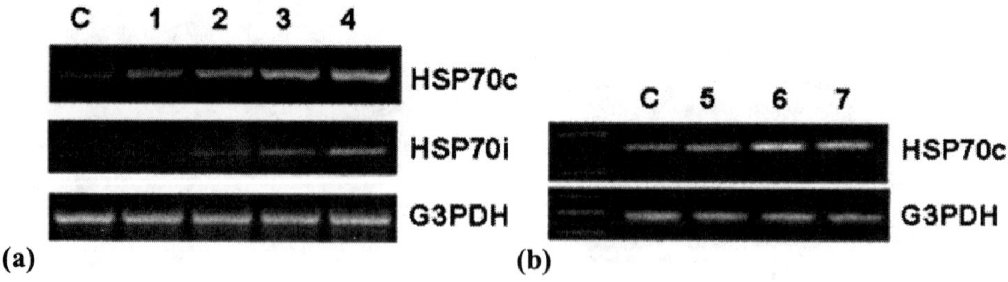

Figure 3. Effect of EGB (a) and terpenoids (b) on myocytes HSP70c and HSP70i expression using RT-PCR analysis. C (control), 1 (EGB 100 µg/ml), 2 (EGB 200 µg/ml), 3 (EGB 300 µg/ml), 4 (EGB 500 µg/ml), 5 (ginkgolide A 30 µg/ml), 6 (ginkgolide B 30 µg/ml), 7 (bilobalide 30 µg/ml).

Figure 4. Effect of EGB on HO-1 gene expression in myocytes. C (control), 1 (EGB100 µg/ml), 2 (EGB200 µg/ml), 3 (EGB300 µg/ml), 4 (EGB500 µg/ml).

INDUCTION OF HO-1 GENE EXPRESSION AND HO ACTIVITY BY EGB PROTECTS CARDIOMYOCYTES AGAINST LPC-INDUCED INJURY

While HO-1 gene expression was undetectable in control myocytes, pretreatment with EGB (100–500 µg/ml) for 24 h dramatically increased HO-1 gene expression (Fig. 4) and heme oxygenase activity in a dose-dependent manner (Fig. 5A). However, pretreatment with individual components, ginkgolide B 30 µg/ml and bilobalide 30 µg/ml showed no significantly effect on expression of HO-1 or HO activity. This finding indicated that the large increase of HO activity was mainly due to elevate HO-1 gene expression. It has been shown that the activity of HO-2 was two-fold more than that of HO-1 in the basal state, but after induction, HO-1 activity increased up to 100-fold and HO-2 was unchanged. The dramatic elevation of heme oxygenase activity (approximately ten-fold) and HO-1 gene expression by EGB suggested that HO-1 was probably the major contributor to up-regulating HO activity.

To verify whether induction of the HO-1 pathway could provide protection against oxidative stress, we examined the effect of HO inhibitor ZnPPIX on EGB's protective effect against LPC-induced damage. We used ZnPPIX instead of other metalloporphyrins to inhibit HO activity because ZnPPIX has been shown to be selective for HO over other microsomal enzyme (Maines 1981). Pretreatment with

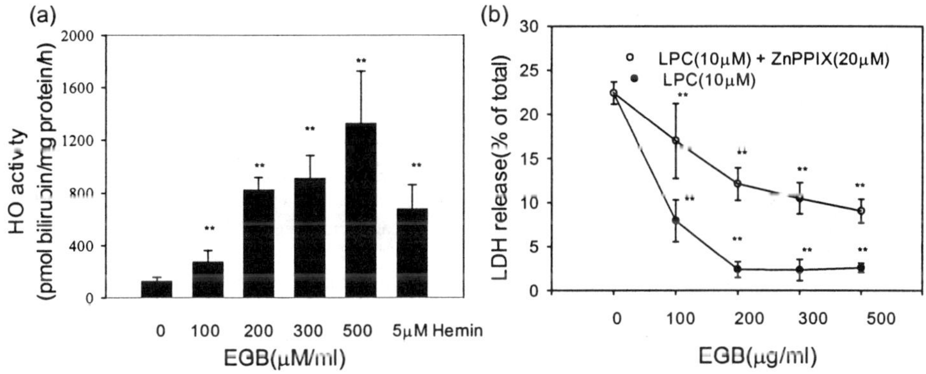

Figure 5. (a) Effect of EGB and Hemin on HO activity: (b) HO-1 specific inhibitor on EGB's protective effects on LPC-induced LDH release (Mean ± S.D., n = 5 – 8, **P < 0.01).

EGB for 24 h resulted in a dose-dependent decline of LDH release in LPC-treated cells (Fig. 5B). Myocytes was exposed to EGB and ZnPPIX (20 µM) for 24 h followed by LPC challenge. ZnPPIX (20 µM) partially suppressed the protective effect of EGB against LPC-induced damage (Fig. 5B), while it had little effect on ginkgolide B and bilobalide protective effects (data not shown). Our data showed that ZnPPIX partially suppressed the protective effect of EGB against LPC indicating that induction of HO-1 may mediate the EGB's powerful anti-oxidative effects. These studies revealed that stimulation of the heme oxygenase-bilirubin pathway by EGB conferred the cell protection against further oxidative assaults.

SUMMARY

This review has highlighted the emerging importance of the HO-bilirubin pathway in oxidative stress in the cardiovascular system. Our results suggest that EGB is a potent inducer of HO-1 in myocytes, and that upregulation of the heme oxygenase gene and its activity are important in EGB mediated cytoprotection against oxidative stress. Our data indicated that increased bilirubin and biliverdin, as a consequence of HO-1 induction, confer cytoprotection against oxidant-mediated damage in cardiomyocytes. This review proposes the hypothesis that heme, as a substrate for augmented HO-1 levels and generation of bilirubin, increases resistance to stressful stimuli. The established cardioprotective role of HO-1 induction in adaptive oxidative stress implicates its potential therapeutic value, and may lead to the development of pharmacological agents or gene therapy for cardioprotection. Development of pharmacological agents or gene therapy strategies to safely augment HO-1 expression in heart and prevent cardiomycytes against ischemial reperfusion injury will be a major area of investigation in the future.

ACKNOWLEDGMENT

The author thanks Dr. Xiang Gao for helpful discussion and for assistance in preparation of this manuscript.

REFERENCES

Abraham, N.G., Lavrovsky, Y., Schwartzman, M.L., Stoltz, R.A., Levere, R.D., Gerritsen, M.E., Shibahara, S., and Kappas, A. (1995) Transfection of the human heme oxygenase gene into rabbit coronary microvessel endothelial cells: protective effect against heme and hemoglobin toxicity. *Proc Natl Acad Sci U S A*, **92**, 6798–6802.

Abraham, N.G., Lin, J.H., Schwartzman, M.L., Levere, R.D., and Shibahara, S. (1988) The physiological significance of heme oxygenase. *Int J Biochem*, **20**, 543–558.

Chen, J.X., Chen, W.Z., Huang, H.L., Chen, L.X., Xie, Z.Z., and Zhu, B.Y. (1998) Protective effects of Ginkgo biloba extract against lysophosphatidylcholine-induced vascular endothelial cell damage. *Zhongguo Yao Li Xue Bao*, **19**, 359–363.

Chong, K.Y., Lai, C.C., Lille, S., Chang, C., and Su, C.Y. (1998) Stable overexpression of the constitutive form of heat shock protein 70 confers oxidative protection. *J Mol Cell Cardiol*, **30**, 599–608.

Clark, J.E., Foresti, R., Sarathchandra, P., Kaur, H., Green, C.J., and Motterlini, R. (2000) Heme oxygenase-1-derived bilirubin ameliorates postischemic myocardial dysfunction. *Am J Physiol Heart Circ Physiol*, **278**, H643–H651.

Durante, W. and Schafer, A.I. (1998) Carbon monoxide and vascular cell function (Review). *Int J Mol Med*, **2**, 255–262.

Foresti, R., Sarathchandra, P., Clark, J.E., Green, C.J., and Motterlini, R. (1999) Peroxynitrite induces haem oxygenase-1 in vascular endothelial cells: a link to apoptosis. *Biochem J*, **339**, 729–736.

Ishikawa, K., Navab, M., Leitinger, N., Fogelman, A.M., and Lusis, A.J. (1997) Induction of heme oxygenase-1 inhibits the monocyte transmigration induced by mildly oxidized LDL. *J Clin Invest*, **100**, 1209–1216.

Johnson, R.A., Kozma, F., and Colombari, E. (1999) Carbon monoxide: from toxin to endogenous modulator of cardiovascular functions. *Braz J Med Biol Res*, **32**, 1–14.

Johnson, R.A., Lavesa, M., DeSeyn, K., Scholer, M.J., and Nasjletti, A. (1996) Heme oxygenase substrates acutely lower blood pressure in hypertensive rats. *Am J Physiol*, **271**, H1132–H1138.

Keyse, S.M. and Tyrrell, R.M. (1989) Heme oxygenase is the major 32-kDa stress protein induced in human skin fibroblasts by UVA radiation, hydrogen peroxide, and sodium arsenite. *Proc Natl Acad Sci U S A*, **86**, 99–103.

Kwak, J.Y., Takeshige, K., Cheung, B.S., and Minakami, S. (1991) Bilirubin inhibits the activation of super-oxide-producing NADPH oxidase in a neutrophil cell-free system. *Biochim Biophys Acta*, **1076**, 369–373.

Lautier, D., Luscher, P., and Tyrrell, R.M. (1992) Endogenous glutathione levels modulate both constitutive and UVA radiation/hydrogen peroxide inducible expression of the human heme oxygenase gene. *Carcinogenesis*, **13**, 227–232.

Lee, P.J., Alam, J., Wiegand, G.W., and Choi, A.M. (1996) Overexpression of heme oxygenase-1 in human pulmonary epithelial cells results in cell growth arrest and increased resistance to hyperoxia. *Proc Natl Acad Sci U S A*, **93**, 10393–10398.

Levere, R.D., Martasek, P., Escalante, B., Schwartzman, M.L., and Abraham, N.G. (1990) Effect of heme arginate administration on blood pressure in spontaneously hypertensive rats. *J Clin Invest*, **86**, 213–219.

Llesuy, S.F. and Tomaro, M.L. (1994) Heme oxygenase and oxidative stress. Evidence of involvement of bilirubin as physiological protector against oxidative damage. *Biochim Biophys Acta*, **1223**, 9–14.

Maines, M.D. (1981) Zinc protoporphyrin is a selective inhibitor of heme oxygenase activity in the neonatal rat. *Biochim Biophys Acta*, **673**, 339–350.

Maines, M.D. (1988) Heme oxygenase: function, multiplicity, regulatory mechanisms, and clinical applications. *Faseb J*, **2**, 2557–2568.

Maines, M.D. (1997) The heme oxygenase system: a regulator of second messenger gases. *Annu Rev Pharmacol Toxicol*, **37**, 517–554.

Maines, M.D., Trakshel, G.M., and Kutty, R.K. (1986) Characterization of two constitutive forms of rat liver microsomal heme oxygenase. Only one molecular species of the enzyme is inducible. *J Biol Chem*, **261**, 411–419.

Marber, M.S., Mestril, R., Chi, S.H., Sayen, M.R., Yellon, D.M., and Dillmann, W.H. (1995) Overexpression of the rat inducible 70-kD heat stress protein in a transgenic mouse increases the resistance of the heart to ischemic injury. *J Clin Invest*, **95**, 1446–1456.

Maulik, N., Sharma, H.S., and Das, D.K. (1996) Induction of the haem oxygenase gene expression during the reperfusion of ischemic rat myocardium. *J Mol Cell Cardiol*, **28**, 1261–1270.

Murry, C.E., Jennings, R.B., and Reimer, K.A. (1986) Preconditioning with ischemia: a delay of lethal cell injury in ischemic myocardium. *Circulation*, **74**, 1124–1136.

Nath, K.A., Balla, G., Vercellotti, G.M., Balla, J., Jacob, H.S., Levitt, M.D., and Rosenberg, M.E. (1992) Induction of heme oxygenase is a rapid, protective response in rhabdomyolysis in the rat. *J Clin Invest*, **90**, 267–270.

Neuzil, J. and Stocker, R. (1993) Bilirubin attenuates radical-mediated damage to serum albumin. *FEBS Lett*, **331**, 281–284.

Neuzil, J. and Stocker, R. (1994) Free and albumin-bound bilirubin are efficient co-antioxidants for alpha-tocopherol, inhibiting plasma and low density lipoprotein lipid peroxidation. *J Biol Chem*, **269**, 16712–16719.

Otterbein, L.E. and Choi, A.M. (2000) Heme oxygenase: colors of defense against cellular stress. *Am J Physiol Lung Cell Mol Physiol*, **279**, L1029–L1037.

Parratt, J.R. (1994) Protection of the heart by ischaemic preconditioning: mechanisms and possibilities for pharmacological exploitation. *Trends Pharmacol Sci*, **15**, 19–25.

Pietri, S., Maurelli, E., Drieu, K., and Culcasi, M. (1997) Cardioprotective and anti-oxidant effects of the terpenoid constituents of Ginkgo biloba extract (EGb 761). *J Mol Cell Cardiol*, **29**, 733–742.

Plumier, J.C., Ross, B.M., Currie, R.W., Angelidis, C.E., Kazlaris, H., Kollias, G., and Pagoulatos, G.N. (1995) Transgenic mice expressing the human heat shock protein 70 have improved post-ischemic myocardial recovery. *J Clin Invest*, **95**, 1854–1860.

Poss, K.D. and Tonegawa, S. (1997) Reduced stress defense in heme oxygenase 1-deficient cells. *Proc Natl Acad Sci U S A*, **94**, 10925–10930.

Sano, K., Nakamura, H., and Matsuo, T. (1985) Mode of inhibitory action of bilirubin on protein kinase C. *Pediatr Res*, **19**, 587–590.

Shen, J.G. and Zhou, D.Y. (1995) Efficiency of Ginkgo biloba extract (EGb 761) in antioxidant protection against myocardial ischemia and reperfusion injury. *Biochem Mol Biol Int*, **35**, 125–134.

Shibahara, S., Muller, R.M., and Taguchi, H. (1987) Transcriptional control of rat heme oxygenase by heat shock. *J Biol Chem*, **262**, 12889–12892.

Siow, R.C., Sato, H., and Mann, G.E. (1999) Heme oxygenase-carbon monoxide signalling pathway in atherosclerosis: anti-atherogenic actions of bilirubin and carbon monoxide? *Cardiovasc Res*, **41**, 385–394.

Stocker, R., Yamamoto, Y., McDonagh, A.F., Glazer, A.N., and Ames, B.N. (1987) Bilirubin is an antioxidant of possible physiological importance. *Science*, **235**, 1043–1046.

Taketani, S., Kohno, H., Yoshinaga, T., and Tokunaga, R. (1989) The human 32-kDa stress protein induced by exposure to arsenite and cadmium ions is heme oxygenase. *FEBS Lett*, **245**, 173–176.

Tosaki, A., Droy-Lefaix, M.T., Pali, T., and Das, D.K. (1993) Effects of SOD, catalase, and a novel antiarrhythmic drug, EGB 761, on reperfusion-induced arrhythmias in isolated rat hearts. *Free Radic Biol Med*, **14**, 361–370.

Trakshel, G.M. and Maines, M.D. (1989) Multiplicity of heme oxygenase isozymes. HO-1 and HO-2 are different molecular species in rat and rabbit. *J Biol Chem*, **264**, 1323–1328.

Wu, T.W., Fung, K.P., and Yang, C.C. (1994) Unconjugated bilirubin inhibits the oxidation of human low density lipoprotein better than Trolox. *Life Sci*, **54**, 477–481.

Yellon, D.M. and Baxter, G.F. (1995) A "second window of protection" or delayed preconditioning phenomenon: future horizons for myocardial protection? *J Mol Cell Cardiol*, **27**, 1023–1034.

Yet, S.F., Tian, R., Layne, M.D., Wang, Z.Y., Maemura, K., Solovyeva, M., Ith, B., Melo, L.G., Zhang, L., Ingwall, J.S., Dzau, V.J., Lee, M.E., and Perrella, M.A. (2001) Cardiac-specific expression of heme oxygenase-1 protects against ischemia and reperfusion injury in transgenic mice. *Circ Res*, **89**, 168–173.

Yoshida, T., Biro, P., Cohen, T., Muller, R.M., and Shibahara, S. (1988) Human heme oxygenase cDNA and induction of its mRNA by hemin. *Eur J Biochem*, **171**, 457–461.

36

TOBACCO-SMOKE-INDUCIBLE HUMAN HEME OXYGENASE-1 GENE EXPRESSION

Role of Distinct Transcription Factors and Reactive Oxygen Intermediates

Florence Favatier[a] and Barbara S. Polla[b]

[a]Laboratoire de Physiologie Respiratoire
UFR Cochin Port-Royal 24 rue du Faubourg Saint-Jacques 750144
Paris
France
[b]INSERM U332, ICGM 22 rue Méchain 75014 Paris
France

1. INTRODUCTION

Expression of stress-related genes are particularly induced following exposure to thiol-reactive compounds (Heikkila et al., 1982), reactive oxygen intermediates (ROI) (Keyse et al., 1989a, 1989b, 1990) (Alam et al., 1994), heavy metal cations (Alam et al., 1994; Takeda et al., 1994; kimberly et al., 1998), UV radiation (Lautier et al., 1992; Vile et al., 1994) and other chemical initiators of cellular stress responses such as lipopolysaccharides or hydrogen peroxide (H2O2) (Kurata et al., 1996; Camhi et al., 1995). These stimuli all induce the expression of a 32–34 kDa protein identified as the microsomal, inducible, heme oxygenase-1 (HO-1), whose regulation is related to heat shock proteins (HSP) in rodents, but not in humans (Shibahara et al., 1985; Mitani et al., 1991; Clerget et al., 1990; Pizursky et al., 1990); a review showed (Stocker et al., 1990) that bile pigments produced by heme degradation catalyzed by HO-1 can function as anti-oxidants, while Keyse and Tyrrell (1989a) demonstrated that HO-1 is involved in the general cellular defence mechanisms against oxidative injury.

Tobacco smoke (TS) contains numerous toxic substances including heavy metals and is also a potent source of oxidants (Church et al., 1985; Müller et al., 1994). TS, in aqueous solutions, induces lipid peroxidation (Churg et al., 1993; Frei et al., 1991), DNA single-strand breaks (Nakayama et al., 1985; Nakayama et al., 1986) and decreases the levels of reduced glutathione and protein thiols (Frei et al., 1991). Although the precise mechanisms by which cells sense and respond to a specific environmental stress to activate the transcription of appropriate target genes are still not completely understood, stimulation of HO-1 expression is controlled primarly at the level of gene transcription. More generally, the activation of gene transcription plays a vital role in the ability of all organisms to induce a protective response to environmental stresses.

Upon exposure to a specific stress, cells activate the transcription of genes encoding proteins that either offer protection from stress, or repair stress-induced intracellular damage. In the case of HO-1, a growing number of transcription factors and their cognate binding sites that mediate transcriptional activation of the gene in response to distinct agents has been described. Alam et al. (1992, 1994) demonstrated that activation of the mouse HO-1 gene by phorbol esters, heme and cadmium is mediated by a fragment, located 4 kb upstream of the transcription initiation site and composed of two elements that interact with the *Fos/Jun* AP1 family of transcription factors, while a cis-acting element responsible for cadmium-mediated induction of the human HO-1 gene called cadmium responsive element (CdRE) has been identified by Takeda et al. (1994). In contrast, heme induction of human HO-1 gene has been shown to be mediated by NF-B and AP-2 transcription factor activation (Shibahara et al., 1989).

In this study, we investigated the mechanisms through which TS-induced HO-1 expression is modulated, and the different transcription factors involved in response to TS exposure.

2. HO-1 STIMULATION BY OXIDATIVE INDUCERS IS TRANSCRIPTIONNALLY REGULATED

Accumulated evidence indicates a role for HO-1 protein in cellular and tissue damage and suggests that HO-1 induction is a protective response against oxidative stress (Choi et al., 1996). Several studies on the regulation of HO-1 gene expression have focused on the induction mediated by thermal stress (Okinaga et al., 1996) metal ions (Takeda et al., 1994) or heme (Inamdar et al., 1996). In this chapter, we have examined the effects of TS, Cd and H2O2 on HO-1 expression. Exposure to TS is done by a peristaltic pump-smoke machine (Heinr. Borgwaldt RM1/G, Hamburg, Germany) used to generate TS-bubbled phosphate buffered saline (PBS), from mainstream smoke of standard research cigarettes (reference 2R1, University of Kentucky, USA) through a puffing mechanism related to the human smoking pattern: (1 puff/minute; 1 puff = 35 ml; each puff of 2 seconds duration). In our experiments, the aqueous smoke fraction of one cigarette corresponded to 10 puffs (350 ml) bubbled into 5 ml of PBS. The final dilutions are expressed as puffs/ml of culture medium. Dose-response experiments were realized between 0.03 and 0.48 puffs/ml. U937 cells were exposed to TS-bubbled PBS overnight, then recovered by pipetting and washed twice in PBS before nuclear extract preparation. HO-1 mRNA accumulation was measured by northern blot analysis. A ^{32}P-labeled oligonucleotide probe was generated by a reverse transcriptase reaction using sequence specific oligonucleotide primers:

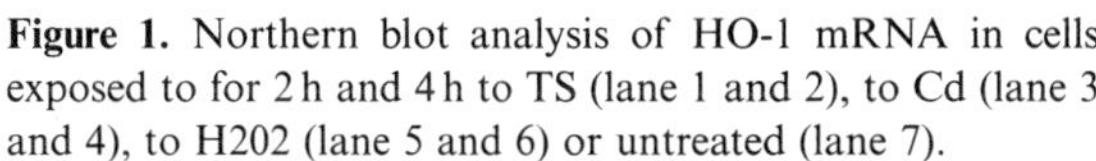

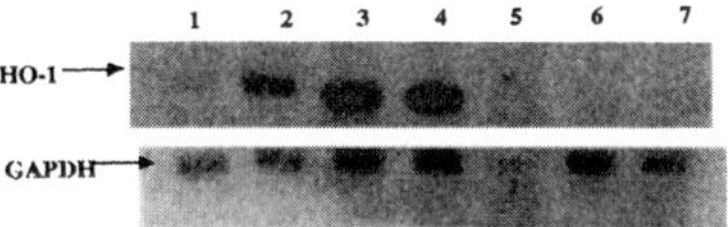

Figure 1. Northern blot analysis of HO-1 mRNA in cells exposed to for 2 h and 4 h to TS (lane 1 and 2), to Cd (lane 3 and 4), to H2O2 (lane 5 and 6) or untreated (lane 7).

5′ CCGCAACCCGACAGCATGCC 3′ (nucleotide position 90) and 5′CCGCTTCACATAGCGCTGC3′ (nucleotide position 414) from the 5′ translated region of cDNA HO-1 (Yoshida et al., 1988). This probe was hybridized to a membrane that contained RNA prepared from control U937 cell line and U937 exposed to 0.24 puffs/ml of TS, 100 ΓM of CdCl2, 2 mM of H2O2 for 2 h and 4 h (Fig. 1). We noted that the TS induced HO-1 mRNA expression is time dependant, 4 hours of incubation give greater mRNA accumulation (Fig. 1 lane 1 and 2). For CdCl2, the induced HO-1 mRNA expression is practically the same at 2 h as at 4 h (Fig. 1 lane 3 and 4), the conditions of time and dose of H2O2 failed to express any HO-1 induced mRNA (Fig. 1 lane 5 and 6). Untreated cells are shown (Fig. 1 lane 7). The human HO-1 mRNA is detected at 1.9 Kb. The membrane was reprobed with a ^{32}P labelled DNA fragment homologous to GAPDH mRNA (detected at 1.27 Kb) in order to take into account variations in the initial amounts of mRNA present in each sample.

3. EFFECTS OF TS AND CD ON HO-1 PROTEIN EXPRESSION

U937 cells were exposed to different concentrations of TS (0.06 to 0.48 puffs/ml) or to Cd (1–10 ΓM) during 16 h, and HO-1 expression analyzed by flow cytometry. Permeabilized cells were incubated with anti-human HO-1 monoclonal antibody as previously described in details (Bachelet et al., 1998). Analysis was performed using an Epics Elite flow cytometer (Coulter, Miami, FL, USA). Exposure of U937 cells to TS resulted in an increased expression of HO-1 protein as compared to control cells, confirming previous reports from our (Pinot et al., 1997) and other laboratories (Müller et al., 1994) using other techniques. HO-1 expression was maximal with 0.06 puffs/ml, while at higher concentrations of TS, HO-1 induction was inhibited reaching extinction at 0.48 puffs/ml (Fig. 2A). Figure 2B shows that HO-1 was also induced by Cd (10 M) for 16 h.

This TS-dependent dual increase and decrease was also observed at shorter incubation times with higher concentrations of TS; we tested several doses of TS varying from one to six hours (0.48 puffs/ml for 1 h, 0.24 puffs/ml for 2 h and 0.12 puffs/ml for 6 h. The time point for which HO-1 begins to be induced is 0.24 puffs/ml for 2 h and the maximum is reached at 0.12 puffs/ml for 6 h. The dose of 100 ΓM CdCl2 for 2 h showed the same curve of HO-1 induction than that obtained at 10 ΓM for 16 h.

4. EFFECTS OF TS ON U937 CELLS GSH CONTENT AND MITOCHONDRIAL MEMBRANE POTENTIAL (Ə∵m)

Multiple ROI derived from TS are the most likely to be responsible for GSH depletion and HO-1 expression in TS-exposed cells. This hypothesis is in good

A

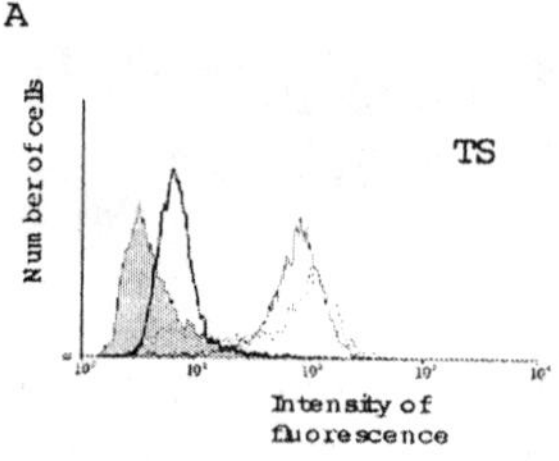

B

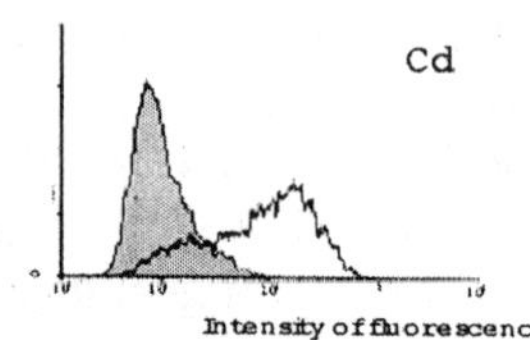

Figure 2. Effects of TS and Cd on HO-1 expression. In A, HO-1 is increased after exposure to 0.06 puffs/ml of TS (dashed curve) and 0.12 puffs (plain curve) but decreased at 0.48 puffs (bold curve). Control (hatched curve); in B, cells are exposed to Cd (bold curve).

agreement with previous work from our laboratory on the role of ROI in TS with respect to the induction of either HSP expression or apoptosis. The depletion of intracellular thiols such as GSH (Fig. 3) resulting from TS exposure and protein thiol groups alterations was also suggested by Kato et al. (1997) and Maranzana et al. (1998). As shown in Fig. 3, both parameters were altered as early as 2 h after exposure to tobacco smoke, as revealed by confocal image.

The non fluorescent sulfydryl probe monobromobimane (MBBr) obtained by Molecular probes was used as described (Hedley et al., 1994). This compound binds avidly to -SH groups to form blue fluorescent conjugates. MBBr was made up as a 4 mM stock solution in 100% ethanol. Under the staining conditions used, fluorescence is highly correlated with the cellular content of GSH (Show et al., 1995). In parallel to GSH content, we also analyzed by confocal microscopy the $\partial\therefore_m$ of

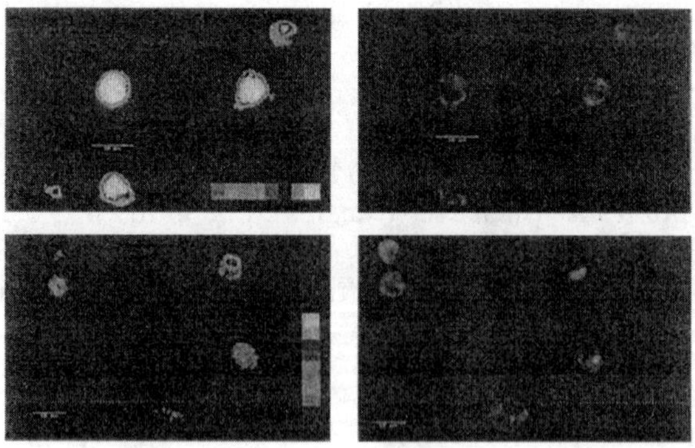

Figure 3. Effects of TS on U937-cell GSH content and $\Delta\chi_m$. C Cells were double stained with MBBr and JC-1. Confocal appearance of cells before and after TS revealed changes of colour reflecting: a depletion in GSH content (A to B) and a decrease in $\Delta\psi_m$ (C to D).

untreated and TS treated U937 cells using 5,5′,6,6′-tetrachloro-1,1′,3,3′,-tetraethyl-benzimidazolcarbocyanine iodide (JC-1) probe as described (Salvioli et al., 1997; Banzet et al., 1999). In these experiments, mitochondrial membrane depolarization is associated with a large shift in JC-1 fluorescence emission from red to green. Depletion of GSH by TS treatment is symbolized by changes in the fluorecence: the cells are colorized in blue after TS treatment (panel A versus panel B). JC-1 emitted red fluorescence in control cells and green fluorescence in cells with depolarized membrane mitochondria after TS exposure (panel C versus panel D). The effect of TS in GSH content have been previously described in different cell types (Maranzana et al., 1998).

5. EFFECTS OF TS ON HO-1 REGULATORY ELEMENTS

Various cis-acting elements in the HO-1 promoter indicated that HO-1 might be induced by distinct mechanisms depending on the induction stimulus. The involvement of different cis-acting elements on the HO-1 promoter and their respective TS-inducible binding activity was performed by electrophoretic gel mobility shift assays (EMSA). Among these elements, we explored HSF, NF-B and AP1 (Fig. 4). Using first a synthetic oligonucleotide containing the radiolabelled consensus HSE-binding site HSE [5′-GCCTCGAATGTTCGCGAAGTT-3′], incubated with nuclear extracts from TS-exposed cells (0.24 puffs/ml, 2h), we observed typical HSF-HSE complexes (Fig. 4A lane 2) and a relevant TS induction of HSF activity. Under our experimental conditions, both the constitutive HSE binding activity (cba) and the inducible forms of DNA binding activity (iba) were suppressed by an excess of cold HSE

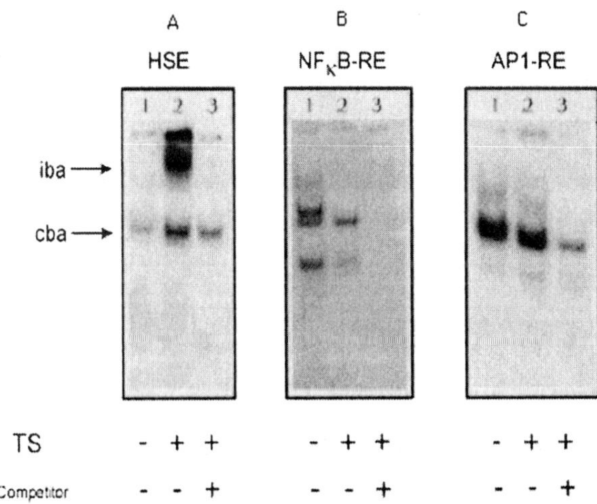

Figure 4. Effects of TS on HSF, NFκB and AP-1 binding activity. In (A), a major inducible HSF binding activity (iba) was revealed (lane 2) as compared to control cells (lane 1). cba indicates constitutive binding activity. Specificity was controlled with a 100-fold molar excess of of unlabelled HSE (lane 3). In B, basal NFκB binding activity (three specific bands) was present in the control (lane 1); TS exposition inhibits the upper band (lane 2). Control specificity (lane 3). In (C), no iba was found with the binding site for AP-1.

(Fig. 4A lane 3). Okinaga et al. (1996) who studied the induction of HO-1 by HS, demonstrated that the HSE sequence on human HO-1 promoter is potentially functional *in vitro* though its function is repressed *in vivo* because of the presence of a purine-rich sequence on HO-1 promoter downstream from the HSE which confers it a silencing effect. We thus deduced that the observed TS-inducible HO-1 expression is likely to be mediated by pathways distinct from HSF.

Binding of NF-NB was next examined using a synthetic oligonucleotide containing the consensus NF-N binding site [5'-AGTTGAGGGGACTTTCCCAGGC-3'] in TS-exposed cells. We observed an inhibition of the complexes and an extinction of the upper band in TS-exposed cells (Fig. 4B lane 2) as compared to control cells (Fig. 4B lane 1). Specificity of NF-NB binding was tested by competition experiments with a molar excess of 100 fold non-radioactive NF-NB (Fig. 4B lane 3). Inhibition of NF-NB activation along with HSF activation has been reported (Vayssier et al., 1998) and is not specific of TS. Other stresses, including HS, arsenite, and cyclopentanone prostaglandins also inhibit NF-B (Rossi et al., 1997). A mediation of the observed TS inducibility of HO-1 by NF-B must thus, be excluded.

Functional AP1 binding sites within the HO-1 promoter have been identified, suggesting the involvement of the *Jun/Fos* family of transcription factors in mediating the induction of HO-1 gene transcription by multiple agents (Alam et al., 1992; Mendelson et al., 1996; Numazawa et al., 1997). We also analyzed nuclear protein extracts from TS-exposed and control cells with a specific oligonucleotide probe containing the binding site for AP1; [5'-GCTGAGTCACCAGTGCC-3'], no modification of the pattern was observed under the same conditions, as described above (Fig. 4C lane 1 and 2).

These results strenghen those published by Takeda et al. (1994) indicating that AP-1 binding site is not directly involved in Cd-mediated activation of the human HO-1 gene. They showed that relative luciferase activities were induced by Cd in the THP-1 cell line transfected with plasmid containing CdRE. In contrast, no Cd mediated increases were detected in the cells transfected with the plasmid containing AP-1 binding site but lacking CdRE fragment. It is still possible that treatment of cells exposed to stressors with longer or shorter TS incubation times should raise AP-1 levels as it has been reported in early genes (Amstad et al., 1992) where increased binding of *fos* and *jun* to AP-1 oligonucleotide was observed only after 30 minutes of treatment with active oxygen.

6. EFFECTS OF TS ON CdRE-BINDING ACTIVITY

We next examined the possibility that TS-induced HO-1 expression was mediated by CdRE, localized about 4 kb upstream from the transcription initiation site of the human HO-1 gene specifically responsible for Cd-mediated induction (Takeda et al., 1994). We examined whether nuclear factors specifically bind to this CdRE after exposure to TS. Cells were exposed to TS (0.24 puffs/ml during 2 hours), nuclear proteins extracted, the 20 bp oligonucleotide containing the radiolabelled core CdRE [5'-AGGCGGATTTTGCTAGATTT-3'] added, and the binding activity was determined. The data are presented in Fig. 5; increasing concentrations of nuclear proteins

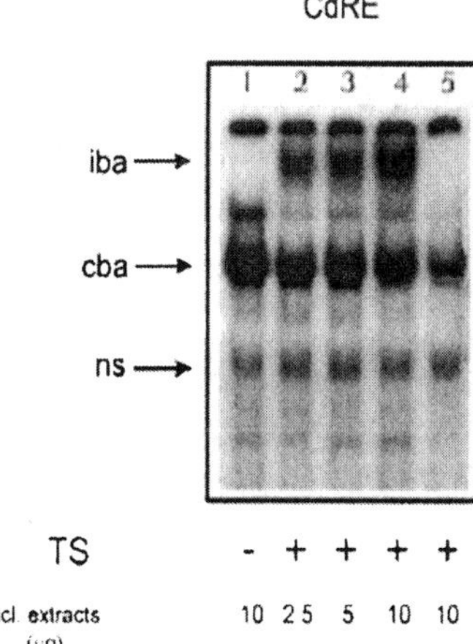

Figure 5. Effects of TS on CdRE-binding activity, iba was increased with increasing concentrations of nuclear extracts (lanes 2,3,4, upper bands) as compared to control (lane 1). Control specificity (lane 5).

induced a dose-dependent binding complex (see lanes 2,3,4 as compared to control, lane1); competition experiments confirmed this specificity (Fig. 5 lane 5).

7. CdRE-BINDING ACTIVITY IS INDUCED BY Cd

Our next aim was to confirm the Cd-dependency of CdRE binding activity. Takeda et al. (1994) have reported in EMSA experiments the presence of two constitutive retarded bands and have shown that CdRE binding activity was unchanged by Cd treatment in their conditions. However, in our experiments, we observed a specific inducible CdRE-binding activity when cells were treated with 100 ΓM Cd for 2 h (Fig. 6 lane 3) as compared to 50 ΓM (lane 2) and non-treated cells (lane 1). This high concentration of Cd did not affect however cell viability as estimated by the percentage

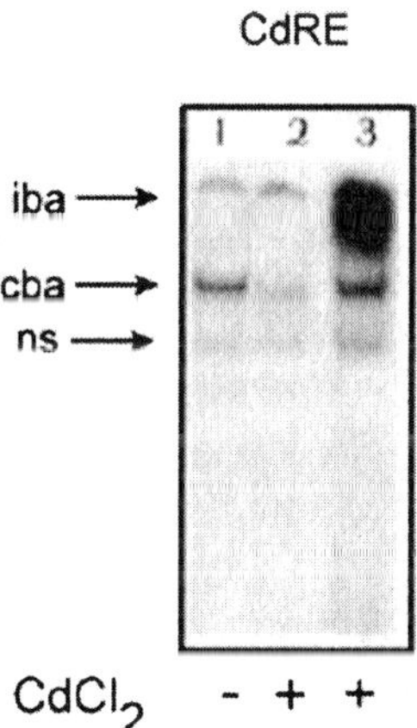

Figure 6. Effects of Cdcl2 on CdRE binding activity. A major iba was revealed with 100 μM of Cd as compared with 50 μM (lane 2); Control (lane 1).

of lactate dehydrogenase (LDH) release (data not shown). In other reported experiments (Liao et al., 1998) where cells were were exposed to Cd, the medium was also supplemented with 100 ΓM CdCl2 for much longer periods of time (8 or 24 h).

Different incubation times and different doses of inducers may differ other elements of transcription and signalling cascades as recently described for rat brain tumor cells cells (Huang et al., 1998). The effects of Cd itself on CdRE-binding activity might be at least in part mediated by ROI, as it is now generally recognized that ROI are the essential mediators of Cd toxicity (Horiguchi et al., 1993). Since Cd has a high affinity for sulfur, some toxic effects of Cd are results of reactions with essential sulfhydryl groups in proteins (Jacobson et al., 1980). Cd shows chemical similarities to zinc and may affect the properties of Zn-containing nuclear transcription factors and enzymes such as Cu/Zn superoxide dismutase, which is inhibited by the substitution of Zn by Cd (Bauer et al., 1980; Karin et al., 1983), thus leading to an increase in ROI production and toxicity.

8. DIFFERENTIAL MODULATION OF HO-1 EXPRESSION BY Cd AND TS

Flow cytometry indicated that Cd-mediated HO-1 protein expression was abolished when using the Cd-chelating resin chelex-100 (Fig. 8A), while depletion of Cd from TS by the same chelating agent did not affect HO-1 induction (Fig. 8B). This suggests that Cd contained in TS was not the only responsible agent for HO-1 induction.

9. EFFECTS OF H2O2 ON CdRE BINDING ACTIVITY

As it has previously been shown that ROI are mainly responsible for TS-dependent DNA damage (Müller et al., 1994; Nakayama et al., 1985; Fielding et al., 1989), we postulated that ROI, and in particular H2O2, might also represent causative agents in TS-dependent CdRE-binding activity. We thus, examined the binding

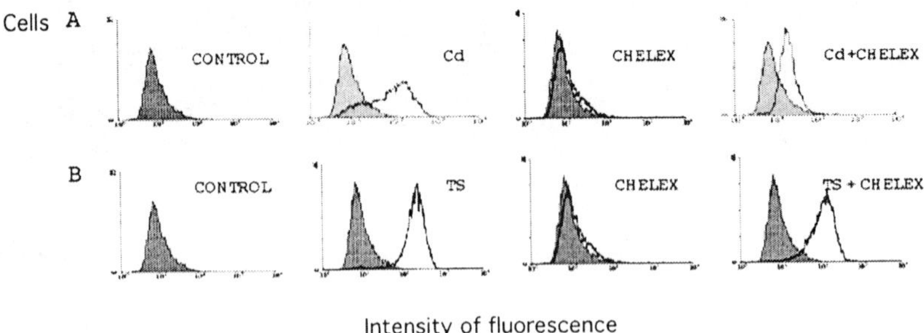

Figure 7. Effects of chelex on TS-induced HO-1 expression. In A, inhibition is practically complete when cells were pre-incubated with chelex 100 before being exposed to Cd (10 μM); In B, there was practically no inhibition before being exposed to TS.

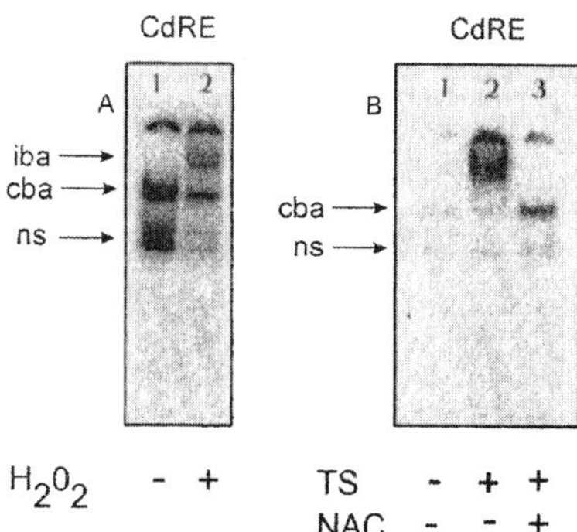

Figure 8. Effects of H2O2 and NAC on CdRe-binding activity. In A, H2O2 induced CdRE-binding activity (lane 2) as compared to control (lane 1). In B, cells are exposed to TS (lane 2); when pre-incubated with NAC before exposed to TS, iba was abolished.

specificity to CdRE of nuclear extracts (Figure 7) of cells treated with H2O2 (2 mM, 2 h). Incubation of nuclear extracts from H2O2-treated cells with the radiolabeled CdRE resulted in a specific inducible binding activity (Fig. 8A lane 2) as compared to control cells (Fig. 8A lane 1), further supporting the hypothesis that CdRE-binding activity might be redox dependent. Interestingly also, the effects of other inducers of HO-1, such as hemine, cobalt protoporphyrin and arsenite, whose effects are not mediated by ROI, do not activate CdRE-binding activity (Takeda et al., 1994).

Activation of binding activity does not always result in activation of corresponding transcription genes as it is shown in Fig. 1; we didn't observe mRNA accumulation after H2O2 treatment (Fig. 1) although an activation of CdRE binding activity is shown. This phenomenon is reported in NIH-3T3 cells (Bruce et al., 1993) where oxidative injury activates the HSF but does not increase levels of HSP.

10. EFFECTS OF NAC ON TS-INDUCED CdRE BINDING ACTIVITY AND HO-1 EXPRESSION

NAC, a known anti-oxidant and a free radical scavenger (Moldeus et al., 1986) increases intracellular GSH stores, which is of particular importance in the case of GSH depletion caused by TS exposure (Maranzana et al., 1998). NAC, as GSH, is able to detoxify reactive electrophiles and free radicals either through conjugation or reduction reactions, and reduces ROI to less reactive products. Pre-incubation of nuclear extracts with 20 mM NAC before exposure to TS resulted in the extinction of CdRE-binding activity (Fig. 8B lane 3) as compared to nuclear extracts from cells exposed to TS but not pre-incubated with NAC (Fig. 8B lane 2). Analysis of HO-1 expression by flow cytometry confirmed the inhibition of HO-1 induction when cells were pre-incubated with NAC before exposure to TS (Fig. 9).

11. DISCUSSION

Depletion of intracellular thiols such as GSH resulting from TS exposure might favor the activation of a redox sensitive component of the CdRE-binding protein

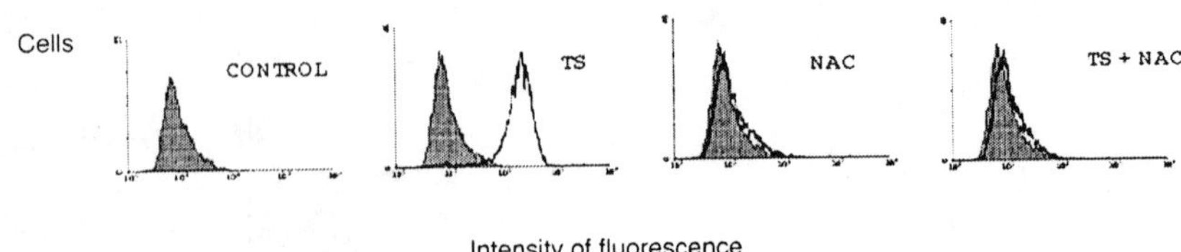

Figure 9. Effects of NAC on TS-induced HO-1 expression. Complete inhibition is realized by preincubation of cells with NAC.

(CdRE-BP) pathway. However Cd, included in TS may also mimic or modify a physiological mediator that would be involved in a signal transduction system leading to phosphorylation or dephosphorylation of the putative CdRE-BP. This putative CdRE-BP may relate to the antioxidant responsive element binding protein (ARE)-BP (Wasserman et al., 1997a). A model of the ARE as a composite regulatory site, where multiple transcription factors interact was presented by the same authors (Wasserman et al., 1997b) in which the ARE "core" sequence RTGACnnnGC is necessary to mediate an anti-oxidant response. Interestingly, CdRE also contains an ARE-like core sequence and it is possible that the as yet unidentified CdRE-BP corresponds to the ARE-BP1 defined by Wasserman and Fahl (1997a, 1997b).

12. SUMMARY

Exposure of eukaryotic cells to a variety of reactive oxygen intermediates (ROI)-mediated sources of cellular injury, including heavy metals and UV radiation, induces the expression of heat shock (HS) and stress-related genes among which a 32–34 kDa protein identified as the inducible heme oxygenase-1 (HO-1). Here, we investigated the induction mechanisms of human HO-1 gene expression by TS in the human pre-monocytic line U937. Northern blot and Flow cytometry revealed a dose- and time-dependent induction of HO-1 mRNA and protein by TS. In order to clarify the role of transacting factors in this induction, electrophoretic mobility shift analysis was performed with nuclear extracts from control, TS, cadmium, or hydrogen peroxide (H2O2)-exposed cells, incubated with consensus elements and binding sites of the promoter region of HO-1: HSE, NF-B, AP-1, and CdRE (isolated by Takeda et al., 1994). These results suggest that TS induced HO-1 expression is modulated by cellular redox state in U937 cell line and that CdRE is involved in the regulation mechanism.

REFERENCES

Alam, J. 1994. Multiple elements within the 5' distal enhancer of the mouse HO gene mediate induction by heavy metals. *J. Biol. Chem.* **269**:25049.

Alam, J. and Den, Z. 1992. Distal AP-1 binding sites mediates basal level enhancement and TPA induction of the mouse heme oxygenase-1 gene. *J. Biol. Chem.* **267**:21894.

Bachelet, M., Mariethoz, E., Banzet, N., Souil, E., Pinot, F., Polla, CZ., Durand, P., Bouchaert, I., and Polla, B.S. 1998. Flow cytometryis a rapid and reliable method for evaluating heat shock protein70 expression in human monocyte. *Cell Stress and Chap.* 3:168–170.

Banzet, N., François, D., and Polla, B.S. 1999. Tobacco smoke induces mitochondrial depolarization along with cell death: effects of antioxidants. *redox report* 4(5):229–236.

Bauer, R., Demeter, I., Haseman, V., and Johansen, J.T. 1980. Structural properties of the zinc site in Cu, Zn-superoxide dismutase perturbed angular correlation of gamma ray spectroscopy on the Cu, 111Cd-superoxide dismutase. *Biochem. Biophys. Res. Commun.* 94:1296.

Bruce, J., Price, B.D., Coleman, C.N., and Calderwood, S.K. 1993. Oxidative injury rapidly activates the heat shock transcription factor but fails to increase levels of heat shock proteins. *Cancer Res.* 53:12.

Camhi, S.L., Alam, J., Otterbein, L., Sylvester, S.L., and Choi, A.M.K. 1995. Induction of heme oxygenase-1 gene expression by lipopolysaccharide is mediated by AP-1 activation. *Am. J. Respir. Cell Miol. Biol.* 13:387.

Choi, A.M.K. and Alam, J. 1996. HO-1 function, regulation and implication of a novel stress-inducible protein in oxidant-induced lung injury. *Am.J. Respir. Cell Mol. Biol.* 15:9.

Church. and Pryor, W.A. 1985. Free radical chemistry of cigarette smoke and its toxicological implications. *Environ. Health Perspect.* 64:11.

Chomcynski, P. and Sacchi, N. 1987. Single-step method of RNA isolation by acid guanidium thiocyanate-phenol-chloroform extraction. *Anal. Biochem.* 162:156.

Churg, A. and Cherukupalli, K. 1993. Cigarette smoke causes rapid peroxidation of rat tracheal epithelium. *Int. J. Exp. Path.* 74:127.

Clerget, M. and Polla, B.S. 1990. Erythrophagocytosis induces heat shock protein synthesis by human monocytes macrophages. *Proc. Natl. Acad. Sci. USA* 87:1081.

Dignam, J.D., Lebovitz, R.M., and Roeder, R.G. 1983. Accurate transcription initiation by RNA polymeraseII in a soluble extract from isolated mammalian nuclei. *Nucleic Acids Res.* 11:1475.

Fielding, S., Short, C., Davies, K., Wald, N., Bridges, B.A., and Waters, R. 1989. Studies on the ability of smoke from differents types of cigarettes to induce DNA single strands breaks in cultured human cells. *Mutat. Res.* 214:147.

Frei, B., Forte, T.M., Ames, B.N., and Cross, C.E. 1991. Gas phase oxidants of cigarette smoke induce lipid peroxidation and changes in lipoprotein properties in human blood plasma. *Biochem. J.* 277:133.

Hedley, D.W. and Show, S. 1994. Evaluation of methods for measuring cellular glutathione content using flow cytometry. *Cytometry* 15:349.

Heikkila, J.J., Schultz, G.A., Iatrou, K., and Gedamu, L. 1982. Expression of a set of fish genes following heat or metal ion exposure. *J. Biol. Chem.* 257:12000.

Horiguchi, H., Mukaida, N., Okamoto, S., Teranishi, H., Kasuya, M., and Matsushima, K. 1993. Cadmium induces interleukin-8 production in human peripheral blood mononuclear concomitant generation of superoxide radicals. *Lymphokine cytokine Res.* 12(6):421.

Hung, J.J., Cheng, T.J., Lai, Y.K., and Chang M. DT. 1998. Differential activation of p38 mitogen-activated protein kinase and extracellular signal-regulated protein kinases confers cadmium-induced HSP70 expression in 9L rat brain tumor cells. *J. Biol. Chem.* 273:31924.

Inamdar, N.M., Ahn, Y.I., and Alam, J. 1996. The heme responsive element of the mouse HO-1 gene is an extended AP1 binding site that resembles the recognition sequences for MAF and NF-E2 transcription factors. *Biochem. Biophys Res. Commun.* 221:570.

Jacobson, K.D. and Turner, J.E. 1980. The interaction of cadmium and certain other metal ions with proteins and nucleic acids. *Toxicology* 16:1.

Karin, M., Cathala, G., and Nguyen-Huu, M.C. 1983. Expression and regulation of a human metallthionein gene carried on an autonomouslly replicating shuttle vector. *Proc. Natl. Acad. Sci. USA* 80:4040.

Kato, K., Ito, H., and Okamato, K. 1997. Modulation of the arsenite-induced expression of stress proteins by reducing agents. *Cell Stress and Chap* 2(3):199.

Keyse, S.M. and Tyrrell, R.M. 1989a. Induction of the heme oxygenase gene by hydrogen peroxide and UVA: evidence for the involvement of the hydroxyl radical. *Cacinogenesis* 11:787.

Keyse, S.M. and Tyrrell, R.M. 1989b. Heme oxygenase is the major 32 KDa stress protein induced in human skin fibroblasts by UVA radiation, hydrogen peroxyde and sodium arsenite. *Proc. Natl. Acad. Sci. USA* 86:99.

Keyse, S.M., Applegate, L.A., Tromvoukis, Y., and Tyrrell, R.M. 1990. Oxidant stress leads to transcriptional activation of the human HO gene in cultured skin fibroblasts. *Mol. Cell. Biol.* **10**:4967.

Kimberly, K.E., Whitmarsh, A.J., Davis, R.J., and Bonkovsky, H.L. 1998. Mechanism of sodium arsenite-mediated induction of HO in hepatoma cells. *J. Biol. Chem.* **273**:8922.

Kurata, S.I., Matsumoto, M., Tsuji, Y., and Nakajima, H. 1996. LPS activates transcription of the HO gene in mouse M1 cells through oxidative activation of nuclear factor. *Eur. J. Biochem.* **239**:566.

Lautier, D., Luscher, P., and Tyrrell, R.M. 1992. Endogenous glutathione levels modulate both constitutive and UVA radiation/hydrogen peroxide inducible expression of the human HO gene. *Carcinogenesis* **13**:227.

Lavrosky, Y., Schwartzman, M.L., Levere, R.D., Kappas, A., and Abraham, N.G. 1994. Identification of binding sites for transcription factors NF-B and AP-2 in the promoter region of the HO1 gene. *Proc. Natl. Acad. Sci. USA* **91**:5987.

Liao, V. HC. and Friedman, J. 1998. Cadmium-regulated genes from the nematode *caenorhabditis elegans*. *J. Biol. Chem.* **273**:31962.

Maranzana, A. and Mehlhorn, R.J. 1998. Loss of Glutathione, ascorbate recycling, and free radical scavenging in human erythrocytes exposed to filtered cigarette smoke. *Arch. of Biochem. and Biophys.* **350**:169.

Mendelson, K.G., Contois, L.R., Tevosian, S.G., Davis, R.J., and Paulson, K.E. 1996. Independent regulation of JNK/p38 mitogen-activated protein kinases by metabolic oxidative stress in the liver. *Proc. Natl. Acad. Sci. USA* **93**:12908.

Mitani, K., Fujita, H., Sassa, S., and Kappas, A. 1991. A heat inducible nuclear factor that binds to the to the HSE of the human HO gene. *Biochem. J.* **277**:895.

Moldeus, P., Cotgreave, I.A., and Berggren, M. 1986. Lung protection by a thiol-containing antioxidant: N-acetyl cysteine. *Respiration* **50**:31.

Müller, T. and Gebel, S. 1994. Heme oxygenase expression in Swiss 3T3 cells following exposure to aqueous cigarette smoke fractions. *Carcinogenesis* **15**:67.

Nakayama, T., Kaneko, M., Kodama, M., and Nagata, C. 1985. Cigarette smoke induces DNA single strands breaks in human cells. *Nature* **314**:462.

Nakayama, T., Kaneko, M., and Kodama, M. 1986. Volatile Gas components contribute to cigarette smoke-induced DNA single strands breaks in cultured human cells. *Agric. Biol. Chem.* **12**:3219.

Numazawa, S., Yamada, H., Furusho, A., Nakaharaz, T., Oguro, T., and Yoshida, T. 1997. Cooperative induction of c-fos and HO gene products under oxidative stress in human fibroblastic cells. *Exp. Cell Res.* **237**:434.

Okinaga, S., Takahashi, K., Takeda, K., Yoshisawa, M., Fujita, H., Sasaki, H., and Shibahara, S. 1996. Regulation of human HO1 gene expression under thermal stress. *Blood* **87**:5074.

Pinot, F., El Yaagoubi, A., Christie, P., Dinh-Xuan, A.T., and Polla, B.S. 1997. Induction of stress proteins by tobacco smoke in human monocytes: modulation by antioxidants. *Cell Stress and Chap.* **2**:156.

Pizurki, L. and Polla, B.S. 1990. cAMP modulates stress protein synthesis in human monocytes-macrophages. *J. Cell Physiol.* **161**:169.

Rossi, A., Elia, G., and Santoro, G. 1997. Inhibition of nuclear factor B by prostaglandin A1: an effect associated with heat shock transcription factor activation. *Proc. Natl. Acad. Sci. USA* **94**:746.

Salvioli, S., Ardizzoni, A., Franchesci, C., and Cossarizza. 1997. JC-1 but not DIOC6(3) or rhodamine 123, is a reliable fluorescent probe to assess ∋∴. changes in intact cells: implications for studies on mitochondrial functionality during apoptosis. *FEBS Letters* **411**:77.

Shibahara, S., Möller, R., Tagushi, H., and Tadashi, Y. 1985. Cloning and expression of cDNA for rat HO *Proc. Natl. Acad. Sci. USA* **82**:7865.

Shibahara, S., Sato, M., Muller, M.R., and Yoshida, T. 1989. Structural organization of the human heme-oxygenase gene and the function of its promoter. *Eur. J. Biochem.* **179**:55.

Show, S. and Hedley, D.W. 1995. Flow cytometric determination of glutathione in clinical samples. *Cytometry* **21**:68.

Stocker, R. 1990. Induction of Haem oxygenase as a defence against oxidative stress. *Free radic. Res. Commun.* **9**:101–112: Review.

Takeda, K., Ishizawa, S., Sato, M., Yoshida, T., and Shibahara, S. 1994. Identification of a cis-acting element that is responsible for Cd mediated induction of the human heme oxygenase gene. *J. Biol. Chem.* **269**:22858.

Vayssier, M., Favatier, F., Pinot, F., Bachelet, M., and Polla, B.S. 1998. Tobacco smoke induces coordinate activation of HSF and inhibition of NFkB in human monocytes: effects on TNF. *Biochem. Biophys. Res. Commun.* **252**:249.

Vile, G.F., Basu-Modak, S., Waltner, C., and Tyrrell, R.M. 1994. Heme oxygenase-1 mediates an adaptive response to oxidative stress in human skin fibroblasts. *Proc. Natl. Acad. Sci. USA* **91**:2607.

Wasserman, W.W. and Fahl, W.E. 1997a. Comprehensive analysis of proteins which interact with the antioxidant responsive element: correlation of ARE-BP1 with the chemoprotective induction response. *Arch. of biochem. and Biophys.* **344**:387.

Wasserman, W.W. and Fahl, W.E. 1997b. Functional antioxidant responsive elements. *Proc. Natl. Acad. Sci. USA* **94**:5361.

Yoshida, T., Biro, P., Cohen, T., Müller, R.M., and Shibahara, S. 1988. Human heme oxygenase cDNA and induction of its mRNA by hemin. *Eur. J. Biochem.* **171**:457.

37

DOPPEL PROTEIN EXPRESSION CORRELATES WITH HEME OXYGENASE 1 AND NITRIC OXIDE SYNTHASE INDUCTION

Boon-Seng Wong[a], Man-Sun Sy[a], and David R. Brown[b]

[a]Institute of Pathology
Case Western Reserve University
Cleveland, Ohio 44106, USA
[b]Department of Biology and Biochemistry
University of Bath, Bath BA2 7AY
UK.

1. INTRODUCTION

In the early 1990's discrepant phenotypes were observed in four reported prion protein gene (*Prnp*)-ablated transgenic mouse lines (*Prnp$^{0/0}$*) (Bueler et al., 1992; Manson et al., 1994; Moore et al., 1995; Sakaguchi et al., 1996). Two of them, *Ngsk* and *Rcm0 Prnp$^{0/0}$* mice (Moore et al., 1999; Sakaguchi et al., 1996), displayed normal early development but revealed progressive ataxia accompanied by the degeneration of cerebellar Purkinje cells at ~70 weeks of age. This phenotype was not observed in the other mouse lines, *Zrch1* and *Npu Prnp$^{0/0}$* mice (Bueler et al., 1992; Manson et al., 1994). Further investigation on the *Rcm0 Prnp$^{0/0}$* mice lead to the discovery of a prion protein (PrP)-like protein named Doppel (Dpl) (Moore et al., 1999).

Doppel is an acronym derived from *downstream prion protein-like gene* (Mastrangelo and Westaway, 2001) because the mouse Dpl gene (*Prnd*) lies ~16 kb downstream of the mouse *Prnp* coding region (Moore et al., 1999). Unlike PrP, Dpl was suggested to be produced in the brains by an unusual mechanism involving exon-skipping and intergenic splicing and they are ectopically expressed (Li et al., 2000a). Although there was no nucleic acid homology between the two genes, the two pro teins do exhibit ~24% sequence identity (Silverman et al., 2000). This similarity in protein sequence and structure (Mo et al., 2001) lies within the C-terminal two-thirds of PrP (Moore et al., 1999). Increasing evidence has demonstrated a role for PrP in

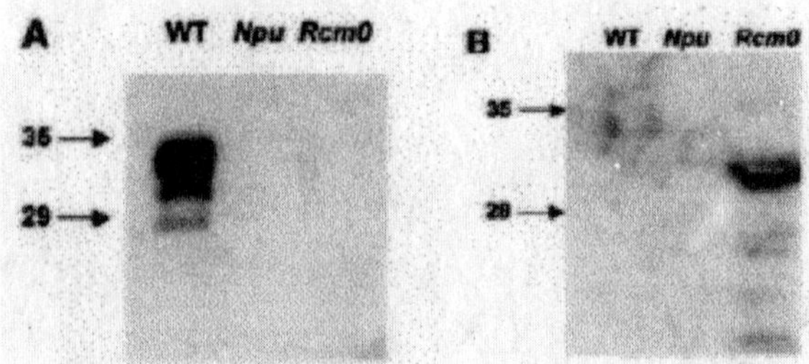

Figure 1. Expression of (A) prion and (B) doppel proteins, modified with permission from (Wong et al., 2001b). Molecular weight markers in kDa. Prion protein was detected using anti-PrP monoclonal 8H4 antibody (Li et al., 2000b), while doppel protein was detected using rabbit anti-Dpl antiserum (Wong et al., 2001b).

ameliorating oxidative stress *in vitro* (Brown et al., 1997; Brown et al., 1999) and *in vivo* (Wong et al., 2001a). This protective function is associated with the octarepeats along the N-terminal of PrP, a region that was missing on Dpl. Therefore, we proposed that Dpl could have an antithetical role in oxidative homeostasis that leads to the abnormalities observed during aging. We decided to test this hypothesis by performing a comparative analysis using two $Prnp^{0/0}$ mice of similar genetic background (129/Ola). $Rcm0$ $Prnp^{0/0}$ mice have been shown to express Dpl (Moore et al., 1999; Silverman et al., 2000), while Npu $Prnp^{0/0}$ mice do not display Dpl-associated Purkinje cell toxicity (Manson et al., 1994) (Fig. 1). Non-transgenic wild-type (WT) mice of similar genetic background which expresses only PrP were used as controls (Fig. 1).

2. HEME OXYGENASE EXPRESSION

Extensive studies have implicated the heme oxygenase (HO) system in the cellular defence against oxidative stress in the brain and therefore, have been widely adopted as a biological marker for oxidative impairment (Dwyer et al., 1998). Three HO isoforms (HO-1, HO-2 and HO-3) have been identified to date. Under normal physiological conditions, HO-1 is barely detected as observed in the brains of wild-type (WT) mice (Maines, 2000) shown in Fig. 2A. While in Npu $Prnp^{0/0}$ mice, a modest increase of ~35% in HO-1 expression was observed. However, the level of HO-1 expression was dramatically increased by ~150% in $Rcm0$ $Prnp^{0/0}$ mice (Fig. 2A). Unlike HO-1, HO-2 is the predominant isozyme constitutively expressed in the brains, and its expression was not significantly altered in the three mice (Fig. 2B).

3. NITRIC OXIDE SYNTHASE SYSTEM

There are increasing evidence indicating an intimate linkage between the HO system and nitric oxide (NO) (Baranano and Snyder, 2001; Maines, 2000). We

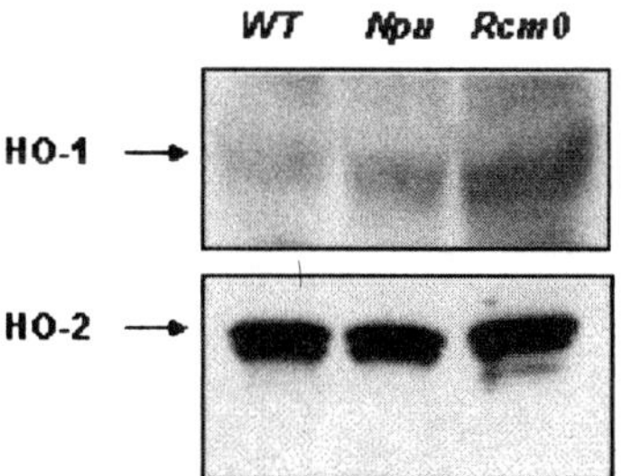

Figure 2. The expression of HO-1 and HO-2 in the brains of wild-type, *Npu Prnp^{0/0}* and *Rcm0 Prnp^{0/0}* mice. Modified with permission from (Wong et al., 2001b).

therefore decided to examine the expression levels of both neuronal nitric oxide synthase (nNOS) and inducible nitric oxide synthase (iNOS) in the brains of the three mice. nNOS is the major NOS isoform present in the brain, and as such, it probably accounts for the majority of the physiologic processes attributed to NO in the nervous system (Dawson and Dawson, 1996). In both WT and *Npu Prnp^{0/0}* mice, the level of nNOS is almost identical (Fig. 3), in line with an earlier observation on another non-Dpl expressing *Zurch1 Prnp^{0/0}* mice (Keshet et al., 1999). However, in the brain of the Dpl-expressing *Rcm0 Prnp^{0/0}* mice, nNOS expression was greatly enhanced (Fig. 3). Although nNOS was believed to be express constitutively, emerging evidence indicates that it can also be regulated by various pathological conditions, including cellular stress (Dawson and Dawson, 1996).

Another component of the NOS system is the iNOS. Under normal physiological conditions, it is not expressed in the nervous system, as demonstrated in both WT and *Npu Prnp^{0/0}* mice (Fig. 3). However, iNOS was activated in the brains of *Rcm0 Prnp^{0/0}* mice (Fig. 3). This is another indication of increased cellular stress in the brains of *Rcm0 Prnp^{0/0}* mice. Interestingly, iNOS activation has been implicated in several neurodegenerative diseases (Dawson and Dawson, 1996), including prion diseases (Ju et al., 1998).

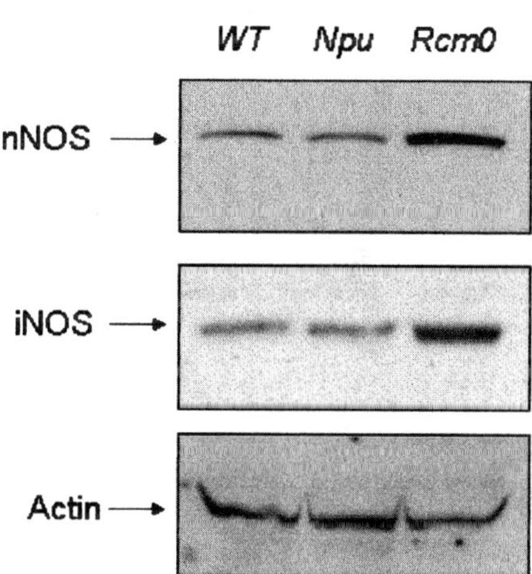

Figure 3. Detection of nNOS and iNOS in the brains of WT, *Npu* and *Rcm0 Prnp^{0/0}* mice. Modified with permission from (Wong et al., 2001b).

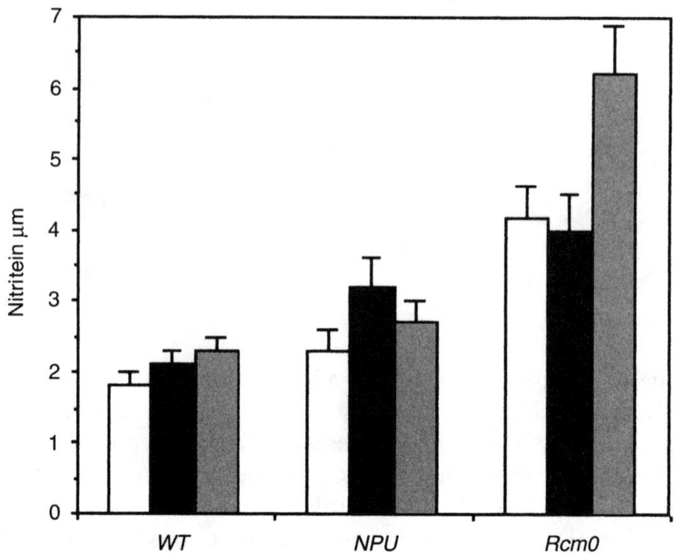

Figure 4. Detection of nitrite before (white bar) and after (black bar) incubation with hydrozxylammo-nium chloride (HAC). The levels of both nitrite and nitrate (grey bar) was determined by incubating the reaction mixture with nitrate reducatase which reduces nitrate to nitrite. Modified with permission from (Wong et al., 2001b).

Although nitric oxide is short-lived, they can react to produce stable nitrite and nitrate. These stable compounds can be assayed using the Griess reaction. The level of nitrite in the brains of *Npu Prnp$^{0/0}$* mice only increased modestly as compared to WT mice, while a dramatic elevation of nitrite was detected in the brains of *Rcm0 Prnp$^{0/0}$* mice (white bars in Fig. 4). To ascertain whether other significant oxidants are present in the brains, the reaction mixtures were pre-incubated with hydroxylammo-nium chloride (HAC) (black bars in Fig. 4), which will converts superoxide and other reactive oxygen intermediates to nitrite. In WT mice, the addition of HAC caused only a slight increase in nitrite levels, while a greater increase was observed in *Npu Prnp$^{0/0}$* mice, indicating an observable presence of other reactive oxygen intermedi-ates. In contrast, no observable increased in nitrite levels was observed after HAC treatment in *Rcm0 Prnp$^{0/0}$* mice, suggestive a lack of observable oxygen intermediates. In addition, the levels of both nitrite and nitrate were also determined by incubating the reaction mixture with nitrate reductase which reduces all nitrate to nitrite (grey bar in Fig. 4). Only *Rcm0 Prnp$^{0/0}$* mice experience significant increases. These results are a clear indication that the destructive oxidants in *Rcm0 Prnp$^{0/0}$* mice are predom-inantly NO-related.

4. PROTEIN CARBONYLATION AND NITROSYLATION

Although enhanced induction of both HO-1 and NOS have been observed, it is important to demonstrate whether these responses are a direct result of increased oxidative stress in the brains of *RcmO Prnp*[0/0] mice. To determine this, we used two established cellular markers; protein carbonylation (Butterfield and Kanski, 2001) and nitrosylation of tyrosine residues (Beckman, 1996).

Biological processes involving the generation of free radicals usually lead to oxidative post-translational modification of cellular proteins. An example is the introduction of site-specific reactive carbonyls groups, and the level of formation is taken to reflect the intensity of oxidative stress (Sayre et al., 1999). Total brain homogenates were derivatised with 2,4-dinitrophenylhydrazine (DNPH) followed by separation of the DNP-derivatised protein samples by SDS-PAGE. The separated proteins were electrotransferred and immunoblotted using a primary antibody against the DNP moiety attached to the protein. Only a modest increased in the level of carbonyl groups was detected in the derivatised brain homogenates from *Npu Prnp*[0/0] mice, while a dramatic elevation of more than 2 folds was observed in the brains of the *RcmO Prnp*[0/0] mice (Figure 5). These observations support our notion that a higher degree of oxidative stress was present in the nervous system of the *RcmO Prnp*[0/0] mice.

Similarly, the levels of nitrotyrosine formation were also examined by using an antibody specific for the nitrated-tyrosine in cellular proteins. This antibody predominantly react with brain proteins from the *RcmO Prnp*[0/0] mice (Fig. 6). In comparison, the levels in both wild-type and *Npu Prnp*[0/0] mice were almost undetectable. These

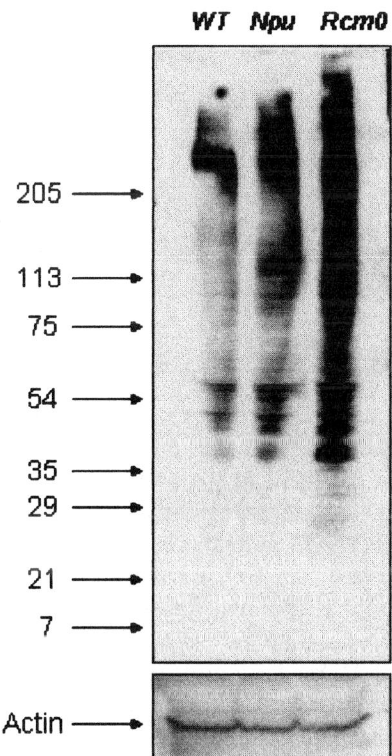

Figure 5. Differential levels of protein oxidation (reactive carbonyl groups) in the total brain homogenates of wild-type, *Npu* and *RcmO Prnp*[0/0] mice. Modified with permission from (Wong et al., 2001b).

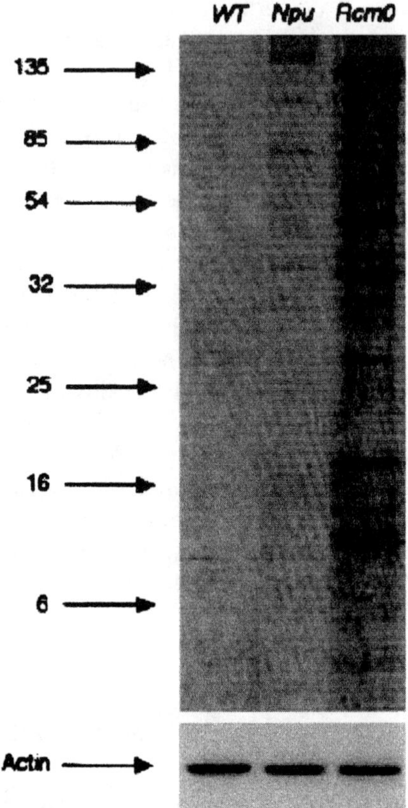

Figure 6. Detection of nitrotyrosine in total brain homogenates from wild-type, *Npu* and *Rcm0 Prnp*$^{0/0}$ mice. Modified with permission from (Wong et al., 2001b).

observations verified our earlier suggestion that the brains of *Rcm0 Prnp*$^{0/0}$ mice are subject to NO-mediated oxidative damage.

5. SUMMARY

In summary, this review has presented a correlation between Dpl expression and the induction of the HO/NOS system. This is likely a response to increase oxidative impairment in the brains of the Dpl-expressing *Rcm0 Prnp*$^{0/0}$ mice. In comparison, the increased oxidative stress in *Npu Prnp*$^{0/0}$ mice did not appear to involve the HO/NOS system. On the contrary, they are likely affected by destructive oxidants other than NO. Furthermore, the lack of severe abnormalities in *Npu Prnp*$^{0/0}$ mice implies the involvement of some compensatory mechanism, such as the ubiquitin-proteasome pathway (Wong et al., 2001a). Our results clearly demonstrate that PrP and Dpl have antithetical roles in modulating the cellular oxidative environment. Indeed, PrP is hardly detected when Dpl is expressed (Li et al., 2000a), suggesting that Dpl could be acting as an antagonist to PrP.

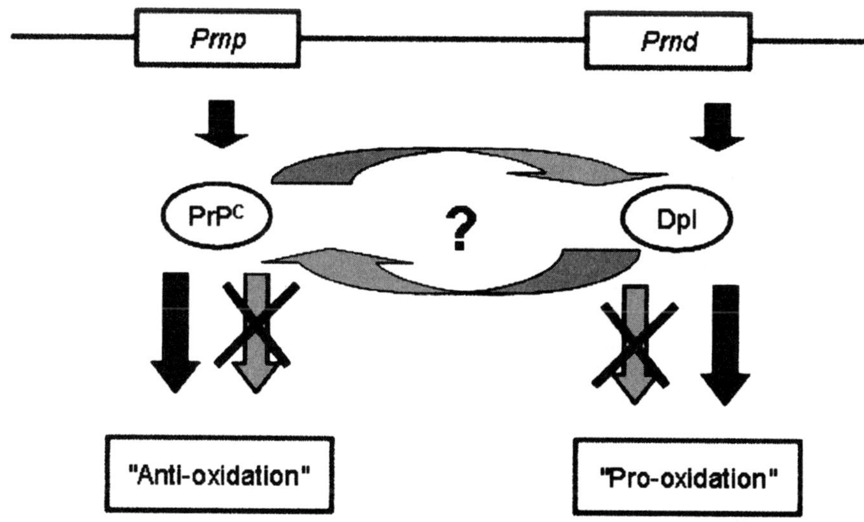

Figure 7. Physiological functions and possible "interaction" (?) associated with PrP and Dpl.

REFERENCES

Baranano, D.E. and Snyder, S.H. (2001) Neural roles for heme oxygenase: Contrasts to nitric oxide synthase. *Proc Natl Acad Sci USA*, **98**, 10996–11002.

Beckman, J.S. (1996) Oxidative damage and tyrosine nitration from peroxynitrite. *Chem Res Toxicol*, **9**, 836–844.

Brown, D.R., Schulz-Schaeffer, W.J., Schmidt, B., and Kretzschmar, H.A. (1997) Prion protein-deficient cells show altered response to oxidative stress due to decreased SOD-1 activity. *Exp Neurol*, **146**, 104–112.

Brown, D.R., Wong, B.S., Hafiz, F., Clive, C., Haswell, S., and Jones, I.M. (1999) Normal prion protein has an activity like that of superoxide dismutase. *Biochem J*, **344**, 1–5.

Bueler, H., Fischer, M., Lang, Y., Bluethmann, H., Lipp, H.P., DeArmond, S.J., Prusiner, S.B., Aguet, M., and Weissmann, C. (1992) Normal development and behaviour of mice lacking the neuronal cell-surface PrP protein. *Nature*, **356**, 577–582.

Butterfield, D.A. and Kanski, J. (2001) Brain protein oxidation in age-related neurodegenerative disorders that are associated with aggregated proteins. *Mech Ageing Dev*, **122**, 945–962.

Dawson, T.M. and Dawson, V.L. (1996) Nitric oxide synthase: role as a transmitter/mediator in the brain and endocrine system. *Annu Rev Med*, **47**, 219–227.

Dwyer, B.E., Lu, S.Y., and Nishimura, R.N. (1998) Heme oxygenase in the experimental ALS mouse. *Exp Neurol*, **150**, 206–212.

Ju, W.K., Park, K.J., Choi, E.K., Kim, J., Carp, R.I., Wisniewski, H.M., and Kim, Y.S. (1998) Expression of inducible nitric oxide synthase in the brains of scrapie-infected mice. *J Neurovirol*, **4**, 445–450.

Keshet, G.I., Ovadia, H., Taraboulos, A., and Gabizon, R. (1999) Scrapie-infected mice and PrP knock-out mice share abnormal localization and activity of neuronal nitric oxide synthase. *J Neurochem*, **72**, 1224–1231.

Li, A., Sakaguchi, S., Atarashi, R., Roy, B.C., Nakaoke, R., Arima, K., Okimura, N., Kopacek, J., and Shigematsu, K. (2000a) Identification of a novel gene encoding a PrP-like protein expressed as chimeric transcripts fused to PrP exon 1/2 in ataxic mouse line with a disrupted PrP gene. *Cell Mol Neurobiol*, **20**, 553–567.

Li, R., Liu, T., Wong, B.S., Pan, T., Morillas, M., Swietnicki, W., O'Rourke, K., Gambetti, P., Surewicz, W.K., and Sy, M.S. (2000b) Identification of an epitope in the C terminus of normal prion protein whose expression is modulated by binding events in the N terminus. *J Mol Biol*, **301**, 567–573.

Maines, M.D. (2000) The heme oxygenase system and its functions in the brain. *Cell Mol Biol (Noisy-Le-Grand)*, **46**, 573–585.

Manson, J.C., Clarke, A.R., Hooper, M.L., Aitchison, L., McConnell, I., and Hope, J. (1994) 129/Ola mice carrying a null mutation in PrP that abolishes mRNA production are developmentally normal. *Mol Neurobiol*, **8**, 121–127.

Mastrangelo, P. and Westaway, D. (2001) The prion gene complex encoding PrP(C) and Doppel: insights from mutational analysis. *Gene*, **275**, 1–18.

Mo, H., Moore, R.C., Cohen, F.E., Westaway, D., Prusiner, S.B., Wright, P.E., and Dyson, H.J. (2001) Two different neurodegenerative diseases caused by proteins with similar structures. *Proc Natl Acad Sci USA*, **98**, 2352–2357.

Moore, R.C., Lee, I.Y., Silverman, G.L., Harrison, P.M., Strome, R., Heinrich, C., Karunaratne, A., Pasternak, S.H., Chishti, M.A., Liang, Y., Mastrangelo, P., Wang, K., Smit, A.F., Katamine, S., Carlson, G.A., Cohen, F.E., Prusiner, S.B., Melton, D.W., Tremblay, P., Hood, L.E., and Westaway, D. (1999) Ataxia in prion protein (PrP)-deficient mice is associated with upregulation of the novel PrP-like protein doppel. *J Mol Biol*, **292**, 797–817.

Moore, R.C., Redhead, N.J., Selfridge, J., Hope, J., Manson, J.C., and Melton, D.W. (1995) Double replacement gene targeting for the production of a series of mouse strains with different prion protein gene alterations. *Biotechnology (NY)*, **13**, 999–1004.

Sakaguchi, S., Katamine, S., Nishida, N., Moriuchi, R., Shigematsu, K., Sugimoto, T., Nakatani, A., Kataoka, Y., Houtani, T., Shirabe, S., Okada, H., Hasegawa, S., Miyamoto, T., and Noda, T. (1996) Loss of cerebellar Purkinje cells in aged mice homozygous for a disrupted PrP gene. *Nature*, **380**, 528–531.

Sayre, L.M., Perry, G., and Smith, M.A. (1999) In situ methods for detection and localization of markers of oxidative stress: application in neurodegenerative disorders. *Methods Enzymol*, **309**, 133–152.

Silverman, G.L., Qin, K., Moore, R.C., Yang, Y., Mastrangelo, P., Tremblay, P., Prusiner, S.B., Cohen, F.E., and Westaway, D. (2000) Doppel is an N-glycosylated, GPI-anchored protein: expression in testis and ectopic production in the brains of Prnpo/o mice predisposed to Purkinje cell loss. *J Biol Chem*, **275**, 26834–26841.

Wong, B.S., Liu, T., Li, R.L., Pan, T., Petersen, R.B., Smith, M.A., Gambetti, P., Perry, G., Manson, J.C., Brown, D.R., and Sy, M.S. (2001a) Increased levels of oxidative stress markers detected in the brains of prion knock-out mice. *J Neurochem*, **76**, 565–572.

Wong, B.S., Liu, T., Paisley, D., Li, R.L., Pan, T., Gunaratne, R.S., Chen, S.G., Perry, G., Petersen, R.B., Smith, M.A., Melton, D.W., Gambetti, P., Brown, D.R., and Sy, M.S. (2001b) Induction of HO-1 and NOS in doppel-expressing mice devoid of prion protein: Implications for doppel function. *Mol Cell Neurosci*, **17**, 768–775.

38

GENERATION OF NITRIC OXIDE AND CARBON MONOXIDE PROVIDE PROTECTION AGAINST CARDIAC ANAPHYLAXIS

Alfredo Vannacci, Cosimo Marzocca, Giovanni Zagli,
Simone Pierpaoli, Daniele Bani[a], Emanuela Masini,
and Pier Francesco Mannaioni

Department of Preclinical and Clinical Pharmacology
[a]Department of Anatomy, Histology and Forensic Medicine
University of Florence
Viale G. Pieraccini n° 6, 50139 Florence, Italy

INTRODUCTION

Cardiac anaphylaxis has been described as the increase in rate and strength of contraction and the onset of arrhythmias in isolated heart preparations from sensitised animals challenged *in vitro* with specific antigen. These changes in myocardial functions were explained by the release of histamine as the sole mediator,[1] or by the combination of histamine release with the production of vasoactive products from the arachidonic acid cascade.[2] Cardiac anaphylaxis is widely recognised as an example of type I hypersensitivity in which the release of histamine participates in myocardial damage.[2] Endogenous autacoids modulate cardiac anaphylaxis. In the isolated guinea pig heart, catecholamine levels and the availability of catecholamine receptors regulate the release of histamine from sensitised hearts: depletion of catecholamines and the blockade of β-receptors decrease anaphylactic histamine release, which is enhanced by noradrenaline.[1] Histamine down-regulates the anaphylactic release of histamine from sensitised guinea pig hearts: H_2-receptor agonists decrease the amount of histamine released, whereas cimetidine increases it.[3] It has been reported that nitric oxide (NO) modulates cardiac anaphylaxis.[4] The NO donor, sodium nitroprusside, decreases anaphylactic histamine release from sensitised guinea pig hearts, which is increased by blocking nitric oxide synthase.[4] Relaxin (RLX) is a peptide hormone secreted mainly by the corpus luteum during pregnancy,[5] with well established effects on the female reproductive organs. Cardiomyocytes from rat atria have been shown to secrete detectable amounts of RLX suggesting the cardiovascular system as a physiological target.[6] In fact, RLX has a powerful vasodilatatory action on rat uterus and mesocaecum, mouse mammary gland and pigeon crop sac,[7] and decreases blood pressure in spontaneous hypertensive rats.[8] In some vascular districts, such as the coronary arteries of the guinea pig, the RLX induced-increase in coronary flow is accounted for by the generation of nitric oxide (NO).[9] Sodium nitroprusside and glyceryl trinitrate, down regulate the immunological and non-immunological release of histamine from sensitised tissues of the guinea pig.[10] RLX shares, with the NO donors the ability to down regulate the immunological and non-immunological release of histamine from isolated rat and guinea pig mast cells.[11] A novel gaseous mediator, carbon monoxide (CO), also has anti-anaphylactic actions in guinea pig mast cells from actively sensitised animals.[12] In this chapter we report the effect of relaxin (a NO generator) and hemin (an inducer of HO-1) on cardiac anaphylaxis.

Guinea pig hearts, taken 18–25 days after sensitisation by intraperitoneal and subcutaneous injections of egg albumin (1 ml of 1% solution), were perfused with Tyrode's solution at 37°C in a Langendorff apparatus and challenged with egg albumin (0.1 ml of 1% solution into the aortic cannula). Records were taken of the strength of contraction, rate and coronary outflow. Control animals were injected intraperitoneally with saline and treated animals were injected with hemin (4 mg/kg i.p.) 18 h before antigen challenge. RLX was added to the perfusion fluid 60 min before antigen challenge. Histamine released in the perfusates and retained in the heart was detected fluorimetrically. In cardiac homogenates, cyclic GMP levels were detected by radioimmunoassay using ^{125}I-labelled cyclic GMP. Tissue Ca^{2+} levels were determined by means of atomic absorption and heme oxygenase (HO) activity by spectrophotometric assay of bilirubin in the presence of biliverdin reductase.[12]

EFFECT OF RELAXIN ON CARDIAC ANAPHYLAXIS

We have previously shown that RLX increases both the coronary flow[9] in the isolated heart of the guinea pig and the amount of nitrite that appears in the perfusates. Both effects were significantly reduced by pretreatment with the nitric oxide synthase inhibitor N-monomethyl-L-arginine, and were comparable with those obtained by the endothelium-dependent vasodilator acetylcholine as well as by the endothelium independent vasodilator, sodium nitroprusside.[9] The present experiments show that in the presence of RLX the pathophysiological responses to antigen challenge were fully abated (Fig. 1, panel a). Consistent eg with these observations, the release of histamine induced by antigen was also diminished in the presence of RLX (Fig. 1, panel b).

Moreover, RLX significantly increased cGMP levels and reduced the antigen-induced increase in cardiac calcium. It should be noted that in the presence of RLX, antigen challenge evoked a less sustained coronary constriction and a more pronounced coronary dilatation. The results of these experiments show that RLX, at a concentration as low as that measured in plasma under physiological conditions, provides protection against cardiac anaphylaxis, in that the responses to antigen are

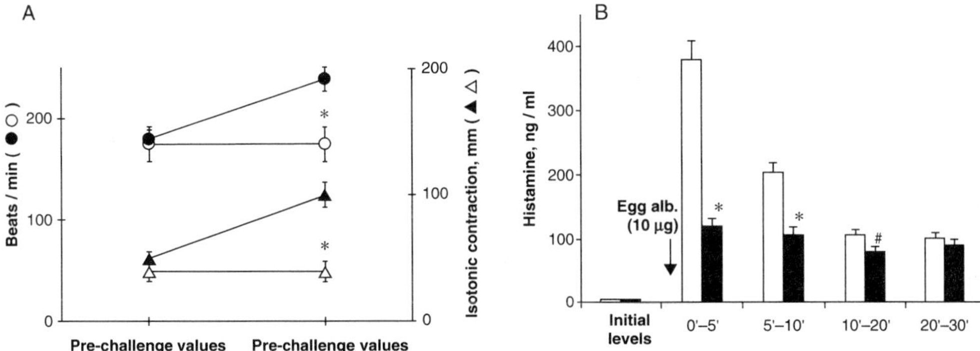

Figure 1. *Panel a:* Effect of Relaxin (RLX, 30 ng/ml) on cardiac anaphylaxis in vitro: control (● π); RLX (ρ ○). *Panel b:* Histamine levels in the perfusates after antigen challenge in either the absence (unshaded columns) or presence of RLX (30 ng/ml) (shaded columns). Values are means ± SEM of 8 experiments. *p < 0.001, #p < 0.05.

fully abated and the release of histamine substantially reduced. These effects are concomitant with the increase in cardiac cGMP and with the corresponding decline in the immunological rise of cardiac calcium. It is conceivable that the protection afforded by RLX on cardiac anaphylaxis could be accounted for by the generation of NO. In fact, the immunological and non-immunological release of histamine from isolated mast cells is similarly reduced by RLX in a NO-dependent fashion. The inhibition of mast cell histamine release by RLX is consistent with the formation of NO, which actually occurs from rat mast cells in the presence of RLX,[11] as well as from isolated guinea pig hearts perfused with RLX.[9] Likewise, the protection afforded by RLX on cardiac anaphylaxis could be related to the generation of NO induced by RLX from endothelial cells, from vascular smooth muscle cells and from perivascular mast cells. The readily diffusable NO could, in turn, down regulate the release of histamine from cardiac mast cells by increasing cGMP and by decreasing calcium levels in a critical way for calcium-dependent granule exocytosis.

EFFECT OF HEMIN ON CARDIAC ANAPHYLAXIS

Guinea pig hearts possess mainly the inducible isoform of HO. Minute amounts of bilirubin, the end product of the HO-catalysed metabolism of heme, are present in cardiac homo-genates of control animals (Fig. 2 panel a). In hemin treated animals, the amount of bilirubin increased significantly (Fig. 2 panel a) showing the presence of the inducible isoform HO-1. Treatment with hemin significantly increases cardiac cGMP levels and reduces the antigen-induced increase in cardiac $(Ca^{2+})_i$. Under these circumstances, the release of histamine induced by antigen is diminished in hemin-pre-treated animals (Table 1). Consistently, the pathophysiological responses to antigen-challenge are fully abated (Fig. 2 panel b).

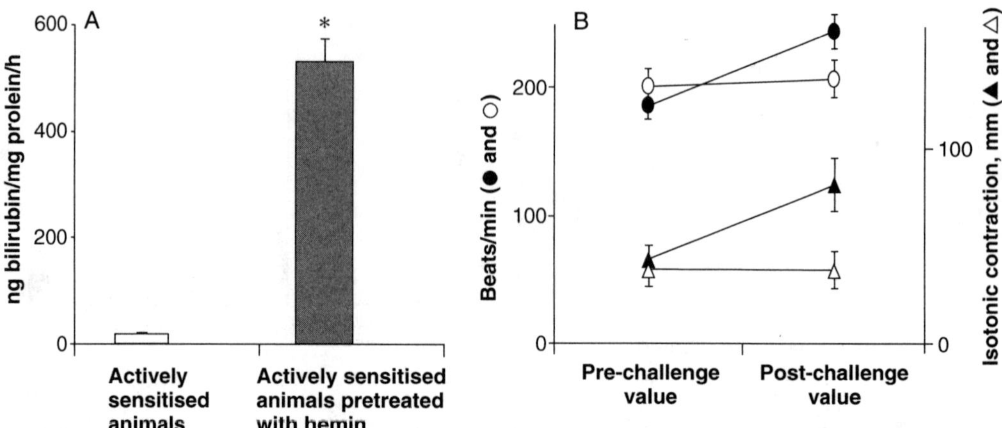

Figure 2. Panel a: Heme oxygenase activity measured as bilirubin production in cardiac homogenates from control animals (n = 4) and from hemin treated animals (n = 6). Panel b: Effect of hemin on cardiac anaphylaxis in vitro: Control (● π), hemin (4 mg/Kg i.p. 18 h before antigen challenge) (ρ ○). Values are means ± ES of 6 experiments. *p < 0.01.

It should be noted that in hemin-pretreated animals, antigen challenge evokes a less sustained coronary constriction and a more pronounced coronary dilatation than in saline-pretreated animals (data not shown). The present experiments show that pretreatment with hemin, an HO-1 inducer, provides protection against cardiac anaphylaxis, in that the responses to antigen are fully abated and the release of histamine significantly reduced. We cannot rule out that the anti-anaphylactic action of hemin is accounted for by the blockade of H1/H2 receptors, or to the direct inhibition of the secretion of histamine from repository cells. However it is possible to hypothesise that the protective effect stems from the hemin-induced increase in HO-1 activity and from the end products of the HO-1 activity, bilirubin and CO. The beneficial role of bilirubin as a physiological antioxidant is widely accepted.[13] However, the net increase of cGMP levels and the corresponding decline in the immunological rise of $[Ca^{2+}]_i$ suggest that hemin actually induces HO-1 activity, increases the generation of CO which enhances the cGMP levels via a guanylyl cyclase dependent mechanism, a

Table 1. Effect of hemin pretreatment and antigen challenge on histamine release, cGMP and total calcium in sensitised guinea pig hearts.

	Number of experiments	Control	Hemin
Histamine release (*)	8	450 ± 73.7	214 ± 27.5
cGMP (**)	6	130 ± 12.7	390 ± 11.6
Total calcium (***)	6	1,250 ± 125.4	398 ± 61.8

(*): Histamine release (ng/ml) in perfusates collected for 10 min. after antigen challenge.
(**): Cardiac cGMP levels (fmoles/mg protein) after antigen challenge.
(***): Cardiac $[Ca^{2+}]_i$ (ng/mg tissue w. wt) after antigen challenge.

common pathway shared with NO.[14] Thus, CO may provide protection against anaphylaxis as it does against oxidative stress.[15]

ACKNOWLEDGMENTS

These investigations were supported by grants from MURST and University of Florence, Italy.

REFERENCES

1. Giotti A., Guidotti A., Mannaioni P.F., and Zilletti L.: The influence of andrenotropic drugs and noradrenaline on the histamine release in cardiac anaphylaxis in vitro. J Physiol Lond 184:924–941, 1966.
2. Capurro N. and Levi R.: The heart as a target organ in systemic allergic reactions: comparison of cardiac anaphylaxis in vivo and in vitro. Circ Res 36:520–528, 1975.
3. Blandina P., Brunelleschi S., Fantozzi R., Giannella E., Mannaioni P.F., and Masini E.: The antianaphylactic action of histamine H2-receptor agonists in the guinea-pig isolated heart. Br J Pharmacol 90:459–466, 1987.
4. Masini E., Pistelli A., Gambassi F., Di Bello M.G., and Mannaioni P.F.: The role of nitric oxide in anaphylactic reaction of isolated guinea pig hearts and mast cells. In: Moncada S., Nisticò G., and Higgs E.A., editors, Nitric Oxide: Brain and Immune System. London and Chapel Hill, Portland Press Ltd., 277–287, 1994.
5. Bryant-Greenwood G.D.: Relaxin as a new hormone. Endocr Rev 3:62–90, 1992.
6. Taylor M.J. and Clark C.L.: Evidence for a novel source of relaxin: atrial cardiocytes. J Endocrinol 143:R5–R8, 1994.
7. Bigazzi M., Bani D., Bani G., and Bani-Sacchi T.: Relaxin and the cardio-circulatory system. In: MacLennan A.H., Tregear R.G., and Bryant-Greenwood, Eds. Progress in Relaxin Research. Singapore, World Scientific Publishing Co., 449–507, 1995.
8. Massicott G., Parent A., and St Louis J.: Blunted response to vasoconstrictors in mesenteric vasculature but not in portal vein of spontaneously hypersensitive rats treated with relaxin. Proc Soc Exp Biol Med, 190:254–259, 1988.
9. Bani Sacchi T., Bigazzi M., Bani D., Mannaioni P.F., and Masini E.: Relaxin-induced increased coronary flow through stimulation of nitric oxide production. Br J Pharmacol, 116:1589–1594, 1995.
10. Masini E., Pistelli A., Gambassi F., Di Bello M.G., and Mannaioni P.F.: The role of nitric oxide in anaphylactic reaction of isolated guinea pig hearts and mast cells. Nitric Oxide: Brain and Immune System. London and Chapel Hill, Moncada S., Nisticò G., and Higgs E.A., Eds., Portland Press Ltd, 277–287, 1994.
11. Masini E., Bani D., Bigazzi M., Mannaioni P.F., and Bani Sacchi T.: Effects of relaxin on mast cells. In vitro and in vivo studies in rats and guinea pigs. J Clin Invest, 94:1974–1980, 1994.
12. Ndisang J.F., Gai P., Berni L., Mirabella C., Baronti R., Mannaioni P.F., and Masini E.: Modulation of the immunological response of guinea pig mast cells by carbon monoxide. Immunopharmacology, 43:65–73, 1999.
13. Stocker R., Yamamoto Y., McDonagh A.F., Glazer A.N., and Ames B.N.: Bilirubin is an antioxidant of possible physiological importance. Science 235:1043–1046, 1987.
14. Deinum G., Stone J.R., Babcock G.T., and Marletta M.A.: Binding of nitric oxide and carbon monoxide to soluble guanylate cyclase as observed with Resonance Raman spectroscopy. Biochemistry 35:1540–1547, 1996.
15. Otterbein L.O., Mantell L.L., and Choi A.M.K.: Carbon monoxide provides protection against hyperoxic lung injury. Lung Cellular and Mole-cular Physiol 276:4 L688 L694, 1999.

THE NETWORK OF HEME OXYGENASE AND PROGRAM CELL GROWTH AND DEATH

39

HEME OXYGENASE-1 INHIBITS VASCULAR SMOOTH MUSCLE CELL PROLIFERATION

Xiao-Ming Liu, MD, Kelly J. Peyton, MA,
and William Durante, PhD

VA Medical Center
Building 109, Room 130
2002 Holcombe Blvd
Houston, TX 77030 and
Departments of Medicine and Pharmacology
Baylor College of Medicine
Houston, TX 77030.

ABSTRACT

In the present study, we overexpressed heme oxygenase-1 (HO-1) in rat aortic smooth muscle cells (SMC) by generating a recombinant defective adenovirus containing the rat HO-1 gene (AdHO-1) and examined the effect on SMC proliferation. Infection of SMC with AdHO-1 resulted in a dose-dependent increase in the expression of HO-1 protein. AdHO-1 inhibited growth factor-stimulated SMC proliferation and DNA synthesis. In contrast, the control adenovirus expressing the green fluorescent protein (AdGFP) failed to induce HO-1 expression and had minimal effects on SMC growth. In addition, the exogenous administration of carbon monoxide (CO) mimicked the antiproliferative effect of AdHO-1, whereas biliverdin and iron had no effect. These results demonstrate that HO-1 inhibits vascular SMC proliferation via the production of CO. Adenovirus-mediated transfer of the HO-1 gene may provide an important therapeutic approach in treating fibroproliferative disorders of the vessel wall.

INTRODUCTION

Heme oxygenase-1 (HO-1) catalyzes the degradation of heme into equal molar amounts of biliverdin, iron, and carbon monoxide (CO).[1] HO-1 is expressed in arterial and venous blood vessels and has been localized to both vascular endothe-lium and smooth muscle.[2] HO-1 can be induced in vascular cells by physiologically and pathophysiologically relevant stimuli. In an early report, Morita et al.[3] demon-strated that hypoxia is a potent inducer of HO-1 gene expression in vascular smooth muscle cells (SMC). In addition, our laboratory found that hemin, sodium arsenite, and heavy metals are also inducers of SMC HO-1 expression.[4] Subsequent studies indicated that inflammatory cytokines, endotoxin, growth factors, cyclic nucleotides, oxidized low density lipoproteins, and nitric oxide are all capable of stimulating vas-cular cell HO-1 gene expression.[2,5] In addition to biochemical stimuli, specific biophy-sical forces such as fluid shear stress and cyclic strain induce the expression of HO-1.[6]

Recent studies indicate that HO-1 plays an important role in the circulation. Induction of HO-1 causes a marked decrease in blood pressure in hypertensive rats, whereas HO-1 inhibition increases blood pressure, suggesting a critical vasoregula-tory role for HO-1.[7–9] HO-1 also modulates platelet vessel wall interactions. Utilizing a platelet:SMC co-incubation system, we found that the induction of HO-1 in SMC markedly attenuates platelet aggregation, indicating a potentially important antithrombotic role for this enzyme.[6] In addition, HO-1 provides significant cytopro-tective effects in the vasculature. The induction of HO-1 in vascular cells leads to increased resistance to oxidative stress, whereas HO-1 deficiency results in enhanced cell injury.[10,11] While the HO-1-mediated inhibitory effect on vascular tone and platelet aggregation appears to be mediated via the release of CO, the antioxidant property of HO-1 is likely due to the formation of biliverdin and its subsequent metabolism to bilirubin by the enzyme biliverdin reductase.[2,7–11]

The recent observation that growth factors are potent inducers of HO-1 gene expression raises the possibility that HO-1 may also regulate SMC growth.[12–14] Indeed, preliminary studies in our laboratory found that HO-1 inhibitors potentiate platelet-derived growth factor-stimulated DNA synthesis in SMC.[15] Interestingly, the mito-genic action of angiotensin II, endothelin, and hypoxia is also augmented by HO-1 inhibition.[16,17] These results suggest that HO-1 may function in a negative feedback fashion to block SMC proliferation. However, inducers and inhibitors of HO-1 gene expression and activity may have actions that are not specific for HO-1,[18–20] poten-tially complicating the interpretation of results of experiments using these com-pounds. In order to circumvent these problems, we generated a recombinant defective adenovirus containing the HO-1 gene to overexpress HO-1 in vascular SMC. We now report that overexpression of HO-1 inhibits vascular SMC proliferation and DNA synthesis via the generation of CO. Thus, the HO-1/CO system represents a poten-tially new therapeutic target for treating vascular occlusive disorders.

METHODS

Materials

Collagenase, elastase, serum, antibiotics, sodium dodecyl sulfate (SDS), acry-lamide, Tween 20, Tes, Hepes, trypsin, and EDTA were from Sigma Chemical Co. (St. Louis, MO); a polyclonal HO-1 antibody was from StressGen (Victoria, Canada); the adenoviral vectors pAdTrack-CMV and pAdEasy-1 were kindly provided by Dr. Bert Vogelstein (John Hopkins Oncology Center, Baltimore, MD); human α-thrombin was from US Biochemicals (Cleveland, OH); [^{3}H]thymidine (85 Ci/mmol) was from NEN-Dupont (Boston, MA).

Cell Culture

SMC were isolated by elastase and collagenase digestion of rat thoracic aorta and were characterized according to morphological and immunological criteria.[21] Cells were serially cultured in minimum essential media containing 10% serum, Earle's balanced salts, 5.6 mM glucose, 2 mM glutamine, 20 mM Tes, 20 mM Hepes, and 100 Units/ml of penicillin, streptomycin, and neomycin.

Construction of Recombinant HO-1 Adenovirus and Gene Transfer

HO-1 cDNA was cloned and amplified by polymerase chain reaction (PCR) from reverse-transcribed rat aortic SMC mRNA. Oligonucleotide primers (sense 5'-CAGTCGCCTCCAGAGTTCC-3'; antisense, 5'-GAGAGCCAGGCAAGATTCTC-3') were designed spanning the translation initiation and termination codons.[22] The PCR product was ligated into pPCRII (Invitrogen, San Diego, CA) and its nucleotide sequence determined to ensure that no PCR-induced mutations were present. The *Hind III/Xba I* restriction fragment of HO-1 was ligated into the *Hind III/Xba I* site of the shuttle vector pAdTrack-CMV. The resultant plasmid was linearized with *Pme I* and co-transformed with the adenoviral backbone vector pAdEasy-1 into electro-competent *E Coli* BJ5183 cells. Homologous recombinants containing HO-1 cDNA (AdHO-1) were detected by restriction endonuclease digestion and agarose gel electrophoresis, and transformed into *E coli* DH10B cells for large scale amplification. The *Pac-I*-digested pAdHO-1 was transfected into mammalian 293 adenoviral packaging cells and the AdHO-1 adenovirus was expanded, purified, and titered, as previously described.[23] The recombinant adenovirus encoding the green fluorescent protein (AdGFP) was used as a control. This adenovirus was similar to AdHO-1 but lacked the HO-1 cDNA.

For adenoviral infection, subconfluent SMC were incubated with AdHO-1 or AdGFP at various multiplicities of infection (MOI) in serum-free media. Following one hour of incubation, media were removed and cells incubated in appropriate fresh media for various periods of time.

HO-1 Protein Analysis

SMC were lysed in electrophoresis buffer [125 mM Tris-HCl (pH 6.8), 12.5% glycerol, and 2% SDS], boiled, and sonicated. The lysate was centrifuged at 13,000 × g for 10 minutes at 4 °C, the supernatant collected and SDS-polyacrylamide gel electrophoresis was performed using 20 µg of protein. The separated blots were electrophoretically transferred to nitrocellulose membranes and blocked for one hour in PBS containing Tween 20 (0.1%) and non-fat milk (5%). Blots were then incubated with the HO-1 (1 : 500 dilution) antibody for one hour. Membranes were then washed in PBS and incubated for one hour with horseradish peroxidase conjugated goat anti-rabbit antibody (1:7,500 dilution). After further washing with PBS, blots were developed using the ECL method (Amersham, Arlington Heights, IL). Relative protein levels were quantified by scanning densitometry (LKB Ultrascan XL laser densitometer, Bromma, Sweden).

SMC Proliferation and DNA Synthesis

For proliferation studies, SMC were seeded at a density of 5×10^4 cells/well in 12-well plates and incubated in serum-free media for 48 hours. SMC were incubated in culture media containing 0 or 5% serum for 4 days. SMC were then harvested with trypsin (0.025%)/EDTA (1 mM) and counted in a calibrated Coulter Counter (model ZF, Coulter Electronics, Hileah, FL). Cell number determinations were corroborated by hemocytometer counting. For DNA synthesis, near confluent SMC were incubated in serum-free media and then treated with serum (5%) or thrombin (1 U/ml). Four hours prior to the termination of the experiment, [^{3}H]thymidine (1 μCi/ml) was added to the cells. SMC were then washed with ice-cold PBS, fixed with 10% trichloroacetic acid for 30 minutes at 4°C, and DNA extracted with 0.2% SDS/0.2 N NaOH. Radioactivity was then determined by liquid scintillation spectrophotometry (Tricarb liquid scintillation analzyer, Model 1900, Packard, Meriden, CT).

CO Exposure

SMC were exposed to CO via a previously described exposure chamber.[24] CO at a concentration of 1% in air was mixed with air containing 5% CO_2 in a stainless steel mixing cylinder prior to delivery into the exposure chamber. Flow into the humidified chamber was at 1 L/minute and the temperature was maintained at 37°C. CO levels in the chamber were continuously monitored by electrochemical detection.[24]

Statistics

Results are expressed as the means ± SEM. Statistical analysis was performed with the use of a Student's two-tailed t-test and *p* values less than 0.05 were considered statistically significant.

RESULTS AND DISCUSSION

Treatment of vascular SMC with AdHO-1 stimulates a dose-dependent increase in HO-1 protein (Fig. 1). Incubation of SMC with AdHO-1 (20 MOI) results in a nearly 60-fold increase in HO-1 protein (Fig. 1). Increased HO-1 exposure following exposure to AdHO-1 is accompanied by an increase in HO-1 activity, as reflected by the rise in bilirubin synthesis (data not shown). In contrast, AdGFP fails to induce the expression of HO-1 (Fig. 1) or the production of bilirubin (data not shown).

Treatment of vascular SMC with serum (5%) for 4 days results in a marked increase in the number of SMC (Fig. 2A). However, infection of rat aortic SMC with

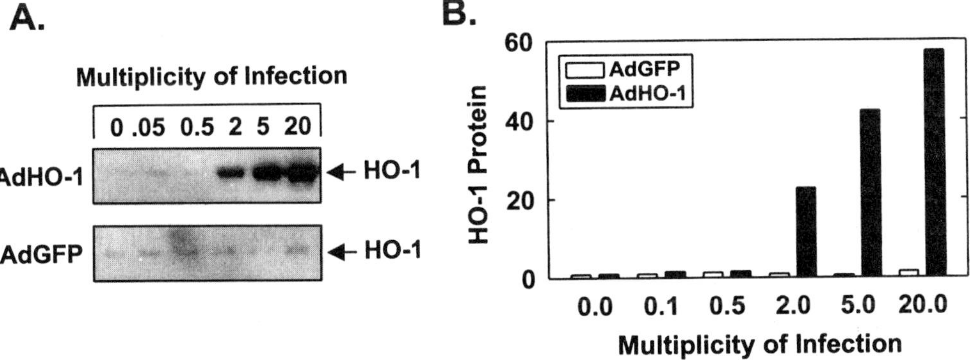

Figure 1. Effect of AdHO-1 infection on the expression of HO-1 protein in vascular SMC. (A) Western blot of HO-1 protein after 48 hours of infection with AdHO-1 or AdGFP. (B) Quantitation of HO-1 protein expression by laser densitometry.

AdHO-1 (20 MOI) inhibits the proliferative response to serum (Fig. 2A). In addition, infection of SMC with AdHO-1 (20 MOI) blocks serum-stimulated DNA synthesis (Fig. 2B). In contrast, the control adenovirus AdGFP (20 MOI) minimally affects SMC number or DNA synthesis (Fig. 2B).

Our finding that adenoviral-mediated HO-1 gene transfer inhibits rat aortic SMC proliferation is consistent with the recent report by Duckers et al.,[25] showing that overexpression of HO-1 inhibits the proliferation of porcine aortic SMC. This study also found that inhibition of SMC growth is dependent on HO-1 activity, since it can be reversed by zinc protoporphyrin-IX. Moreover, inhibition of soluble

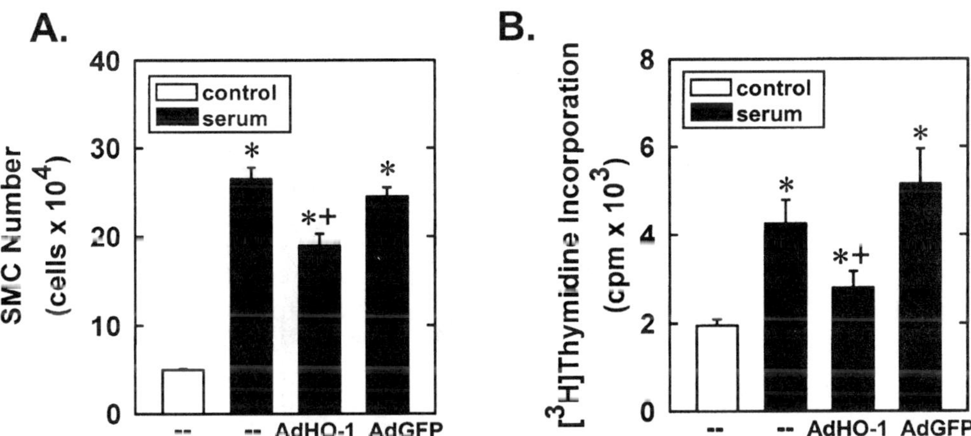

Figure 2. Effect of AdHO-1 on vascular SMC proliferation (A) and DNA synthesis (B). *SMC were treated with serum (5%) in the presence of AdHO-1 (20 MOI) or AdGFP (20 MOI) for 4 days (A) or 24 hours (B). Statistically significant effect of serum. +Statistically significant effect of AdHO-1.

guanylate cyclase by 1H-[1,2,4]oxadiazolo[4,3a]quinoxaline-1-one (ODQ) or by the cGMP antagonist, Rp-8-BrcGMP, was observed to restore DNA synthesis to control levels in SMC overexpressing HO-1. These results suggest that HO-1 attenuates SMC proliferation through the CO-mediated activation of the cGMP signaling pathway. Flow cytometric analysis further revealed that HO-1 arrests SMC in the G_1/G_0 phase of the cell cycle.[24] The cell cycle arrest appears to be mediated, in part, by the cyclin-dependent kinase inhibitor, p21. Both growth inhibition and cell cycle arrest are associated with the induction of p21 expression. In addition, the antiproliferative action of HO-1 is significantly reduced in SMC obtained from p21 null mice. Furthermore, SMC obtained from HO-1 knockout animals exhibit enhanced growth and DNA synthesis with a corresponding reduction in p21 levels compared with wild-type mice, further corroborating a link between HO-1 and p21 in growth regulation.[25]

To directly confirm that CO exerts an antiproliferative effect, vascular SMC were exposed to CO via a previously described environmental chamber.[24] Exposure of vascular SMC to CO (100 ppm) attenuates thrombin-stimulated DNA synthesis (Fig. 3). This concentration of exogeneous CO is estimated to be comparable to that produced by HO-1 activity in cultured vascular cells,[24,26] indicating that physiologically relevant levels of CO are able to inhibit vascular SMC growth. In contrast, the exogenous administration of biliverdin or iron failed to block SMC proliferation (data not shown). These results further support the hypothesis that CO is the HO-1 product responsible for inhibiting SMC growth.

Recent studies suggest that HO-1 is an important modulator of the vascular response to injury. Utilizing the rat carotid artery model of balloon angioplasty, we demonstrated that two-day pretreatment of rats with the HO-1 inducer, hemin, followed by daily injections of hemin for 14 days post-injury results in a significantly diminished development of neointima. Neointimal area and neointimal thickness is

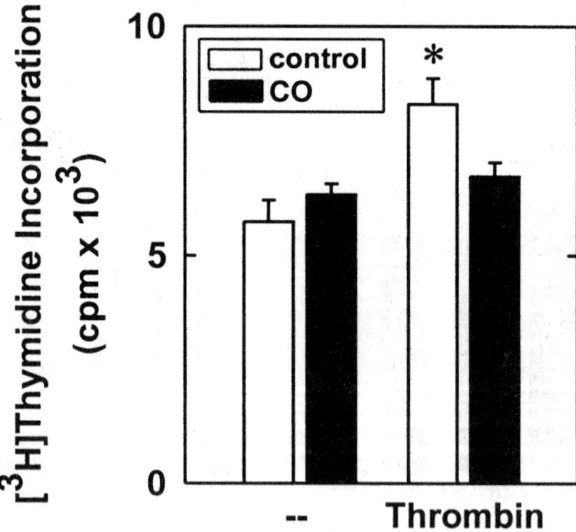

Figure 3. Effect of CO on vascular SMC DNA synthesis. SMC were treated with thrombin (1 U/mL) for 48 hours in the presence or absence of CO (100 ppm). *Statistically significant effect of thrombin.

reduced by nearly 60% while the ratio of neointimal area to medial wall area is decreased by 40%.[27] Concomitant treatment with SnPP-IX, however, completely restores the neointima back to control levels, demonstrating that HO-1 activity is required for this response.[27] More recently, we found that adenovirus-mediated HO-1 gene transfer to carotid arteries immediately following vascular injury also inhibits neointima formation.[28] The diminution in neointimal thickening is associated with a significant decrease in PCNA-labeling of SMC in the arteries exposed to AdHO-1.[28] Interestingly, a significant increase in TUNEL-labeling is also observed in AdHO-1-treated arteries.[28] This latter finding is consistent with preliminary studies in our laboratory and others demonstrating that HO-1 overexpression induces apoptosis in vascular SMC.[29,30] These results suggest that HO-1 may limit the accumulation of SMC in the intima by inhibiting their proliferation and stimulating their death. Our demonstration that HO-1 ameliorates neointimal development following balloon angioplasty is consistent with other studies[17,25,31] and further supports the hypothesis that HO-1 is an important pathophysiologic determinant of post-interventional stenosis and remodeling.

In conclusion, the present study demonstrates that adenovirus-mediated transfer of HO-1 inhibits proliferation in vascular SMC by generating CO. Genetic approaches targeting HO-1 to the vessel wall may provide an effective strategy in alleviating vascular disease.

ACKNOWLEDGMENTS

This study was supported by the National Heart, Lung, and Blood Institute grant HL59976 and by an Established Investigator Grant from the American Heart Association.

REFERENCES

1. Tenhunen R., Marver H.S., and Schmidt R. The enzymatic conversion of heme to bilirubin by microsomal heme oxygenase. *Proc. Natl. Acad. Sci. U.S.A.* 1968;244:6388–6394.
2. Durante W. and Schafer A.I. Carbon monoxide and vascular cell function. *Int. J. Mol. Med.* 1998;2:255–262.
3. Morita T., Perella M.A., Lee M.E., and Kourembanas S. Smooth muscle cell-derived carbon monoxide is a regulator of vascular cGMP. *Proc. Natl. Acad. Sci. U.S.A.* 1995;92:1475–1479.
4. Christodoulides N., Durante W., Kroll M.H., and Schafer A.I. Vascular smooth muscle cell heme oxygenases generate guanylyl cyclase-stimulatory carbon monoxide. *Circulation* 1995;91:2306–2309.
5. Ishikawa K., Navab M., Leitinger N., Fogelman A.M., and Lusis A.J. Induction of heme oxygenase-1 inhibits the monocyte transmigration induced by mildly oxidized LDL. *J. Clin. Invest.* 1997;100:1209–1216.
6. Wagner C.T., Durante W., Christodoulides N., Hellums J.D., and Schafer A.I. Hemodynamic forces induce the expression of heme oxygenase in cultured vascular smooth muscle cells. *J. Clin. Invest.* 1997;100:589–596.
7. Levere R.D., Martasek P., Escalante B., Schwartzman M.L., and Abraham N.G. Effect of heme arginate administration on blood pressure in spontaneously hypertensive rats. *J. Clin. Invest.* 1990;86:213–219.

8. Sabaawy H.E., Zhang F., Nguyen X., ElHosseiny A., Nasjletti A., Schwartzman M., Dennery P., Kappas A., and Abraham N.G. Human heme oxygenase-1 gene transfer lowers blood pressure and promotes growth in spontaneously hypertensive rats. 2001;38:210–115.

9. Johnson R.A., Lavesa M., Askari B., Abraham N.G., and Nasjletti A. A heme oxygenase product, presumably carbon monoxide, mediates a vasodepressor function in rats. *Hypertension* 1995;25:166–169.

10. Abraham N.G., Lavrovsky Y., Schwartzman M.L., Stoltz R.A., Levere R.D., Gerritsen M.E., Shibahara S., and Kappas S. Transfection of human heme oxygenase gene into rabbit coronary microvessel endothelial cells: protective effect against heme and hemoglobin toxicity. *Proc. Natl. Acad. Sci. U.S.A.* 1995;92:6798–6802.

11. Yachie A., Niida Y., Wada T., Igarashi N., Kaneda H., Toma T., Ohta K., Kasahara Y., and Koizumi S. Oxidative stress causes enhanced endothelial cell injury in human heme oxygenase-1 deficiency. *J. Clin. Invest.* 1999:103:129–135.

12. Durante W., Peyton K.J., and Schafer A.I. Platelet-derived growth factor stimulates heme oxygenase-1 gene expression and carbon monoxide production in vascular smooth muscle cells. *Arteriosler. Thromb. Vasc. Biol.* 1999;19:2666–2672.

13. Kutty R.K., Nagineni C.N., Kutty G., Hooks J.J., Chader G.J., and Wiggert B. Increased expression of heme oxygenase-1 in human retinal pigment epithelial cells by transforming growth factor-beta. *J. Cell. Physiol.* 1994;159:371–378.

14. Ishizaka N., Aizawa T., Mori I., Taguchi J., Yazaki Y., Nagai R., and Ohno M. Heme oxygenase-1 is regulated in the rat heart in response to chronic administration of angiotensin II. *Am. J. Physiol.* 2000;279:H672–H678.

15. Durante W., Peyton K.J., and Schafer A.I. Heme oxygenase-1 is an autocrine inhibitor of vascular smooth muscle cell growth. *Circulation* 2000;102:II-298.

16. Morita T., Mitsialis S.A., Hoke H., Liu Y., and Kourembanas S. Carbon monoxide controls the proliferation of hypoxic smooth muscle cells. *J. Biol. Chem.*1997;272:32804–32809.

17. Togane Y., Morita T., Suematsu M., Ishimura Y., Yamazaki J.-I., and Katayama S. Protective roles of endogenous carbon monoxide in neointimal development elicited by arterial injury. *Am. J. Physiol.* 2000;278:H623–H632.

18. Grundemar L. and Ny L. Pitfalls using metalloporphyrins in carbon monoxide research. *Trends Pharmacol. Sci.* 1997;18:193–195.

19. Ignarro L.J., Woods K.S., and Wolin M.S. Regulation of soluble guanylate cyclase by porphyrins and metalloporphyrins. *J. Biol. Chem.* 1984;259:6201–6207.

20. Meffert M.K., Haley J.E., Schuman E.M., Schulman H., and Madison D.V. Inhibition of hippocampal heme oxygenase, nitric oxide synthase, and long-term potentiation by metalloporphyrins. *Neuron* 1994;13:1225–1233.

21. Durante W., Schini V.B., Catovsky S., Kroll M.H., Vanhoutte P.M., and Schafer A.I. Plasmin potentiates the induction of nitric oxide synthase by interleukin-1β in vascular smooth muscle cells. *Am. J. Physiol.* 1993;264:H617–H623.

22. Durante W., Kroll M.H., Christodoulides N., Peyton K.J., and Schafer A.I. Nitric oxide induces heme oxygenase-1 gene expression and carbon monoxide production in vascular smooth muscle cells. *Circulation* 1997;80:557–564.

23. He T.-C., Zhou S., Da Costa L.T., Yu J., Kinzler K.W., and Vogelstein B. A simpified system for generating recombinant adenoviruses. *Proc. Natl. Acad. Sci. U.S.A.* 1998;95:2509–2514.

24. Otterbein L.E., Bach F.H., Alam J., Soares M., Lu H.T., Wysk M., Davis R.J., Favell R.A., and Choi A.M.K. Carbon monoxide has anti-inflammatory effects involving the mitogen-activated protein kinase pathway. *Nature Med.* 2000;6:422–428.

25. Duckers H.J., Boehm M., True A.L., Yet S.-F., San H., Park J.L., Webb R.C., Lee M.-E., Nabel G.J., and Nabel E.G. Heme oxygenase-1 protects against vascular constriction and proliferation. *Nature Med.* 2001;7:693–698.

26. Brouard S., Otterbein L.E., Anrather J., Tobiasch E., Bach F.H., Choi A.M.K., and Soares M.P. Carbon monoxide generated by heme oxygenase-1 suppresses endothelial apoptosis. *J. Exp. Med.* 2000;192:1015–1025.

27. Tulis D.A., Durante W., Peyton K.J., Evans A.J., and Schafer A.I. Heme oxygenase-1 attenuates vascular remodeling following balloon injury in rat carotid arteries. *Atherosclerosis* 2001;155:113–122.

28. Tulis D.A., Durante W., Liu X., Evans A.J., Peyton K.J., and Schafer A.I. Adenovirus-mediated heme oxygenase-1 gene delivery inhibits injury-induced vascular neointima formation. *Circulation 2001*; In press.
29. Liu X., Chapman G.B., Peyton K.J., and Durante W. Adenovirus-mediated heme oxygenase-1 gene expression stimulates apoptosis in vascular smooth muscle cells. *FASEB J.* 2001;15:476.
30. Chau L.-Y. and Juan S.-H. Induction of apoptosis in vascular smooth muscle cells by heme oxygenase-derived carbon monoxide. *FASEB J.* 2001;15:773.
31. Aizawa T., Ishizaki N., Taguchi J., Kimura S., Kurokawa K., and Ohno M. Balloon injury does not induce heme oxygenase-1 expression, but administration of hemin inhibits neointimal formation in balloon injured rat carotid arteries. *Biochem. Biophys. Res. Commun.* 1999;261:302–307.

40

INDUCTION OF APOPTOSIS IN VASCULAR SMOOTH MUSCLE CELLS BY HEME OXYGENASE-1-DERIVED CARBON MONOXIDE

Shu-Hui Juan and Lee-Young Chau

Institute of Biomedical Sciences
Academia Sinica, Taipei 11529
Taiwan, ROC

INTRODUCTION

Heme oxygenase (HO) is the rate-limiting enzyme to catalyze the heme degradation, leading to the generation of biliverdin, which is subsequently converted to antioxidant bilirubin by biliverdin reductase,[1,2] free iron, and carbon monoxide (CO). Three HO isozymes derived from distinct genes were identified.[3] Among them, HO-1 is a stress-responsive protein and can be induced by a variety of oxidative-inducing agents, including heme, heavy metals, UV radiation, cytokines, and endotoxin.[4,5] Recently, numerous in vitro and in vivo studies have demonstrated that the induction of HO-1 represents an important cellular protective mechanism against oxidative injury.[4,5] In addition to the antioxidant effect of bilirubin,[6–8] increasing interest has been drawn on the potential effects of CO. It has been shown that CO shares similar properties with nitric oxide (NO) to act as a putative neural transmitter in central nervous system and a potent vasodilator to regulate the vascular tone through the activation of soluble guanylate-cyclase-cGMP pathway.[9–11] Both gases also exert potent anti-proliferative effect on vascular smooth muscle cells (VSMCs).[12–15] Very recently, there are studies demonstrating that low concentration of CO provides protection against hyperoxic lung injury in animal[16] and apoptotic death in fibroblasts and endothelial cells induced by TNF-α.[17,18] Furthermore, a study on macrophages has shown that CO exhibits anti-inflammatory effect through a pathway involving the mitogen-activated protein kinases.[19] It is apparent that CO mediates some of the beneficial effects of HO-1 induction under certain circumstances. Nevertheless, it has been documented that CO can poison cellular hemoproteins according to the level and duration of exposure and leads to the generation of reactive oxygen species, which seems to be secondary to the inhibition of mitochondrial respiration.[20,21] The potential cytotoxicity of the massive production of endogenous CO upon overexpression of HO-1 in cells, however, was less explored. In this study, we were able to demonstrate that adenovirus-mediated HO-1 overexpression in vascular smooth muscle cells (VSMCs) significantly reduced cell viability. The DNA fragmentation, which was preceded by the induction of p53, decrease in the ratio of Bcl-2/Bax, increase of cytosolic cytochrome C, and activation of caspase-3, was evident in VSMCs overexpressing HO-1, indicating the occurrence of apoptotic death.[22] Furthermore, hemoglobin (a CO scavenger), but not desferrioxamine (an iron chelator), effectively inhibited the apoptotic events, implicating that CO, but not free iron, derived from heme degradation mediates the cytotoxic effect.

CYTOTOXICITY INDUCED BY ADENOVIRUS-MEDIATED HO-1 OVEREXPRESSION IN VSMCS

Although HO-1 induction is generally considered to provide protection against oxidative injury, two recent studies demonstrated that the cytoprotective effects is within a narrow range of expression.[23,24] To test the effect of HO-1 overexpression on the growth of VSMCs, we constructed an adenoviral vector carrying human HO-1 gene (Adv-HO-1) and infected rat VSMCs with the recombinant virus. As shown in Fig. 1A, the cell growth, as assessed by MTT assay for a period of 3 days, was markedly reduced in VSMCs[25] infected with Adv-HO-1 in a dose-dependent manner. In contrast, VSMCs infected with the same doses of empty adenovirus lacking trangene (Adv) proliferated to the similar extent as control cells without infection.

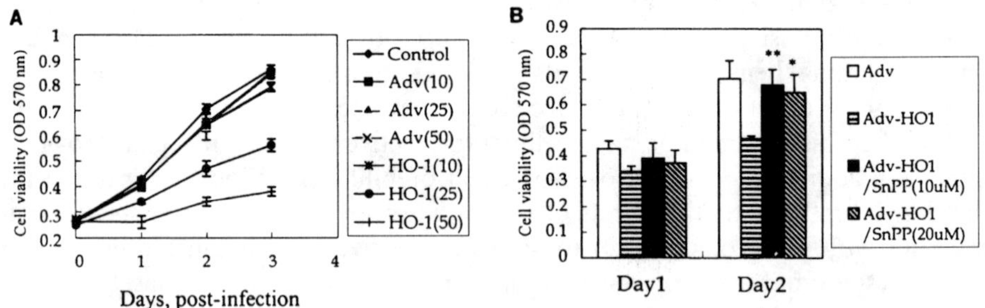

Figure 1. Effect of adenovirus-mediated HO-1 gene transfer on viability of VSMCs. A. VSMCs plated at 2×10^4/well were infected with empty adenovirus (Adv) or adenovirus carrying human HO-1 gene (HO-1) at indicated MOIs. Cell viability was determined by MTT assay at given time points. B. VSMCs at a density of 5×10^4 cells/well were infected with indicated Adv constructs at MOI of 25 and cultured in the absence or presence of HO-1 inhibitor, SnPP. Cell viability was determined at 1 day and 2 days post-infection. Data shown are mean ±SD of three independent experiments done in triplicates. Significant differences vs cells infected with Adv-HO-1 without treatment with SnPP. *p < 0.01; **<0.005.

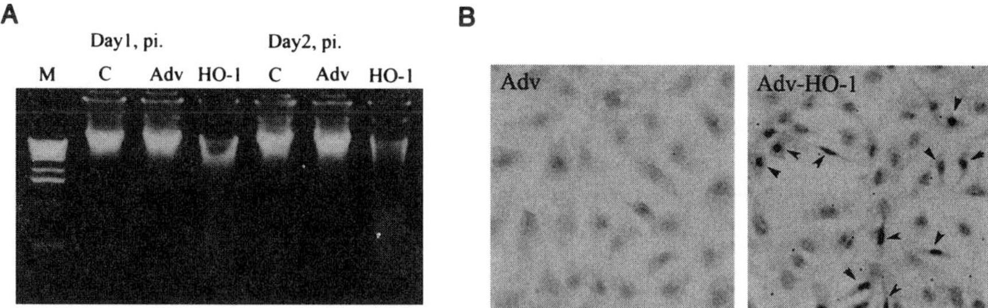

Figure 2. DNA fragmentation in VSMCs after infection with Adv-HO-1. A. VSMCs were infected with Adv or Adv-HO-1 at titer of 50 MOI. Chromosomal DNAs were then isolated at 1 and 2 days post infection and analyzed by 2% agarose gel electrophoresis. B. Cells grown on cover slips were infected with 50 MOI of indicated virus constructs for 2 days, followed by TUNEL assay. Nuclei with positive stains are indicated by arrows.

HO-1 inhibitor, tin-protoporphyrin (SnPP) effectively reversed the decrease in cell growth of VSMCs following infection with Adv-HO-1 (Fig. 1B), indicating that the inhibitory effect was resulted from HO-1 activity. When we examined cells by light microscopy, it was found that infection of VSMCs with Adv-HO-1 resulted in the appearance of floating dead cells after 24 hr (data not shown). To characterize whether the cell death induced by Adv-HO-1 was resulted from apoptosis, we examined the integrity of chrosomal DNA of infected cells by electrophoretic analysis and TUNEL staining. As shown in Fig. 2A, agarose gel electrophoresis revealed the appearance of DNA laddering in DNA preparations extracted from cells infected with Adv-HO-1, but not with Adv, for 1 and 2 days. The DNA fragmentation was further confirmed by the in situ TUNEL staining, which revealed dark, condensed nuclei in Adv-HO-1-infected cells, but not in Adv-infected control cells (Fig. 2B).

EFFECT OF ADV-HO-1 INFECTION ON THE EXPRESSION OF APOPTOSIS-RELATED PROTEINS AND THE ACTIVATION OF CASPASE-3

To elucidate the cellular mechanism underlying the apoptosis induced by HO-1 overexpression, Western blot analysis was performed to examine the expression profile of apoptosis-related proteins in VSMCs infected with Adv-HO-1. As shown in Fig. 3A, the level of HO-1 protein was highly expressed in VSMCs infected with Adv-HO-1 in a dose-dependent manner. Densitometric quantification revealed approximately 5.0 ± 0.6, 12.0 ± 1.0, and 23.0 ± 1.3-fold increases in HO-1 expression level in cells infected with Adv-HO-1 at MOI of 10, 25 and 50, respectively. The increase in HO-1 was coincided with elevated levels of p53 and Bax proteins, which were also in proportion to the doses of Adv-HO-1 used. Conversely, the expression level of antiapoptotic protein, Bcl-2, was down-regulated concurrently. Since the decrease in the ratio of Bcl-2/Bax may lead to the efflux of cytochrome C from

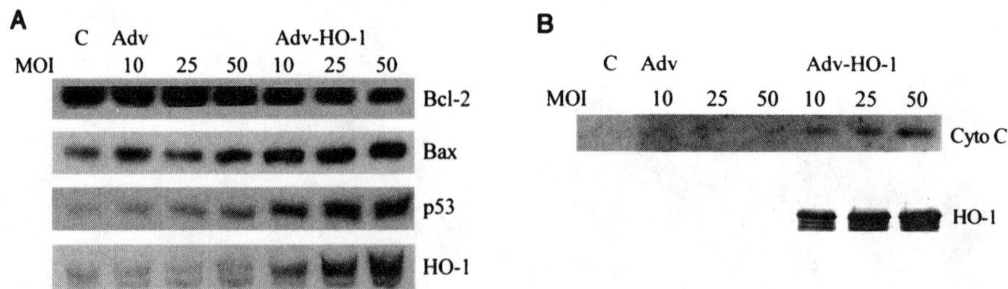

Figure 3. Changes in apoptosis-associated proteins (A) and cytosolic level of cytochrome C (B) in VSMCs infected with increasing MOIs of Adv or Adv-HO-1. Whole cell lysates or cytosolic fractions of VSMCs infected with adenovirus constructs for 24 hr were prepared. Levels of p53, Bax, Bcl-2, HO-1 and cytochrome C were assessed by Western blot analysis. Results shown are the representative of at least 3 independent experiments.

mitochondria,[22] the cytosolic level of cytochrome C was also examined. As shown in Fig. 3B, the level of cytosolic cytochrome C was increased in parallel to the overexpression of HO-1. When the time course experiment was performed, it was shown that the overexpression of HO-1 was detected as early as 18 hr post infection, reached a maximum at 30 hr, and sustained up to 42 hr, the longest time point examined (Fig. 4A). The alteration in the levels of apoptotic-related proteins was also observed along with the increment in HO-1 expression (Figs. 4A and 4B). Experiment was further conducted to assess the activity of caspase-3, which is a major effector caspase responsible for the proteolytic events in cell-death pathway,[22] in VSMCs following infection with Adv-HO-1. In concordant with the time course of HO-1 expression, the activation of caspase-3 was evident as early as 18 hr, reached a maximum at 30 hr, and sustained up to 42 hr after infection with Adv-HO-1 (Fig. 5).

EFFECTS OF HEMOGLOBIN, N-ACETYLCYSTEINE, DESFERRIOXAMINE, AND GUANYLATE CYCLASE INHIBITORS

The apoptotic death induced by HO-1 overexpression in VSMCs may be caused by the mass production of CO or the release of free iron from HO reaction. When

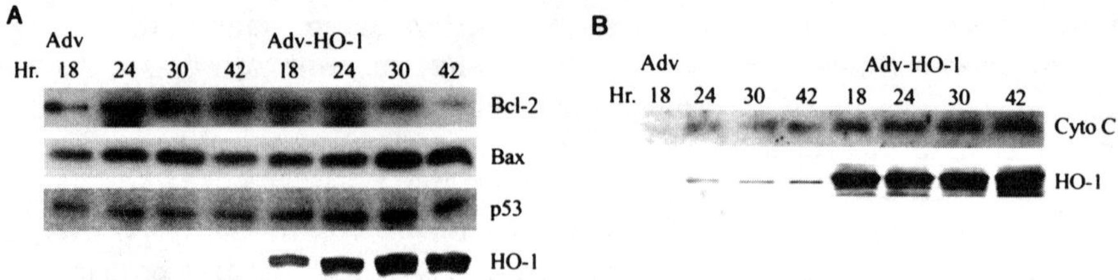

Figure 4. Time course of expression profiles of HO-1, p53, Bax, Bcl-2 and cytochrome C in VSMCs following infection with Adv-HO-1. VSMCs were infected with 25 MOI of Adv or Adv-HO-1. Cell lysates or cytosolic fractions were prepared at indicated times post infection. Expression levels of HO-1, p53, Bax, Bcl-2 and cytosolic cytochrome C were examined by Western blot analysis.

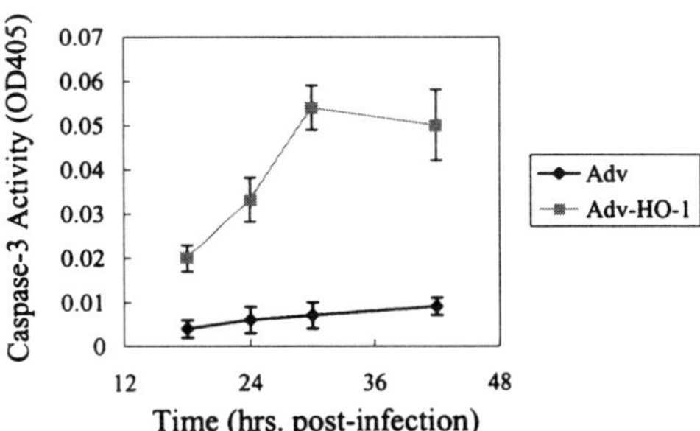

Figure 5. Time course of caspase-3 activation in VSMCs following infection with Adv-HO-1. Cells were infected with 25 MOI of Adv or Adv-HO-1 for indicated times. Cell lysates were prepared and the changes in caspase-3 activity were determined.

we examined the effect of hemoglobin (Hb), a CO scavenger, on the cell viability of Adv-HO-1-infected cells, it was clearly shown that Hb at concentrations of 1–2 µM was able to reverse the decrease in cell viability in a dose-dependent manner (Fig. 6A). Consistently, Hb inhibited the induction of p53 protein as well as the activation of caspase-3 in VSMCs overexpressing HO-1 (Figs. 6B & 6C), suggesting that the CO derived from HO-1 reaction mediates the cytotoxic effect. Desferrioxamine (DEF), an iron chelator, and N-acetylcysteine (NAC), a glutathione precursor, were also tested for their effects on the induction of apoptotic events. As shown in Fig. 7, NAC treatment partially inhibited the induction of p53 and caspase-3 activity; whereas DEF did not exhibit significant inhibitory effect. To assess whether the increased cGMP is involved in triggering the apoptotic pathway, the effects of soluble guanylate cyclase inhibitors, methylene blue (MB) and 1H-oxadiazolo-1,2-4-[4,3-α] quinoxaline-1-one (ODQ), on the Adv-HO-1-induced apoptosis were examined. As illustrated in Fig. 8, both MB and ODQ did not significantly affect the expression of p53 as well as activation of caspase-3 in VSMCs overexpressing HO-1.

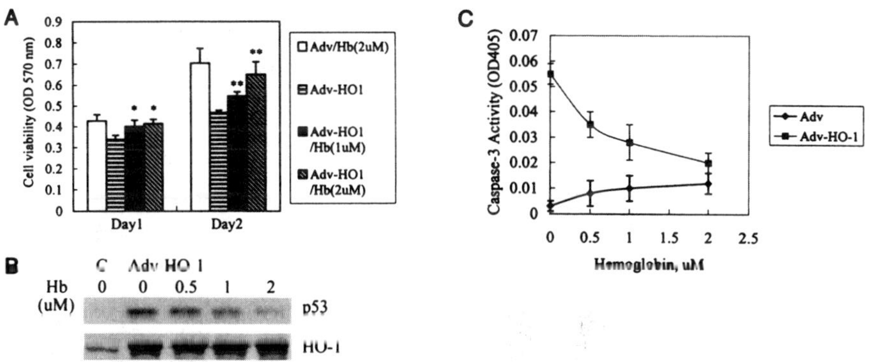

Figure 6. Effect of hemoglobin on cell viability, expression of p53, and caspase-3 activity in VSMCs overexpressing HO-1. VSMCs infected with 25 MOI of Adv or Adv-HO-1 were cultured in the absence or presence of indicated concentrations of hemoglobin (Hb). A. Cell viability was determined by MTT assay at 1 and 2 days post infection. Data shown are mean ±SD of three independent experiments in triplicates. Significant differences vs cells infected with Adv-HO-1 without treatment with Hb. *p < 0.01; **p < 0.005. B. Expression levels of p53 and HO-1 were examined by Western blot analysis at 1 day post infection. C. Caspase-3 activity was determined at 1 day post infection. Data shown are mean ±SD of three independent experiments in triplicates.

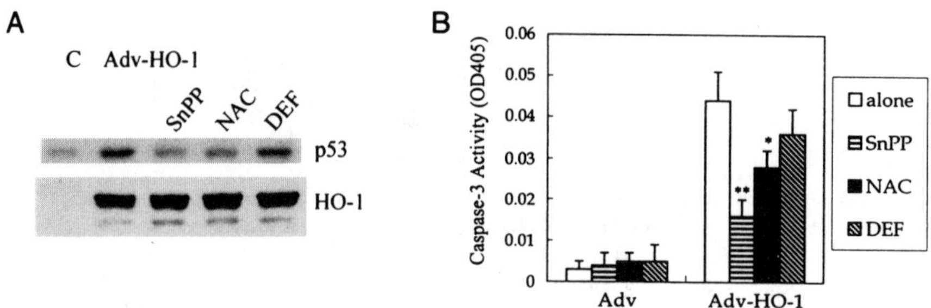

Figure 7. Effects of tin-protoporphyrin, N-acetylcysteine and desferrioxamine on p53 expression and caspase-3 activation in VSMCs overexpressing HO-1. Cells infected with 25 MOI of Adv or Adv-HO-1 were treated with 10 µM tin-protoporphyrin (SnPP), 5 mM N-acetylcysteine (NAC), or 10 µM desferrioxamine (DEF) for 24 hr. Cell lysates were prepared and the expression of p53 (A) and caspase-3 activity (B) were examined. Data shown in B are mean ±SD of three independent experiments done in triplicates. Significant difference vs cells infected with Adv-HO-1 alone. *p < 0.01; **p < 0.005.

Apparently, the present finding is contradictory to recent reports showing that CO mediates the anti-apoptotic effect of HO-1 in TNF-induced cell death in fibroblasts and endothelial cells.[17,18] Although the levels of endogenous CO produced under the condition of HO-1 overexpression in various cell types are largely unknown, it is very likely that the susceptibility to CO toxicity is variable in different cell types. Nevertheless, it is well known that CO binds tightly to heme centers of hemoglobin and respiratory proteins, such as cytochrome C oxidase in mitochondria, leading to the reduction in the oxygen carrying capacity of the red blood cells and oxygen utilization by the mitochondria.[20,21] Toxic exposure to CO can induce oxidative stress, which is resulted from the inhibition of mitochondrial respiration and the generation of free radicals, including hydroxyl, nitric oxide, and peroxynitrite, and cause cellular damage.[26,27] In vivo studies on rats have shown that exposure to high levels of CO leads to the production of NO and NO-derived oxidants, which mediate CO toxicity in brain, lung and vasculature.[27–29] In vitro, CO-induced apoptosis has been reported in

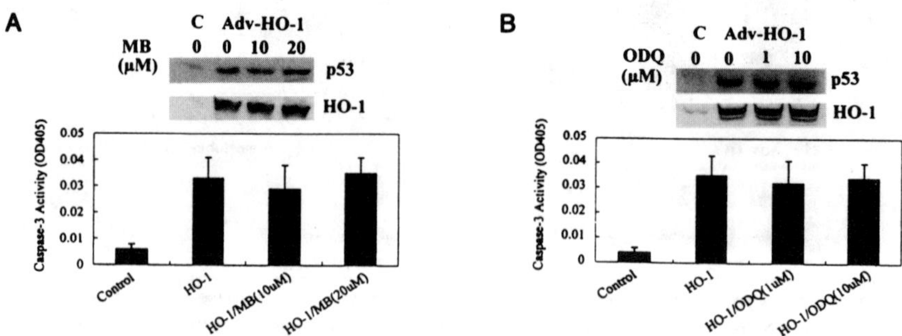

Figure 8. Effects of guanylate cyclase inhibitors on p53 expression and caspase-3 activity in VSMCs overexpressing HO-1. Cells infected with Adv-HO-1 were treated without or with indicated concentrations of methyl blue (MB) (A) or 1H-oxadiazolo-1,2,4-[4,3,-α]quinoxaline-1-one (ODQ) (B) in culture for 24 hr. The expression level of p53 and caspase-3 activity were examined.

thymocytes and endothelial cells, although mediated by differential pathways.[30,31] It has been shown that the death event in thymocytes is mediated by a free-radical mechanism; whereas the cytotoxicity in endothelial cells is mediated by the production of NO as well as peroxynitrite. To test whether the production of NO is involved in the apoptosis in VSMCs induced by HO-1 overexpression, the effect of nitric oxide synthase inhibitors was assessed. However, we did not observe significant inhibition on cell death and caspase-3 activity induced in VSMCs overexpressing HO-1 by treatment with N-nitro-L-arginine methyl ester or S-isopropylisothiourea (data not shown). Furthermore, the level of nitrite detected in the culture medium of VSMCs overexpressing HO-1 was not significantly higher than that of control cells (data not shown), indicating that NO is unlikely involved. Nevertheless, it was noted that antioxidant NAC inhibited the p53 induction as well as caspase-3 activation to some extent, suggesting the possible involvement of reactive oxygen species in triggering the death signaling.

It is interesting to learn that CO and NO share similar proapoptotic activities in VSMCs. A recent study by Iwashina et al. demonstrated that NO produced by overexpression of inducible nitric oxide synthase induces apoptosis in VSMCs,[32] although the molecular mechanism is not necessarily same as that underlying the CO-induced cell death. Nevertheless, p53 appears to be the common death transducer in both NO- and CO-induced apoptosis.[32] It appears that CO exerts differential effects in different cells or tissues through distinct mechanisms. For example, in a recent report By Otterbein et al,[19] it was noted that the intracellular cGMP level was not significantly increased in macrophages exposed to CO, and it was not responsible for the anti-inflammatory effect observed in these cells. In VSMCs CO acts as a vasodilator[9–11] and exhibits anti-proliferative activity[12–15] through guanylate cyclase-cGMP-dependent pathway. However we found that specific inhibitors of guanylyl cyclase were not able to inhibit the cell death induced by HO-1 overexpression in VSMCs, indicating the CO-induced apoptosis is cGMP-independent. The signaling pathway mediating the CO-induced death event remains to be established.

Previous work from our laboratory has shown that HO-1 is induced in atherosclerotic lesions of humans and experimental animal.[33] Although the direct evidence is lacking, it is envisioned that endogenous CO produced may have a role in the compensatory response to induce apoptosis and inhibit proliferation of VSMCs in the course of disease development. Very recently, there are studies showing that increased HO-1 expression in arteries inhibits neointimal formation after balloon injury in rats.[14,34] The apoptotic effect of CO on SMCs seems to contribute, at least in part, to the reduction of hyperplasia in the process of vascular remodeling. In view of the potential diverse effects of the products derived from heme degradation by HO, it is important to fully understand the physiological significance of HO-1 induction in vascular biology and pathology.

ACKNOWLEDGMENT

This work was supported by research grants from the Institute of Biomedical Sciences, Academia Sinica, and the Department of Health (DOH 89-TD-1139, DOH 90-TD-1036), Taiwan, ROC.

REFERENCES

1. M.D. Maines, Heme oxygenase: function, multiplicity, regulatory mechanisms and clinical applications, *FASEB J.* **2** (10), 2557–2568 (1988).
2. P. Ponka, Cell biology of heme, *Am. J. Med. Sci.* **318** (4), 241–256 (1999).
3. M.D. Maines, The oxygenase system: A regulator of second messenger gases, *Annu. Rev. Pharmacol. Toxicol.* **37**, 517–554 (1997).
4. J.L. Platt and K.A. Nath, Heme oxygenase: protective gene or Trojan horse, *Nature Med.* **4** (12), 1364–1365 (1998).
5. L.E. Otterbein and A.M.K. Choi, Heme oxygenase: colors of defense against cellular stress, *Am. J. Physiol. Lung Cell Mol. Physiol.* **279** (6), L1029–L1037 (2000).
6. R. Stocker, Y. Yamamoto, A.F. McDonagh, A.N. Glazer, and B.N. Ames, Bilirubin is an antioxidant of possible physiological importance, *Science* **235** (4792), 1043–1046 (1987).
7. S.F. Llesuy, S.F., and M.L. Tomaro, Heme oxygenase and oxidative stress: evidence of involvement of bilirubin as physiological protector against oxidative damage, *Biochim. Biophys. Acta* **1223** (1), 9–14 (1994).
8. R. Stocker, Induction of haem oxygenase as a defense against oxidative stress, *Free Radical Res. Commun.* **9** (2), 101–112 (1990).
9. H.H. Schmidt, NO, CO, and HO endogenous soluble guanylyl cyclase-activating factors, *FEBS Lett.* **307** (1), 102–107 (1992).
10. A. Verma, D.J. Hirsch, C.E. Glatt, G.V. Ronnett, and S.H. Snyder, Carbon monoxide: a putative neural messenger, *Science* **259** (5093), 381–384 (1993).
11. T. Morita, M.A. Perrella, M.E. Lee, and S. Kourembanas, Smooth muscle cell-derived carbon monoxide is a regulator of vascular cGMP, *Proc. Natl. Acad. Sci. USA* **92** (5), 1475–1479 (1995).
12. D.M. Lloyd-Jones and K.D. Bloch, The vascular biology of nitric oxide and its role in atherosclerosis, *Annu. Rev. Med.* **47**, 365–375 (1996).
13. M. Kibbe, T. Billiar, and E. Tzeng, Inducible nitric oxide synthase and vascular injury, *Cardiovas. Res.* **43** (3), 650–657 (1999).
14. Y. Togane, T. Morita, M. Suematsu, Y. Ishimura, J.I. Yamazaki, and S. Katayama, Protective roles of endogenous carbon monoxide in neointimal development elicited by arterial injury, *Am. J. Physiol. Heart Circ. Physiol.* **278** (2), H623–H632 (2000).
15. T. Morita, S.A. Mitsialis, H. Koike, Y. Liu, and S. Kourembanas, Carbon monoxide controls the proliferation of hypoxic vascular smooth muscle cells. *J. Biol. Chem.* **272** (52), 32804–32809 (1997).
16. L.E. Otterbein, L.L. Mantell, and A.M.K. Choi, Carbon monoxide provides protection against hyperoxic lung injury, *Am. J. Physiol.* **276** (4 Pt1), L688–L694 (1999).
17. I. Petrache, L.E. Otterbein, J. Alam, G.W. Wiegand, and A.M.K. Choi, Heme oxygenase-1 inhibits TNF-α-induced apoptosis in cultured fibroblasts, *Am. J. Physiol. Lung Cell Mol. Physiol.* **278** (2), L312–L319 (2000).
18. S. Brouard, L.E. Otterbein, J. Anrather, E. Tobiasch, F.H. Bach, A.M.K. Choi, and M.P. Soares, Carbon monoxide generated by heme oxygenase-1 suppresses endothelial cell apoptosis, *J. Exp. Med.* **192** (7), 1015–1025 (2000).
19. L.E. Otterbein, F.H. Bach, J. Alam, M. Soares, H.T. Lu, M. Wysk, R.J. Davis, R.A. Flavell, and A.M.K. Choi, Carbon monoxide has anti-inflammatory effects involving the mitogen-activated protein kinase pathway, *Nature Med.* **6** (4), 422–428 (2000).
20. A. Ernst and J.D. Zibrak, Carbon monoxide poisoning, *N. Engl. J. Med.* **339** (22), 1603–1608 (1998).
21. J.S. Stamler and C.A. Piantadosi, O=O NO: It's CO., *J. Clin. Invest.* **97** (10), 2260–2267 (1996).
22. M.O. Hengartner, The biochemistry of apoptosis, *Nature* **407** (6805), 770–776 (2000).
23. P.A. Dennery, K.J. Sridhar, C.S. Lee, H.E. Wong, V. Shokoohi, P.A. Rodgers, and D.R. Spitz, Heme oxygenase-mediated resistance to oxygen toxicity in Hamster fibroblasts, *J. Biol. Chem.* **272** (23), 14937–14942 (1997).
24. D.M. Suttner and P.A. Dennery, Reversal of HO-1 related cytoprotection with increased expression is due to reactive iron, *FASEB J.* **13** (13), 1800–1809 (1999).

25. K. Fisher-Dzoga, R.M. Jones, D. Vesselinbovitch, and R.W. Wissler, Ultrastructural and immuno-histochemical studies of primary cultures of aortic medial cells, *Exp. Mol. Pathol.* **18** (2), 162–176 (1973).

26. J. Zhang and C.A. Piantadosi, Mitochondrial oxidative stress after carbon monoxide hypoxia in the rat brain, *J. Clin. Invest.* **90** (4), 1193–1199 (1992).

27. H. Ischizopoulos, M.F. Beers, S.T. Ohnishi, D. Fisher, S.E. Garner, and S.R. Thom, Nitric oxide production and perivascular tyrosine nitration in brain after carbon monoxide poisoning in the rat, *J. Clin. Invest.* **97** (10), 2260–2267 (1996).

28. S.R. Thom, S.T. Ohnishi, D. Fisher, Y.A. Xu, and H. Ischiropoulos, Pulmonary vascular stress from carbon monoxide, *Toxicol. Appl. Pharmacol.* **154** (1), 12–19 (1999).

29. S.R. Thom, D. Fisher, Y.A. Xu, S. Garner, and H. Ischiropoulos, Role of nitric oxide-derived oxidants in vascular injury from carbon monoxide in the rat, *Am. J. Physiol. Heart Circ. Physiol.* **276** (3 Pt2), H984–H992 (1999).

30. V. Turcanu, M. Dhouib, J-L. Gendrault, and P. Poindron, Carbon monoxide induces murine thymocyte apoptosis by a free-radical mediated mechanism, *Cell Biol. & Toxicol.* **14** (1), 47–54 (1998).

31. S.R. Thom, D. Fisher, Y.A. Xu, K. Notarfrancesco, and H. Ischiropoulos, Adaptive responses and apoptosis in endothelial cells exposed to carbon monoxide, *Proc. Natl. Acad. Sci. USA* **97** (3), 1305–1310 (2000).

32. M. Iwashina, M. Shichiri, F. Marumo, and Y. Hirata, Transfection of inducible nitric oxide synthase gene causes apoptosis in vascular smooth muscle cells, *Circulation* **98** (12), 1212–1218 (1998).

33. L.-J. Wang, T.-S. Lee, F.-Y. Lee, R.-C. Pai, and L.-Y. Chau, Expression of heme oxygenase-1 in atherosclerotic lesions, *Am. J. Pathol.* **152** (3), 711–720 (1998).

34. T. Aizawa, N. Ishizaka, J. Taguchi, S. Kimura, K. Kurokawa, and M. Ohno, Balloon injury does not induce heme oxygenase-1 expression, but a dministration of hemin inhibits neointimal formation in balloon-injured rat carotid artery, *Biochem. Biophys. Res. Commun.* **261** (2), 302–307 (1999).

41

ROLE OF HEME OXYGENASE IN ANGIOGENESIS AND RENAL CARCINOMA

Alvin I. Goodman[a], Giovanni Li Volti[b], Nader G. Abraham[b], Lucia Malaguarnera[b,c]

Department of Medicine[a] and
Department of Pharmacology[b]
New York Medical College, Valhalla, NY, and
Department of Biomedical Sciences[c]
University of Catania, Italy
Telfax: ++39/095/320267

1. INTRODUCTION

1.1. Heme Oxygenase, Angiogenesis and Cancer

Angiogenesis is a fundamental process by which new blood vessels are formed.[1] Angiogenesis is increased during embryogenesis, and in pathological events, such as hypoxia, ischemia, inflammation, tumor growth and wound healing, in response to angiogenic factors. These factors increase vascular permeability, endothelial cell (EC) activation, migration, proliferation, and capillary formation[1] and lead to the simultaneous production of a number of soluble mediators, including cytokines and acute-phase proteins, such as heme oxygenase -1 (HO-1).[2] Several studies have demonstrated the relationship between cancer and increases of HO-1. Goodman, et al.[3] found elevated HO activity in renal adenocarcinoma compared with juxtatumor or normal renal tissues. An elevation of HO activity was found in lymphosarcoma-bearing rats.[4] This up-regulation of HO may be the result of local or circulating factors responding to oxidative stress and causing uncontrolled and uncoordinated cell growth. Induction of HO-1 biosynthesis during tumor growth has been investigated in an experimental solid tumor model (AH136B hepatoma) in rats.[5] HO-1 was found only in tumor cells. Strong induction of HO-1 was also observed in solid tumors after occlusion or embolization of the tumor-feeding artery, indicating that ischaemic stress, which may involve oxidative stress, triggers HO-1 induction in the solid tumor.[5] Others have shown that an HO inhibitor, zinc protoporphyrin IX, injected intra-arterially into a solid tumor suppressed tumor growth to a great extent and HO-1 expression in the solid tumor may confer resistance of tumor cells to hypoxic stress.[5] Elevation of HO-1 was observed in the hepatoma stage of the Long-Evans (LEC) rat.[6] On the other hand, Tozer et al.,[7] state that the HO/CO pathway does not play a major vasodilatory role in rat P22 carcinosarcoma. An inhibitor of HO, such as zinc protoporphyrin and copper protoporphyrin, was useful for inducing a relatively selective decrease in tumor blood flow by mechanisms unrelated to HO inhibition, especially when combined with NOS inhibition.[7]

HO-1 plays a role not only in tumor growth but also in tumor metastasis. Tsuji et al.[8] showed that high levels of HO-1 were detected in oral squamous cell carcinoma without lymph node metastasis and in well-differentiated cells. They concluded that high HO-1 expression in oral squamous cell carcinoma might be useful in identifying patients at risk for lymph node metastasis. The HO-1 gene is expressed selectively in brain tumors[9] and also in reactive astrocytes and macrophage-like cells.[10] Moreover, Kurata et al.[11] have reported expression of the HO-1 gene in macrophages in brain tumors. HO-1 gene expression appears to be related to the number of activated macrophages in the stromal compartment of tumors. Correlation of HO-1 expression with vascular density supports an association of macrophages with neovascularization.[14] It has been demonstrated that HO-1 gene expression could be a useful marker for macrophage infiltration as well as neovascularization in human gliomas[12] and melanomas.[13]

Angiogenesis is necessary for the continued growth of solid tumors, and acceleration of tumor growth accompanies neovascularization.[1] Development of blood vessels within tumor tissue is closely correlated with invasion and metastasis, as well as growth, as has been demonstrated in malignant melanoma and cancers in breast, lung, prostate and other organs.[14] Endogenous angiogenic factors as well as angiogenesis inhibitors released by tumor cells and other type cells are thought to regulate tumor angiogenesis.

We recently characterized a model of hemoglobin-induced endothelial (EC) injury and demonstrated that adenovirus or liposome-cDNA-mediated HO-1 gene transfer markedly reduced heme/hemoglobin toxicity.[15] However, the link between overexpression of HO in EC and blood vessel formation was not established. In the present study, we examined the relationship of HO-1 gene expression in EC and the formation of blood vessels.

1.2. *In Vitro* Angiogenesis and Capillary EC Formation

Endothelial cells were isolated from the midportion of the rabbit myocardium by collagenase digestion, filteration homogenization, and centrifugation as previously described.[16] Permanent transfection with human HO-1 gene was performed.[2] Cell proliferation in nontransfected and transfected cells was determined[17,18] Assessment of in vitro capillary formation used growth factor reduced basement membrane Matrigel matrix.

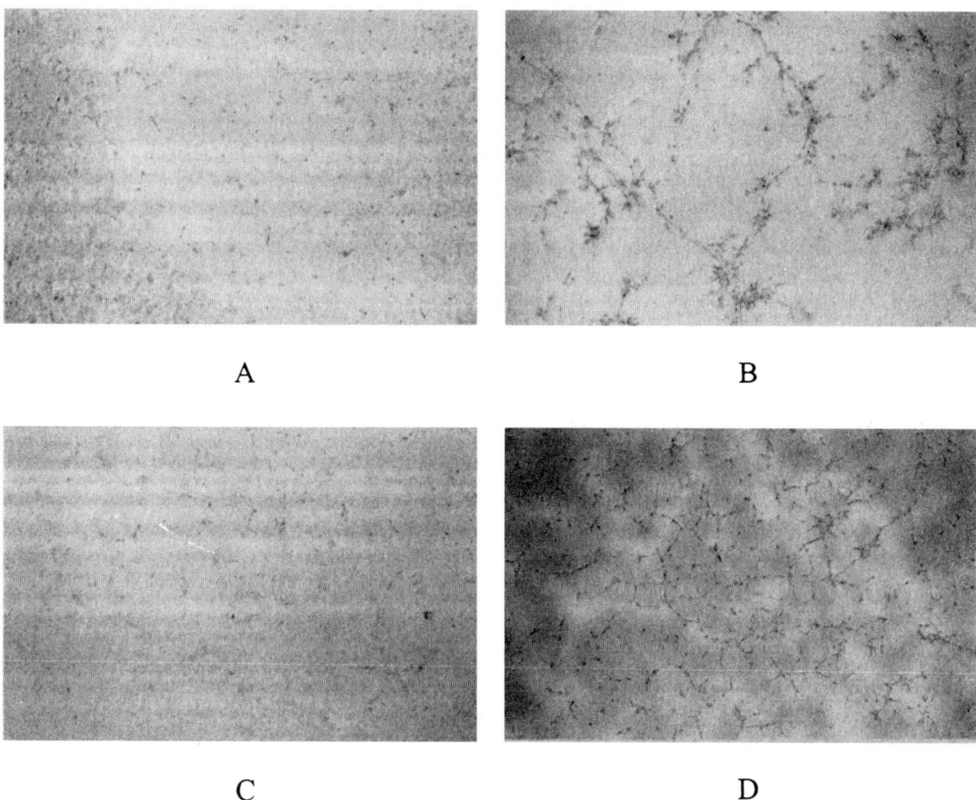

A   B

C   D

Figure 1. In vitro angiogenesis of endothelial cells (EC). **A:** Control cells. **B:** Cells treated with heme 10 μM) for 24 h. **C:** Cells treated with $SnCl_2$ (10 μM). **D:** Cells transfected with human HO-1 gene. Photos are representative of three separate experiments, each performed in quantrate. Quantitative analysis of tubular network area were performed with an OPTIMAS video images software analyzer. Data plotted are given in the lower panel. Results are presented as percentage increase in vessels tube size over control. (*) Significantly different (p < 0.01) from control.

Although transfected EC with human HO-1 gene exhibited similar AC-LDL uptake and factor VIII antigen to nontransfected cells, transfected EC seemed to grow faster and to stay closer to each other. The ability of the human HO-1 gene transfer to stimulate cell growth (increase in cell number) would implicate this gene expression in the regulation of cell cycle and in the modulation of angiogenesis. We, therefore, examined the rate of cell proliferation in both nontransfected and transfected cells Quiescent EC were plated at the same cell number, and cell proliferation was measured as previously described.[17,18] As shown in Table 1, transfection of the human HO-1 gene significantly stimulated cell growth of quiescent EC at 24 h (p < 0.001). The observed effect required a background serum concentration of 0.5% FBS Figure 1 and suggested that one of the HO products may be involved in the regulation of the cell cycle. The observed effect was not seen in the pCMV-transfected cells without human HO-1 (data not shown).

Table 1. Influence of human HO-1 gene transfer on endothelial cell proliferation. Nontransfected and transfected cell numbers were determined after plating the same number of cells in 96-well flasks in the presence of bFGF and 0.5% FBS. Results are average ± SEM; n = 3 in triplicates. P < 0.001

Conditions	Cell number ×10	% of control
Control	264.705 ± 83.256	100%
HO-1 Cells	441.176 ± 103.15	150%

2. HEME OXYGENASE AND HEME-MEDIATED ARACHIDONIC ACID METABOLISM

The rate-limiting enzyme in heme catabolism, heme oxygenase (HO), is a stress protein and its induction has been suggested to represent an important protective response against oxidative stress produced by heme and hemoprotein.[19–23] Induction of HO may specifically decrease cellular heme (prooxidant) and elevate bilirubin (antioxidant) levels.[15,24] Elevation of HO in human tumor and partial hepatectomy rats[3,25,26] resulted in a decrease in renal and liver hemoproteins such as cytochrome P450. Two HO isozymes, the products of distinct genes, have been described.[27,28] HO is ubiquitously distributed in mammalian tissues and is strongly and rapidly induced by many compounds that elicit cell injury. The natural substrate of HO, heme, is itself a potent inducer of the enzyme.[19] HO-2, which is believed to be constitutively expressed, is present in high concentrations in such tissues and in the brain and testis.[29]

Endotoxin, interleukin-1 and other stress agents cause a rapid (within 5–10 min.) activation of the HO gene and subsequent accumulation of HO mRNA.[30,31] This process involves transcriptional activation of several regulatory sites in the human HO promoter region.[22,32] A previous study from this laboratory demonstrated that the proximal promoter region of the human HO gene contains NFκβ and AP-2 binding sequences.[32] The finding of these binding sites on the HO promoter suggests the importance of HO in the stress response, when these transcriptional factors are known to be activated.[32] Substantial evidence indicates that induction of renal HO resulted in a decrease in hemoproteins and thromboxane synthesis[33] and in cytochrome P450 arachidonic acid-dependent metabolites.[34,35] Human renal and liver cytochrome P450 metabolizes arachidonic acid epoxygenase and ω-hydroxylase to novel metabolites, which have been shown to influence renal function.[36,37]

Several studies by our laboratory and others demonstrated that cytochrome P450 epoxygenase and ω-hydroxylase metabolite epoxyeicosatrienoic acid (EETs), and 20-HETE, respectively, possess a wide range of biological functions.[33,36,38] These include stimulation of peptide hormone release, modulation of Na^+-K^+-ATPase, vasodilation and inhibition of platelet aggregation (for review, see references).[36,39]

3. HEME OXYGENASE IN CELL GROWTH

It has been demonstrated that rabbit coronary microvessel endothelial cells (EC) transfected with human HO-1 gene increased in cell number by 45% as compared to control cells[17] The same authors showed that the transfected cells exhibited a twofold increase in blood vessel formation. Thus HO-1 may participate in the regulation of EC proliferation and angiogenesis. Retrovirus-mediated HO gene transfer into EC has been shown to protect EC against oxidant-induced injury.[40] Human HO-1 gene transduced ED acquired substantial resistance to toxicity produced by exposure to heme and H_2O_2 compared with that in nontransduced cells. The protective effect of enhancement of HO-1 activity against heme and H_2O_2 was reversed by pretreatment with stannic mesoporphyrin, a competitive inhibitor of HO. It has been suggested that HO-1 plays an important role in protecting human renal tubules from oxidative injury. HO-1 inhibits the production of 20-hydroxyeicosatetraenoic acid (20-HETE), a major cytochrome P450 arachidonic acid metabolite found in the rat and human kidney.[38] It has been shown that epidermal growth factor (EGF) stimulates proximal tubular 20-HETE production and the latter is a potent mitogen to these cells.[41]

We have recently shown that rat lung microvessel endothelial cells (RLMV) transfected with a retroviral mediated human HO-1 gene exhibited a 4-fold increase in HO activity with a significant increase in the steady-state levels of heme and cGMP.[42] We have also recently shown that tumor necrosis factor mediated cell death is attenuated by retrovirus-mediated human heme oxygenase gene transfer in human microvessel endothelial cells (unpublished data). It is thus clear that HO-1 has a significant regulatory role in cell growth.

4. HEME OXYGENASE AND RENAL CELL CARCINOMA

Heme oxygenase (HO) activity has been implicated in the regulation of renal function and cell growth in normal and disease states. Expression of HO genes has been shown to regulate important hemoprotein(s), such as cytochrome P450. In a recent study, HO activity was measured in samples of human adenocarcinoma, juxtatumor, and normal renal tissue.[3] The samples were histologically examined to verify the malignant and normal nature. Biopsies of tumor and normal tissues were taken from the adenocarcinoma and juxtatumor tissue sites, respectively, of the same kidney just after nephrectomy. Each sample was processed for histological and morphological examination. HO activity was measured in adenocarcinoma and juxtatumor tissue biopsies from six patients undergoing nephrectomy. As seen in Fig 2, in all patients, the activity of HO in the adenocarcinoma tissue was 3- to 6-fold higher than the average value in the normal tissue adjacent to the tumor of the same kidney.

The adenocarcinoma tissue demonstrated an activity that ranged from 1.26 to 5.62 nmol bilirubin/mg/hr (2.66 ± 0.64 [mean + SEM]; n = 6), while in the juxtatumor tissue, HO activities ranged from 0.38 to 1.04 nmol bilirubin/mg/hr (0.66 ± 0.11; n = 6). In normal and tissues, HO activity was expressed at the same level as that seen in

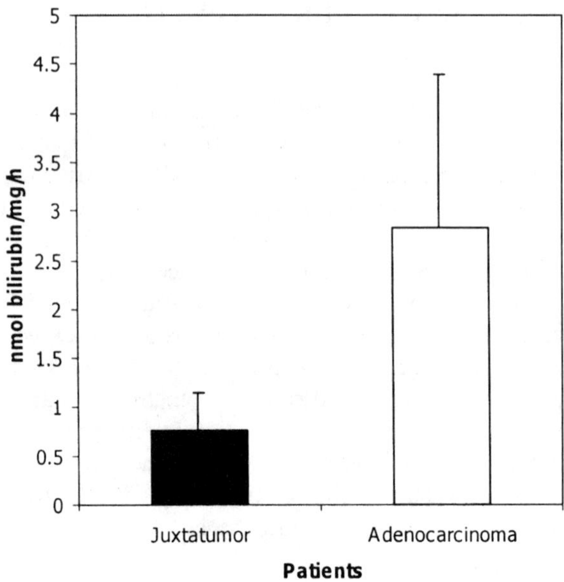

Figure 2. HO activities in tumor and juxtatumor tissues of the hypernephroma kidneys. Microsomes were prepared by different centrifugations and microsomal HO activity was determined by measuring bilirubin formation.

juxtatumor tissues. These results indicated a four-fold difference (p < 0.01) between the carcinoma and juxtatumor tissues. Ho activity in the juxtatumor tissues demonstrated a level comparable tot hat described previously for normal human kidney.[43]

Northern blot analysis using cytochrome P450 cDNA probe 4A2 cDNA for the ω-hydroxylase gene family revealed that mRNA levels for ω-hydroxylase transcripts were significantly decreased in the adenocarcinoma compared with juxtatumor. The decrease in cytochrome P450 4A2 mRNA levels correlated with a decrease in the arachidonic acid ω-hydroxylation metabolite 20-HETE. The production of 20-HETE was significantly higher in juxtatumor in agreement with ω-hydroxylase mRNA. Higher levels of HO-1 may be a contributing factor for the undetectable levels of cytochrome P450 arachidonic acid metabolite, 20-HETE, in the adenocarcinoma. Our results suggest that increased generation of mitogenic activities by ω-hydroxylase and 20-HETE in the juxtatumor may be a contributing factor in the development and growth of neoplastic tissues, an the induction of HO in the tumor tissue may be and attempt to limit oxidative injury caused by the cytochrome P450 metabolites and other oxidative stress.

4.1. Northern Blot Analysis of HO-1 Expression In Normal Kidney, Adenocarcinoma and Juxtatumor

The RNA preparations from the different samples were also subjected to Northern blot analysis and probed for HO-1, HO-2 and G3PDH. As can be seen in

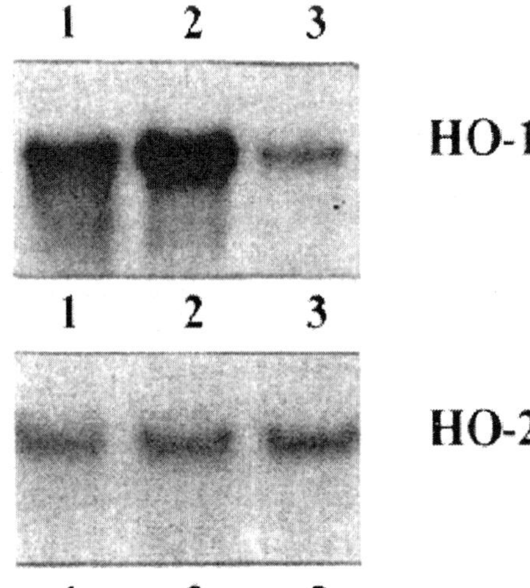

Figure 3. Northern blot analysis of RNA from human kidney. Total RNA (10 µg/lane) was separated on a 1% agarose gel containing 1 M formaldehyde. After blotting to Hybond N$^+$ membranes, the blot was probed with a ^{32}P-labeled cDNA for human HO-1, as described in Materials and Methods. Lane 1: normal kidney, Lane 2: renal adenocarcinoma, Lane 3: juxtatumor.

Fig. 3, Lane 2, the level of HO-1 mRNA in the adenocarcinoma was elevated compared with the transcript levels in normal or juxtatumor tissues (Lanes 1 and 3, respectively), In contrast, HO-2 transcript levels did not change in any of the three renal tissues. Evaluation of the intensity of the bands by scanning densitometry indicate that adenocarcinoma RNA contains severalfold higher levels of HO-1 mRNA compared with normal tissues. On the other hand, HO-2 mRNA levels did not change in all three tissues (Fig. 3, middle panel). Hybridization of the filters with a radiolabeled G3PDH probe confirmed that a similar amount of total RNA was transferred to the filters in each lane of the paired experiment (Fig. 3, lower panel).

4.2. Cytochrome P450 Arachidonic Acid Epoxygenase and ω-Hydroxylase Activity

Cytochrome P450 arachidonic acid epoxygenase and ω-hydroxylase activities were markedly reduced in the tumor tissues, whereas, in the juxtatumor tissue, cytochrome P450 ω-hydroxylase activity was significantly increased (Table 2).

Table 2. Activity of ω/ω-1 hydroxylase and expoxygenase activity were obtained by reverse-phase HPLC chromatogram of arachidonic acid metabolites formed by hypernephroma microsomes

	ω/ω-1 hydroxylase activity pmol/mg/30 min	Epoxygenase activity pmol/mg/30 min
Juxtatumor tissues	304.7 ± 95.6	161.86 ± 44.49
Carcinoma tissues	12.15 ± 7.49	14.68 ± 9.49

5. SUMMARY

The reciprocal relationship between HO-1 and cytochrome P450 may dictate the state of tumor growth and development, as well as tumor resistance to treatment and the ability of the surrounding normal tissue to metabolize endogenous substrates such as arachidonic acid, the products of which may possess the ability to modulate tumor growth. These speculations suggest the need to examine further the role of HO-1 levels and cytochrome P450-arachidonic acid metabolites in tumor growth and development. Our present studies provide direct evidence involving heme oxygenase in angiogenesis and cell growth. At this time, it is unknown whether HO-1 is up-regulated to counter the mitogenic propensities of metabolite, e.g., 20-HETE or whether the HO-1 increase itself is ultimately responsible for uncontrolled cell growth.

REFERENCES

1. Folkman, J. and Shing, Y. (1992) *J. Biol. Chem.* **267**, 10931–10934.
2. Abraham, N.G., Drummond, G.S., Lutton, J.D., and Kappas, A. (1996) *Cell. Physiol. Biochem.* **6**, 129–168.
3. Goodman, A.I., Choudhury, M., da-Silva, J.L., Schwartzman, M.L., and Abraham, N.G. (1997) *Proc. Soc. Exp. Biol. Med.* **214**, 54–61.
4. Matsumoto, A., Hanayama, R., Nakamura, M., Suzuki, K., Fujii, J., Tatsumi, H., and Taniguchi, N. (1998) *Free. Radic. Res.* **28**, 383–391.
5. Nishie, A., Ono, M., Shono, T., Fukushi, J., Otsubo, M., Onoue, H., Ito, Y., Inamura, T., Ikezaki, K., Fukui, M., Iwaki, T., and Kuwano, M. (1999) *Clin. Cancer. Res.* **5**, 1107–1113.
6. Tozer, G.M., Prise, V.E., Motterlini, R., Poole, B.A., Wilson, J., and Chaplin, D.J. (1998) *Int. J. Radiat. Oncol. Biol. Phys.* **42**, 849–853.
7. Tsuji, M.H., Yanagawa, T., Iwasa, S., Tabuchi, K., Onizawa, K., Bannai, S., Toyooka, H., and Yoshida, H. (1999) *Cancer. Lett.* **138**, 53–59.
8. Dwyer, B.E., Nishimura, R.N., Lu, S.Y., and Alcaraz, A. (1996) *Brain. Res. Mol. Brain. Res.* **38**, 251–259.
9. Hara, E., Takahashi, K., Tominaga, T., Kumabe, T., Kayama, T., Suzuki, H., Fujita, H., Yoshimoto, T., Shirato, K., and Shibahara, S. (1996) *Biochem. Biophys. Res. Commun.* **224**, 153–158.
10. Torisu-Itakura, H., Furue, M., Kuwano, M., and Ono, M. (2000) *Jpn. J. Cancer. Res.* **91**, 906–910.
11. Kurata, S.I., Matsumoto, M., and Nakajima, H. (1996) *J. Cell. Biochem.* **62**, 314–324.
12. Schacter, B.A. and Kurz, P. (1986) *Clin. Invest. Med.* **9**, 150–155.
13. Weidner, N., Semple, J.P., Welch, W.R., and Folkman, J. (1991) *N. Engl. J. Med.* **324**, 1–8.
14. Maines, M.D. and Abrahamsson, P.A. (1996) *Urology.* **47**, 727–733.
15. Abraham, N.G., Lavrovsky, Y., Schwartzman, M.L., Stoltz, R.A., Levere, R.D., Gerritsen, M.E., Shibahara, S., and Kappas, A. (1995) *Proc. Natl. Acad. Sci. USA* **92**, 6798–6802.
16. Gerritsen, M.E., Carley, W., and Milici, A.J. (1988) *Adv. Cell. Cult.* **6**, 35–67.
17. Deramaudt, B.M., Braunstein, S., Remy, P., and Abraham, N.G. (1998) *J. Cell. Biochem.* **68**, 121–127.
18. Stoltz, R.A. and Abraham, N.G. (1996) *Proc. Natl. Acad. Sci. USA* **93**, 2832–2837.
19. Abraham, N.G., Lin, J.H., Schwartzman, M.L., Levere, R.D., and Shibahara, S. (1988) *Int. J. Biochem.* **20**, 543–558.
20. Maines, M.D. (1988) *FASEB J.* **2**, 2557–2568.
21. Marks, G.S., Brien, J.F., Nakatsu, K., and McLaughlin, B.E. (1991) *Trends. Pharmacol. Sci.* **12**, 185–188.
22. Shibahara, S., Muller, M., and Taguchi, H. (1987) *J. Biol. Chem.* **262**, 12889–12892.
23. Mitani, K., Fujita, H., Sassa, S., and Kappas, A. (1989) *Biochem. Biophys. Res. Commun.* **165**, 437–441.

24. Nath, K.A., Balla, J., Jacob, H.S., Vercellotti, G.M., Levitt, M., and Rosenberg, M.E. (1992) *J. Clin. Invest.* **90**, 267–270.

25. Solangi, K., Sacerdoti, D., Goodman, A.I., Schwartzman, M.L., Abraham, N.G., and Levere, R.D. (1988) *Am. J. Med. Sci.* **296**, 387–391.

26. Goodman, A.I., Choudhury, M., da Silva, J.L., Jiang, S., and Abraham, N.G. (1996) *J. Cell. Biochem.* **63**, 342–348.

27. McCoubrey, W.K., Jr., Ewing, J.F., and Maines, M.D. (1992) *Arch. Biochem. Biophys.* **295**, 13–20.

28. Shibahara, S., Yoshizawa, M., Suzuki, H., Takeda, K., Meguro, K., and Endo, K. (1993) *J. Biochem. Tokyo* **113**, 214–218.

29. Hibbs, J.B., Westenfelder, C., Taintor, R., Vavrin, Z., Kablitz, C., Babanowski, J.P., Ward, J.H., and Menlove, R.L. (1992) *J. Clin. Invest.* **89**, 867–877.

30. Cantoni, L., Rossi, C., Rizzardini, M., Gadina, M., and Ghezzi, P. (1991) *Biochem.J.* **279**, 891–894.

31. Lutton, J.D., da Silva, J.-L., Moqattash, S., Brown, A.C., Levere, R.D., and Abraham, N.G. (1992) *J. Cell. Biochem.* **49**, 259–265.

32. Lavrovsky, Y., Schwartzman, M.L., Levere, R.D., Kappas, A., and Abraham, N.G. (1994) *Proc. Natl. Acad. Sci. USA* **91**, 5987–5991.

33. Sessa, W.C., Abraham, N.G., Escalante, B., and Schwartzman, M.L. (1989) *J. Hypertens.* **7**, 37–42.

34. Pinto, A., Abraham, N.G., and Mullane, K.M. (1986) *J. Pharmacol. Exp. Ther.* **236**, 445–451.

35. Sacerdoti, D., Escalante, B., Abraham, N.G., McGiff, J.C., Levere, R.D., and Schwartzman, M.L. (1989) *Science* **243**, 388–390.

36. Laniado-Schwartzman, M. and Abraham, N.G. (1992) *Pediatr. Nephrol.* **6**, 490–498.

37. Laniado-Schwartzman, M., Abraham, N.G., Sacerdoti, D., Escalante, B., and McGiff, J.C. (1992) *Tohoku J Exp. Med.* **166**, 85–91.

38. Schwartzman, M.L., Martasek, P., Rios, A.R., Levere, R.D., Solangi, K., Goodman, A.I., and Abraham, N.G. (1990) *Kidney Int* **37**, 94–99.

39. Laniado-Schwartzman, M., Lavrovsky, Y., Stoltz, R.A., Conners, M.S., Falck, J.R., Chauhan, K, and Abraham, N.G. (1994) *J. Biol. Chem.* **269**, 24321–24327.

40. Yang, L., Quan, S., and Abraham, N.G. (1999) *Am. J. Physiol.* **277**, L127-L133.

41. Lin, F., Rios, A., Falck, J.R., Belosludtsev, Y., and Schwartzman, M.L. (1995) *Am. J. Physiol.* **269**, F806–F816.

42. Abraham, N.G., Mieyal, P.A., Quan, S., Yang, L., Burke-Wolin, T., Mingone, C.J., Goodman, A.I., Nasjletti, A., and Wolin, M.S. (2001) *Am. J. Physiol.* **submitted**.

43. Schwartzman, M.L., Martasek, P., Rios, A.R., Levere, R.D., Solangi, K., Goodman, A.I., and Abraham, N.G. (1990) *Kidney. Int.* **37**, 94–99.

42

ON THE PROMOTING ACTION OF TAMOXIFEN IN THE P-DIMETHYLAMINOAZOBENZENE INDUCED HEPATOCARCINOGENESIS CF1 MICE MODEL AND THE CYTOPROTECTIVE ROLE OF HEME OXYGENASE

Elba Vázquez[a], Esther Gerez[a], Fabiana Caballero[a], Leda Oliveri[a], Nora Falcoff[b], María Lujan Tomaro[c], and Alcira Batlle[a,*]

[a]Centro de Investigaciones sobre Porfirinas y Porfirias (CIPYP)
(CONICET-Facultad de Ciencias Exactas y Naturales, UBA)
Ciudad Universitaria, Pabellón II
2do Piso, 1428 Buenos Aires
Argentina
[b]Departamento de Patología
Hospital "Dr. Bernardo Houssay"
Vicente López, Buenos Aires
Argentina
[c]Departamento de Química Biológica
Facultad de Farmacia y Bioquímica, UBA
Buenos Aires
Argentina

* Corresponding author: Vlamonte 1881 10A-1056 Buenos Aires, Argentina.
Fax: +54 11 4811 7447, Email: batlle@mail.retina.ar

Part of the text, Fig. 1 and Fig. 3 were reprinted from The International Journal of Biochemistry and Cell Biology, Vol 33, N° 7, Caballero F., Gerez E., Oliveri L., Falcoff N., Batlle A., Vazquez E.: "On the promoting action of tamoxifen in a model of hepatocarcinogenesis induced by p-dimethylaminoazobenzene in CF1 mice", pp. 681–690, Copyright (2001), with permission from Elsevier Science.

1. INTRODUCTION

Carcinogenesis is a multistage, multimechanism process, involving the irreversible alteration of an initiated cell, followed by its clonal proliferation, from which the acquisition of the invasive and metastatic phenotypes are generated (Trosko and Chang, 2001). Chemically induced and spontaneous liver tumors share some metabolic alterations. We have developed an experimental mice model, by administrating p-dimethylaminoazobenzene (DAB) in the diet, to study the onset of hepatocarcinogenesis, that includes a primary activating liver stage, provoking biochemical alterations that lead to the true initiation step (Gerez et al., 1997; Vazquez et al., 1999). Tamoxifen (TMX) has proven to be an effective palliative treatment for advanced breast cancer with low reported incidence of side effects (Forbes, 1997). It has been observed, that TMX is a carcinogen capable of both initiating and promoting liver carcinogenesis in female rats (Dragan et al., 1996). Recently, we have demonstrated that TMX produced changes in hepatic enzyme activities which may be relevant to the metabolism and disposition of this and/or other drugs. We have found that long-term treatment with TMX enhances hepatocarcinogenesis induced by DAB (Caballero et al., 2001).

While drugs intervention to develop cancer could occur at each step, in this chapter we will focus on the analysis of the response of hemo oxygenase (HO) activity against oxidative stress and liver damage induced by the carcinogen during the initiation and promotion steps.

We will report data from our laboratory on the effects of DAB and/or TMX on hepatic HO in an animal model of chemically induced liver carcinogenesis. We also propose that the inducible isoform of this enzyme has a protective role in the initiation phase of carcinogenesis.

2. CHEMICAL CARCINOGENESIS AND OXIDATIVE STRESS

In a pluricellular organism, carcinogenesis is the result of a chemical, physical, biological or genetic insult or a combination of them. Generally, exposition to some carcinogens involves changes occuring in the transformed cell genome and these changes include several types of mutations (Harris, 1991; Pitot, 1993; Loeb and Loeb, 2001). Chronic administration of chemical carcinogens can induce one or more mutations, as frequently observed in experimental models. The potential carcinogenic effect of several agents can only be shown when they are administrated for long periods, and in high doses (Yuspa and Shields, 1997). Three stages are accepted to occur in the carcinogenic process: initiation, promotion and progression (Pitot, 1991; Pitot and Dragan, 1991). The first and the last stages involve cell genetic alterations. The promotion step is characterized by an altered expression of the initiated cell genome, and changes occurring are probably the result of the interaction between the genetic alterations which were induced during the initiation step and microenvironmental factors called promoters. Such alterations in the genetic expression, include a selective increase in the replication of the initiated cells (Dragan and Pitot, 1992). The initiation stage is not easy to study due to difficulties to identify, isolate and specifically characterize the initiated cells. Nevertheless, liver rodent is the most appropriate experimental model, where the early progenie of initiated cells give rise to the hepatic altered enzyme foci.

Most of the initiated cells never progress to the following steps, and remain quiescent in the organism. Therefore, the first step in the development of cancer is a common event that can occur spontaneously and can be induced in experimental models (Pitot and Dragan, 1991). We have developed an experimental mice model to study the onset of hepatocarcinogenesis that includes a primary activating liver stage, which involves the whole organ, provoking biochemical alterations that lead to the true initiation step (Gerez et al., 1997; Vazquez et al., 1999). The carcinogen used is p-dimethylaminoazobenzene (DAB).

DAB metabolization through the microsomal hepatic fraction generates free radicals. The reduction of azodyes depends on the electron transport from NADPH to P450-azodyes complex, through NADPH-P450 reductase, generating an intermediate radical which can react with oxygen to form a superoxide anion. However, intermediates with a reduced electron can be temporarily stable, even in the presence of oxygen and proceed to the addition of a second electron, followed by protonation producing unstable hidrazo derivates, which spontaneously form a primary amine (Zbaida, 1995).

Oxidative stress occurs when the balance between the production of reactive oxygen species (ROS) overrides the antioxidant capability of the cell. ROS may interact with and modify macromolecules resulting in altered cell function (Halliwell and Aruoma, 1993; Klaunig et al., 1998). Oxidative injury may produce selective cell death and a compensatory increase in cell proliferation. This stimulus leads to unrepaired DNA damage resulting in the formation of a new mutation and, potentially, new initiated cells and/or enhancement of the selective clonal expansion of latent initiated cells (Klaunig et al., 1998).

ROS induce damage to DNA, including breakage of chains, base modifications and crosslinking of proteins and DNA (Halliwell and Aruoma, 1993).

Cerruti et al. (1994) showed that during oxidative stress "early genes", such as c-fos and c-myc, are activated altering the cell cycle progression.

Under our experimental model, the activating liver status was indicated by a an important and sustained P450 levels increase, enhanced lipid peroxidation and diminished levels of the natural antioxidant defense system (Gerez et al., 1997, 1998; Vazquez et al., 1999). In the model, the alterations provoked by the carcinogen would be fixed through cell proliferation, reflected by hepatomegalia and in response to necrosis (Caballero et al., 2001).

3. HEME OXYGENASE AGAINST OXIDATIVE DAMAGE IN CANCER

Recently, its was proposed that the heme oxygenase inducible isoform (HO1) could play an effective role to counteract oxidative damage (Maines, 1997). HO1 is induced by host oxidative stress stimuli, and activation of HO1 gene expression is considered to be an adaptive cellular response to environmental oxidative stress.

Heme oxygenase the rate limiting step in the heme degradation pathway generates free iron, biliverdine and carbon monoxide (Abraham et al., 1995; Maines, 1997; Tomaro and Batlle, 2001). Three isoforms: HO1, HO2, HO3 have been described in mammals, products of different genes (Elbirt and Bonkovsky, 1999).

HO1 is inducible in response to its own substrate heme, heavy metals, stress, UV radiation, chemical agents, hypoxia, hyperoxia, hyperthermia, ischemia, drugs and many other stressful stimuli and HO is identical to heat shock protein 32 (Maines et al., 1986; Abraham et al., 1996; Maines, 1999). HO2 is chemically an structurally different to HO1 and is constitutively expressed in most tissues (Maines et al., 1986; Maines, 1999). HO3 is similar to HO2 and dissimilar to HO1 in structure, but it is a poor catalyst (McCoubrey et al., 1997). Due to its inducibility, HO1 is the best known isoform. This gene includes structural moieties for termic shock and metal dependent transcriptional factors (Galbraith, 1999).

For a long time, since its identification in 1974 (Maines and Kappas, 1974), the HO system was always associated to the degradation of the heme moiety of hemoglobin to toxic waste compounds, including biliverdin and bilirubin. However in the late 1980s, a physiological role for bilirubin as a potent scavenger and antioxidant was proposed (Stocker et al., 1987).

Which could be the reason why the degradative heme pathway is activated?. One probable answer is, to deplete heme as a protective regulatory response. However, heme is very stable and it is frequently bound to transport proteins.

HO1 also plays a regulatory role in iron metabolism and might be involved in the chaperoning of iron to ferritin (Poss and Tonegawa, 1997; Yachie et al., 1999). Iron catalyzes the production of HO Leading to biological macromolecules damage (McCord, 1998). Metalloporphyrins, potent inhibitors of HO, protect cells from various insults (Wagner et al., 2000). The antioxidant role of HO1 is supported by crucial experimental observations: 1) the gene expression is very sensible to oxidative stress in several mammal tissues (Maines, 1997), 2) the induction of the protein or the transfection with HO1 gene protect tissues against oxidative damage (Abraham et al., 1995) and 3) HO knock out mice show a very low defense to oxidative stress (Poss and Tonegawa, 1997). In addition, it was reported that HO1 derived bilurubine decreased apoptosis mediated by peroxinitrite in endothelian cells (Clark et al., 2000).

Yan et al. (1998) demonstrated that DRH rats, a strain derived from DONRYU, resistant to carcinogens, showed a very low incidence of hepatic tumors induced by diet 3′-methyl-DAB administration. The induction of HO1 and hepatocyte growth factor (HGF) mRNA was higher in DONRYU, confirming the cytotoxic sensitivity of this strain.

Other suggested function for HO1 is related to neovascularization. Macrophages are key participants in the angiogenis process and its infiltration is correlated to the appearance of histologic malignancies. Nishie et al. (1999) proposed HO1 as a marker of activated macrophages infiltration and of neovascularization in human gliomas.

It was also suggested that HO1 expression in oral carcinoma can be used to identify patients with low risk of limph nodules metastases (Tsuji et al., 1999). Other authors have detected HO1 expression only in the tumoral cells but not in macrophages and intra arterial Zn protoporphyrin administration in rat hepatomas, considerably suppressed tumor growth (Doi et al., 1999).

4. POTENTIAL PROTECTIVE ROLE OF HO1 IN CHEMICAL CARCINOGENESIS

The present study examines the modulation of HO in response to oxidative damage during the initiation and promotion of chemical hepatocarcinogenesis. Male CF1 mice (n = 50) were placed on dietary p-dimethylaminoazobenzene (DAB, 0.5%, w/w) during a whole period of 28 weeks. Control animals (n = 30) were fed with a standard laboratory diet for the same period. Other two groups of non-treated animals (n = 40) and DAB treated animals (n = 50), received TMX citrate (Gador Laboratories, 0.025% w/w) in the diet since day 20.

The toxicity effect of the drugs was evaluated by the ratio of liver to body weight along the treatment. Under our protocol, studies were undertaken for over 6 months during which no variation, when compared with their respective control groups, was detected in the weight ratio in CF1 mice, either in animals treated with TMX alone or in those receiving DAB+TMX, reflecting that TMX by itself would not evoke any additional liver damage (Caballero et al., 2001). The toxicity of DAB was confirmed by the significant increase in the weight ratio observed in all the animals that received the carcinogen.

In all groups, HO activity was significantly diminished when compared with controls on weeks 9 and 12 (40% for DAB groups and 15% for TMX group). At longer periods, HO activity showed a progressive induction for animals fed with DAB. Moreover, we have found that the activity of HO, in the group receiving DAB + TMX was not further affected by TMX when compared to the DAB group. (Fig. 1A).

As expected GST activity, considered as a marker of liver damage (Gerez et al., 1997; Vazquez et al., 1999), increased about 100% along the 20 weeks of DAB treatment, but then the enzyme activity diminished on week 28 with a tendency of reaching its lowest values in the DAB+TMX group. TMX alone produced a 30% and a 75% enhancement of GST activity between weeks 12 and 20 and then on week 28 respectively (Fig. 1B).

DAB treatment greatly increased TBARS content (Gerez et al., 1998), reaching its highest values (DAB group: 325%; DAB+TMX: 230%) on week 12, remaining unchanged for the rest of the period examined. In the TMX treated group, TBARS were unmodified during the first 12 weeks and then increased to 60% from week 20 onwards (Fig. 1C).

Adaptive compensatory responses seem to be the most likely explanation for the majority of the changes observed after TMX administration in the drug metabolizing system and in the oxidation status. TMX is a mixed inducer of phase I enzymes with a weak phenobarbital-like pattern of induction (White et al., 1993), TMX has also significant effect on hepatic phase II enzymes expression which may have implications for the carcinogenicity and/or therapeutic activity of the drug (Nuwaysir et al., 1996).

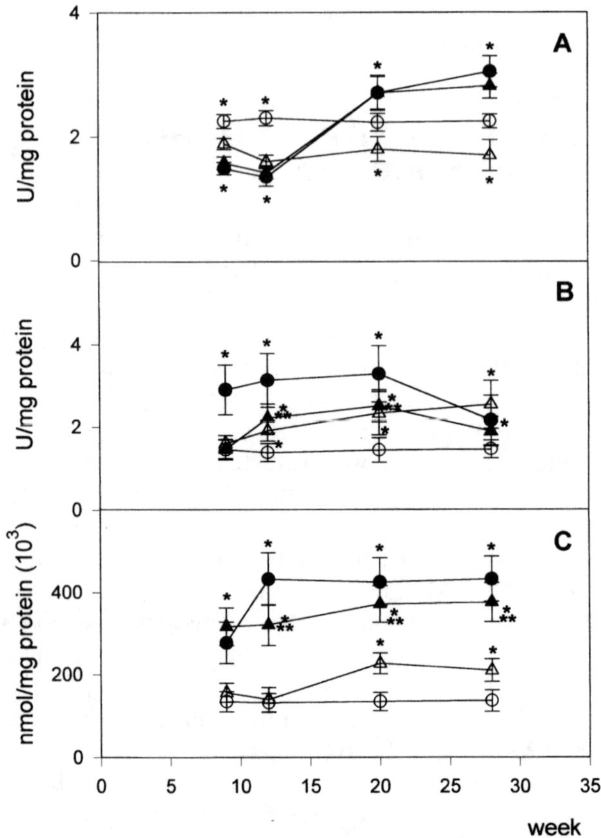

Figure 1. Long-term effect of DAB and TMX on hepatic HO (A) and GST (B) activities and peroxidation index (TBARS content, C). The control groups of animals received a standard laboratory diet (○), during 28 weeks. Other group were fed with DAB (0.5% w/w) during the same period (●). Non treated animals and DAB treated animals received TMX citrate (0.025% w/w) in the diet since day 20 (TMX group △; DAB+TMX: ▲). Mice were killed at 9, 12, 20 and 28 weeks of treatment The data represent mean values ± SD. Other experimental conditions are as indicated in Materials and methods. Significantly different (p < 0.05) from control group * and DAB group **.

It is worth noting the great increase in GST activity towards the end of the assay, reflecting liver damage. This could be of particular importance, since TMX required metabolic activation for the subsequent formation of covalent bound protein adducts (Pathak and Badell, 1994). Other properties of TMX may cause alteration in liver cell homeostasis and thereby lead to liver damage.

The discrepancy between the low HO activity and the high oxidative stress provoked by the carcinogen during the first 10 weeks of treatment, did not match with the cytoprotective role proposed for HO1. Various agents can induce HO1 via the stress response elements (SER) on the mouse HO1 gene and this is the best described mechanism of HO1 induction (Alam et al., 1995). Therefore, we decided to analize HO1 mRNA expression using the pRHO1 (kindly provided by Dr. Shigeki

Shibahara, Tohoku University, Sendai, Japan), during the initiation stage of liver carcinogenesis. Northern blot analysis clearly revealed a significant induction (170–190%) in DAB treated animals from week 6 to week 28 (Fig. 2). These results suggest that HO1 might have a protective role against the damaging effect of ROS inducing the synthesis of HO as an adaptive cellular response, whereas the enzyme activity would be modulated by post transcriptional events.

Histological liver sections of 6/6 DAB treated animals (28 weeks) showed normal architecture with cell hyperplasia, determined by morphological features. The presence of bile thrombosis and abundant hemosiderosis in hepatocytes and in Kupffer cells was also detected. Foci of necrosis, irregularly distributed and surrounded by leucocytes, were observed with a frequency of 3/6.

Grossly, at the end point of the assay hepatic neoplastic soft nodules (0.3–0.4 cm in diameter) at the periphery of the liver left lobe, were observed in four of six of the DAB+TMX treated mice. This pathology was not found in mice receiving either DAB or TMX alone.

Microscopically, histological sections of the nodules showed atypical hepatocytes with solid, trabecular and pseudoglandular pattern of growth. Cytologically, the tumor cells were polygonal, had an abundant granular eosinophilic cytoplasm and more than one prominent nucleoli. Neoplastic cells showed intracytoplasmic bile pigment (Fig. 3).

Neighbouring parenchymal cells were dysplasic, with intense cholestasis and intracytoplasmic bile pigment in Kupffer cells and canaliculi, most of the cells showed fatty change microvesicular steatosis.

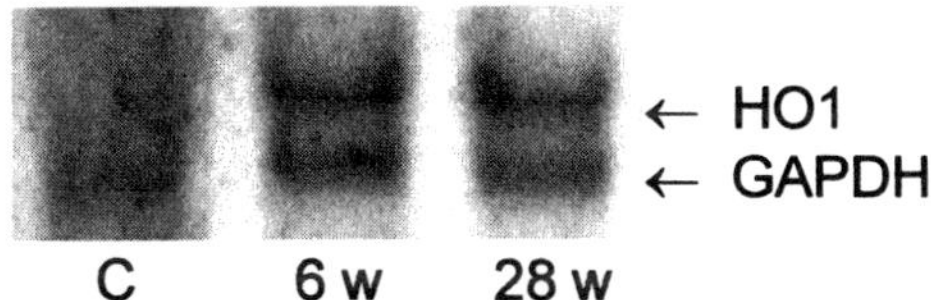

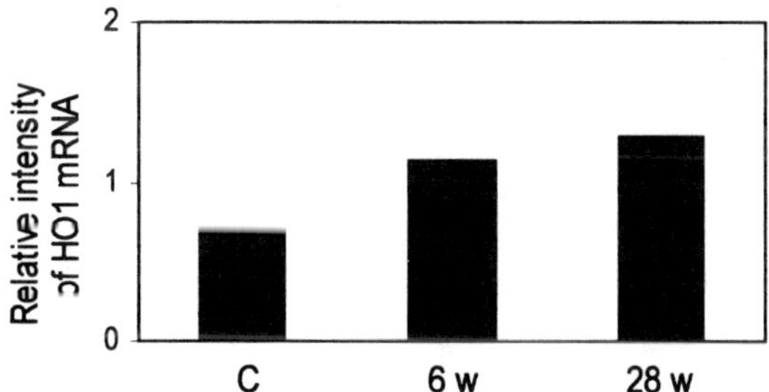

Figure 2. Northern blot analysis from HO1 mRNA expression in controls and DAB treated animals during 6 or 28 weeks. Northern hybridization was performed with use of pRHO1 (provided by Dr. Shigeki Shibahara, Tohoku University, Sendai, Japan). In the lower panel, the HO1 mRNA signals were quantified by densitometric analysis, after normalization with GAPDH mRNA signals as a standard mRNA expression, using an Image Scanner and the ImageMaster software (Amersham Pharmacia, Biotech AB, SE-751 84, Uppsala, Sweden).

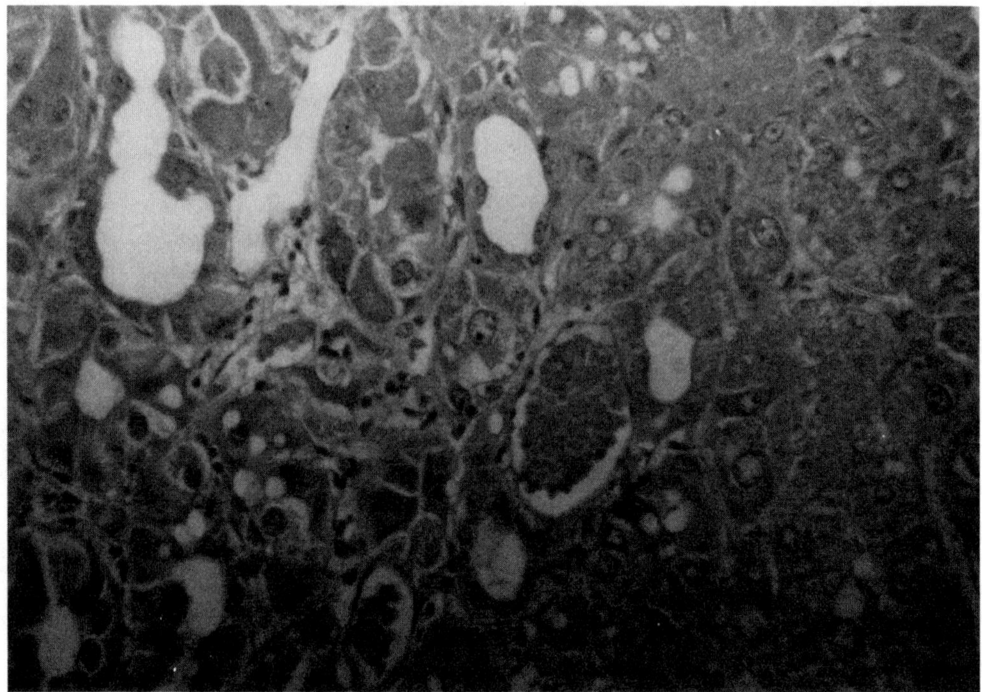

Figure 3. Histological features of liver nodules. Mice were treated with DAB (0.5% w/w) and TMX (0.025% w/w) for 28 weeks. Liver was sectioned and stained with hematoxylin-eosin (magnification × 160). The figure shows hepatocellular carcinoma solid, trabecular and acinar type. Other experimental conditions are as indicated in Material and methods.

The present study demonstrates that the long-term treatment with TMX promotes hepatocarcinogenesis initiated with DAB in mice. It is worth noting that the same dose of TMX in rat increases the incidence of hepatocellular carcinoma in diethylnitrosamine-initiated lesions (Wogan, 1997) and that this dose is comparable to the levels observed in patients under usual treatment protocols (Dragan et al., 1996). In contrast to previously published reports with our model, we have found that TMX fails to initiate hepatocarcinogenesis. This apparent disagreement may be due to the use of different animal species and to the length of treatment (Hirsimäki et al., 1993; Greaves et al., 1993).

5. SUMMARY

This report has focused on the response of HO against liver damage produced by oxidative stress during the carcinogenic process and the possible cytoprotective role of HO1. We have shown that HO1 mRNA levels were greatly induced in the liver of DAB treated animals, and that this level of induction persisted until the end of the intoxication protocol and as long as the oxidative damage lasted. We have demonstrated that the profile of HO activity is the same in animals treated with both the initiator agent alone and the initiator and the promoter and that this enzyme activity would be modulated by post transcriptional events.

ACKNOWLEDGMENTS

We are very grateful to Dr. Shigeki Shibahara, Tohoku University, Sendai, Japan for providing the pRHO1. We also thank to Mrs. B. Corvalan and the staff of the Department of Pathology of Dr. Bernardo Houssay Hospital for valuable technical assistance during this study. A. Batlle and E. Vazquez are members of the Career of Scientific Researcher at the Argentine National Research Council (CONICET). F. Caballero and E. Gerez hold the post of Research Assistant at the CONICET. N. Falcoff is head of the Department of Pathology at the Hospital "Dr. Bernardo Houssay", Vicente López. This work has been supported by grants from the CONICET; the University of Buenos Aires, Argentina; AICR, UK; and Science and Technology Argentine Agency.

REFERENCES

Abraham N.G., Lavrovsky Y., Schwartzman M., Stoltz R., Levere R., Gerritsen M., Shibahara S., and Kappas A., 1995. Transfection of the human heme oxygenase gene into rabbit coronary microvessel endothelial cells: protective effect against heme and hemoglobin toxicity. *Proc Natl Acad Sci USA* **92**:6798.

Abraham N.G., Drummed J.D., Lutton A., and Kappas A., 1996. The biological significance and physiological role of heme oxygenase. *Cell Physiol Biochem* **6**:129.

Alam J., Camhi S., and Choi A.M., 1995. Identification of a second region upstream of the mouse heme oxygenase-1 gene that functions as a basal level and inducer-dependent transcription enhancer. *J Biol Chem* **270**:11977.

Caballero F., Gerez E., Oliveri L., Falcoff N., Batlle A., and Vazquez E., 2001. On the promoting action of tamoxifen in a model of hepatocarcinogenesis induced by p-dimethylaminoazobenzene in CF1 mice. *Int J Biochem Cell Biol* **33**:681.

Cerutti P., Ghosh R., Oya Y., and Amstad P., 1994. The role of the cellular antioxidant defense in oxidant carcinogenesis. *Env Health Perspect* **102**:123.

Clark J., Foresti R., Green C., and Motterlini R., 2000. Dynamics of haem oxygenase-1 expression and bilirubin production in cellular protection against oxidative stress. *Biochem J* **348**:615.

Doi K., Akaike T., Fujii S., Tanaka S., Ikebe N., Beppu T., Shibahara S., Ogawa M., and Maeda H., 1999. Induction of haem oxygenase-1 nitric oxide and ischaemia in experimental solid tumours and implications for tumour growth. *Br J Cancer* **80**:1945.

Dragan Y. and Pitot H., 1992. The role of the stages of initiation and promotion in phenotypic diversity during hepatocarcinogenesis in the rat. *Carcinogenesis* **13**:739.

Dragan Y., Fahey S., Nuwaysir E., Sattler C., Babcock K., Vaughan J., McCague R., and Pitot H., 1996. The effect of tamoxifen and two of its non-isomerizable fixed-ring analogues on multistage rat hepatocarcinogenesis. *Carcinogenesis* **17**:585.

Elbirt K. and Bonkovsky H., 1999. Heme oxygenase: recent advances in understanding its regulation and role. *Proc Assoc Am Physicians* **111**:438.

Forbes J. 1997. Scientific review of tamoxifen. *Semin Oncol* **24**:S1–5.

Galbraith R., 1999. Heme oxygenase: who needs it?. *Proc Soc Exp Biol Med* **222**:299.

Gerez E., Vazquez E., Caballero F., Polo C., and Batlle A., 1997. Altered heme pathway regulation and drug metabolizing enzyme system in a mouse model of hepatocarcinogenesis. Effect of veronal. *Gen Pharmac* **29**:569.

Gerez E., Caballero F., Vazquez E., Polo C., and Batlle A., 1998. Hepatic enzymatic metabolism alteration and oxidative stress during the onset of carcinogenesis: protective role of alpha tocopherol. *Eur J Cancer Prevention* **7**:69.

Greaves P., Goonetilleke R., Nunn G., Topham J., and Orton T., 1993. Two-year carcinogenicity study of tamoxifen in Alderley Park Wistar-derived rats. *Cancer Res* **53**:3919.

Halliwell B. and Aruoma O., 1993. In: DNA and free Radicals (Ed. Ellis Horwood) Chichester, England.

Harris C., 1991. Chemical and physical carcinogenesis: advances and perspectives for the 1990s. *Cancer Res* **51**:5023.

Hirsimäki P., Hirsimäki Y., Nieminen L., and Payne B., 1993. Tamoxifen induces hepatocellular carcinoma in rat liver: a one year study with two carcinogens. *Arch Toxicol* **67**:49.

Klaunig J., Xu Y., Isenberg J., Bachowski S., Kolaja K., Jiang J., Stevenson D., and Walborg E., 1998. *Environ Health Perspect* **106**:289.

Loeb K. and Loeb L., 2001. Significance of multiple mutations in cancer. *Carcinogenesis* **21**:379.

Maines M.D., 1997. The heme oxygenase system: a regulator of second messenger gases. *Annu Rev Pharmacol Toxicol* **37**:517.

Maines M.D., 1999. In: Heme Oxygenase: Clinical Applications and Functions. (Ed. Maines MD.) Boca Ratón, Fl, CRC Press, pp. 109.

Maines M.D. and Kappas A., 1974. Cobalt induction of hepatic heme oxygenase, with evidence that cytochrome P-450 is not essential for this enzyme activity. *Proc. Natl. Acad. Sci. U.S.A.* **71**:4293.

Maines M.D., Traskhel G.M., and Kutty R.K., 1986. Characterization of two constitutive forms of microsomal heme oxygenase ony one molecular species of the enzyme is inducible. *J. Biol. Chem* **261**:411–419.

McCord J.M., 1998. Iron, free radicals, and oxidative injury. *Semin Hematol* **35**:5.

McCoubrey W.K., Huang T., and Maines M.D., 1997. Isolation and characterization of a cDNA from the rat brain that encodes hemoprotein heme oxygenase-3. *Eur J Biochem* **247**:725.

Nishie A., Ono M., Shono T., Fukushi J., Otsubo M., Onoue H., Ito Y., Inamura T., Ikezaki K., Fukui M., Iwaki T., and Kuwano M., 1999. Macrophage infiltration and heme oxygenase-1 expression correlate with angiogenesis in human gliomas. *Clin Cancer Res* **5**:1107.

Nuwaysir E.F., Daggett D.A., Jordan V.C., and Pitot H.C., 1996. Phase II enzyme expression in rat liver in response to the antiestrogen tamoxifen. *Cancer Res.* **56**:3704.

Pathak D.N. and Badell W.Y., 1994. DNA adducts formation by tamoxifen with rat and human liver microsomal activating systems. *Carcinogenesis* **15**:529.

Pitot H., 1991. Endogenous carcinogenesis: the role of tumor promotion. *Proc Soc Exp Biol Med* **198**:661.

Pitot H., 1993. The molecular biology of carcinogenesis. *Cancer* **72**:962.

Pitot H. and Dragan Y., 1991. Facts and theories concerning the mechanisms of carcinogenesis. *FASEB J* **5**:2280.

Poss K. and Tonegawa S., 1997. Heme oxygenase 1 is required for mammalian iron reutilization. *Proc Natl Acad Sci USA* **94**:10919.

Stocker R., Yamamoto Y., Ms Donagh AF., Glazer AN., and Ames BN., 1987. Bilirubin is an antioxidant of possible physiological importance. *Science* **235**:1043–1046.

Tomaro M.L. and Batlle A., 2001. Bilirubin: its role in cytoprotection against oxidative stress. *Int. J. Biochem. Cell Biol.* **33**:1.

Tsuji M., Yanagawa T., Iwasa S., Tabuchi K., Onizawa K., Bannai S., Toyooka H., and Yoshida H., 1999. Heme oxygenase-1 expression in oral squamous cell carcinoma as involved in lymph node metastasis. *Cancer Lett* **138**:53.

Trosko J. and Chang C., 2001. Mechanism of up-regulated gap junctional intercellular communication during chemoprevention and chemotherapy of cancer. *Mutat Res* **480–481**:219.

Vazquez E., Gerez E., Caballero F., Polo C., and Batlle A., 1999. Drug metabolizing enzyme system and heme pathway in hepatocarcinogenesis. *Cancer Biochem Biophys* **17**:25.

Wagner K., Hua Y., De Courten-Myers G., Broderick J., Nishimura R., Lu S., and Dwyer B., 2000. Tin-mesoporphyrin, a potent heme oxygenase inhibitor, for treatment of intracerebral hemorrhage: in vivo and in vitro studies. *Cell Mol Biol* **46**:597.

White I.N.H., Davies A., Smith L.L., Dawson S., and De Matteis F., 1993. Induction of CYP2B1 and 3A1 and associated monooxygenase activities by tamoxifen and certain analogues in the liver of female rats and mice. *Biochem Pharmacol* **45**:21.

Wogan G., 1997. Review of the toxicology of tamoxifen. *Seminars in Oncology* **24**:S1–87.

Yachie A., Niida Y., Wada T., Igarashi N., Kaneda H., Toma T., Ohta K., Kasahara Y., and Koizumi S., 1999. Oxidative stress causes enhanced endothelial cell injury in human heme oxygenase-1 deficiency. *J Clin Invest* **103**:129.

Yan Y., Higashi K., Yamamura K., Fukamachi Y., Abe T., Gotoh S., Sugiura T., Hirano T., Higashi T., and Ichiba M., 1998. Different responses other than the formation of DNA-adducts between the livers of carcinogen-resistant rats (DRH) and carcinogen-sensitive rats (Donryu) to 3′-methyl-4-dimethylaminoazobenzene administration. *Jpn J Cancer Res* **89**:806.

Yuspa S. and Shields P., 1997. In: Cancer: Principles & Practice of Oncology (Ed. V. DeVita, S. Hellman, S. Rosenberg) Lippincott-Raven Publishers, Philadelphia, pp. 185.

Zbaida S., 1995. The mechanism of microsomal azoreduction: predictions based on electronic aspects of structure-activity relationships. *Drug Metab Rev* **27**:497.

43

HUMAN HEME OXYGENASE GENE TRANSFER PROMOTES BODY GROWTH AND NORMALIZES BLOOD PRESSURE IN SPONTANEOUSLY HYPERTENSIVE RATS WITHOUT AFFECTING SPRAGUE-DAWLEY RATS

Liming Yang, Shuo Quan, and Nader G. Abraham[a]

[a]Director of Gene Therapy
New York Medical College
Valhalla, New York 10595
USA
Tel: 914.594.4132
Fax: 914.594.4119
E-mail: nader_abraham@nymc.edu

1. INTRODUCTION

1.1. Heme Oxygenase and the Cardiovascular System

HO catalyzes the conversion of heme to bilirubin, free iron and CO. As the key enzyme in heme degradation, HO activity governs cellular heme concentration. Heme is ubiquitous and of vital importance in eukaryotes, functioning as the prosthetic moiety of various heme proteins including CYP, COX, thromboxane and prostacyclin synthases, NOS, catalase, peroxidase, and hemoglobin. Excess release of heme from myoglobin and hemoglobin has been shown to be highly toxic to several organs including the kidney. In view of the widespread utilization of heme and its potential toxicity,[1] it is not surprising that tissues that normally process relatively large amounts of heme possess the biochemical means to efficiently and safely degrade the heme ring and regulate heme availability for essential functions. The kidney has a high capacity

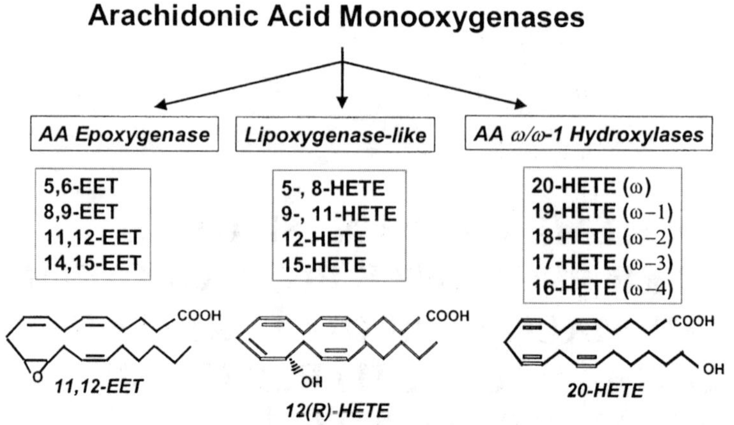

Arachidonic Acid Monooxygenases

Figure 1. The three metabolic pathways of arachidonic acid. HETE-hydroxyeicosatraenoic acid; EET-epoxyeicosatrienoic acid; DHET-dihydroxyeicosatrienoic acid.

to metabolize heme.[2] The importance of HO thus resides in degrading heme and regulating the activity of heme proteins (Fig. 1). To date, three isoforms, the products of three distinct genes, have been identified. HO-2 is constitutively expressed and localized primarily in the brain, testis and the vascular endothelium. The newly described HO-3 exhibits 90% homology to HO-2 and its mRNA has been detected in many tissues including the kidney.[3] However, it lacks significant catalytic activity and may only function as a heme regulatory protein through its capacity to bind heme.[3] HO-1 is an inducible isoform with widespread tissue distribution including the liver, kidney and lung;[4] however, its induction is also dependent in the cellular heme pool. Cellular heme is regulated by the rate limiting enzyme of heme synthesis, aminolevalinic acid (ALAS) and the rate limiting enzyme in heme degradation, HO-1.

Most of our knowledge with regard to kidney HO relates to HO-1 and its role in cardiovascular and inflammatory conditions. HO-1 is induced in cardiovascular tissues not only in response to its substrate heme, but also by a wide array of stimuli that include H_2O_2, heavy metals (such as Sn, Au, Cd), endotoxin, cytokines (TNF, IL-1, TGFβ), and glutathione depletion. The induction of HO-1 by these stimuli, which all share the capacity to exert oxidative stress, has led to the suggestion that induction of HO-1 in cells may be a beneficial response in such inimical settings. Preinduction of HO-1 by endotoxin improved renal function following the administration of glycerol (Balla's group). Renal failure in glycerol treated rats was greatly exacerbated by tin protoporphyrin (SnPP), an inhibitor of HO,[5] but prevented by treatment with an inducer of HO. Additional studies have shown that HO-1 is induced in a time- and dose-dependent fashion in the kidney following the administration of cisplatin, a chemotherapeutic agent associated with nephrotoxicity.[6] Inhibition of HO activity worsened renal injury in this disease model.[7] Together, these studies documented the presence and inducibility of HO-1 in the kidney and indicated its importance as a major defense mechanism against injury. The protective property of HO-1 was mainly attributed to the production of biliverdin/bilirubin which are potent antioxidants and to the removal of prooxidant heme and free iron via rapid induction of ferritin.[8] However, there is another side to the heme-HO system with regard to renal function in which the differential significance of HO-1 and HO-2 has not been explored. Both heme and products derived from its metabolism potentially influence blood pressure, As mentioned above, heme is required for the synthesis and activity of heme proteins that can affect vascular and renal function including guanylate cyclase, NOS, CYP and COX (Fig. 1). HO not only controls the availability of heme for the full expression of these activities but also is solely responsible for the generation of CO which, by itself, can bind to the heme moiety of these heme proteins causing either enzyme activation or inhibition (vide infra).

1.2. Carbon Monoxide and the Cardiovascular System

CO has recently been reported to play an important role in the regulation of vascular tone and blood pressure. Furchgott and others have concluded that CO arising from heme via metabolism by HO exerts a vasodilatory effect[9;10] and that inhibition of HO by tin mesoporphyrin (SnMP) causes vascular constriction.[11] More recently, heme has been shown to elicit vasodilation in the ductus arteriosus[12] and inhibition of HO by chromium mesoporphyrin (CrMP) resulted in elevation of arterial pressure. Kozma et al.[13] showed that HO inhibitors magnify myogenic tone in the gracilis muscle arterioles and suggested that endogenous HO-derived carbon monoxide plays a role in the regulation of basal tone in resistance vessels. Recent studies using isoform-specific HO antisense oligonucleotides demonstrated that HO-2-derived CO is an inhibitory regulator of small renal artery reactivity to phenylephrine further emphasizing the potential role of HO isoforms in the regulation of renal hemodynamics.[14] Snyder and coworkers found that an HO inhibitor elicited vasodilation that was prevented by the inhibition of NOS with L-NAME.[15] NOS is a heme-dependent enzyme which can be inhibited by CO.[16;17] On the other hand, NO donors caused a noticeable increase in the expression of HO-1 mRNA.[18] Given the widely accepted importance of NOS, and the observation that CO and NO cross-talk between the HO and NOS genes, it is likely that this cross talk will impact on a number of biological processes related to COX and CYP. The current application does not address HO cross talk with NOS per se. Yet, the possibility of such an interaction must be considered in terms of its functional significance as ignorance of this interaction may confound interpreting key experiments in this proposal. We intend to focus our interests on the physiological significance of HO isoforms in terms of regulating CYP and COX activities within kidney structures. Relevant to this proposal then, is a consideration of renal CYP and COX activities and possible interaction between HO isoforms and their expression.

1.3. CYP-AA Metabolism and HO

Capdevila et al.[19] have identified arachidonic acid (AA) as a substrate for liver CYP. McGiff and colleagues have uncovered the significant role of renal CYP-AA metabolites that elicited their effects on blood vessels and renal tubules by modulating vascular tone and ion transport.[20] CYP enzymes catalyze the metabolism of AA to four epoxyeicosatrienoic acids (EETs), ω/ω-1 alcohols (20-HETE and 19-HETE), and six regioisomeric cis-trans conjugated monohydroxyeicosatetraenoic acids (HETEs)[19;20]. Some of these metabolites (e.g., 5,6 EET and 20-HETE) can be processed further by COX to products having biological activities[21;22] (Fig. 1).

The renal synthesis of EETs is increased in rats fed a diet with a high content of NaCl.[23] The renal excretion of 20-HETE is increased in SHR[24] as is the excretion of EETs in patients with pregnancy-induced hypertension. Metabolites of AA via the CYP pathway are endowed with biological activities most relevant to the vascular and renal mechanisms of blood pressure regulation. For example, 20-HETE can stimulate contraction of vascular smooth muscle[21;25], inhibit Na^+-K^+-ATPase,[26] and reduce the activity of K-channels in arterial smooth muscle and renal tubular cells.[27;28] On the other hand, EETs can relax vascular smooth muscle[29], increase K-channel activity in arterial smooth muscle,[30] and inhibit ion transport in the renal tubules. 20-HETE affects movement of ions, constricts blood vessels, participates in tubuloglomerular feedback, and acts as a mitogen[31;32] effects that are prohypertensive. In contrast, 20-HETE may also act in an antihypertensive mechanism as 20-HETE inhibits sodium chloride reabsorption in the TALH. A deficiency in 20-HETE formation is postulated to be responsible for increased chloride reabsorption, the underlying abnormality in the Dahl salt-sensitive hypertensive rat. 20-HETE has been implicated in mediation of renal vasoconstriction;[22] and EETs in mediation of renal vasodilatory responses to AA and bradykinin[33] and inhibition of renin secretion. Oxygenation of AA via the CYP pathway yields products capable of affecting blood pressure that act in renal and vascular mechanisms. Generally speaking, the EETs are involved in antihypertensive mechanisms while 20-HETE is prohypertensive. Accordingly, the development of hypertension in the SHR has been linked to increased expression of a prohypertensive mechanism mediated by a metabolite of AA via the CYP pathway, presumably, 20-HETE.[34;35]

Since the publication of our first reports characterizing rat and human kidney HO and demonstrating its role in modulating renal and liver CYP-AA metabolism[36-41] over 90 publications have confirmed that induction of HO activity affects CYP-AA metabolism (reviewed in ref. 7). Sacerdoti et al. used stannous chloride, a specific inducer of renal HO, and demonstrated that increased HO activity resulted in depletion of CYP-AA metabolites, 20-HETE and 19-HETE, associated with reduction of blood pressure.[37] For review of AA metabolism and hypertension, see Quilley et al.[42] We have shown that administration of heme arginate caused a rapid decrease in blood pressure in the young SHR[43-45] and pretreatment with zinc 2,4 deuteroprotoporpyrin-X-bisglycol (ZnDPBG), a potent HO inhibitor developed in our laboratory,[46] greatly attenuated the antihypertensive response to heme arginate.[41;42] We further explored the action of $SnCl_2$ in the kidney by comparing its time-dependent activation of HO with inhibition of CYP-AA synthesis of 20-HETE, reduction in blood pressure and urinary electrolyte excretion.[45] The findings suggested a reciprocal relationship between 20-HETE levels and blood pressure. However, it has yet to be determined if these antihypertensive effects are mediated by HO in renal extra vascular tissues or HO in the renal vasculature and whether they are linked to HO-dependent changes in CYP-AA activity and expression. The present study was undertaken to establish

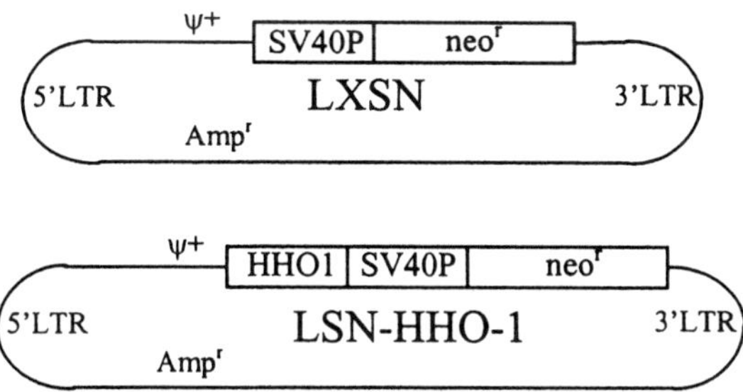

Figure 2. Schematic diagram of retroviral vectors LSN-HHO-1 and LXSN. HHO-1, human heme oxygenase-1 cDNA; neo^r, neomycin resistant gene; SV40P, SV40 promoter; LTR, long termial repeat.

chimeric rats expressing the human HO-1 (HHO-1) gene with the use of a retroviral vector and to investigate the impact of augmentation of HO activity on the development of hypertension in spontaneously hypertensive rats (SHR) and Sprague-Dawley (SD) rats.

2. GENERAL METHODS

2.1. Construction of the Retroviral Recombinant LSN-HHO-1

The HHO-1–expressing replication-deficient retrovirus vector LSN-HHO-1 was constructed with the use of the backbone of LXSN[47] vector (Fig. 2), as previously described.[48]

2.2. Virus Production and HHO-1 Gene Expression in Transduced Cells

PA317 retroviral packaging cells were transfected with retroviral vectors (LXSN and LSN-HHO-1) using lipofectamine reagent (GIBCO-BRL). The cells were selected for neomycin resistance (neo^r) in medium containing G418 (600 µg/ml) (GIBCO-BRL). For each isolated clone, the viral titer was determined by infection of NIH3T3 fibroblasts.[19] A clone of packaging cell line PA317/LSN-hHO (PA317/hHO) producing the highest viral titer of 1.4×10^6 colony-forming unit (cfu)/ml was employed in the experiments described below. Simultaneously, we got NIH3T3/LXSN and NIH3T3/LSN-hHO (NIH3T3/hHO) cell lines. Using supernatants of the PA317/LXSN and PA317/LSN-hHO cells, we infected RLMV cells. After selection with G418, transduced endothelial cells RLMV/LXSN and RLMV/LSN-hHO (RLMV/hHO) were obtained. HHO-1 and neo^r gene expression were confirmed by RT-PCR (Fig. 3).

2.3. Animal Treatment

Pregnant SHR and SD mothers were purchased from Taconic laboratories (Germantown, NY). Five-day-old SHR and SD from the same litter were divided into

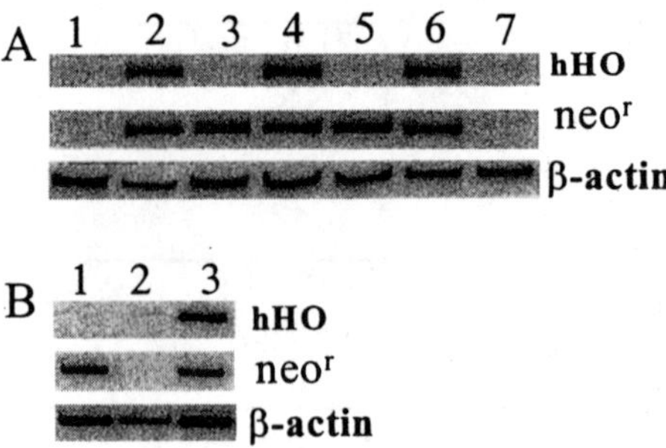

Figure 3. **(A)** Detection of human HO-1, neo[r] and β-actin transcripts by RT-PCR in RNA extracted from PA317 and NIH3T3 cells transduced or transduced with retroviral vectors LSN-hHO or LXSN. RT-PCR products amplified in the RNA extracted from NIH3T3 fibroblasts (lane 1), NIH3T3/hHO (lanes 2 and 4), NIH3T3/LXSN (lane 3), PA317/LXSN (lane 5), PA317/hHO (lane 6) and PA317 (lane 7) cells. α-[^{32}P]-dCTP-incorporated RT-PCR was performed to amplify hHO, neo[r] and β-actin transcripts, respectively. After electrophoresis in 8% non-denaturing polyacrylamide gel, the dried gel was exposed to an X-ray film. **(B)** Detection of human HO-1 transcripts by RT-PCR analysis in the RNA isolated from RLMV endothelial cells transduced or transduced with retroviral vectors (LSN-hHO or LXSN). Lane 1, RLMV/LXSN; Lane 2, RLMV; Lane 3, RLMV/hHO.

3 treatment groups: vehicle (HBSS), LXSN (viral control), and LSN-HHO-1 (experimental). Treatments were administered by bolus injection of 10 µL HBSS with or without LXSN or LSN-HHO-1 concentrated viral particles (1×10^{10} CFU/ml) directly into the left ventricle under methoxyflurane anesthesia with 96% survival, as previously described.[49] Animals were weaned at 21-days of age; males were separated and utilized for all experiments. Systolic blood pressure was measured by tail cuff sphygmomanometry twice weekly, starting at 4 weeks of age. Total body weight gain and average daily food intake were measured for all treatment groups. The daily food intake was estimated by measuring the weight of food utilized by each cage divided by the number of animals in that cage. At 12 weeks of age, animals were subjected to radiography (n = 6) and measurements of nose-to-tail length and fibula length. At different times (4, 8, 12, 16 and 20 weeks), rats were sacrificed and tissues, including kidney, liver, lung, brain and aorta were isolated for determination of HHO-1 expression and HO activity.

3. RESULTS

3.1. Enhancement of HO Activity in Endothelial Cells Tranduced with Human HO-1 Gene

We examined the expression of total human HO-1 protein and HO activity in endothelial cells nontransduced or transduced with retroviral vector LSN-hHO or control vector LXSN, repectively (Fig. 4A and 4B). The basal levels of HO activity in nontransduced or LXSN-transduced endothelial cells were not significantly different (P > 0.05). HO activity in human HO-1 gene-transduced endothelial cells was increased by 2.3-fold as compared to transduced cells (P < 0.05) (Fig. 4B). To ascertain the characteristics of the expressed protein, we tested the effect of stannic mesoporphyrin (SnMP), a potent inhibitor of HO activity,[5] on cultured endothelial cells. Addition of 10 µM SnMP to cell cultures inhibited the enzyme activity (data not shown). To assess that the increase in human HO-1 mRNA and HO activity was associated with the elevation of HO-1 protein, Western blot analysis was performed on cell lysates obtained from nontransduced and HO transduced cells. Using HO antibodies which recognize both human and rat HO-1 protein, we found that human HO-1 gene-transduced endothelial cells exhibited a strong signal for HO-1 protein. The quantitative evaluation of HO-1 protein was measured by scanning densitometry. The results showed that human HO-1 protein was increased by 2.1 fold (Fig. 4A). Comparison of the levels of HO-1 protein with HO activity showed that HO activity was increased at a comparable level to HO-1 protein. These findings indicated that LSN-HHO-1 transduced endothelial cells generate increased levels of HO-1 protein and activity as compared to control cells.

3.2. Effect of HHO-1 Gene Transduction on Blood Pressure and Body Weight in Rats

Blood pressure increased as a function of time in all treatment groups; however, up to 20 weeks of age, the blood pressure of LSN-HHO-1 treated SHR was significantly lower than that of vehicle-treated or LXSN-treated SHR (Fig. 5).

The fact that hypertension is attenuated in SHR expressing the HHO-1 gene implies that a mechanism dependent on the function of this human gene lowers blood pressure in SHR. Thus, the attenuating influence of LSN-HHO-1 treatment on the development of hypertension in SHR is most likely the functional manifestation of increased HO activity consequent to HHO-1 expression. This is in agreement with

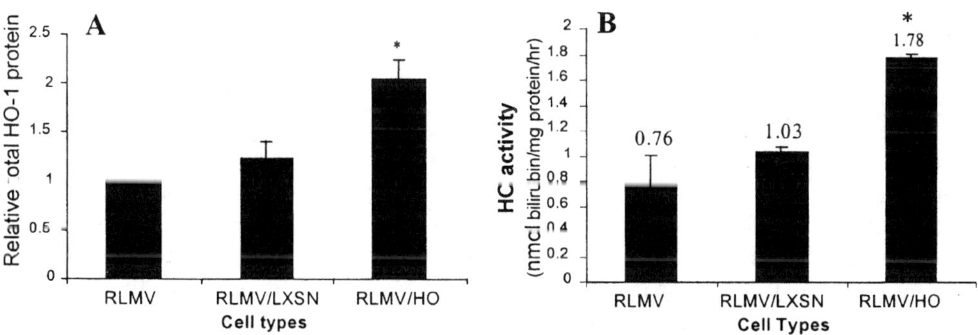

Figure 4. (A) Relatively immunoreactive HO 1 protein in RLMV endothelial cells nontransduced or transduced with LSN-hHO or LXSN by Western blot-ECL analysis. * P < 0.05 vs control RLMV endothelial cells. Values are expressed as mean + SD of three experiments. (B) HO activity in RLMV endothelial cells nontransduced or transduced with retroviral vectors LSN-hHO or LXSN. HO activity (nmol bilirubin/mg protein) is expressed as mean ± SD of three experiments. * P < 0.05 vs control RLMV cells.

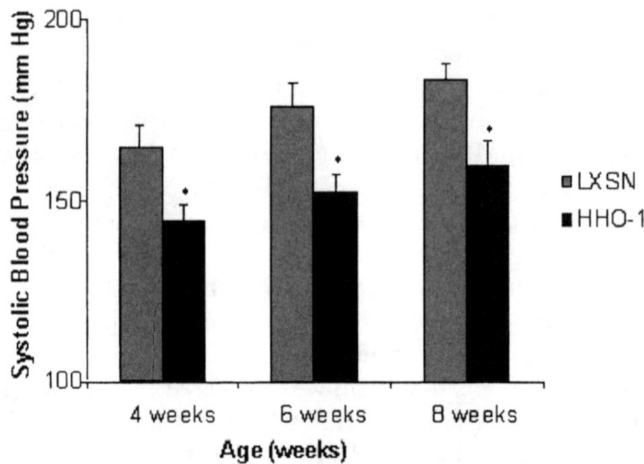

Figure 5. Effect of LSN-HHO-1 treatment on systolic blood pressure in SHR. At 5 days of age, animals were injected with LXSN or LSN-HHO-1 (1×10^{10} CFU/ml). Blood pressure was determined twice weekly by the tail cuff method at 4, 6 and 8 weeks of age. Results are mean ± SE. Blood pressure of LSN-HHO-1 treated rats (n = 32) was significantly different (p < 0.01) from vehicle-treated SHR (n = 10) or LXSN-treated rats (n = 32) by ANOVA and the Student's t test.

reports that other interventions that increase HO activity in SHR as a result of inducing HO-1 expression also attenuate the development of hypertension.[50]

In this study, 24-hour urinary excretion of 20-HETE in 7-week-old SHR treated with LSN-HHO-1 was lower ($P < 0.01$) than that of vehicle or LXSN treated SHR (1.29 ± 0.06, 2.34 ± 0.09 and 2.2 ± 0.06 ng 20-HETE/ml in LSN-HHO-1, LXSN and vehicle-treated, respectively; n = 6). Hence, the renal excretion of 20-HETE, a vasoconstrictor eicosanoid, is reciprocally related to the expression of HO-1. Reduction in 20-HETE production may favor, at least in part, the lowering of blood pressure in SHR since 20-HETE promotes vasoconstriction at renal and extrarenal sites and, consequently, may be a mediator of prohypertensive mechanisms in SHR, an experimental model in which 20-HETE production was reported to increase.[51;52]

Endogenous CO was proposed to inhibit myogenic vascular tone,[13] which may explain, in some way, the lower blood pressure of LSN-HHO-1 treated SHR. We studied pressure-diameter relationships in isolated gracilis muscle arterioles of 12-week-old SHR treated with LXSN or LSN-HHO-1 viral particles. Stepwise elevation of intraluminal pressure over the range of 40–100 mmHg elicited pressure-dependent reductions in arteriolar diameter expressed as a percentage of the passive diameter in the absence of calcium. The pressure-induced constrictor response at both 80 and 100 mmHg was less intense ($P < 0.01$) in arterioles of SHR treated with LSN-HHO-1 than in arterioles of SHR treated with LXSN viral particles (Fig. 6). Importantly, after treatment of the vessels with chromium mesoporphyrin (CrMP, 15 μmol/L), an inhibitor of HO, the intensity of the pressure-induced reduction in arteriolar diameter significantly increased (Fig. 6).

Beginning at 4 weeks of age, the body weight of SHR treated with LSN-HHO-1 viral particles surpassed that of SHR treated with vehicle alone or with LXSN viral control (Fig. 7A); the nose-to-tail length and fibula length of SHR treated with LSN-HHO-1 also exceeded the corresponding values in SHR treated with vehicle or LXSN (Fig. 7B), however, food intake was similar in all treatment groups (data not shown). SHR expressing the HHO-1 gene grew faster than SHR lacking the HHO-1 gene, particularly during the first 12 weeks.

Importantly, the increase in somatic growth associated with HHO-1 expression in SHR was both proportionate and not associated with an increase in food intake. The latter observation, striking and most unexpected, implies that SHR expressing the HHO-1 gene are, in metabolic terms, more efficient than their counterparts lacking the HHO-1 gene and thus can develop somatically at a faster pace without consuming greater amounts of food.

Recent reports indicate that both human[53] and mice[54] lacking the HO-1 gene display severe growth retardation. HO-1 gene expression has been shown to play a role in cell proliferation and cell death; indeed, previous studies demonstrated that elevation of HO-1 activity by gene transfer to rabbit coronary microvessel endothelial cells enhances cell proliferation and increases angiogenesis.[55] In contrast, Lee et al.[56] demonstrated that overexpression of HO-1 in pulmonary epithelial human cell line results in cell growth arrest, highlighting the cell specific effects of HO-1 on cellular proliferation. A priori, a consequence of HO activity may impact directly on somatic growth by influencing the production and/or cellular actions of hormones and factors that stimulate or inhibit growth. Cheriathundam et al.[57] have found a significant correlation between hepatic levels of HO-1 and growth hormone in transgenic mice. Others showed that in the rat, hormones as thyroid hormone and insulin increase hepatic HO. Moreover, consensus binding sites for nuclear factor-$\kappa\beta$, activator protein-1, activator protein-2 and interleukin-6 responsive elements, as well as

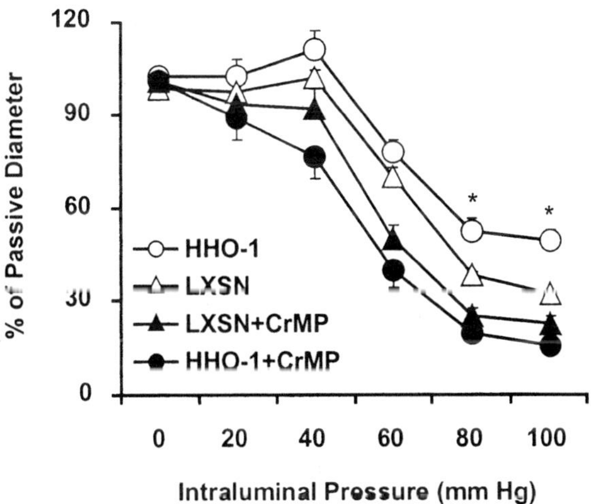

Figure 6. Pressure-diameter relationship in gracilis muscle arterioles isolated from SHR treated with LSN-HHO-1 or LXSN in (15 µmol/L). Results are mean ± SE; n = 6. *Significantly different (p < 0.01) fromLXSN-treated or LSN-HHO-1 treated rats in the presence of CrMP.

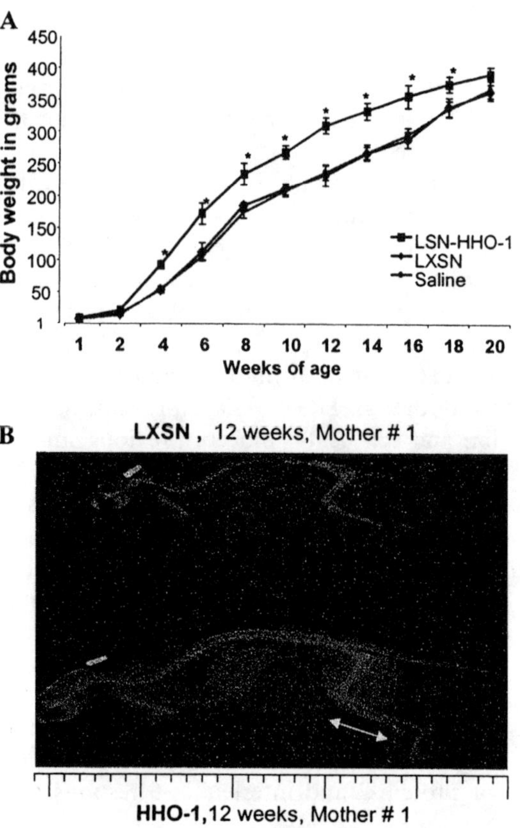

Figure 7. Growth promoting activity of HHO-1 gene transfer in LSN-HHO-1 treated SHR. **(A)** Total body weight gain of LSN-HHO-1 treated SHR rats (n = 32) and LXSN treated SHR rats (n = 23). *(P < 0.01) significantly different from vehicle or LXSN-treated rats. **(B)** A representative x-ray radiogram of 12-week-old LSN-HHO-1 (lower) and LXSN (upper panel) treated SHR from the same mother. The arrows indicate that the fibula length of SHR treated with LSN-HHO-1 significantly greater than the corresponding values in SHR treated with LXSN (31.9 ± 1.3 and 23.8 ± 0.7 mm, respectively, P < 0.01), the fibula length of SHR treated with LXSN vector alone were comparable to those of vehicle-treated SHR.

other transcription factors have been reported in the promoter region of the HO-1 gene.[58] Whether these transcription factors activate certain elements involved in promoting SHR growth remains to be investigated. Our study offers no information on the mechanism(s) responsible for the observed growth-promoting effect of HHO-1 expression in SHR. This does not detract from the importance of our finding, which, for the first time, link the heme-HO system to the regulation of somatic growth in SHR. Similar experiments were performed to inject the concentrated retroviruses (expressing HHO-1) to 5-day-old SD rats via cardiac route, but did not modulate blood pressure or body growth (data not shown).

Taken together, this study demonstrates that delivery of the human HO-1 gene to SHR and SD by means of a recombinant retrovirus vector attenuates the development of hypertension and accelerates somatic growth. These findings support the notion that one or more consequences of HO activity subserve vasodepressor and body growth promoting functions in SHR without affecting SD. The study also highlights the usefulness of gene transfer approaches to the investigation of the functional tasks of the heme-HO system in situations such as hypertension.

4. SUMMARY

Heme oxygenase (HO) catalyzes the conversion of heme to biliverdin, with release of free iron and carbon monoxide. Both heme and carbon monoxide have been implicated in the regulation of vascular tone. A retroviral vector containing human HO-1 cDNA (LSN-HHO-1) was constructed, subjected to purification and concentration of the viral particles to achieve $5 \times 10^9 - 1 \times 10^{10}$ CFU/ml. The ability of concentrated infectious viral particles to express human HO-1 *in vivo* was tested. A single intra-cardiac injection of the concentrated infectious viral particles (expressing HHO-1) to 5-day-old spontaneously hypertensive rats (SHR) resulted in functional expression of the human HO-1 gene and attenuation of the development of hypertension. Rats expressing HHO-1 showed a significant decrease in urinary excretion of a vasoconstrictor arachidonic acid metabolite and a reduction in myogenic responses to increased intraluminal pressure in isolated arterioles. Similarly, a single intra-cardiac injection of the concentrated infectious viral particles (expressing HHO-1) to 5-day-old Sprague-Dawley (SD) rats did not modulate blood pressure or body growth. Unexpectedly, HHO-1 chimeric rats showed a simultaneous significant proportionate increase in somatic growth only in SHR but not SD. Thus, delivery of human HO-1 gene by retroviral vector attenuates the development of hypertension and promotes body growth in SHR without modulation in normal rats.

REFERENCES

1. Wagener, F.A.D.T.G., E. Feldman, T. de-Witte, and N.G. Abraham. Heme induces the expression of adhesion molecules ICAM-1, VCAM-1, and E selectin in vascular endothelial cells. *Proc. Soc. Exp. Biol. Med.* 216:456–463, 1997.
2. Hebbel, R.P., W.T. Morgan, J.W. Eaton, and B.E. Hedlund. Accelerated autooxidation and heme loss due to instability of sickle hemoglobin. *Proc. Natl. Acad. Sci. USA* 85:237–241, 1988.
3. McCoubrey, W.K., Jr., T.J. Huang, and M.D. Maines. Isolation and characterization of a cDNA from the rat brain that encodes hemoprotein heme oxygenase-3. *Eur. J. Biochem.* 247:725–732, 1997.
4. Abraham, N.G., G.S. Drummond, J.D. Lutton, and A. Kappas. The biological significance and physiological role of heme oxygenase. *Cell. Physiol. Biochem.* 6:129–168, 1996.
5. Nath, K.A., J. Balla, H.S. Jacob, G.M. Vercellotti, M. Levitt, and M.E. Rosenberg. Induction of heme oxygenase is a rapid protective response in rhabdomyolysis in the rat. *J. Clin. Invest.* 90:267–270, 1992.
6. Agarwal, A., J. Balla, J. Alam, A.J. Croatt, and K.A. Nath. Induction of heme oxygenase in toxic renal injury: a protective role in cisplatin nephrotoxicity in the rat. *Kidney Int.* 48:1298–1307, 1995.
7. Nath, K.A., J. Balla, A.J. Croatt, and G.M. Vercellotti. Heme protein-mediated renal injury: a protective role for 21 aminosteroids in vitro and in vivo. *Kidney Int.* 47:592–602, 1995
8. Eisenstein, R.S., D. Garcia-Mayol, W. Pettingell, and H.N. Munro. Regulation of ferritin and heme oxygenase synthesis in rat fibroblasts by different forms of iron. *Proc. Natl. Acad. Sci. U.S.A.* 88:688–692, 1991.

9. Furchgott, R.F. and D. Jothianandan. Endothelium-dependent and -independent vasodilation involving cyclic GMP: relaxation induced by nitric oxide, carbon monoxide and light. *Blood Vessels* 28:52–61, 1991.

10. Johnson, R.A., M. Lavesa, B. Askari, N.G. Abraham, and A. Nasjletti. A heme oxygenase product, presumably carbon monoxide, mediates a vasodepressor function in rats. *Hypertension* 25:166–169, 1995.

11. Johnson, R.A., M. Lavesa, K. DeSeyn, M.J. Scholer, and A. Nasjletti. Heme oxygenase substrates acutely lower blood pressure in hypertensive rats. *Am. J. Physiol.* 271:H1132–H1138, 1996.

12. Coceani, F., L. Kelsey, E. Seidlitz, G.S. Marks, B.E. McLaughlin, H.J. Vreman, D.K. Stevenson, M. Rabinovitch, and C. Ackerley. Carbon monoxide formation in the ductus arteriosus in the lamb: implications for the regulation of muscle tone. *Br. J. Pharmacol.* 120:599–608, 1997.

13. Kozma, F., R.A. Johnson, F. Zhang, C. Yu, X. Tong, and A. Nasjletti. Contribution of endogenous carbon monoxide to regulation of diameter in resistance vessels. *Am. J. Physiol.* 276:R1087–R1094, 1999.

14. Kaide, J.-I., F. Zhang, C. Yu, N.G. Abraham, and A. Nasjletti. Heme oxygenase (HO)-2-derived carbon monoxide (CO) is an inhibitory regulator of small renal artery reactivity to phenylephrine (PE). *Hypertension* 34:P151, 1999.

15. Zakhary, R., S.P. Gaine, J.L. Dinerman, M. Ruat, N.A. Flavahan, and S.H. Snyder. Heme oxyge-nase 2: endothelial and neuronal localization and role in endothelium-dependent relaxation. *Proc. Natl. Acad. Sci. U.S.A.* 93:795–798, 1996.

16. Chakder, S., S. Rathi, X.L. Ma, and S. Rattan. Heme oxygenase inhibitor zinc protoporphyrin IX causes an activation of nitric oxide synthase in the rabbit internal anal sphincter. *J. Pharmacol. Exp. Ther.* 277:1376–1382, 1996.

17. Klatt, P., K. Schmidt, and B. Mayer. Brain nitric oxide synthase is a haemoprotein. *Biochem. J* 288 (Pt 1):15–17, 1992.

18. Wang, J.L., H.F. Cheng, M.Z. Zhang, J.A. McKanna, and R.C. Harris. Selective increase of cyclooxygenase-2 expression in a model of renal ablation. *Am. J. Physiol.* 275:F613–F622, 1998.

19. Capdevila, J., N. Chacos, J. Werringloer, R.A. Prough, and R.W. Estabrook. Liver microsomal cytochrome P-450 and the oxidative metabolism of arachidonic acid. *Proc. Natl. Acad. Sci. U.S.A.* 78:5362–5366, 1981.

20. McGiff, J.C. Cytochrome P-450 metabolism of arachidonic acid. *Annu. Rev. Pharmacol. Toxicol.* 31:339–369, 1991.

21. Schwartzman, M., M.A. Carroll, D. Sacerdoti, N.G. Abraham, and J.C. McGiff. The renal cytochrome P450 system generates novel arachidonic acid metabolites. *Adv. Exp. Med. Biol.* 259:109–129, 1989.

22. Carroll, M.A., M.P. Garcia, J.R. Falck, and J.C. McGiff. Cyclooxygenase dependency of the renovascular actions of cytochrome P450-derived arachidonate metabolites. *J. Pharmacol. Exp. Ther.* 260:104–109, 1992.

23. Capdevila, J.H., J.R. Falck, and R.W. Estabrook. Cytochrome P450 and the arachidonate cascade. *FASEB J.* 6:731–736, 1992.

24. Schwartzman, M.L., K. Omata, F.M. Lin, R.K. Bhatt, J.R. Falck, and N.G. Abraham. Detection of 20-hydroxyeicosatetraenoic acid in rat urine. *Biochem. Biophys. Res. Commun.* 180:445–449, 1991.

25. Abraham, N.G., J.L. Chertkov, and J. Harrison. Gene transfer into hematopoietic stem cell: Role of adherent stromal cell layer. In Carella, A.G. ed., ed. Acute Leukemias and Chronic Myelogenous Leukemia New Developments. Comm-Tur, Genova, 1993, 94–103.

26. Schwartzman, M.L., N.G. Abraham, J. Masferrer, M.W. Dunn, and J.C. McGiff. Cytochrome P450 dependent metabolism of arachidonic acid in bovine corneal epithelium. *Biochem. Biophys. Res. Commun.* 132:343–351, 1985.

27. Zou, A.P., J.T. Fleming, J.R. Falck, E.R. Jacobs, D. Gebremedhin, D.R. Harder, and R.J. Roman. 20-HETE is an endogenous inhibitor of the large-conductance Ca(2+)-activated K+ channel in renal arterioles. *Am. J. Physiol* 270:R228–R237, 1996.

28. Zou, A.P., H.A. Drummond, and R.J. Roman. Role of 20-HETE in elevating loop chloride reabsorption in Dahl SS/Jr rats. *Hypertension* 27:631–635, 1996.

29. Carroll, M.A., M. Schwartzman, M. Baba, N.G. Abraham, and J.C. McGiff. Formation of biologically active cytochrome P450-arachidonate metabolites in renomedullary cells. *Adv.Prost.Thromb. Leukot.Res.* 17B:714–718, 1987.

30. Campbell, W.B., D. Gebremedhin, P.F. Pratt, and D.R. Harder. Identification of epoxyeicosatrienoic acids as endothelium-derived hyperpolarizing factors. *Circ.Res.* 78:415–423, 1996.

31. Zou, A.P., J.D. Imig, P.R. Ortiz-de-Montellano, Z. Sui, J.R. Falck, and R.J. Roman. Effect of P-450 omega-hydroxylase metabolites of arachidonic acid on tubuloglomerular feedback. *Am.J.Physiol.* 266:F934–F941, 1994.

32. Lin, F., A. Rios, J.R. Falck, Y. Belosludtsev, and M.L. Schwartzman. 20-Hydroxyeicosatetraenoic acid is formed in response to EGF and is a mitogen in rat proximal tubule. *Am.J.Physiol.* 269:F806–F816, 1995.

33. Fulton, D., J.C. McGiff, and J. Quilley. Contribution of NO and cytochrome P450 to the vasodilator effect of bradykinin in the rat kidney. *Br.J. Pharmacol.* 107:722–725, 1992.

34. Laniado-Schwartzman, M. and N.G. Abraham. The renal cytochrome P-450 arachidonic acid system. *Pediatr.Nephrol.* 6:490–498, 1992.

35. Stec, D.E., M.R. Trolliet, J.E. Krieger, H.J. Jacob, and R.J. Roman. Renal cytochrome P4504A activity and salt sensitivity in spontaneously hypertensive rats. *Hypertension* 27:1329–1336, 1996.

36. Schwartzman, M.L., N.G. Abraham, M.A. Carroll, R.D. Levere, and J.C. McGiff. Regulation of arachidonic acid metabolism by cytochrome P-450 in rabbit kidney. *Biochem.J.* 238:283–290, 1986.

37. Sacerdoti, D., B. Escalante, N.G. Abraham, J.C. McGiff, R.D. Levere, and M.L. Schwartzman. Treatment with tin prevents the development of hypertension in spontaneously hypertensive rats. *Science* 243:388–390, 1989.

38. Schwartzman, M.L., P. Martasek, A.R. Rios, R.D. Levere, K. Solangi, A.I. Goodman, and N.G. Abraham. Cytochrome P450-dependent arachidonic acid metabolism in human kidney. *Kidney Int.* 37:94–99, 1990.

39. Sessa, W.C., N.G. Abraham, B. Escalante, and M.L. Schwartzman. Manipulation of cytochrome P-450 dependent renal thromboxane synthase activity in spontaneously hypertensive rats. *J.Hypertens.* 7:37–42, 1989.

40. Martasek, P., K. Solangi, A.I. Goodman, R.D. Levere, R.J. Chernick, and N.G. Abraham. Properties of human kidney heme oxygenase: inhibition by synthetic heme analogues and metalloporphyrins. *Biochem.Biophys.Res.Commun.* 157:480–487, 1988.

41. Lin, J.H., P. Villalon, P. Martasek, and N.G. Abraham. Regulation of heme oxygenase gene expression by cobalt in rat liver and kidney. *Eur.J. Biochem.* 192:577–582, 1990.

42. Quilley, J., C.P. Bell-Quilley, and J.C. McGiff. Eicosanoids and hypertension. *Hypertension* Second Edition: 1995.

43. Martasek, P., M.L. Schwartzman, A.I. Goodman, K.B. Solangi, R.D. Levere, and N.G. Abraham. Hemin and L-arginine regulation of blood pressure in spontaneous hypertensive rats. *J.Am.Soc.Nephrol.* 2:1078–1084, 1991.

44. Levere, R.D., P. Martasek, B. Escalante, M.L. Schwartzman, and N.G. Abraham. Effect of heme arginate administration on blood pressure in spontaneously hypertensive rats. *J.Clin.Invest* 86:213–219, 1990.

45. da-Silva, J.L., M. Tiefenthaler, E. Park, B. Escalante, M.L. Schwartzman, R.D. Levere, and N.G. Abraham. Tin-mediated heme oxygenase gene activation and cytochrome P450 arachidonate hydroxylase inhibition in spontaneously hypertensive rats [published erratum appears in Am J Med Sci 1994 Aug;308(2):138]. *Am.J.Med.Sci.* 307:173–181, 1994.

46. Chernick, R.J., P. Martasek, R.D. Levere, R. Margreiter, and N.G. Abraham. Sensitivity of human tissue heme oxygenase to a new synthetic metalloporphyrin. *Hepatology* 10:365–369, 1989.

47. Miller, A.D. and G.J. Rosman. Improved retroviral vectors for gene transfer and expression. *Biotechniques* 7:980–986, 989, 1989.

48. Yang, L., S. Quan, and N.G. Abraham. Retrovirus-mediated HO gene transfer into endothelial cells protects against oxidant induced injury. *Am.J.Physiol.* 277:L127–L133, 1999.

49. Iyer, S.N., D. Lu, M.J. Katovich, and M.K. Raizada. Chronic control of high blood pressure in the spontaneously hypertensive rat by delivery of angiotensin type 1 receptor antisense. *Proc.Natl.Acad.Sci.U.S.A.* 93:9960–9965, 1996.

50. Bakken, A.F., M.M. Thaler, and R. Schmid. Metabolic regulation of heme catabolism and bilirubin production. I. Hormonal control of hepatic heme oxygenase activity. *J. Clin.Invest.* 51:530–536, 1972.

51. Omata, K., N.G. Abraham, B. Escalante, and M.L. Schwartzman. Age-related changes in renal cytochrome P-450 arachidonic acid metabolism in spontaneously hypertensive rats. *Am.J.Physiol.* 262:F8–16, 1992.

52. Imig, J.D., J.R. Falck, D. Gebremedhin, D.R. Harder, and R.J. Roman. Elevated renovascular tone in young spontaneously hypertensive rats. Role of cytochrome P-450. *Hypertension* 22:357–364, 1993.

53. Yachie, A., Y. Niida, T. Wada, N. Igarashi, H. Kaneda, T. Toma, K. Ohta, Y. Kasahara, and S. Koizumi. Oxidative stress causes enhanced endothelial cell injury in human heme oxygenase-1 deficiency. *J.Clin.Invest.* 103:129–135, 1999.

54. Poss, K.D. and S. Tonegawa. Reduced stress defense in heme oxygenase-1 deficient cells. *Proc.Natl.Acad.Sci.U.S.A.* 94:10925–10930, 1997.

55. Deramaudt, B.M., S. Braunstein, P. Remy, and N.G. Abraham. Gene transfer of human heme oxygenase into coronary endothelial cells potentially promotes angiogenesis. *J.Cell Biochem.* 68:121–127, 1998.

56. Lee, P.J., J. Alam, G.W. Wiegand, and A.M.K. Choi. Overexpression of heme oxygenase-1 in human pulmonary epithelial cells results in cell growth arrest and increased resistance to hyperoxia. *Proc.Natl.Acad.Sci.U.S.A.* 93:10393–10398, 1996.

57. Cheriathundam, E., S.Q. Doi, J.R. Knapp, M.Z. Jasser, J.J. Kopchick, and A.P. Alvares. Consequences of overexpression of growth hormone in transgenic mice on liver cytochrome P450 enzymes. *Biochem.Pharmacol.* 55:1481–1487, 1998.

58. Lavrovsky, Y., M.L. Schwartzman, R.D. Levere, A. Kappas, and N.G. Abraham. Identification of binding sites for transcription factors NF-kappa B and AP-2 in the promoter region of the human heme oxygenase 1 gene. *Proc.Natl.Acad.Sci.U.S.A.* 91:5987–5991, 1994.

INDEX